Differential Equations on the World Wide Web

The Boston University
Ordinary Differential Equations Project
http://math.bu.edu/odes

The Brooks/Cole DiffEQ Resource Center
http://diffeq.brookscole.com

DIFFERENTIAL EQUATIONS

ABOUT THE AUTHORS

Paul Blanchard

Paul Blanchard was raised in Sutton, Massachusetts, spent his undergraduate years at Brown University, and received his Ph.D. from Yale University. He has taught college mathematics for twenty years, mostly at Boston University. He has coauthored or contributed chapters to four different textbooks. His main area of mathematical research is complex analytic dynamical systems and the related point sets—Julia sets and the Mandelbrot set. His recent efforts have focused on reforming the traditional differential equations course, and he currently heads the Boston University Differential Equations Project and leads workshops in this innovative approach to teaching differential equations. When he becomes exhausted fixing the errors made by his two coauthors, he usually closes up his CD store and heads for the golf course with his caddy, Glen Hall.

Robert L. Devaney

Robert L. Devaney was raised in Methuen, Massachusetts. He received his undergraduate degree from Holy Cross College and his Ph.D. from the University of California, Berkeley. He has taught at Boston University since 1980. His main area of research is complex dynamical systems, and he has lectured extensively throughout the world on this topic. In 1996 he received the National Excellence in Teaching Award from the Mathematical Association of America. When he gets sick of arguing with his coauthors over which topics to include in the differential equations course, he either turns up the volume of his opera CDs, or heads for waters off New England for a long-distance sail.

Glen R. Hall

Glen R. Hall spent most of his youth in Denver, Colorado. His undergraduate degree comes from Carleton College and his Ph.D. from the University of Minnesota. His research interests are mainly in low-dimensional dynamics and celestial mechanics. He has published numerous articles on the dynamics of circle and annulus maps. For his research he has been awarded both National Science Foundation Postdoctoral and Sloan Foundation Fellowships. He has no plans to open a CD store since he is busy raising his two young sons. He is an untalented, but earnest, trumpet player and golfer. He once bicycled 148 miles in a single day.

DIFFERENTIAL EQUATIONS

Paul Blanchard

Robert L. Devaney

Glen R. Hall

Boston University

 Brooks/Cole Publishing Company

I(T)P® *An International Thomson Publishing Company*

Pacific Grove • Albany • Belmont • Bonn • Boston • Cincinnati • Detroit • Johannesburg • London
Madrid • Melbourne • Mexico City • New York • Paris • Singapore • Tokyo • Toronto • Washington

A GARY W. OSTEDT BOOK

Publisher: *Gary W. Ostedt*
Marketing Team: *Caroline Croley, Margaret Parks, and Laura Caldwell*
Editorial Associate: *Carol Ann Benedict*
Production Editor: *Jamie Sue Brooks*
Manuscript Editor: *Barbara Kimmel*

Interior Design: *John Edeen*
Cover Design: *Vernon T. Boes*
Cover Photos: *AP/World Wide Photos and The Image Bank*
Art Editor: *Kathy Joneson*
Cover Printing: *Phoenix Color Corp.*
Printing and Binding: *R. R. Donnelley & Sons, Crawfordsville*

For more information, contact:

BROOKS/COLE PUBLISHING COMPANY
511 Forest Lodge Road
Pacific Grove, CA 93950
USA

International Thomson Publishing Europe
Berkshire House 168-173
High Holborn
London WC1V 7AA
England

Thomas Nelson Australia
102 Dodds Street
South Melbourne, 3205
Victoria, Australia

Nelson Canada
1120 Birchmount Road
Scarborough, Ontario
Canada M1K 5G4

International Thomson Editores
Seneca 53
Col. Polanco
11560 México, D. F., México

International Thomson Publishing GmbH
Königswinterer Strasse 418
53227 Bonn
Germany

International Thomson Publishing Asia
221 Henderson Road
#05-10 Henderson Building
Singapore 0315

International Thomson Publishing Japan
Hirakawacho Kyowa Building, 3F
2-2-1 Hirakawacho
Chiyoda-ku, Tokyo 102
Japan

Printed in the United States of America

10   9   8   7   6   5   4

Library of Congress Cataloging-in-Publication Data

Blanchard, Paul
    Differential equations / Paul Blanchard, Robert L. Devaney
Glen R. Hall
       p.   cm
    Includes index
    ISBN 0-534-34550-6 (alk. paper)
    1. Differential equations.   I. Devaney, Robert L.   II. Hall,
Glen R.   III. Title.
QA371.B63   1997
515 . 35—dc21   97-34431

PREFACE

The study of differential equations is a beautiful application of the ideas and techniques of calculus to our everyday lives. Indeed, it could be said that calculus was developed mainly so that the fundamental principles that govern many phenomena could be expressed in the language of differential equations. Unfortunately, it was difficult to convey the beauty of the subject in the traditional first course on differential equations because the number of equations that can be treated by analytic techniques is very limited. Consequently, the course tended to focus on technique rather than on concept.

This book is an outgrowth of our opinion that we are now able to effect a radical revision, and we approach our updated course with several goals in mind. First, the traditional emphasis on specialized tricks and techniques for solving differential equations is no longer appropriate given the technology that is readily available. Second, many of the most important differential equations are nonlinear, and numerical and qualitative techniques are more effective than analytic techniques in this setting. Finally, the differential equations course is one of the few undergraduate courses where it is possible to give students a glimpse of the nature of contemporary mathematical research.

The Qualitative, Numeric, and Analytic Approaches

Accordingly, this book is a radical departure from the typical "cookbook" differential equations text. We have eliminated most specialized techniques for deriving formulas for solutions, and we have replaced them with topics that focus on the formulation of differential equations and the interpretation of their solutions. To obtain an understanding of the solutions, we generally attack a given equation from three different points of view.

One major approach we adopt is qualitative. We expect students to be able to visualize differential equations and their solutions in many geometric ways. For example, we readily use slope fields, graphs of solutions, vector fields, and solution curves in the phase plane as tools to gain a better understanding of solutions. We also ask students to become adept at moving among these geometric representations and more traditional analytic representations.

Since differential equations are readily studied using the computer, we also emphasize numerical techniques. We assume that students have access to some sort of technology that approximates solutions and graphs these solutions easily. Even if we can find an explicit formula for a solution, we often work with the equation both numerically and qualitatively to understand the geometry and the long-term behavior of solutions. When we can find explicit solutions easily (such as in the case of separable first-order equations or constant-coefficient, linear systems), we do the calculations. But we never fail to examine the resulting formulas we obtain using qualitative and numerical points of view as well.

Specific Changes

There are several specific ways in which this book differs from other books at this level. First, we incorporate modeling throughout. We expect students to understand the meaning of the variables and parameters in a differential equation and to be able to interpret this meaning in terms of a particular model. Certain models reappear often as running themes, and they are drawn from a variety of disciplines so that students with various backgrounds will find something familiar.

We also adopt a dynamical systems point of view. Thus, we are always concerned with the long-term behavior of solutions of an equation, and using all of the appropriate approaches outlined above, we ask students to predict this long-term behavior of solutions. In addition, we emphasize the role of parameters in many of our examples, and we specifically address the manner in which the behavior of solutions changes as these parameters are varied.

Like other texts, we begin with first-order equations, but the only analytic technique we use to find closed-form solutions is separation of variables (and, at the end of the chapter, an integrating factor or two to handle certain linear equations). Instead, we emphasize the meaning of a differential equation and its solutions in terms of its slope field and the graphs of its solutions. If the differential equation is autonomous, we also discuss its phase line. This discussion of the phase line serves as an elementary introduction to the idea of a phase plane, which plays a fundamental role in subsequent chapters.

We then move directly from first-order equations to systems of first-order differential equations. Rather than consider second-order equations separately, we convert these equations to first-order systems. When these equations are viewed as systems, we are able to use qualitative and numerical techniques more readily. Of course, we then use the information about these systems gleaned from these techniques to recover information about the solutions of the original equation.

We also begin the treatment of systems with a general approach. We do not immediately restrict our attention to linear systems. Qualitative and numerical techniques work just as easily when a system is nonlinear, and one can proceed a long way toward understanding systems without resorting to algebraic techniques. However, qualitative ideas do not tell the whole story, and we are led naturally to the idea of linearization. With this background in the fundamental geometric and qualitative concepts, we then discuss linear systems in detail. As always, we not only emphasize the formula for the general solution of a linear system but also the geometry of its solution curves and of the related eigenvectors and eigenvalues.

While our study of systems requires the minimal use of some linear algebra, it is definitely not a prerequisite. As we deal primarily with two-dimensional systems, we easily develop all of the necessary algebraic techniques as we proceed. In the process, we give considerable insight into the geometry of such topics as eigenvectors and eigenvalues.

These topics form the core of our approach. However, there are many additional topics that one would like to cover in the course. Consequently, we have included discussions of forced second-order equations, nonlinear systems, Laplace transforms, numerical methods, and discrete dynamical systems. Although some of these topics are quite traditional, we always present them in a manner that is consistent with the philosophy developed in the first half of the text.

At the end of each chapter, we have included several "labs." Doing detailed numerical experimentation and writing reports has been our most successful modification of the traditional course at Boston University. Good labs are tough to write and to grade, but we feel that the benefit to students is extraordinary.

Pathways Through This Book

There are a number of possible tracks that instructors can follow in using this book. We feel that Chapters 1–3 form the core (with the possible exception of Sections 2.5 and 3.8, which cover systems in three dimensions). Most of the later chapters assume familiarity with this material. Certain sections such as Section 1.7 (bifurcations) and Section 1.9 (changing variables) may be skipped if some care is taken in choosing material from subsequent sections. However, the material on phase lines and phase planes, qualitative analysis, and solutions of linear systems is central.

A typical track for an engineering-oriented course would follow Chapters 1–3 (perhaps skipping Sections 1.9, 2.5, and 3.8). These chapters will take roughly two-thirds of a semester. The final third of the course might cover Sections 4.1–4.3 (forced, second-order linear equations and resonance), Section 5.1 (linearization of nonlinear systems), and Chapter 6 (Laplace transforms). Chapters 4 and 5 are independent of each other and can be covered in either order. In particular, Section 5.1 on linearization of nonlinear systems near equilibrium points forms an excellent capstone for the material on linear systems in Chapter 3.

Incidentally, it is possible to cover Sections 6.1 and 6.2 (Laplace transforms for first-order equations) immediately after Chapter 1. As we have learned from our colleagues in the College of Engineering at Boston University, some engineering programs teach a circuit theory course that uses the Laplace transform at an earlier point than is typically the case. Consequently, Sections 6.1 and 6.2 are written so that the differential equations course and such a circuits course could proceed in parallel. However, if possible, we recommend waiting to cover Chapter 6 entirely until after the material in Sections 4.1–4.3 has been discussed.

Instructors may wish to substitute material on discrete dynamics (Chapter 8) for Laplace transforms. A course for students with a strong background in physics might involve more of Chapter 5, including a treatment of Hamiltonian (Section 5.3) and gradient systems (Section 5.4). A course geared toward applied mathematics might include a more detailed discussion of numerical methods (Chapter 7).

Changes in the First Edition

We have been quite pleased with the reception that the preliminary edition of this book has enjoyed since its publication in 1995. We are especially indebted to the large number of readers and instructors who made comments about various points in the earlier edition. Accordingly, we have made some changes in this edition. The most significant changes include more thorough treatments of forcing and resonance for second-order equations and a revised treatment of Laplace transforms. The material in Chapter 2 has been extensively rewritten to follow more closely our intent to introduce analytic, qualitative, and numerical methods for systems at an early stage. Two appendices have been added. The first is an alternate treatment of first-order linear equations and can be used in place of Section of 1.8. The second appendix is a review of complex numbers and Euler's formula.

Most of the other changes involve only minor rearrangements of topics so that most instructors can avoid skipping sections within a chapter. As with any significant revision of an existing course, we anticipate that this book will continue to evolve in future editions. We encourage comments, suggestions, and criticism. The best way to comment is to send e-mail to **odes@math.bu.edu**. We'll do our best to acknowledge the e-mail, but we will definitely read and consider every comment.

Our Website and Ancillaries

Readers and instructors are invited to make extensive use of our web site

http://math.bu.edu/odes

At this site we have posted an on-line instructor's guide that includes discussions of how we use the text. We have also posted sample syllabi contributed by users at various institutions as well as information about workshops and seminars dealing with the teaching of differential equations. We also maintain a list of errata at this site. The *Instructor's Guide with Solutions*, available to instructors who have adopted the text for class use, contains a hardcopy of the on-line guide along with the solutions to all the problems. The *Student Solutions Manual* contains the solutions to all odd-numbered problems in the text.

Our publisher, Brooks/Cole, also maintains the DiffEQ Resource Center at

http://diffeq.brookscole.com

This site contains a wealth of information about the teaching and learning of differential equations, including an extensive array of laboratory and project ideas and links to a number of other sites related to the teaching of differential equations.

The Boston University Differential Equations Project

This book is a product of the now complete National Science Foundation Boston University Differential Equations Project sponsored by the National Science Foundation (NSF Grant DUE-9352833) and Boston University. The goal of that project was to rethink the traditional, sophomore-level differential equations course. We are especially thankful for that support.

Paul Blanchard
Robert L. Devaney
Glen R. Hall
Boston University

ACKNOWLEDGMENTS

As we move from the preliminary to the first edition, the list of people we are privileged to thank has grown exponentially. For this edition, we owe the greatest debt to **Gareth Roberts**. As project manager, he has overseen the production of the text and pictures. As mathematician and teacher he has been an invaluable critic and able assistant. As with his predecessor Sam Kaplan, the project manager for the preliminary edition, Gareth has left his mark on this text in many positive ways. Thanks, Gareth.

With the exception of a few professionally drawn figures, this book was entirely produced at Boston University's Department of Mathematics using Alex Kasman's ASTEX macro package in concert with LATEX2$_\epsilon$. Alex is a true TEX wizard, and anyone who is writing a textbook should consider his package. In fact, Alex's graphics macros are extremely useful in many contexts (see Alex's Web page available from the site **http://math.bu.edu**).

Much of the production work, solutions to exercises, accuracy checking and rendering of pictures was done by our team of graduate students: Bill Basener, Lee DeVille, and Stephanie Ruggiano. They spent many long days and nights in an alternately too-hot-or-too-cold windowless computer lab to bring this book to completion. We still rely on much of the work done by Adrian Iovita, Kinya Ono, Adrian Vajiac, and Nuria Fagella during the production of the preliminary edition.

Many other individuals at Boston University have made important contributions. In particular, our teaching assistants Duff Campbell, Michael Hayes, Eileen Lee, and Clara Bodelon had to put up with the headaches associated with our experimentation.

We received support from many of our colleagues at Boston University and at other institutions. Our chair, Marvin Freedman, encouraged us throughout. It was a special pleasure for us to work closely with colleagues in the College of Engineering—Michael Ruane (who coordinates the circuits course), Moe Wasserman (who permitted one of the authors to audit his course), and John Baillieul (a member of our advisory board). We also thank Donna Molinek (Davidson College), Carolyn Narasimhan (DePaul University), and James Walsh (Oberlin College) for organizing workshops for faculty on their campuses.

As is mentioned in the Preface, this book would not exist if our project had not received support from the National Science Foundation's Division of Undergraduate Education, and we thank the program directors at NSF/DUE for their enthusiastic support. We also thank the members of the advisory board—John Baillieul, Morton Brown, John Franks, Deborah Hughes Hallett, Philip Holmes, and Nancy Kopell. All have contributed their scarce time during workshops and trips to Boston University.

We were pleased that so many of our colleagues outside of Boston University were willing to help us with this project. Bill Krohn gave us valuable advice regarding our exposition of Laplace transforms, and Bruce Elenbogen did a thorough reading of early drafts of the beginning chapters. Preliminary drafts of our original notes were class tested in a number of different settings by Gregory Buck, Scott Sutherland, Kathleen Alligood, Diego Benedette, Jack Dockery, Mako Haruta, Jim Henle, Ed Packel, and Ben Pollina.

We have been pleased with the reception given to the preliminary edition of this text and are particularly grateful for the patience with which students and teachers alike have accepted our first attempt. Many have written us with excellent comments and suggestions—a few even caught the typo that we made. We thank them all. An updated list can be found at the web page address in the preface.

Thoughtful and insightful reviews have also been a tremendous help in the creation of both editions of the text. Reviews for the preliminary edition were done by Charles Boncelet, University of Delaware; Dean R. Brown, Youngstown State University; Michael Colvin, California Polytechnic State University; Peter Colwell, Iowa State University; James P. Fink, Gettysburg College; Michael Frame, Union College; Donnie Hallstone, Green River Community College; Stephen J. Merrill, Marquette University; LTC Joe Myers, U. S. Military Academy; Carolyn C. Narasimhan, DePaul University; Roger Pinkham, Stevens Institute of Technology; T. G. Proctor, Clemson University; Tim Sauer, George Mason University; Monty J. Strauss, Texas Tech University; and Paul Williams, Austin Community College.

Reviewers for this edition were David Arnold, College of the Redwoods; Steven H. Izen, Case Western Reserve University; Joe Marlin, North Carolina State University; Kenneth Meyer, University of Cincinnati; Joel Robbin, University of Wisconsin at Madison; Clark Robinson, Northwestern University; and Jim Walsh, Oberlin College.

Finally, as any author knows, writing a book requires significant sacrifices from one's family. Extra special thanks go to Lori, Kathy, and Dottie.

G.R.H., R.L.D., P.B.

A NOTE TO THE STUDENT

This book is probably different from most of your previous mathematics texts. If you thumb through it, you will see that there are few "boxed" formulas, no margin notes, and very few n-step procedures. We've written the book this way because we think that you are now at a point in your education where you should be learning to identify and work effectively with the mathematics inherent in everyday life. As you pursue your careers, no one is going to ask you to do all of the odd exercises at the end of some employee manual. They are going to give you some problem whose mathematical component may be difficult to identify and ask you to do your best with it. One of our goals in this book is to start preparing you for this type of work by avoiding artificial algorithmic exercises.

Our intention is that you will read this book as you would any other text, then work on the exercises, rereading sections and examples as necessary. Even though there are no template examples, you will find the discussions full of examples. Since one of our main goals is to demonstrate how differential equations are used to model physical systems, we often start with the description of a physical system, build a model, and then study the model to make conclusions and predictions about the original system. Many of the exercises ask you to produce or modify a model of a physical system, analyze it, and explain your conclusions. This is hard stuff, and you will need to practice. Since the days when you could make a living plugging and chugging through computations are over (computers do that now), you will need to learn these skills, and we hope that this book helps you develop them.

Another way in which this book may differ from your previous texts is that we expect you to make judicious use of a graphing calculator or a computer as you work the exercises and labs. The computer won't do the thinking for you, but it will provide you with numerical evidence that is essentially impossible for you to get in any other way. One of our goals is to give you practice as a sophisticated consumer of computer cycles as well as a skeptic of computer results.

Incidentally, one of the authors is known to have made a mistake or two in his life that the other two authors have overlooked. So we maintain a very short list of errata at our web site **http://math.bu.edu/odes**. Please check this page if you think that something you have read is not quite right.

Finally, you should know that the authors take the study of differential equations very seriously. However, we don't take ourselves very seriously (and we certainly don't take the other two authors seriously). We have tried to express both the beauty of the mathematics and some of the fun we have doing mathematics. If you think (some of?) the jokes are old or stupid, you're probably right.

All of us who worked on this book have learned something about differential equations along the way, and we hope that we are able to communicate our appreciation for the subject's beauty and range of application. We would enjoy hearing your comments. Feel free to send us e-mail at **odes@math.bu.edu**. We sometimes get busy and cannot always respond, but we do read and appreciate your feedback.

We had fun writing this book. We hope you have fun reading it.

G.R.H., R.L.D., P.B.

CONTENTS

1 FIRST-ORDER DIFFERENTIAL EQUATIONS

This book is about how to predict the future. To do so, all we have is a knowledge of how things are and an understanding of the rules that govern the changes that will occur. From calculus we know that change is measured by the derivative, and using the derivative to describe how a quantity changes is what the subject of differential equations is all about.

Turning the rules that govern the evolution of a quantity into a differential equation is called modeling, and in this chapter we study many models. Our goal is to use the differential equation to predict the future value of the quantity being modeled.

There are three basic types of techniques for making these predictions. Analytical techniques involve finding formulas for the future values of the quantity. Qualitative techniques involve obtaining a rough sketch of the graph of the quantity as a function of time as well as a description of its long-term behavior. Numerical techniques involve doing arithmetic (or having a computer do arithmetic) that yields approximations of the future values of the quantity. We introduce and use all three of these approaches in this chapter.

1.1 MODELING VIA DIFFERENTIAL EQUATIONS

The hardest part of using mathematics to study an application is the translation from real life into mathematical formalism. This translation is usually difficult because it involves the conversion of imprecise assumptions into very precise formulas. There is no way to avoid it. Modeling is difficult, and the best way to get good at it is the same way you get to play Carnegie Hall—practice, practice, practice.

What Is a Model?

It is important to remember that mathematical models are like other types of models. The goal is not to produce an exact copy of the "real" object but rather to give a representation of some aspect of the real thing. For example, a portrait of a person, a store mannequin, and a pig can all be models of a human being. None is a perfect copy of a human, but each has certain aspects in common with a human. The painting gives a description of what a particular person looks like; the mannequin wears clothes as a person does; and the pig is alive. Which of the three models is "best" depends on how we use the model—to remember old friends, to buy clothes, or to study biology.

The mathematical models we study are systems that evolve over time, but they often depend on other variables as well. In fact, real-world systems can be notoriously complicated—the population of rabbits in Wyoming depends on the number of coyotes, the number of bobcats, the number of mountain lions, the number of mice (alternative food for the predators), farming practices, the weather, any number of rabbit diseases, etc. We can make a model of the rabbit population simple enough to understand only by making simplifying assumptions and lumping together effects that may or may not belong together.

Once we've built the model, we should compare predictions of the model with data from the system. If the model and the system agree, then we gain confidence that the assumptions we made in creating the model are reasonable, and we can use the model to make predictions. If the system and the model disagree, then we must study and improve our assumptions. In either case we learn more about the system by comparing it to the model.

The types of predictions that are reasonable depend on our assumptions. If our model is based on precise rules such as Newton's laws of motion or the rules of compound interest, then we can use the model to make very accurate quantitative predictions. If the assumptions are less precise or if the model is a simplified version of the system, then precise quantitative predictions would be silly. In this case we would use the model to make qualitative predictions such as "the population of rabbits in Wyoming will increase ...." The dividing line between qualitative and quantitative prediction is itself imprecise, but we will see that it is frequently better and easier to make qualitative use of even the most precise models.

Some hints for model building

The basic steps in creating the model are

Step 1 Clearly state the assumptions on which the model will be based. These assumptions should describe the relationships among the quantities to be studied.

Step 2 Completely describe the variables and parameters to be used in the model — "you can't tell the players without a program."

Step 3 Use the assumptions formulated in Step 1 to derive equations relating the quantities in Step 2.

Step 1 is the "science" step. In Step 1, we describe how we think the physical system works or, at least, what the most important aspects of the system are. In some cases these assumptions are fairly speculative, as, for example, "rabbits don't mind being overcrowded." In other cases the assumptions are quite precise and well accepted, such as "force is equal to the product of mass and acceleration." The quality of the assumptions determines the validity of the model and the situations to which the model is relevant. For example, some population models apply only to small populations in large environments, whereas others consider limited space and resources. Most important, we must avoid "hidden assumptions" that make the model seem mysterious or magical.

Step 2 is where we name the quantities to be studied and, if necessary, describe the units and scales involved. Leaving this step out is like deciding you will speak your own language without telling anyone what the words mean.

The quantities in our models fall into three basic categories: the **independent variable**, the **dependent variables**, and the **parameters**. In this book the independent variable is (almost) always time. Time is "independent" of any other quantity in the model. On the other hand, the dependent variables are quantities that are functions of the independent variable. For example, if we say that "position is a function of time," we mean that position is a variable that depends on time. We can vaguely state the goal of a model expressed in terms of a differential equation as "Describe the behavior of the dependent variable as the independent variable changes." For example, we may ask whether the dependent variable increases or decreases, or whether it oscillates or tends to a limit.

Parameters are quantities that don't change with time (or with the independent variable) but that can be adjusted (by natural causes or by a scientist running the experiment). For example, if we are studying the motion of a rocket, the initial mass of the rocket is a parameter. If we are studying the amount of ozone in the upper atmosphere, then the rate of release of fluorocarbons from refrigerators is a parameter. Determining how the behavior of the dependent variables changes when we adjust the parameters can be the most important aspect of the study of a model.

In Step 3 we create the equations. Most of the models we consider are expressed as differential equations. In other words, we expect to find derivatives in our equations. Look for phrases such as "rate of change of . . ." or "rate of increase of . . .," since rate of change is synonymous with derivative. Of course, also watch for "velocity" (derivative of position) and "acceleration" (derivative of velocity) in models from physics. The word *is* means "equals" and indicates where the equality lies. The phrase "*A* is proportional to *B*" means $A = kB$, where k is a proportionality constant (often a parameter in the model).

An important rule of thumb we use when formulating models is: *Always make the algebra as simple as possible.* For example, when modeling the velocity v of a cat falling from a tall building, we could assume:

• Air resistance increases as the cat's velocity increases.

This assumption says that air resistance provides a force that counteracts the force of gravity and that this force increases as the velocity v of the cat increases. We could choose kv or kv^2 for the air resistance term, where k is the friction coefficient, a parameter. Both expressions increase as v increases, so they satisfy the assumption. However, we most likely would try kv first because it is the simplest expression that satisfies the assumption. In fact, it turns out that kv yields a good model for falling bodies with low densities like snowflakes, but kv^2 is a more appropriate model for dense objects like raindrops.

Now we turn to a series of models of population growth based on various assumptions about the species involved. Our goal here is to study how to go from a set of assumptions to a model. These examples are not "state-of-the-art" models from population ecology, but they are good ones to consider initially. We also begin to describe the analytic, qualitative, and numerical techniques that we use to make predictions based on these models. Our approach is meant to be illustrative only; we discuss these mathematical techniques in much more detail throughout the entire book.

Unlimited Population Growth

An elementary model of population growth is based on the assumption that

• The rate of growth of the population is proportional to the size of the population.

Note that the rate of change of a population depends on only the size of the population and nothing else. In particular, limitations of space or resources have no effect. This assumption is reasonable for small populations in large environments — for example, the first few spots of mold on a piece of bread or the first European settlers in the United States.

Because the assumption is so simple, we expect the model to be simple as well. The quantities involved are

$$t = \text{time (independent variable)},$$
$$P = \text{population (dependent variable), and}$$
$$k = \text{proportionality constant (parameter) between the rate}$$
$$\text{of growth of the population and the size of the population.}$$

The parameter k is often called the "growth-rate coefficient."

The units for these quantities depend on the application. If we are modeling the growth of mold on bread, then t might be measured in days and $P(t)$ might be either the area of bread covered by the mold or the weight of the mold. If we are talking about the European population of the United States, then t probably should be measured in years and $P(t)$ in millions of people. In this case we could let $t = 0$ correspond to any time we wanted. The year 1790 (the year of the first census) is a convenient choice.

Now let's express our assumption using this notation. The rate of growth of the population P is the derivative dP/dt. Being proportional to the population is expressed as the product, kP, of the population P and the proportionality constant k. Hence our assumption is expressed as the differential equation

$$\frac{dP}{dt} = kP.$$

In other words, the rate of change of P is proportional to P.

This equation is our first example of a differential equation. Associated with it are a number of adjectives that describe the type of differential equation that we are considering. In particular, it is a **first-order** equation because it contains only first derivatives of the dependent variable, and it is an **ordinary differential equation** because it does not contain partial derivatives. In this book we deal only with ordinary differential equations.

We have written this differential equation using the dP/dt Leibniz notation — the notation that we tend to use. However, there are many other ways to express the same differential equation. In particular, we could also write this equation as $P' = kP$ or as $\dot{P} = kP$. The "dot" notation is often used when the independent variable is time t.

What does the model predict?

More important than the adjectives or how the equation is written is what the equation tells us about the situation being modeled. Since $dP/dt = kP$ for some constant k, $dP/dt = 0$ if $P = 0$. Thus the constant function $P(t) = 0$ is a solution of the differential equation. This special type of solution is called an **equilibrium solution** because it is constant forever. In terms of the population model, it corresponds to a species that is nonexistent.

If $P(t_0) \neq 0$ at some time t_0, then at time $t = t_0$

$$\frac{dP}{dt} = k\,P(t_0) \neq 0.$$

As a consequence, the population is not constant. If $k > 0$ and $P(t_0) > 0$, we have

$$\frac{dP}{dt} = kP(t_0) > 0,$$

at time $t = t_0$ and the population is increasing (as one would expect). As t increases, $P(t)$ becomes larger, so dP/dt becomes larger. In turn, $P(t)$ increases even faster. That is, the rate of growth increases as the population increases. We therefore expect that the graph of the function $P(t)$ might look like Figure 1.1.

The value of $P(t)$ at $t = 0$ is called an **initial condition**. If we start with a different initial condition we get a different function $P(t)$ as is indicated in Figure 1.2. If

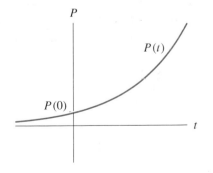

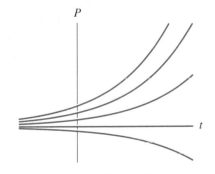

Figure 1.1
The graph of a function that satisfies the differential equation

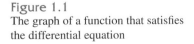

Figure 1.2
The graphs of several different functions that satisfy the differential equation $dP/dt = kP$. Each has a different value at $t = 0$.

$P(0)$ is negative (remembering $k > 0$), we then have $dP/dt < 0$ for $t = 0$, so $P(t)$ is initially decreasing. As t increases, $P(t)$ becomes more negative. The picture below the t-axis is the flip of the picture above, although this isn't "physically meaningful" because a negative population doesn't make much sense.

Our analysis of the way in which $P(t)$ increases as t increases is called a **qualitative analysis** of the differential equation. If all we care about is whether the model predicts "population explosions," then we can answer "yes, as long as $P(0) > 0$."

Analytic solutions of the differential equation

If, on the other hand, we know the exact value P_0 of $P(0)$ and we want to predict the value of $P(10)$ or $P(100)$, then we need more precise information about the function $P(t)$. The pair of equations

$$\frac{dP}{dt} = kP, \quad P(0) = P_0,$$

is called an **initial-value problem**. A **solution** to the initial-value problem is a function $P(t)$ that satisfies both equations. That is,

$$\frac{dP}{dt} = kP \text{ for all } t \quad \text{and} \quad P(0) = P_0.$$

Consequently, to find a solution to this differential equation we must find a function $P(t)$ whose derivative is the product of k with $P(t)$. One (not very subtle) way to find such a function is to guess. In this case, it is relatively easy to guess the right form for $P(t)$ because we know that the derivative of an exponential function is essentially itself. (We can eliminate this guesswork by using the method of separation of variables, which we describe in the next section. But for now, let's just try the exponential and see where that leads us.) After a couple of tries with various forms of the exponential, we see that

$$P(t) = e^{kt}$$

is a function whose derivative, $dP/dt = ke^{kt}$, is the product of k with $P(t)$. But there are other possible solutions, since $P(t) = ce^{kt}$ (where c is a constant) yields $dP/dt = c(ke^{kt}) = k(ce^{kt}) = kP(t)$. Thus $dP/dt = kP$ for all t for any value of the constant c.

We have infinitely many solutions to the differential equation, one for each value of c. To determine which of these solutions is the correct one for the situation at hand, we use the given initial condition. We have

$$P_0 = P(0) = c \cdot e^{k \cdot 0} = c \cdot e^0 = c \cdot 1 = c.$$

Consequently, we should choose $c = P_0$, so a solution to the initial-value problem is

$$P(t) = P_0 e^{kt}.$$

We have obtained an actual formula for our solution, not just a qualitative picture of its graph.

The function $P(t)$ is called the solution to the initial-value problem as well as a **particular solution** of the differential equation. The collection of functions $P(t) = ce^{kt}$

is called the **general solution** of the differential equation because we can use it to find the particular solution corresponding to any initial-value problem. Figure 1.2 consists of the graphs of exponential functions of the form $P(t) = ce^{kt}$ with various values of the constant c — that is, with different initial values. In other words, it is a picture of the general solution to the differential equation.

The U.S. Population

As an example of how this model can be used, consider the U.S. census figures since 1790 given in Table 1.1.

Let's see how well the unlimited growth model fits this data. We measure time in years and the population $P(t)$ in millions of people. We also let $t = 0$ be the year 1790, so the initial condition is $P(0) = 3.9$. The corresponding initial-value problem

$$\frac{dP}{dt} = kP, \quad P(0) = 3.9,$$

has $P(t) = 3.9e^{kt}$ as a solution. We cannot use this model to make predictions yet because we don't know the value of k. However, we do know that the population in the year 1800 was 5.3 million, and we can use this value to determine k. If we set

$$5.3 = P(10) = 3.9e^{k \cdot 10},$$

then we have

$$e^{k \cdot 10} = \frac{5.3}{3.9}$$

$$10k = \ln\left(\frac{5.3}{3.9}\right)$$

$$k \approx 0.03067.$$

Table 1.1

U.S. census figures, in millions of people (see Funk and Wagnalls, *1994 World Almanac*)

Year	t	Actual	$P(t) = 3.9e^{0.03067t}$	Year	t	Actual	$P(t) = 3.9e^{0.03067t}$
1790	0	3.9	3.9	1930	140	122	286
1800	10	5.3	5.3	1940	150	131	388
1810	20	7.2	7.2	1950	160	151	528
1820	30	9.6	9.8	1960	170	179	717
1830	40	12	13	1970	180	203	975
1840	50	17	18	1980	190	226	1,320
1850	60	23	25	1990	200	249	1,800
1860	70	31	33	2000	210		2,450
1870	80	38	45	2010	220		3,320
1880	90	50	62	2020	230		4,520
1890	100	62	84	2030	240		6,140
1900	110	75	114	2040	250		8,340
1910	120	91	155	2050	260		11,300
1920	130	105	210				

Thus our model predicts that the United States population is given by

$$P(t) = 3.9e^{0.03067t}.$$

As we see from Figure 1.3, this model of $P(t)$ does a decent job of predicting the population until roughly 1860, but after 1860 the prediction is much too large. (Table 1.1 includes a comparison of the predicted values to the actual data.)

Our model is fairly good provided the population is relatively small. However, as time goes on, the model predicts that the population will continue to grow without any limits, and obviously, this cannot happen in the real world. Consequently, if we want a model that is accurate over a large time scale, we should account for the fact that populations exist in a finite amount of space and with limited resources.

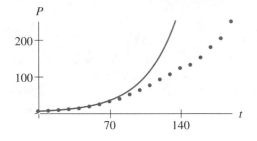

Figure 1.3
The dots represent actual census data and the solid line is the solution of the exponential growth model

$$\frac{dP}{dt} = 0.03067P.$$

Time t is measured in years since the Year 1790.

Logistic Population Model

To adjust the exponential growth population model to account for a limited environment and limited resources, we add the assumptions:

• If the population is small, the rate of growth of the population is proportional to its size.
• If the population is too large to be supported by its environment and resources, the population will decrease. That is, the rate of growth is negative.

For this model, we again use

$$t = \text{time (independent variable)},$$
$$P = \text{population (dependent variable)},$$
$$k = \text{growth-rate coefficient for small}$$
$$\text{populations (parameter)}.$$

However, our assumption about limited resources introduces another quantity, the size of the population that corresponds to being "too large." This quantity is a second parameter, denoted by N, that we call the "carrying capacity" of the environment. In terms of the carrying capacity, we are assuming that $P(t)$ is increasing if $P(t) < N$. However, if $P(t) > N$, we assume that $P(t)$ is decreasing.

Using this notation, we can restate our assumptions as:

• $\dfrac{dP}{dt} \approx kP$ if P is small (first assumption).

• If $P > N$, $\dfrac{dP}{dt} < 0$ (second assumption).

We also want the model to be "algebraically simple," or at least as simple as possible, so we try to modify the exponential model as little as possible. For instance, we might look for an expression of the form

$$\frac{dP}{dt} = k \cdot (\text{something}) \cdot P.$$

We want the "something" factor to be close to 1 if P is small, but if $P > N$ we want "something" to be negative. The simplest expression that has these properties is the function

$$(\text{something}) = \left(1 - \frac{P}{N}\right).$$

Note that this expression equals 1 if $P = 0$, and it is negative if $P > N$. Thus our model is

$$\frac{dP}{dt} = k\left(1 - \frac{P}{N}\right)P.$$

This is called the **logistic population model** with growth rate k and carrying capacity N. It is another first-order differential equation. This equation is said to be **nonlinear** because its right-hand side is not a linear function of P as it was in the exponential growth model.

Qualitative analysis of the logistic model

Although the logistic differential equation is just slightly more complicated than the exponential growth model, there is no way that we can just guess solutions. The method of separation of variables discussed in the next section produces a formula for the solution of this particular differential equation. But for now, we rely solely on qualitative methods to see what this model predicts over the long term.

First, let

$$f(P) = k\left(1 - \frac{P}{N}\right)P$$

denote the right-hand side of the differential equation. In other words, the differential equation can be written as

$$\frac{dP}{dt} = f(P) = k\left(1 - \frac{P}{N}\right)P.$$

We can derive qualitative information about the solutions to the differential equation from a knowledge of where dP/dt is zero, where it is positive, and where it is negative.

If we sketch the graph of the quadratic function f (see Figure 1.4), we see that it crosses the P-axis at exactly two points, $P = 0$ and $P = N$. In either case we have

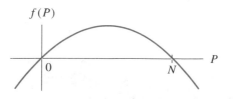

$f(P)$

Figure 1.4

Graph of the right-hand side

$$f(P) = k\left(1 - \frac{P}{N}\right)P$$

of the logistic differential equation.

$dP/dt = 0$. Since the derivative of P vanishes for all t, the population remains constant if $P = 0$ or $P = N$. That is, the constant functions $P(t) = 0$ and $P(t) = N$ are solutions of the differential equation. These two constant solutions make perfect sense: If the population is zero, the population remains zero indefinitely; if the population is exactly at the carrying capacity, it neither increases nor decreases. As before, we say that $P = 0$ and $P = N$ are *equilibria*. The constant functions $P(t) = 0$ and $P(t) = N$ are called *equilibrium solutions* (see Figure 1.5).

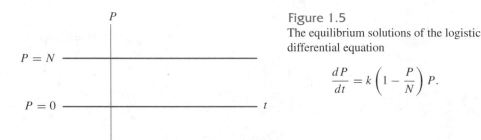

Figure 1.5
The equilibrium solutions of the logistic differential equation

$$\frac{dP}{dt} = k\left(1 - \frac{P}{N}\right)P.$$

The long-term behavior of the population is very different for other values of the population. If the initial population lies between 0 and N, then we have $f(P) > 0$. In this case the rate of growth $dP/dt = f(P)$ is positive, and consequently the population $P(t)$ is increasing. As long as $P(t)$ lies between 0 and N, the population continues to increase. However, as the population approaches the carrying capacity N, $dP/dt = f(P)$ approaches zero, so we expect that the population might level off as it approaches N (see Figure 1.6).

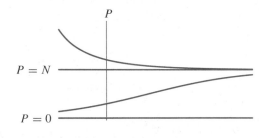

Figure 1.6
Solutions of the logistic differential equation

$$\frac{dP}{dt} = k\left(1 - \frac{P}{N}\right)P$$

approaching the equilibrium solution $P = N$.

If $P(0) > N$, then $dP/dt = f(P) < 0$, and the population is decreasing. As above, when the population approaches the carrying capacity N, dP/dt approaches zero, and we again expect the population to level off at N.

Finally, if $P(0) < 0$ (which does not make much sense in terms of populations), we also have $dP/dt = f(P) < 0$. Again we see that $P(t)$ decreases, but this time it does not level off at any particular value since dP/dt becomes more and more negative as $P(t)$ decreases.

Thus, just from a knowledge of the graph of f, we can sketch a number of different solutions with different initial conditions, all on the same axes. The only information that we need is the fact that $P = 0$ and $P = N$ are equilibrium solutions, $P(t)$ increases if $0 < P < N$ and $P(t)$ decreases if $P > N$ or $P < 0$. Of course the exact values of $P(t)$ at any given time t depend on the values of $P(0)$, k, and N (see Figure 1.7).

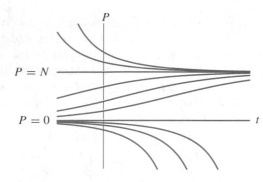

$P = N$

$P = 0$

Figure 1.7
Solutions of the logistic differential equation

$$\frac{dP}{dt} = k\left(1 - \frac{P}{N}\right)P$$

approaching the equilibrium solution $P = N$ and moving away from the equilibrium solution $P = 0$.

Predator-Prey Systems

No species lives in isolation, and the interactions among species give some of the most interesting models to study. We conclude this section by introducing a simple predator-prey system of differential equations where one species "eats" another. The most obvious difference between the model here and previous models is that we have *two* quantities that depend on time. Thus our model has two dependent variables that are both functions of time. Since both predator and prey begin with "p," we call the prey "rabbits" and the predators "foxes," and we denote the prey by R and the predators by F. The assumptions for our model are:

- If no foxes are present, the rabbits reproduce at a rate proportional to their population, and they are not affected by overcrowding.
- The foxes eat the rabbits, and the rate at which the rabbits are eaten is proportional to the rate at which the foxes and rabbits interact.
- Without rabbits to eat, the fox population declines at a rate proportional to itself.
- The rate at which foxes are born is proportional to the number of rabbits eaten by foxes which, by the second assumption, is proportional to the rate at which the foxes and rabbits interact.

To formulate this model in mathematical terms, we need four parameters in addition to our independent variable t and our two dependent variables F and R. The parameters are

$\alpha =$ growth-rate coefficient of rabbits,

$\beta =$ constant of proportionality that measures the number of rabbit-fox interactions in which the rabbit is eaten,

$\gamma =$ death-rate coefficient of foxes,

$\delta =$ constant of proportionality that measures the benefit to the fox population of an eaten rabbit.

When we formulate our model, we follow the convention that α, β, γ, and δ are all positive.

Our first and third assumptions above are similar to the assumption in the unlimited growth model discussed earlier in this section. Consequently, they give terms of the form αR in the equation for dR/dt and $-\gamma F$ (since the fox population declines) in the equation for dF/dt.

The rate at which the rabbits are eaten is proportional to the rate at which the foxes and rabbits interact, so we need a term that models the rate of interaction of the two populations. We want a term that increases if either R or F increases, but it should vanish if either $R = 0$ or $F = 0$. A simple term that incorporates these assumptions is RF. Thus we model the effects of rabbit-fox interactions on dR/dt by a term of the form $-\beta RF$. The fourth assumption gives a similar term in the equation for dF/dt. In this case, eating rabbits helps the foxes, so we add a term of the form δRF.

Given these assumptions, we obtain the model

$$\frac{dR}{dt} = \alpha R - \beta RF$$
$$\frac{dF}{dt} = -\gamma F + \delta RF.$$

Considered together, this pair of equations is called a **first-order system** (only first derivatives, but more than one dependent variable) of ordinary differential equations. The system is said to be *coupled* because the rates of change of R and F depend on both R and F.

It is important to note the signs of the terms in this system. Because $\beta > 0$, the term "$-\beta RF$" is nonpositive, so an increase in the number of foxes decreases the growth rate of the rabbit population. Also, since $\delta > 0$, the term "δRF" is nonnegative. Consequently, an increase in the number of rabbits increases the growth rate of the fox population.

Although this model may seem relatively simpleminded, it has been the basis of some interesting ecological studies. In particular, Volterra and D'Ancona successfully used the model to explain the increase in the population of sharks in the Mediterranean during World War I when the fishing of "prey" species decreased. The model can also be used as the basis for studying the effects of pesticides on the populations of predator and prey insects.

A *solution* to this system of equations is, unlike our previous models, a pair of functions, $R(t)$ and $F(t)$, that describe the populations of rabbits and foxes as functions of time. Since the system is coupled, we cannot simply determine one of these functions first and then the other. Rather, we must solve both differential equations simultaneously. Unfortunately, for most values of the parameters, it is impossible to determine explicit formulas for $R(t)$ and $F(t)$. These functions cannot be expressed in terms of known functions such as polynomials, sines, cosines, exponentials, and the like. However, as we will see in Chapter 2, these solutions do exist, although we have no hope of ever finding them exactly. Since analytic methods for solving this system are destined to fail, we must use either qualitative or numerical methods to "find" $R(t)$ and $F(t)$.

The Analytic, Qualitative, and Numerical Approaches

Our discussion of the three population models in this section illustrates three different approaches to the study of the solutions of differential equations. The **analytic** approach searches for explicit formulas that describe the behavior of the solutions. Here we saw that exponential functions give us explicit solutions to the exponential growth model. Unfortunately, a large number of important equations cannot be handled with the analytic approach; there simply is no way to find an exact formula that describes the situation. We are therefore forced to turn to alternative methods.

One particularly powerful method of describing the behavior of solutions is the **qualitative** approach. This method involves using geometry to give an overview of the behavior of the model, just as we did with the logistic population growth model. We do not use this method to give precise values of the solution at specific times, but we are often able to use this method to determine the long-term behavior of the solutions. Frequently, this is just the kind of information we need.

The third approach to solving differential equations is **numerical**. The computer approximates the solution we seek. Although we did not illustrate any numerical techniques in this section, we will soon see that numerical approximation techniques are a powerful tool for giving us intuition regarding the solutions we desire.

All three of the methods we use have certain advantages, and all have drawbacks. Sometimes certain methods are useful while others are not. One of our main tasks as we study the solutions to differential equations will be to determine which method or combination of methods works in each specific case. In the next three sections, we elaborate on these three techniques.

EXERCISES FOR SECTION 1.1

1. Consider the population model

$$\frac{dP}{dt} = 0.4P\left(1 - \frac{P}{230}\right),$$

where $P(t)$ is the population at time t.

   (a) For what values of P is the population in equilibrium?
   (b) For what values of P is the population increasing?
   (c) For what values of P is the population decreasing?

2. Consider the population model

$$\frac{dP}{dt} = 0.3\left(1 - \frac{P}{200}\right)\left(\frac{P}{50} - 1\right)P,$$

where $P(t)$ is the population at time t.

   (a) For what values of P is the population in equilibrium?
   (b) For what values of P is the population increasing?
   (c) For what values of P is the population decreasing?

3. Consider the differential equation

$$\frac{dy}{dt} = y^3 - y^2 - 12y.$$

   (a) For what values of y is $y(t)$ in equilibrium?
   (b) For what values of y is $y(t)$ increasing?
   (c) For what values of y is $y(t)$ decreasing?

4. The following table provides the land area in Australia colonized by the American marine toad (*Bufo marinis*) every five years from 1939–1974. Model the migration of this toad using an exponential growth model

$$\frac{dA}{dt} = kA,$$

where in this case $A(t)$ is the land area occupied at time t. Make predictions about the land area occupied in the years 2010, 2050, and 2100. You should do this by

(a) solving the initial-value problem,

(b) determining the constant k,

(c) computing the predicted areas, and

(d) comparing your solution to the actual data. Do you believe your prediction?

Year	Cumulative area occupied (km^2)
1939	32,800
1944	55,800
1949	73,600
1954	138,000
1959	202,000
1964	257,000
1969	301,000
1974	584,000

(Note that there are many exponential growth models that you can form using this data. Is one a more reasonable model than the others? Note also that the area of Queensland is 1,728,000 km^2 and the area of Australia is 7,619,000 km^2.)*

Remark: The American marine toad was introduced to Australia to control sugar cane beetles and, in the words of J. W. Hedgpath (see *Science*, July 1993 and *The New York Times*, July 6, 1993),

> Unfortunately the toads are nocturnal feeders and the beetles are abroad by day, while the toads sleep under rocks, boards and burrows. By night the toads flourish, reproduce phenomenally well and eat up everything they can find. The cane growers were warned by Walter W. Froggart, president of the New South Wales Naturalist Society, that the introduction was not a good idea and that the toads would eat the native ground fauna. He was immediately denounced as an ignorant meddlesome crank. He was also dead right.

---

*All data taken from "Cumulative Geographical Range of *Bufo Marinis* in Queensland, Australia from 1935 to 1974," by Michael D. Sabath, Walter C. Boughton, and Simon Easteal, in *Copeia*, No. 3, 1981, pp. 676–680.

Exercises 5–7 consider an elementary model of the learning process: Although human learning is an extremely complicated process, it is possible to build models of certain simple types of memorization. For example, consider a person presented with a list to be studied. The subject is given periodic quizzes to determine exactly how much of the list has been memorized. (The lists are usually things like nonsense syllables, randomly generated three-digit numbers, or entries from tables of integrals.) If we let $L(t)$ be the fraction of the list learned at time t, where $L = 0$ corresponds to knowing nothing and $L = 1$ corresponds to knowing the entire list, then we can form a simple model of this type of learning based on the assumption:

- The rate dL/dt is proportional to the fraction of the list left to be learned.

Since $L = 1$ corresponds to knowing the entire list, the model is

$$\frac{dL}{dt} = k(1 - L),$$

where k is the constant of proportionality.

5. For what value of L, $0 \le L \le 1$, does learning occur most rapidly?

6. Suppose two students memorize lists according to the same model:

$$\frac{dL}{dt} = 2(1 - L).$$

(a) If one of the students knows one-half of the list at time $t = 0$ and the other knows none of the list, which student is learning most rapidly at this instant?

(b) Will the student who starts out knowing none of the list ever catch up to the student who starts out knowing one-half of the list?

7. Consider the following two differential equations that model two students' rates of memorizing a poem. Jillian's rate is proportional to the amount to be learned with proportionality constant $k = 2$. Beth's rate is proportional to the square of the amount to be learned with proportionality constant 3. The corresponding differential equations are

$$\frac{dL_J}{dt} = 2(1 - L_J) \quad \text{and} \quad \frac{dL_B}{dt} = 3(1 - L_B)^2,$$

where $L_J(t)$ and $L_B(t)$ are the fractions of the poem learned at time t by Jillian and Beth, respectively.

(a) Which student has a faster rate of learning at $t = 0$ if they both start memorizing together having never seen the poem before?

(b) Which student has a faster rate of learning at $t = 0$ if they both start memorizing together having already learned half the poem?

(c) Which student has a faster rate of learning at $t = 0$ if they both start memorizing together having already learned one-third of the poem?

In Exercises 8–12, we consider the phenomenon of radioactive decay which, from experimentation, we know behaves according to the law:

> The rate at which a quantity of a radioactive isotope decays is proportional to the amount of the isotope present. The proportionality constant depends only on which radioactive isotope is used.

8. Model radioactive decay using the notation

$$t = \text{time (independent variable)},$$
$$r(t) = \text{amount of particular radioactive isotope}$$
$$\text{present at time } t \text{ (dependent variable)},$$
$$-\lambda = \text{decay rate (parameter)}.$$

Note that the minus sign is used so that $\lambda > 0$.

(a) Using this notation, write a model for the decay of a particular radioactive isotope.

(b) If the amount of the isotope present at $t = 0$ is r_0, state the corresponding initial-value problem for the model in part (a).

9. The **half-life** of a radioactive isotope is the amount of time it takes for a quantity of radioactive material to decay to one-half of its original amount.

(a) The half-life of Carbon 14 (C-14) is 5230 years. Determine the decay-rate parameter λ for C-14.

(b) The half-life of Iodine 131 (I-131) is 8 days. Determine the decay-rate parameter for I-131.

(c) What are the units of the decay-rate parameters in parts (a) and (b)?

(d) To determine the half-life of an isotope, we could start with 1000 atoms of the isotope and measure the amount of time it takes 500 of them to decay, or we could start with 10,000 atoms of the isotope and measure the amount of time it takes 5000 of them to decay. Will we get the same answer? Why?

10. Carbon dating is a method of determining the time elapsed since the death of organic material. The assumptions implicit in carbon dating are that

- Carbon 14 (C-14) makes up a constant proportion of the carbon that living matter ingests on a regular basis, and
- once the matter dies, the C-14 present decays, but no new carbon is added to the matter.

Hence, by measuring the amount of C-14 still in the organic matter and comparing it to the amount of C-14 typically found in living matter, a "time-since-death" can be approximated. Using the decay-rate parameter you computed in Exercise 9, determine the time-since-death if

(a) 88% of the original C-14 is still in the material.

(b) 12% of the original C-14 is still in the material.

(c) 2% of the original C-14 is still in the material.

(d) 98% of the original C-14 is still in the material.

Remark: There has been speculation that the amount of C-14 available to living creatures has not been exactly constant over long periods (thousands of years). This makes accurate dates much trickier to determine.

11. In order to apply the carbon dating technique of Exercise 10, we must measure the amount of C-14 in a sample. Chemically, radioactive Carbon 14 (C-14) and regular carbon behave identically. How can we determine the amount of C-14 in a sample? [*Hint*: See Exercise 8.]

12. The radioactive isotope I-131 is used in the treatment of hyperthyroid. I-131 administered to a patient naturally collects in the thyroid gland, where it decays and kills part of the gland.

   (a) Suppose that it takes 72 hours to ship I-131 from the producer to the hospital. What percentage of the original amount shipped actually arrives at the hospital? (See Exercise 9.)

   (b) If the I-131 is stored at the hospital for an additional 48 hours before it is used, how much of the original amount shipped from the producer is left when it is used?

   (c) How long will it take for the I-131 to *completely* decay so that the remnants can be thrown away without special precautions?

13. Suppose a species of fish in a particular lake has a population that is modeled by the logistic population model with growth rate k, carrying capacity N, and time t measured in years. Adjust the model to account for each of the following situations.

   (a) 100 fish are harvested each year.

   (b) One-third of the fish population is harvested annually.

   (c) The number of fish harvested each year is proportional to the square root of the number of fish in the lake.

14. Suppose that the growth-rate parameter $k = 0.3$ and the carrying capacity $N = 2500$ in the logistic population model of Exercise 13. Suppose $P(0) = 2500$.

   (a) If 100 fish are harvested each year, what does the model predict for the long-term behavior of fish population? In other words, what does a qualitative analysis of the model yield?

   (b) If one-third of the fish are harvested each year, what does the model predict for the long-term behavior of fish population?

15. The rhinoceros is now extremely rare. Suppose enough game preserve land is set aside so that there is sufficient room for many more rhinoceros territories than there are rhinoceroses. Consequently, there will be no danger of overcrowding. However, if the population is too small, fertile adults have difficulty finding each other when it is time to mate. Write a differential equation that models the rhinoceros population based on these assumptions. (Note that there is more than one reasonable model that fits these assumptions.)

16. Consider the following assumptions concerning the fraction of a piece of bread covered by mold.

   - Mold spores fall on the bread at a constant rate.
   - When the proportion covered is small, the fraction of the bread covered by mold increases at a rate proportional to the amount of bread covered.
   - When the fraction of bread covered by mold is large, the growth rate decreases.
   - In order to survive, mold must be in contact with the bread.

   Using these assumptions, write a differential equation that models the proportion of a piece of bread covered by mold. (Note that there is more than one reasonable model that fits these assumptions.)

17. The following table contains data for the population of tawny owls in Wyman Woods, Oxford England (collected by Southern).*

   (a) What population model would you use to model this population?

   (b) Can you approximate (or even make reasonable guesses for) the parameter values?

   (c) What does your model predict for the population today?

Year	Population	Year	Population
1947	34	1954	52
1948	40	1955	60
1949	40	1956	64
1950	40	1957	64
1951	42	1958	62
1952	48	1959	64
1953	48		

18. For the following predator-prey systems, identify which dependent variable, x or y, is the prey population and which is the predator population. Is the growth of the prey limited by any factors other than the number of predators? Do the predators have sources of food other than the prey? (Assume that the parameters α, β, γ, δ, and N are all positive.)

   (a)
   $$\frac{dx}{dt} = -\alpha x + \beta xy$$
   $$\frac{dy}{dt} = \gamma y - \delta xy$$

   (b)
   $$\frac{dx}{dt} = \alpha x - \alpha \frac{x^2}{N} - \beta xy$$
   $$\frac{dy}{dt} = \gamma y + \delta xy$$

19. In the following predator-prey population models, x represents the prey, and y represents the predators.

   (i)
   $$\frac{dx}{dt} = 5x - 3xy$$
   $$\frac{dy}{dt} = -2y + \tfrac{1}{2}xy$$

   (ii)
   $$\frac{dx}{dt} = x - 8xy$$
   $$\frac{dy}{dt} = -2y + 6xy$$

*See J. P. Dempster, *Animal Population Ecology*, Academic Press, 1975, p. 99.

(a) In which system does the prey reproduce more quickly when there are no predators (when $y = 0$) and equal numbers of prey?

(b) In which system are the predators more successful at catching prey? In other words, if the number of predators and prey are equal for the two systems, in which system do the predators have a greater effect on the rate of change of the prey?

(c) Which system requires more prey for the predators to achieve a given growth rate (assuming identical numbers of predators in both cases)?

20. The system

$$\frac{dx}{dt} = ax - by\sqrt{x}$$

$$\frac{dy}{dt} = cy\sqrt{x}$$

has been proposed as a model for a predator-prey system of two particular species of microorganisms (where a, b, and c are positive parameters).

(a) Which variable, x or y, represents the predator population? Which variable represents the prey population?

(b) What happens to the predator population if the prey is extinct?

21. The following systems are models of the populations of pairs of species that either *compete* for resources (an increase in one species decreases the growth rate of the other) or *cooperate* (an increase in one species increases the growth rate of the other). For each system, identify the variables (independent and dependent) and the parameters (carrying capacity, measures of interaction between species, etc.) Do the species compete or cooperate? (Assume all parameters are positive.)

(a)
$$\frac{dx}{dt} = \alpha x - \alpha \frac{x^2}{N} + \beta xy$$
$$\frac{dy}{dt} = \gamma y + \delta xy$$

(b)
$$\frac{dx}{dt} = \gamma x - \delta xy$$
$$\frac{dy}{dt} = \alpha y - \beta xy$$

1.2 ANALYTIC TECHNIQUE: SEPARATION OF VARIABLES

What Is a Differential Equation and What Is a Solution?

A first-order differential equation is an equation for an unknown function in terms of its derivative. As we saw in the previous section, there are three types of "variables" in differential equations — the independent variable (almost always t for time in our examples), one or more dependent variables (which are functions of the independent variable), and the parameters. This terminology is standard but a bit confusing. The dependent variable is actually a function, so technically it should be called the dependent function.

The standard form for a first-order differential equation is

$$\frac{dy}{dt} = f(t, y).$$

Here the right-hand side typically depends on both the dependent and independent variables, although we often encounter cases where either t or y is missing.

A **solution** of the differential equation is a function of the independent variable that, when substituted into the equation as the dependent variable, satisfies the equation for all values of the independent variable. That is, a function $y(t)$ is a solution if it satisfies $dy/dt = y'(t) = f(t, y(t))$. This terminology doesn't tell us how to find solutions, but it does tell us how to check whether a candidate function is or is not a solution. For example, consider the simple differential equation

$$\frac{dy}{dt} = y.$$

We can easily check that the function $y_1(t) = 3e^t$ is a solution, whereas $y_2(t) = \sin t$ is not a solution. The function $y_1(t)$ is a solution because

$$\frac{dy_1}{dt} = \frac{d(3e^t)}{dt} = 3e^t = y_1 \quad \text{for all } t.$$

On the other hand, $y_2(t)$ is not a solution since

$$\frac{dy_2}{dt} = \frac{d(\sin t)}{dt} = \cos t,$$

and certainly the function $\cos t$ is not the same function as $y_2(t) = \sin t$.

Checking that a given function is a solution to a given equation

If we look at a more complicated equation such as

$$\frac{dy}{dt} = \frac{y^2 - 1}{t^2 + 2t},$$

then we have considerably more trouble finding a solution. On the other hand, if somebody hands us a function $y(t)$, then we know how to check whether or not it is a solution.

For example, suppose we meet three differential equations textbook authors — say Paul, Bob, and Glen — at our local espresso bar, and we ask them to find solutions of this differential equation. After a few minutes of furious calculation, Paul says that

$$y_1(t) = 1 + t$$

is a solution. Glen then says that

$$y_2(t) = 1 + 2t$$

is a solution. After several more minutes, Bob says that

$$y_3(t) = 1$$

is a solution. Which of these functions is a solution? Let's see who is right by substituting each function into the differential equation.

First we test Paul's function. We compute the left-hand side by differentiating $y_1(t)$. We have

$$\frac{dy_1}{dt} = \frac{d(1+t)}{dt} = 1.$$

Substituting $y_1(t)$ into the right-hand side, we find

$$\frac{(y_1(t))^2 - 1}{t^2 + 2t} = \frac{(1+t)^2 - 1}{t^2 + 2t} = \frac{t^2 + 2t}{t^2 + 2t} = 1.$$

The left-hand side and the right-hand side of the differential equation are identical, so Paul is correct.

To check Glen's function, we again compute the derivative

$$\frac{dy_2}{dt} = \frac{d(1+2t)}{dt} = 2.$$

With $y_2(t)$, the right-hand side of the differential equation is

$$\frac{(y_2(t))^2 - 1}{t^2 + 2t} = \frac{(1+2t)^2 - 1}{t^2 + 2t} = \frac{4t^2 + 4t}{t^2 + 2t} = \frac{4(t+1)}{t+2}.$$

The left-hand side of the differential equation does not equal the right-hand side for all t since the right-hand side is not the constant function 2. Glen's function is *not* a solution.

Finally, we check Bob's function the same way. The left-hand side is

$$\frac{dy_3}{dt} = \frac{d(1)}{dt} = 0$$

because $y_3(t) = 1$ is a constant. The right-hand side is

$$\frac{y_3(t)^2 - 1}{t^2 + t} = \frac{1 - 1}{t^2 + t} = 0.$$

Both the left-hand side and the right-hand side of the differential equation vanish for all t. Hence, Bob's function *is* a solution of the differential equation.

The lessons we learn from this example are that a differential equation may have solutions that look very different from each other algebraically and that (of course) not every function is a solution. Given a function, we can test to see whether it is a solution by just substituting it into the differential equation and checking to see whether the left-hand side is identical to the right-hand side. This is a very nice aspect of differential equations: *We can always check our answers.* So we should never be wrong.

Initial-Value Problems and the General Solution

When we encounter differential equations in practice, they often come with **initial conditions**. We seek a solution of the given equation that assumes a given value at a particular time. A differential equation along with an initial condition is called an **initial-value problem**. Thus the usual form of an initial-value problem is

$$\frac{dy}{dt} = f(t, y), \quad y(t_0) = y_0.$$

Here we are looking for a function $y(t)$ that is a solution of the differential equation *and* assumes the value y_0 at time t_0. Often, the particular time in question is $t = 0$ (hence the name *initial condition*), but any other time could be specified.

For example,

$$\frac{dy}{dt} = t^3 - 2\sin t, \quad y(0) = 3.$$

is an initial-value problem. To solve this problem, note that the right-hand side of the differential equation depends only on t, not on y. We are looking for a function whose derivative is $t^3 - 2\sin t$. This is a typical antidifferentiation problem from calculus, so all we need to do is to integrate this expression. We find

$$\int (t^3 - 2\sin t)\, dt = \frac{t^4}{4} + 2\cos t + c,$$

where c is a constant of integration. Thus the solution of the differential equation must be of the form

$$y(t) = \frac{t^4}{4} + 2\cos t + c.$$

We now use the initial condition $y(0) = 3$ to determine c by

$$3 = y(0) = \frac{0^4}{4} + 2\cos 0 + c = 0 + 2 \cdot 1 + c = 2 + c.$$

Thus $c = 1$, and the solution to this initial-value problem is

$$y(t) = \frac{t^4}{4} + 2\cos t + 1.$$

The expression

$$y(t) = \frac{t^4}{4} + 2\cos t + c$$

is called the **general solution** of the differential equation because we can use it to solve any initial-value problem whatsoever. For example, if the initial condition is $y(0) = \pi$, then we would choose $c = \pi - 2$ to solve the initial-value problem $dy/dt = t^3 - 2\sin t$, $y(0) = \pi$.

Separable Equations

Now that we know how to check that a given function is a solution to a differential equation, the question is: How can we get our hands on a solution in the first place? Unfortunately, it is rarely the case that we can find explicit solutions of a differential equation. Many differential equations have solutions that cannot be expressed in terms of known functions such as polynomials, exponentials, or trigonometric functions. However, there are a few special types of differential equations for which we can derive explicit solutions, and in this section we discuss one of these types of differential equations.

The typical first-order differential equation is given in the form

$$\frac{dy}{dt} = f(t, y).$$

The right-hand side of this equation generally involves both the independent variable t and the dependent variable y (although there are many important examples where either the t or the y is missing). A differential equation is called **separable** if the function $f(t, y)$ can be written as the product of two functions: one that depends on t alone and another that depends only on y. That is, a differential equation is separable if it can be written in the form

$$\frac{dy}{dt} = g(t)h(y).$$

For example, the differential equation

$$\frac{dy}{dt} = yt$$

is clearly separable, and the equation

$$\frac{dy}{dt} = y + t$$

is not. We might have to do a little work to see that an equation is separable. For instance,

$$\frac{dy}{dt} = \frac{t+1}{ty+t}$$

is separable since we can rewrite the equation as

$$\frac{dy}{dt} = \frac{(t+1)}{t(y+1)} = \left(\frac{t+1}{t}\right)\left(\frac{1}{y+1}\right).$$

Two important types of separable equations occur if either t or y is missing from the right-hand side of the equation. The differential equation

$$\frac{dy}{dt} = g(t)$$

is separable since we may regard the right-hand side as $g(t) \cdot 1$, where we consider 1 as a (very simple) function of y. Similarly,

$$\frac{dy}{dt} = h(y)$$

is also separable. This last type of differential equation is said to be **autonomous**. Many of the most important first-order differential equations that arise in applications (including all of our models in the previous section) are autonomous. For example, the right-hand side of the logistic equation

$$\frac{dP}{dt} = kP\left(1 - \frac{P}{N}\right)$$

depends on the dependent variable P alone, so this equation is autonomous.

How to solve separable differential equations

To find explicit solutions of separable differential equations, we use a technique familiar from calculus. To illustrate the method, consider the differential equation

$$\frac{dy}{dt} = \frac{t}{y^2}.$$

There is a temptation to solve this equation by simply integrating both sides of the equation with respect to t. This yields

$$\int \frac{dy}{dt}\, dt = \int \frac{t}{y^2}\, dt,$$

and, consequently,

$$y(t) = \int \frac{t}{y^2}\, dt.$$

Now we are stuck. We can't evaluate the integral on the right-hand side because we don't know the function $y(t)$. In fact, that is precisely the function we wish to find. We have simply replaced the differential equation with an *integral equation*.

We need to do something to this equation *before* we try to integrate. Returning to the original differential equation

$$\frac{dy}{dt} = \frac{t}{y^2},$$

we first do some "informal" algebra and rewrite this equation in the form

$$y^2\, dy = t\, dt.$$

That is, we multiply both sides by $y^2\, dt$. Of course, it makes no sense to split up dy/dt by multiplying by dt. However, this should remind you of the technique of integration known as u-substitution in calculus. We will soon see that substitution is exactly what we are doing here.

We now integrate both sides: the left with respect to y and the right with respect to t. We have

$$\int y^2\, dy = \int t\, dt,$$

which yields

$$\frac{y^3}{3} = \frac{t^2}{2} + c.$$

Technically there is a constant of integration on both sides of this equation, but we can lump them together as a single constant c on the right. We may rewrite this expression as

$$y(t) = \left(\frac{3t^2}{2} + 3c \right)^{1/3};$$

or since c is an arbitrary constant, we may write this even more compactly as

$$y(t) = \left(\frac{3t^2}{2} + k \right)^{1/3},$$

where k is an arbitrary constant. As usual, we can check that this expression really is a solution of the differential equation, so despite the questionable separation we just performed, we do obtain a solution in the end.

Note that this process yields many solutions of the differential equation. Each choice of the constant k gives a different solution.

What is really going on in our informal algebra

If you read the previous example closely, you probably became nervous at one point. Treating dt as a variable is a tip-off that something a little more complicated is actually going on. Here is the real story.

We began with a separable equation

$$\frac{dy}{dt} = g(t)h(y),$$

and then rewrote it as

$$\frac{1}{h(y)} \frac{dy}{dt} = g(t).$$

This equation actually has a function of t on each side of the equals sign because y is a function of t. So we really should write it as

$$\frac{1}{h(y(t))} \frac{dy}{dt} = g(t).$$

In this form, we can integrate both sides with respect to t to get

$$\int \frac{1}{h(y(t))} \frac{dy}{dt}\, dt = \int g(t)\, dt.$$

Now for the important step: We make a "u-substitution" just as in calculus by replacing the function $y(t)$ by the new variable, say y. (In this case, the substitution is actually a y-substitution.) Of course, we must also replace the expression $(dy/dt)\, dt$ by dy. The method of substitution from calculus tells us that

$$\int \frac{1}{h(y(t))} \frac{dy}{dt}\, dt = \int \frac{1}{h(y)}\, dy,$$

and therefore we can combine the last two equations to obtain

$$\int \frac{1}{h(y)}\, dy = \int g(t)\, dt.$$

Hence, we can integrate the left-hand side with respect to y and the right-hand side with respect to t.

Separating variables and multiplying both sides of the differential equation by dt is simply a notational convention that helps us remember the method. It is justified by the argument above.

Missing Solutions

If it is possible to separate variables in a differential equation, it appears that solving the equation reduces to a matter of computing several integrals. This is true, but there are some hidden pitfalls, as the following example shows. Consider the differential equation

$$\frac{dy}{dt} = y^2.$$

This is an autonomous and hence separable equation, and its solution looks straightforward. If we separate and integrate as usual, we obtain

$$\int \frac{dy}{y^2} = \int dt$$

$$-\frac{1}{y} = t + c$$

$$y(t) = -\frac{1}{t+c}.$$

We are tempted to say that this expression for $y(t)$ is the general solution. However, we cannot solve all initial-value problems with solutions of this form. In fact, we have $y(0) = -1/c$, so we cannot use this expression to solve the initial-value problem $y(0) = 0$.

What's wrong? Note that the right-hand side of the differential equation vanishes if $y = 0$. So the constant function $y(t) = 0$ is a solution to this differential equation. In other words, in addition to those solutions that we derived using the method of separation of variables, this differential equation possesses the equilibrium solution $y(t) = 0$ for all t, and it is this equilibrium solution that satisfies the initial-value problem $y(0) = 0$. Even though it is "missing" from the family of solutions that we obtain by separating variables, it is a solution that we need if we want to solve every initial-value problem for this differential equation. Thus the general solution consists of functions of the form $y(t) = -1/(t + c)$ together with the equilibrium solution $y(t) = 0$.

Getting Stuck

As another example, consider the differential equation

$$\frac{dy}{dt} = \frac{y}{1 + y^2}.$$

As before, this equation is autonomous. So we first separate to obtain

$$\left(\frac{1 + y^2}{y}\right) dy = dt.$$

Then we integrate

$$\int \left(\frac{1}{y} + y\right) dy = \int dt,$$

which yields

$$\ln|y| + \frac{y^2}{2} = t + c.$$

But now we are stuck; there is no way to solve the equation

$$\ln|y| + \frac{y^2}{2} = t + c$$

for y alone. Thus we cannot generate an explicit formula for y. We do, however, have an **implicit form** for the solution which, for many purposes, is perfectly acceptable.

Even though we don't obtain explicit solutions by separating variables for this equation, we can find one explicit solution. The right-hand side vanishes if $y = 0$. Thus the constant function $y(t) = 0$ is an equilibrium solution. Note that this equilibrium solution does not appear in the implicit solution we derived from the method of separation of variables.

There is another problem that arises with this method. It is often impossible to perform the required integrations. For example, the differential equation

$$\frac{dy}{dt} = \sec(y^2)$$

is autonomous. Separating variables and integrating we get

$$\int \frac{1}{\sec(y^2)} \, dy = \int dt,$$

or equivalently,

$$\int \cos(y^2) \, dy = \int dt.$$

The integral on the left-hand side is difficult, to say the least. (In fact, there is a special function that was defined just to give us a name for this integral.) The lesson is that, even for special equations of the form

$$\frac{dy}{dt} = f(y),$$

carrying out the required algebra or integration is frequently impossible. We will not be able to rely solely on analytic tools and explicit solutions when studying differential equations, even if we can separate variables.

A Savings Model

Suppose we deposit \$5000 in a savings account with interest accruing at the rate of 5% compounded continuously. If we let $A(t)$ denote the amount of money in the account at time t, then the differential equation for A is

$$\frac{dA}{dt} = 0.05A.$$

As we saw in the previous section, the general solution to this equation is the exponential function

$$A(t) = ce^{0.05t},$$

where $c = A(0)$. Thus

$$A(t) = 5000e^{0.05t}$$

is our particular solution.

Assuming interest rates never change, after 10 years we will have

$$A(10) = 5000e^{0.5} \approx 8244$$

dollars in this account. That is a nice little nest egg, so we decide we should have some fun in life. We decide to withdraw \$1000 (mad money) from the account each year in a continuous way beginning in year 10. How long will this money last? Will we ever go broke?

The differential equation for $A(t)$ must change, but only beginning in year 10. For $0 \le t \le 10$, our previous model works fine. However, for $t > 10$, the differential equation becomes

$$\frac{dA}{dt} = 0.05A - 1000.$$

Thus we really have a differential equation of the form

$$\frac{dA}{dt} = \begin{cases} 0.05A & \text{for } t < 10; \\ 0.05A - 1000 & \text{for } t > 10, \end{cases}$$

whose right-hand side consists of two pieces.

To solve this two-part equation, we solve the first part and determine $A(10)$. We just did that and obtained $A(10) \approx 8244$. Then we solve the second equation using $A(10) \approx 8244$ as the initial value. This equation is also separable, and we have

$$\int \frac{dA}{0.05A - 1000} = \int dt.$$

We calculate this integral using substitution and the natural logarithm function. Let $u = 0.05A - 1000$. Then $du = 0.05 \, dA$, or $20 \, du = dA$ since $0.05 = 1/20$. We obtain

$$\int \frac{20 \, du}{u} = t + c_1$$

$$20 \ln |u| = t + c_1$$

$$20 \ln |0.05A - 1000| = t + c_1,$$

for some constant c_1.

At $t = 10$, we know that $A \approx 8244$. Thus at $t = 10$,

$$\frac{dA}{dt} = 0.05A - 1000 \approx -587.8 < 0.$$

In other words, we are withdrawing at a rate that exceeds the rate at which we are earning interest. Since dA/dt at $t = 10$ is negative, A will decrease and $0.05A - 1000$ remains negative for all $t > 10$. If $0.05A - 1000 < 0$, then

$$|0.05A - 1000| = -(0.05A - 1000) = 1000 - 0.05A.$$

Consequently, we have

$$20 \ln(1000 - 0.05A) = t + c_1,$$

or

$$\frac{\ln(1000 - 0.05A)}{0.05} = t + c_1.$$

Multiplying both sides by 0.05 and exponentiating yields

$$1000 - 0.05A = e^{0.05(t + c_1)}$$
$$1000 - 0.05A = c_2 e^{0.05t},$$

where $c_2 = e^{0.05c_1}$. Solving for A, we obtain

$$A = \frac{1000 - c_2 e^{0.05t}}{0.05}$$
$$= 20 \left(1000 - c_2 e^{0.05t}\right) = 20000 - c_3 e^{0.05t},$$

where $c_3 = 20c_2$. (Although we have been careful to spell out the relationships among the constants c_1, c_2, and c_3, we need only remember that c_3 is a constant that is determined from the initial condition.)

Now we use the initial condition to determine c_3. We know that

$$8244 \approx A(10) = 20000 - c_3 e^{0.05(10)} \approx 20000 - c_3(1.6487).$$

Solving for c_3, we obtain $c_3 \approx 7130$. Our solution for $t \geq 10$ is

$$A(t) \approx 20000 - 7130 e^{0.05t}.$$

We see that

$$A(11) \approx 7641$$
$$A(12) \approx 7008$$

and so forth. Our account is being depleted, but not by that much. In fact, we can find out just how long the good times will last by asking when our money will run out. In other words, we solve the equation $A(t) = 0$ for t. We have

$$0 = 20000 - 7130 e^{0.05t},$$

which yields

$$t = 20 \ln \left(\frac{20000}{7130}\right) \approx 20.63.$$

After letting the $5000 accumulate interest for ten years, we can withdraw $1000 per year for more than ten years.

A Mixing Problem

The name *mixing problem* refers to a large collection of different problems where two or more substances are mixed together at various rates. Examples range from the mixing of pollutants in a lake to the mixing of chemicals in a vat to the diffusion of cigar smoke in the air in a room to the blending of spices in a serving of curry.

Mixing in a vat

Consider a large vat containing sugar-water that is to be made into soft drinks (see Figure 1.8). Suppose:

- The vat contains 100 gallons of liquid. Moreover, the amount flowing in is the same as the amount flowing out, so there are always 100 gallons in the vat.
- The vat is kept well mixed, so the concentration of sugar is uniform throughout the vat.
- Sugar-water containing 5 tablespoons of sugar per gallon enters the vat through pipe A at a rate of 2 gallons per minute.
- Sugar-water containing 10 tablespoons of sugar per gallon enters the vat through pipe B at a rate of 1 gallon per minute.
- Sugar-water leaves the vat through pipe C at a rate of 3 gallons per minute.

To make the model, we let t be time measured in minutes (the independent variable). For the dependent variable, we have two choices. We could choose either the total amount of sugar, $S(t)$, in the vat at time t measured in tablespoons, or $C(t)$, the concentration of sugar in the vat at time t measured in tablespoons per gallon. We will develop the model for S, leaving the model for C as an exercise.

Using the total sugar $S(t)$ in the vat as the dependent variable, the rate of change of S is the difference between the amount of sugar being added and the amount of sugar being removed. The sugar entering the vat comes from pipes A and B and can be easily computed by multiplying the number of gallons per minute of sugar mixture entering the vat by the amount of sugar per gallon. The amount of sugar leaving the vat through pipe C at any given moment depends on the concentration of sugar in the vat at that moment. The concentration is given by $S/100$, so the sugar leaving the vat is the product of the number of gallons leaving per minute (3 gallons per minute) and the concentration ($S/100$). The model is

$$\frac{dS}{dt} = \underbrace{2 \cdot 5}_{\substack{\text{sugar in} \\ \text{from pipe A}}} + \underbrace{1 \cdot 10}_{\substack{\text{sugar in} \\ \text{from pipe B}}} - \underbrace{3 \cdot \frac{S}{100}}_{\substack{\text{sugar out} \\ \text{from pipe C}}}.$$

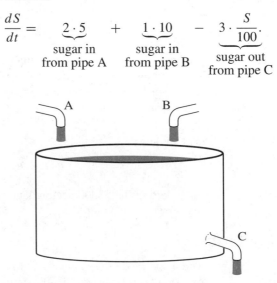

Figure 1.8
Mixing vat.

That is,

$$\frac{dS}{dt} = 20 - \frac{3S}{100} = \frac{2000 - 3S}{100}.$$

To solve this equation analytically, we separate and integrate. We find

$$\frac{dS}{2000 - 3S} = \frac{dt}{100}$$

$$\frac{\ln|2000 - 3S|}{-3} = \frac{t}{100} + c_1$$

$$\ln|2000 - 3S| = -\frac{3t}{100} - 3c_1$$

$$\ln|2000 - 3S| = -0.03t + c_2,$$

where $c_2 = -3c_1$. Exponentiating we obtain

$$|2000 - 3S| = e^{(-0.03t + c_2)} = c_3 e^{-0.03t},$$

where $c_3 = e^{c_2}$. Note that this means that c_3 is a positive constant. Now we must be careful. Removing the absolute value signs yields

$$2000 - 3S = \pm c_3 e^{-0.03t},$$

where we choose the plus sign if $S(t) < 2000/3$ and the minus sign if $S(t) > 2000/3$. Therefore we may write this equation more simply as

$$2000 - 3S = c_4 e^{-0.03t},$$

where c_4 is an arbitrary constant (positive, negative, or zero). Solving for S yields the general solution

$$S(t) = ce^{-0.03t} + \frac{2000}{3},$$

where $c = -c_4/3$ is an arbitrary constant. We can determine the precise value of c if we know the exact amount of sugar that is initially in the vat. Note that, if $c = 0$, the solution is simply $S(t) = 2000/3$, an equilibrium solution.

EXERCISES FOR SECTION 1.2

1. Bob, Glen, and Paul are once again sitting around having a nice, cold glass of iced cappucino when one of their students asks them to come up with solutions to the differential equation

$$\frac{dy}{dt} = \frac{y + 1}{t + 1}.$$

After much discussion, Bob says $y(t) = t$, Glen says $y(t) = 2t + 1$, and Paul says $y(t) = t^2 - 2$.

(a) Who is right?

(b) What solution should they have seen right away?

2. Make up a differential equation of the form

$$\frac{dy}{dt} = 2y - t + g(y)$$

that has the function $y(t) = e^{2t}$ as a solution.

3. Make up a differential equation of the form

$$\frac{dy}{dt} = f(t, y)$$

that has $y(t) = e^{t^3}$ as a solution. (Try to come up with one whose right-hand side $f(t, y)$ depends explicitly on both t and y.)

4. Pick some relatively simple function and produce a differential equation that has that function as a solution. Then ask a classmate to guess a solution to the resulting differential equation without telling them the solution you have in mind. (Be kind. Your classmate will ask the same thing of you.)

(a) Did your classmate come up with a solution?

(b) Was it the same one you had in mind?

(c) Why is it easier to write differential equations problems than it is to solve them?

In Exercises 5–24, find the general solution of the differential equation specified. (You may not be able to reach the ideal answer of an equation with only the dependent variable on the left and only the independent variable on the right, but get as far as you can.)

5. $\dfrac{dy}{dt} = ty$

6. $\dfrac{dy}{dt} = t^4 y$

7. $\dfrac{dy}{dt} = 2y + 1$

8. $\dfrac{dy}{dt} = 2 - y$

9. $\dfrac{dy}{dt} = e^{-y}$

10. $\dfrac{dy}{dt} = (ty)^2$

11. $\dfrac{dy}{dt} = \dfrac{t}{t^2 y + y}$

12. $\dfrac{dy}{dt} = t\sqrt[3]{y}$

13. $\dfrac{dy}{dt} = \dfrac{1}{2y + 1}$

14. $\dfrac{dy}{dt} = \dfrac{t}{1 + y^2}$

15. $\dfrac{dy}{dt} = y(1 - y)$

16. $\dfrac{dy}{dt} = \dfrac{t}{y}$

17. $\dfrac{dy}{dt} = t^2 y + 1 + y + t^2$

18. $\dfrac{dy}{dt} = \dfrac{1}{ty + t + y + 1}$

19. $\dfrac{dy}{dt} = \dfrac{e^t y}{1 + y^2}$

20. $\dfrac{dy}{ds} = \sec y$

21. $\dfrac{dw}{dt} = \dfrac{w}{t}$

22. $\dfrac{dy}{dt} = \dfrac{2y + 1}{t}$

23. $\dfrac{dy}{dt} = \dfrac{t^2 + 1}{y^4 + 3y}$

24. $\dfrac{dy}{dt} = 1 + \dfrac{1}{y^2}$

In Exercises 25–34, solve the given initial-value problem.

25. $\dfrac{dy}{dt} = 2y + 1, \quad y(0) = 3$

26. $\dfrac{dy}{dt} = ty^2 + 2y^2, \quad y(0) = 1$

27. $\dfrac{dy}{dt} = -y^2, \quad y(0) = 1/2$

28. $\dfrac{dx}{dt} = -xt, \quad x(0) = 1/\sqrt{\pi}$

29. $\dfrac{dy}{dt} = -y^2, \quad y(0) = 0$

30. $\dfrac{dy}{dt} = \dfrac{t}{y - t^2y}, \quad y(0) = 4$

31. $\dfrac{dy}{dt} = \dfrac{t^2}{y + t^3y}, \quad y(0) = -2$

32. $\dfrac{dy}{dt} = 2ty^2 + 3t^2y^2, \quad y(1) = -1$

33. $\dfrac{dy}{dt} = (y^2 + 1)t, \quad y(0) = 1$

34. $\dfrac{dy}{dt} = \dfrac{1}{2y + 3}, \quad y(0) = 1$

35. A 5-gallon bucket is full of pure water. Suppose we begin dumping salt into the bucket at a rate of 1/4 pounds per minute. Also, we open the spigot so that 1/2 gallons per minute leaves the bucket, and we add pure water to keep the bucket full. If the saltwater solution is always well mixed, what is the amount of salt in the bucket after

(a) 1 minute?   **(b)** 10 minutes?   **(c)** 60 minutes?
(d) 1000 minutes?   **(e)** a very, very long time?

36. Consider the following very simple model of blood cholesterol levels based on the fact that cholesterol is manufactured by the body for use in the construction of cell walls and is absorbed from foods containing cholesterol: Let $C(t)$ be the amount of cholesterol in the blood of a particular person at time t (in milligrams per deciliter). Then

$$\frac{dC}{dt} = k_1(C_0 - C) + k_2E,$$

where

$$C_0 = \text{the person's natural cholesterol level,}$$
$$k_1 = \text{production parameter,}$$
$$E = \text{daily rate at which cholesterol is eaten, and}$$
$$k_2 = \text{absorption parameter.}$$

(a) Suppose $C_0 = 200$, $k_1 = 0.1$, $k_2 = 0.1$, $E = 400$, and $C(0) = 150$. What will the person's cholesterol level be after 2 days on this diet?

(b) With the initial conditions as above, what will the person's cholesterol level be after 5 days on this diet?

(c) What will the person's cholesterol level be after a long time on this diet?

(d) High levels of cholesterol in the blood are known to be a risk factor for heart disease. Suppose that, after a long time on the high cholesterol diet described above, the person goes on a very low cholesterol diet, so E changes to $E = 100$. (The initial cholesterol level at the starting time of this diet is the result of part (c).) What will the person's cholesterol level be after 1 day on the new diet, after 5 days on the new diet, and after a very long time on the new diet?

(e) Suppose the person stays on the high cholesterol diet but takes drugs that block some of the uptake of cholesterol from food, so k_2 changes to $k_2 = 0.075$. With the cholesterol level from part (c), what will the person's cholesterol level be after 1 day, after 5 days, and after a very long time?

37. A cup of hot chocolate is initially 170° F and is left in a room with an ambient temperature of 70° F. Suppose that at time $t = 0$ it is cooling at a rate of 20° per minute.

    (a) Assume that Newton's law of cooling applies: The rate of cooling is proportional to the difference between the current temperature and the ambient temperature. Write an initial-value problem that models the temperature of the hot chocolate.

    (b) How long does it take the hot chocolate to cool to a temperature of 110° F?

38. In the mixing problem in this section, we had to make a choice of dependent variable. We used the amount of sugar as the dependent variable, but we could have used the concentration of sugar as the dependent variable. If $S(t)$ is the total amount of sugar in the vat at time t, then the concentration at time t is given by $C(t) = S(t)/100$ and is measured in tablespoons per gallon. Write a differential equation modeling the assumptions in the section using $C(t)$ as the dependent variable.

39. Use the techniques of this section to solve the differential equation in Exercise 38. Are there any equilibrium solutions for this differential equation?

40. Suppose you are having a dinner party for a large group of people, and you decide to make 2 gallons of chili. The recipe calls for 2 teaspoons of hot sauce per gallon, but you misread the instructions and put in 2 tablespoons of hot sauce per gallon. (Since each tablespoon is 3 teaspoons, you have put in 6 teaspoons per gallon, which is a total of 12 teaspoons of hot sauce in the chili.) You don't want to throw the chili out because there isn't much else to eat (and some people like hot chili), so you serve the chili anyway. However, as each person takes some chili, you fill up the pot with beans and tomatoes without hot sauce until the concentration of hot sauce agrees with the recipe. Suppose the guests take 1 cup of chili per minute from the pot (there are 16 cups in a gallon), how long will it take to get the chili back to the recipe's concentration of hot sauce? How many cups of chili will have been taken from the pot?

41. Suppose Ms. Lee is buying a new house and must borrow \$150,000. She wants a 20-year mortgage and she has two choices. She can either borrow money at 7% per year with no points, or she can borrow the money at 6.85% per year with a charge of 3 points. (A "point" is a fee of 1% of the loan amount that the borrower pays the lender at the beginning of the loan. For example, a mortgage with 3 points requires Ms. Lee to pay \$4,500 extra to get the loan.) As an approximation, we assume that interest is compounded and payments are made continuously. Let

$$M(t) = \text{amount owed at time } t \text{ (measured in years)},$$
$$i = \text{annual interest rate, and}$$
$$p = \text{annual payment.}$$

Then the model for the amount owed is

$$\frac{dM}{dt} = iM - p.$$

(a) How much does Ms. Lee pay in each case?

(b) Which is a better deal over the entire time of the loan (assuming Ms. Lee does not invest the money she would have paid in points)?

(c) If Ms. Lee can invest the $4,500 she would have paid in points for the second mortgage at 5% compounded continuously, which is the better deal?

1.3 QUALITATIVE TECHNIQUE: SLOPE FIELDS

Finding an analytic expression (in other words, finding a formula) for a solution to a differential equation is often a useful way to describe a solution of a differential equation. However, there are other ways to describe solutions, and these alternative representations are frequently easier to understand and use. In this section we focus on geometric techniques for representing solutions, and we develop a method for visualizing the graphs of the solutions to the differential equation

$$\frac{dy}{dt} = f(t, y).$$

The Geometry of $dy/dt = f(t, y)$

If the function $y(t)$ is a solution of the equation $dy/dt = f(t, y)$ and if its graph passes through the point (t_1, y_1) where $y_1 = y(t_1)$, then the differential equation says that the derivative dy/dt at $t = t_1$ is given by the number $f(t_1, y_1)$. Geometrically, this equality of dy/dt at $t = t_1$ with $f(t_1, y_1)$ means that the slope of the tangent line to the graph of $y(t)$ at the point (t_1, y_1) is $f(t_1, y_1)$ (see Figure 1.9). Note that there is nothing special about the point (t_1, y_1) other than the fact that it is a point on the graph of the solution $y(t)$. The equality of dy/dt and $f(t, y)$ must hold for all t for which $y(t)$ satisfies the differential equation. In other words, the values of the right-hand side of the differential equation yield the slopes of the tangents at all points on the graph of $y(t)$ (see Figure 1.10).

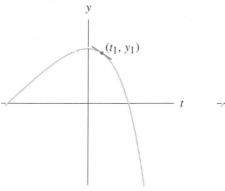

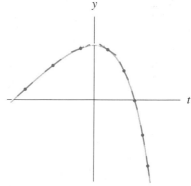

Figure 1.9
Slope of the tangent at the point (t_1, y_1) is given by the value of $f(t_1, y_1)$.

Figure 1.10
If $y = y(t)$ is a solution, then the slope of any tangent must equal $f(t, y)$.

Slope Fields

This simple geometric observation leads to our main device for the visualization of the solutions to a first-order differential equation

$$\frac{dy}{dt} = f(t, y).$$

If we are given the function $f(t, y)$, we obtain a rough idea of the graphs of the solutions to the differential equation by sketching its corresponding **slope field**. We make this sketch by selecting points in the ty-plane and computing the numbers $f(t, y)$ at these points. At each point (t, y) selected, we use $f(t, y)$ to draw a minitangent line whose slope is $f(t, y)$ (see Figure 1.11). These minitangent lines are also called **slope marks**. Once we have a lot of slope marks, we can visualize the graphs of the solutions. For example, consider the differential equation

$$\frac{dy}{dt} = y - t.$$

In other words, the right-hand side of the differential equation is given by the function $f(t, y) = y - t$. To get some practice with the idea of a slope field, we sketch its slope field by hand at a small number of points. Then we discuss a computer-generated version of this slope field.

Generating slope fields by hand is tedious, so we consider only the nine points in the ty-plane. For example, at the point $(t, y) = (1, -1)$, we have $f(t, y) = f(1, -1) = -1 - 1 = -2$. Therefore we sketch a "small" line segment with slope -2 centered at the point $(1, -1)$ (see Figure 1.12). To sketch the slope field for all nine points, we use the function $f(t, y)$ to compute the appropriate slopes. The results are summarized in Table 1.2. Once we have these values, we use them to give a sparse sketch of the slope field for this equation (see Figure 1.12).

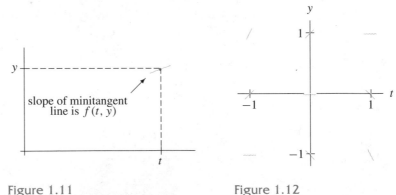

Figure 1.11
The slope of the minitangent at the point (t, y) is determined by the right-hand side $f(t, y)$ of the differential equation.

Figure 1.12
A "sparse" slope field generated from Table 1.2.

Sketching slope fields is best done using a computer. Figure 1.13 is a sketch of the slope field for this equation over the region $-3 \leq t \leq 3$ and $-3 \leq y \leq 3$ in the ty-plane. We calculated values of the function $f(t, y)$ over 25×25 points (625 points) in that region.

Table 1.2

Selected slopes corresponding to the differential equation $dy/dt = y - t$

(t, y)	$f(t, y)$	(t, y)	$f(t, y)$	(t, y)	$f(t, y)$
$(-1, 1)$	2	$(0, 1)$	1	$(1, 1)$	0
$(-1, 0)$	1	$(0, 0)$	0	$(1, 0)$	-1
$(-1, -1)$	0	$(0, -1)$	-1	$(1, -1)$	-2

A glance at this slope field suggests that the graph of one solution is a diagonal line passing through the points $(-1, 0)$ and $(0, 1)$. Solutions corresponding to initial conditions that are below this line seem to increase until they reach an absolute maximum. Solutions corresponding to initial conditions that are above the line seem to increase more and more rapidly.

In fact, in Section 1.8 we will learn an analytic technique for finding solutions of this equation. We will see that the general solution consists of the family of functions

$$y(t) = t + 1 + ce^t,$$

where c is an arbitrary constant. (At this point it is important to emphasize that, even though we have not studied the technique that gives us these solutions, we can still check to see whether these functions are indeed solutions. If $y(t) = t + 1 + ce^t$, then $dy/dt = 1 + ce^t$. Also $f(t, y) = y - t = (t + 1 + ce^t) - t = 1 + ce^t$. Hence all these functions are solutions.)

In Figure 1.14 we sketch the graphs of these functions with $c = -2, -1, 0, 1, 2, 3$. Note that each of these graphs is tangent to the slope field. Also note that, if $c = 0$, the graph is a straight line whose slope is 1. It goes through the points $(-1, 0)$ and $(0, 1)$.

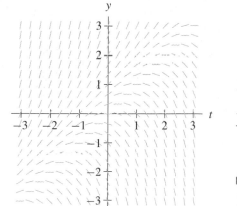

Figure 1.13
A computer-generated version of the slope field for $dy/dt = y - t$.

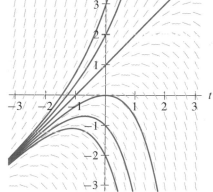

Figure 1.14
The graphs of six solutions to $dy/dt = y - t$ superimposed on its slope field.

Important Special Cases

From an analytic point of view, differential equations of the forms

$$\frac{dy}{dt} = f(t) \quad \text{and} \quad \frac{dy}{dt} = f(y)$$

are somewhat easier to consider than more complicated equations because they are separable. The geometry of their slope fields is equally special.

Slope fields for $dy/dt = f(t)$

If the right-hand side of the differential equation in question is solely a function of t, or in other words, if

$$\frac{dy}{dt} = f(t),$$

the slope at any point is the same as the slope of any other point with the same t-coordinate (see Figure 1.15).

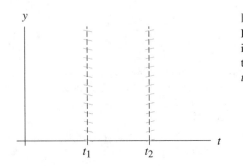

Figure 1.15
If the right-hand side of the differential equation is a function of t alone, then the slope marks in the slope field are determined solely by their t-coordinate.

Geometrically, this implies that all of the slope marks on each vertical line are parallel. Whenever a slope field has this geometric property for all vertical lines throughout the domain in question, we know that the corresponding differential equation is really an equation of the form

$$\frac{dy}{dt} = f(t).$$

(Note that finding solutions to this type of differential equation is the same thing as finding an antiderivative of $f(t)$ in calculus.)

For example, consider the slope field shown in Figure 1.16. We generated this slope field from the equation

$$\frac{dy}{dt} = 2t,$$

and from calculus we know that

$$y(t) = \int 2t \, dt = t^2 + c,$$

where c is the constant of integration. Hence the general solution of the differential equation consists of functions of the form

$$y(t) = t^2 + c.$$

In Figure 1.17 we have superimposed graphs of such solutions on this field. Note that all of these graphs simply differ by a vertical translation. If one graph is tangent to the slope field, we can get infinitely many graphs — all tangent to the slope field — by translating the original graph either up or down.

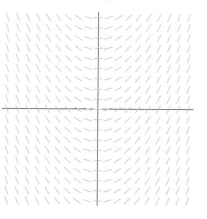

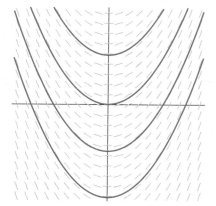

Figure 1.16
A slope field with parallel slopes along vertical lines.

Figure 1.17
Graphs of solutions to $dy/dt = 2t$ superimposed on its slope field.

Slope fields for autonomous equations
In the case of an autonomous differential equation

$$\frac{dy}{dt} = f(y),$$

the right-hand side of the equation does not depend on the independent variable t. The slope field in this case is also somewhat special. Here, the slopes that correspond to two different points with the same y-coordinate are equal. That is, $f(t_1, y) = f(t_2, y) = f(y)$ since the right-hand side of the differential equation depends only on y. In other words, the slope field of an autonomous equation is parallel along each horizontal line (see Figure 1.18).

For example, the slope field for the autonomous equation

$$\frac{dy}{dt} = 4y(1 - y)$$

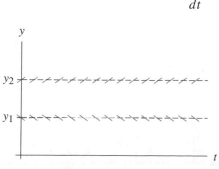

Figure 1.18
If the right-hand side of the differential equation is a function of y alone, then the slope marks in the slope field are determined solely by their y-coordinate.

is given in Figure 1.19. Note that, along each horizontal line, the slope marks are parallel. In fact, if $0 < y < 1$, then dy/dt is positive, and the tangents suggest that a solution with $0 < y < 1$ is increasing. On the other hand, if $y < -1$ or if $y > 1$, then dy/dt is negative and any solution with either $y < -1$ or $y > 1$ is decreasing.

We have equilibrium solutions at $y = 0$ and at $y = 1$ since the right-hand side of the differential equation vanishes along these lines. The slope field is horizontal all along these two horizontal lines, and therefore we know that these lines are the graphs of solutions. Solutions whose graphs are between these two lines are increasing. Solutions that are above the line $y = 1$ or that are below the line $y = 0$ are decreasing (see Figure 1.20).

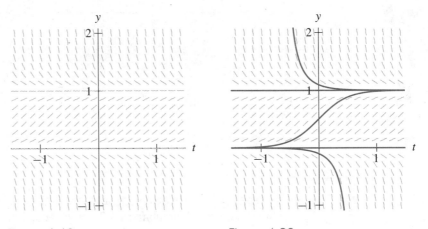

Figure 1.19
The slope field for $dy/dt = 4y(1 - y)$.

Figure 1.20
The graphs of five solutions superimposed on the slope field for $dy/dt = 4y(1 - y)$.

The fact that autonomous equations produce slope fields that are parallel along horizontal lines indicates that we can get infinitely many solutions from one solution simply by translating the graph of the given solution left or right (see Figure 1.21). We will make extensive use of this simple geometric observation about the solutions to autonomous equations in Section 1.6.

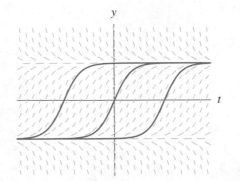

Figure 1.21
The graphs of three solutions to an autonomous equation. Note that each is a horizontal translate of the others.

Analytic versus Qualitative Analysis

For the autonomous equation

$$\frac{dy}{dt} = 4y(1 - y),$$

we could have used the analytic techniques of the previous section to find explicit formulas for the solutions. In fact, we can perform all of the required integrations to determine the general solution (see Exercise 15 on page 32). However, these integrations are complicated, and the formulas that result are by no means easy to interpret. This points out the power of geometric and qualitative methods for solving differential equations. With very little work, we gain a lot of insight into the behavior of solutions. Although we cannot use qualitative methods to answer specific questions, such as what the exact value of the solution is at any given time, we can use these methods to understand the long-term behavior of a solution.

These ideas are especially important if the differential equation in question cannot be handled by analytic techniques. As an example, consider the differential equation

$$\frac{dy}{dt} = e^{y^2/10} \sin^2 y.$$

This equation is autonomous and hence separable. To solve this equation analytically, we must evaluate the integrals

$$\int \frac{dy}{e^{y^2/10} \sin^2 y} = \int dt.$$

However, the integral on the left-hand side cannot be evaluated so easily. Thus we resort to qualitative methods. The right-hand side of this differential equation is positive except if $y = n\pi$ for any integer n. These special lines correspond to equilibrium solutions of the equation. Between these equilibria, solutions must always increase. From the slope field, we expect that their graphs either lie on one of the horizontal lines $y = n\pi$ or increase from one of these lines to the next higher as $t \to \infty$ (see Figure 1.22). Hence we can predict the long-term behavior of the solutions even though we cannot explicitly solve the equation.

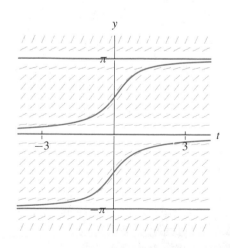

Figure 1.22

Slope field and graphs of solutions for

$$\frac{dy}{dt} = e^{y^2/10} \sin^2 y.$$

The lines $y = n\pi$ correspond to equilibrium solutions, and between these equilibria, solutions are increasing.

Although the computer pictures of solutions of this differential equation are convincing, some subtle questions remain. For example, how do we *really* know that these pictures are correct? In particular, for $dy/dt = e^{y^2/10} \sin^2 y$, how do we know that the graphs of solutions do not cross the horizontal lines that are the graphs of the equilibrium solutions (see Figure 1.22)? Such a solution could not cross these lines at a nonzero angle since we know that the tangent line to the solution must be horizontal. But what prevents certain solutions from crossing these lines tangentially and then continuing to increase?

In the differential equation

$$\frac{dy}{dt} = 4y(1 - y)$$

we can eliminate these questions because we can evaluate all of the integrals and check the accuracy of the pictures using analytic techniques. But using analytic techniques to check our qualitative analysis does not work if we cannot find explicit solutions. Besides, having to resort to analytic techniques to check the qualitative results defeats the purpose of using these methods in the first place. In Section 1.5 we discuss powerful theorems that answer many of these questions without undue effort.

The Mixing Problem Revisited

Recall that in the previous section (page 31) we found precise analytic solutions for the differential equation

$$\frac{dS}{dt} = \frac{2000 - 3S}{100},$$

where S describes the amount of sugar in a vat at time t. We found that the general solution of this equation was

$$S(t) = ce^{-0.03t} + \frac{2000}{3},$$

where c is an arbitrary constant.

Using the slope field of this equation, we can easily derive a qualitative description of these solutions. In Figure 1.23, we display the slope field and graphs of selected solutions. Note that, as expected, the slope field is horizontal if $S = 2000/3$, the equilibrium solution. Slopes are positive if $S < 2000/3$ and negative if $S > 2000/3$. So we expect solutions to tend toward the equilibrium solution as t increases. This qualitative analysis indicates that, no matter what the initial amount of sugar, the amount of sugar in the vat tends to $2000/3$ as $t \to \infty$. Of course, we obtain the same information by taking the limit of the general solution as $t \to \infty$, but it is nice to see the same result in a geometric setting. Furthermore, in other examples, taking such a limit may not be as easy as in this case, but qualitative methods may still be used to determine the long-term behavior of the solutions.

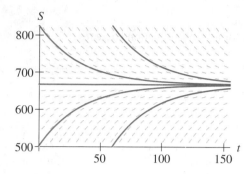

Figure 1.23
The slope field and graphs of a few solutions of

$$\frac{dS}{dt} = \frac{2000 - 3S}{100}.$$

An RC Circuit

The simple electric circuit pictured in Figure 1.24 contains a capacitor, a resistor, and a voltage source. The behavior of the resistor is specified by a positive parameter R (the "resistance"), and the behavior of the capacitor is specified by a positive parameter C (the "capacitance"). The input voltage across the voltage source at time t is denoted by $V(t)$. This voltage source could be a constant source such as a battery, or it could be a source that varies with time such as alternating current. In any case, we consider $V(t)$ to be a function that is specified by the circuit designer. In other words, it is part of the design of the circuit.

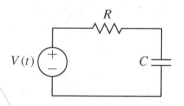

Figure 1.24
Circuit diagram with resistor, capacitor, and voltage source.

The quantities that specify the behavior of the circuit at a particular time t are the current $i(t)$ and the voltage across the capacitor $v_c(t)$. In this example we are interested in the voltage $v_c(t)$ across the capacitor. From the theory of electric circuits, we know that $v_c(t)$ satisfies the differential equation

$$RC\frac{dv_c}{dt} + v_c = V(t).$$

If we rewrite this in our standard form $dv_c/dt = f(t, v_c)$, we have

$$\frac{dv_c}{dt} = \frac{V(t) - v_c}{RC}.$$

We use slope fields to visualize solutions for four different types of voltage sources $V(t)$. (If you don't know anything about electric circuits, don't worry; Paul, Bob, and Glen don't either. In examples like this, all we need to do is accept the differential equation and "go with it.")

Zero input

If $V(t) = 0$ for all t, the equation becomes

$$\frac{dv_c}{dt} = \frac{-v_c}{RC}.$$

A sample slope field for a particular choice of R and C is given in Figure 1.25. We see clearly that all solutions "decay" toward $v_c = 0$ as t increases. If there is no voltage source, the voltage across the capacitor $v_c(t)$ decays to zero. This prediction for the voltage agrees with what we obtain analytically since the general solution of this equation is $v_c(t) = v_0 e^{-t/RC}$, where v_0 is the initial voltage across the capacitor. (Note that this equation is essentially the same as the exponential growth model that we studied in Section 1.1, and consequently, we can solve it analytically by either guessing the correct form of a solution or by separating variables.)

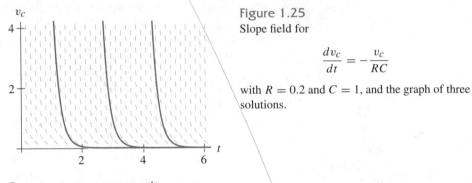

Figure 1.25
Slope field for

$$\frac{dv_c}{dt} = -\frac{v_c}{RC}$$

with $R = 0.2$ and $C = 1$, and the graph of three solutions.

Constant nonzero voltage source

Suppose $V(t)$ is a nonzero constant K for all t. The equation for voltage across the capacitor becomes

$$\frac{dv_c}{dt} = \frac{K - v_c}{RC}.$$

This equation is autonomous with one equilibrium solution at $v_c = K$. The slope field for this equation shows that all solutions tend toward this equilibrium as t increases (see Figure 1.26). Given any initial voltage $v_c(0)$ across the capacitor, the voltage $v_c(t)$ tends to the value $v = K$ as time increases.

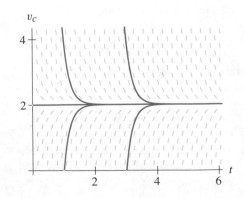

Figure 1.26
Slope field for

$$\frac{dv_c}{dt} = \frac{K - v_c}{RC}$$

for $R = 0.2$, $C = 1$, and $K = 2$, and the graphs of several solutions with different initial conditions.

We could find a formula for the general solution by separating variables and integrating, but we leave this as an exercise.

On-off voltage source

Suppose $V(t) = K > 0$ for $0 \le t < 3$, but at $t = 3$, this voltage is "turned off." Then $V(t) = 0$ for $t > 3$. Our differential equation is

$$\frac{dv_c}{dt} = \frac{V(t) - v_c}{RC} = \begin{cases} \dfrac{K - v_c}{RC} & \text{for } 0 \le t < 3; \\[2mm] \dfrac{-v_c}{RC} & \text{for } t > 3. \end{cases}$$

The right-hand side is given by two different formulas depending on the value of t. We can see this in the slope field for this equation (see Figure 1.27). It resembles Figures 1.25 and 1.26 pasted together along $t = 3$. Since the differential equation is not defined at $t = 3$, we must add an additional assumption to our model. We assume that the voltage $v_c(t)$ is a continuous function at $t = 3$.

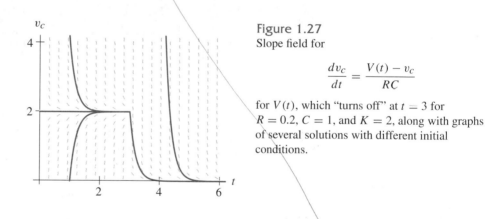

Figure 1.27
Slope field for

$$\frac{dv_c}{dt} = \frac{V(t) - v_c}{RC}$$

for $V(t)$, which "turns off" at $t = 3$ for $R = 0.2$, $C = 1$, and $K = 2$, along with graphs of several solutions with different initial conditions.

The particular solution with the initial condition $v_c(0) = K$ is constant for $t < 3$, but for $t > 3$ it decays exponentially. Solutions with $v_c(0) \ne K$ move toward K for $t < 3$, but then decay toward zero for $t > 3$. We could find formulas for the solutions by first finding the solution for $t < 3$, then starting over for $t > 3$ (see Section 1.2), but we again leave this as an exercise.

Periodic on-off voltage source

Suppose $V(t)$ alternates periodically between the values K and zero every three seconds. That is,

$$V(t) = \begin{cases} K & \text{for } 0 \le t < 3; \\ 0 & \text{for } 3 < t < 6; \\ K & \text{for } 6 < t < 9; \\ \quad \vdots \end{cases}$$

This corresponds to someone switching the source voltage off every three seconds and back on three seconds later. The slope field for the differential equation

$$\frac{dv_c}{dt} = \frac{V(t) - v_c}{RC}$$

is given in Figure 1.28. Parts of the slope fields in Figures 1.25 and 1.26 are patched together every three seconds. The solutions are also patched together from these two equations. When $V(t) = K$, the solution approaches the equilibrium value $v_c = K$, and when $V(t) = 0$, the solution decays toward zero.

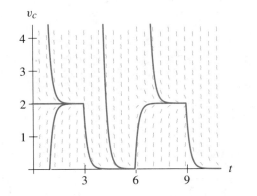

Figure 1.28
Slope field for

$$\frac{dv_c}{dt} = \frac{V(t) - v_c}{RC}$$

where $V(t)$ alternates between K and zero every three seconds for $R = 0.2$, $C = 1$, and $K = 2$, along with the graphs of several solutions with different initial conditions.

Combining Qualitative with Quantitative Results

When only knowledge of the qualitative behavior of the solution is required, sketches of solutions obtained from slope fields can sometimes suffice. In other applications it is necessary to know the exact value (or almost exact value) of the solution with a given initial condition. In these situations analytic and/or numerical methods can't be avoided. But even then, it is nice to have a picture of what solutions look like.

EXERCISES FOR SECTION 1.3

In Exercises 1–6, sketch the slope fields for the given differential equation. (You may use a computer or compute the slopes by hand. If you compute by hand, start with a grid of points (t, y) with $t = -2, -1, 0, 1, 2$ and $y = -2, -1, 0, 1, 2$. Then add more slope marks at other important points in the ty-plane. If you use a computer, you should compute the slope field at many more points to get a more accurate picture.)

1. $\dfrac{dy}{dt} = t^2 - t$

2. $\dfrac{dy}{dt} = 1 - y$

3. $\dfrac{dy}{dt} = y + t + 1$

4. $\dfrac{dy}{dt} = t^2 + 1$

5. $\dfrac{dy}{dt} = 2y(1 - y)$

6. $\dfrac{dy}{dt} = 4y^2$

In Exercises 7–10, a differential equation and its associated slope field are given. For each equation,

(a) sketch a number of different solutions on the slope field, and

(b) describe briefly the behavior of the solution with $y(0) = 1/2$ as t increases.

7. $\dfrac{dy}{dt} = 3y(1 - y)$

8. $\dfrac{dy}{dt} = 2y - t$

9. $\dfrac{dy}{dt} = \left(y + \tfrac{1}{2}\right)(y + t)$

10. $\dfrac{dy}{dt} = (t + 1)y$

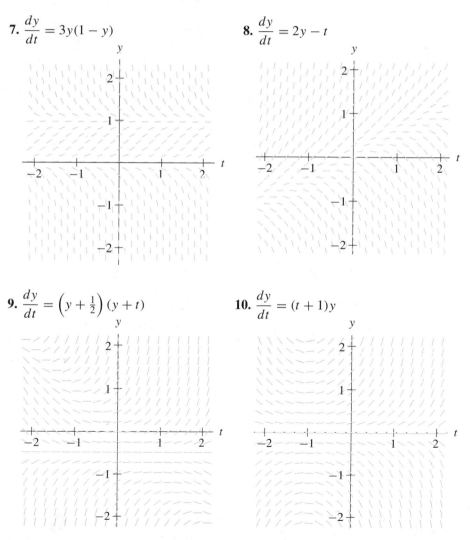

11. Eight differential equations and four slope fields are given below. Determine the equation that corresponds to each slope field and state briefly how you know your choice is correct. You should do this exercise without using technology.

(i) $\dfrac{dy}{dt} = t - 1$ (ii) $\dfrac{dy}{dt} = 1 - y^2$ (iii) $\dfrac{dy}{dt} = y^2 - t^2$ (iv) $\dfrac{dy}{dt} = 1 - t$

(v) $\dfrac{dy}{dt} = 1 - y$ (vi) $\dfrac{dy}{dt} = t^2 - y^2$ (vii) $\dfrac{dy}{dt} = 1 + y$ (viii) $\dfrac{dy}{dt} = y^2 - 1$

(a)   **(b)**

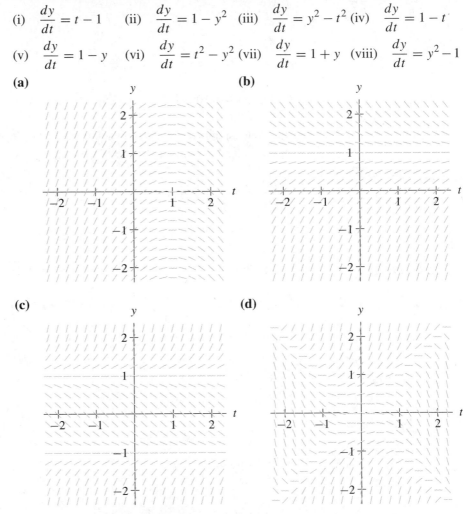

(c)   **(d)**

12. Consider the autonomous differential equation

$$\frac{dS}{dt} = S^3 - 2S^2 + S.$$

(a) By hand, give a rough sketch of the slope field.

(b) Using this sketch, sketch the graphs of the solutions $S(t)$ with the initial conditions $S(0) = 0$, $S(0) = 1/2$, $S(1) = 1/2$, $S(0) = 3/2$, and $S(0) = -1/2$.

13. Suppose we know that the function $f(t, y)$ is continuous and that $f(t, 3) = -1$ for all t.

(a) What does this information tell us about the slope field for the differential equation $dy/dt = f(t, y)$?

(b) What can we conclude about solutions $y(t)$ of $dy/dt = f(t, y)$? For example, if $y(0) < 3$, can $y(t) \to \infty$ as t increases?

14. Suppose we know that the graph below is the graph of the right-hand side $f(t)$ of the differential equation $dy/dt = f(t)$. Give a rough sketch of the slope field that corresponds to this differential equation.

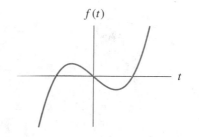

15. Suppose we know that the graph below is the graph of the right-hand side $f(y)$ of the differential equation $dy/dt = f(y)$. Give a rough sketch of the slope field that corresponds to this differential equation.

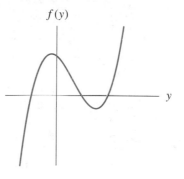

16. Suppose we know that the graph below is the graph of a solution to $dy/dt = f(t)$.

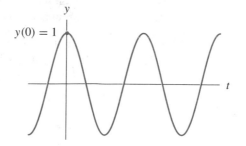

(a) How much of the slope field can you sketch from this information? [*Hint*: Note that the differential equation depends only on t.]

(b) What can you say about the solution with $y(0) = 2$? (For example, can you sketch the graph of this solution?)

17. Suppose we know that the graph below is the graph of a solution to $dy/dt = f(y)$.

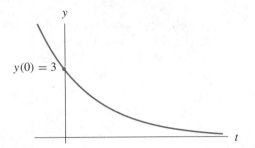

(a) How much of the slope field can you sketch from this information? [*Hint*: Note that the equation is autonomous.]

(b) What can you say about the solution with $y(0) = 2$? Sketch this solution.

18. Suppose the constant function $y(t) = 2$ for all t is a solution of the differential equation

$$\frac{dy}{dt} = f(t, y).$$

(a) What does this tell you about the slope field? In other words, how much of the slope field can you sketch using this information?

(b) What does this tell you about solutions?

Exercises 19–23 refer to the RC circuit discussed in this section. The differential equation for the voltage v_c across the capacitor is

$$\frac{dv_c}{dt} = \frac{V(t) - v_c}{RC}.$$

19. Find the formula for the general solution of the RC circuit equation above if the voltage source is constant for all time. In other words, $V(t) = K$ for all t. (Your solution will contain the three parameters, R, C, and K, along with a constant that depends on the initial condition.)

20. Find the solution for the voltage $v_c(t)$ with initial value $v_c(0) = 1$ in the RC circuit equation given above if the voltage source $V(t)$ is the step function given by

$$V(t) = \begin{cases} K & \text{for } 0 \leq t < 3; \\ 0 & \text{for } t > 3. \end{cases}$$

Your answer should contain the three parameters R, C, and K.

21. Given the source voltage $V(t) = 2t$ and the parameter values $R = 0.2$ and $C = 1$,

(a) sketch the slope field,

(b) sketch the graph of the solution with the initial condition $v_c(0) = 0$, and

(c) sketch the graph of the solution with the initial condition $v_c(0) = 3$.

22. Given the source voltage

$$V(t) = \begin{cases} 0 & \text{for } 0 \le t < 1; \\ 2 & \text{for } t \ge 1; \end{cases}$$

and the parameter values $R = 0.2$ and $C = 1$,

(a) sketch the slope field,

(b) sketch the graph of the solution with the initial condition $v_c(0) = 0$, and

(c) sketch the graph of the solution with the initial condition $v_c(0) = 3$.

23. Given the source voltage

$$V(t) = \begin{cases} 2t & \text{for } 0 \le t < 1; \\ 2 & \text{for } t \ge 1; \end{cases}$$

and parameter values $R = 0.2$ and $C = 1$,

(a) sketch the slope field,

(b) sketch the graph of the solution with the initial condition $v_c(0) = 0$,

(c) sketch the graph of the solution with the initial condition $v_c(0) = 3$, and

(d) discuss in a few sentences the differences between the solutions for this differential equation and the solutions for the differential equations in Exercises 21 and 22.

24. Suppose that a population can be accurately modeled by the logistic equation

$$\frac{dp}{dt} = 0.4p \left(1 - \frac{p}{30}\right).$$

(Note that the growth-rate parameter is 0.4 and the carrying capacity is 30.) Suppose that, at time $t = 5$, a disease is introduced into the population that kills 25% of the population per year. To adjust the model, we change the differential equation to

$$\frac{dp}{dt} = \begin{cases} 0.4p \left(1 - \dfrac{p}{30}\right) & \text{for } 0 \le t < 5; \\ 0.4p \left(1 - \dfrac{p}{30}\right) - 0.25p & \text{for } t > 5. \end{cases}$$

(a) Sketch the slope field for this equation.

(b) Using the slope field, sketch the graphs of a few representative solutions to this equation.

(c) Find formulas for the solutions of this equation for initial conditions $p(0) = 30$ and $p(0) = 20$.

(d) In a few sentences, describe the behavior of the solutions with initial conditions $p(0) = 30$ and $p(0) = 20$. (You can use either the sketches from the slope field or the formulas, but give a qualitative description of the solutions.)

1.4 NUMERICAL TECHNIQUE: EULER'S METHOD

The geometric concept of a slope field as discussed in the previous section is closely related to a fundamental numerical method for approximating solutions to a differential equation. Given an initial-value problem

$$\frac{dy}{dt} = f(t, y), \quad y(t_0) = y_0,$$

we can get a rough idea of the graph of its solution by first sketching the slope field in the ty-plane and then, starting at the initial value (t_0, y_0), sketching the solution by drawing a graph that is tangent to the slope field at each point along the graph. In this section we describe a numerical procedure that automates this idea. Using a computer or a calculator, we obtain numbers and graphs that approximate solutions to initial-value problems.

Numerical methods provide quantitative information about solutions even if we cannot find their formulas. There is also the advantage that most of the work can be done by machine. The disadvantage is that we obtain only approximations, not precise solutions. If we remain aware of this fact and are prudent, numerical methods become powerful tools for the study of differential equations. It is not uncommon to turn to numerical methods even when it is possible to find formulas for solutions. (Most of the graphs of solutions of differential equations in this text were drawn using numerical approximations even when formulas were available.)

The numerical technique that we discuss in this section is called *Euler's method*. A more detailed discussion of the accuracy of Euler's method as well as other numerical methods is given in Chapter 7.

Stepping along the Slope Field

To describe Euler's method, we begin with the initial-value problem

$$\frac{dy}{dt} = f(t, y), \quad y(t_0) = y_0.$$

Since we are given $f(t, y)$, we can plot its slope field in the ty-plane. The idea of the method is to start at the point (t_0, y_0) in the slope field and take tiny steps dictated by the tangents in the slope field.

We begin by choosing a (small) **step size** Δt. The slope of the approximate solution is updated every Δt units of t. In other words, for each step, we move Δt units along the t-axis. The size of Δt determines the accuracy of the approximate solution as well as the number of computations that are necessary to obtain the approximation.

Starting at (t_0, y_0), our first step is to the point (t_1, y_1) where $t_1 = t_0 + \Delta t$ and (t_1, y_1) is the point on the line through (t_0, y_0) with slope given by the slope field at (t_0, y_0) (see Figure 1.29). At (t_1, y_1) we repeat the procedure. Taking a step whose size along the t-axis is Δt and whose direction is determined by the slope field at (t_1, y_1), we reach the new point (t_2, y_2). The new time is given by $t_2 = t_1 + \Delta t$ and (t_2, y_2) is on the line segment that starts at (t_1, y_1) and has slope $f(t_1, y_1)$. Continuing, we use the slope field at the point (t_k, y_k) to determine the next point (t_{k+1}, y_{k+1}). The sequence of values $y_0, y_1, y_2, \ldots$ serves as an approximation to the solution at the times

$t_0, t_1, t_2, \ldots$. Geometrically, we think of the method as producing a sequence of tiny line segments connecting (t_k, y_k) to (t_{k+1}, y_{k+1}) (see Figure 1.30). Basically, we are stitching together little pieces of the slope field to form a graph that approximates our solution curve.

This method uses tangent line segments, given by the slope field, to approximate the graph of the solution. Consequently, at each stage we make a slight error (see Figure 1.30). Hopefully, if the step size is sufficiently small, these errors do not get out of hand as we continue to step, and the resulting graph is close to the desired solution.

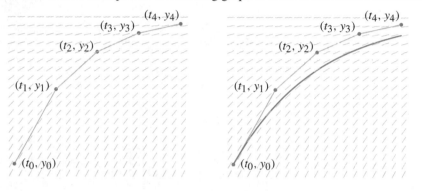

Figure 1.29
Stepping along the slope field.

Figure 1.30
The graph of a solution and its approximation obtained using Euler's method.

Euler's Method

To put Euler's method into practice, we need a formula for determining (t_{k+1}, y_{k+1}) from (t_k, y_k). Finding t_{k+1} is easy. We specify the step size Δt at the outset, so

$$t_{k+1} = t_k + \Delta t.$$

To obtain y_{k+1} from (t_k, y_k), we use the differential equation. We know that the slope of the solution to the equation $dy/dt = f(t, y)$ at the point (t_k, y_k) is $f(t_k, y_k)$, and Euler's method uses this slope to determine y_{k+1}. In fact, the method determines the point (t_{k+1}, y_{k+1}) by assuming that it lies on the line through (t_k, y_k) with slope $f(t_k, y_k)$ (see Figure 1.31).

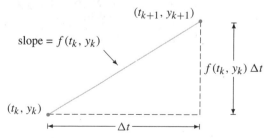

Figure 1.31
Euler's method uses the slope at the point (t_k, y_k) to approximate the solution for $t_k \leq t \leq t_{k+1}$.

Now we can use our basic knowledge of slopes to determine y_{k+1}. The formula for the slope of a line gives

$$\frac{y_{k+1} - y_k}{t_{k+1} - t_k} = f(t_k, y_k).$$

Since $t_{k+1} = t_k + \Delta t$, the denominator $t_{k+1} - t_k$ is just Δt, and therefore we have

$$\frac{y_{k+1} - y_k}{\Delta t} = f(t_k, y_k)$$

$$y_{k+1} - y_k = f(t_k, y_k)\,\Delta t$$

$$y_{k+1} = y_k + f(t_k, y_k)\,\Delta t.$$

This is the formula for Euler's method (see Figures 1.31 and 1.32).

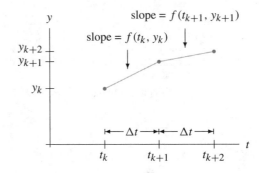

Figure 1.32
Two successive steps of Euler's method.

Euler's method for $\dfrac{dy}{dt} = f(t, y)$

Given the initial condition $y(t_0) = y_0$ and the step size Δt, compute the point (t_{k+1}, y_{k+1}) from the preceding point (t_k, y_k) as follows:

1. Use the differential equation to compute the slope $f(t_k, y_k)$.

2. Calculate the next point (t_{k+1}, y_{k+1}) using the formulas

$$t_{k+1} = t_k + \Delta t$$

and

$$y_{k+1} = y_k + f(t_k, y_k)\Delta t.$$

Approximating an Autonomous Equation

To illustrate Euler's method, we first use it to approximate the solution to a differential equation whose solution we already know. In this way, we are able to compare the approximation we obtain to the known solution. Consequently, we are able to gain some insight into the effectiveness of the method in addition to seeing how it is implemented.

Consider the initial-value problem

$$\frac{dy}{dt} = 2y - 1, \quad y(0) = 1.$$

This equation is separable, and by separating and integrating we obtain the solution

$$y(t) = \frac{e^{2t} + 1}{2}.$$

In this example, $f(t, y) = 2y - 1$, so Euler's method is given by

$$y_{k+1} = y_k + (2y_k - 1)\Delta t.$$

To illustrate the method, we start with a relatively large step size of $\Delta t = 0.1$ and approximate the solution over the interval $0 \le t \le 1$. In order to approximate the solution over an interval whose length is 1 with a step size of 0.1, we must compute ten iterations of the method. The initial condition $y(0) = 1$ provides the initial value $y_0 = 1$. Given $\Delta t = 0.1$, we have $t_1 = t_0 + 0.1 = 0 + 0.1 = 0.1$. We compute the y-coordinate for the first step by

$$y_1 = y_0 + (2y_0 - 1)\Delta t = 1 + (1)\,0.1 = 1.1.$$

Thus the first point (t_1, y_1) on the graph of the approximate solution is $(0.1, 1.1)$.

To compute the y-coordinate y_2 for the second step, we now use y_1 rather than y_0. That is,

$$y_2 = y_1 + (2y_1 - 1)\Delta t = 1.1 + (1.2)\,0.1 = 1.22,$$

and the second point for our approximate solution is $(t_2, y_2) = (0.2, 1.22)$.

Continuing this procedure, we obtain the results given in Table 1.3. After ten steps, we obtain the approximation of $y(1)$ by $y_{10} = 3.596$. (Different machines use different algorithms for rounding numbers, so you may get slightly different results on your computer or calculator. Keep this fact in mind whenever you compare the numerical results presented in this book with the results of your calculation.) Since we know that

$$y(1) = \frac{e^2 + 1}{2} \approx 4.195,$$

the approximation y_{10} is off by slightly less than 0.6. This is not a very good approximation, but we'll soon see how to avoid this (usually). The reason for the error can be seen by looking at the graph of the solution and its approximation. The slope field for

Table 1.3
Euler's method (to three decimal places) for $dy/dt = 2y - 1$, $y(0) = 1$ with $\Delta t = 0.1$

k	t_k	y_k	$f(t_k, y_k)$
0	0	1	1
1	0.1	1.100	1.20
2	0.2	1.220	1.44
3	0.3	1.364	1.73
4	0.4	1.537	2.07
5	0.5	1.744	2.49
6	0.6	1.993	2.98
7	0.7	2.292	3.58
8	0.8	2.650	4.30
9	0.9	3.080	5.16
10	1.0	3.596	

this differential equation always lies below the graph (see Figure 1.33), so we expect our approximation to come up short.

Using a smaller step size usually reduces the error, but more computations must be done to approximate the solution over the same interval. For example, if we halve the step size in this example ($\Delta t = 0.05$), then we must calculate twice as many steps, since $t_1 = 0.05$, $t_2 = 0.1, \ldots, t_{20} = 1.0$. Again we start with $(t_0, y_0) = (0, 1)$ as specified by the initial condition. However, with $\Delta t = 0.05$, we obtain

$$y_1 = y_0 + (2y_0 - 1)\Delta t = 1 + (1)\,0.05 = 1.05.$$

This step yields the point $(t_1, y_1) = (0.05, 1.05)$ on the graph of our approximate solution. For the next step, we compute

$$y_2 = y_1 + (2y_1 - 1)\Delta t = 1.05 + (1.1)\,0.05 = 1.105.$$

Now we have the point $(t_2, y_2) = (1.1, 1.105)$. This type of calculation gets tedious fairly quickly, but luckily calculations such as these are perfect for a computer or a calculator. For $\Delta t = 0.05$, the results of Euler's method are given in Table 1.4.

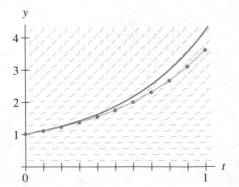

Figure 1.33
The graph of the solution to

$$\frac{dy}{dt} = 2y - 1$$

with $y(0) = 1$ and the approximation produced by Euler's method with $\Delta t = 0.1$.

Table 1.4

Euler's method (to three decimal places) for $dy/dt = 2y - 1$, $y(0) = 1$ with $\Delta t = 0.05$

k	t_k	y_k	$f(t_k, y_k)$
0	0	1	1
1	0.05	1.050	1.100
2	0.10	1.105	1.210
3	0.15	1.166	1.331
⋮	⋮	⋮	⋮
19	0.95	3.558	6.116
20	1.00	3.864	

If we carefully compare the final results of our two computations, we see that, with $\Delta t = 0.1$, we approximate $y(1) \approx 4.195$ with $y_{10} = 3.596$. With $\Delta t = 0.05$, we approximate $y(1)$ with $y_{20} = 3.864$. The error in the first approximation is slightly less than 0.6, whereas the error in the second approximation is 0.331. Roughly speaking we halve the error by halving the step size. This type of improvement is typical of Euler's method. (We will be much more precise about how the error in Euler's method is related to the step size in Chapter 7.)

With the even smaller step size of $\Delta t = 0.01$, we must do much more work since we need 100 steps to go from $t = 0$ to $t = 1$. However, in the end, we obtain a much better approximation to the solution (see Table 1.5).

This example illustrates the typical trade off that occurs with numerical methods. There are always decisions to be made such as the choice of the step size Δt. Lowering Δt often results in a better approximation — at the expense of more computation.

Table 1.5

Euler's method (to four decimal places) for $dy/dt = 2y - 1$, $y(0) = 1$ with $\Delta t = 0.01$

k	t_k	y_k	$f(t_k, y_k)$
0	0	1	1
1	0.01	1.0100	1.0200
2	0.02	1.0202	1.0404
3	0.03	1.0306	1.0612
⋮	⋮	⋮	⋮
98	0.98	3.9817	6.9633
99	0.99	4.0513	7.1026
100	1.00	4.1223	

A Nonautonomous Example

Note that it is the value $f(t_k, y_k)$ of the right-hand side of the differential equation at (t_k, y_k) that determines the next point (t_{k+1}, y_{k+1}). The last example was an autonomous differential equation, so the right-hand side $f(t_k, y_k)$ depended only on y_k.

Table 1.6
Euler's method for $dy/dt = -2ty^2$, $y(0) = 1$ with $\Delta t = 1/2$

k	t_k	y_k	$f(t_k, y_k)$
0	0	1	0
1	1/2	1	−1
2	1	1/2	−1/2
3	3/2	1/4	−3/16
4	2	5/32	

However, if the differential equation is nonautonomous, the value of t_k also plays a role in the computations.

To illustrate Euler's method applied to a nonautonomous equation, we consider the initial-value problem

$$\frac{dy}{dt} = -2ty^2, \quad y(0) = 1.$$

This differential equation is also separable, and we can separate variables to obtain the solution

$$y(t) = \frac{1}{1+t^2}.$$

We use Euler's method to approximate this solution over the interval $0 \le t \le 2$. The value of the solution at $t = 2$ is $y(2) = 1/5$. Again, it is interesting to see how close we come to this value with various choices of Δt. The formula for Euler's method is

$$y_{k+1} = y_k + f(t_k, y_k)\,\Delta t = y_k - (2t_k y_k^2)\Delta t$$

with $t_0 = 0$ and $y_0 = 1$. We begin by approximating the solution from $t = 0$ to $t = 2$ using just four steps. This involves so few computations that we can perform the arithmetic "by hand." To cover an interval of length 2 in four steps, we must use $\Delta t = 2/4 = 1/2$. The entire calculation is displayed in Table 1.6. Note that we end up approximating the exact value $y(2) = 1/5 = 0.2$ by $y_4 = 5/32 = 0.15625$. Figure 1.34 shows the graph of the solution as compared to the results of Euler's method over this interval.

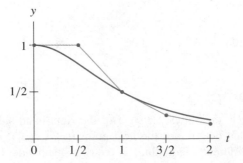

Figure 1.34
The graph of the solution to the initial-value problem

$$\frac{dy}{dt} = -2ty^2, \quad y(0) = 1,$$

and the approximation produced by Euler's method with $\Delta t = 1/2$.

Table 1.7

Euler's method (to four decimal places) for $dy/dt = -2ty^2$, $y(0) = 1$ with $\Delta t = 0.1$

k	t_k	y_k
0	0	1
1	0.1	1.0000
2	0.2	0.9800
3	0.3	0.9416
$\vdots$	$\vdots$	$\vdots$
19	1.9	0.2101
20	2.0	0.1933

Table 1.8

Euler's method (to six decimal places) for $dy/dt = -2ty^2$, $y(0) = 1$ with $\Delta t = 0.001$

k	t_k	y_k
0	0	1
1	0.001	1.000000
2	0.002	0.999998
3	0.003	0.999994
$\vdots$	$\vdots$	$\vdots$
1999	1.999	0.200097
2000	2	0.199937

As before, choosing smaller values of Δt yields better approximations. For example, if $\Delta t = 0.1$, the Euler approximation of the exact value $y(2) = 0.2$ is $y_{20} = 0.1933$. If $\Delta t = 0.001$, we need to compute 2000 steps, but the approximation improves to $y_{2000} = 0.199937$ (see Tables 1.7 and 1.8).

Note that the convergence of the approximation to the actual value is slow. We computed 2000 steps and obtained an answer that is only accurate to three decimal places. In Chapter 7, we present more complicated algorithms for numerical approximation of solutions. Although the algorithms are more complicated from a conceptual point of view, they obtain better accuracy with less computation.

An RC Circuit with Periodic Input

Recall from Section 1.3 that the voltage v_c across the capacitor in the simple circuit shown in Figure 1.35 is given by the differential equation

$$\frac{dv_c}{dt} = \frac{V(t) - v_c}{RC}$$

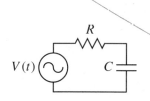

Figure 1.35

Circuit diagram with resistor, capacitor, and voltage source.

where R is the resistance, C is the capacitance, and $V(t)$ is the source or input voltage. We have seen how we can use slope fields to give a qualitative sketch of solutions. Using Euler's method we can also obtain numerical approximations of the solutions.

Suppose we consider a circuit where $R = 0.5$ and $C = 1$. (The usual units are "ohms" for resistance and "farads" for capacitance. We choose these numbers so that the numbers in the solution work out nicely. A 1 farad capacitor would be extremely large.) Then the differential equation is

$$\frac{dv_c}{dt} = \frac{V(t) - v_c}{0.5} = 2(V(t) - v_c).$$

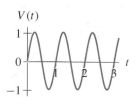

Figure 1.36

Graph of $V(t) = \sin(2\pi t)$, the input voltage.

To understand how the voltage v_c varies if the voltage source $V(t)$ is periodic in time, we consider the case where $V(t) = \sin(2\pi t)$. Consequently, the voltage oscillates between -1 and 1 once each unit of time (see Figure 1.36). The differential equation is now

$$\frac{dv_c}{dt} = -2v_c + 2\sin(2\pi t).$$

From the slope field for this equation (see Figure 1.37), we might predict that the solutions oscillate. Using Euler's method applied to this equation for several different initial conditions, we see that the solutions do indeed oscillate. In addition, we see that they also approach each other and collect around a single solution (see Figure 1.38). This uniformity of long-term behavior is not so easily predicted from the slope field alone.

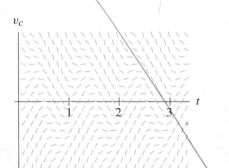

Figure 1.37
Slope field for
$dv_c/dt = -2v_c + 2\sin(2\pi t)$.

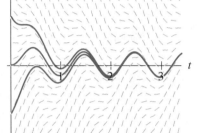

Figure 1.38
Graphs of approximate solutions to
$dv_c/dt = -2v_c + 2\sin(2\pi t)$ obtained
using Euler's method.

Errors in Numerical Methods

By its very nature, any numerical approximation scheme is inaccurate. For instance, in each step of Euler's method, we almost always make an error of some sort. These errors can accumulate and sometimes lead to disastrously wrong approximations. As an example, consider the differential equation

$$\frac{dy}{dt} = e^t \sin y.$$

There are equilibrium solutions for this equation if $\sin y = 0$. In other words, any constant function of the form $y(t) = n\pi$ for any integer n is a solution.

Using the initial value $y(0) = 5$ and a step size $\Delta t = 0.1$, Euler's method yields the approximation graphed in Figure 1.39. It seems that something must be wrong. At first, the solution tends toward the equilibrium solution $y(t) = \pi$, but then just before $t = 5$ something strange happens. The graph of the approximation jumps dramatically. If we lower Δt to 0.05, we still find erratic behavior, although t is slightly greater than 5 before this happens (see Figure 1.40).

The difficulty arises in Euler's method for this equation because of the term e^t on the right-hand side. It becomes very large as t increases, and consequently slopes in the slope field are quite large for large t. Even a very small step in the t-direction throws us far from the actual solution.

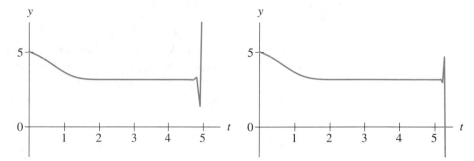

Figure 1.39
Euler's method applied to $dy/dt = e^t \sin y$ with $\Delta t = 0.1$

Figure 1.40
Euler's method applied to $dy/dt = e^t \sin y$ with $\Delta t = 0.05$.

This problem is typical of the use of numerics in the study of differential equations. Numerical methods, when they work, work beautifully. But they sometimes fail. We must always be aware of this possibility and be ready with an alternate approach. In the next section we present theoretical results that help identify when numerical approximations have gone awry.

The Big Three

We have now introduced examples of all three of the fundamental methods for attacking differential equations — the analytic, the numeric, and the qualitative approaches. Which method is the best depends both on the differential equation in question and on what we want to know about the solutions. Often all three methods "work," but a great deal of labor can be saved if we think first about which method gives the most direct route to the information we need.

Stephen Smale (1930–) is one of the founders of modern day dynamical systems theory. In the mid 1960s, Smale began to advocate taking a more qualitative approach to the study of differential equations, as we do in this book. Using this approach, he was among the first mathematicians to encounter and analyze a "chaotic" dynamical system. Since this discovery, scientists have found that many important physical systems exhibit chaos.

Smale's research has spanned many disciplines, including economics, theoretical computer science, mathematical biology, as well as many subareas of mathematics. In 1966 he was awarded the Fields Medal, the equivalent of the Nobel Prize in mathematics. After retiring from the University of California, Berkeley in 1995, he assumed a Research Professorship at the City University of Hong Kong, where he currently teaches.

EXERCISES FOR SECTION 1.4

In Exercises 1–8, use Euler's method with the given step size Δt to approximate the solution to the given initial-value problem over the time interval specified. Your answer should include a table of the approximate values of the dependent variable. It should also include a sketch of the graph of the approximate solution.

1. $\dfrac{dy}{dt} = 2y + 1, \quad y(0) = 3, \quad 0 \le t \le 2, \quad \Delta t = 0.5$

2. $\dfrac{dy}{dt} = t - y^2, \quad y(0) = 1, \quad 0 \le t \le 1, \quad \Delta t = 0.25$

3. $\dfrac{dy}{dt} = y^2 - 2y + 1, \quad y(0) = 2, \quad 0 \le t \le 2, \quad \Delta t = 0.5$

4. $\dfrac{dy}{dt} = \sin y, \quad y(0) = 1, \quad 0 \le t \le 3, \quad \Delta t = 0.5$

5. $\dfrac{dw}{dt} = (3 - w)(w + 1), \quad w(0) = 4, \quad 0 \le t \le 5, \quad \Delta t = 1.0$

6. $\dfrac{dw}{dt} = (3 - w)(w + 1), \quad w(0) = 0, \quad 0 \le t \le 5, \quad \Delta t = 0.5$

7. $\dfrac{dy}{dt} = e^{(2/y)}, \quad y(0) = 2, \quad 0 \le t \le 2, \quad \Delta t = 0.5$

8. $\dfrac{dy}{dt} = e^{(2/y)}, \quad y(1) = 2, \quad 1 \le t \le 3, \quad \Delta t = 0.5$

9. Compare your answers to Exercises 7 and 8 and explain your observations.

10. Compare your answers to Exercises 5 and 6. Is Euler's method doing a good job in this case? What would you do to avoid the difficulties that arise in this case?

11. Do a qualitative analysis of the solution of the initial-value problem in Exercise 6 and compare your conclusions with your results in Exercise 6. What's wrong with the approximate solutions given by Euler's method?

12. Consider the initial-value problem

$$\frac{dy}{dt} = \sqrt{y}, \quad y(0) = 1.$$

Using Euler's method, compute three different approximate solutions corresponding to $\Delta t = 1.0$, 0.5, and 0.25 over the interval $0 \le t \le 4$. Graph all three solutions. What predictions do you make about the actual solution to the initial-value problem?

13. Consider the initial-value problem

$$\frac{dy}{dt} = 2 - y, \quad y(0) = 1.$$

Using Euler's method, compute three different approximate solutions corresponding to $\Delta t = 1.0$, 0.5, and 0.25 over the interval $0 \le t \le 4$. Graph all three solutions.

What predictions do you make about the actual solution to the initial-value problem? How do the graphs of these approximate solutions relate to the graph of the actual solution? Why?

In Exercises 14–17, we consider the RC circuit model equation from the text

$$\frac{dv_c}{dt} = \frac{V(t) - v_c}{RC}.$$

Suppose $V(t) = e^{-0.1t}$ (the voltage source $V(t)$ is decaying exponentially). If $R = 0.2$ and $C = 1$, use Euler's method to compute values of the solutions with the given initial conditions over the interval $0 \leq t \leq 10$.

14. $v_c(0) = 0$ **15.** $v_c(0) = 2$ **16.** $v_c(0) = -2$ **17.** $v_c(0) = 4$

18. Consider the polynomial $p(y) = -y^3 - 2y + 2$. Using appropriate technology,
    (a) sketch the slope field for $dy/dt = p(y)$,
    (b) sketch the graphs of some of the solutions using the slope field,
    (c) describe the relationship between the roots of $p(y)$ and the solutions of the differential equation, and
    (d) using Euler's method, approximate the real root(s) of $p(y)$ to three decimal places.

19. Consider the polynomial $p(y) = -y^3 + 4y + 1$. Using appropriate technology,
    (a) sketch the slope field for $dy/dt = p(y)$,
    (b) sketch the graphs of some of the solutions using the slope field,
    (c) describe the relationship between the roots of $p(y)$ and solutions of the differential equation, and
    (d) using Euler's method, approximate the real root(s) of $p(y)$ to three decimal places. [*Hint*: Euler's method also works with a negative Δt.]

1.5 EXISTENCE AND UNIQUENESS OF SOLUTIONS

What Does It Mean to Say Solutions Exist?

We have seen analytic, qualitative, and numerical techniques for studying solutions of differential equations. One problem we have not considered is: How do we know there are solutions? Although this may seem to be a subtle and abstract question, it is also a question of great importance. If solutions to the differential do not exist, then there is no use trying to find or approximate them. More important, if a differential equation is supposed to model a physical system but the solutions of the differential equation do not exist, then we should have serious doubts about the validity of the model.

To get an idea of what is meant by the existence of solutions, consider the algebraic equation

$$2x^5 - 10x + 5 = 0.$$

A solution to this equation is a value of x for which the left-hand side is zero. In other words, it is a root of the fifth-degree polynomial $2x^5 - 10x + 5$. We can easily compute that the value of $2x^5 - 10x + 5$ is -3 if $x = 1$ and 13 if $x = -1$. Since polynomials are continuous, there must therefore be a value of x between -1 and 1 for which the left-hand side is zero.

So we have established the existence of at least one solution of this equation between -1 and 1. We did not construct this value of x or approximate it (other than to say it is between -1 and 1). Unfortunately, there is no "quadratic formula" for finding roots of fifth-degree polynomials, so there is no way to write down the exact values of the solutions of this equation. But this does not make us any less sure of the existence of this solution. The point here is that we can discuss the existence of solutions without having to compute them.

It is also possible that there is more than one solution between -1 and 1. In other words, the solution may not be unique.

In the same way, if we are given an initial-value problem

$$\frac{dy}{dt} = f(t, y), \quad y(0) = y_0,$$

we can ask if there is a solution. This is a different question than asking what the solution is or what its graph looks like. We can say there is a solution without having any knowledge of a formula for the solution, just as we can say that the algebraic equation above has a solution between -1 and 1 without knowing its exact or even approximate value.

Existence

Luckily, the question of existence of solutions for differential equations has been extensively studied and some very good results have been established. For our purposes, we will use the standard existence theorem.

EXISTENCE THEOREM   Suppose $f(t, y)$ is a continuous function in a rectangle of the form $\{(t, y) \mid a < t < b, c < y < d\}$ in the ty-plane. If (t_0, y_0) is a point in this rectangle, then there exists an $\epsilon > 0$ and a function $y(t)$ defined for $t_0 - \epsilon < t < t_0 + \epsilon$ that solves the initial-value problem

$$\frac{dy}{dt} = f(t, y), \quad y(t_0) = y_0. \quad \blacksquare$$

This theorem says that as long as the function on the right-hand side of the differential equation is reasonable, solutions exist. (It does not rule out the possibility that solutions exist even if $f(t, y)$ is not a nice function, but it doesn't guarantee it either.) This is reassuring. When we are studying the solutions of a reasonable initial-value problem, there is something there to study.

Extendability

Given an initial-value problem

$$\frac{dy}{dt} = f(t, y), \quad y(t_0) = y_0,$$

the Existence Theorem guarantees that there is a solution. If you read the theorem very closely (with a lawyer's eye for loopholes), you will see that the solution may have a very small domain of definition. The theorem says that there exists an $\epsilon > 0$ and that the solution has a domain that includes the open interval $(t_0 - \epsilon, t_0 + \epsilon)$. The ϵ may be very, very small, so although the theorem guarantees that a solution exists, it may be defined for only a very short interval of time.

Unfortunately, this is a serious but necessary restriction. Consider the initial-value problem

$$\frac{dy}{dt} = 1 + y^2, \quad y(0) = 0.$$

The slopes in the slope field for this equation increase in steepness very rapidly as y increases (see Figure 1.41). Hence, dy/dt increases more and more rapidly as $y(t)$ increases. There is a danger that solutions "blow up" (tend to infinity very quickly) as t increases. By looking at solutions sketched by the slope field, we can't really tell if the solutions blow up in finite time or if they stay finite for all time, so we try analytic methods.

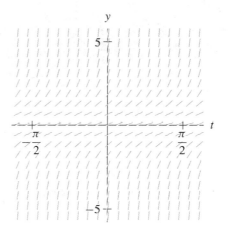

Figure 1.41
The slope field for the equation
$dy/dt = 1 + y^2$.

This is an autonomous equation, so we can separate variables and integrate as usual. We have

$$\int \frac{1}{1 + y^2} \, dy = \int dt.$$

Integration yields

$$\arctan y = t + c,$$

where c is an arbitrary constant. Therefore

$$y(t) = \tan(t + c),$$

which is the general solution of the differential equation. Using the initial value

$$0 = y(0) = \tan(0 + c),$$

we find $c = 0$ (or $c = n\pi$ for any integer n). Thus, the particular solution is $y(t) = \tan t$, and the domain of definition for this particular solution is $-\pi/2 < t < \pi/2$.

As we see from Figure 1.42, our fears were well founded. The graph of this particular solution has vertical asymptotes at $t = \pm\pi/2$. As t approaches $\pi/2$ from the left and $-\pi/2$ from the right, the solution blows up. If this differential equation were a model of a physical system, then we would expect the system to break as t approaches $\pi/2$.

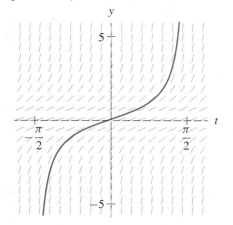

Figure 1.42
The graph of the solution $y(t) = \tan t$ with initial condition $y(0) = 0$ along with the slope field for

$$\frac{dy}{dt} = 1 + y^2.$$

As t approaches $\pi/2$ from the left, $y(t) = \tan t \to \infty$. As t approaches $-\pi/2$ from the right, $y(t) = \tan t \to -\infty$.

Uniqueness

When dealing with initial-value problems of the form

$$\frac{dy}{dt} = f(t, y), \quad y(t_0) = y_0,$$

we have always said "consider *the* solution." By the Existence Theorem we know there is a solution, but how do we know there is only one? Why don't we have to say "consider *a* solution" instead of "consider *the* solution?" In other words, how do we know the solution is unique?

Knowing that the solution to an initial-value problem is unique is very valuable from both theoretical and practical standpoints. If solutions weren't unique, then we would have to worry about all possible solutions, even when we were doing numerical or qualitative work. Different solutions could give completely different predictions for how the system would work. Fortunately, there is a very good theorem that guarantees that solutions of initial-value problems are unique.

UNIQUENESS THEOREM   Suppose $f(t, y)$ and $\partial f/\partial y$ are continuous functions in a rectangle of the form $\{(t, y) \mid a < t < b, \ c < y < d\}$ in the ty-plane. If (t_0, y_0) is a point in this rectangle and if $y_1(t)$ and $y_2(t)$ are two functions that solve the initial-value problem

$$\frac{dy}{dt} = f(t, y), \quad y(t_0) = y_0$$

for all t in the interval $t_0 - \epsilon < t < t_0 + \epsilon$ (where ϵ is some positive number), then

$$y_1(t) = y_2(t)$$

for $t_0 - \epsilon < t < t_0 + \epsilon$. That is, the solution to the initial-value problem is *unique*. ∎

Before giving applications of the Uniqueness Theorem we should emphasize that both the Existence and the Uniqueness Theorems have *hypotheses* — conditions that must hold before we can use these theorems. Before we say that the solution of an initial-value problem

$$\frac{dy}{dt} = f(t, y), \quad y(t_0) = y_0$$

exists and is unique, we must check that $f(t, y)$ satisfies the necessary hypotheses.

Often we lump these two theorems together (using the more restrictive hypotheses of the Uniqueness Theorem) and refer to the combination as the **Existence and Uniqueness Theorem**.

Lack of Uniqueness

It is pretty difficult to construct an example of a reasonable differential equation that does not have solutions. However, it is not so hard to find examples where $f(t, y)$ is a reasonable function but where uniqueness fails. (Of course, in these examples, either $f(t, y)$ or $\partial f / \partial y$ is not continuous.)

For example, consider the differential equation

$$\frac{dy}{dt} = 3y^{2/3}.$$

The right-hand side is a continuous function on the entire ty-plane. Unfortunately, the partial derivative of $y^{2/3}$ with respect to y fails to exist if $y = 0$, so the Uniqueness Theorem does not tell us anything about the number of solutions to an initial-value problem of the form $y(t_0) = 0$.

Let's apply the qualitative and analytic techniques that we have already discussed. First, if we look for equilibrium solutions, we see that the function $y(t) = 0$ for all t is a solution. Second, we note that this equation is separable, so we separate variables and obtain

$$\int y^{-2/3} \, dy = \int 3 \, dt.$$

Integrating, we obtain the solutions

$$y(t) = (t + c)^3$$

where c is an arbitrary constant.

Now consider the initial-value problem

$$\frac{dy}{dt} = 3y^{2/3}, \quad y(0) = 0.$$

One solution is the equilibrium solution $y_1(t) = 0$ for all t. However, a second solution is obtained by setting $c = 0$ after we separate variables. We have $y_2(t) = t^3$.

Consequently, we have two solutions, $y_1(t) = 0$ and $y_2(t) = t^3$, to the same initial-value problem (see Figure 1.43).

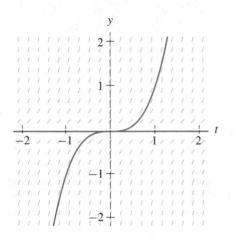

Figure 1.43
The slope field and the graphs of two solutions to the initial-value problem

$$\frac{dy}{dt} = 3y^{2/3}, \quad y(0) = 0.$$

This differential equation does not satisfy the hypothesis of the Uniqueness Theorem if $y = 0$. Note that we have two different solutions whose graphs intersect at $(0, 0)$.

Applications of the Uniqueness Theorem

The Uniqueness Theorem says that two solutions to the same initial-value problem are identical. This result is reassuring, but it may not sound useful in a practical sense. Here we discuss a few examples to illustrate why this theorem is, in fact, very useful.

Suppose $y_1(t)$ and $y_2(t)$ are both solutions of a differential equation

$$\frac{dy}{dt} = f(t, y),$$

where $f(t, y)$ satisfies the hypotheses of the Uniqueness Theorem. If for some t_0 we have $y_1(t_0) = y_2(t_0)$, then both of these functions are solutions of the same initial-value problem

$$\frac{dy}{dt} = f(t, y), \quad y(t_0) = y_1(t_0) = y_2(t_0).$$

The Uniqueness Theorem guarantees that $y_1(t) = y_2(t)$, at least for all t for which both solutions are defined. We can paraphrase the Uniqueness Theorem as:

"If two solutions are ever in the same place at the same time, then they are the same function."

This form of Uniqueness Theorem is very valuable as the following examples show.

Role of equilibrium solutions
Consider the initial-value problem

$$\frac{dy}{dt} = \frac{(y^2 - 4)(\sin^2 y^3 + \cos y - 2)}{2}, \quad y(0) = \frac{1}{2}.$$

Finding the explicit solution to this equation is not easy because, even though the equation is autonomous and hence separable, the integrals involved are very difficult (try them). On the other hand, if $y = 2$, the right-hand side of the equation vanishes. Thus the constant function $y_1(t) = 2$ is an equilibrium solution for this equation.

Suppose $y_2(t)$ is the solution to the differential equation that satisfies the initial condition $y_2(0) = 1/2$. The Uniqueness Theorem implies that $y_2(t) < 2$ for all t since the graph of $y_2(t)$ cannot touch the line $y = 2$, which is the graph of the constant solution $y_1(t)$ (see Figure 1.44).

This observation is not a lot of information about the solution of the initial-value problem with $y(0) = 1/2$. On the other hand, we didn't have to do a lot of work to get this information. Identifying $y_1(t) = 2$ as a solution is pretty easy, and the rest follows from the Uniqueness Theorem. By doing a little bit of work, we get some information. If all we care about is how large the solution of the original initial-value problem can possibly become, then the fact that it is bounded above by $y = 2$ may suffice. If we need more detailed information, we must look more carefully at the equation.

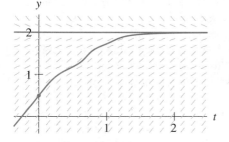

Figure 1.44
The slope field and the graphs of two solutions of

$$\frac{dy}{dt} = \frac{(y^2 - 4)(\sin^2 y^3 + \cos y - 2)}{2}.$$

Although it looks as if these two graphs agree for $t > 2$, the Uniqueness Theorem tells us that there is always a little space between them.

Comparing solutions

We can also use this technique to obtain information about solutions by comparing them to "known" solutions. For example, consider the differential equation

$$\frac{dy}{dt} = \frac{(1 + t)^2}{(1 + y)^2}.$$

It is easy to check that $y_1(t) = t$ is a solution to the differential equation with the initial condition $y_1(0) = 0$. If $y_2(t)$ is the solution satisfying the initial condition $y(0) = -0.1$, then $y_2(0) < y_1(0)$, so $y_2(t) < y_1(t)$ for all t. Thus $y_2(t) < t$ for all t (see Figure 1.45). Again, this is only a little bit of information about the solution of the initial-value problem, but then we only did a little work.

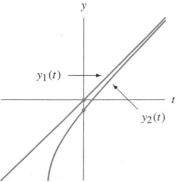

Figure 1.45
The graphs of two solutions $y_1(t)$ and $y_2(t)$ of

$$\frac{dy}{dt} = \frac{(1 + t)^2}{(1 + y)^2}.$$

The graph of the solution $y_1(t)$ that satisfies the initial condition $y_1(0) = 0$ is the diagonal line, and the graph of the solution that satisfies the initial condition $y_2(0) = -0.1$ must lie below the line.

Uniqueness and qualitative analysis

In some cases we can use the Uniqueness Theorem and some qualitative information to give more exact information about solutions. For example, consider the differential equation

$$\frac{dy}{dt} = (y-2)(y+1).$$

The right-hand side of this autonomous equation is the function $f(y) = (y-2)(y+1)$. Note that $f(2) = f(-1) = 0$. Thus $y = 2$ and $y = -1$ are equilibrium solutions (see the slope field in Figure 1.46). By the Existence and Uniqueness Theorem, any solution $y(t)$ with an initial condition $y(0)$ that satisfies $-1 < y(0) < 2$ must also satisfy $-1 < y(t) < 2$ for all t.

In this case we can say even more about these solutions. For example, consider the solution with the initial condition $y(0) = 0.5$. Not only do we know that $-1 < y(t) < 2$ for all t, but because this equation is autonomous, the sign of dy/dt depends only on the value of y. For $-1 < y < 2$, $dy/dt = f(y) < 0$. Hence the solution $y(t)$ with the initial condition $y(0) = 0.5$ satisfies $dy/dt < 0$ for all t. Consequently this solution is decreasing for all t.

Since the solution is decreasing for all t and since it always remains above $y = -1$, it is reasonable to guess that, as $t \to \infty$, $y(t) \to -1$. In fact this is precisely what happens. If $y(t)$ were to limit to any value y_0 larger than -1 as $t \to \infty$, then when t is very large, $y(t)$ must be close to y_0. But $f(y_0)$ is negative because $-1 < y_0 < 2$. So when $y(t)$ is close to y_0, we have dy/dt close to $f(y_0)$, which is negative, so the solution must continue to decrease past y_0. That is, solutions of this differential equation can be asymptotic only to the equilibrium solutions.

We can sketch the solution of this initial-value problem. For all t the graph is between the lines $y = -1$ and $y = 2$, and for all t it decreases (see Figure 1.47).

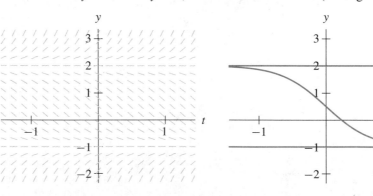

Figure 1.46
The slope field for $dy/dt = (y-2)(y+1)$.

Figure 1.47
Graphs of the equilibrium solutions and the solution with initial condition $y(0) = 0.5$ for $dy/dt = (y-2)(y+1)$.

Uniqueness and Numerical Approximation

As the preceding examples show, the Uniqueness Theorem gives us qualitative information concerning the behavior of solutions. We can use this information to check the behavior of numerical approximations of solutions. If numerical approximations of solutions violate the Uniqueness Theorem, then we are certain that something is wrong.

The graph of the Euler approximation to the solution of the initial-value problem

$$\frac{dy}{dt} = e^t \sin y, \quad y(0) = 5,$$

with $\Delta t = 0.05$ is shown in Figure 1.48. As noted in Section 1.4, the behavior seems erratic, and hence we are suspicious.

We can easily check that the constant function $y(t) = n\pi$ is a solution for any integer n and, hence by the Uniqueness Theorem, each solution is trapped between $y = n\pi$ and $y = (n + 1)\pi$ for some integer n. The approximations in Figure 1.48 violate this requirement. This confirms our suspicions that the numerical results in this case are not to be believed.

This equation is unusual because of the e^t term on the right-hand side. When t is large, the slopes of solutions become gigantic and hence Euler's method overshoots the true solution for even a very small step size.

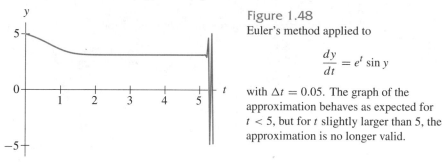

Figure 1.48
Euler's method applied to

$$\frac{dy}{dt} = e^t \sin y$$

with $\Delta t = 0.05$. The graph of the approximation behaves as expected for $t < 5$, but for t slightly larger than 5, the approximation is no longer valid.

EXERCISES FOR SECTION 1.5

In Exercises 1–4, we refer to a function f, but we do not provide its formula. However, we do assume that f satisfies the hypotheses of the Uniqueness Theorem in the entire ty-plane, and we do provide various solutions to the given differential equation. Finally, we specify an initial condition. Using the Uniqueness Theorem, what can you conclude about the solution to the equation with the given initial condition?

1. $\dfrac{dy}{dt} = f(t, y)$

$y_1(t) = 3$ for all t is a solution,
initial condition $y(0) = 1$

2. $\dfrac{dy}{dt} = f(y)$

$y_1(t) = 4$ for all t is a solution,
$y_2(t) = 2$ for all t is a solution,
$y_3(t) = 0$ for all t is a solution,
initial condition $y(0) = 1$

3. $\dfrac{dy}{dt} = f(t, y)$

$y_1(t) = t + 2$ for all t is a solution,
$y_2(t) = -t^2$ for all t is a solution,
initial condition $y(0) = 1$

4. $\dfrac{dy}{dt} = f(t, y)$

$y_1(t) = -1$ for all t is a solution,
$y_2(t) = 1 + t^2$ for all t is a solution,
initial condition $y(0) = 0$

In Exercises 5–8, an initial condition for the differential equation

$$\frac{dy}{dt} = (y-2)(y-3)y$$

is given. What does the Existence and Uniqueness Theorem say about the corresponding solution?

5. $y(0) = 4$ **6.** $y(0) = 3$ **7.** $y(0) = 1$ **8.** $y(0) = -1$

9. (a) Show that $y_1(t) = t^2$ and $y_2(t) = t^2 + 1$ are solutions to

$$\frac{dy}{dt} = -y^2 + y + 2yt^2 + 2t - t^2 - t^4.$$

    (b) Show that if $y(t)$ is a solution to the differential equation in part (a) and if $0 < y(0) < 1$, then $t^2 < y(t) < t^2 + 1$ for all t.

10. Consider the differential equation

$$\frac{dy}{dt} = y^{2/3}.$$

    (a) Show that $y_1(t) = 0$ for all t is a solution.

    (b) Show that $y_2(t) = t^3/27$ is a solution.

    (c) Verify that $y_1(0) = y_2(0)$ but that $y_1(t) \neq y_2(t)$ for all t. Why doesn't this example contradict the Uniqueness Theorem?

11. Consider a differential equation of the form $dy/dt = f(y)$ — an autonomous equation — and assume that the function $f(y)$ is continuously differentiable.

    (a) Suppose $y_1(t)$ is a solution and $y_1(t)$ has a local maximum at $t = t_0$. Let $y_0 = y_1(t_0)$. Show that $f(y_0) = 0$.

    (b) Use the information of part (a) to sketch the slope field along the line $y = y_0$ in the ty-plane.

    (c) Show that the constant function $y_2(t) = y_0$ is a solution (in other words, $y_2(t)$ is an equilibrium solution).

    (d) Show that $y_1(t) = y_0$ for all t.

    (e) Show that if a solution of $dy/dt = f(y)$ has a local minimum, then it is a constant function; that is, it also corresponds to an equilibrium solution.

12. (a) Show that

$$y_1(t) = \frac{1}{t-1} \quad \text{and} \quad y_2(t) = \frac{1}{t-2}$$

    are solutions of $dy/dt = -y^2$.

    (b) What can you say about solutions of $dy/dt = -y^2$ for which the initial condition $y(0)$ satisfies the inequality $-1 < y(0) < -1/2$? [*Hint:* You could find the general solution, but what information can you get from your answer to part (a) alone?]

13. Consider the differential equation

$$\frac{dy}{dt} = \frac{y}{t^2}.$$

(a) Show that the constant function $y_1(t) = 0$ is a solution.

(b) Show that there are infinitely many other functions that satisfy the differential equation, that agree with this solution when $t \leq 0$, but that are nonzero when $t > 0$. [*Hint*: You need to define these functions using language like "$y(t) = \ldots$ when $t \leq 0$ and $y(t) = \ldots$ when $t > 0$."]

(c) Why doesn't this example contradict the Uniqueness Theorem?

In Exercises 14–17, an initial-value problem is given.

(a) Find a formula for the solution.

(b) State the domain of definition of the solution.

(c) Describe what happens to the solution as it approachs the limits of its domain of definition. Why can't the solution be extended for more time?

14. $\dfrac{dy}{dt} = y^3, \quad y(0) = 1$

15. $\dfrac{dy}{dt} = \dfrac{1}{(y+1)(t-2)}, \quad y(0) = 0$

16. $\dfrac{dy}{dt} = \dfrac{1}{(y+2)^2}, \quad y(0) = 1$

17. $\dfrac{dy}{dt} = \dfrac{t}{y-2}, \quad y(-1) = 0$

18. We have emphasized that the Uniqueness Theorem does not apply to every differential equation. There are hypotheses that must be verified before we can apply the theorem. However, there is a temptation to think that, since models of "real-world" problems must obviously have solutions, we don't need to worry about the hypotheses of the Uniqueness Theorem when we are working with differential equations modeling the physical world. The following model illustrates the flaw in this assumption:

Suppose we wish to study the formation of raindrops in the atmosphere. We make the reasonable assumption that raindrops are approximately spherical. We also assume that the rate of growth of the volume of a raindrop is proportional to its surface area.

Let $v(t)$ be the volume of the raindrop at time t, and let $r(t)$ be its radius. We have

$$v = \tfrac{4}{3}\pi r^3$$

by the usual formula for the volume of a sphere. Therefore

$$r = \left(\frac{3v}{4\pi}\right)^{1/3}.$$

The surface area of the drop is given by $4\pi r^2$, which is therefore $3^{2/3}(4\pi)^{1/3}v^{2/3}$. Hence the differential equation that models the volume of the raindrop is

$$\frac{dv}{dt} = kv^{2/3},$$

where k is the product of the proportionality constant and $3^{2/3}(4\pi)^{1/3}$ (see Exercise 10 in this section).

(a) Why doesn't this equation satisfy the hypotheses of the Uniqueness Theorem?

(b) Give a physical interpretation of the fact that solutions to this equation with the initial condition $v(0) = 0$ are not unique. Does this model say anything about the way raindrops begin to form?

1.6 EQUILIBRIA AND THE PHASE LINE

Given a differential equation

$$\frac{dy}{dt} = f(t, y),$$

we can get an idea of how solutions behave by drawing slope fields and sketching their graphs or by using Euler's method and computing approximate solutions. Sometimes we can even derive explicit formulas for solutions and plot the results. All of these techniques require quite a bit of work, either numerical (computation of slopes or Euler's method) or analytic (integration).

In this section we consider differential equations where the right-hand side is independent of t — **autonomous** equations. For these differential equations, there are qualitative techniques that help us sketch the graphs of the solutions with less arithmetic than with other methods.

Autonomous Equations

Autonomous equations are differential equations of the form $dy/dt = f(y)$. In other words, the rate of change of the dependent variable can be expressed as a function of the dependent variable alone. Autonomous equations appear frequently as models for two reasons. First, many physical systems work the same way at any time. For example, a spring compressed the same amount at 10:00 AM and at 3:00 PM provides the same force. Second, for many systems, the time dependence "averages out" over the time scales being considered. For example, if we are studying how wolves and field mice interact, we might find that wolves eat many more field mice during the day than they do at night. However, if we are interested in how the wolf and mouse populations behave over a period of years or decades, then we can average the number of mice eaten by each wolf per week. We ignore the daily fluctuations.

We have already noticed that autonomous equations have slope fields that have a special form (see page 39 in Section 1.3). Because the right-hand side of the equation does not depend on t, the slope marks are parallel along horizontal lines in the ty-plane. That is, for an autonomous equation, two points with the same y-coordinate but different t-coordinates have the same slope marks (see Figure 1.49).

Hence there is a great deal of redundancy in the slope field of an autonomous equation. If we know the slope field along a single vertical line $t = t_0$, then we know the slope field in the entire ty-plane. So instead of drawing the entire slope field, we should be able to draw just one line containing the same information. This line is called the **phase line** for the autonomous equation.

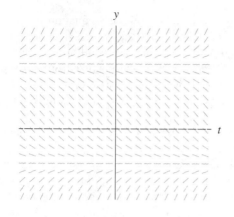

Figure 1.49
Slope field for the autonomous differential equation

$$\frac{dy}{dt} = (y - 2)(y + 1).$$

The slopes are parallel along horizontal lines.

Metaphor of the rope

Suppose you are given an autonomous differential equation

$$\frac{dy}{dt} = f(y).$$

Think of a rope hanging vertically and stretching infinitely far up and infinitely far down. The dependent variable y tells you a position on the rope (the rope is the y-axis). The function $f(y)$ gives a number for each position on the rope. Suppose the number $f(y)$ is actually printed on the rope at height y for every value of y. For example, at the height $y = 2.17$, the value $f(2.17)$ is printed on the rope.

Suppose that you are placed on the rope at height y_0 at time $t = 0$ and given the following instructions: Read the number that is printed on the rope and climb up or down the rope with velocity equal to that number. Climb up the rope if the number is positive or down the rope if the number is negative. (A large positive number means you climb up very quickly, whereas a negative number near zero means you climb down slowly.) As you move, continue to read the numbers on the rope and adjust your velocity so that it always agrees with the number printed on the rope.

If you follow this rather bizarre set of instructions, you will generate a function $y(t)$ that gives your position on the rope at time t. Your position at time $t = 0$ is $y(0) = y_0$ because that is where you were placed initially. The velocity of your motion dy/dt at time t will be given by the number on the rope, so $dy/dt = f(y(t))$ for all t. Hence, your position function $y(t)$ is a solution to the initial-value problem

$$\frac{dy}{dt} = f(y), \quad y(0) = y_0.$$

The phase line is a picture of this rope. Because it is tedious to record the numerical values of all the velocities, we only mark the phase line with the numbers where the velocity is zero and indicate the sign of the velocity on the intervals in between. The phase line provides qualitative information about the solutions.

Phase Line of a Logistic Equation

For example, consider the differential equation

$$\frac{dy}{dt} = (1 - y)y.$$

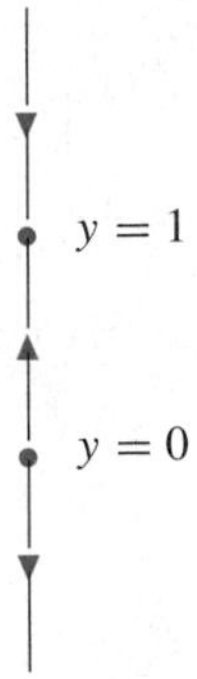

The right-hand side of the differential equation is $f(y) = (1 - y)y$. In this case, $f(y) = 0$ precisely when $y = 0$ and $y = 1$. Therefore the constant function $y_1(t) = 0$ for all t and $y_2(t) = 1$ for all t are equilibrium solutions for this equation. We call the points $y = 0$ and $y = 1$ on the y-axis **equilibrium points**. Also note that $f(y)$ is positive if $0 < y < 1$, whereas $f(y)$ is negative if $y < 0$ or $y > 1$. We can draw the phase line (or "rope") by placing dots at the equilibrium points $y = 0$ and $y = 1$. For $0 < y < 1$, we put arrows pointing up because $f(y) > 0$ means you climb up; and for $y < 0$ or $y > 1$, we put arrows pointing down because $f(y) < 0$ means you climb down (see Figure 1.50).

If we compare the phase line to the slope field, we see that the phase line contains all the information about the equilibrium solutions and whether the solutions are increasing or decreasing. Information about the *speed* of increase or decrease of solutions is lost (see Figure 1.51), But we can give rough sketches of the graphs of solutions using the phase line alone. These sketches will not be quite as accurate as the sketches from the slope field, but they will contain all the information about the behavior of solutions as t gets large (see Figure 1.52).

Figure 1.50
Phase line for
$dy/dt = (1 - y)y$.

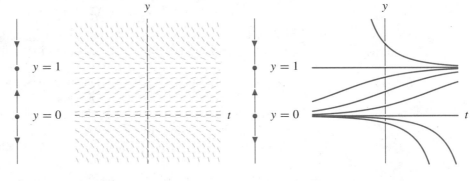

Figure 1.51
Phase line and slope field of
$dy/dt = (1 - y)y$.

Figure 1.52
Phase line and sketches of the graphs of
solutions for $dy/dt = (1 - y)y$.

How to Draw Phase Lines

We can give a more precise definition of the phase line by giving the steps required to draw it. For the autonomous equation $dy/dt = f(y)$:

- Draw the y-line.
- Find the equilibrium points (the numbers such that $f(y) = 0$), and mark them on the line.
- Find the intervals of y-values for which $f(y) > 0$, and draw arrows pointing up in these intervals.
- Find the intervals of y-values for which $f(y) < 0$, and draw arrows pointing down in these intervals.

We sketch several examples of phase lines in Figure 1.53. When looking at the phase line, you should remember the metaphor of the rope and think of solutions of the differential equation "dynamically" — people climbing up and down the rope as time increases.

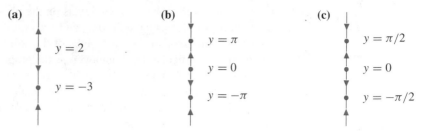

Figure 1.53
Phase lines for   **(a)** $dy/dt = (y - 2)(y + 3)$,   **(b)** $dy/dt = \sin y$, and
(c) $dy/dt = y \cos y$.

How to Use Phase Lines to Sketch Solutions

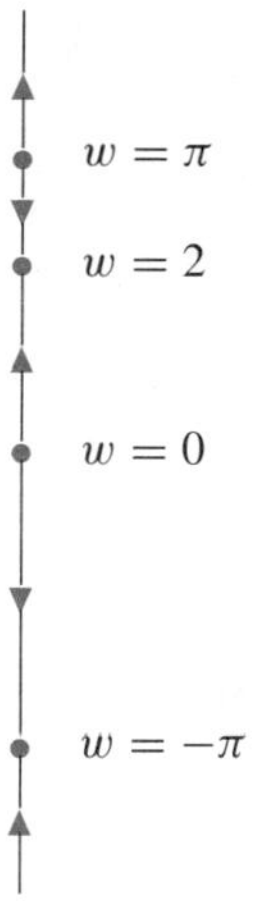

We can obtain rough sketches of the graphs of solutions directly from the phase lines, provided we are careful in interpreting these sketches. The sort of information that phase lines are very good at predicting is the limiting behavior of solutions as t increases or decreases.

Consider the equation

$$\frac{dw}{dt} = (2 - w) \sin w.$$

The phase line for this differential equation is given in Figure 1.54. Note that the equilibrium points are $w = 2$ and $w = k\pi$ for any integer k. Suppose we want to sketch the graph of the solution $w(t)$ with the initial value $w(0) = 0.4$. Because $w = 0$ and $w = 2$ are equilibrium points of this equation and $0 < 0.4 < 2$, we know from the Existence and Uniqueness Theorem that $0 < w(t) < 2$ for all t. Moreover, because $(2 - w) \sin w > 0$ for $0 < w < 2$, the solution is always increasing. Because the velocity of the solution is small only when $(2 - w) \sin w$ is close to zero and because this happens only near equilibrium points, we know that the solution $w(t)$ increases toward $w = 2$ as $t \to \infty$ (see Section 1.5).

Similarly, if we run the clock backward, the solution $w(t)$ decreases. It always remains above $w = 0$ and cannot stop, since $0 < w < 2$. Thus as $t \to -\infty$, the solution tends toward $w = 0$. We can draw a qualitative picture of the graph of the solution with the initial condition $w(0) = 0.4$ (see Figure 1.55).

Likewise, we can sketch other solutions in the tw-plane from the information on the phase line. The equilibrium solutions are easy to find and draw because they are

Figure 1.54
Phase line for
$dw/dt = (2 - w) \sin w$.

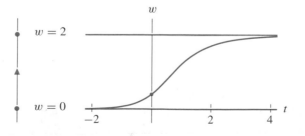

Figure 1.55
Graph of the solution to the initial-value problem

$$\frac{dw}{dt} = (2-w) \sin w, \quad w(0) = 0.4.$$

marked on the phase line. The intervals on the phase line with upward-pointing arrows correspond to increasing solutions, and those with downward-pointing arrows correspond to decreasing solutions. Graphs of the solutions do not cross by the Uniqueness Theorem. In particular, they cannot cross the graphs of the equilibrium solutions. Also, solutions must continue to increase or decrease until they come close to an equilibrium solution. Hence we can sketch many solutions with different initial conditions quite easily. The only information that we do not have is how quickly the solutions increase or decrease (see Figure 1.56).

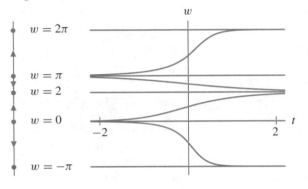

Figure 1.56
Graphs of many solutions to $dw/dt = (2 - w)\sin w$.

These observations lead to some general statements that can be made for all solutions of autonomous equations. Suppose $y(t)$ is a solution to an autonomous equation

$$\frac{dy}{dt} = f(y).$$

- If $f(y(0)) = 0$, then $y(0)$ is an equilibrium point and $y(t) = y(0)$ for all t.
- If $f(y(0)) > 0$, then $y(t)$ is increasing for all t and either $y(t) \to \infty$ as t increases or $y(t)$ tends to the first equilibrium point larger than $y(0)$.
- If $f(y(0)) < 0$, then $y(t)$ is decreasing for all t and either $y(t) \to -\infty$ as t increases or $y(t)$ tends to the first equilibrium point smaller than $y(0)$.

Similar results hold as t decreases (as time runs backward). If $f(y(0)) > 0$, then $y(t)$ either tends (in negative time) to $-\infty$ or to the next smaller equilibrium point. If $f(y(0)) < 0$, then $y(t)$ either tends (in negative time) to $+\infty$ or the next larger equilibrium point.

An example with three equilibrium points

For example, consider the differential equation

$$\frac{dP}{dt} = \left(1 - \frac{P}{20}\right)^3 \left(\frac{P}{5} - 1\right) P^7.$$

If the initial condition is given by $P(0) = 8$, what happens as t becomes very large? First we draw the phase line for this equation. Let

$$f(P) = \left(1 - \frac{P}{20}\right)^3 \left(\frac{P}{5} - 1\right) P^7.$$

We find the equilibrium points by solving $f(P) = 0$. Thus $P = 0$, $P = 5$, and $P = 20$ are the equilibrium points.

If $0 < P < 5$, $f(P)$ is negative; if $P < 0$ or $5 < P < 20$, $f(P)$ is positive; and if $P > 20$, $f(P)$ is negative. We can place the arrows on the phase line appropriately (see Figure 1.57). Note that we only have to check the value of $f(P)$ at one point in each of these intervals to determine the sign of $f(P)$ in the entire interval.

The solution $P(t)$ with initial condition $P(0) = 8$ is in the region between the equilibrium points $P = 5$ and $P = 20$, so $5 < P(t) < 20$ for all t. The arrows point up in this interval, so $P(t)$ is increasing for all t. As $t \to \infty$, $P(t)$ tends toward the equilibrium point $P = 20$.

As $t \to -\infty$, the solution with initial condition $P(0) = 8$ decreases toward the next smaller equilibrium point, which is $P = 5$. Hence $P(t)$ is always greater than $P = 5$. If we compute the solution $P(t)$ numerically, we see that it increases from $P(0) = 8$ to close to $P = 20$ very quickly (see Figure 1.58). From the phase line alone, we cannot tell how quickly the solution increases.

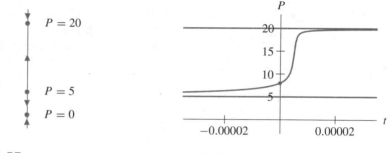

Figure 1.57
Phase line for $dP/dt = f(P) = (1 - P/20)^3 ((P/5) - 1) P^7$.

Figure 1.58
Graph of the solution to the initial-value problem $dP/dt = (1 - P/20)^3 ((P/5) - 1) P^7$, $P(0) = 8$.

Warning: Not All Solutions Exist for All Time

Suppose y_0 is an equilibrium point for the equation $dy/dt = f(y)$. Then $f(y_0) = 0$. We are assuming $f(y)$ is continuous, so if solutions are close to y_0, the value of f is small. Thus solutions move slowly when they are close to equilibrium points. A solution that approaches an equilibrium point as t increases (or decreases) moves more and more slowly as it approaches the equilibrium point. By the Existence and Uniqueness Theorem, a solution that approaches an equilibrium point never actually gets there. It is *asymptotic* to the equilibrium point, and the graph of the solution in the ty-plane has a horizontal asymptote.

On the other hand, unbounded solutions often speed up as they move. For example, the equation

$$\frac{dy}{dt} = (1 + y)^2$$

has one equilibrium point at $y = -1$ and $dy/dt > 0$ for all other values of y (see Figure 1.59).

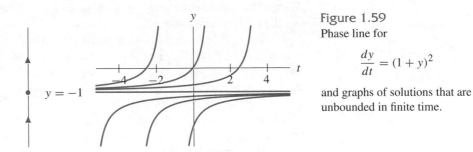

Figure 1.59

Phase line for

$$\frac{dy}{dt} = (1+y)^2$$

and graphs of solutions that are unbounded in finite time.

The phase line indicates that solutions with initial conditions that are greater than -1 increase for all t and tend to $+\infty$ as t increases.

If we separate variables and compute the explicit form of the solution, we can determine that these solutions actually blow up in finite time. In fact, the explicit form of any nonconstant solution is given by

$$y(t) = -1 - \frac{1}{t+c}$$

for some constant c. Since we are assuming that $y(0) > -1$, we must have

$$y(0) = -1 - \frac{1}{c} > -1,$$

which implies that $c < 0$. Therefore these solutions are defined only for $t < -c$, and they tend to ∞ as $t \to -c$ from below (see Figure 1.59). We cannot tell if solutions blow up in finite time like this simply by looking at the phase line.

The solutions with initial conditions $y(0) < -1$ are asymptotic to the equilibrium point $y = -1$ as t increases, so they are defined for all $t > 0$. However, these solutions tend to $-\infty$ in finite time as t decreases.

Another dangerous example is

$$\frac{dy}{dt} = \frac{1}{1-y}.$$

If $y > 1$, dy/dt is negative, and if $y < 1$, dy/dt is positive. If $y = 1$, dy/dt does not exist. The phase line has a hole in it. There is no standard way to denote such points on the phase line, but we will use a small empty circle to mark them (see Figure 1.60).

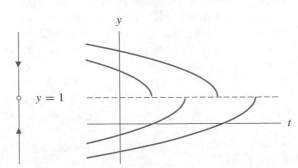

Figure 1.60

Phase line for

$$\frac{dy}{dt} = \frac{1}{1-y}.$$

Note that dy/dt is not defined for $y = 1$. Also, the graphs of solutions reach the "hole" at $y = 1$ in finite time.

All solutions tend toward $y = 1$ as t increases. Because the value of dy/dt is large if y is close to 1, solutions speed up as they get close to $y = 1$, and solutions reach $y = 1$ in a finite amount of time. Once a solution reaches $y = 1$, it cannot be continued because it has left the domain of definition of the differential equation. It has fallen into a hole in the phase line.

Drawing Phase Lines from Qualitative Information Alone

To draw the phase line for the differential equation $dy/dt = f(y)$, we need to know the location of the equilibrium points and the intervals over which the solutions are increasing or decreasing. That is, we need to know the points where $f(y) = 0$, the intervals where $f(y) > 0$, and the intervals where $f(y) < 0$. Consequently, we can draw the phase line for the differential equation with only qualitative information about the function $f(y)$.

For example, suppose we do not know a formula for $f(y)$, but we do have its graph (see Figure 1.61). From the graph we can determine the values of y for which $f(y) = 0$ and decide on which intervals $f(y) > 0$ and $f(y) < 0$. With this information we can draw the phase line (see Figure 1.62). From the phase line we can then get qualitative sketches of solutions (see Figure 1.63). Thus we can go from qualitative information about $f(y)$ to graphs of solutions of the differential equation $dy/dt = f(y)$ without ever writing down a formula. For models where the information available is completely qualitative, this approach is very appropriate.

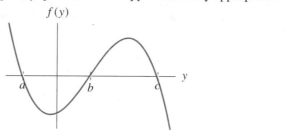

Figure 1.61
Graph of $f(y)$.

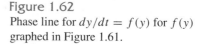

Figure 1.62
Phase line for $dy/dt = f(y)$ for $f(y)$ graphed in Figure 1.61.

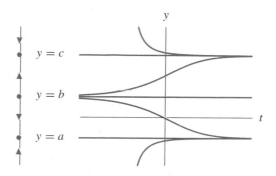

Figure 1.63
Sketch of solutions for $dy/dt = f(y)$ for $f(y)$ graphed in Figure 1.61.

The Role of Equilibrium Points

We have already determined that every solution to an autonomous differential equation $dy/dt = f(y)$ either tends to $+\infty$ or $-\infty$ as t increases (perhaps becoming infinite in finite time) or tends asymptotically to an equilibrium point as t increases. Hence the equilibrium points are extremely important in understanding the long-term behavior of solutions.

Also we have seen that, when drawing a phase line, we need to find the equilibrium points, the intervals on which $f(y)$ is positive, and the intervals on which $f(y)$ is negative. If f is continuous, it can switch from positive to negative only at points y_0 when $f(y_0) = 0$, that is, at equilibrium points. Hence the equilibrium points also play a crucial role in sketching the phase line.

In fact the equilibrium points are the key to understanding the entire phase line. For example, suppose we have an autonomous differential equation $dy/dt = g(y)$. Suppose all we know about this differential equation is that it has exactly two equilibrium points, at $y = 2$ and $y = 7$, and that the phase line near $y = 2$ and $y = 7$ is as shown in Figure 1.64. We can use this information to sketch the entire phase line. We know that the sign of $g(y)$ can change only at an equilibrium point. Hence the sign of $g(y)$ does not change for $2 < y < 7$, for $y < 2$, or for $y > 7$. Thus if we know the direction of the arrows anywhere in these intervals (say near the equilibrium points), then we know the directions on the entire phase line (see Figure 1.65). Consequently if we understand the equilibrium points for an autonomous differential equation, we should be able to understand (at least qualitatively) any solution of the equation.

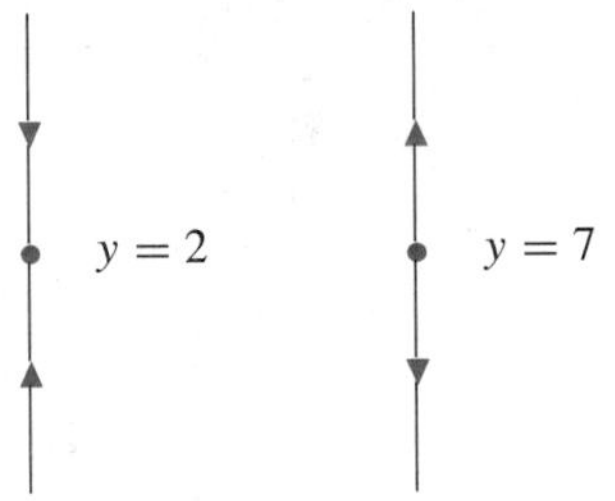

Figure 1.64
Pieces of the phase line for $dy/dt = g(y)$ near $y = 2$ and $y = 7$.

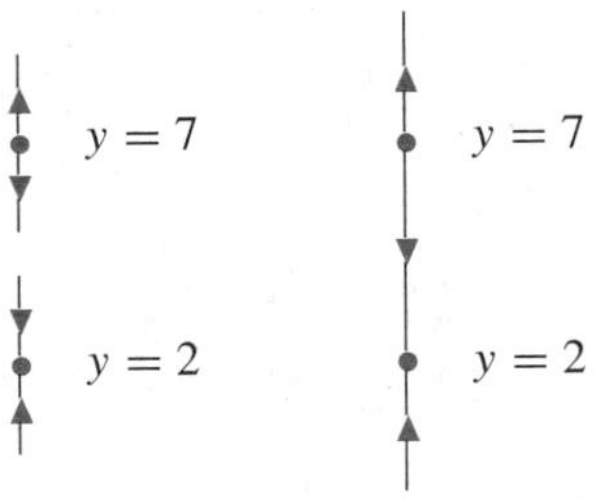

Figure 1.65
Entire phase line of $dy/dt = g(y)$ constructed from the phase line near the equilibrium points.

Classification of Equilibrium Points

Given their significance, it is useful to name the different types of equilibrium points and to classify them according to the behavior of nearby solutions. Consider an equilibrium point $y = y_0$, as shown in Figure 1.66. For y slightly less than y_0, the arrows point up, and for y slightly larger than y_0, the arrows point down. A solution with initial condition close to y_0 is asymptotic to y_0 as $t \to \infty$.

We say an equilibrium point y_0 is a **sink** if any solution with initial condition sufficiently close to y_0 is asymptotic to y_0 as t increases. (The name *sink* is supposed to bring to mind a kitchen sink with the equilibrium point as the drain. If water starts close enough to the drain, it will run toward it.)

Another possible phase line near an equilibrium point y_0 is shown in Figure 1.67. Here, the arrows point up for values of y just above y_0 and down for values of y just below y_0. A solution that has an initial value near y_0 tends away from y_0 as t increases. If time is run backward, solutions that start near y_0 tend toward y_0.

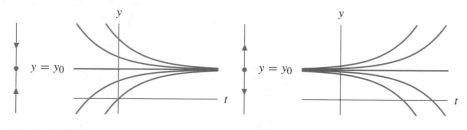

Figure 1.66
Phase line at a sink and graphs of solutions near a sink.

Figure 1.67
Phase line at a source and graphs of solutions near a source.

We say an equilibrium point y_0 is a **source** if all solutions that start sufficiently close to y_0 tend toward y_0 as t decreases. This means that all solutions that start close to y_0 (but not at y_0) will tend away from y_0 as t increases. So a source is a sink if time is run backward. (The name *source* is supposed to help you picture solutions flowing out of or away from a point.)

Sinks and sources are the two major types of equilibrium points. Every equilibrium point that is neither a source nor a sink is called a **node**. Two possible phase line pictures near nodes are shown in Figure 1.68.

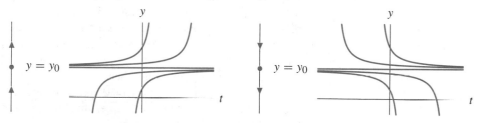

Figure 1.68
Examples of node equilibrium points and graphs of nearby solutions.

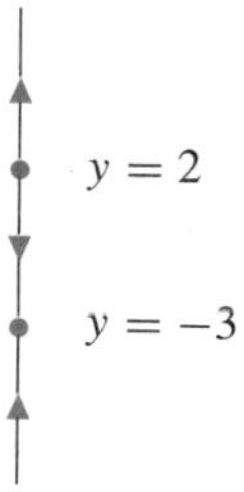

Given a differential equation, we can classify the equilibrium points as sinks, sources, or nodes from the phase line. For example, consider

$$\frac{dy}{dt} = y^2 + y - 6 = (y+3)(y-2).$$

The equilibrium points are $y = -3$ and $y = 2$. Also $dy/dt < 0$ for $-3 < y < 2$, and $dy/dt > 0$ for $y < -3$ and $y > 2$. Given this information, we can draw the phase line, and from the phase line we see that $y = -3$ is a sink and $y = 2$ is a source (see Figure 1.69).

Figure 1.69
Phase line for
$dy/dt = y^2 + y - 6$.

Suppose we are given a differential equation $dw/dt = g(w)$, where the right-hand side $g(w)$ is specified in terms of a graph rather than in terms of a formula. Then we can still sketch the phase line. For example, suppose that $g(w)$ is the function graphed in Figure 1.70. The corresponding differential equation has three equilibrium points, $w = -0.5$, $w = 1$, and $w = 2.5$; and $g(w) > 0$ if $w < -0.5$, $1 < w < 2.5$, and $w > 2.5$. For $-0.5 < w < 1$, $g(w) < 0$. Using this information we can draw the phase line (see Figure 1.71) and classify the equilibrium points. The point $w = -0.5$ is a sink, the point $w = 1$ is a source, and the point $w = 2.5$ is a node.

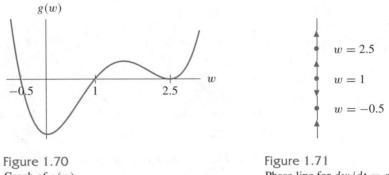

Figure 1.70
Graph of $g(w)$.

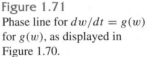

Figure 1.71
Phase line for $dw/dt = g(w)$
for $g(w)$, as displayed in
Figure 1.70.

Identifying the type of an equilibrium point and "linearization"

From the previous examples we know that we can determine the phase line and classify the equilibrium points for an autonomous differential equation $dy/dt = f(y)$ from the graph of $f(y)$ alone. Since the classification of an equilibrium point depends only on the phase line near the equilibrium point, then we should be able to determine the type of an equilibrium point y_0 from the graph of $f(y)$ near y_0.

If y_0 is a sink, then the arrows on the phase line just below y_0 point up and the arrows just above y_0 point down. Hence $f(y)$ must be positive for y just smaller than y_0 and negative for y just larger than y_0 (see Figure 1.72). So f must be decreasing for y near y_0. Conversely, if $f(y_0) = 0$ and f is decreasing for all y near y_0, then $f(y)$ is positive just to the left of y_0 and negative just to the right of y_0. Hence, y_0 is a sink. Similarly, the equilibrium point y_0 is a source if and only if f is increasing for all y near y_0 (see Figure 1.73).

From calculus we have a powerful tool for telling whether a function is increasing or decreasing at a particular point — the derivative. Using the derivative of $f(y)$ combined with the geometric observations above, we can give criteria that specify the type of the equilibrium point.

LINEARIZATION THEOREM   Suppose y_0 is an equilibrium point of the differential equation $dy/dt = f(y)$ where f is a continuously differentiable function. Then,

- if $f'(y_0) < 0$, then y_0 is a sink;
- if $f'(y_0) > 0$, then y_0 is a source; or
- if $f'(y_0) = 0$ or if $f'(y_0)$ does not exist, then we need additional information to determine the type of y_0.   ■

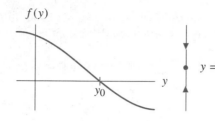

Figure 1.72
Phase line near a sink at $y = y_0$ for $dy/dt = f(y)$ and graph of $f(y)$ near $y = y_0$.

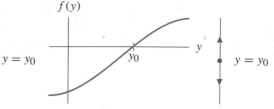

Figure 1.73
Phase line near a source at $y = y_0$ for $dy/dt = f(y)$ and graph of $f(y)$ near $y = y_0$.

This theorem follows immediately from the discussion prior to its statement once we recall that if $f'(y_0) < 0$, then f is decreasing near y_0, and if $f'(y_0) > 0$, then f is increasing near y_0. This analysis and these conclusions are an example of **linearization**, a technique that we will often find useful. The derivative $f'(y_0)$ tells us the behavior of the best linear approximation to f near y_0. If we replace f with its best linear approximation, then the differential equation we obtain is very close to the original differential equation for y near y_0.

We cannot make any conclusion about the classification of y_0 if $f'(y_0) = 0$, because all three possibilities can occur (see Figure 1.74).

As another example, consider the differential equation

$$\frac{dy}{dt} = h(y) = y(\cos(y^5 + 2y) - 27\pi y^4).$$

What does the phase line look like near $y = 0$? Drawing the phase line for this equation would be a very complicated affair. We would have to find the equilibrium points

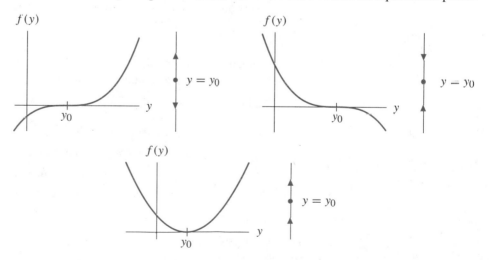

Figure 1.74
Graphs of various functions f along with the corresponding phase lines for the differential equation $dy/dt = f(y)$. In all cases, $y_0 = 0$ is an equilibrium point and $f'(y_0) = 0$.

and determine the sign of $h(y)$. On the other hand, it is easy to see that $y = 0$ is an equilibrium point because $h(0) = 0$. We compute

$$h'(y) = (\cos(y^5 + 2y) - 27\pi y^4) + y\frac{d}{dy}(\cos(y^5 + 2y) - 27\pi y^4).$$

Thus $h'(0) = (\cos(0) - 0) + 0 = 1$. By the Linearization Theorem, we conclude that $y = 0$ is a source. Solutions that start sufficiently close to $y = 0$ move away from $y = 0$ as t increases. Of course, there is the dangerous loophole clause "sufficiently close." Initial conditions might have to be very, very close to $y = 0$ for the above to apply. Again we did a little work and got a little information. To get more information, we would need to study the function $h(y)$ more carefully.

Modified Logistic Model

As an application of these ideas, we use the techniques of this section to discuss a modification of the logistic population model we introduced in Section 1.1.

The fox squirrel is a small mammal native to the Rocky Mountains. These squirrels are very territorial, so if their population is large, their rate of growth decreases and may even become negative. On the other hand, if the population is too small, fertile adults run the risk of not being able to find suitable mates, so again the rate of growth is negative.

The model

We can restate these assumptions succinctly:

- If the population is too big, the rate of growth is negative.
- If the population is too small, the rate of growth is negative.

So the population grows only if it is between "too big" and "too small." Also, it is reasonable to assume that, if the population is zero, it will stay zero. Thus we also assume:

- If the population is zero, the growth rate is zero. (Compare these assumptions with those of the logistic population model of Section 1.1.)

We let

$$t = \text{time (independent variable)},$$
$$S(t) = \text{population of squirrels at time } t \text{ (dependent variable)},$$
$$k = \text{growth-rate coefficient (parameter)},$$
$$N = \text{carrying capacity (parameter), and}$$
$$M = \text{"sparsity" constant (parameter)}.$$

The carrying capacity N indicates what population is "too big," and the sparsity parameter M indicates what population is "too small."

Now we want a model of the form $dS/dt = g(S)$ that conforms to the assumptions. We can think of the assumptions as determining the shape of the graph of $g(S)$, in particular where $g(S)$ is positive and where it is negative. Note that $dS/dt = g(S) < 0$

Nancy Kopell (1942–) received her doctorate in mathematics at the University of California, Berkeley, where she wrote her thesis under the direction of Stephen Smale. She is one of the leading figures in the world in the use of differential equations to model natural phenomena. Kopell has employed techniques similar to those that we study in this book to tackle such diverse problems as spontaneous pattern formation in chemical systems and the networks of neurons that govern rhythmic motion in animals and other oscillations in the central nervous system.

For her work, she has received numerous awards, including a MacArthur Fellowship "genius grant" in 1990. In 1996, she was elected to the National Academy of Sciences. She is currently Professor of Mathematics and Director of the Center for Biodynamics (and the authors' colleague) at Boston University.

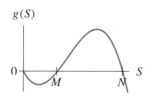

Figure 1.75
Graph of $g(S)$.

if $S > N$ because the population decreases if it is too big. Also $g(S) < 0$ when $S < M$ because the population decreases if it is too small. Finally, $g(S) > 0$ when $M < S < N$ and $g(0) = 0$. That is, we want $g(S)$ to have a graph shaped like Figure 1.75. The graph of g for $S < 0$ does not matter because a negative number of squirrels (anti-squirrels?) is meaningless.

The logistic model would give "correct" behavior for populations near the carrying capacity, but for small populations (below the "sparsity" level M), the solutions of the logistic model do not agree with the assumptions. Hence we will need to modify the logistic model to include the behavior of small populations and to include the parameter M. We make a model of the form

$$\frac{dS}{dt} = g(S) = kS \left(1 - \frac{S}{N}\right) \text{(something)}.$$

The "something" term must be positive if $S > M$ and negative if $S < M$. The simplest choice that satisfies these conditions is

$$\text{(something)} = \left(\frac{S}{M} - 1\right).$$

Hence our model is

$$\frac{dS}{dt} = kS \left(1 - \frac{S}{N}\right) \left(\frac{S}{M} - 1\right).$$

This is the logistic model with the extra term

$$\left(\frac{S}{M} - 1\right).$$

We call it the modified logistic population model. (Other models might also be called the modified logistic, but modified in a different way.)

Analysis of the model

To analyze solutions of this differential equation, we could use analytic techniques, since the equation is separable. However, qualitative techniques provide a lot of information about the solutions with a lot less work. The differential equation is

$$\frac{dS}{dt} = g(S) = kS\left(1 - \frac{S}{N}\right)\left(\frac{S}{M} - 1\right),$$

with $0 < M < N$ and $k > 0$. There are three equilibrium points — $S = 0$, $S = M$, and $S = N$. If $0 < S < M$, we have $g(S) < 0$, so solutions with initial conditions between 0 and M decrease. Similarly, if $S > N$, $g(S) < 0$, solutions with initial conditions larger than N also decrease. For $M < S < N$, we have $g(S) > 0$. Consequently, solutions with initial conditions between M and N increase. Thus we conclude that the equilibria at 0 and N are sinks, and the equilibrium point at M is a source. The phase line and graphs of typical solutions are shown in Figure 1.76.

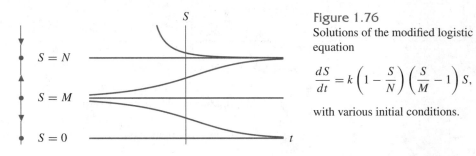

Figure 1.76
Solutions of the modified logistic equation

$$\frac{dS}{dt} = k\left(1 - \frac{S}{N}\right)\left(\frac{S}{M} - 1\right)S,$$

with various initial conditions.

EXERCISES FOR SECTION 1.6

In Exercises 1–8, sketch the phase lines for the given differential equation. Identify the equilibrium points as sinks, sources, or nodes.

1. $\dfrac{dy}{dt} = 3y(1 - y)$ **2.** $\dfrac{dy}{dt} = y^2 - 6y - 16$ **3.** $\dfrac{dy}{dt} = \cos y$

4. $\dfrac{dw}{dt} = w \cos w$ **5.** $\dfrac{dw}{dt} = (w - 2)\sin w$ **6.** $\dfrac{dy}{dt} = \dfrac{1}{y - 2}$

7. $\dfrac{dw}{dt} = w^2 + 2w + 10$ **8.** $\dfrac{dy}{dt} = \tan y$

In Exercises 9–15, a differential equation and various initial conditions are specified. Sketch the graphs of the solutions satisfying these initial conditions. For each exercise, put all your graphs on one pair of axes.

9. Equation from Exercise 1; $y(0) = 1$, $y(2) = -1$, $y(0) = 1/2$, $y(0) = 2$.

10. Equation from Exercise 2; $y(0) = 1$, $y(1) = 0$, $y(0) = -10$, $y(0) = 5$.

11. Equation from Exercise 3; $y(0) = 0$, $y(-1) = 1$, $y(0) = -\pi/2$, $y(0) = \pi$.

12. Equation from Exercise 4; $w(0) = 0$, $w(3) = 1$, $w(0) = 2$, $w(0) = -1$.

13. Equation from Exercise 5; $w(0) = 1$, $w(0) = 7/4$, $w(0) = -1$, $w(0) = 3$.

14. Equation from Exercise 6; $y(0) = 0$, $y(1) = 3$, $y(0) = 2$ (trick question).

15. Equation from Exercise 7; $w(0) = 0$, $w(1/2) = 1$, $w(0) = 2$.

In Exercises 16–21, describe the long-term behavior of the solution to the differential equation

$$\frac{dy}{dt} = y^2 - 4y + 2$$

with the given initial condition.

16. $y(0) = 0$ **17.** $y(0) = 1$ **18.** $y(0) = -1$

19. $y(0) = -10$ **20.** $y(0) = 10$ **21.** $y(3) = 1$

22. Consider the autonomous equation $dy/dt = f(y)$. Suppose we know that $f(-1) = f(2) = 0$.

    (a) Describe all the possible behaviors of the solution $y(t)$ that satisfies the initial condition $y(0) = 1$.

    (b) Suppose also that $f(y) > 0$ for $-1 < y < 2$. Describe all the possible behaviors of the solution $y(t)$ that satisfies the initial condition $y(0) = 1$.

In Exercises 23–26, the graph of a function $f(y)$ is given. Sketch the phase line for the autonomous differential equation $dy/dt = f(y)$.

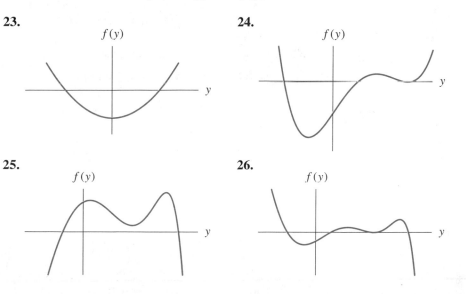

23.

$f(y)$

y

24.

$f(y)$

y

25.

$f(y)$

y

26.

$f(y)$

y

In Exercises 27–30, a phase line for an autonomous equation $dy/dt = f(y)$ is shown. Make a rough sketch of the graph of the corresponding function $f(y)$. (Assume $y = 0$ is in the middle of the segment shown in each case.)

27.   **28.**   **29.**   **30.**

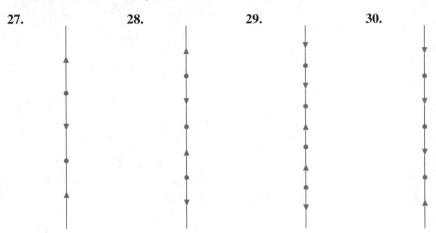

31. Suppose you wish to model a population with a differential equation of the form $dP/dt = f(P)$, where $P(t)$ is the population at time t. Experiments have been performed on the population that give the following information:

- The only equilibrium points in the population are $P = 0$, $P = 10$, and $P = 50$.
- If the population is 100, the population decreases.
- If the population is 25, the population increases.

(a) Sketch the possible phase lines for this system for $P > 0$ (there are two).

(b) Give a rough sketch of the corresponding functions $f(P)$ for each of your phase lines.

(c) Give a formula for functions $f(P)$ whose graph agrees (qualitatively) with the rough sketches in part (b) for each of your phase lines.

32. Suppose the experimental information in Exercise 31 is changed as follows:

- The population $P = 0$ remains constant.
- A population close to 0 will decrease.
- A population of $P = 20$ will increase.
- A population of $P > 100$ will decrease.

(a) Sketch the simplest possible phase line that agrees with the experimental information above.

(b) Give a rough sketch of the function $f(P)$ for the phase line of part (a).

(c) What other phase lines are possible?

33. Let $f(y)$ be a continuous function.

(a) Suppose that $f(-10) > 0$ and $f(10) < 0$. Show that there is an equilibrium point for $dy/dt = f(y)$ between $y = -10$ and $y = 10$.

(b) Suppose that $f(-10) > 0$, that $f(10) < 0$, and that there are finitely many equilibrium points between $y = -10$ and $y = 10$. If $y = 1$ is a source, show that $dy/dt = f(y)$ must have at least two sinks between $y = -10$ and $y = 10$. (Can you say where they are located?)

34. Consider the differential equation $dy/dt = f(y)$. Suppose y_0 is an isolated equilibrium point; that is, y_0 is an equilibrium point and there is an open interval containing y_0 that contains no other equilibrium points. We define the **index** of y_0 to be

$$\text{index}(y_0) = \begin{cases} +1 & \text{if } y_0 \text{ is a source;} \\ -1 & \text{if } y_0 \text{ is a sink;} \\ 0 & \text{if } y_0 \text{ is a node.} \end{cases}$$

(a) Suppose that there are finitely many equilibrium points in the interval $-100 < y < 100$ and that $f(-100) < 0$ and $f(100) > 0$. Show that, if $y_1, y_2, \ldots, y_n$ are the equilibrium points between -100 and 100, then

$$\text{index}(y_1) + \text{index}(y_2) + \cdots + \text{index}(y_n) = 1.$$

(b) Suppose that there are finitely many equilibrium points in the interval $-100 < y < 100$ and that $f(-100) > 0$ and $f(100) < 0$. Show that

$$\text{index}(y_1) + \text{index}(y_2) + \cdots + \text{index}(y_n) = -1.$$

(c) Suppose that there are finitely many equilibrium points in the interval $-100 < y < 100$ and that $f(-100) < 0$ and $f(100) < 0$. Show that

$$\text{index}(y_1) + \text{index}(y_2) + \cdots + \text{index}(y_n) = 0.$$

35. Suppose $dy/dt = f(y)$ has an equilibrium point at $y = y_0$ and

(a) $f'(y_0) = 0$, $f''(y_0) = 0$, and $f'''(y_0) > 0$: Is y_0 a source, a sink, or a node?
(b) $f'(y_0) = 0$, $f''(y_0) = 0$, and $f'''(y_0) < 0$: Is y_0 a source, a sink, or a node?
(c) $f'(y_0) = 0$ and $f''(y_0) > 0$: Is y_0 a source, a sink, or a node?

36. **(a)** Sketch the phase line for the differential equation

$$\frac{dy}{dt} = \frac{1}{(y-2)(y+1)},$$

and discuss the behavior of the solution with initial condition $y(0) = 1/2$.
(b) Apply analytic techniques to the initial-value problem

$$\frac{dy}{dt} = \frac{1}{(y-2)(y+1)}, \quad y(0) = \frac{1}{2},$$

and compare your results with your discussion in part (a).

The proper scheduling of city bus and train systems is a difficult problem, which the City of Boston seems to ignore. It is not uncommon in Boston to wait a long time for the trolley, only to have several trolleys arrive simultaneously. In Exercises 37–40 we study a very simple model of the behavior of trolley cars.

Consider two trolley cars on the same track moving toward downtown Boston. Let $x(t)$ denote the amount of time between the two cars at time t. That is, if the first car arrives at a particular stop at time t, then the other car will arrive at the stop $x(t)$ time units later. We assume that the first car runs at a constant average speed (not a bad assumption for a car running before rush hour). We wish to model how $x(t)$ changes as t increases.

We first assume that, if no passengers are waiting for the second train, then it has an average speed greater than the first train and hence will catch up to the first train. Thus the time between trains $x(t)$ will decrease at a constant rate if no people are waiting for the second train. However, the speed of the second train decreases if there are passengers to pick up. We assume that the speed of the second train decreases at a rate proportional to the number of passengers it picks up and that the passengers arrive at the stops at a constant rate. Hence the number of passengers waiting for the second train is proportional to the time between trains.

37. Let $x(t)$ be the amount of time between two consecutive trolley cars as described above. We claim that a reasonable model for $x(t)$ is

$$\frac{dx}{dt} = \beta x - \alpha.$$

Which term represents the rate of decrease of the time between the trains if no people are waiting, and which term represents the effect of the people waiting for the second train? (Justify your answer.) Should the parameters α and β be positive or negative?

38. For the model in Exercise 37:

  (a) Find the equilibrium points.

  (b) Classify the equilibrium points (source, sink, or node).

  (c) Sketch the phase line.

  (d) Sketch the graphs of solutions.

  (e) Find the formula for the general solution.

39. Use the model in Exercise 37 to predict what happens to $x(t)$ as t increases. Include the effect of the initial value $x(0)$. Is it possible for the trains to run at regular intervals? Given that there are always slight variations in the number of passengers waiting at each stop, is it likely that a regular interval can be maintained? Write two brief reports (of one or two paragraphs):

  (a) The first report is addressed to other students in the class (hence you may use technical language we use in class).

  (b) The second report is addressed to the Mayor of Boston.

40. Assuming the model for $x(t)$ from Exercise 37, what happens if trolley cars leave the station at fixed intervals? Can you use the model to predict what will happen for a whole sequence of trains? Will it help to increase the number of trains so that they leave the station more frequently?

1.7 BIFURCATIONS

Equations with Parameters

In many of our models, a common feature is the presence of **parameters** along with the other variables involved. Parameters are quantities that do not depend on time (the independent variable) but that assume different values depending on the specifics of the application at hand. For instance, the exponential growth model for population

$$\frac{dP}{dt} = kP$$

contains the parameter k, the constant of proportionality for the growth rate dP/dt versus the total population P. One of the underlying assumptions of this model is that the growth rate dP/dt is a constant multiple of the total population. However, when we apply this model to different species, we expect to use different values for the constant of proportionality. For example, the value of k that we would use for rabbits would be significantly larger than the value for humans.

How the behavior of solutions changes as the parameters vary is a particularly important aspect of the study of differential equations. For some models, we must study the behavior of solutions for all parameter values in a certain range. As an example, consider a model for the motion of a bridge over time. In this case, the number of cars on the bridge may affect how the bridge reacts to wind, and a model for the motion of the bridge might contain a parameter for the total mass of the cars on the bridge. In that case, we would want to know the behavior of various solutions of the model for a variety of different values of the mass.

In many models we know only approximate values for the parameters. However, in order for the model to be useful to us, we must know the effect of slight variations in the values of the parameters on the behavior of the solutions. Also there may be effects that we have not included in our model that make the parameters vary in unexpected ways. In many complicated physical systems, the long-term effect of these intentional or unintentional adjustments in the parameters can be very dramatic.

In this section we study how solutions of a differential equation change as a parameter is varied. We study autonomous equations with one parameter. We find that a small change in the parameter usually results in only a small change in the nature of the solutions. However, occasionally a small change in the parameter can lead to a drastic change in the long-term behavior of solutions. Such a change is called a **bifurcation**. We say that a differential equation that depends on a parameter *bifurcates* if there is a qualitative change in the behavior of solutions as the parameter changes.

Notation for differential equations depending on a parameter

An example of an autonomous differential equation that depends on a parameter is

$$\frac{dy}{dt} = y^2 - 2y + \mu.$$

The parameter is μ. The independent variable is t and the dependent variable is y, as usual. Note that this equation really represents infinitely many different equations, one for each value of μ. We think of the value of μ as a constant in each equation, but different values of μ yield different differential equations, each with a different set of solutions. Because of their different roles in the differential equation, we use a notation that distinguishes the dependence of the right-hand side on y and μ. We let

$$f_\mu(y) = y^2 - 2y + \mu.$$

The parameter μ appears in the subscript, and the dependent variable y is the argument of the function f_μ. If we want to specify a particular value of μ, say $\mu = 3$, then we write

$$f_3(y) = y^2 - 2y + 3.$$

With $\mu = 3$, we obtain the corresponding differential equation

$$\frac{dy}{dt} = f_3(y) = y^2 - 2y + 3.$$

We use this notation in general. A function of the dependent variable y, which also depends on a parameter μ, is denoted by $f_\mu(y)$. The corresponding differential equation with dependent variable y and parameter μ is

$$\frac{dy}{dt} = f_\mu(y).$$

Since such a differential equation really refers to a collection of different equations, one for each value of μ, we call such an equation a **one-parameter family** of differential equations.

A One-Parameter Family with One Bifurcation

Let's consider the one-parameter family

$$\frac{dy}{dt} = f_\mu(y) = y^2 - 2y + \mu$$

more closely. For each value of μ we have an autonomous differential equation, and we can draw its phase line and analyze it using the techniques of the previous section. We begin our study of this family by studying the differential equations obtained from particular choices of μ. Since we do not yet know the most interesting values of μ, we just pick integer values, say $\mu = -4$, $\mu = -2$, $\mu = 0$, $\mu = 2$, and $\mu = 4$, for starters. This gives us five different autonomous differential equations with five different phase lines. One equation is

$$\frac{dy}{dt} = f_{-2}(y) = y^2 - 2y - 2.$$

This differential equation has equilibrium points at values of y for which

$$f_{-2}(y) = y^2 - 2y - 2 = 0.$$

The equilibrium points are $y = 1 - \sqrt{3}$ and $y = 1 + \sqrt{3}$. Between the equilibrium points, the function f_{-2} is negative, and above and below the equilibrium points, f_{-2} is positive. Hence $y = 1 - \sqrt{3}$ is a sink and $y = 1 + \sqrt{3}$ is a source. With this information we can draw the phase line. For the other values of μ we follow a similar procedure and draw the phase lines. All these phase lines are shown in Figure 1.77.

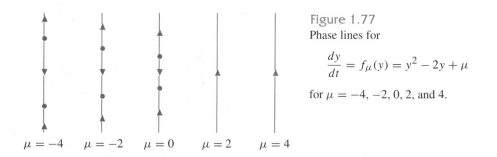

Figure 1.77
Phase lines for

$$\frac{dy}{dt} = f_\mu(y) = y^2 - 2y + \mu$$

for $\mu = -4, -2, 0, 2,$ and 4.

Each of the phase lines is somewhat different from the others. However, the basic description of the phase lines for $\mu = -4$, $\mu = -2$, and $\mu = 0$ is the same: There are exactly two equilibrium points; the smaller one is a sink and the larger one is a source. Although the exact position of these equilibrium points changes as μ increases, their relative position and type do not change. Solutions of these equations with large initial values blow up in finite time as t increases and tend to an equilibrium point as t decreases. Solutions with very negative initial conditions tend to an equilibrium point as t increases and to $-\infty$ as t decreases. Solutions with initial values between the equilibrium points tend to the smaller equilibrium point as t increases and to the larger equilibrium point as t decreases (see Figure 1.78).

If $\mu = 2$ and $\mu = 4$, we have something very different. There are no equilibrium points. All solutions tend to $+\infty$ as t increases and to $-\infty$ as t decreases. Because there is a significant change in the nature of the solutions, we say that a bifurcation has occurred somewhere between $\mu = 0$ and $\mu = 2$.

To investigate the nature of this bifurcation, we draw the graphs of f_μ for the μ-values above (see Figure 1.79). For $\mu = -4, -2,$ and 0, $f_\mu(y)$ has 2 roots, but for $\mu = 2$ and 4, the graph of $f_\mu(y)$ does not cross the y-axis. Somewhere between $\mu = 0$ and $\mu = 2$ the graph of $f_\mu(y)$ must be tangent to the y-axis.

The roots of the quadratic equation

$$y^2 - 2y + \mu = 0$$

are $y = 1 \pm \sqrt{1 - \mu}$. If $\mu < 1$, this quadratic has two real roots; if $\mu = 1$, it has only one root; and if $\mu > 1$, it has no real roots. The corresponding differential equations have two equilibrium points if $\mu < 1$, one equilibrium point if $\mu = 1$, and no equilibrium points if $\mu > 1$. Hence the qualitative nature of the phase lines changes when $\mu = 1$. We say that a bifurcation occurs at $\mu = 1$ and that $\mu = 1$ is a **bifurcation value**.

The graph of $f_1(y)$ and the phase line for $dy/dt = f_1(y)$ are shown in Figures 1.80 and 1.81. The phase line has one equilibrium point (which is a node), and everywhere else solutions increase. The fact that the bifurcation occurs at the parameter value for which the equilibrium point is a node is no coincidence. In fact, this entire bifurcation scenario is quite common.

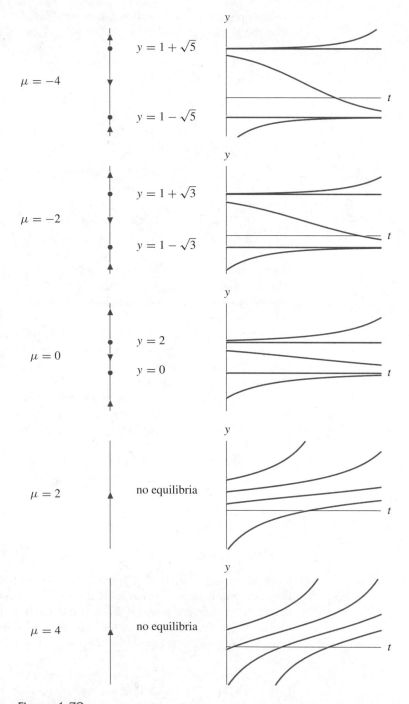

Figure 1.78
Phase lines and sketches of solutions for $dy/dt = f_\mu(y) = y^2 - 2y + \mu$ for $\mu = -4, -2, 0, 2, 4$.

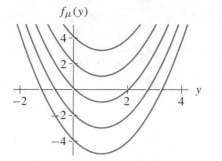

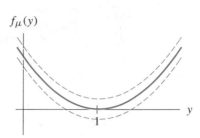

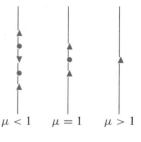

$\mu < 1 \quad \mu = 1 \quad \mu > 1$

Figure 1.79
Graphs of $f_\mu(y) = y^2 - 2y + \mu$ for $\mu = -4, -2, 0, 2,$ and 4.

Figure 1.80
Graphs of $f_\mu(y) = y^2 - 2y + \mu$ for μ slightly less than 1, equal to 1, and slightly greater than 1.

Figure 1.81
Corresponding phase lines for $dy/dt = f_\mu(y) = y^2 - 2y + \mu$.

The Bifurcation Diagram

An extremely helpful way to understand the qualitative behavior of solutions is through the **bifurcation diagram**. This diagram is a picture (in the μy-plane) of the phase lines near a bifurcation value. It highlights the changes that the phase lines undergo as the parameter passes through this value.

To plot the bifurcation diagram, we plot the parameter values along the horizontal axis. For each μ-value (not just integers), we draw the phase line corresponding to μ on the vertical line through μ. We think of the bifurcation diagram as a movie: As our eye scans the picture from left to right, we see the phase lines evolve through the bifurcation. Figure 1.82 shows the bifurcation diagram for $f_\mu(y) = y^2 - 2y + \mu$.

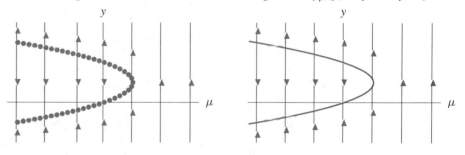

Figure 1.82
Bifurcation diagram for the differential equation $dy/dt = f_\mu(y) = y^2 - 2y + \mu$. The horizontal axis is the μ-value and the vertical lines are the phase lines for the differential equations with the corresponding μ-values.

A bifurcation from one to three equilibria

Let's look now at another one-parameter family of differential equations

$$\frac{dy}{dt} = g_\alpha(y) = y^3 - \alpha y = y(y^2 - \alpha).$$

In this equation, α is the parameter. There are three equilibria if $\alpha > 0$ ($y = 0, \pm\sqrt{\alpha}$), but there is only one equilibrium point ($y = 0$) if $\alpha \le 0$. Therefore a bifurcation occurs when $\alpha = 0$. To understand this bifurcation, we plot the bifurcation diagram.

First, if $\alpha < 0$, the term $y^2 - \alpha$ is always positive. Thus $g_\alpha(y) = y(y^2 - \alpha)$ has the same sign as y. Solutions tend to ∞ if $y(0) > 0$ and to $-\infty$ if $y(0) < 0$. If $\alpha > 0$, the situation is different. The graph of $g_\alpha(y)$ shows that $g_\alpha(y) > 0$ in the intervals $\sqrt{\alpha} < y < \infty$ and $-\sqrt{\alpha} < y < 0$ (see Figure 1.83). Thus solutions increase in these intervals. In the other intervals, $g_\alpha(y) < 0$, so solutions decrease. The bifurcation diagram is depicted in Figure 1.84.

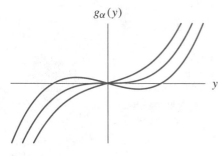

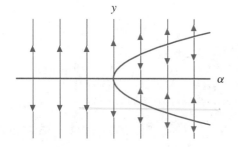

Figure 1.83
Graphs of $g_\alpha(y)$ for $\alpha > 0$, $\alpha = 0$, and $\alpha < 0$. Note that for $\alpha \leq 0$ the graph crosses the y-axis once, whereas if $\alpha > 0$, the graph crosses the y-axis three times.

Figure 1.84
Bifurcation diagram for the one-parameter family $dy/dt = g_\alpha(y) = y^3 - \alpha y$.

Bifurcations of Equilibrium Points

Throughout the rest of this section, we assume that all the one-parameter families of differential equations that we consider depend smoothly on the parameter. That is, for a one-parameter family

$$\frac{dy}{dt} = f_\mu(y),$$

the partial derivatives of $f_\mu(y)$ with respect to y and μ exist and are continuous. So changing μ a little changes the graph of $f_\mu(y)$ only slightly.

When bifurcations do not happen

The most important fact about bifurcations is that they usually do not happen. A small change in the parameter usually leads to only a small change in the behavior of solutions. This is very reassuring. For example, suppose we have a one-parameter family

$$\frac{dy}{dt} = f_\mu(y),$$

and the differential equation for $\mu = \mu_0$ has an equilibrium point at $y = y_0$. Also suppose that $f'_{\mu_0}(y_0) < 0$, so the equilibrium point is a sink. We sketch the phase line and the graph of $f_{\mu_0}(y)$ near $y = y_0$ in Figure 1.85.

Now if we change μ just a little bit, say from μ_0 to μ_1, then the graph of $f_{\mu_1}(y)$ is very close to the graph of $f_{\mu_0}(y)$ (see Figure 1.86). So the graph of $f_{\mu_1}(y)$ is strictly

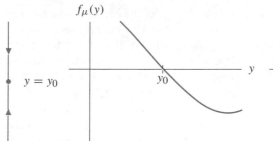

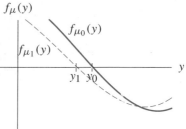

Figure 1.85
Graph of $f_{\mu_0}(y)$ near the sink y_0 and the phase line for the differential equation $dy/dt = f_{\mu_0}(y)$ near y_0.

Figure 1.86
Graphs of $f_{\mu_0}(y)$ and $f_{\mu_1}(y)$ for μ_1 close to μ_0. Note that $f_{\mu_1}(y)$ decreases across the y-axis at $y = y_1$ near y_0, so $dy/dt = f_{\mu_1}(y)$ has a sink at $y = y_1$.

decreasing near y_0, passing through the horizontal axis near $y = y_0$. The corresponding differential equation

$$\frac{dy}{dt} = f_{\mu_1}(y)$$

has a sink at some point $y = y_1$ very near y_0.

We can make this more precise: If y_0 is a sink for a differential equation

$$\frac{dy}{dt} = f_{\mu_0}(y)$$

with $f'_{\mu_0}(y_0) < 0$, then for all μ_1 sufficiently close to μ_0, the differential equation

$$\frac{dy}{dt} = f_{\mu_1}(y)$$

has a sink at a point $y = y_1$ very near y_0 (and no other equilibrium points near y_0). A similar statement holds if y_0 is a source and $f'_{\mu_0}(y_0) > 0$. These are the situations in which we can say for sure that no bifurcation occurs, at least not near y_0.

With these observations in mind, we see that bifurcations occur only if the above conditions do not hold. Consequently, given a one-parameter family of differential equations

$$\frac{dy}{dt} = f_\mu(y),$$

we look for values $\mu = \mu_0$ and $y = y_0$ for which $f_{\mu_0}(y_0) = 0$ and $f'_{\mu_0}(y_0) = 0$.

Determining bifurcation values
Consider the one-parameter family of differential equations given by

$$\frac{dy}{dt} = f_\mu(y) = y(1 - y)^2 + \mu.$$

If $\mu = 0$, the equilibrium points are $y = 0$ and $y = 1$. Also $f'_0(0) = 1$. Hence $y = 0$ is a source for the differential equation $dy/dt = f_0(y)$. Thus for all μ sufficiently close to zero, the differential equation $dy/dt = f_\mu(y)$ has a source near $y = 0$.

On the other hand, for the equilibrium point $y = 1$, $f_0'(1) = 0$. The Linearization Theorem from Section 1.6 says nothing about what happens in this case. To see what is going on, we sketch the graph of $f_\mu(y)$ for several μ-values near $\mu = 0$ (see Figure 1.87). If $\mu = 0$, the graph of f_μ is tangent to the horizontal axis at $y = 1$. Since $f_0(y) > 0$ for all $y > 0$ except $y = 1$, it follows that the equilibrium point at $y = 1$ is a node for this parameter value. Changing μ moves the graph of $f_\mu(y)$ up (if μ is positive) or down (if μ is negative). If we make μ slightly positive, $f_\mu(y)$ does not touch the horizontal axis near $y = 1$. So the equilibrium point at $y = 1$ for $\mu = 0$ disappears. A bifurcation occurs at $\mu = 0$. For μ slightly negative, the corresponding differential equation has two equilibrium points near $y = 1$. Since f_μ is decreasing at one of these equilibria and increasing at the other, one of these equilibria is a source and the other is a sink.

There is a second bifurcation in this one-parameter family. To see this, note what happens as μ decreases. There is a value of μ for which the graph of $f_\mu(y)$ again has a tangency with the horizontal axis (see Figure 1.88). For larger μ-values, the graph crosses the horizontal axis three times, but for lower μ-values, the graph crosses only once. Thus a second bifurcation occurs at this μ-value.

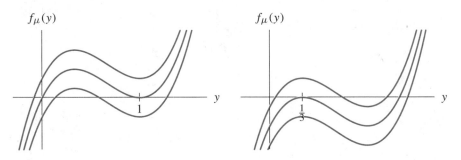

Figure 1.87
Graphs of $f_\mu(y) = y(1 - y)^2 + \mu$ for μ slightly greater than zero, μ equal to zero, and μ slightly less than zero.

Figure 1.88
Graphs of $f_\mu(y) = y(1 - y)^2 + \mu$ for μ slightly greater than $-4/27$, for μ equal to $-4/27$, and for μ slightly less than $-4/27$.

To find this bifurcation value exactly, we must find the μ-values for which the graph of f_μ is tangent to the horizontal axis. That is, we must find the μ-values for which, at some equilibrium point y, we have $f_\mu'(y) = 0$. Since

$$f_\mu'(y) = (1 - y)^2 - 2y(1 - y) = (1 - y)(1 - 3y),$$

it follows that the graph of $f_\mu(y)$ is horizontal at the two points $y = 1$ and $y = 1/3$. We know that the graph of $f_0(y)$ is tangent to the horizontal axis $y = 1$, so let's look at $y = 1/3$. We have $f_\mu(1/3) = \mu + 4/27$, so the graph is also tangent to the horizontal axis if $\mu = -4/27$. This is our second bifurcation value. Using analogous arguments to those above, we find that f_μ has three equilibria for $-4/27 < \mu < 0$ and only one equilibrium point when $\mu < -4/27$. The bifurcation diagram summarizes all this information in one picture (see Figure 1.89).

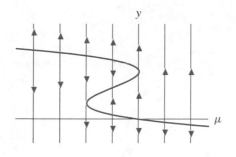

Figure 1.89
Bifurcation diagram for

$$\frac{dy}{dt} = f_\mu(y) = y(1-y)^2 + \mu.$$

Note the two bifurcation values of μ,
$\mu = -4/27$ and $\mu = 0$.

Harvesting of Natural Resources

When harvesting a natural resource, it is important to control the amount harvested so that the resource is not completely depleted. To accomplish this, we must study the particular species involved and pay close attention to the possible changes that may occur if the harvesting level is increased.

Suppose we model the population $P(t)$ of a particular species of fish with a logistic model

$$\frac{dP}{dt} = kP\left(1 - \frac{P}{N}\right),$$

where k is the growth-rate parameter and N is the carrying capacity of the habitat. Suppose that fishing removes a certain constant number C (for catch) of fish per season from the population. Then a modification of the model that takes fishing into account is

$$\frac{dP}{dt} = k\left(1 - \frac{P}{N}\right)P - C.$$

How does the population of fish vary as C is increased?

This model has three parameters, k, N, and C; but we are concerned only with what happens if C is varied. Therefore we think of k and N as fixed constants determined by the type of fish and their habitat. Our predictions involve the values of k and N. For example, if $C = 0$, we know from Section 1.1 that all positive initial conditions yield solutions that tend toward the equilibrium point $P = N$. So if fishing is prohibited, we expect the population to be close to $P = N$.

Let

$$f_C(P) = k\left(1 - \frac{P}{N}\right)P - C.$$

As C increases, the graph of $f_C(P)$ slides down (see Figure 1.90). The points where $f_C(P)$ crosses the P-axis tend toward each other. In other words, the equilibrium points for the corresponding differential equations slide together.

$f_C(P)$

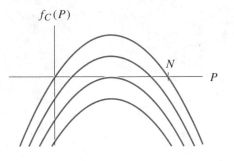

Figure 1.90
Graphs of

$$f_C(P) = k\left(1 - \frac{P}{N}\right)P - C$$

for several values of C. Note that, as C increases, the graph of $f_C(P)$ slides down the vertical axis.

We can compute the equilibrium points by solving $f_C(P) = 0$, which yields

$$k\left(1 - \frac{P}{N}\right)P - C = 0$$

or, equivalently,

$$-kP^2 + kNP - CN = 0.$$

This quadratic equation has solutions

$$P = \frac{N}{2} \pm \sqrt{\frac{N^2}{4} - \frac{CN}{k}}.$$

As long as the term under the square root (the discriminant of the quadratic) is positive, the function crosses the horizontal axis twice and the corresponding differential equation has two equilibrium points — a source and a sink. Thus, for small values of C, the phase line has two equilibrium points (see Figure 1.90).

If

$$\frac{N^2}{4} - \frac{CN}{k} < 0,$$

then the graph of $f_C(P)$ does not cross the P-axis and the corresponding differential equation has no equilibrium points. Thus, if

$$\frac{N^2}{4} < \frac{CN}{k}$$

or equivalently if

$$C > \frac{kN}{4},$$

then there are no equilibria. For these values of C, the function $f_C(P)$ is negative for all values of P and the solutions of the corresponding differential equation tend toward $-\infty$. Since negative populations do not make any sense, we say that the species has become extinct when the population reaches zero.

With this information, we can sketch the bifurcation diagram for this system (see Figure 1.91). A bifurcation occurs as we increase C. The bifurcation value for the parameter C is $kN/4$ because, at this value, the graph of $f_C(P)$ is tangent to the P-axis. The corresponding differential equation has a node at $P = N/2$. If C is slightly less

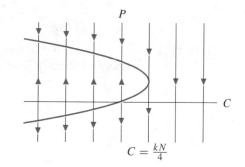

Figure 1.91
Bifurcation diagram for

$$\frac{dP}{dt} = f_C(P) = k\left(1 - \frac{P}{N}\right)P - C.$$

Note that if $C < kN/4$, the phase line has two equilibrium points, whereas if $C > kN/4$, the phase line has no equilibrium points and all solutions decrease.

than $kN/4$, the corresponding differential equation has two equilibrium points, a source and a sink, near $P = N/2$. If C is slightly greater than $kN/4$, the corresponding differential equation has no equilibrium points (see Figure 1.91).

It is interesting to consider what happens to the fish population as the parameter C is slowly increased. If $C = 0$, the population tends to the sink at $P = N$. Then, if there is a relatively small amount of fishing, the fish population is close to $P = N$. That is, if C is slightly positive, the sink for $C = 0$ at $P = N$ moves to the slightly smaller value

$$P = \frac{N}{2} + \sqrt{\frac{N^2}{4} - \frac{CN}{k}}.$$

For somewhat larger values of C, the value of the sink continues to decrease, and the fish population adjusts to stay close to this sink. We observe a gradual decrease in the fish population. When C is close to $kN/4$, the fish population is close to the sink for the corresponding differential equation, which is close to $P = N/2$. If C increases just a little more so that $C > kN/4$, then the corresponding differential equation has no equilibrium points and all solutions decrease. If C is slightly larger than $kN/4$, $f_C(P)$ is slightly negative near $P = N/2$, so the population decreases slowly at first. As P decreases, $f_C(P)$ becomes more negative and the rate of decrease of P accelerates. The population reaches zero in a finite amount of time, and the fish species becomes extinct.

So as the number of fish removed by fishing increases gradually, we initially expect a gradual decline in the fish population. This decline continues until the fishing parameter C reaches the bifurcation value $C = kN/4$. At this point, if we allow even slightly more fishing, the fish population decreases slowly at first and then collapses, and the fish become extinct in the area. This is a pretty frightening scenario. The fact that a little fishing causes only a small population decline over the long term does not necessarily imply that a little more fishing causes only a little more population decline. Once the bifurcation value is passed, the fish population tends to zero.

This model is a very simple one, and as such it should not be taken too seriously. The lesson to be learned is that, if this sort of behavior can be observed in simple models, we would expect that the same (and even more surprising behavior) occurs in more complicated models and in the actual populations. To properly manage resources, we need to have accurate models and to be aware of possible bifurcations.

EXERCISES FOR SECTION 1.7

In Exercises 1–4, locate the bifurcation values for the one-parameter family and draw the phase lines for values of the parameter slightly smaller than, slightly larger than, and at the bifurcation value.

1. $\dfrac{dy}{dt} = y^2 + a$

2. $\dfrac{dy}{dt} = y^2 + 3y + a$

3. $\dfrac{dy}{dt} = y^2 - ay + 1$

4. $\dfrac{dy}{dt} = \cos y + a$

5. For the one-parameter family

$$\frac{dy}{dt} = y^6 - 2y^3 + \alpha,$$

identify the bifurcation values of α and describe the bifurcations that take place as α increases. [*Hint*: Rewrite y^6 as $(y^3)^2$ and use the quadratic equation to find the equilibrium points.]

6. For the one-parameter family

$$\frac{dy}{dt} = y^6 - 2y^4 + \alpha,$$

identify the bifurcation values of α and describe the bifurcations that take place as α increases. [*Hint*: It might be useful to look at the graph of the right-hand side of the equation for various α.]

7. Consider the population model

$$\frac{dP}{dt} = -\frac{P^2}{50} + 2P,$$

for a species of fish in a lake. Suppose it is decided that fishing will be allowed, but it is unclear how many fishing licenses should be issued. Suppose the average catch of a fisherman with a license is 3 fish per year (these are hard fish to catch).

(a) What is the largest number of licenses that can be issued if the fish are to have a chance to survive in the lake?

(b) Suppose the number of fishing licenses in part (a) is issued. What will happen to the fish population — that is, how does the behavior of the population depend on the initial population?

(c) The simple population model above can be thought of as a model of an ideal fish population that is not subject to many of the environmental problems of an actual lake. For the actual fish population, there will be occasional changes in the population that were not considered when this model was constructed. For example, if the water level increases due to a heavy rainstorm, a few extra fish might be able to swim down a usually dry stream bed to reach the lake, or the extra water might wash toxic waste into the lake, killing a few fish. Given the possibility of unexpected perturbations of the population not included in the model, what do you think will happen to the actual fish population if we allow fishing at the level determined in part (b)?

8. Consider our model

$$\frac{dS}{dt} = f(S) = kS\left(1 - \frac{S}{N}\right)\left(\frac{S}{M} - 1\right)$$

of a fox squirrel population from the previous section. Suppose that the parameters M and k remain relatively constant over the long term but as more people move into the area, the parameter N (the carrying capacity) decreases.

   (a) Assuming that $M \leq N$, sketch the graph of the function $f(S)$ for fixed values of k and M and several values of N.

   (b) At what value of N does a bifurcation occur?

   (c) How does the population of fox squirrels behave if the parameter N slowly and continuously decreases toward the bifurcation value?

9. Suppose that an increase in the human population of a region makes that region less desirable to fox squirrels and they emigrate. We can adjust the fox squirrel population model to reflect this emigration by subtracting a fixed rate E of emigration. We have

$$\frac{dS}{dt} = kS\left(1 - \frac{S}{N}\right)\left(\frac{S}{M} - 1\right) - E,$$

where S, k, M, and N are as in Exercise 8. Suppose that the parameters k, M, and N are fixed.

   (a) Sketch the graph of the function

$$f_E(S) = kS\left(1 - \frac{S}{N}\right)\left(\frac{S}{M} - 1\right) - E$$

   for several values of E.

   (b) What bifurcation takes place as E varies? How does the population of fox squirrels behave if the parameter E is slowly and continuously increased past the bifurcation value?

   (c) What is the bifurcation value of E? (Your answer will be in terms of the parameters k, M, and N.)

10. For the differential equation that models fish populations with harvesting,

$$\frac{dP}{dt} = f_C(P) = kP\left(1 - \frac{P}{N}\right) - C,$$

we saw that if $C > kN/4$ the fish population will become extinct. If the fish population falls to near zero because the fishing level C is slightly greater than $kN/4$, why must fishing be banned completely in order for the population to recover? That is, if a level of fishing just above $C = kN/4$ causes a collapse of the population, why can't the population be restored by reducing the fishing level to just below $C = kN/4$?

The differential equations in Exercises 11 and 12 depend on two parameters, α and β, and the qualitative aspects of the phase line depend on the values of both parameters. Determine all of the possible qualitatively distinct phase lines and the corresponding regions in the $\alpha\beta$-plane.

11. $\dfrac{dy}{dt} = y^2 + \alpha y + \beta$ **12.** $\dfrac{dy}{dt} = y^4 + \alpha y^2 + \beta$

In Exercise 34 in Section 1.6, we defined the index of an equilibrium point for an autonomous differential equation. A source has index $+1$, a sink has index -1, and nodes have index 0. Consider the one-parameter family of differential equations $dy/dt = f_\alpha(y)$, where f depends continuously on α and y. In Exercises 13–15 we suppose that this differential equation has finitely many equilibrium points in the interval $0 \leq y \leq 1$ for all α.

13. Suppose that for $\alpha = 0$, the sum of the indices of the equilibrium points in the interval $0 \leq y \leq 1$ is -1. If $f_\alpha(0) \neq 0$ and $f_\alpha(1) \neq 0$ for all α, show that $dy/dt = f_\alpha(y)$ must have at least one sink in $0 \leq y \leq 1$ for all α.

14. Suppose that for $\alpha = 0$, the sum of the indices of the equilibrium points in the interval $0 \leq y \leq 1$ is $+1$, but for $\alpha = 1$, the sum of the indices of the equilibrium points in the interval $0 \leq y \leq 1$ is 0. Show that $f_\alpha(0) = 0$ or $f_\alpha(1) = 0$ for some α between 0 and 1.

15. Suppose that for $\alpha = 0$, the sum of the indices of the equilibrium points in the interval $0 \leq y \leq 1$ is 0. If for each α, $f_\alpha(0) > 0$ and $f_\alpha(1) \neq 0$, show that there is a positive number M_α depending on α such that $dy/dt = f_\alpha(y) + M_\alpha$ has no equilibrium points in the interval $0 \leq y \leq 1$.

Suppose that n is a positive, even integer. Then the differential equation

$$\frac{dy}{dt} = y^n,$$

has a node at $y = 0$ and no other equilibrium points. One might be tempted to try to classify all the different bifurcations that can occur. In Exercises 16–19, we consider one-parameter families that include this differential equation.

16. For the one-parameter family

$$\frac{dy}{dt} = y^3 + \alpha y^2,$$

how many equilibrium points are near $y = 0$ if α is just slightly greater than zero? How many are there if α is just slightly less than zero?

17. For the one-parameter family

$$\frac{dy}{dt} = y^4 + \alpha y^2,$$

how many equilibrium points are near $y = 0$ if α is just slightly greater than zero? How many equilibrium points are near zero if α is just slightly less than zero?

18. For the one-parameter family

$$\frac{dy}{dt} = y^6 + \alpha(y^2 - 1000y^4),$$

how many equilibrium points are near $y = 0$ if α is just slightly greater than zero? How many equilibrium points are near $y = 0$ if α is just slightly less than zero?

19. Find a one-parameter family of differential equations with parameter α that has exactly one equilibrium point if $\alpha = 0$ and exactly six equilibrium points if $\alpha < 0$.

20. For the one-parameter family of differential equations

$$\frac{dy}{dt} = \frac{y}{y^2 + 1} + \alpha,$$

locate the equilibrium points if α is slightly greater than zero, if $\alpha = 0$, and if α is slightly less than zero. Describe the bifurcation that occurs at $\alpha = 0$.

1.8 LINEAR DIFFERENTIAL EQUATIONS

In Section 1.2 we developed an analytic method for finding the explicit solutions to separable differential equations. Although many interesting problems lead to separable differential equations, most differential equations are not separable. The qualitative and numerical techniques we developed in Sections 1.5 and 1.6 apply to a much wider range of problems. It would be nice if we could also extend our analytic methods by developing ways of finding explicit solutions of equations that are not separable.

Unfortunately, there is no general way to compute explicit solutions that works for all differential equations, and in fact the problem is even worse. Although we know from the Existence Theorem that every reasonable differential equation has solutions, there is no guarantee that these solutions are at all familiar. Usually they are not. Over the centuries, this dilemma led to the development of many solution techniques for special differential equations, and today these techniques are available to us as one-line commands in sophisticated computer packages such as Maple and *Mathematica*. Nevertheless, a student of the subject should be familiar with a few of the standard analytic techniques, and in this section we develop one of the standard techniques for solving *linear* differential equations.

Linear Differential Equations

A first-order differential equation is **linear** if it can be written in the form

$$\frac{dy}{dt} = g(t)y + r(t),$$

where $g(t)$ and $r(t)$ are arbitrary functions of t. Examples of linear equations include

$$\frac{dy}{dt} = t^2 y + \cos t,$$

where $g(t) = t^2$ and $r(t) = \cos t$, and

$$\frac{dy}{dt} = \frac{e^{4\sin t}}{t^3 + 7t}y + 23t^3 - 7t^2 + 3,$$

where $g(t) = e^{4\sin t}/(t^3 + 7t)$ and $r(t) = 23t^3 - 7t^2 + 3$. Sometimes it is necessary to do a little algebra in order to see that an equation is linear. For example, the differential equation

$$ty + 2 = \frac{dy}{dt} - 3y$$

can be rewritten as

$$\frac{dy}{dt} = (t + 3)y + 2.$$

In this form we see that the equation is linear with $g(t) = t + 3$ and $r(t) = 2$. Some differential equations fit into several categories. For example, the equation

$$\frac{dy}{dt} = -2y + 8$$

is linear with $g(t) = -2$ and $r(t) = 8$. (Both g and r are very simple functions of t.) It is also separable because it is autonomous.

The word *linear* in the name of these equations refers to the fact that the dependent variable y appears in the equation only to the first power. The equation

$$\frac{dy}{dt} = y^2$$

is not linear because the y^2 term cannot be rewritten in the form $g(t)y + r(t)$, no matter how $g(t)$ and $r(t)$ are chosen. Of course, there is nothing magical about the names of the variables. The equation

$$\frac{dP}{dt} = e^{2t}P - \sin t$$

is linear ($g(t) = e^{2t}$ and $r(t) = -\sin t$). Also,

$$\frac{dw}{dt} = (\sin t)w$$

is both linear (where $g(t) = \sin t$ and $r(t) = 0$) and separable. But

$$\frac{dz}{dt} = t \sin z$$

is not linear.

Solving Linear Differential Equations

Given a linear differential equation

$$\frac{dy}{dt} = g(t)y + r(t),$$

how can we go about finding the general solution? There is a clever trick that turns an equation of this form into a differential equation that can be solved by integration. As with many techniques in mathematics, the cleverness of this trick might leave you feeling a little depressed—the "how could I ever think of something like that?" feeling. The thing to remember is that differential equations have been around for more than 300 years. If you were given 300 years to work on it, you would certainly come up with a method for solving these equations.

Integrating factors

To see how to solve a linear differential equation, we first rewrite it as

$$\frac{dy}{dt} + a(t)y = r(t),$$

where $a(t) = -g(t)$. After staring at this equation for a while (a couple of decades or so), we notice that, with poor eyesight, the left-hand side looks somewhat like what we get when we differentiate using the Product Rule. The Product Rule says that the derivative of the product of $y(t)$ with a function $\mu(t)$ is

$$\frac{d(\mu(t) \cdot y(t))}{dt} = \mu(t)\frac{dy}{dt} + \frac{d\mu}{dt}y(t).$$

Now here's the clever part. Let's multiply both sides of the differential equation by an (as yet unspecified) function $\mu(t)$. We obtain

$$\mu(t)\frac{dy}{dt} + \mu(t)a(t)y = \mu(t)r(t).$$

The left-hand side looks even more like the derivative of a product of two functions. For the moment, let's *assume* that we have found a function $\mu(t)$ so that the left-hand side actually is the derivative of $\mu(t) \cdot y(t)$. That is, suppose we have found a function $\mu(t)$ that satisfies

$$\frac{d(\mu(t) \cdot y(t))}{dt} = \mu(t)\frac{dy}{dt} + \mu(t)a(t)y.$$

Then the differential equation becomes

$$\frac{d(\mu(t) \cdot y(t))}{dt} = \mu(t)r(t).$$

How does this help? We can now integrate both sides of this equation with respect to t to obtain

$$\mu(t)\,y(t) = \int \mu(t)\,r(t)\,dt$$

so that

$$y(t) = \frac{1}{\mu(t)}\int \mu(t)\,r(t)\,dt.$$

That is, assuming we have $\mu(t)$ and can evaluate this integral, we can compute our solution $y(t)$.

Finding the integrating factor

Therefore, the question is: How do we find such a function $\mu(t)$ in the first place? The Product Rule says that

$$\frac{d(\mu(t) \cdot y(t))}{dt} = \mu(t)\frac{dy}{dt} + \frac{d\mu}{dt}y(t),$$

but given the previous assumptions on $\mu(t)$, we know that we also want

$$\frac{d(\mu(t) \cdot y(t))}{dt} = \mu(t)\frac{dy}{dt} + \mu(t)a(t)y(t).$$

Because we want $\mu(t)$ to satisfy both of these equations, we have

$$\mu(t)\frac{dy}{dt} + \frac{d\mu}{dt}y(t) = \mu(t)\frac{dy}{dt} + \mu(t)a(t)y(t).$$

Canceling the first term on both sides yields

$$\frac{d\mu}{dt}y(t) = \mu(t)\,a(t)\,y(t).$$

Dividing by $y(t)$, we see that we need to find a function $\mu(t)$ that satisfies the equation

$$\frac{d\mu}{dt} = \mu(t)\,a(t).$$

This equation is another differential equation — an equation for the unknown function $\mu(t)$, and it is a separable equation!

We can separate variables and obtain

$$\frac{1}{\mu}\frac{d\mu}{dt} = a(t),$$

which leads to the solution

$$\mu(t) = e^{\int a(t)\,dt}.$$

Given this formula for $\mu(t)$, we now know that this strategy is going to work and we can solve a first-order linear differential equation. The function $\mu(t)$ is called an **integrating factor** for the original differential equation because, if we multiply the original equation by $\mu(t)$, we can then solve the new equation by integration. (When we calculate $\mu(t)$, there is an arbitrary constant of integration in the exponent. Since we need only one integrating factor $\mu(t)$ to solve the equation, we choose the constant to be whatever is most convenient. That choice is usually zero.)

To summarize, whenever we want to determine an explicit solution to a linear differential equation of the form

$$\frac{dy}{dt} + a(t)y = r(t),$$

we first compute the integrating factor

$$\mu(t) = e^{\int a(t)\,dt}.$$

Then we can determine the solutions by multiplying both sides of the linear differential equation by $\mu(t)$ and integrating. To see this method at work, let's look at some examples. The method looks very general. However, because there are two integrals to calculate, it is quite easy to get stuck before we obtain an explicit solution.

Complete success

Consider the differential equation

$$\frac{dy}{dt} + \frac{2}{t}y = t - 1.$$

This equation is linear with $a(t) = 2/t$ and $r(t) = t - 1$. We first compute the integrating factor

$$\mu(t) = e^{\int a(t)\,dt} = e^{\int (2/t)\,dt} = e^{2\ln t} = e^{\ln(t^2)} = t^2.$$

Remember that the idea behind this method is multiply both sides by $\mu(t)$ so that the left-hand side of the new equation is the result of the Product Rule. In this case, multiplying by $\mu(t) = t^2$ yields

$$t^2 \frac{dy}{dt} + 2ty = t^2(t - 1).$$

Note that the left-hand side is the derivative of the product of t^2 and $y(t)$. Thus this equation is the same as

$$\frac{d}{dt}(t^2 y) = t^2(t - 1) = t^3 - t^2.$$

Integrating both sides with respect to t yields

$$t^2 y = \frac{t^4}{4} - \frac{t^3}{3} + c,$$

where c is an arbitrary constant. The general solution is

$$y(t) = \frac{t^2}{4} - \frac{t}{3} + \frac{c}{t^2}.$$

Of course, we can check that these functions satisfy the differential equation by substituting them back into the equation.

Problems with the integration

The previous example was rather carefully chosen. Another example which doesn't look any more difficult is

$$\frac{dy}{dt} = t^2 y + (t - 1),$$

which is a linear differential equation with $g(t) = t^2$ and $r(t) = t - 1$. Again we rewrite the differential equation as

$$\frac{dy}{dt} - t^2 y = (t - 1)$$

and compute the integrating factor

$$\mu(t) = e^{\int -t^2\,dt} = e^{-t^3/3}.$$

Now we multiply both sides by $\mu(t)$ and obtain

$$e^{-t^3/3}\frac{dy}{dt} - e^{-t^3/3}t^2\,y = e^{-t^3/3}(t-1).$$

Note that the left-hand side is the derivative of the product of $e^{-t^3/3}$ and $y(t)$, so we have

$$\frac{d}{dt}\left(e^{-t^3/3}\,y\right) = e^{-t^3/3}(t-1).$$

Integrating both sides yields

$$e^{-t^3/3}\,y = \int e^{-t^3/3}(t-1)\,dt,$$

but then we are stuck. It turns out that the integral on the right-hand side of this equation is not expressible in terms of the familiar functions (sin, cos, ln, and so on), so we can't obtain explicit formulas for the solutions.

This example indicates what can go wrong with techniques that involve the calculation of explicit integrals. Even reasonable-looking functions can quickly lead to complicated integrating factors and very complicated integrals. On the other hand, we can express the solution in terms of integrals with respect to t (which is some progress), and although many integrals are impossible to calculate explicitly, many others are possible. Indeed, as we mentioned at the beginning of this section, there are now a number of computer programs that are quite good at calculating the indefinite integrals involved in this technique.

Mixing Problems Revisited

In Section 1.2 we considered a model of the concentration of a substance in solution. Typically in these problems we have a container in which there is a certain amount of fluid (such as water or air) and to which a contaminant is added at some rate. The fluid is kept well mixed at all times. If the total volume of fluid is kept fixed, then the resulting differential equation for the amount of contaminant is an autonomous differential equation. If the total volume of fluid changes with time, then the differential equation is nonautonomous. If the problem requires finding an exact value for the amount or concentration of the contaminant at a particular time, then we need to use techniques like those of this section.

A polluted pond

Consider a pond that has an initial volume of 10,000 cubic meters. Suppose that at time $t = 0$, the water in the pond is clean and that the pond has two streams flowing into it, stream A and stream B, and one stream flowing out, stream C (see Figure 1.92). Suppose 500 cubic meters per day of water flow into the pond from stream A, 750 cubic meters per day flow into the pond from stream B, and 1250 cubic meters flow out of the pond via stream C.

At time $t = 0$, the water flowing into the pond from stream A becomes contaminated with road salt at a concentration of 5 kilograms per 1000 cubic meters. Suppose the water in the pond is well mixed so the concentration of salt at any given time is constant. To make matters worse, suppose also that at time $t = 0$ someone begins dumping trash into the pond at a rate of 50 cubic meters a day. The trash settles to the bottom of the pond, reducing the volume by 50 cubic meters per day. To adjust for the incoming trash, the rate that water flows out via stream C increases to 1300 cubic meters per day and the banks of the pond do not overflow.

The description looks very much like the mixing problems we have already considered (where "pond" replaces "vat" and "stream" replaces "pipe"). The new element here is that the total volume is not constant. Because of the dumping of trash, the volume decreases by 50 cubic meters per day.

If we let $S(t)$ be the amount of salt (in kilograms) in the pond at time t, then dS/dt is the difference between the rate that salt enters the pond and the rate that salt leaves the pond. Salt enters the pond from stream A only, and the rate at which it enters is the product of its concentration in the water and the rate at which the water flows in through stream A. Since the concentration is 5 kilograms per 1000 cubic meters and the rate that water flows into the pond from stream A is 500 cubic meters per day, the rate at which salt enters the pond is $(500)(5/1000) = 5/2$ kilograms per day. The rate at which the salt leaves via stream C is the product of its concentration *in the pond* and the rate at which water flows out of the pond. The rate at which water flows out is 1300 cubic meters per day. To determine the concentration, we note that it is the quotient of the amount S of salt in the pond by the volume V. Because the volume is initially 10,000 cubic meters and it decreases by 50 cubic meters per day, we know that $V(t) = 10,000 - 50t$. Hence, the concentration is $S/(10,000 - 50t)$, and the rate at which salt flows out of the pond is

$$1300 \left(\frac{S}{10,000 - 50t} \right) = \frac{26S}{200 - t}.$$

The differential equation that models the amount of salt in the pond is therefore

$$\frac{dS}{dt} = \frac{5}{2} - \frac{26S}{200 - t}.$$

This model is valid only as long as there is water in the pond — that is, as long as the volume $V(t) = 10,000 - 50t$ is positive. So the differential equation is valid for $0 \le t < 200$. Because the water is clean at time $t = 0$, the initial condition is $S(0) = 0$.

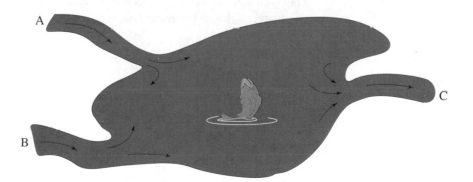

Figure 1.92
Schematic of the pond with three streams.

The differential equation for salt in the pond is nonautonomous. Its slope field is given in Figure 1.93. From the slope field or by using Euler's method, we could approximate the solution with initial value $S(0) = 0$. Because the equation is linear, we can also find a formula for the solution.

Rewriting the differential equation as

$$\frac{dS}{dt} + \frac{26S}{200 - t} = \frac{5}{2}$$

indicates that the integrating factor is

$$\mu(t) = e^{\int \frac{26}{200-t} \, dt} = e^{-26 \ln(200-t)} = e^{\ln((200-t)^{-26})} = (200 - t)^{-26}.$$

Multiplying both sides by $\mu(t)$ gives

$$(200 - t)^{-26} \frac{dS}{dt} + 26(200 - t)^{-27} S = \frac{5}{2}(200 - t)^{-26}.$$

By the Product Rule, this equation is the same as the differential equation

$$\frac{d}{dt}\left((200 - t)^{-26} S\right) = \frac{5}{2}(200 - t)^{-26}.$$

Integrating both sides yields

$$(200 - t)^{-26} S = \frac{5}{2} \int (200 - t)^{-26} \, dt$$

$$= \frac{5}{2} \frac{(200 - t)^{-25}}{25} + c,$$

where c is an arbitrary constant. Solving for S, we obtain the general solution

$$S = \frac{200 - t}{10} + c(200 - t)^{26}.$$

Using the initial condition $S(0) = 0$, we find that $c = -20/200^{26}$ and the particular solution for the initial-value problem is

$$S = \frac{200 - t}{10} - 20\left(\frac{200 - t}{200}\right)^{26}.$$

This is an unusual-looking expression because of the large number 200^{26}. However, the graph reveals that its behavior is not at all unusual (see Figure 1.94). The amount of salt

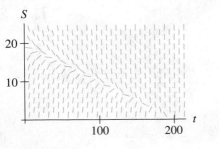

Figure 1.93
Slope field for the equation
$dS/dt = 5/2 - 26S/(200 - t)$.

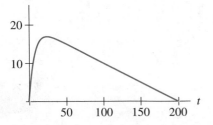

Figure 1.94
Graph of solution of $dS/dt = 5/2 - 26S/(200 - t)$, with $S(0) = 0$.

in the pond rises fairly quickly, reaching a maximum close to $S = 20$ at $t \approx 25$. After that time, the amount of salt decreases almost linearly, reaching zero at $t = 200$.

The behavior of this solution is quite reasonable if we recall that the pond starts out containing no salt and that eventually it is completely filled with trash. (It contains no salt or water at time $t = 200$.) As we mentioned above, the concentration of salt in the pond water is given by $C(t) = S(t)/V(t) = S(t)/(10,000 - 50t)$. Graphing $C(t)$, we see that it increases asymptotically toward 0.002 kilograms per cubic meter even as the water level decreases (see Figure 1.95).

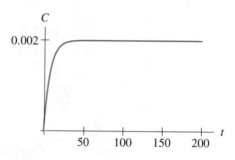

Figure 1.95
Graph of concentration of salt versus time for the solution graphed in Figure 1.94.

EXERCISES FOR SECTION 1.8

In Exercises 1–8, find the general solution of the differential equation specified.

1. $\dfrac{dy}{dt} = -\dfrac{y}{t} + 2$

2. $\dfrac{dy}{dt} = \dfrac{3}{t}y + t^5$

3. $\dfrac{dy}{dt} = y - 3e^{-t}$

4. $\dfrac{dy}{dt} = -2ty + 4e^{-t^2}$

5. $\dfrac{dy}{dt} = \dfrac{2t}{1+t^2}y + \dfrac{2}{1+t^2}$

6. $\dfrac{dy}{dt} = -2y - 5t$

7. $\dfrac{dy}{dt} = -y + t^2$

8. $\dfrac{dy}{dt} = -2y + \sin t$

In Exercises 9–14, find the particular solution with the given initial condition.

9. $\dfrac{dy}{dt} = -\dfrac{y}{1+t} + 2, \quad y(0) = 3$

10. $\dfrac{dy}{dt} = -y + e^t, \quad y(0) = 0.4$

11. $\dfrac{dy}{dt} = -\dfrac{y}{t} + 2, \quad y(1) = 3$

12. $\dfrac{dy}{dt} = \dfrac{2t}{1+t^2}y + \dfrac{2}{1+t^2}, \quad y(0) = -2$

13. $\dfrac{dy}{dt} = \dfrac{2y}{t} + 2t^2, \quad y(-2) = 4$

14. $\dfrac{dy}{dt} = -5y + \sin t, \quad y(0) = 1$

In Exercises 15–18, the differential equations are linear, so in theory we can find their general solutions. However, since finding the general solution of a linear equation involves computing at least two integrals, in practice it is frequently impossible to reach a formula for the solution that is free of integrals. For these exercises, determine the general solution to the equation and express it with as few integrals as possible.

15. $\dfrac{dy}{dt} = (\sin t)y + 4$

16. $\dfrac{dy}{dt} = t^2 y + 4$

17. $\dfrac{dy}{dt} = \dfrac{y}{t^2} + 4\cos t$

18. $\dfrac{dy}{dt} = y + 4\cos t^2$

19. For what value(s) of the parameter a is it possible to find explicit formulas (without integrals) for the solutions to

$$\frac{dy}{dt} = aty + 4e^{-t^2}?$$

20. For what value(s) of the parameter r is it possible to find explicit formulas (without integrals) for the solutions to

$$\frac{dy}{dt} = t^r y + 4?$$

21. A person initially places $500 in a savings account that pays interest at the rate of 4% per year compounded continuously. Suppose the person arranges for $10 per week to be deposited automatically into the savings account.

(a) Write a differential equation for $P(t)$, the amount on deposit after t years (assume that "weekly deposits" is close enough to "continuous deposits" so that we may model the balance with a differential equation.)

(b) Find the amount on deposit after 5 years.

22. A student has saved $30,000 for her college tuition. When she starts college, she invests the money in a savings account that pays 6% interest per year, compounded continuously. Suppose her college tuition is $10,000 per year and she arranges with the college that the money will be deducted from her savings account in small payments. In other words, we assume that she is paying continuously. How long will she be able to stay in school before she runs out of money?

23. A college professor contributes $1200 per year into her retirement fund by making many small deposits throughout the year. The fund grows at a rate of 8% per year compounded continuously. After 30 years, she retires and begins withdrawing from her fund at a rate of $3000 per month. If she does not make any deposits after retirement, how long will the money last? [*Hint*: Solve this in two steps, before retirement and after retirement.]

24. A 30-gallon tank initially contains 15 gallons of saltwater containing 6 pounds of salt. Suppose saltwater containing 1 pound of salt per gallon is pumped into the top of the tank at the rate of 2 gallons per minute, while a well-mixed solution leaves the bottom of the tank at a rate of 1 gallon per minute. How much salt is in the tank when the tank is full?

25. A 400-gallon tank initially contains 200 gallons of water containing 2 parts per billion by weight of dioxin, an extremely potent carcinogen. Suppose water containing 5 parts per billion of dioxin flows into the top of the tank at a rate of 4 gallons per minute. The water in the tank is kept well mixed, and 2 gallons per minute are removed from the bottom of the tank. How much dioxin is in the tank when the tank is full?

26. A 100-gallon tank initially contains 100 gallons of sugar-water at a concentration of 0.25 pounds of sugar per gallon. Suppose that sugar is added to the tank at a rate of p pounds per minute, that sugar-water is removed at a rate of 1 gallon per minute, and that the water in the tank is kept well mixed.

    (a) What value of p should we pick so that, when 5 gallons of sugar solution is left in the tank, the concentration is 0.5 pounds of sugar per gallon?

    (b) Is it possible to choose p so that the last drop of water out of the bucket has a concentration of 0.75 pounds of sugar per gallon?

27. Suppose a 50-gallon tank contains a volume V_0 of clean water at time $t = 0$. At time $t = 0$, we begin dumping 2 gallons per minute of salt solution containing 0.25 pounds of salt per gallon into the tank. Also at time $t = 0$, we begin removing 1 gallon per minute of saltwater from the tank. As usual, suppose the water in the tank is well mixed so that the salt concentration at any given time is constant throughout the tank.

    (a) Set up the initial-value problem for the amount of salt in the tank. [*Hint*: The initial value of V_0 will appear in the differential equation.]

    (b) What is your model equation when $V_0 = 0$ (when the tank is initially empty)? Comment on the validity of the model in this situation. What will be the amount of salt in the tank at time t for this situation?

1.9 CHANGING VARIABLES

Many of the techniques we have dealt with in this chapter concern particular types of equations. For example, the analytic techniques we have studied apply only to separable and linear equations, and the concept of the phase line applies only to autonomous equations.

    Given a differential equation, it is unlikely that it will immediately be in a form that is appropriate for any particular technique, but it would be nice if we could transform it into a new equation that has a standard form. This type of transformation is the idea behind changing variables. By rewriting the equation in terms of a new variable, we may be able to put the equation into a form that is appropriate for a particular technique. In this section we give a method for transforming differential equations by changing variables into (hopefully) simpler forms.

u-Substitution

The idea behind changing variables is not new. We use it in calculus when we do "*u*-substitutions" to compute antiderivatives. For example, given the integral

$$\int t \sin t^2 \, dt,$$

we define a new variable u by $u = t^2$ and rewrite the integral in terms of u instead of t. Because $du/dt = 2t$ (sometimes informally written as $du = 2t \, dt$), we obtain

$$\int t \sin t^2 \, dt = \int \frac{1}{2} \sin u \, du.$$

This integral is the same one written in terms of the new variable u. The variable u was chosen so that the new integral is easy to evaluate. We have

$$\frac{1}{2} \int \sin u \, du = -\frac{\cos u}{2} + c.$$

We now find the solution to the original problem by replacing the new variable u with the original variable t, and we obtain

$$\int t \sin t^2 \, dt = \int \frac{\sin u}{2} \, du$$

$$= -\frac{\cos u}{2} + c$$

$$= -\frac{\cos t^2}{2} + c.$$

Changing variables from t to u converts a difficult computation into an easy one. After computing the antiderivative in terms of the new variable, we can recover the desired antiderivative by replacing the new variable with the old variable.

Changing Variables in Differential Equations

A similar *u*-substitution method works for differential equations. The idea is the same as for integrals: Define a new variable in terms of the old variable(s) and then rewrite the equation in terms of the new variable. This gives us the "same" differential equation expressed in terms of a new variable, but from an algebraic point of view, the new equation can be very different from the original equation. If we make a wise (or clever) choice of the new variable, then the new equation may be much easier to study than the old equation.

An example
Consider the differential equation

$$\frac{dy}{dt} = -\frac{y}{t} + \frac{t-1}{2y}.$$

This looks like a pretty complicated equation. It is neither autonomous, separable, nor linear, so if we had to find the general solution we would be at a loss using the techniques that we have already studied. But of course we don't give up. We try a little algebraic simplification first.

The most annoying term in this equation is the second fraction on the right. The fact that the dependent variable y appears in the denominator is what makes the equation nonlinear. If the y were not there, then the equation would be linear. So let's try to simplify by multiplying both sides by y to obtain

$$y\frac{dy}{dt} = -\frac{y^2}{t} + \frac{t-1}{2}.$$

This equation is still not linear, and there is a y on the left-hand side. However, the right-hand side of the equation would be the right-hand side of a linear equation if the y^2 term were only y. This motivates our substitution. We define a new dependent variable u by

$$u = y^2,$$

and we rewrite the equation in terms of u. This means eliminating all the y's from the equation, including dy/dt, and replacing them with u's (and du/dt). To express dy/dt in terms of u, we differentiate the equation $u = y^2$ to obtain

$$\frac{du}{dt} = 2y\frac{dy}{dt},$$

so

$$\frac{1}{2}\frac{du}{dt} = y\frac{dy}{dt}.$$

Substituting this expression for $y(dy/dt)$ on the left-hand side and replacing the y^2 with u on the right-hand side yields the new equation

$$\frac{1}{2}\frac{du}{dt} = -\frac{u}{t} + \frac{t-1}{2}.$$

Multiplying through by 2 gives

$$\frac{du}{dt} = -\frac{2}{t}u + (t-1).$$

This is a linear equation. In fact, it is one of the examples that we worked out in complete detail in Section 1.8. The general solution is

$$u(t) = \frac{t^2}{4} - \frac{t}{3} + \frac{c}{t^2}.$$

To find the general solution for $y(t)$ in the original equation, we just recall that $u = y^2$. Thus $y = \pm\sqrt{u}$ and

$$y(t) = \pm\sqrt{\frac{t^2}{4} - \frac{t}{3} + \frac{c}{t^2}},$$

where the choice of sign and the value of c depend on the initial condition.

This technique is pretty remarkable. By defining a new dependent variable and rewriting the differential equation in terms of that variable, we obtained a linear differential equation that we could solve. Of course, we had to figure out (or guess) that the "right" dependent variable or the right **change of variables** is $u = y^2$. Just as with u-substitution, it sometimes takes several attempts to find the best substitution, and often there just isn't a good substitution.

Changing Variables and Qualitative Analysis

The technique of changing variables is not a "magic bullet" with which all differential equations can be solved. Finding a change of variables that makes the differential equation manageable depends both on the form of the equation and on the goal of the analysis. In the following examples we consider various types of changes of variables.

An example

We can use the method of changing variables in concert with qualitative and numerical methods. For example, consider the very complicated equation

$$\frac{dy}{dt} = y^2 - 4ty + 4t^2 - 4y + 8t - 3.$$

Of course, we can draw the slope field for this equation and use Euler's method to sketch solutions (see Figure 1.96).

This equation is neither linear nor separable, so we might try to look for a change of variables that simplifies the equation algebraically. To make an intelligent guess of a new dependent variable, we first rewrite the equation using some algebra. After staring at the right-hand side for a while, we see that we can make it look a little simpler by collecting terms and factoring, that is,

$$\frac{dy}{dt} = y^2 - 4ty + 4t^2 - 4y + 8t - 3$$
$$= (y - 2t)^2 - 4(y - 2t) - 3.$$

This new form of the right-hand side of the equation suggests a possible choice of a new dependent variable. Let

$$u = y - 2t.$$

This new dependent variable u is a combination of the old dependent variable y and the

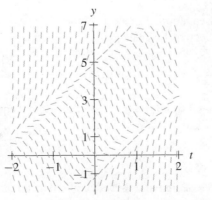

Figure 1.96
Slope field for

$$\frac{dy}{dt} = y^2 - 4ty + 4t^2 - 4y + 8t - 3.$$

Solutions seem to be increasing in two regions and decreasing in a third region.

independent variable t. To replace all the occurrences of the old variable y with the new dependent variable u, we compute dy/dt in terms of u by differentiating $u = y - 2t$. We get

$$\frac{du}{dt} = \frac{dy}{dt} - 2 \text{ or } \frac{dy}{dt} = \frac{du}{dt} + 2.$$

Thus the equation

$$\frac{dy}{dt} = (y - 2t)^2 - 4(y - 2t) - 3$$

becomes

$$\frac{du}{dt} + 2 = u^2 - 4u - 3,$$

which is equivalent to

$$\frac{du}{dt} = u^2 - 4u - 5.$$

This is an autonomous equation, so we can study it by drawing its phase line. Because $u^2 - 4u - 5 = (u - 5)(u + 1)$, the equilibrium points are $u = 5$ and $u = -1$. The phase line and a sketch of several solutions of this equation are given in Figure 1.97. The equilibrium point $u = -1$ is a sink, and the equilibrium point $u = 5$ is a source.

What does this tell us about the original differential equation? Because the new equation is separable, we can find explicit solutions by integration. However, we can also obtain information about the solutions $y(t)$ even more directly. We know that $u_1(t) = -1$ is an equilibrium solution for the new equation. But any solution of the new equation corresponds to a solution of the original equation. Hence

$$y_1(t) = u_1(t) + 2t = -1 + 2t$$

is a solution of the original equation.

Similarly, because $u_2(t) = 5$ is a solution of the new equation,

$$y_2(t) = u_2(t) + 2t = 5 + 2t$$

is a solution of the original equation. Every solution $u(t)$ of the new equation corresponds to a solution of the original equation by the change of variables $y(t) = u(t) + 2t$. Thus the graphs of solutions of the new equation in the tu-plane and the solutions of the

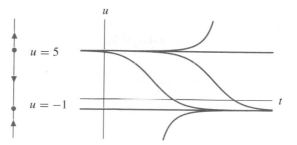

Figure 1.97
Phase line and graphs of solutions for $du/dt = u^2 - 4u - 5$.

original equation in the ty-plane are closely related. The ty-plane can be obtained from the tu-plane by adding $2t$ to every solution. In effect, this is the same as taking the tu-plane and shearing it upward with slope 2. The equilibrium solutions in u correspond to solutions whose graphs are lines with slope 2 in the ty-plane. Also, because $u = -1$ is a sink, solutions tend toward the solution $y_1(t) = -1 + 2t$ as t increases. Similarly, solutions tend away from the solution $y_2(t) = 5 + 2t$ because $u = 5$ is a source. So we may graph solutions in the ty-plane just from the corresponding graphs in the tu-plane (see Figure 1.98).

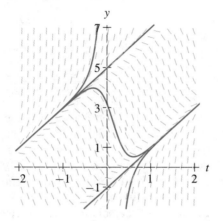

Figure 1.98

The slope field and graphs of solutions for the original equation

$$\frac{dy}{dt} = y^2 - 4ty + 4t^2 - 4y + 8t - 3.$$

Solutions tend toward the solution $y_1(t) = -1 + 2t$ as t increases. Similarly, solutions tend away from the solution $y_2(t) = 5 + 2t$.

Another example

For another example, consider the equation

$$\frac{dv}{dt} = v - \frac{v^2}{t} + \frac{v}{t}.$$

Although there might be several reasonable choices of new variables for this equation, we try to simplify the equation by getting rid of the fractions on the right-hand side of the equation. We try

$$u = \frac{v}{t},$$

so $tu = v$. The Product Rule yields

$$\frac{dv}{dt} = t\frac{du}{dt} + u,$$

and the new equation is

$$t\frac{du}{dt} + u = tu - \frac{(tu)^2}{t} + \frac{tu}{t}.$$

Simplifying, we find

$$t\frac{du}{dt} + u = tu - tu^2 + u,$$

which is equivalent to

$$\frac{du}{dt} = (1 - u)u.$$

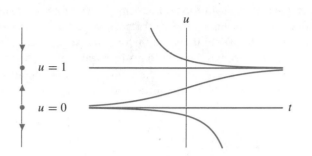

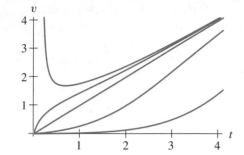

Figure 1.99
Phase line and graphs of solutions for the logistic
equation $du/dt = (1 - u)u$.

Figure 1.100
Graphs of solutions for the equation
$dv/dt = v - v^2/t + v/t$.

Note that this equation is our old friend, the logistic. We sketch the phase line and
solutions for this equation (probably from memory) (see Figure 1.99). The equilibrium
points $u = 0$ and $u = 1$ correspond to the constant solutions $u_0(t) = 0$ and $u_1(t) = 1$,
respectively, and the corresponding solutions in t and v are $v_0(t) = tu_0(t) = t \cdot 0 = 0$
and $v_1(t) = tu_1(t) = t \cdot 1 = t$. Moreover, because $u = 0$ is a source, solutions in the
tv-plane tend to move away from the solution $v_0(t)$ as t increases. Similarly, because
$u = 1$ is a sink, solutions in the tv-plane tend toward the solution $v_1(t)$ as t increases
(see Figure 1.100).

The graphs of solutions $v_0(t) = 0$ and $v_1(t) = t$ intersect at $t = 0$. This makes us
very nervous because the Existence and Uniqueness Theorem says that solutions whose
graphs intersect are actually the same. However, there is no contradiction here because
the original equation is not defined when $t = 0$. The change of variables $u = v/t$ is
also undefined when $t = 0$. When t is close to 0, a small value of v yields a big value
of u, so the region near $t = 0$ on the tv-plane is "blown up" by the change of variables.

A Mixing Problem Revisited

Suppose a 10-gallon vat contains 4 gallons of clean water, and we start pouring sugar
into the vat at time $t = 0$ at a rate of 0.25 pounds per minute. We also pour clean water
into the vat at a rate of 2 gallons per minute. Finally, suppose that the water is kept well
mixed so the concentration of sugar is uniform in the vat and that we remove 1 gallon
per minute of sugar-water from the bottom of the vat (see Figure 1.101). What will the
concentration of sugar be at the moment when the tank starts to overflow?

Figure 1.101
Schematic of the vat with sugar and clean water
being added and sugar-water being removed. Sugar
enters the vat at a rate of 0.25 pounds per minute,
and clean water enters the vat at a rate of 2 gallons
per minute. Sugar-water leaves the vat at the rate of
1 gallon per minute.

Let $S(t)$ be the amount of sugar in the tank at time t. The rate of change of S is determined by the rate that sugar enters the vat (a positive term) and the rate that sugar leaves the vat (a negative term). In this case the rate that sugar enters the vat is a constant (0.25 pounds per minute), but the rate that sugar leaves the vat depends on the concentration. At time t, the concentration is the ratio of the total amount $S(t)$ of sugar in the vat and the total volume of liquid in the vat. Hence the differential equation governing the change of the amount of sugar in the vat is

$$\frac{dS}{dt} = 0.25 - \frac{S}{4+t}.$$

The vat will overflow when $4 + t = 10$, that is, when $t = 6$. We wish to determine the concentration $S(6)/10$ corresponding to the initial condition $S(0) = 0$. The equation is linear, so we could use the technique of Section 1.8: We calculate an integrating factor and do the integrals. But we can also change variables, which makes the equation separable.

In this case our choice of the new variable depends on physical considerations. Since our goal is to determine the concentration at time $t = 6$, we let the new variable be the concentration of sugar in the vat at time t. Let $C(t)$ denote the concentration at time t in pounds of sugar per gallon, so

$$C = \frac{S}{4+t},$$

which can be expressed as $(4+t)C = S$. To find the new differential equation in terms of the concentration, we differentiate the equation $(4+t)C = S$ and obtain

$$(4+t)\frac{dC}{dt} + C = \frac{dS}{dt}.$$

Using this relationship between dC/dt and dS/dt, we have

$$(4+t)\frac{dC}{dt} + C = \frac{dS}{dt}$$

$$= 0.25 - \frac{S}{4+t}$$

$$= 0.25 - C.$$

In terms of the concentration, the differential equation is

$$(4+t)\frac{dC}{dt} + C = 0.25 - C$$

$$\frac{dC}{dt} = \frac{0.25 - 2C}{4+t}.$$

This equation describes the same physical situation as the equation for dS/dt. It is really the same equation expressed in terms of the concentration C instead of the amount of sugar S.

It is somewhat surprising that this new equation is separable. Separating variables, we find that

$$\frac{1}{0.25 - 2C}\, dC = \frac{1}{4+t}\, dt,$$

and integrating both sides yields

$$-\tfrac{1}{2}\ln(0.25 - 2C) = \ln(4 + t) + c,$$

where c is the constant of integration. Since the water in the vat is free of sugar at $t = 0$, the initial condition is $C(0) = 0$, and therefore we can determine the value of c by

$$-\tfrac{1}{2}\ln(0.25) = \ln(4) + c$$

which is the same as

$$\tfrac{1}{2}\ln(4) = \ln(4) + c.$$

Thus $c = -(\tfrac{1}{2})\ln(4) = \ln(4^{-1/2}) = \ln(\tfrac{1}{2})$. Now that we have determined c, we can solve for the concentration C. We have

$$-\frac{1}{2}\ln(0.25 - 2C) = \ln(4 + t) + \ln\left(\frac{1}{2}\right) = \ln\left(\frac{1}{2}(4 + t)\right) = \ln\left(2 + \frac{t}{2}\right).$$

Multiplying both sides by -2 yields

$$\ln\left(\frac{1}{4} - 2C\right) = -2\ln\left(2 + \frac{t}{2}\right) = \ln\left[\left(2 + \frac{t}{2}\right)^{-2}\right],$$

and exponentiating we obtain

$$\frac{1}{4} - 2C = \left(2 + \frac{t}{2}\right)^{-2}.$$

Solving for C gives

$$C = \frac{1}{8} - \frac{1}{2}\left(2 + \frac{t}{2}\right)^{-2}.$$

The concentration at the time of overflow is $C(6) = 0.105$ pounds of sugar per gallon.

Whether this equation was easier or harder to solve than the dS/dt equation depends on the person doing the arithmetic. Having two ways to find the solution is useful for any problem, since it is unlikely that an arithmetic error will give the same wrong answer with both methods. From a theoretical point of view, it is interesting that changing from the total amount of sugar to the concentration makes such a large difference in the type of differential equation obtained and the technique of solution used.

A Linearization Problem

As we have noted, changing variables is not just a method for helping to find analytic solutions. It can also help to determine the qualitative behavior of solutions. For example, consider the logistic population model

$$\frac{dP}{dt} = 0.06P\left(1 - \frac{P}{500}\right),$$

where $P(t)$ is a population at time t measured in hundreds or thousands of individuals. Note that the growth-rate parameter is 0.06 and the carrying capacity is 500.

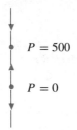

Figure 1.102

Phase line for the familiar equation

$$\frac{dP}{dt} = 0.06P\left(1 - \frac{P}{500}\right).$$

The equilibrium points are a source at $P = 0$ and a sink at $P = 500$. All solutions with positive initial conditions approach the sink $P = 500$ as $t \to \infty$ (see Figure 1.102). We know that if P is small, the term $(1 - P/500)$ is close to 1 and solutions of this logistic equation are very close to solutions of the exponential growth model

$$\frac{dP}{dt} = 0.06P,$$

which have the form $P(t) = ke^{0.06t}$. So, for P small, the population grows at an exponential rate with an exponent of $0.06t$.

After a long time, we expect that the population will be near $P = 500$. Suppose that some unexpected event (not included in the model) pushes the population away from $P = 500$. Such an event could be a period of uncharacteristically harsh weather, unlawful poaching, or the unexpected immigration of a small number of individuals. Any of these events will give a population near, but not at, $P = 500$. We know that, if conditions return to those of the model, then the population will again approach the carrying capacity of $P = 500$. A natural question is: How long will this recovery take? In other words, for $P(0)$ near 500, how will the solution behave?

The sort of answer we want is similar to the description of the behavior of small populations. We change variables, moving the point $P = 500$ to the origin. In the new variables, the behavior of solutions near the equilibrium point are easier to discover. Let

$$u = P - 500$$

so that $P = u + 500$.

The new differential equation in terms of u is given by

$$\frac{du}{dt} = \frac{dP}{dt}$$

$$= 0.06P\left(1 - \frac{P}{500}\right)$$

$$= 0.06(u + 500)\left(1 - \frac{u + 500}{500}\right)$$

$$= 0.06(u + 500)\left(-\frac{u}{500}\right).$$

Collecting terms, this equation becomes

$$\frac{du}{dt} = -0.06u - 0.06\frac{u^2}{500}.$$

The equilibrium point $P = 500$ corresponds to the equilibrium point $u = 0$ in the new variable. For u very close to 0, the term containing u^2 is the square of a very small number. Hence, for u near 0, the behavior of solutions of this equation will be very close to the behavior of solutions of the equation where we neglect the u^2 term, that is

$$\frac{du}{dt} = -0.06u.$$

Solutions of this equation are of the form $u(t) = ke^{-0.06t}$, so they decay toward $u = 0$ at an exponential rate with an exponent of $-0.06t$ (see Figure 1.103). Because the new dependent variable u is just a translation of the original variable P, solutions for the logistic equation that have initial condition near $P = 500$ approach $P = 500$ at an exponential rate with an exponent of $-0.06t$ (see Figure 1.104).

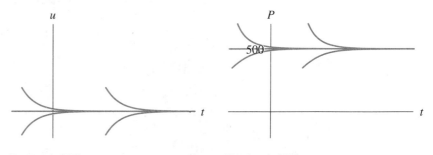

Figure 1.103
Solutions of the equation
$du/dt = -0.06u$.

Figure 1.104
Solutions of the equation
$dP/dt = 0.06P(1 - P/500)$ with initial
condition near $P = 500$.

Linearization

In general, to find the behavior of a solution near an equilibrium point, we first change variables moving that equilibrium point to the origin. Next we neglect all the "nonlinear" terms in the equation. Since we are only concerned with values of the new variable near 0, any higher power of that variable will be so close to 0 that it can be safely neglected, at least on first approximation. The equation that results is linear and hence easy to analyze.

This method involves an approximation and hence is valid only for values of the variable where that approximation is reasonable. Once the new variable becomes large, the nonlinear terms in the equation become significant and cannot be ignored. Exactly how small is small and how large is large depends on the particular equation considered (see the exercises).

This technique is called **linearization**. The idea is to approximate a complicated equation with a simpler, linear equation. Hopefully, the linear equation will be simple enough that we understand the behavior of solutions. Linearization is an important tool in our study of systems of differential equations, and we will discuss it at length in Section 5.1.

EXERCISES FOR SECTION 1.9

In Exercises 1–4, change the dependent variable from y to u using the change of variables indicated. Describe the equation in the new variable (such as separable, linear, ...).

1. $\dfrac{dy}{dt} = y - 4t + y^2 - 8yt + 16t^2 + 4,$ let $u = y - 4t$

2. $\dfrac{dy}{dt} = \dfrac{y^2 + ty}{y^2 + 3t^2},$ let $u = \dfrac{y}{t}$

3. $\dfrac{dy}{dt} = t(y + ty^2) + \cos ty,$ let $u = ty$

4. $\dfrac{dy}{dt} = e^y + \dfrac{t^2}{e^y},$ let $u = e^y$

In Exercises 5–7, find a change of variables that transforms the equation into an autonomous equation. Sketch the phase line for the equation that you obtain and use it to sketch the graphs of solutions for the original equation.

5. $\dfrac{dy}{dt} = (y - t)^2 - (y - t) - 1$

6. $\dfrac{dy}{dt} = \dfrac{y^2}{t} + 2y - 4t + \dfrac{y}{t}$

7. $\dfrac{dy}{dt} = y \cos(ty) - \dfrac{y}{t}$

In Exercises 8–10, change variables as indicated and then find the general solution of the resulting equation. Use this information to give the general solution of the original equation.

8. $\dfrac{dy}{dt} = \dfrac{ty}{2} + \dfrac{e^{t^2/2}}{2y},$ let $y = \sqrt{u}$

9. $\dfrac{dy}{dt} = \dfrac{y}{1+t} - \dfrac{y}{t} + t^2(1+t),$ let $u = \dfrac{y}{1+t}$

10. $\dfrac{dy}{dt} = y^2 - 2yt + t^2 + y - t + 1,$ let $u = y - t$

In Exercises 11–13, create your own change of variables problems. Starting with the simple equation $du/dt = (1 - u)u$, change variables as indicated. Sketch the graphs of solutions of the new equations using what you know about the solutions of $du/dt = (1 - u)u$.

11. $y = u + 3t$ **12.** $y = \sqrt{u}$ **13.** $y = u^2$

14. Consider a 20-gallon vat that at time $t = 0$ contains 5 gallons of clean water. Suppose water enters the vat from two pipes. From the first pipe, saltwater containing 2 pounds of salt per gallon enters the vat at a rate of 3 gallons per minute. From the second pipe, saltwater containing 0.5 pounds of salt per gallon enters the vat at a rate of 4 gallons per minute. Suppose the water is kept well mixed and saltwater is removed from the vat at a rate of 2 gallons per minute.

   (a) Derive a differential equation for the rate of change of the total amount of salt $S(t)$ in the vat at time t.

   (b) Convert this equation to a differential equation for the concentration $C(t)$ of salt in the vat at time t.

   (c) Find the concentration of salt in the vat at the instant when the vat first starts to overflow.

15. Consider a very large vat that initially contains 10 gallons of clean water. Suppose water enters the vat from two pipes. From the first pipe, saltwater containing 2 pounds of salt per gallon enters the vat at a rate of 1 gallon per minute. From the second pipe, saltwater containing 0.2 pounds of salt per gallon enters the vat at a rate of 5 gallons per minute. Suppose the liquid is kept well mixed and saltwater is removed from the vat at a rate of 3 gallons per minute.

    (a) Derive a differential equation for the total amount of salt $S(t)$ in the vat at time t.

    (b) Convert this equation to a differential equation for the concentration $C(t)$ of salt in the vat at time t.

    (c) What will the concentration of salt in the vat be after a very long time?

    (d) Find the concentration of salt in the vat at the time $t = 5$.

16. Suppose a 20-gallon vat contains 10 gallons of clean water at time $t = 0$. Water containing 0.3 pounds per gallon of salt enters the vat at a rate of 2 gallons per minute. Suppose clean water enters the vat at a rate of 1 gallon per minute. Finally, suppose the vat is well mixed and saltwater exits the vat at a rate of 0.5 gallons per minute.

    (a) Develop a differential equation for the concentration $C(t)$ of salt in the vat at time t. Be sure to justify each term in your equation carefully. [*Hint*: Consider how each pipe changes the concentration. In particular, does the removal of saltwater affect the concentration?]

    (b) Take the equation for the rate of change of the concentration of salt in the vat and convert it into an equation for the amount of salt in the vat at time t. [*Hint*: This should provide a check of your differential equation for the concentration.]

In Exercises 17–20,

  (a) Identify the equilibrium points.

  (b) For each equilibrium point, give the change of variables that moves the equilibrium point to the origin and give the linear approximation of the new system near the origin. Identify the equilibrium point as a source, a sink, or a node.

  (c) For each equilibrium point, describe quantitatively the behavior of solutions with initial values that are very close to the equilibrium point. (For example, how quickly does a solution that starts close to the equilibrium point approach or recede from the equilibrium point as t increases?) In particular, approximate the amount of time it takes a point to halve or double its distance from the equilibrium point as t increases.

17. $\dfrac{dy}{dt} = 10y^3 - 1$

18. $\dfrac{dy}{dt} = (y+1)(y-3)$

19. $\dfrac{dy}{dt} = (y+1)(3-y)$

20. $\dfrac{dy}{dt} = y^3 - 3y^2 + y$

21. Consider the differential equation

$$\frac{dy}{dt} = -3 \sin y.$$

(a) What is the third-degree Taylor polynomial about $y = 0$ for $\sin y$?

(b) What is the linearization of this differential equation near the equilibrium point at $y = 0$?

(c) Use the result of part (b) to approximate $y(2)$ for the solution that satisfies the initial condition $y(0) = 0.04$.

22. Consider the population model (see Section 1.6, page 87)

$$\frac{dS}{dt} = 0.05S \left(\frac{S}{20} - 1 \right) \left(1 - \frac{S}{100} \right),$$

where S is measured in thousands of individuals. Suppose the population is near the carrying capacity of the system, and a tornado suddenly reduces the population by 2000 individuals.

(a) Approximately how long will it take for the population to return to within 1000 individuals of the carrying capacity?

(b) Approximately how long will it take for the population to return to within 100 individuals of the carrying capacity?

23. Consider the logistic population model with a small amount of periodic harvesting

$$\frac{dP}{dt} = 0.05P \left(1 - \frac{P}{100} \right) - 0.02(1 + \sin t).$$

(a) Change variables using $u = P - 100$ to move the carrying capacity to 0.

(b) Find a linear equation that is close to your equation from part (a) if u is close to zero (that is, P is close to 100).

(c) Solve the equation you found in part (b).

(d) What do you conclude about the behavior of the population with small periodic harvesting given by the equation above? (For example, how far do you predict the population will vary from the value $P = 100$?)

24. Consider the differential equation

$$\frac{dy}{dt} = -y + \frac{y^2}{4}.$$

(a) What is the linear approximation of this equation near the equilibrium point $y = 0$?

(b) If $y(0) = y_0$ is very close to zero, approximately how long will it take the solution to reach $y_0/2$?

(c) For what interval of y's near $y = 0$ is $0.9|y| \leq |-y + y^2/4| \leq 1.1|y|$?

(d) For how large a value of $y(0) = y_0$ will your answer to part (b) apply to the nonlinear equation? Justify your answer.

25. Consider the differential equation $dy/dt = f(y)$, where $f(y)$ is a smooth function (it can be differentiated as many times as we like). Suppose $y = y_0$ is an equilibrium point for this equation.

    (a) Let $u = y - y_0$ and write the differential equation in terms of the new dependent variable u.

    (b) Show that the linear approximation of the differential equation for du/dt near the equilibrium point at $u = 0$ is given by

$$\frac{du}{dt} = f'(y_0)u.$$

26. So far we have changed only the dependent variable. It is also possible to change the independent variable to obtain a new equation. This is analogous to changing the way we measure time. Consider the differential equation

$$\frac{dy}{dt} = 3t^2 y + 3t^5.$$

Suppose we define a new time independent variable using the formula $s = t^3$. Thus $t = \sqrt[3]{s}$.

    (a) Show that

$$\frac{dy}{ds} = y + s.$$

    (b) Find the general solution of the differential equation with the variable s.

    (c) What is the general solution of the original differential equation?

LAB 1.1 Logistic Population Models with Harvesting

In this lab we consider logistic models of population growth that have been modified to include terms that account for "harvesting." In particular, you should imagine a fish population subject to various degrees and types of fishing. The differential equation models are given below. (Your instructor will indicate the values of the parameters k, N, a_1, and a_2 you should use. Several possible choices are listed in Table 1.9.) In your report, you should include a discussion of the meaning of each variable and parameter and an explanation of why the equation is written the way it is.

We have discussed three general approaches that can be employed to study a differential equation: numerical techniques yield graphs of approximate solutions, geometric/qualitative techniques provide predictions of the long-term behavior of the solution and in special cases analytic techniques provide explicit formulas for the solution. In your report, you should employ as many of these techniques as is appropriate to help understand the models.

Your report should consider the following equations:

1. (Logistic growth with constant harvesting) The equation

$$\frac{dp}{dt} = kp\left(1 - \frac{p}{N}\right) - a$$

represents a logistic model of population growth with constant harvesting at a rate a. For $a = a_1$, what will happen to the fish population for various initial conditions? (Note: This equation is autonomous, so you can take advantage of the special techniques that are available for autonomous equations.)

2. (Logistic growth with periodic harvesting) The equation

$$\frac{dp}{dt} = kp\left(1 - \frac{p}{N}\right) - a(1 + \sin(bt))$$

is a nonautonomous equation that considers periodic harvesting. What do the parameters a and b represent? Let $b = 1$. If $a = a_1$, what will happen to the fish population for various initial conditions?

3. Consider the same equation as in part 2 above, but let $a = a_2$. What will happen to the fish population for various initial conditions with this value of a?

Your report: In your report you should address these three questions, one at a time, in the form of a short essay. Begin questions 1 and 2 with a description of the meaning of each of the variables and parameters and an explanation of why the differential equation is the way it is. You should include pictures and graphs of data and of solutions of your models *as appropriate*. (Remember that one carefully chosen picture can be worth a thousand words, but a thousand pictures aren't worth anything.)

Table 1.9
Possible choices for the parameters

Choice	k	N	a_1	a_2
1	0.25	4	0.16	0.25
2	0.50	2	0.21	0.25
3	0.20	5	0.21	0.25
4	0.20	5	0.16	0.25
5	0.25	4	0.09	0.25
6	0.20	5	0.09	0.25
7	0.50	2	0.16	0.25
8	0.20	5	0.24	0.25
9	0.25	4	0.21	0.25
10	0.50	2	0.09	0.25

LAB 1.2   Rate of Memorization Model

Human learning is, to say the least, an extremely complicated process. The biology and chemistry of learning is far from understood. While simple models of learning cannot hope to encompass this complexity, they can illuminate limited aspects of the learning process. In this lab we study a simple model of the process of memorization of lists (lists of nonsense syllables or entries from tables of integrals).

The model is based on the assumption that the rate of learning is proportional to the amount left to be learned. We let $L(t)$ be the fraction of the list already committed to memory at time t. So $L = 0$ corresponds to knowing none of the list, and $L = 1$ corresponds to knowing the entire list. The differential equation is

$$\frac{dL}{dt} = k(1 - L).$$

Different people take different amounts of time to memorize a list. According to the model this means that each person has his or her own personal value of k. The value of k for a given individual must be determined by experiment.
Carry out the following steps:

1. Four lists of three-digit numbers are given in Table 1.10, and additional lists can be generated by a random number generator on a computer. Collect the data necessary to determine your personal k value as follows:

    (a) Spend one minute studying one of the lists of numbers in table Table 1.10. (Measure the time carefully. A friend can help.)
    (b) Quiz yourself on how many of the numbers you have memorized by writing down as many of the numbers as you remember in their correct order. (You may skip over numbers you don't remember and obtain "credit" for numbers you remember later in the list.) Put your quiz aside to be graded later.

(c) Spend another minute studying the same list.

(d) Quiz yourself again.

Repeat the process ten times (or until you have learned the entire list). Grade your quizzes (a correct answer is having a correct number in its correct position in the list). Compile your data in a graph with t, the amount of time spent studying, on the horizontal axis, and L, the fraction of the list learned, on the vertical axis.

2. Use this data to approximate your personal k-value and compare your data with the predictions of the model. You may use numeric or analytic methods, but be sure to carefully explain your work. Estimate how long it would take you to learn a list of 50 and 100 three-digit numbers.

3. Repeat the process in part 1 on two of the other lists and compute your k-value on these lists. Is your personal k-value really constant, or does it improve with practice? If k does improve with practice, how would you modify the model to include this?

Your report: In your report you should give your data in parts 1 and 3 neatly and clearly. Your answer to the questions in parts 2 and 3 should be in the form of short essays. You should include graphs (hand or computer drawn) of your data and solutions of the model *as appropriate*. (Remember that one carefully chosen picture can be worth a thousand words, but a thousand pictures aren't worth anything.)

Table 1.10
Four lists of random three-digit numbers

	List 1	List 2	List 3	List 4
1	457	167	733	240
2	938	603	297	897
3	363	980	184	935
4	246	326	784	105
5	219	189	277	679
6	538	846	274	011
7	790	040	516	020
8	895	891	051	013
9	073	519	925	144
10	951	306	102	209
11	777	424	826	419
12	300	559	937	191
13	048	911	182	551
14	918	439	951	282
15	524	140	643	587
16	203	155	434	609
17	719	847	921	391
18	518	245	820	364
19	130	752	017	733
20	874	552	389	735

LAB 1.3 Exponential and Logistic Population Models

In the text we modeled the U.S. population over the last 200 years using both an exponential growth model and a logistic growth model. For this lab project, we ask that you model the population growth of a particular state. Population data for several states are given in Table 1.11. (Your instructor will assign the state(s) you are to consider.)

We have also discussed three general approaches that can be employed to study a differential equation: numerical techniques yield graphs of approximate solutions, geometric/qualitative techniques provide predictions of the long-term behavior of the solution, and in special cases analytic techniques provide explicit formulas for the solution. In your report, you should employ as many of these techniques as is appropriate to help understand the models.

Your report should address the following items:

1. Using an exponential growth model, determine as accurate a prediction as possible for the population of your state in the Year 2000. How much does your prediction differ from the prediction that comes from linear extrapolation using the populations in 1980 and 1990? To what extent do solutions of your model agree with the historical data?

2. Produce a logistic growth model for the population of your state. What is the carrying capacity for your model? Using Euler's method, predict the population in the Years 2000 and 2050. Using analytic techniques, obtain a formula for the population function $P(t)$ that satisfies your model. To what extent do solutions of your model agree with the historical data?

3. Comment on how much confidence you have in your predictions of the future populations. Discuss which model, exponential or logistic growth, is better for your data and why (and if neither is very good, suggest alternatives).

Your report: The body of your report should address all three items, one at a time, in the form of a short essay. For each model, you must choose specific values for certain parameters (the growth-rate parameter and the carrying capacity). Be sure to give a complete justification of why you made the choices that you did. You should include pictures and graphs of data and of solutions of your models *as appropriate*. (Remember that one carefully chosen picture can be worth a thousand words, but a thousand pictures aren't worth anything.)

Table 1.11
Population (in thousands) of selected states (Data from *1994 World Almanac*)

Year	Massachusetts	New York	North Carolina	Alabama	Florida	California	Montana	Hawaii
1790	379	340	394					
1800	423	589	478	1				
1810	472	959	556	9				
1820	523	1373	638	127				
1830	610	1919	738	309	35			
1840	738	2429	753	591	54			
1850	995	3097	869	772	87	93		
1860	1231	3881	993	964	140	380		
1870	1457	4383	1071	996	188	560	20	
1880	1783	5083	1399	1262	269	865	39	
1890	2239	6003	1618	1513	391	1213	143	
1900	2805	7269	1893	1829	529	1485	243	154
1910	3366	9114	2206	2138	753	2378	376	192
1920	3852	10385	2559	2348	968	3427	549	256
1930	4250	12588	3170	2646	1468	5677	538	368
1940	4317	13479	3571	2832	1897	6907	559	423
1950	4691	14830	4061	3062	2771	10586	591	500
1960	5149	16782	4556	3267	4952	15717	675	633
1970	5689	18241	5084	3444	6791	19971	694	770
1980	5737	17558	5880	3894	9747	23668	787	965
1990	6016	17990	6628	4040	12938	29760	799	1108

LAB 1.4   Growth of a Population of Mold

In the text we modeled the U.S. population using both an exponential growth model and a logistic growth model. The assumptions we used to create the models are easy to state. For the exponential model we assumed only that the growth of the population is proportional to the size of the population. For the logistic model we added the assumption that the ratio of the population to the growth rate decreases as the population increases. In this lab we apply these same principles to model the colonization of a piece of bread by mold.

Place a piece of mold-free bread in a plastic bag with a small amount of water and leave the bread in a warm place. Each day, record the area of the bread that is covered with mold. (One way to do this is to trace the grid from a piece of graph paper onto a clear piece of plastic. Hold the plastic over the bread and count the number of squares that are mostly covered by mold.)

Warning: It takes at least two weeks to accumulate a reasonable amount of data. Some types of bread seem to be resistant to mold growth, and the bread just dries out. If the mold grows, then after about a week the bread will look pretty disgusting. Take precautions to make sure your assignment isn't thrown out.

In your report, address the following questions:

1. Model the growth of mold using an exponential growth model. How accurately does the model fit the data? Be sure to explain carefully how you obtained the value for the growth-rate parameter.

2. Model the growth of mold using a logistic growth model. How accurately does the model fit the data? Be sure to explain carefully how you obtained the value of the growth-rate parameter and carrying capacity.

3. Discuss the models for mold growth population. Were there any surprises? Does it matter that we are measuring the area covered by the mold rather than the total weight of the mold? To what extent would you believe predictions of future mold populations based on these models?

Your report: You should include in your report the details of the type of bread used, where it was kept, and how and how often the mold was measured. Your analysis of the models may include qualitative, numerical, and analytic arguments, and graphs of data and solutions of your models as appropriate. (Remember that a well-chosen picture can be worth a thousand words, but a thousand pictures aren't worth anything.) **Do not hand in the piece of bread.**

2 | FIRST-ORDER SYSTEMS

Fe w phenomena are completely described by a single number. For example, the size of a population of rabbits can be represented using one number, but to know its rate of change, we should consider other quantities such as the size of predator populations and the availability of food.

In this chapter we begin the study of systems of differential equations — equations that involve more than one dependent variable. As with first-order equations, the techniques for studying these systems fall into three general categories: analytic, qualitative, and numeric. Only special systems of differential equations can be attacked using analytic methods, so we focus primarily on qualitative and numerical methods. The main class of systems that can be studied analytically — linear systems — are the subject of Chapter 3.

We continue to study models involving differential equations by discussing models that have more than one dependent variable. Included is a system known as the harmonic oscillator. This particular model has numerous applications in many branches of science such as mechanics, electronics, and physics.

2.1   MODELING VIA SYSTEMS

In this section we discuss models of two very different phenomena — the evolution of the two populations in a predator-prey system, and the motion of a mass-spring system. Initially these models seem quite different, but from the right point of view, they possess a number of similarities.

The Predator-Prey System Revisited

We begin our study of systems of differential equations by considering two versions of the predator-prey model discussed briefly in Section 1.1. Recall that $R(t)$ denotes the population (in thousands, or millions, or whatever) of prey present at time t and that $F(t)$ denotes the population of predators. We assume that both $R(t)$ and $F(t)$ are nonnegative. One system of differential equations that might govern the changes in the population of these two species is

$$\frac{dR}{dt} = 2R - 1.2RF$$
$$\frac{dF}{dt} = -F + 0.9RF.$$

The $2R$ term in the equation for dR/dt represents exponential growth of the prey in the absence of predators, and the $-1.2RF$ term corresponds to the negative effect on the prey of predator-prey interaction. The $-F$ term in dF/dt corresponds to the assumption that the predators die off if there are no prey to eat, and the $0.9RF$ term corresponds to the positive effect on the predators of predator-prey interaction.

The coefficients 2, -1.2, -1, and 0.9 depend on the species involved. Similar systems with different coefficients are considered in the exercises. (We choose these values of the parameters in this example solely for convenience.)*

The presence of the RF terms in these equations makes this system difficult to solve. It is impossible to derive explicit formulas for the general solution, but there are some initial conditions that do yield simple solutions. For instance, suppose that both $R = 0$ and $F = 0$. Then the right-hand sides of both equations vanish ($dR/dt = dF/dt = 0$) for all t, and consequently the pair of constant functions $R(t) = 0$ and $F(t) = 0$ form a solution to the system. By analogy to first-order equations, we call such a pair of constant functions an **equilibrium solution** to the system. This equilibrium solution makes perfect sense: If both the predator and prey populations vanish, we certainly do not expect the populations to grow at any later time.

We can also look for other values of R and F that correspond to constant solutions. We rewrite the system as

$$\frac{dR}{dt} = (2 - 1.2F)R$$
$$\frac{dF}{dt} = (-1 + 0.9R)F$$

---

*For more details on the development, use, and limitations of this system as a model of predator-prey interactions in the wild, we refer the reader to the excellent discussions in J.P. Dempster, *Animal Population Ecology* (New York: Academic Press, 1975) and M. Braun, *Differential Equations and Their Applications* (New York: Springer-Verlag, 1993).

and note that both equations vanish if $R = 1/0.9 \approx 1.11$ and $F = 2/1.2 \approx 1.67$. Thus the pair of constant functions $R(t) \approx 1.11$ and $F(t) \approx 1.67$ together form another equilibrium solution. This solution says that, if the prey population is 1.11 and the predator population is 1.67, the system is in perfect balance. There are just enough prey to support a constant predator population of 1.67, and similarly there are neither too many predators (which would cause the population of prey to fall) nor too few (in which case the number of prey would rise). Each species' birth rate is exactly equal to its death rate, and these populations are maintained indefinitely. The system is in *equilibrium*.

For certain initial conditions, we can use the techniques that we have already developed for first-order equations to study systems. For example, if $R = 0$, the first equation in this system vanishes. Therefore the constant function $R(t) = 0$ satisfies this differential equation no matter what initial condition we choose for F. In this case the second differential equation reduces to

$$\frac{dF}{dt} = -F,$$

which we recognize as the exponential decay model for the predator population — a familiar and very simple differential equation. From this equation we know that the population of predators tends exponentially to zero. This entire scenario for $R = 0$ is reasonable since, if there are no prey at some time, then there never will be any prey no matter how many predators there are. Moreover, without a food supply, the predators must die out.

In similar fashion, note that the equation for dF/dt vanishes if $F = 0$, and the equation for dR/dt reduces to

$$\frac{dR}{dt} = 2R$$

— an exponential growth model. As we saw in Section 1.1, any nonzero prey population grows without bound under these assumptions. Again these conclusions make sense because there are no predators to control the growth of the prey population. On the other hand, we could make the more realistic assumption that the prey population obeys a logistic growth law. Our second example in this section incorporates this additional assumption.

$R(t)$- and $F(t)$-graphs

In order to understand all solutions of this predator-prey system

$$\frac{dR}{dt} = 2R - 1.2RF$$
$$\frac{dF}{dt} = -F + 0.9RF,$$

it is important to note that the rate of change of either population depends both on R and on F. Hence we need two numbers, an initial value R_0 of R and an initial value F_0 for F, to determine the manner in which these populations evolve over time. In other words, an **initial condition** which determines a solution to this system of equations is a pair of numbers, R_0 and F_0, which are then used to determine the initial values of

dR/dt and dF/dt. This initial condition yields a **solution** of the system which consists of two functions $R(t)$ and $F(t)$ that, taken together, satisfy the system of equations.

For the study of solutions to systems of differential equations, there is good news and bad news. The bad news is that for many systems there are few analytic techniques that yield formulas for the solutions. The good news is that there are numerical and qualitative methods that give us a good understanding of the solutions even if we cannot find analytic representations for them. For example, if we specify the initial conditions $R_0 = 1$ and $F_0 = 0.5$, we can use a numerical method akin to Euler's method to obtain approximate values for the corresponding solutions $R(t)$ and $F(t)$. (We will develop this method in Section 2.4.)

In Figures 2.1 and 2.2 we graph the solutions $R(t)$ and $F(t)$ that correspond to the initial condition $R_0 = 1$ and $F_0 = 0.5$, and we see that both $R(t)$ and $F(t)$ rise and fall in a *periodic* fashion.

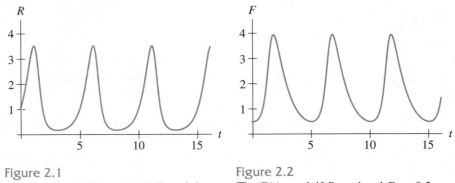

Figure 2.1

The $R(t)$-graph if $R_0 = 1$ and $F_0 = 0.5$.

Figure 2.2

The $F(t)$-graph if $R_0 = 1$ and $F_0 = 0.5$.

In Figure 2.3 we graph both $R(t)$ and $F(t)$ on the same set of axes. Although this graph is somewhat misleading because there are really two scales on the vertical axis — one corresponding to the units of $R(t)$ and the other corresponding to the units of $F(t)$, it does provide information that is hard to read from the individual $R(t)$- and $F(t)$-graphs. For example, for this particular solution we see that the increases in the

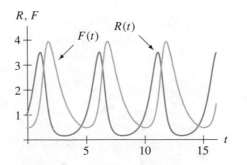

Figure 2.3

The $R(t)$- and $F(t)$-graphs given by the initial condition $R_0 = 1$ and $F_0 = 0.5$.

predator population lag the increases in the prey population and that the predator population continues to increase for a short amount of time after the prey population starts to decline. Perhaps the most important observation that we can make from this graph is that both $R(t)$ and $F(t)$ seem to repeat with the same period (roughly five time units). Although we could reach the same conclusion by closely studying Figures 2.1 and 2.2, this fact is much easier to observe if both the $R(t)$- and $F(t)$-graphs are drawn on the same pair of axes.

The phase portrait for this system

There is another way to graph the solution of the system that corresponds to the initial condition $(R_0, F_0) = (1, 0.5)$. Given $R(t)$, $F(t)$, and a value of t, we can form the pair $(R(t), F(t))$ and think of it as a point in the RF-plane. In other words, the co-ordinates of the point are the values of the two populations at time t. As t varies, the pair $(R(t), F(t))$ sweeps out a curve in the RF-plane. This curve is the **solution curve** determined by the original initial condition. The coordinates of each point on the curve are the prey and predator populations at the associated time t, and the point (R_0, F_0) that corresponds to the initial condition for the solution is often referred to as the **initial point** of this solution curve.

It is often helpful to view a solution curve for a system of differential equations not merely as a set of points in the plane but, rather in a more dynamic fashion, as a point following a curve that is determined by the solution to the differential equation. In Figure 2.4 we show the solution curve corresponding to the solution with initial conditions $R_0 = 1$ and $F_0 = 0.5$ in the RF-plane. This curve starts at the point $P = (1, 0.5)$. As t increases, the corresponding point on the curve moves to the right. This motion implies that $R(t)$ is increasing but that $F(t)$ initially stays relatively constant. Near $R = 3$, the solution curve turns significantly upward. Thus the predator population $F(t)$ starts increasing significantly. As $F(t)$ nears $F = 2$, the curve starts heading to the left. Thus $R(t)$ has reached a maximum and is starting to decrease. As t increases, the values of $R(t)$ and $F(t)$ change as indicated by the shape of the solution curve. Eventually the solution curve returns to its starting point P and begins its cycle again.

The RF-plane is called the **phase plane**, and it is analogous to the phase line for an autonomous first-order differential equation. Just as the phase line has a point

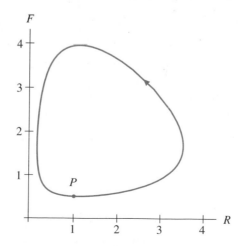

Figure 2.4
The solution curve for the predator-prey system

$$\frac{dR}{dt} = 2R - 1.2RF$$

$$\frac{dF}{dt} = -F + 0.9RF,$$

corresponding to the initial condition $P = (R_0, F_0) = (1, 0.5)$.

for each value of the dependent variable but does not explicitly show the corresponding value of time, the phase plane has a point for each ordered pair (R, F) of populations. The dependence of a solution on the independent variable t can only be imagined as a point moving along the solution curve as t evolves.

We can plot many solutions curves on the phase plane simultaneously. In Figure 2.5 we see the complete **phase portrait** for our predator-prey system. Of course, since negative populations do not make sense for this model, we restrict our attention to the first quadrant of the RF-plane.

Equilibrium solutions are solutions that are constant, and consequently they produce solution curves $(R(t), F(t))$, where $R(t)$ and $F(t)$ never vary. In other words, the solution curves that correspond to equilibrium solutions are really just points, and we refer to them as **equilibrium points**. Just as with the phase line, the equilibrium points in the phase plane are especially important parts of the phase portrait, and therefore we usually mark them with large dots. (Note the dots at the equilibrium points at $(0, 0)$ and $(1.11, 1.67)$ in Figure 2.5.)

In this predator-prey system, all other solutions for which $R_0 > 0$ and $F_0 > 0$ yield solution curves that loop around the equilibrium point $(1.11, 1.67)$ in a counter-clockwise fashion. Ultimately, they return to their initial points, and hence, this model predicts that, except for the equilibrium solution, both $R(t)$ and $F(t)$ rise and fall in a periodic fashion.

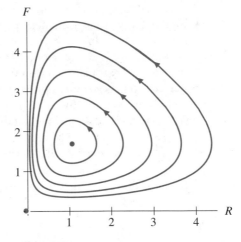

Figure 2.5
The phase plane for the predator-prey system.

A Modified Predator-Prey Model

We now consider a modification of the predator-prey model in which we assume that, in the absence of predators, the prey population obeys a logistic rather than an exponential growth model. One such model for this situation is the system

$$\frac{dR}{dt} = 2R \left(1 - \frac{R}{2} \right) - 1.2RF$$

$$\frac{dF}{dt} = -F + 0.9RF.$$

In this system, when the predators are not present (that is, $F = 0$), the prey population obeys a logistic growth model with carrying capacity 2. Once again, with the use of numerical methods we see that the behavior of solution curves (and therefore the predictions made by this model) are quite different from those made by the previous predator-prey model.

First, let's find the equilibrium solutions for this system. Recall that these solutions occur at points (R, F) where the right-hand sides of both of the differential equations vanish. As before, $(R, F) = (0, 0)$ is one equilibrium solution. There are two other equilibria — $(R, F) = (2, 0)$ and $(R, F) = (10/9, 20/27) \approx (1.11, 0.74)$.

As in our first predator-prey model, if there are no prey present, the predator population declines exponentially. In other words, if $R = 0$, then $dR/dt = 0$ for all t, so $R(t) = 0$. Then the equation for dF/dt reduces to the familiar exponential decay model

$$\frac{dF}{dt} = -F.$$

In the absence of predators the situation is somewhat different. If $F = 0$, we have $dF/dt = 0$ for all t, and the equation for R simplifies to the familiar logistic model

$$\frac{dR}{dt} = 2R\left(1 - \frac{R}{2}\right).$$

From this equation we see that the growth coefficient for low populations of prey is 2 and the carrying capacity is 2. Thus, if $F = 0$, we expect any nonzero initial population of prey to approach 2 eventually.

When both R and F are nonzero, the evolution of the two populations is more complicated. In Figure 2.6 we plot three solution curves for $t \geq 0$. Note that, in all cases, the solutions tend to the equilibrium point A, which has coordinates $(R, F) = (1.11, 0.74)$.

Once we have the solution curve that corresponds to a given initial condition, we know what the model predicts for the solution that satisfies this initial condition. For example, we see that the initial condition B in Figure 2.6 corresponds to an overabundance of both predators and prey. Following the solution curve we see that the predator population initially rises while the prey population declines. However, once the supply of prey is sufficiently low, the predator population declines and eventually approaches

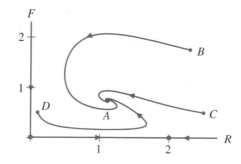

Figure 2.6
The equilibria and three solution curves for
the logistic predator-prey model.

the equilibrium value $F = 0.74$. On the other hand, the prey population eventually re-covers, and this population also tends to stabilize at the equilibrium value $R = 1.11$. This evolution of $R(t)$ and $F(t)$ is exactly what we see if we plot the corresponding $R(t)$- and $F(t)$-graphs (see Figure 2.7).

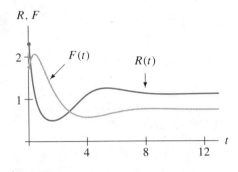

Figure 2.7
The $R(t)$- and $F(t)$-graphs for the solution curve B in Figure 2.6.

The other two solution curves that are shown in Figure 2.6 can be interpreted in a similar fashion (see Figures 2.8 and 2.9). Note that the graphs of both $F(t)$ and $R(t)$ tend to the equilibrium values $R = 1.11$ and $F = 0.74$. We can predict this from the solution curves in the phase plane (see Figure 2.6).

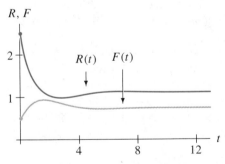

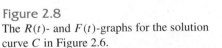

Figure 2.8
The $R(t)$- and $F(t)$-graphs for the solution curve C in Figure 2.6.

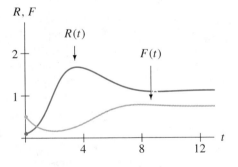

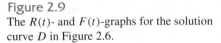

Figure 2.9
The $R(t)$- and $F(t)$-graphs for the solution curve D in Figure 2.6.

The Motion of a Mass Attached to a Spring

At first glance, the standard model of the motion of an undamped mass-spring system seems quite different from the population models that we have just discussed, but there are some important similarities in the corresponding mathematical models.

Consider a mass that is attached to a spring and that slides on a frictionless ta-ble (see Figure 2.10). We wish to understand its horizontal motion when the spring is

Figure 2.10
A mass-spring system.

stretched (or compressed) and then released. In order to keep the model as simple as possible, we assume that the only force acting on the mass is the force of the spring. In particular, we ignore air resistance and other forces that would dampen the motion of the mass.

There are two key quantities in this model—a quantity that measures the displacement of the mass from its natural rest position and the restoring force on the mass caused by the spring. We wish to determine the position of the mass as a function of time, so we let $y(t)$ denote the position of the mass at time t. It is convenient to let $y = 0$ represent the rest position of the mass (see Figure 2.11). At the rest position the spring is neither stretched nor compressed, and it exerts no force on the mass. We adopt the convention that $y(t) < 0$ if the spring is compressed and $y(t) > 0$ if the spring is stretched using whatever units are convenient (see Figures 2.11–2.13).

The main idea from physics needed to derive the differential equation that models

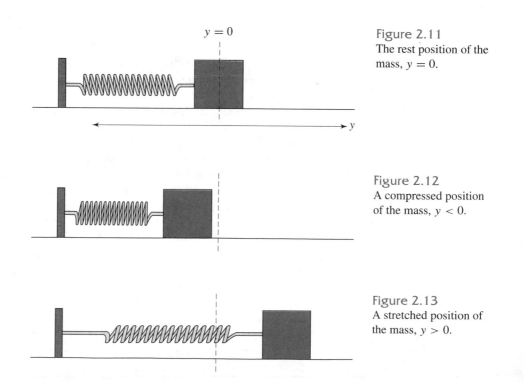

Figure 2.11
The rest position of the mass, $y = 0$.

Figure 2.12
A compressed position of the mass, $y < 0$.

Figure 2.13
A stretched position of the mass, $y > 0$.

this motion is Newton's second law,

$$\text{Force } F = \text{mass} \times \text{acceleration.}$$

Since the displacement is $y(t)$, the acceleration is d^2y/dt^2. If we let m denote the mass, Newton's law becomes

$$F = m\frac{d^2y}{dt^2}.$$

To complete the model we must specify an expression for the force that the spring exerts on the mass. We use **Hooke's law of springs** as our model for the restoring force F_s of the spring:

The restoring force exerted by a spring is linearly proportional to the spring's displacement from its rest position and is directed toward the rest position.

Therefore we have

$$F_s = -ky,$$

where $k > 0$ is a constant of proportionality called the **spring constant** — a parameter we can adjust by changing springs. Combining this expression for the force with Newton's law, we obtain the differential equation

$$F_s = -ky = m\frac{d^2y}{dt^2},$$

which models the motion of the mass. It is traditional to rewrite this equation in the form

$$\frac{d^2y}{dt^2} + \frac{k}{m}y = 0.$$

This equation is the differential equation for what is often called a simple (or undamped) **harmonic oscillator**. Since the equation contains the second derivative of the dependent variable y, it is a **second-order** differential equation. The coefficients m and k are parameters that are determined by the particular mass and spring involved.

From a notational point of view, this second-order equation seems to have little in common with the first-order predator-prey systems that we discussed earlier in this section. In particular, the equation contains only a single dependent variable, and it involves a second derivative rather than two first derivatives.

However, once we attempt to use this second-order equation to describe the motion of a particular mass-spring system, the similarities start to emerge. For example, suppose that we want to describe the motion of the mass. What do we need for initial conditions? Certainly we need an initial condition y_0 that corresponds to the initial displacement of the mass, but does y_0 alone determine the subsequent motion of the mass? The answer is no because we cannot ignore the initial velocity v_0 of the mass. For example, the motion that results from extending the mass-spring system by 1 foot and releasing it is different than the motion that results from extending the system by 1 foot and then pushing with an initial velocity of 1 foot/second. There is a theory of existence and uniqueness for solutions to this equation just as with first-order equations (see Section 2.4), and this theory tells us that we need two numbers, y_0 and v_0, to determine the motion of the simple harmonic oscillator.

Now that the velocity of the motion has been identified as a key part of the overall picture, we are only one step away from completing the analogy between first-order systems such as the predator-prey system and second-order equations such as the equation for the simple harmonic oscillator. If we let $v(t)$ denote the velocity of the mass at time t, then we know from calculus that $v = dy/dt$. Therefore, the acceleration d^2y/dt^2 is the derivative dv/dt of the velocity, and we can rewrite our second-order equation

$$\frac{d^2y}{dt^2} = -\frac{k}{m}y$$

as

$$\frac{dv}{dt} = -\frac{k}{m}y.$$

In other words, we can rewrite the second-order equation as the first-order system

$$\frac{dy}{dt} = v$$

$$\frac{dv}{dt} = -\frac{k}{m}y.$$

This technique of reducing the order of the system by increasing the number of dependent variables gives us two ways of representing the same model for the motion of the mass. Each representation has its advantages and disadvantages. The representation of the mass-spring system as a second-order equation involving one variable is more convenient for certain analytic techniques, whereas the representation as a first-order system is much better for numerical and qualitative analysis.

An initial-value problem

To demonstrate the connections between these two points of view, we consider a very specific initial-value problem. Suppose m and k are fixed so that $k/m = 1$. Then the second order equation simplifies to

$$\frac{d^2y}{dt^2} = -y.$$

In other words, the second derivative of $y(t)$ is $-y(t)$. Two such functions, sine and cosine, come to mind immediately. As we will see in Chapter 3, there are many other functions that also satisfy this differential equation, but for the purposes of this discussion, we focus on the initial-value problem $(y(0), v(0)) = (y_0, v_0) = (1, 0)$. In this case the function $y(t) = \cos t$ satisfies this initial condition since $y(0) = \cos 0 = 1$ and $y'(0) = -\sin 0 = 0$.

If we convert this second-order equation to a first-order system where $v = dy/dt$, we obtain

$$\frac{dy}{dt} = v$$

$$\frac{dv}{dt} = -y.$$

In this context the same initial condition yields a solution that consists of the pair of functions $y(t) = \cos t$ and $v(t) = -\sin t$. Their $y(t)$- and $v(t)$-graphs are shown in Figure 2.14.

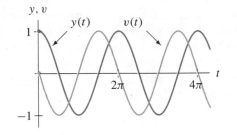

Figure 2.14

Graphs of the solutions $y(t)$ and $v(t)$ for the initial-value problem

$$\frac{d^2y}{dt^2} + y = 0, \quad y(0) = 1, \quad v(0) = 0.$$

In the yv-phase plane the corresponding solution curve is

$$(y(t), v(t)) = (\cos t, -\sin t).$$

With the help of a little trigonometry we see that

$$y^2 + v^2 = (\cos t)^2 + (-\sin t)^2 = 1,$$

and therefore this curve sweeps out the unit circle centered at the origin. Due to the minus sign in $v(t) = -\sin t$, the unit circle is swept out in a clockwise direction (as indicated by the arrowhead on the circle in Figure 2.15).

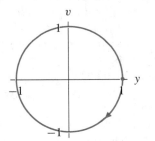

Figure 2.15

Graph of the solution curve in the yv-phase plane for the solution to the initial-value problem

$$\frac{d^2y}{dt^2} + y = 0, \quad y(0) = 1, \quad v(0) = 0.$$

Either the periodic $y(t)$- and $v(t)$-graphs (Figure 2.14) or the parameterization of the unit circle in the yv-plane (Figure 2.15) indicates that the solution is periodic, with $y(t)$ and $v(t)$ alternately increasing and decreasing, repeating the same cycle again and again. The mass oscillates back and forth across its rest position, $y = 0$, forever. Of course, this phenomenon is possible only because we have neglected damping.

Taken together, Figures 2.14 and 2.15 give a complete picture of the solution. It would be nice if we could make one picture that included all of the information in both Figures 2.14 and 2.15. Such a picture must be three dimensional since three important variables — t, y, and v — are involved. Due to the fact that we are so familiar with the functions that arise in this example, we can be successful for this equation (see Figure 2.16). Note that Figure 2.14 comes from the projections of Figure 2.16 into both the ty- and tv-planes and that Figure 2.14 is the projection of Figure 2.16 into the yv-phase plane.

Drawing these types of three-dimensional figures requires considerable graphical skill even when the solution $y(t)$ is the very familiar cosine function. In addition, interpreting these pictures requires an even greater skill in visualization. We therefore generally avoid graphs that involve all three variables at once. We restrict our attention to the graphs of the solutions, the $y(t)$- and $v(t)$-graphs, and the solution curve in the yv-phase plane.

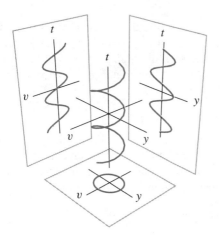

Figure 2.16
The graph of a solution of

$$\frac{d^2y}{dt^2} + y = 0$$

in tyv-space and its projections onto the ty-, tv-, and yv-coordinate planes. Note that the $y(t)$-graph is the graph of $\cos t$, the $v(t)$-graph is the graph of $-\sin t$, and the solution curve in the yv-phase plane is the unit circle.

The Study of Systems of Differential Equations

In Chapter 1 we learned that there were three basic ways to understand the solutions of a differential equation — with the use of analytic, geometric (or qualitative), and numeric techniques. In the next three sections of this chapter, we will concentrate on analogous approaches for systems and second-order equations. In the next section we introduce vector notation in order to provide a geometric approach. In Section 2.3 we discuss analytic techniques that we can use to find explicit formulas for solutions in somewhat specialized situations, and in Section 2.4 we use our vector notation to generalize Euler's method.

EXERCISES FOR SECTION 2.1

Exercises 1–6 refer to the following systems of equations:

(i)
$$\frac{dx}{dt} = 10x\left(1 - \frac{x}{10}\right) - 20xy$$
$$\frac{dy}{dt} = -5y + \frac{xy}{20}$$

(ii)
$$\frac{dx}{dt} = 0.3x - \frac{xy}{100}$$
$$\frac{dy}{dt} = 15y\left(1 - \frac{y}{15}\right) + 25xy.$$

1. In one of these systems, the prey are very large animals and the predators are very small animals, such as elephants and mosquitoes. Thus it takes many predators to eat one prey, but each prey eaten is a tremendous benefit for the predator population. The other system has very large predators and very small prey. Determine which system is which and provide a justification for your answer.

2. Find all equilibrium points for the two systems. Explain the significance of these points in terms of the predator and prey populations.

3. Suppose that the predators are extinct at time $t_0 = 0$. For each system, verify that the predators remain extinct for all time.

4. For each system, describe the behavior of the prey population if the predators are extinct. (Sketch the phase line for the prey population assuming that the predators are extinct, and sketch the graphs of the prey population as a function of time for several solutions. Then interpret these graphs for the prey population.)

5. For each system, suppose that the prey are extinct at time $t_0 = 0$. Verify that the prey remain extinct for all time.

6. For each system, describe the behavior of the predator population if the prey are extinct. (Sketch the phase line for the predator population assuming that the prey are extinct, and sketch the graphs of the predator population as a function of time for several solutions. Then interpret these graphs for the predator population.)

7. Consider the predator-prey system

$$\frac{dR}{dt} = 2\left(1 - \frac{R}{3}\right)R - RF$$

$$\frac{dF}{dt} = -2F + 4RF.$$

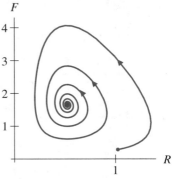

The figure to the right shows a computer-generated plot of a solution curve for this system in the RF-plane. What can you say about the fate of the prey (R) and predator (F) populations based on this image?

8. Consider the predator-prey system

$$\frac{dR}{dt} = 2R\left(1 - \frac{R}{2.5}\right) - 1.5RF$$

$$\frac{dF}{dt} = -F + 0.8RF$$

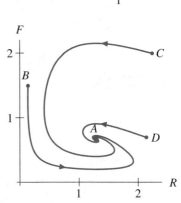

and the solution curves in the phase plane on the right.

(a) Sketch the $R(t)$- and $F(t)$-graphs for the solutions with initial points A, B, C, and D.

(b) Interpret each of these solution curves in terms of the behavior of the populations over time.

Exercises 9–14 refer to the predator-prey and the modified predator-prey systems discussed in the text (repeated here for convenience):

(i)
$$\frac{dR}{dt} = 2R - 1.2RF$$
$$\frac{dF}{dt} = -F + 0.9RF$$

(ii)
$$\frac{dR}{dt} = 2R\left(1 - \frac{R}{2}\right) - 1.2RF$$
$$\frac{dF}{dt} = -F + 0.9RF.$$

9. How would you modify these systems to include the effect of hunting of the prey at a rate of α units of prey per unit of time?

10. How would you modify these systems to include the effect of hunting of the predators at a rate proportional to the number of predators?

11. Suppose the predators discover a second, unlimited source of food, but they still prefer to eat prey when they can catch them. How would you modify these systems to include this assumption?

12. Suppose the predators found a second food source that is limited in supply. How would you modify these systems to include this fact?

13. Suppose predators migrate to an area if there are five times as many prey as predators in that area (that is, if $R > 5F$), and they move away if there are fewer than five times as many prey as predators. How would you modify these systems to take this into account?

14. Suppose prey move out of an area at a rate proportional to the number of predators in the area. How would you modify these systems to take this into account?

15. Consider the two systems of differential equations

(i)
$$\frac{dx}{dt} = 0.3x - 0.1xy$$
$$\frac{dy}{dt} = -0.1y + 2xy$$

(ii)
$$\frac{dx}{dt} = 0.3x - 3xy$$
$$\frac{dy}{dt} = -2y + 0.1xy.$$

One of these systems refers to a predator-prey system with very lethargic predators—predators who seldom catch prey but who can live for a long time on a single prey (for example, boa constrictors). The other system refers to a very active predator that requires many prey to stay healthy (such as a small cat). The prey in each case is the same. Identify which system is which and justify your answer.

16. Consider the system of predator-prey equations

$$\frac{dR}{dt} = 2\left(1 - \frac{R}{3}\right)R - RF$$
$$\frac{dF}{dt} = \quad 16F + 4RF.$$

The figure below shows a computer-generated plot of a solution curve for this system in the RF-plane. What can you say about the fate of the rabbit R and fox F populations based on this image?

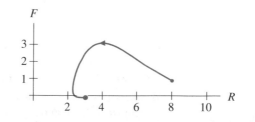

17. Pesticides that kill all insect species are not only bad for the environment, but they can also be inefficient at controlling pest species. Suppose a pest insect species in a particular field has population $R(t)$ at time t, and suppose its primary predator is another insect species with population $F(t)$ at time t. Suppose the populations of these species are accurately modeled by the system

$$\frac{dR}{dt} = 2R - 1.2RF$$

$$\frac{dF}{dt} = -F + 0.9RF$$

studied in this section. Finally, suppose that at time $t = 0$ a pesticide is applied to the field that reduces both the pest and predator populations to very small but nonzero numbers.

(a) Using Figures 2.3 and 2.5, predict what will happen as t increases to the population of the pest species.

(b) Write a short essay, in nontechnical language, warning of the possibility of the "paradoxical" effect that pesticide application can have on pest populations.

18. Some predator species seldom capture healthy adult prey, eating only injured or weak prey. Because weak prey consume resources but are not as successful at reproduction, the harsh reality is that their removal from the population increases prey population. Discuss how you would modify a predator-prey system to model this sort of interaction.

19. Consider the initial-value problem

$$\frac{d^2y}{dt^2} + y = 0$$

with $y(0) = 0$ and $y'(0) = v(0) = 1$.

(a) Show that the function $y(t) = \sin t$ is a solution to this initial-value problem.

(b) Plot the solution curve corresponding to this solution in the yv-plane.

(c) In what ways is this solution curve the same as the one shown in Figure 2.15?

(d) How is this curve different from the one shown in Figure 2.15?

20. Consider the equation

$$\frac{d^2y}{dt^2} + \frac{k}{m}y = 0$$

for the motion of a simple harmonic oscillator.

(a) Consider the function $y(t) = \cos \beta t$. Under what conditions on β is $y(t)$ a solution?

(b) What is the initial point at $t = 0$ that corresponds to this solution?

(c) In terms of k and m, what is the period of this solution?

(d) Sketch the solution curve (in the yv-plane) associated to this solution. [*Hint:* Consider the quantity $y^2 + (v/\beta)^2$.]

21. A mass weighing 12 pounds stretches a spring 3 inches. What is the spring constant for this spring?

22. A mass weighing 4 pounds stretches a spring 4 inches.

   (a) Formulate an initial-value problem that corresponds to the motion of this un-damped mass-spring system if the mass is extended 1 foot from its rest position and released (with no initial velocity).

   (b) Using the result of Exercise 20, find the solution of this initial-value problem.

23. Do the springs in an "extra firm" mattress have a large spring constant or a small spring constant?

24. Consider a mass-spring system as shown in the figure below.

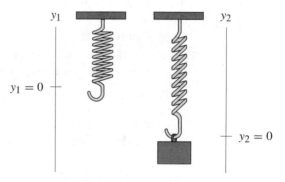

Figure 2.17
Vertical mass-spring system

Before the mass is placed on the end of the spring, the spring has a natural length. After the mass is placed on the end of the spring, the system has a new equilibrium position, which corresponds to the position where the force on the mass due to grav-ity is equal to the force on the mass due to the spring.

   (a) Assuming that the only forces acting on the spring are the force due to gravity and the force of the spring, formulate two different (but related) second-order differential equations that describe the motion of the spring. For one equation, let the position $y_1(t)$ be measured from the point at the end of the spring when it hangs without the mass attached. For the other equation, let $y_2(t)$ be mea-sured from the equilibrium position once the mass is attached to the spring.

   (b) Rewrite these two second-order equations as first-order systems and calculate their equilibrium points. Interpret your results in terms of the mass-spring sys-tem.

   (c) Given a solution $y_1(t)$ to one system, how can you produce a solution $y_2(t)$ to the second system?

   (d) Which choice of coordinate system, y_1 or y_2, do you prefer? Why?

Exercises 25–30 refer to a situation in which models similar to the predator-prey population models arise. Suppose A and B represent two substances that can combine to form a new substance C (chemists would write A + B → C). Suppose we have a container with a solution containing low concentrations of substances A and B, and A and B molecules react only when they happen to come close to each other. If $a(t)$ and $b(t)$ represent the amount of A and B in the solution, respectively, then the chance that a molecule of A is close to a molecule of B at time t is proportional to the product $a(t) \cdot b(t)$. Hence the rate of reaction of A and B to form C is proportional to ab. Suppose C precipitates out of the solution as soon as it is formed, and the solution is always kept well mixed.

25. Write a system of differential equations that models the evolution of $a(t)$ and $b(t)$. Be sure to identify and describe any parameters you introduce.

26. Describe an experiment you could perform to determine an approximate value for the parameter(s) in the system you developed in Exercise 25. Include the calculations you would perform using the data from your experiment to determine the parameter(s).

27. Suppose substances A and B are added to the solution at constant (perhaps unequal) rates. How would you modify your system to include this assumption?

28. Suppose A and B are being added to the solution at constant (perhaps unequal) rates, and, in addition to the A + B → C reaction, a reaction A + A → D also can occur when two molecules of A are close and substance D precipitates out of the solution. How would you modify your system of equations to include these assumptions?

29. Suppose A and B are being added to the solutions at constant (perhaps unequal) rates, and, in addition to the A + B → C reaction, a reaction B + B → A can also occur when two molecules of B are close. How would you modify your system of equations to include these assumptions?

30. Suppose A and B are being added to the solution at constant (perhaps unequal) rates, and, in addition to the A + B → C reaction, a reaction A + 2B → D can occur when two B and one A molecules are close. Suppose substance D precipitates out of the solution. How would you modify your system of equations to include these assumptions?

2.2 THE GEOMETRY OF SYSTEMS

In Section 2.1 we displayed $R(t)$- and $F(t)$-graphs of solutions to two different predator-prey systems, but we did not describe how we generated these graphs. We will ultimately answer this question in Section 2.4 in which we generalize Euler's method to produce numerical approximations to solutions of systems. But first we must introduce some vector notation. This notation provides a convenient shorthand for writing systems of differential equations, but it is also important for a more fundamental reason. Using vectors, we build a geometric representation of a system of differential equations. As we saw when we used slope fields in Chapter 1, having a geometric representation of a differential equation gives us a convenient way to understand the corresponding solutions.

The Predator-Prey Vector Field

Recall that the predator-prey system

$$\frac{dR}{dt} = 2R - 1.2RF$$

$$\frac{dF}{dt} = -F + 0.9RF$$

models the evolution of two populations, R and F, over time. In the previous section we studied two different (but related) ways to visualize this evolution. We can plot the graphs of $R(t)$ and $F(t)$ as functions of t, or we can plot the solution curve $(R(t), F(t))$ in the RF-plane. Although we can think of $(R(t), F(t))$ as simply a combination of the two scalar-valued functions $R(t)$ and $F(t)$, there are advantages if we take a different approach. We consider the pair $(R(t), F(t))$ as a vector-valued function in the RF-plane.

For each t, we let $\mathbf{P}(t)$ denote the vector

$$\mathbf{P}(t) = \left(\begin{array}{c} R(t) \\ F(t) \end{array} \right).$$

Then the vector-valued function $\mathbf{P}(t)$ corresponds to the solution curve $(R(t), F(t))$ in the RF-plane.

To compute the derivative of the vector-valued function $\mathbf{P}(t)$, we compute the derivatives of each component. That is,

$$\frac{d\mathbf{P}}{dt} = \left(\begin{array}{c} \frac{dR}{dt} \\ \frac{dF}{dt} \end{array} \right).$$

Using this notation, we can rewrite the predator-prey system as the single vector equation

$$\frac{d\mathbf{P}}{dt} = \left(\begin{array}{c} \frac{dR}{dt} \\ \frac{dF}{dt} \end{array} \right) = \left(\begin{array}{c} 2R - 1.2RF \\ -F + 0.9RF \end{array} \right).$$

So far we have only introduced more notation. We have converted our first-order system consisting of two scalar equations into a single vector equation involving vectors with two components.

The advantages of the vector notation start to become evident once we consider the right-hand side of this system as a *vector field*. The right-hand side of the predator-prey system is a function that assigns a vector to each point in the RF-plane. If we denote this function using the vector $\mathbf{V}$, we have

$$\mathbf{V} \left(\begin{array}{c} R \\ F \end{array} \right) = \left(\begin{array}{c} 2R - 1.2RF \\ -F + 0.9RF \end{array} \right).$$

For example, at the point $(R, F) = (2, 1)$,

$$\mathbf{V}\begin{pmatrix} 2 \\ 1 \end{pmatrix} = \begin{pmatrix} 2(2) - 1.2(2)(1) \\ -(1) + 0.9(2)(1) \end{pmatrix} = \begin{pmatrix} 1.6 \\ 0.8 \end{pmatrix}.$$

To save paper, we will sometimes write vectors vertically (as "column" vectors) and at other times horizontally (as "row" vectors). The vertical notation is more consistent with how we have written systems up to now, whereas the horizontal notation is easier on trees. In any case, we always write vectors in boldface type to distinguish them from scalars. Written as a row vector, the predator-prey vector field is expressed as

$$\mathbf{V}(R, F) = (2R - 1.2RF, -F + 0.9RF),$$

and $\mathbf{V}(2, 1) = (1.6, 0.8)$.

In the previous computation there was nothing special about the point $(R, F) = (2, 1)$. Similarly, we have $\mathbf{V}(1, 1) = (0.8, -0.1)$, $\mathbf{V}(0.5, 2.2) = (-0.32, -1.21)$, and so forth. The function $\mathbf{V}(R, F)$ can be evaluated at any point in the RF-plane.

The use of vectors enables us to simplify the notation considerably. We can now write the predator-prey system very economically as

$$\frac{d\mathbf{P}}{dt} = \mathbf{V}(\mathbf{P}).$$

The vector notation is much more than just a way to save ink. It also gives us a new way to think about and to visualize systems of differential equations. We can sketch the vector field $\mathbf{V}$ by attaching the vector $\mathbf{V}(\mathbf{P})$ to the corresponding point $\mathbf{P}$ in the plane. Computing $\mathbf{V}(\mathbf{P})$ for many different values of $\mathbf{P}$ and carefully sketching these vectors in the plane is tedious work for a human, but it is just the sort of job that computers and calculators are good at. A few vectors in the predator-prey vector field $\mathbf{V}$ are shown in Figure 2.18. In general we visualize this vector field as a "field" of arrows, one based at each point in the RF-plane.

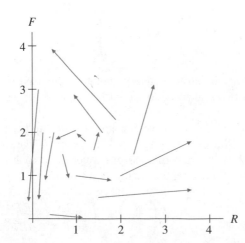

Figure 2.18
Selected vectors in the vector field $\mathbf{V}(R, F)$ for the predator-prey equations

$$\frac{dR}{dt} = 2R - 1.2RF$$

$$\frac{dF}{dt} = -F + 0.9RF.$$

At each point (R, F), the vector $\mathbf{V}(R, F)$ is drawn starting at (R, F).

The Vector Field for a Simple Harmonic Oscillator

In Section 2.1 we modeled the motion of an undamped mass-spring system by a second-order differential equation of the form

$$\frac{d^2y}{dt^2} + \frac{k}{m}y = 0,$$

where k is the spring constant and m is the mass. We also saw that this mass-spring system can be written as the first-order system

$$\frac{dy}{dt} = v$$

$$\frac{dv}{dt} = -\frac{k}{m}y,$$

where $v = dy/dt$ is the velocity of the mass.

In the special case where $k/m = 1$, we obtained the especially nice system

$$\frac{dy}{dt} = v$$

$$\frac{dv}{dt} = -y.$$

One reason this system is so nice is that its vector field $\mathbf{F}(y, v) = (v, -y)$ in the yv-plane is relatively easy to understand. After plotting a few vectors in the vector field, it is natural to wonder if all of the vectors are tangent to circles centered at the origin and in fact, they are (see Figure 2.19 and Exercise 32).

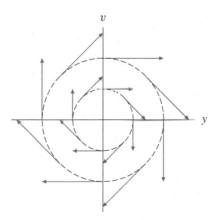

Figure 2.19
Selected vectors in the vector field $\mathbf{F}(y, v) = (v, -y)$. Note that all of these vectors are tangent to circles centered at the origin.

Although computers can take the tedium out of the process of plotting vector fields, there is one aspect of vector fields that make them much harder to plot than slope fields. By definition, the vectors in a vector field have various lengths as determined by the system of equations. Some of the vectors can be quite short while others can be quite long. Therefore if we plot a vector field by evaluating it over a regular grid in the plane, we often get overlapping vectors. For example, Figure 2.20 is a plot of the vector field $\mathbf{F}(y, v) = (v, -y)$ for the simple harmonic oscillator. We don't need to take many points before we end up with a picture that is basically useless.

To avoid the confusion of overlapping vectors in our pictures of vector fields, we often scale the vectors so they all have the same (short) length. The resulting picture is called the **direction field** associated to the original vector field. Figure 2.21 is a plot of the direction field associated to the vector field $\mathbf{F}(y, v) = (v, -y)$ for a simple harmonic oscillator.

While the direction field gives a picture that is much easier to visualize than the vector field, there is some loss of information. The lengths of a vector in the vector field give the speed of the solution as it passes through the associated point in the plane. In the direction field all information about the speed of the solution is lost. Because of the artistic advantages of using the direction field, we are almost always willing to live with this loss.

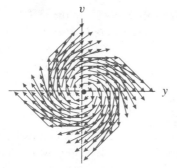

Figure 2.20
Vector field for $\mathbf{F}(y, v) = (v, -y)$.

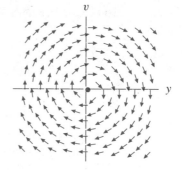

Figure 2.21
Direction field for $\mathbf{F}(y, v) = (v, -y)$.

Examples of Systems and Vector Fields

In general, for a system with two dependent variables of the form

$$\frac{dx}{dt} = f(x, y)$$
$$\frac{dy}{dt} = g(x, y),$$

we introduce the vector $\mathbf{Y}(t) = (x(t), y(t))$ and the **vector field**

$$\mathbf{F}(\mathbf{Y}) = \mathbf{F}(x, y) = (f(x, y), g(x, y)).$$

With this notation the system of two equations may be written in the compact form

$$\frac{d\mathbf{Y}}{dt} = \begin{pmatrix} \dfrac{dx}{dt} \\ \dfrac{dy}{dt} \end{pmatrix} = \begin{pmatrix} f(x, y) \\ g(x, y) \end{pmatrix} = \mathbf{F}(\mathbf{Y}),$$

or even more economically as

$$\frac{d\mathbf{Y}}{dt} = \mathbf{F}(\mathbf{Y}).$$

Elementary (but important) examples

The system

$$\frac{dx}{dt} = x$$

$$\frac{dy}{dt} = y$$

yields the vector field $\mathbf{F}(x, y) = (x, y)$, and the vectors in the vector field always point directly away from the origin (see Figure 2.22). On the other hand, the system

$$\frac{dx}{dt} = -x$$

$$\frac{dy}{dt} = -y$$

yields the vector field $\mathbf{G}(x, y) = (-x, -y)$, and the vectors in the vector field always point toward the origin (see Figure 2.23).

The system

$$\frac{dx}{dt} = -x$$

$$\frac{dy}{dt} = -2y$$

also yields a vector field $\mathbf{H}(x, y) = (-x, -2y)$ which (more or less) points toward the origin (see Figure 2.24). We will soon see that the trained eye can distinguish important differences between the vector field $\mathbf{G}(x, y)$ in Figure 2.23 and the vector field $\mathbf{H}(x, y)$ in Figure 2.24.

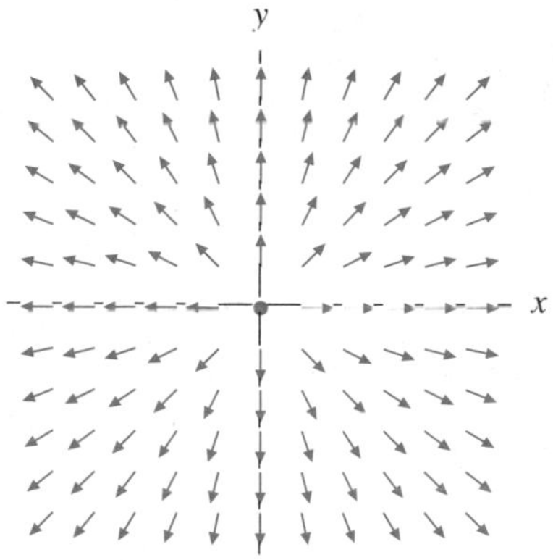

Figure 2.22
Direction field for $\mathbf{F}(x, y) = (x, y)$.

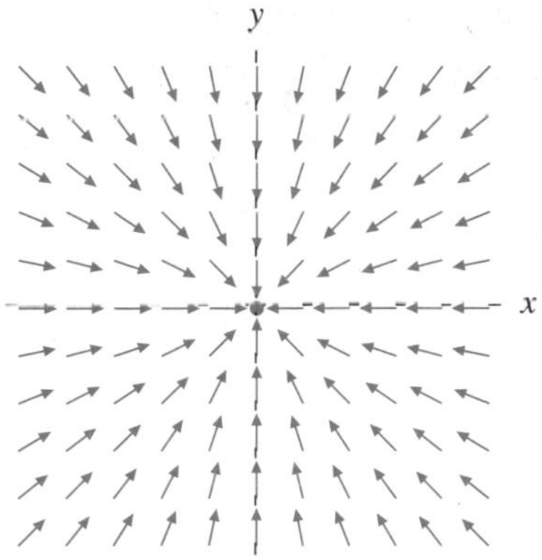

Figure 2.23
Direction field for $\mathbf{G}(x, y) = (-x, -y)$.

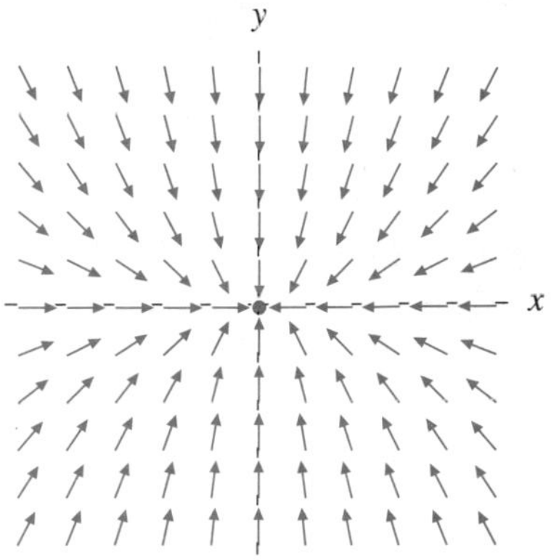

Figure 2.24
Direction field for $\mathbf{H}(x, y) = (-x, -2y)$.

The Geometry of Solutions

We can think of the picture of a vector field or a direction field as a picture of a system of differential equations, and we can use this picture to sketch solution curves of the system. To be more precise let's consider a system of the form

$$\frac{dx}{dt} = f(x, y)$$

$$\frac{dy}{dt} = g(x, y).$$

As we have seen, this system yields the vector field $\mathbf{F}(x, y) = (f(x, y), g(x, y))$. Letting $\mathbf{Y}(t) = (x(t), y(t))$, the system can be written in terms of the vector equation

$$\frac{d\mathbf{Y}}{dt} = \mathbf{F}(\mathbf{Y}).$$

Interpreting this vector equation geometrically is the key to a geometric understanding of this system of differential equations. If we think of a solution $\mathbf{Y}(t) = (x(t), y(t))$ as a parameterization of a curve in the xy-plane, then $d\mathbf{Y}/dt$ yields the tangent vectors of the curve. Therefore the equation $d\mathbf{Y}/dt = \mathbf{F}(\mathbf{Y})$ says that the tangent vectors for the solution curves are given by the vectors in the vector field.

One consequence of this geometric interpretation is that we can go directly from a sketch of a vector field $\mathbf{F}$ (or its direction field) to a sketch of the solution curves of the equation $d\mathbf{Y}/dt = \mathbf{F}(\mathbf{Y})$ without ever knowing a formula for $\mathbf{F}$ (see Figures 2.25 and 2.26).

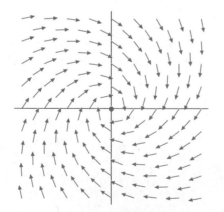

Figure 2.25
A direction field that spirals about the origin.

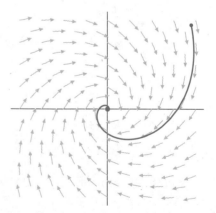

Figure 2.26
A solution curve for the solution corresponding to the initial condition indicated.

Metaphor of the parking lot

To help visualize solution curves of a system from this point of view, imagine an infinite, perfectly flat parking lot. At each point in the lot, an arrow is painted on the pavement. These arrows come from the vector field $\mathbf{F}(\mathbf{Y})$. As you drive through the parking lot, your instructions are to look out your window at the ground and drive so that your velocity vector always agrees with the arrow on the ground. (Imagine you are a professional driver in a closed parking lot.) You steer so that your car goes in the direction given by the direction of the arrow, and you go as fast as the length of the vector indicates. As you move, the arrow outside your window changes, so you must adjust the speed and direction of the car accordingly. The path you follow is the solution curve associated to a solution of the system. In fact as you will soon see, you can use exactly this idea to sketch solution curves of a system using only this interpretation of the vector field (but don't hit anything).

A solution curve of the harmonic oscillator

For example, in Section 2.1 we saw that the functions $y(t) = \cos t$ and $v(t) = -\sin t$ satisfy the simple harmonic oscillator system

$$\frac{dy}{dt} = v$$
$$\frac{dv}{dt} = -y.$$

Since $y^2 + v^2 = 1$, we know that the vector-valued function

$$\mathbf{Y}(t) = (y(t), v(t)) = (\cos t, -\sin t)$$

sweeps out the unit circle centered at the origin of the yv-plane in a clockwise fashion. As we see in Figure 2.27, the velocity vectors for this motion agree precisely with the vectors in the vector field $\mathbf{F}(y, v) = (v, -y)$.

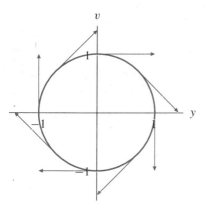

Figure 2.27
The unit circle in the yv-plane is a solution curve for the system

$$\frac{dy}{dt} = v$$
$$\frac{dv}{dt} = -y.$$

Recall that the vectors in this vector field are always tangent to circles centered at the origin.

A solution curve for a predator-prey system

In Section 2.1 we plotted the solution curve to the system

$$\frac{dR}{dt} = 2R - 1.2RF$$
$$\frac{dF}{dt} = -F + 0.9RF$$

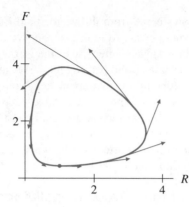

Figure 2.28
The solution curve corresponding to the solution of the predator-prey system

$$\frac{dR}{dt} = 2R - 1.2RF$$

$$\frac{dF}{dt} = -F + 0.9RF$$

with the initial condition $(R_0, F_0) = (1, 0.5)$, along with vectors from the predator-prey vector field.

corresponding to the initial condition $(R_0, F_0) = (1, 0.5)$. In Figure 2.28 we see the relationship between the solution curve and the vectors in the vector field.

Equilibrium Solutions

Just as there are special points—equilibrium points—on the phase line, there are distinguished points in the phase plane of systems of the form

$$\frac{dx}{dt} = f(x, y)$$

$$\frac{dy}{dt} = g(x, y).$$

These points also correspond to constant solutions.

DEFINITION   The point $\mathbf{Y}_0$ is an **equilibrium point** for the system $d\mathbf{Y}/dt = \mathbf{F}(\mathbf{Y})$ if $\mathbf{F}(\mathbf{Y}_0) = \mathbf{0}$. The constant function $\mathbf{Y}(t) = \mathbf{Y}_0$ is an **equilibrium solution**.   ∎

Equilibrium points are simply points at which the right-hand side of the system vanishes. If $\mathbf{Y}_0$ is an equilibrium point, then the constant function

$$\mathbf{Y}(t) = \mathbf{Y}_0 \text{ for all } t$$

is a solution of the system. To verify this claim, note that the constant function has $d\mathbf{Y}/dt = (0, 0)$ for all t. On the other hand, $\mathbf{F}(\mathbf{Y}(t)) = \mathbf{F}(\mathbf{Y}_0) = (0, 0)$ at an equilibrium point. Hence equilibrium points in the vector field correspond to constant solutions.

Computation of equilibrium points
The system

$$\frac{dx}{dt} = 3x + y$$

$$\frac{dy}{dt} = x - y$$

has only one equilibrium point, the origin $(0, 0)$. To see why, we simultaneously solve the two equations

$$\begin{cases} 3x + y = 0 \\ x - y = 0, \end{cases}$$

which are given by the right-hand side of the system. (Add the first equation to the second to see that $x = 0$, then use either equation to conclude that $y = 0$.) If we look at the vector field for this system, we see that the vectors near the origin are relatively short (see Figure 2.29). Thus solution curves move slowly as they pass near the origin. Although all nonzero vectors in the direction field are the same length by definition, we can still tell that there must be an equilibrium point at the origin because the directions of the vectors in the direction field change radically near the origin $(0, 0)$ (see Figure 2.30).

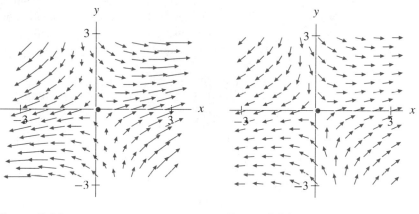

Figure 2.29
Vector field.

Figure 2.30
Direction field.

As a solution passes near an equilibrium point, both dx/dt and dy/dt are close to zero. Therefore, the $x(t)$- and $y(t)$-graphs are nearly flat over the corresponding time interval (see Figure 2.31).

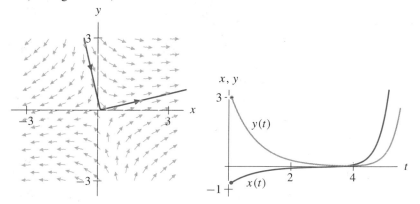

Figure 2.31
As a solution curve travels near an equilibrium point, the $x(t)$- and $y(t)$-graphs are nearly flat.

A Population Model for Two Competing Species

To illustrate all of the concepts introduced in this section, we conclude with an analysis of the system

$$\frac{dx}{dt} = 2x\left(1 - \frac{x}{2}\right) - xy$$

$$\frac{dy}{dt} = 3y\left(1 - \frac{y}{3}\right) - 2xy.$$

We think of x and y as representing the populations of two species that compete for the same resource. Note that, left on its own, each species evolves according to a logistic population growth model. The interaction of the two species is modeled by the xy-terms in both equations. For example, the effect of the population y on the rate of change of x is determined by the term $-xy$ in the dx/dt equation. This term is negative since we are assuming that the two species compete for resources. Similarly, the term $-2xy$ determines the effect of the x population on the rate of change of y. Since x and y represent populations, we focus our attention on the solutions whose initial conditions lie in the first quadrant.

Finding the equilibrium points

First, we find the equilibrium points by setting the right-hand sides of the differential equations to zero and solving for x and y in the resulting system of equations

$$\begin{cases} 2x\left(1 - \dfrac{x}{2}\right) - xy = 0 \\ 3y\left(1 - \dfrac{y}{3}\right) - 2xy = 0. \end{cases}$$

These equations can be rewritten in the form

$$\begin{cases} x(2 - x - y) = 0 \\ y(3 - y - 2x) = 0. \end{cases}$$

The first equation is satisfied if $x = 0$ or if $2 - x - y = 0$, and the second equation is satisfied if either $y = 0$ or $3 - y - 2x = 0$.

Suppose first that $x = 0$. Then the equation $y = 0$ yields an equilibrium point at the origin, and the equation $3 - y - 2x = 0$ yields an equilibrium point at $(0, 3)$.

Now suppose that $2 - x - y = 0$. Then the equation $y = 0$ yields an equilibrium point at $(2, 0)$, and the equation $3 - y - 2x = 0$ yields an equilibrium point at $(1, 1)$. (Solve the equations $2 - x - y = 0$ and $3 - y - 2x = 0$ simultaneously.) Hence the equilibrium points are $(0, 0)$, $(0, 3)$, $(2, 0)$, and $(1, 1)$.

Sketching the phase portrait

Next, we use the direction field to sketch solution curves. To get a good sketch of the phase portrait, we must choose enough solutions to see all the different types of solution curves, but not so many curves that the picture gets messy (see Figure 2.32). It is advisable to make the sketch with the aid of a computer or calculator, and in Section 2.4 we will generalize Euler's method to numerically approximate solution curves. Note that the phase portrait for this competing species model suggests that for most initial conditions, one or the other species dies out and the surviving population stabilizes.

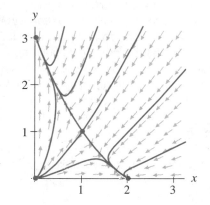

Figure 2.32
Direction field and phase portrait for the competing species model

$$\frac{dx}{dt} = 2x\left(1 - \frac{x}{2}\right) - xy$$

$$\frac{dy}{dt} = 3y\left(1 - \frac{y}{3}\right) - 2xy.$$

Note that this phase portrait suggests that for most initial conditions, one or the other species dies out and the surviving population stabilizes.

Just as we did in Chapter 1 when we started sketching slope fields and graphs of solutions, we should pause and wonder if sketches such as this one represent the true behavior of the solutions. For example, how do we know that distinct solution curves in the phase plane do not cross or even touch? As in Chapter 1 the answer follows from a powerful theorem regarding the uniqueness of solutions. We will study this theorem in Section 2.4, but in the meantime, you should assume that, if the differential equations are sufficiently nice, then distinct solution curves will not cross or even touch.

$x(t)$- and $y(t)$-graphs

As we saw in Section 2.1, the phase portrait is just one way of visualizing the solutions of systems of differential equations. Not all information about a particular solution can be seen by studying its solution curve in the phase plane. In particular, when we look at a picture of a solution curve in the phase plane, we do not see the time variable, so we don't know how fast the solution traverses the curve. The best way to get information about the time variable is to watch a computer sketch the solution curve in "real time." The next best thing is to give the solution in the phase plane, along with the $x(t)$- and $y(t)$-graphs.

In Figure 2.33, we see the $x(t)$- and $y(t)$-graphs for two solutions of the competing species model. For one initial condition the y population is essentially extinct after $t = 8$, but for the other initial condition the x population does not die out until at least $t = 15$.

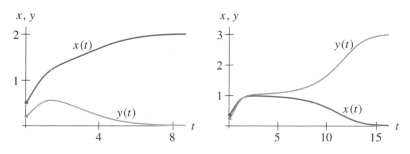

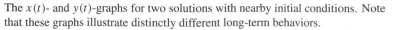

Figure 2.33
The $x(t)$- and $y(t)$-graphs for two solutions with nearby initial conditions. Note that these graphs illustrate distinctly different long-term behaviors.

Even though solution curves and $x(t)$- and $y(t)$-graphs display different information about solutions, it is important to be able to connect the two different representations. Two solution curves that come from these particular initial conditions are shown in Figure 2.34. From the solution curve corresponding to the initial condition on the right, we can conclude that the solution approaches the equilibrium point $(2, 0)$. In particular, for this initial condition the y population becomes extinct. The solution for the left initial condition approaches the equilibrium point $(0, 3)$, so the x population becomes extinct. We observed the same long-term behavior when we plotted the $x(t)$- and $y(t)$-graphs (see Figure 2.33).

In the phase plane we also note that the solution curve for the initial condition on the left crosses the line $y = x$. In other words, from the solution curve in the phase plane, we can see from the phase portrait that there is one time t at which the two populations are equal. However, to determine that particular time, we must consult the corresponding $x(t)$- and $y(t)$-graphs. Similarly, for the other initial condition, we know that the x population is always larger than the y population.

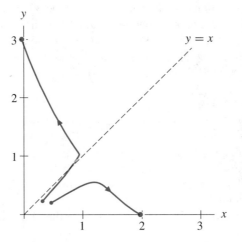

Figure 2.34
Two solution curves for solutions to the system

$$\frac{dx}{dt} = 2x\left(1 - \frac{x}{2}\right) - xy$$
$$\frac{dy}{dt} = 3y\left(1 - \frac{y}{3}\right) - 2xy.$$

These curves correspond to solutions with nearby initial conditions. The long-term behavior of these two solutions is also illustrated in Figure 2.33.

Qualitative Thinking

In all the systems considered so far, the independent variable has not appeared on the right-hand side. Systems with this property are said to be **autonomous**. The word *autonomous* means self-governing, and roughly speaking, an autonomous system is self-governing because it evolves according to differential equations that are determined entirely by the values of the dependent variables. An important geometric consequence is that the vector field associated to an autonomous system depends only on the dependent variables. As a result we need not consider the independent variable when we sketch the vector field (or direction field), the solution curves, and the phase portrait.

Although we will continue to focus on autonomous systems for the remainder of this chapter and throughout Chapter 3, many important systems are nonautonomous. We will first encounter nonautonomous systems in Chapter 4. In the next two sections of this chapter, we complement the geometric approach introduced here with analytic and numerical approaches.

EXERCISES FOR SECTION 2.2

In Exercises 1–6,

  (a) determine the vector field associated to the first-order system specified,

  (b) sketch enough vectors in the vector field to get a sense of its geometric structure,

  (c) sketch an associated direction field, and

  (d) give a rough sketch of the phase portrait of the system.

1. $\dfrac{dx}{dt} = 1$

  $\dfrac{dy}{dt} = 0$

2. $\dfrac{dx}{dt} = x$

  $\dfrac{dy}{dt} = 1$

3. $\dfrac{dy}{dt} = -v$

  $\dfrac{dv}{dt} = y$

4. $\dfrac{du}{dt} = u - 1$

  $\dfrac{dv}{dt} = v - 1$

5. $\dfrac{dx}{dt} = x$

  $\dfrac{dy}{dt} = -y$

6. $\dfrac{dx}{dt} = x$

  $\dfrac{dy}{dt} = 2y$

7. Convert the second-order differential equation

$$\frac{d^2y}{dt^2} - y = 0$$

into a first-order system in terms of y and v, where $v = dy/dt$, and

  (a) determine the vector field associated to the first-order system,

  (b) sketch enough vectors in the vector field to get a sense of its geometric structure,

  (c) sketch an associated direction field, and

  (d) give a rough sketch of the phase portrait of the system.

8. Convert the second-order differential equation

$$\frac{d^2y}{dt^2} + 2y = 0$$

into a first-order system in terms of y and v, where $v = dy/dt$, and

  (a) determine the vector field associated to the first-order system,

  (b) sketch enough vectors in the vector field to get a sense of its geometric structure,

  (c) sketch an associated direction field, and

  (d) give a rough sketch of the phase portrait of the system.

9. Consider the system

$$\frac{dx}{dt} = x + 2y$$

$$\frac{dy}{dt} = -y$$

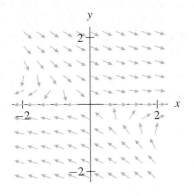

and its corresponding direction field.

(a) Sketch a number of different solution curves on the phase plane.

(b) Describe the behavior of the solution that satisfies the initial condition $(x_0, y_0) = (-2, 2)$.

10. Consider the system

$$\frac{dx}{dt} = -2x + y$$

$$\frac{dy}{dt} = -2y$$

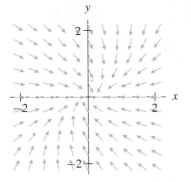

and its corresponding direction field.

(a) Sketch a number of different solution curves on the phase plane.

(b) Describe the behavior of the solution that satisfies the initial condition $(x_0, y_0) = (0, 2)$.

In Exercises 11–16,
   (a) find the equilibrium points of the system,
   (b) using a computer or calculator, sketch the direction field of the system, and
   (c) using the direction field, sketch the phase portrait for the system.

11. $\dfrac{dx}{dt} = 4x - 7y + 2$

$\dfrac{dy}{dt} = 3x + 6y - 1$

12. $\dfrac{dR}{dt} = 4R - 7F - 1$

$\dfrac{dF}{dt} = 3R + 6F - 12$

13. $\dfrac{dz}{dt} = \cos w$

$\dfrac{dw}{dt} = -z + w$

14. $\dfrac{dx}{dt} = y$

$\dfrac{dy}{dt} = x - x^3 - y$

15. $\dfrac{dx}{dt} = y$

$\dfrac{dy}{dt} = -\cos x - y$

16. $\dfrac{dx}{dt} = y(x^2 + y^2 - 1)$

$\dfrac{dy}{dt} = -x(x^2 + y^2 - 1)$

In Exercises 17–20, match the direction field with one of the following eight systems. Provide justification for your choices.

(i) $\dfrac{dx}{dt} = -x$

$\dfrac{dy}{dt} = y - 1$

(ii) $\dfrac{dx}{dt} = x^2 - 1$

$\dfrac{dy}{dt} = y$

(iii) $\dfrac{dx}{dt} = x + 2y$

$\dfrac{dy}{dt} = -y$

(iv) $\dfrac{dx}{dt} = 2x$

$\dfrac{dy}{dt} = y$

(v) $\dfrac{dx}{dt} = x$

$\dfrac{dy}{dt} = 2y$

(vi) $\dfrac{dx}{dt} = x - 1$

$\dfrac{dy}{dt} = -y$

(vii) $\dfrac{dx}{dt} = x^2 - 1$

$\dfrac{dy}{dt} = -y$

(viii) $\dfrac{dx}{dt} = x - 2y$

$\dfrac{dy}{dt} = -y$

17.

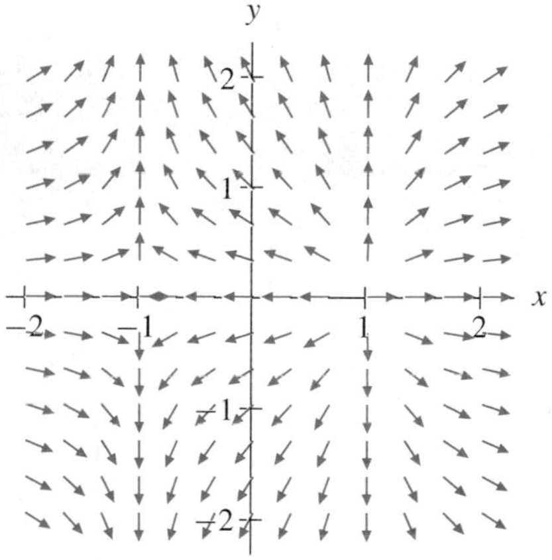

18.

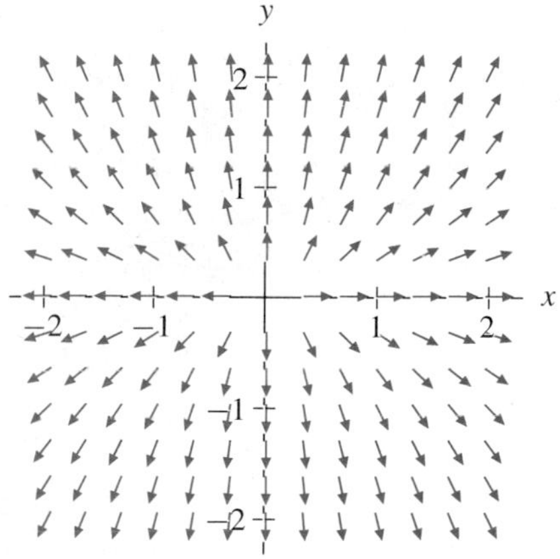

19.

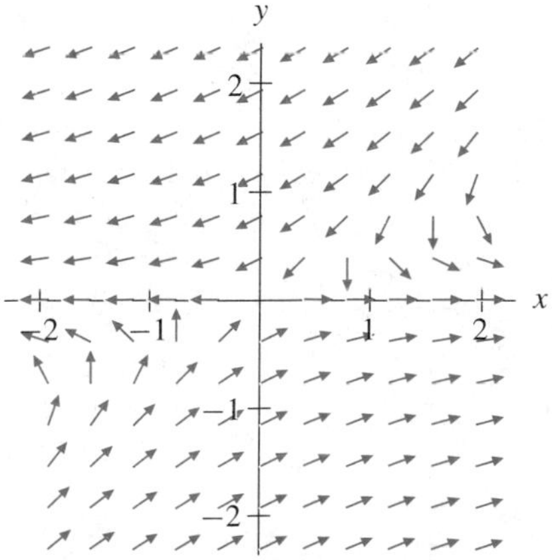

20.

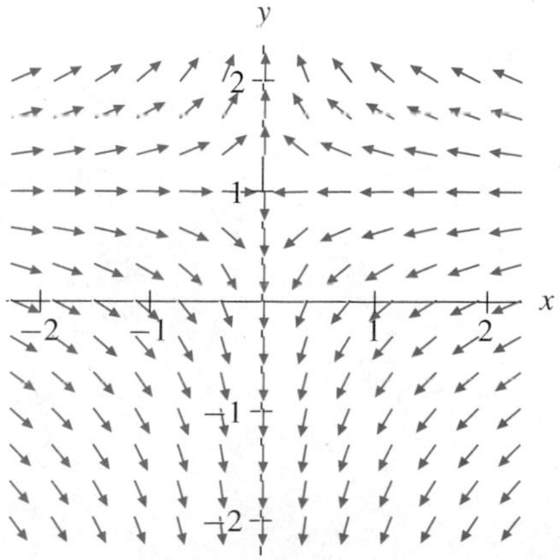

Exercises 21–24 involve the four solution curves in the phase portrait below.

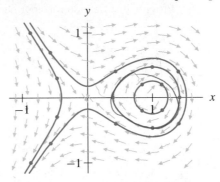

In each exercise, a pair of $x(t)$- and $y(t)$-graphs is shown. Determine which solution curve corresponds to the graphs provided. Then on the t-axis mark the t-values that correspond to the distinguished points along the curve.

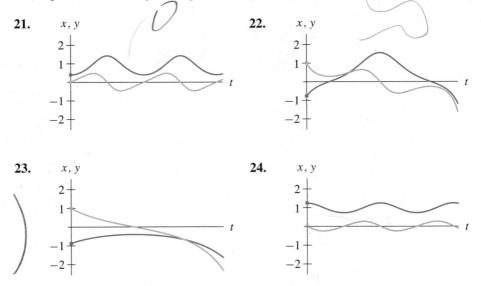

21. x, y

22. x, y

23. x, y

24. x, y

In Exercises 25–28, a solution curve in the xy-plane and an initial condition on that curve are specified. Sketch the $x(t)$- and $y(t)$-graphs for the solution.

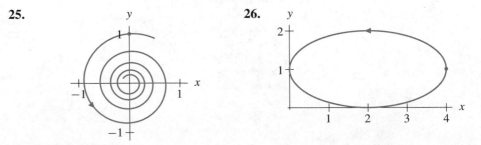

25.

26.

27.

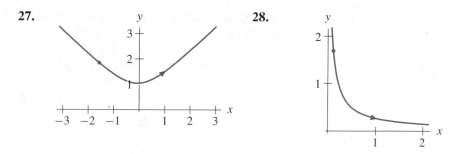

28.

29. The following graphs are the $x(t)$- and the $y(t)$-graphs for a solution curve in the xy-phase plane. Sketch that curve and indicate the direction that the solution travels as time increases.

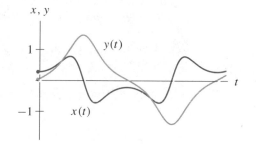

30. Recall the Metaphor of the Parking Lot from this section. Suppose two people, say Gib and Harry, are both driving cars on the parking lot and both are carefully following the rules prescribed in the metaphor. If they start at time $t = 0$ at different points, will they ever collide? (Neglect the width of their cars.)

31. Consider the two drivers, Gib and Harry, from Exercise 30. Suppose that at time $t = 0$ they start at different points in the parking lot, but at time $t = 1$ Gib drives over the point where Harry started. Will they ever collide? What can you say about their paths?

32. Show that all vectors in the vector field $\mathbf{F}(y, v) = (v, -y)$ are tangent to circles centered at the origin (see Figure 2.19). [*Hint*: You can verify this fact using slopes or the dot product of two vectors.]

2.3 ANALYTIC METHODS FOR SPECIAL SYSTEMS

When we studied first-order differential equations in Chapter 1, we saw that we could sometimes derive a formula for the general solution if the differential equation had a special form. When that happened, the analytic techniques for computing the solutions were especially adapted to the form of the differential equation.

For systems of differential equations, the special forms for which we can apply analytic techniques to find explicit solutions are few and far between. Because they are rare, these special systems are very valuable. We can use them to develop intuition (even wisdom) that we then use when studying systems for which analytical techniques are unavailable.

The most important class of systems that we can solve explicitly, the linear systems, is studied at length in Chapter 3. In this section we discuss analytic techniques that apply to very special classes of equations. We use the formulas that we obtain to become more familiar with solution curves and $x(t)$- and $y(t)$-graphs.

Checking Solutions

As noted above, finding formulas for a solution of a system can range from difficult to impossible. However, once we have the formulas, checking that they give a solution is not so bad. This observation is important for two reasons. First, we can double-check the (sometimes daunting) arithmetic we did while calculating the formulas. Second, and more important, many of the "techniques" for solving systems are really just sophisticated guessing schemes. Once we make a guess, we test to see if our guess actually is a solution.

Consider the system

$$\frac{dx}{dt} = -x + y$$
$$\frac{dy}{dt} = -3x - 5y.$$

We can rewrite this system in vector notation as

$$\frac{d\mathbf{Y}}{dt} = \mathbf{F}(\mathbf{Y}),$$

where $\mathbf{Y}(t) = (x(t), y(t))$ and $\mathbf{F}(x, y) = (-x + y, -3x - 5y)$. Now suppose someone says that

$$\mathbf{Y}(t) = (x(t), y(t)) = (e^{-4t} - 3e^{-2t}, -3e^{-4t} + 3e^{-2t})$$

is a solution to this system.

To verify this claim, we compute the derivatives of both $x(t)$ and $y(t)$. We have

$$\frac{dx}{dt} = \frac{d(e^{-4t} - 3e^{-2t})}{dt} = -4e^{-4t} + 6e^{-2t}$$
$$\frac{dy}{dt} = \frac{d(-3e^{-4t} + 3e^{-2t})}{dt} = 12e^{-4t} - 6e^{-2t}.$$

We must also substitute $x(t)$ and $y(t)$ into the right-hand side of the system. We get

$$-x + y = -(e^{-4t} - 3e^{-2t}) + (-3e^{-4t} + 3e^{-2t}) = -4e^{-4t} + 6e^{-2t}$$

and

$$-3x - 5y = -3(e^{-4t} - 3e^{-2t}) - 5(-3e^{-4t} + 3e^{-2t}) = 12e^{-4t} - 6e^{-2t}.$$

Thus dx/dt is equal to $-x + y$ and dy/dt is equal to $-3x - 5y$ for all t. Hence

$$\mathbf{Y}(t) = (x(t), y(t)) = (e^{-4t} - 3e^{-2t}, -3e^{-4t} + 3e^{-2t})$$

is a solution.

Note that $\mathbf{Y}(0) = (-2, 0)$. Consequently we have checked that $\mathbf{Y}(t)$ is a solution of the initial-value problem

$$\frac{d\mathbf{Y}}{dt} = \mathbf{F}(\mathbf{Y}), \quad \mathbf{Y}(0) = (-2, 0).$$

As a second example, consider the system

$$\frac{dx}{dt} = 2x - y$$

$$\frac{dy}{dt} = x - 2y,$$

and suppose we want to see if the function $\mathbf{Y}(t) = (e^{-t}, 3e^{-t})$ is a solution that satisfies the initial condition $\mathbf{Y}(0) = (1, 3)$.

To check that $\mathbf{Y}(t)$ satisfies the initial condition, we evaluate it at $t = 0$. This gives $\mathbf{Y}(0) = (e^{-0}, 3e^{-0}) = (1, 3)$. Next we check to see if the first equation of the system is satisfied. We have

$$\frac{dx}{dt} = \frac{d(e^{-t})}{dt} = -e^{-t}.$$

Substituting $x(t)$ and $y(t)$ into the right-hand side of the equation for dx/dt gives

$$2x - y = 2e^{-t} - 3e^{-t} = -e^{-t}.$$

Thus the first equation holds for all t. Finally, we must check the second equation in the system. We have

$$\frac{dy}{dt} = \frac{d(3e^{-t})}{dt} = -3e^{-t},$$

and

$$x - 2y = e^{-t} - 2(3e^{-t}) = -5e^{-t}.$$

Since the second equation is not satisfied, the function $\mathbf{Y}(t) = (e^{-t}, 3e^{-t})$ is not a solution of the initial-value problem.

The moral of these two examples is very important and often overlooked. Given a formula for a function $\mathbf{Y}(t)$, we can always check to see if that function satisfies the system simply by direct computation. This type of computation is certainly not the most exciting part of the subject, but it is straightforward. We can immediately determine if a given vector-valued function is a solution.

Decoupled Systems

One of the things that makes systems of differential equations so difficult (and so interesting) is that the rate of change of each of the dependent variables often depends on the values of other dependent variables. However, sometimes there is not too much interdependence among the variables and, in that case we can often derive the general solution using techniques from Chapter 1.

A system of differential equations is said to *decouple* if the rate of change of one or more of the dependent variables depends only on its own value.

A completely decoupled example

Consider the system

$$\frac{dx}{dt} = -2x$$
$$\frac{dy}{dt} = -y.$$

Since the equation for dx/dt involves only x and the equation for dy/dt involves only y, we can solve the two equations separately. When this happens, we say the system is **completely decoupled**. The general solution of $dx/dt = -2x$ is $x(t) = k_1 e^{-2t}$, where k_1 is any constant. The general solution of $dy/dt = -y$ is $y(t) = k_2 e^{-t}$, where k_2 is any constant. We can put these together to find the general solution

$$(x(t), y(t)) = (k_1 e^{-2t}, k_2 e^{-t})$$

of the system. This general solution has two undetermined constants, k_1 and k_2. These constants can be adjusted so that any given initial condition can be satisfied. For example, given the initial condition $\mathbf{Y}(0) = (1, 1)$, we let $k_1 = 1$ and $k_2 = 1$ to obtain the solution

$$\mathbf{Y}(t) = \begin{pmatrix} e^{-2t} \\ e^{-t} \end{pmatrix}.$$

In Figure 2.35 we plot this curve along with the direction field associated to the vector field $\mathbf{F}(x, y) = (-2x, -y)$. From the formula for $\mathbf{Y}(t)$, we note that $\mathbf{Y}(t)$ gives a parameterization of the upper half of the curve $x = y^2$ in the plane because

$$(y(t))^2 = (e^{-t})^2 = e^{-2t} = x(t).$$

We only obtain the upper half of this parabola because $y(t) = e^{-t} > 0$ for all t.

The solution curve in the phase plane hides the behavior of our solution with respect to the independent variable t. The solution actually tends exponentially toward the origin. Since we have the formulas for $x(t)$ and $y(t)$, it is not difficult to sketch the $x(t)$- and $y(t)$-graphs (see Figure 2.36).

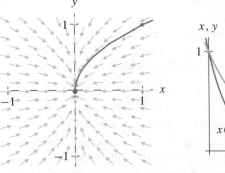

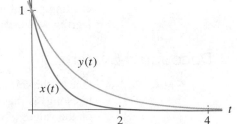

Figure 2.35
The solution curve
$\mathbf{Y}(t) = (e^{-2t}, e^{-t})$.

Figure 2.36
The $x(t)$- and $y(t)$-graphs for the solution
$(x(t), y(t)) = (e^{-2t}, e^{-t})$.

A partially decoupled example

Our next example is the system

$$\frac{dx}{dt} = 2x + 3y$$
$$\frac{dy}{dt} = -4y.$$

For this system the rate of change of x depends on both x and y, but the rate of change of y depends only on y. We say that the dependent variable y decouples from the system and the system is **partially decoupled**.

The general solution of the equation for y is $y(t) = k_2 e^{-4t}$, where k_2 is an arbitrary constant. Substituting this expression for y into the equation for x gives

$$\frac{dx}{dt} = 2x + 3k_2 e^{-4t}.$$

This is a first-order linear equation. Recall from Section 1.8 that we write this equation in the form

$$\frac{dx}{dt} - 2x = 3k_2 e^{-4t}$$

to find the general solution. Multiplying both sides by the integrating factor

$$e^{\int -2\,dt} = e^{-2t},$$

we obtain

$$e^{-2t} \frac{dx}{dt} - 2e^{-2t} x = 3k_2 e^{-6t},$$

which is equivalent to

$$\frac{d(e^{-2t} x)}{dt} = 3k_2 e^{-6t}.$$

Integrating both sides with respect to t yields

$$e^{-2t} x = -\tfrac{1}{2} k_2 e^{-6t} + k_1,$$

where k_1 is any constant. Simplifying, we have

$$x(t) = k_1 e^{2t} - \tfrac{1}{2} k_2 e^{-4t}.$$

Putting this together with the general solution of the equation for y, we obtain the general solution

$$x(t) = k_1 e^{2t} - \tfrac{1}{2} k_2 e^{-4t}$$
$$y(t) = k_2 e^{-4t}.$$

The constants k_1 and k_2 can be adjusted to obtain any desired initial condition. For example, suppose we have $x(0) = 0$ and $y(0) = 1$. To find the appropriate values of k_1 and k_2, we substitute $t = 0$ into the formula for the general solution and solve. That is,

$$x(0) = 0 = k_1 - \tfrac{1}{2} k_2$$
$$y(0) = 1 = k_2,$$

which gives $k_1 = 1/2$ and $k_2 = 1$. So the solution of the initial-value problem is

$$x(t) = \tfrac{1}{2}e^{2t} - \tfrac{1}{2}e^{-4t}$$
$$y(t) = e^{-4t}.$$

For the initial condition $(x(0), y(0)) = (-1/2, 1)$, we can follow the same steps as above, obtaining $k_1 = 0$ and $k_2 = 1$. The formula for this solution is

$$x(t) = -\tfrac{1}{2}e^{-4t}$$
$$y(t) = e^{-4t}.$$

Note that $y(t)/x(t) = -2$ for all t and that the solution tends toward the equilibrium point at the origin as t increases and toward infinity as t decreases. Since the ratio y/x is constant, the solution curve lies on a line through the origin in the phase plane (see Figure 2.37). The fact that this system has a solution curve that lies on a line is an artifact of the simple algebra of the equations. This sort of special geometry will be extensively exploited in Chapter 3.

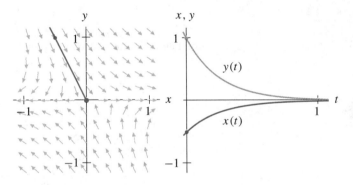

Figure 2.37
Even though the $x(t)$- and $y(t)$-graphs are graphs of exponential functions, the corresponding solution curve lies on a line in the xy-phase plane.

The Damped Harmonic Oscillator

As a final example, we return to the model of the harmonic oscillator that we discussed in Section 2.1 (see Figure 2.38). We let $y(t)$ denote the position of the mass measured from the rest position of the spring. The undamped harmonic oscillator equation is

$$m\frac{d^2y}{dt^2} = -ky,$$

where m is the mass and k is the spring constant.

We saw in Sections 2.1 and 2.2 that this equation has solutions that involve sine and cosine functions. Such solutions oscillate forever with constant amplitude, and therefore they correspond to perpetual motion. To make the model more realistic, we must include some form of friction or damping. A damping force slows the motion,

Figure 2.38
Mass-spring
system.

dissipating energy from the system. A realistic model including air resistance and the frictional forces between the mass and the table is very complicated because friction is a surprisingly subtle phenomenon.* As a first model, we lump together all the damping forces and assume that the strength of this force is proportional to the velocity. Thus the form of the damping force is

$$-b\left(\frac{dy}{dt}\right),$$

where $b > 0$ is called the **coefficient of damping**. The minus sign indicates that the damping pushes against the direction of motion, always reducing the speed. The parameter b can be adjusted by adjusting the viscosity of the medium through which the mass moves (for example, by putting the whole mechanism in the bathtub).

To obtain the new model, we equate the product of the mass and the acceleration with the sum of the spring force and the damping and we get

$$m\frac{d^2y}{dt^2} = -ky - b\frac{dy}{dt},$$

which is typically written

$$m\frac{d^2y}{dt^2} + b\frac{dy}{dt} + ky = 0.$$

This equation is often called the equation for the **damped harmonic oscillator**. To simplify the notation, we often let $p = b/m$ and $q - k/m$, and rewrite the equation as

$$\frac{d^2y}{dt^2} + p\frac{dy}{dt} + qy = 0.$$

We can convert this second-order equation into a system by letting v denote the velocity, so $v = dy/dt$ and we have

$$\frac{dy}{dt} = v$$

$$\frac{dv}{dt} = -qy - pv.$$

Guessing solutions

To get an idea of the behavior of solutions of the damped harmonic oscillator, it would be nice to have some explicit solutions, and we can use the time-honored *guess-and-test*

---

*See Jacqueline Krim, "Friction at the Atomic Scale," *Scientific American*, Vol. 275, No. 4, Oct. 1996 for an interesting discussion of friction.

method to obtain some. The idea of this "method" is to make a reasonable guess of the form of the solution and then to substitute this guess into the differential equation. The hope is that by adjusting the guess, we can obtain a solution.

Consider the equation

$$\frac{d^2y}{dt^2} + 5\frac{dy}{dt} + 6y = 0.$$

A solution $y(t)$ is a function whose second derivative can be expressed in terms of y, dy/dt, and constants. The most familiar function whose derivative is almost exactly itself is the exponential function, so we guess that there is a solution of the form $y(t) = e^{st}$ for some choice of the constant s. To determine which (if any) choices of s make $y(t)$ a solution, we substitute $y(t) = e^{st}$ into the left-hand side of the differential equation, obtaining

$$\frac{d^2y}{dt^2} + 5\frac{dy}{dt} + 6y = \frac{d^2(e^{st})}{dt^2} + 5\frac{d(e^{st})}{dt} + 6(e^{st})$$

$$= s^2e^{st} + 5se^{st} + 6e^{st}$$

$$= (s^2 + 5s + 6)e^{st}$$

In order for $y(t) = e^{st}$ to be a solution, this expression must equal the right-hand side of the differential equation for all t. In other words, we must have

$$(s^2 + 5s + 6)e^{st} = 0$$

for all t. Now, $e^{st} \neq 0$ for all t, so we must choose s so that

$$s^2 + 5s + 6 = 0.$$

This equation is satisfied only if $s = -2$ or $s = -3$. Hence this process yields two solutions, $y_1(t) = e^{-2t}$ and $y_2(t) = e^{-3t}$, of this equation.

These solutions can be converted into solutions of the system by letting $v_1 = dy_1/dt = -2e^{-2t}$ and $v_2 = dy_2/dt = -3e^{-3t}$. So, $\mathbf{Y}_1(t) = (e^{-2t}, -2e^{-2t})$ and $\mathbf{Y}_2(t) = (e^{-3t}, -3e^{-3t})$ are solutions of the associated system.

The solution curves and the $y(t)$- and $v(t)$-graphs for these two solutions are given in Figures 2.39–2.41. The direction field indicates that all solutions tend to the

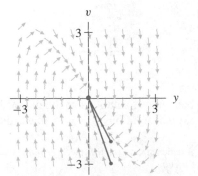

Figure 2.39

The two solution curves that correspond to the solutions $y_1(t) = e^{-2t}$ and $y_2(t) = e^{-3t}$ of the equation

$$\frac{d^2y}{dt^2} + 5\frac{dy}{dt} + 6y = 0.$$

Both curves lie on lines in the yv-plane.

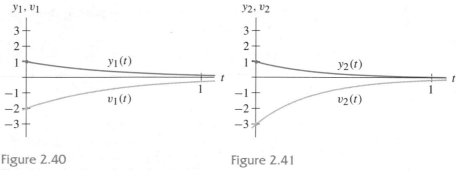

Figure 2.40
The $y(t)$- and $v(t)$-graphs for the solution $y_1(t) = e^{-2t}$.

Figure 2.41
The $y(t)$- and $v(t)$-graphs for the solution $y_2(t) = e^{-3t}$.

origin. This is no surprise because the damping reduces the speed. The two particular solutions we have computed are special because the solution curves lie on lines in the phase plane. From the direction field we can see that most solution curves are not straight lines.

General comment on guess-and-test methods

The guess-and-test "method" of finding explicit solutions is never satisfying. How do we know what to guess if we do not already know the solution? If our first guess does not work, what then? In Chapter 3, we study linear systems (including the damped harmonic oscillator). We will see that our "guess" of e^{st} above is a result of the interplay between the algebraic form of the equations and the geometric structure of the solution curves in the phase plane for these systems.

EXERCISES FOR SECTION 2.3

In Exercises 1–4, we consider the system

$$\frac{dx}{dt} = 2x + 2y$$
$$\frac{dy}{dt} = x + 3y.$$

For the given functions $\mathbf{Y}(t) = (x(t), y(t))$, check to see if $\mathbf{Y}(t)$ is a solution to the system.

1. $(x(t), y(t)) = (2e^t, -e^t)$

2. $(x(t), y(t)) = (3e^{2t} + e^t, -e^t + e^{4t})$

3. $(x(t), y(t)) = (2e^t - e^{4t}, -e^t + e^{4t})$

4. $(x(t), y(t)) = (4e^t + e^{4t}, -2e^t + e^{4t})$

In Exercises 5–12, we consider the partially decoupled system

$$\frac{dx}{dt} = 2x + y$$
$$\frac{dy}{dt} = -y.$$

5. Although we can use the method described in this section to derive the general solution to this system, why should we immediately know that $\mathbf{Y}(t) = (x(t), y(t)) = (e^{2t} - e^{-t}, e^{-2t})$ is *not* a solution to the system?

6. Although we can use the method described in this section to derive the general solution to this system, is there an easier way to show that $\mathbf{Y}(t) = (x(t), y(t)) = (4e^{2t} - e^{-t}, 3e^{-t})$ is a solution to the system?

7. Use the method described in this section to derive the general solution to this system.

8. **(a)** Can you choose constants in the general solution obtained in Exercise 7 that yield the function $\mathbf{Y}(t) = (e^{-t}, 3e^{-t})$?

(b) Suppose that the result of Exercise 7 was not immediately available. How could you tell that $\mathbf{Y}(t) = (e^{-t}, 3e^{-t})$ is not a solution?

9. **(a)** Using the result of Exercise 7, determine the solution that satisfies the initial condition $\mathbf{Y}(0) = (x(0), y(0)) = (1, 0)$.

(b) In the xy-phase plane, plot the solution curve associated to this solution.

(c) Plot the corresponding $x(t)$- and $y(t)$-graphs.

10. **(a)** Using the result of Exercise 7, determine the solution that satisfies the initial condition $\mathbf{Y}(0) = (x(0), y(0)) = (-1, 3)$.

(b) In the xy-phase plane, plot the solution curve associated to this solution.

(c) Plot the corresponding $x(t)$- and $y(t)$-graphs.

11. **(a)** Using the result of Exercise 7, determine the solution that satisfies the initial condition $\mathbf{Y}(0) = (x(0), y(0)) = (0, 1)$.

(b) Using a computer or calculator, plot the corresponding solution curve in the xy-phase plane and compare the result with the curve that you would have drawn directly from the direction field for the system.

(c) Using only the solution curve, sketch the $x(t)$- and $y(t)$-graphs.

(d) Compare your sketch with the $x(t)$- and $y(t)$-graphs that the computer provides.

12. **(a)** Using the result of Exercise 7, determine the solution that satisfies the initial condition $\mathbf{Y}(0) = (x(0), y(0)) = (1, -1)$.

(b) Using a computer or calculator, plot the corresponding solution curve in the xy-phase plane and compare the result with the curve that you would have drawn directly from the direction field for the system.

(c) Using only the solution curve, sketch the $x(t)$- and $y(t)$-graphs.

(d) Compare your sketch with the $x(t)$- and $y(t)$-graphs that the computer provides.

In Exercises 13–16, a second-order equation for $y(t)$ is given.

(a) Plot its direction field in the yv-plane, where $v = dy/dt$.

(b) Using the guess-and-test method described in this section, find two nonzero solutions that are not multiples of one another.

(c) For each solution, plot both its solution curve in the yv-plane and its $y(t)$- and $v(t)$-graphs.

13. $\dfrac{d^2y}{dt^2} + 3\dfrac{dy}{dt} - 10y = 0$

14. $\dfrac{d^2y}{dt^2} + 3\dfrac{dy}{dt} + 2y = 0$

15. $\dfrac{d^2y}{dt^2} + 4\dfrac{dy}{dt} + y = 0$

16. $\dfrac{d^2y}{dt^2} + 5\dfrac{dy}{dt} + 6y = 0$

In Exercises 17 and 18, we consider a mass sliding on a frictionless table between two walls that are 1 unit apart and connected to both walls with springs, as shown below.

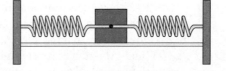

Let k_1 and k_2 be the spring constants of the left and right spring, respectively, let m be the mass, and let b be the damping coefficient of the medium the spring is sliding through. Suppose L_1 and L_2 are the rest lengths of the left and right springs, respectively.

17. Write a second-order differential equation for the position of the mass at time t. [*Hint*: The first step is to pick an origin, that is, a point where the position is 0. Halfway between the walls is a natural choice, but remember that at this point the system may not be in equilibrium. The springs may be exerting forces, depending on their rest lengths.]

18.   (a) Convert the second-order equation of Exercise 17 into a first-order system.

(b) Find the equilibrium point of this system.

(c) Using your result from part (b), pick a new coordinate system and rewrite the system in terms of this new coordinate system.

(d) How does this new system compare to the system for a damped harmonic oscillator?

19. Consider the partially coupled system

$$\frac{dx}{dt} = xy$$

$$\frac{dy}{dt} = y + 1.$$

(a) Derive the general solution.

(b) Find the equilibrium points of the system.

(c) Find the solution that satisfies the initial condition $(x_0, y_0) = (1, 0)$.

(d) With the aid of a computer or calculator, plot the phase portrait for this system, and identify the solution curve that corresponds to the solution with initial condition $(x_0, y_0) = (1, 0)$.

2.4 EULER'S METHOD FOR SYSTEMS

Many of the examples in this chapter include some type of plot of solutions, either as curves in the phase plane or as $x(t)$- or $y(t)$-graphs. In most cases these plots are provided without any indication of how we obtain them. Occasionally the solutions are line segments or circles or ellipses, and we are able to verify this analytically. But more often the solutions do not lie on familiar curves. For example, consider the predator-prey type system

$$\frac{dx}{dt} = 2x - 1.2xy$$

$$\frac{dy}{dt} = -y + 1.2xy$$

and the solution that satisfies the initial condition $(x(0), y(0)) = (1.75, 1.0)$. Figure 2.42 shows this solution in the phase plane, the xy-plane, and Figure 2.43 contains the corresponding $x(t)$- and $y(t)$-graphs. Figure 2.42 suggests that this solution is a closed curve, but the curve is certainly neither circular nor elliptical. Similarly, the $x(t)$- and $y(t)$-graphs appear to be periodic, although they do not seem to be graphs of any of the standard periodic functions (sine, cosine, etc.). So how did we compute these graphs?

The answer to this question is essentially the same as the answer to the analogous question for first-order equations. We use a dependable numerical technique and the aid of a computer. In this section we define Euler's method for first-order systems. Other numerical methods are discussed in Chapter 7.

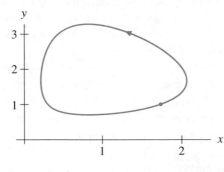

Figure 2.42
A solution curve corresponding to the initial condition $(x_0, y_0) = (1.75, 1.0)$.

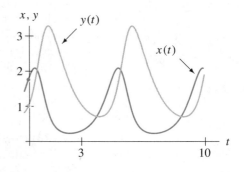

Figure 2.43
The corresponding $x(t)$- and $y(t)$-graphs for the solution curve in Figure 2.42.

Derivation of Euler's Method

Consider the first-order autonomous system

$$\frac{dx}{dt} = f(x, y)$$

$$\frac{dy}{dt} = g(x, y),$$

along with the initial condition $(x(t_0), y(t_0)) = (x_0, y_0)$. We have seen that we can use vector notation to rewrite this system as

$$\frac{d\mathbf{Y}}{dt} = \mathbf{F(Y)},$$

where $\mathbf{Y} = (x, y)$, $d\mathbf{Y}/dt = (dx/dt, dy/dt)$, and $\mathbf{F(Y)} = (f(x, y), g(x, y))$. The vector-valued function $\mathbf{F}$ yields a vector field, and a solution is a curve whose tangent vector at any point on the curve agrees with the vector field (see Figure 2.44). In other words the "velocity" vector for the curve is equal to the vector $\mathbf{F}(x(t), y(t))$.

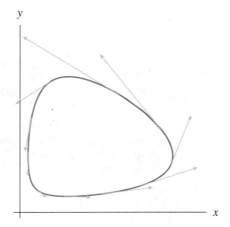

Figure 2.44
A solution curve is a curve that is everywhere tangent to the vector field.

As we saw in Section 1.4, Euler's method for a first-order equation is based on the idea of approximating the graph of a solution by line segments whose slopes are obtained from the differential equation. Euler's approximation scheme for systems is the same basic idea interpreted in a vector framework.

Given an initial condition (x_0, y_0), how can we use the vector field $\mathbf{F}(x, y)$ to approximate the solution curve? Just as for equations, we first pick a step size Δt. The vector $\mathbf{F}(x_0, y_0)$ is the velocity vector of the solution through (x_0, y_0), so we begin our approximate solution by using $\Delta t \, \mathbf{F}(x_0, y_0)$ to form the first "step." In other words we step from (x_0, y_0) to (x_1, y_1), where the point (x_1, y_1) is given by

$$(x_1, y_1) = (x_0, y_0) + \Delta t \, \mathbf{F}(x_0, y_0)$$

$\mathbf{F}(x_0, y_0)$

(x_1, y_1)

(x_0, y_0)

Figure 2.45
The vector at (x_0, y_0) and the point (x_1, y_1) obtained from one step of Euler's method.

(see Figure 2.45). This corresponds to traveling along a straight line for time Δt with velocity $\mathbf{F}(x_0, y_0)$.

Having calculated a point (x_1, y_1) on the approximate solution curve, we calculate the new velocity vector $\mathbf{F}(x_1, y_1)$. The second step in the approximation is

$$(x_2, y_2) = (x_1, y_1) + \Delta t \, \mathbf{F}(x_1, y_1).$$

We repeat this scheme and obtain an approximate solution curve (see Figure 2.46).

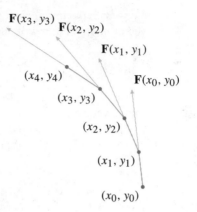

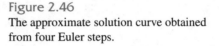

Figure 2.46
The approximate solution curve obtained
from four Euler steps.

In practice we choose a step size Δt that is small enough to provide an accurate solution over the given time interval. (See Chapter 7 for a technical discussion of how small is small and how small is too small.)

Euler's Method for Autonomous Systems

Euler's method for systems can be written without the vector notation as follows. Given the system

$$\frac{dx}{dt} = f(x, y)$$

$$\frac{dy}{dt} = g(x, y),$$

the initial condition (x_0, y_0), and the step size Δt, we calculate the Euler approximation by repeating the calculations:

$$m_k = f(x_k, y_k)$$

$$n_k = g(x_k, y_k),$$

$$x_{k+1} = x_k + m_k \Delta t$$

$$y_{k+1} = y_k + n_k \Delta t.$$

Euler's Method Applied to the Van der Pol Equation

For example, consider the second-order differential equation

$$\frac{d^2x}{dt^2} - (1 - x^2)\frac{dx}{dt} + x = 0.$$

This equation is called the **Van der Pol equation**. To study it numerically, we first convert it into a first-order system by letting $y = dx/dt$. The resulting system is

$$\frac{dx}{dt} = y$$

$$\frac{dy}{dt} = -x + (1 - x^2)y.$$

Suppose we want to find an approximate solution for the initial condition $(x(0), y(0)) = (1, 1)$. We do a few calculations by hand to see how Euler's method works, and then turn to the computer for the repetitive part. The method is best illustrated by doing a calculation with a relatively large step size, although in practice we would never use such a large value for Δt.

Let $\Delta t = 0.25$. Given the initial condition $(x_0, y_0) = (1, 1)$, we compute the vector field

$$\mathbf{F}(x, y) = (y, -x + (1 - x^2)y)$$

at $(1, 1)$. We obtain the vector $\mathbf{F}(1, 1) = (1, -1)$. Thus our first step starts at $(1, 1)$ and ends at

$$(x_1, y_1) = (x_0, y_0) + \Delta t \, \mathbf{F}(x_0, y_0)$$

$$= (1, 1) + 0.25 \, (1, -1)$$

$$= (1.25, 0.75).$$

In other words, since $\Delta t = 0.25$, we obtain (x_1, y_1) from (x_0, y_0) by stepping one-quarter of the way along the displacement vector $(1, -1)$ (see Figure 2.47).

The next step is obtained by computing the vector field at (x_1, y_1). We have $\mathbf{F}(1.25, 0.75) = (0.75, -1.67)$ (to 2 decimal places). Consequently, our next step starts at $(1.25, 0.75)$ and ends at

$$(x_2, y_2) = (1.25, 0.75) + 0.25 \, (0.75, -1.67)$$

$$= (1.44, 0.33)$$

(see Figure 2.47).

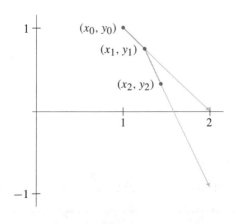

Figure 2.47
Two steps of Euler's method applied to the Van der Pol equation with initial condition $(x_0, y_0) = (1, 1)$ and step size $\Delta t = 0.25$.

Table 2.1
Ten steps of Euler's method.

i	x_i	y_i	m_i	n_i
0	1	1	1	-1
1	1.25	0.75	0.75	-1.671875
2	1.4375	0.332031	0.332031	-1.791580
3	1.520507	-0.115864	-0.115864	-1.368501
4	1.491542	-0.457989	-0.457989	-0.930644
5	1.377045	-0.690650	-0.690650	-0.758048
6	1.204382	-0.880162	-0.880162	-0.807837
7	0.984342	-1.082121	-1.082121	-1.017965
8	0.713811	-1.336613	-1.336613	-1.369384
9	0.379658	-1.678959	-1.678959	-1.816611
10	-0.040082	-2.133112		

Table 2.1 illustrates the computations necessary to calculate ten steps of Euler's method, starting at the initial condition $(x_0, y_0) = (1, 1)$ with $\Delta t = 0.25$. The resulting approximate solution curve is shown in Figure 2.48.

As we mentioned above, $\Delta t = 0.25$ is much larger than the typical step size, so let's repeat our calculations with $\Delta t = 0.1$. Since we will use a computer to do these calculations, we might as well do more steps, too. Figure 2.49 shows the result of this calculation. In this figure we show both the points obtained in the calculation as well as a graph of an approximate solution curve obtained by joining successive points by line segments. Note that the curve is hardly a "standard" shape and that it is almost a closed curve.

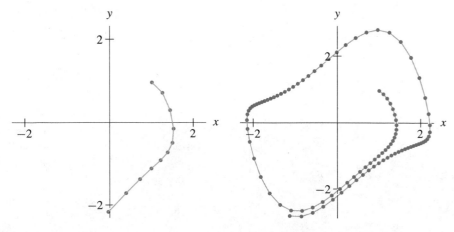

Figure 2.48
Ten steps of Euler's method applied to the Van der Pol equation with initial condition $(x_0, y_0) = (1, 1)$ and step size $\Delta t = 0.25$.

Figure 2.49
One hundred steps of Euler's method applied to the Van der Pol equation with initial condition $(x_0, y_0) = (1, 1)$ and step size $\Delta t = 0.1$.

James H. Curry (1948–) received his Ph.D. in Mathematics at the University of California at Berkeley in 1976. He spent several years as a Postdoctoral Fellow at MIT and the National Center for Atmospheric Research where he met and worked with E. N. Lorenz. He has also taught at Howard University and the University of Colorado where he currently holds the position of Professor of Applied Mathematics and Associate Chair of the Applied Mathematics Department.

Curry's research has focused on qualitative methods in differential equations that model the atmosphere. He has also published extensively on iterative methods for solving nonlinear equations. The methods Curry considers are faster and more advanced methods than the type of numerical methods we describe in this section. We describe some of these more advanced techniques in Chapter 7.

To show the $x(t)$- and $y(t)$-graphs for this approximate solution, we must include information about the independent variable t in our Euler's method table. If we assume that the initial condition $(x_0, y_0) = (1, 1)$ corresponds to the initial time $t_0 = 0$, we can augment that table by adding the corresponding times (see Table 2.2). Thus we are able

Table 2.2
Ten steps of Euler's method with $t_0 = 0$

i	t_i	x_i	y_i	m_i	n_i
0	0	1	1	1	1
1	0.25	1.25	0.75	0.75	−1.671875
2	0.50	1.4375	0.332031	0.332031	−1.791580
3	0.75	1.520507	−0.115864	−0.115864	−1.368501
4	1.00	1.491542	−0.457989	−0.457989	−0.930644
5	1.25	1.377045	−0.690650	−0.690650	−0.758048
6	1.50	1.204382	−0.880162	−0.880162	−0.807837
7	1.75	0.984342	−1.082121	−1.082121	−1.017965
8	2.00	0.713811	−1.336613	−1.336613	−1.369384
9	2.25	0.379658	−1.678959	−1.678959	−1.816611
10	2.50	−0.040082	−2.133112		

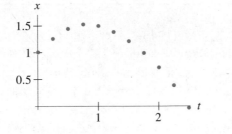

Figure 2.50
The $x(t)$-graph corresponding to the approximate solution curve obtained in Table 2.2.

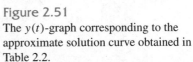

Figure 2.51
The $y(t)$-graph corresponding to the approximate solution curve obtained in Table 2.2.

to produce $x(t)$- and $y(t)$-graphs of approximate solutions (see Figures 2.50 and 2.51). Figures 2.52 and 2.53 illustrate how the "almost" closed solution curve in the phase plane (the xy-plane) corresponds to the functions $x(t)$ and $y(t)$, which are essentially periodic.

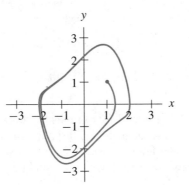

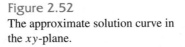

Figure 2.52
The approximate solution curve in the xy-plane.

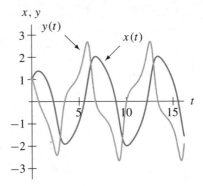

Figure 2.53
The corresponding $x(t)$- and $y(t)$-graphs.

Existence and Uniqueness

Numerical methods, such as Euler's method, give approximations to solutions. Controlling the difference between the numerical approximation and the actual solution is a difficult problem since we usually do not know the actual solution (see Chapter 7). As we saw in Section 1.5, the Existence and Uniqueness Theorem gives us (among other things) qualitative information about solutions, which we can use to check our numerics. The same is true for systems.

EXISTENCE AND UNIQUENESS THEOREM   Let

$$\frac{d\mathbf{Y}}{dt} = \mathbf{F}(t, \mathbf{Y})$$

be a system of differential equations. Suppose that t_0 is an initial time and $\mathbf{Y}_0$ is an initial value. Suppose also that the function $\mathbf{F}$ is continuously differentiable. Then there is an $\epsilon > 0$ and a function $\mathbf{Y}(t)$ defined for $t_0 - \epsilon < t < t_0 + \epsilon$, such that $\mathbf{Y}(t)$ satisfies the initial-value problem $d\mathbf{Y}/dt = \mathbf{F}(t, \mathbf{Y})$ and $\mathbf{Y}(t_0) = \mathbf{Y}_0$. Moreover, for t in this interval, this solution is unique. ∎

We emphasize that, as with equations, we must first verify the hypotheses of this theorem to apply it properly. If the hypotheses do not hold, then the theorem yields no information. In this case solutions may not exist, or they may not be unique.

The existence part of the theorem is mostly just reassuring. If we are studying a certain system, then it is nice to know that what we are studying exists. The uniqueness part is useful in a much more practical way. Roughly speaking, the Uniqueness Theorem says that two different solutions cannot start at the same place at the same time.

All of the systems that we study in this chapter and in Chapter 3 are autonomous. In other words the right-hand sides of the differential equations do not depend on the independent variable t, so they have the form $d\mathbf{Y}/dt = \mathbf{F}(\mathbf{Y})$. For this type of system, the Uniqueness Theorem is particularly useful. Since the vector field $\mathbf{F}(\mathbf{Y})$ does not change with time, we obtain the same solution curves from two different solutions that start at the same point $\mathbf{Y}_0$ at different times. This geometric observation implies that distinct solution curves $\mathbf{Y}_1(t)$ and $\mathbf{Y}_2(t)$ cannot cross.

To verify this conclusion, suppose two solution curves do intersect at the point $\mathbf{Y}_0$. In other words suppose

$$\mathbf{Y}_1(t_1) = \mathbf{Y}_0 = \mathbf{Y}_2(t_2)$$

for two solutions $\mathbf{Y}_1(t)$ and $\mathbf{Y}_2(t)$. Then the solution curve for $\mathbf{Y}_1(t)$ before and after time t_1 is exactly the same as the curve for $\mathbf{Y}_2$ before and after time t_2, that is,

$$\mathbf{Y}_1(t_1 + t) = \mathbf{Y}_2(t_2 + t)$$

for all t. Hence, if two solution curves intersect, then their images are the same curves in the phase plane, and they differ only in their parameterizations (see Exercises 14 and 15).

The consequences of the Uniqueness Theorem are not as strong if the system is nonautonomous. For example, it is possible to have solution curves that cross themselves in the phase plane. We will consider the geometry of nonautonomous systems in Chapter 4, but in the meantime, we can assume that different solution curves do not touch as long as the system satisfies the hypothesis of the Uniqueness Theorem.

A Swaying Skyscraper

As an example of a system for which qualitative analysis (including the Uniqueness Theorem) and numerical analysis are needed to understand the behavior, we consider a model of the swaying motion of a tall building. Modern skyscrapers are built to be flexible. In strong gusts of wind or in earthquakes these buildings tend to sway back and forth to absorb the shocks. Oscillations with an amplitude of a meter and periods on the order of 5 to 10 seconds are common.

As an application of Euler's method, let's see how we can analyze two simple differential equations that model the swaying of a building.

The model

To describe the swaying motion of the skyscraper, let $y(t)$ be a measure of how far the building is bent—the displacement (in meters) of the top of the building with $y = 0$

corresponding to the perfectly vertical position. When y is not zero, the building is bent and the structure applies a strong restoring force back toward the vertical (see Figure 2.54). This is reminiscent of the harmonic oscillator described in Section 2.3. Therefore, as a very crude first approximation of the motion of a swaying building, we use the damped harmonic oscillator equation

$$\frac{d^2y}{dt^2} + p\frac{dy}{dt} + qy = 0.$$

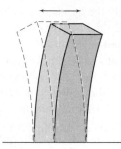

Figure 2.54

Schematic of a skyscraper swaying.

Here the constants q and p are chosen to reflect the characteristics of the particular building being studied. For the sake of definiteness we fix the constants $p = 0.2$ and $q = 0.25$ and consider the second-order equation

$$\frac{d^2y}{dt^2} + 0.2\frac{dy}{dt} + 0.25y = 0$$

and the corresponding system

$$\frac{dy}{dt} = v$$

$$\frac{dv}{dt} = -0.25y - 0.2v.$$

These numbers are chosen for the purpose of demonstrating the behavior of solutions (and do not refer to any building, currently standing or not).

For this system we could use analytic techniques to obtain an exact solution (see Chapter 3). But in practice, even when other techniques are available, we often start by getting an idea of the behavior of solutions using numerical methods. Using Euler's method, we can sketch the phase plane for the system (see Figure 2.55). All solutions spiral toward the origin in the yv-plane. Thus, this model predicts that once displaced from the vertical, the building will sway back and forth, and the amplitude of the oscillation decreases with each oscillation. If we sketch the $y(t)$-graph for several different

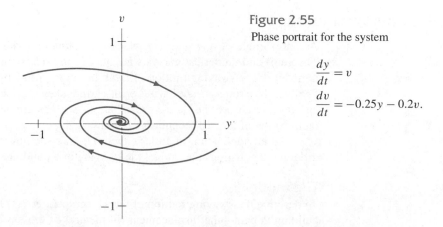

Figure 2.55

Phase portrait for the system

$$\frac{dy}{dt} = v$$

$$\frac{dv}{dt} = -0.25y - 0.2v.$$

initial conditions, we see that these oscillations always have the same frequency independent of their amplitude or initial condition (see Figure 2.56). This frequency is called the *natural frequency* of the harmonic oscillator and is a fundamental characteristic of solutions of these equations (see Section 3.4).

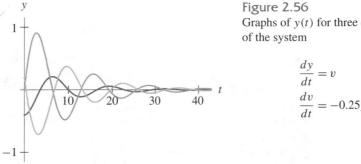

Figure 2.56
Graphs of $y(t)$ for three different solutions of the system

$$\frac{dy}{dt} = v$$

$$\frac{dv}{dt} = -0.25y - 0.2v.$$

The P-Delta Effect

Modeling the swaying building with a harmonic oscillator equation is extremely crude. We do not claim that the forces present in a swaying building are identical to those of a spring. The harmonic oscillator is only a first approximation of a complicated physical system. To extend the usefulness of this model, we must consider other factors that govern the motion of a swaying building.

One aspect of the model of the swaying building that we have not yet included is the effect of gravity. When the building undergoes small oscillations, gravity does not play a very important role. However, if the oscillations become large enough, then gravity can have a significant effect. When $y(t)$ is at its maximum value, a portion of the building is not directly above any other part of the building (see Figure 2.57). Therefore, gravity pulls downward on this portion of the building and this force tends to bend the building farther. This is called the "P-Delta" effect ("Delta" is the overhang distance and "P" is the force of gravity).*

To include this effect in our model in a way that is quantitatively accurate requires knowledge of the density of the building and the flexibility of the construction materials. Without going into specific detail, we can construct a much simplified model that is a caricature of the P-Delta effect.

The P-Delta effect is very small when y is small, much smaller than the restoring force. As y increases, the P-Delta effect becomes quite large. As a first model, we may assume that the force provided by the P-Delta effect is proportional to y^3. Adding this force corresponds to adding a term to the expression for the acceleration of y, that is, adding a term proportional to y^3 to the right-hand side of the second-order differential equation. For the system, this corresponds to adding a term proportional to y^3 to the equation for dv/dt.

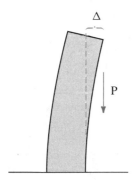

Figure 2.57
The force of gravity on a bent building.

---

*Matthys Levy and Mario Salvadori, *Why Buildings Fall Down*, New York: W.W. Norton and Co., 1992, p. 109, give an excellent description of this model along with the amusing story of the John Hancock Tower in Boston.

To study the qualitative behavior of solutions, we assume that the coefficient of the y^3-term is 1 since we have no particular building in mind. Hence our new model is

$$\frac{d^2y}{dt^2} + 0.2\frac{dy}{dt} + 0.25y = y^3,$$

which converts to the first-order system

$$\frac{dy}{dt} = v$$

$$\frac{dv}{dt} = -0.25y + y^3 - 0.2v.$$

Using Euler's method, we can compute solutions on the phase plane and $y(t)$- and $v(t)$-graphs (see Figures 2.58 and 2.59). The behavior of solutions is somewhat unnerving. If the initial condition is sufficiently close to zero, then the solution spirals toward the origin as in the case of our original model. If the initial condition is sufficiently far from the origin, however, the behavior is quite different. The solution in the phase plane moves away from the origin. Solutions with these two types of behavior are separated by solution curves that tend to the equilibrium points as t increases.

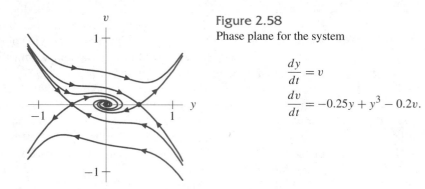

Figure 2.58

Phase plane for the system

$$\frac{dy}{dt} = v$$

$$\frac{dv}{dt} = -0.25y + y^3 - 0.2v.$$

The interpretation of the behavior of these solutions in terms of the behavior of the building yields dramatic results. For small oscillations the building sways with decreasing amplitude and eventually returns to its rest position. However, if the initial displacement exceeds a threshold distance, then the amplitude of the solution quickly moves away from zero. The building sways more and more violently, and the result is a disaster.

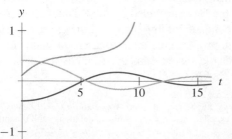

Figure 2.59

Graphs of $y(t)$ for three different solutions of the system

$$\frac{dy}{dt} = v$$

$$\frac{dv}{dt} = -0.25y + y^3 - 0.2v.$$

Reality Check

We must emphasize that this model is only a caricature of the actual dynamics of a swaying building. However, the model does teach an important lesson. Solutions with initial conditions in one region of the phase plane may behave very differently from solutions in another region. The Uniqueness Theorem guarantees that, if an initial condition is in one of these regions, then the corresponding solution stays in the region for all time. The transition between different types of solutions can occur abruptly as the initial conditions are varied. Just because a physical system is "stable" with respect to small initial displacements does not imply that it will be stable with respect to all initial conditions. If this simple model can behave in such a radical way, then we should not be surprised to find such bizarre behavior in an actual building.

EXERCISES FOR SECTION 2.4

1. For the system

$$\frac{dx}{dt} = -y$$
$$\frac{dy}{dt} = x,$$

the curve $\mathbf{Y}(t) = (\cos t, \sin t)$ is a solution. This solution is periodic. Its initial position is $\mathbf{Y}(0) = (1, 0)$, and it returns to this position when $t = 2\pi$. So $\mathbf{Y}(2\pi) = (1, 0)$ and $\mathbf{Y}(t + 2\pi) = \mathbf{Y}(t)$ for all t.

(a) Check that $\mathbf{Y}(t) = (\cos t, \sin t)$ is a solution.

(b) Use Euler's method with step size 0.5 to approximate this solution, and check how close the approximate solution is to the real solution when $t = 4$, $t = 6$, and $t = 10$.

(c) Use Euler's method with step size 0.1 to approximate this solution, and check how close the approximate solution is to the real solution when $t = 4$, $t = 6$, and $t = 10$.

(d) The points on the solution curve $\mathbf{Y}(t)$ are all 1 unit distance from the origin. Is this true of the approximate solutions? Are they too far from the origin or too close to it? What will happen for other step sizes (that is, will approximate solutions formed with other step sizes be too far or too close to the origin)?

[It is well worthwhile to use a computer or calculator to help with the arithmetic of Euler's method.]

2. For the system

$$\frac{dx}{dt} = 2x$$
$$\frac{dy}{dt} = y,$$

we claim that the curve $\mathbf{Y}(t) = (e^{2t}, 3e^t)$ is a solution. Its initial position is $\mathbf{Y}(0) = (1, 3)$.

(a) Check that $\mathbf{Y}(t) = (e^{2t}, 3e^t)$ is a solution.

(b) Use Euler's method with step size $\Delta t = 0.5$ to approximate this solution, and check how close the approximate solution is to the real solution when $t = 2$, $t = 4$, and $t = 6$.

(c) Use Euler's method with step size $\Delta t = 0.1$ to approximate this solution, and check how close the approximate solution is to the real solution when $t = 2$, $t = 4$, and $t = 6$.

(d) Discuss how and why the Euler approximations differ from the solution.

[It is well worthwhile to use a computer or calculator to help with the arithmetic of Euler's method.]

In Exercises 3–6, a system, an initial condition, a step size, and an integer n are given. The direction field for the system is also provided.

(a) Calculate the approximate solution given by Euler's method for the given system with the given initial condition and step size for n steps.

(b) Plot your approximate solution on the direction field. Make sure that your approximate solution is consistent with the direction field.

(c) Using a computer or a graphing calculator, obtain a more detailed sketch of the phase portrait for the system.

3. $\dfrac{dx}{dt} = y$

$\dfrac{dy}{dt} = -2x - 3y$

$\begin{cases} (x_0, y_0) = (1, 1) \\ \Delta t = 0.25 \\ n = 5 \end{cases}$

4. $\dfrac{dx}{dt} = y$

$\dfrac{dy}{dt} = -\sin x$

$\begin{cases} (x_0, y_0) = (0, 2) \\ \Delta t = 0.25 \\ n = 8 \end{cases}$

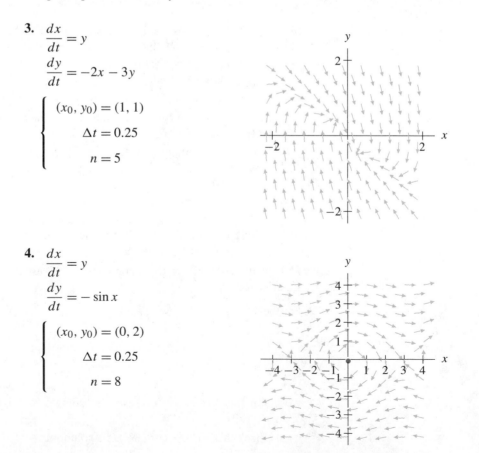

5. $\dfrac{dx}{dt} = y + y^2$

$\dfrac{dy}{dt} = -x + \dfrac{y}{5} - xy + \dfrac{6y^2}{5}$

$\begin{cases} (x_0, y_0) = (1, 1) \\ \quad \Delta t = 0.25 \\ \quad n = 5 \end{cases}$

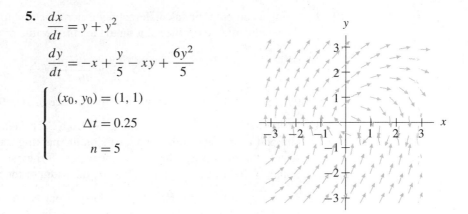

6. $\dfrac{dx}{dt} = y + y^2$

$\dfrac{dy}{dt} = -\dfrac{x}{2} + \dfrac{y}{5} - xy + \dfrac{6y^2}{5}$

$\begin{cases} (x_0, y_0) = (-0.5, 0) \\ \quad \Delta t = 0.25 \\ \quad n = 7 \end{cases}$

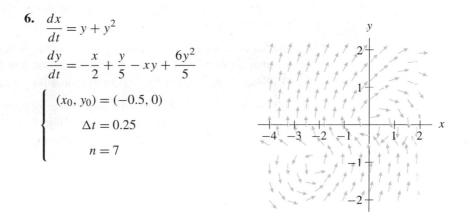

In Exercises 7–9, consider the system

$$\frac{dx}{dt} = -x + 3y$$
$$\frac{dy}{dt} = -3x - y.$$

7. Verify that $\mathbf{Y}_1(t) = (e^{-t} \sin(3t), e^{-t} \cos(3t))$ is a solution of this system.

8. Verify that $\mathbf{Y}_2(t) = (e^{-(t-1)} \sin(3(t-1)), e^{-(t-1)} \cos(3(t-1)))$ is a solution.

9. Sketch the solution curves for $\mathbf{Y}_1(t)$ and $\mathbf{Y}_2(t)$ in the xy-phase plane. Why don't $\mathbf{Y}_1(t)$ and $\mathbf{Y}_2(t)$ contradict the Uniqueness Theorem?

10. Using a computer or calculator, apply Euler's method to sketch an approximation to the solution curve for the solution to the initial-value problem

$$2\frac{d^2y}{dt^2} + \frac{dy}{dt} + 4y = 0,$$

where $(y_0, v_0) = (2, 0)$. How does your choice of Δt affect your result?

11. Using a computer or calculator, apply Euler's method to sketch an approximation to the solution curve for the solution to the initial-value problem

$$5\frac{d^2y}{dt^2} + \frac{dy}{dt} + 5y = 0,$$

where $(y_0, v_0) = (0, 1)$. How does your choice of Δt affect your result?

12. Recall the Metaphor of the Parking Lot from Section 2.2. Suppose two people, say Gib and Harry, are both driving cars on the parking lot and both are carefully following the rules prescribed in the metaphor. If they start at time $t = 0$ at different points, will they ever collide? (Neglect the width of their cars.)

13. Consider the two drivers, Gib and Harry, from Exercise 12. Suppose that at time $t = 0$ they start at different points in the parking lot, but at time $t = 1$ Gib drives over the point where Harry started. Will they ever collide? What can you say about their paths?

14. **(a)** Suppose $\mathbf{Y}_1(t)$ is a solution of an autonomous system $d\mathbf{Y}/dt = \mathbf{F}(\mathbf{Y})$. Show that $\mathbf{Y}_2(t) = \mathbf{Y}_1(t + t_0)$ is also a solution for any constant t_0.

(b) What is the relationship between the solution curves of $\mathbf{Y}_1(t)$ and $\mathbf{Y}_2(t)$?

15. Suppose $\mathbf{Y}_1(t)$ and $\mathbf{Y}_2(t)$ are solutions of an autonomous system $d\mathbf{Y}/dt = \mathbf{F}(\mathbf{Y})$, where $\mathbf{F}(\mathbf{Y})$ satisfies the hypotheses of the Uniqueness Theorem. Suppose also that $\mathbf{Y}_2(1) = \mathbf{Y}_1(0)$. How are $\mathbf{Y}_1(t)$ and $\mathbf{Y}_2(t)$ related?

16. Consider the system

$$\frac{dx}{dt} = x^2 + y$$
$$\frac{dy}{dt} = x^2y^2.$$

Show that, for the solution $(x(t), y(t))$ with initial condition $x(0) = y(0) = 1$, there is a time t_* such that $x(t) \to \infty$ as $t \to t_*$. In other words the solution blows up in finite time. [*Hint*: Note that $dy/dt \geq 0$ for all x and y.]

2.5 THE LORENZ EQUATIONS

We have already seen that the behavior of solutions of autonomous systems of differential equations can be much more interesting and complicated than solutions of single autonomous equations. For autonomous equations with one dependent variable, the solutions live on a phase line, and their behavior is completely governed by the position and nature of the equilibrium points. Solutions of systems with two dependent variables live in two-dimensional phase planes. A plane has much more "room" than a line, so solutions in a phase plane can do many more interesting things. This includes forming loops (periodic solutions) and approaching and retreating from equilibrium points.

However, there are still severe restrictions on the types of phase plane pictures that are possible for systems. The Uniqueness Theorem says that solution curves in the phase plane do not cross. So, for example, if there is a periodic solution that forms a loop in the phase plane, then solutions with initial conditions inside the loop must stay

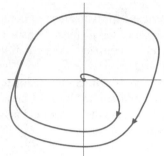

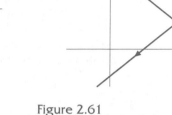

Figure 2.60
Solutions with initial conditions
inside a periodic solution must
stay inside for all time.

Figure 2.61
Two solutions and an equilibrium point cut
the phase plane into regions. Solutions with
initial conditions in one region must stay in
that region for all time.

inside for all time (see Figure 2.60). Also, two or three solutions can fit together to
divide the phase plane, and from the Uniqueness Theorem solutions have to stay in the
same region as their initial point (see Figure 2.61).

If we raise the number of dependent variables to three, the situation becomes
much more complicated. A solution of an autonomous system with three dependent
variables is a curve in a three-dimensional "phase space." The Uniqueness Theorem
still applies, so solution curves do not cross, but in three dimensions this restriction is
not nearly so confining as it is in two dimensions. In Figure 2.62 we see examples of
curves in three dimensions that do not cross. These curves can knot and link in very
complicated ways.

The first to realize the possible complications of three-dimensional systems was
Henri Poincaré. In the 1890s, while working on the Newtonian three-body problem,
Poincaré realized that systems with three dependent variables can have behavior so
complicated that he did not even want to attempt to draw them. With a computer, we

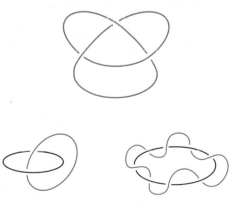

Figure 2.62
A knot and two links in space.

can easily draw numerical approximations of complicated solution curves. The problem now is to make sense of the pictures. This is an active area of current research in dynamical systems, and the complete story for systems with three dependent variables is still far from being written.

In this section we study a three-dimensional system known as the Lorenz equations. This system was first studied by Ed Lorenz in 1963 in an effort to model the weather. It is important because the vector field is formed by very simple equations, yet solutions are very complicated curves.

The Lorenz System

The behavior of a physical system like the weather on Earth is (obviously) extremely complicated. To predict the weather, many mathematical models have been developed. The readings from weather stations and satellites are used as initial conditions, and numerical approximations of solutions are used to obtain predictions.

The success of long-range weather forecasts (that is, more than five days in the future) is very limited. This might be because the model equations are inaccurate in their representation of some aspect of the evolution of the weather. It is also possible that the model is accurate but that some property of the equations makes prediction difficult. Consequently it is important to study these models theoretically as well as numerically.

Since the weather is so complicated, it is necessary to start the theoretical study by looking at simplifications. Through a process of simplification, meteorologist Ed Lorenz came up with the following system,

$$\frac{dx}{dt} = \sigma(y - x)$$
$$\frac{dy}{dt} = \rho x - y - xz$$
$$\frac{dz}{dt} = -\beta z + xy,$$

where x, y, and z are dependent variables and σ, ρ, and β are parameters. This system is so much simpler than the equations used for modeling the weather that they have nothing to tell us about tomorrow's temperature. However, by studying this system, Lorenz helped start a scientific revolution by making scientists and engineers aware of the field of mathematics now called Chaos Theory.

The Vector Field

Lorenz chose to study the system with the parameter values $\sigma = 10$, $\beta = 8/3$, and $\rho = 28$,

$$\frac{dx}{dt} = 10(y - x)$$
$$\frac{dy}{dt} = 28x - y - xz$$
$$\frac{dz}{dt} = -\frac{8}{3}z + xy.$$

The right-hand sides of these equations define a vector field in three-dimensional space

$$\mathbf{F}(x, y, z) = (10(y - x), 28x - y - xz, -\tfrac{8}{3}z + xy),$$

which assigns a three-component vector to each point (x, y, z). Just as in two dimensions, the equilibrium points are the points (x, y, z) for which the vector field is zero, that is, $\mathbf{F}(x, y, z) = (0, 0, 0)$. By direct computation (see Exercise 1) we find that the equilibrium points are $(0, 0, 0)$, $(6\sqrt{2}, 6\sqrt{2}, 27)$, and $(-6\sqrt{2}, -6\sqrt{2}, 27)$.

An initial point for a solution must include values for the three coordinates x, y, and z, so it is a point in space. Given an initial point, we can sketch the corresponding solution by sketching a curve whose velocity vector equals the vector field at each point. For example, suppose we choose the initial point $(x(0), y(0), z(0)) = (10, 7, 7)$. Then the vector field at this point is $\mathbf{F}(10, 7, 7) = (-30, 203, 154/3)$. As we move along the solution, we compute the vector field and adjust the velocity vector accordingly. In Figure 2.63 we sketch a small portion of the solution with initial condition $(10, 7, 7)$ using this method. As we can see from this picture, sketching solutions using the vector field is challenging.

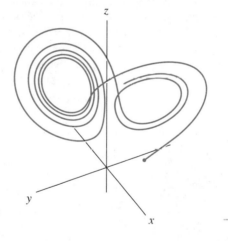

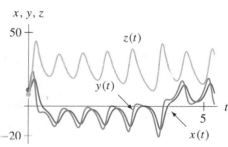

Figure 2.63
A portion ($0 \leq t \leq 5.3$) of the solution curve of the Lorenz system with initial condition $(10, 7, 7)$.

Figure 2.64
The $x(t)$-, $y(t)$-, and $z(t)$- graphs for the solution with initial condition $(10, 7, 7)$.

With the exception of the equilibrium points and solutions with initial conditions on the z-axis, there is little hope of finding a formula for solutions (see Exercises 1 and 3). It is natural to turn to numerical methods. Euler's method for three-dimensional systems works exactly the same as in two dimensions. The approximate solution is constructed by following the vector field for short time steps. Lorenz began his study of this system by finding numerical approximations of solutions, so we follow in his footsteps.

Edward N. Lorenz (1917 –   ) began his career as a mathematics graduate student at Harvard, but turned his attention to meteorology during World War II. In 1961, using a primitive computer by today's standards, Lorenz attempted to solve a much-simplified model for weather prediction. His model seemed to simulate real weather patterns quite well, but it also illustrated something much more important: When Lorenz changed the initial conditions in the model slightly, the resulting weather patterns changed completely after a short time. Lorenz had discovered the fact that simple differential equations can behave "chaotically." We describe some aspects of this important discovery in this and subsequent chapters.

Lorenz is currently Professor Emeritus of Meteorology at MIT.

Numerical Approximation of Solutions

We begin by looking at a numerical approximation of the solution with initial condition $(0, 1, 0)$ (that is, $x(0) = 0$, $y(0) = 1$, $z(0) = 0$—see Figure 2.65). Clearly something interesting is going on. The solution does not seem to have any particular pattern. For example, the x-coordinate jumps from positive to negative values in an "unpredictable" way. Although this is somewhat unnerving, we see something even more interesting if we look at the solution with initial condition $(0, 1.001, 0)$ (see Figure 2.66).

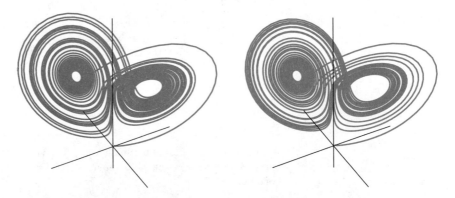

Figure 2.65
Solution of Lorenz system with initial point $(0, 1, 0)$.

Figure 2.66
Solution of Lorenz system with initial point $(0, 1.001, 0)$.

This solution starts out very close to the previous one, with x, y, and z oscillating unpredictably. However, if we compare the two solutions, we see that eventually they become quite far apart (see Figure 2.67). So a very small change in initial condition leads to a big change in the solution.

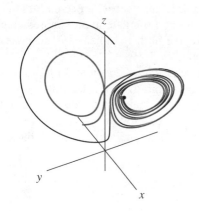

Figure 2.67
Comparison of solutions of Lorenz system with initial points (0, 1, 0) (black curve) and (0, 1.001, 0) (blue curve) for $44 \leq t \leq 47$.

To get a better view of what is going on, we show these solution curves from two different angles to better display their three-dimensional character (see Figures 2.68 and 2.69).

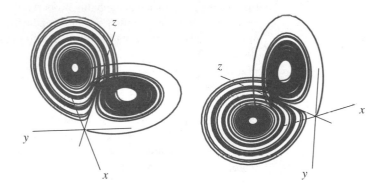

Figure 2.68
Two views of the solution curve with initial point (0, 1, 0) in the phase space.

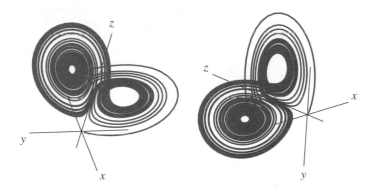

Figure 2.69
Two views of the solution curve with initial point (0, 1.001, 0) in the phase space.

It turns out that this strange behavior occurs for almost every solution curve. The functions $x(t)$, $y(t)$, and $z(t)$ oscillate in an "unpredictable" and unique way, but the graph of the solution curve in three dimensions generates a figure that is roughly the same for every solution.

The solution curves we have drawn seem to loop around the equilibrium points above $z = 0$ in increasing spirals. Once the radius gets too large, the solution passes close to $(0, 0, 0)$ and then gets "reinjected" toward one of the two equilibria. (It is very instructive to watch a movie of a solution moving in real time through the phase space.) In the next chapter we develop the tools to study the behavior near the equilibrium points.

Chaos

The qualitative analysis of this system is a difficult undertaking that must wait until Chapters 5 and 8. However, there is a moral to what we have seen so far that is having an important effect on many different branches of science. The Lorenz system has two important properties. The first is that a small change in initial conditions leads fairly quickly to large differences in the corresponding solutions. If a system as simple as the Lorenz equations can have this property, it is entirely reasonable to think that much more complicated systems (like the weather) might have it as well. This means that any small error in the initial conditions will lead quickly to a big error in prediction of the solution. This might well be why physical systems like the weather are so hard to predict.

The second property of the Lorenz system is that although the details of individual solutions are quite different, the pictures of the solution curves in the three-dimensional phase space look remarkably alike. Many solutions seem to be "filling in" the same region in three dimensions. So the solutions of the Lorenz system still have structure that we can study. We don't have to give up studying the solutions of the Lorenz equations, we just have to ask the right questions.

EXERCISES FOR SECTION 2.5

1. **(a)** Verify that $(0, 0, 0)$, $(6\sqrt{2}, 6\sqrt{2}, 27)$, and $(-6\sqrt{2}, -6\sqrt{2}, 27)$ are equilibrium points of the Lorenz system

$$\frac{dx}{dt} = 10(y - x)$$
$$\frac{dy}{dt} = 28x - y - xz$$
$$\frac{dz}{dt} = -\frac{8}{3}z + xy.$$

   (b) Verify that these are the only equilibrium points of the Lorenz system.

2. Suppose that in the Lorenz system we fix $\sigma = 10$ and $\beta = 8/3$, as in the text, but we

leave ρ a parameter. For the resulting system,

$$\frac{dx}{dt} = 10(y - x)$$

$$\frac{dy}{dt} = \rho x - y - xz$$

$$\frac{dz}{dt} = -\frac{8}{3}z + xy,$$

(a) show that if $\rho > 1$, there are three equilibrium points, and compute their locations; and

(b) show that if $\rho \leq 1$, there is just one equilibrium point, and compute its location.

3. For the Lorenz system with $\sigma = 10$, $\rho = 28$, and $\beta = 8/3$,

(a) verify that if $(x(t), y(t), z(t))$ is a solution with $x(0) = y(0) = 0$, then $x(t) = y(t) = 0$ for all t;

(b) find the solution with initial condition $(0, 0, 1)$; and

(c) find the solution with initial condition $(0, 0, z_0)$, where z_0 is any constant, and sketch these solutions in the three-dimensional phase space.

4. With the help of computing equipment, find numerical approximations of solutions of the Lorenz system with $\sigma = 10$, $\rho = 28$, and $\beta = 8/3$ for the given initial conditions. Comment on how long it takes two solutions with initial conditions close together to separate.

(a) $(1, 0, 0)$	**(b)** $(0, 1.1, 0)$	**(c)** $(0, 1.01, 0)$
(d) $(1.001, 0, 0)$	**(e)** $(0.001, 1, 0)$	**(f)** $(0, 1, -0.001)$

5. Close to the origin, where x, y, and z are very small, the quadratic terms $-xz$ and $+xy$ will be very, very small. So near $(x, y, z) = (0, 0, 0)$ we can approximate the Lorenz system with the system

$$\frac{dx}{dt} = 10(y - x)$$

$$\frac{dy}{dt} = 28x - y$$

$$\frac{dz}{dt} = -\frac{8}{3}z.$$

(This is called *linearization* at the origin and will be studied in detail in Chapter 5.) Notice that z does not appear in the equations for dx/dt and dy/dt, and the equation for dz/dt does not contain x or y; that is, the system decouples into a two-dimensional system and a one-dimensional equation.

(a) Sketch the direction field and the phase plane for the planar system

$$\frac{dx}{dt} = 10(y - x)$$
$$\frac{dy}{dt} = 28x - y.$$

(b) Sketch the phase line for the equation

$$\frac{dz}{dt} = -\frac{8}{3}z.$$

(c) Sketch solutions in the three-dimensional phase space for the system above.

Remark: This picture gives the behavior of the Lorenz system *near* $(0, 0, 0)$.

LAB 2.1 | The Harmonic Oscillator with Modified Damping

Autonomous second-order differential equations are studied numerically by reducing them to first-order systems with two dependent variables. In this lab you will use the computer to analyze three somewhat related second-order equations. In particular, you will analyze phase planes and $y(t)$- and $v(t)$-graphs to describe the long-term behavior of the solutions.

In Sections 2.1 and 2.3, we discuss the most classic of all second-order equations, the harmonic oscillator. The harmonic oscillator is

$$m\frac{d^2y}{dt^2} + b\frac{dy}{dt} + ky = 0.$$

It is an example of a second-order, homogeneous, linear equation with constant coefficients. In the text we explain how this equation is used to model the motion of a spring. The force due to the spring is assumed to obey Hooke's law (the force is proportional to the amount the spring is compressed or stretched). The force due to damping is assumed to be proportional to the velocity. In your report you should describe the motion of the spring assuming certain values of m, b, and k. (A table of values of the parameters is given below. Your instructor will tell you what values of m, b, and k to consider.) Your report should discuss the following:

1. (Harmonic oscillator with no damping) The first equation that you should study is the harmonic oscillator with no damping; that is, $b = 0$ and with $k \neq 0$. Examine solutions using both their graphs and the phase plane. Are the solutions periodic? If so, what does the period seem to be? Describe the behavior of three different solutions that have especially different initial conditions and be specific about the physical interpretation of the different initial conditions. (Analytic methods to answer these questions are discussed in Chapter 3. For now, work numerically.)

2. (Harmonic oscillator with damping) Repeat part 1 using the equation

$$m\frac{d^2y}{dt^2} + b\frac{dy}{dt} + ky = 0.$$

3. (Harmonic oscillator with nonlinear damping) Repeat part 1 using the equation

$$m\frac{d^2y}{dt^2} + b\left|\frac{dy}{dt}\right|\frac{dy}{dt} + ky = 0$$

in place of the usual harmonic oscillator equation. (Note that even with the same value of the parameter b, the drag forces in this equation and the equation in part 2 have the same magnitude only for velocity ± 1. Also, note that the sign of the term

$$\left|\frac{dy}{dt}\right|\frac{dy}{dt}$$

is the same as the sign of dy/dt, hence this damping force is always directed opposite the direction of motion. The difference between this equation and that in part 2 is the size of the damping for small and large velocities. One of the many examples of situations for which this is a better model than linear damping is the drag on airplane tires from wet snow or slush. Drag from only four inches of slush was enough to cause the 1958 crash during take-off of the plane carrying the Manchester United soccer team. Currently, large airplanes are allowed to take off and land in no more than one half inch of wet snow or slush.*

4. (Nonlinear second-order equation) Finally, consider a somewhat related second-order equation where the damping coefficient b is replaced by the factor $(y^2 - \alpha)$; that is,

$$m\frac{d^2y}{dt^2} + (y^2 - \alpha)\frac{dy}{dt} + ky = 0.$$

Is it reasonable to interpret this factor as some type of damping? Provide a complete description of the long-term behavior of the solutions. Are the solutions periodic? If so, what does the period seem to be? Explain why this equation is not a good model for something like a mass-spring system. Give an example of some other type of physical or biological phenomenon that could be modeled by this equation.

Your report: Address the questions in each item above in the form of a short essay. Be particularly sure to describe the behavior of the solution and the corresponding behavior of the mass-spring system. You may use the phase planes and graphs of $y(t)$ to illustrate the points you make in your essay. (However, please remember that, although one good illustration may be worth 1000 words, 1000 illustrations are usually worth nothing.)

Table 2.3
Possible choices for the parameters.

Choice	m	k	b	α
1	2	5	2	3
2	3	5	3	3
3	5	5	4	3
4	2	6	3	5
5	3	6	3	5
6	5	6	3	5
7	5	4	4	2
8	5	5	4	2
9	5	6	4	2
10	5	4	4	2

---

*See Stanley Stewart, *Air Disasters*, Barnes & Noble, 1986.

LAB 2.2 Cooperative and Competitive Species Population Models

In this chapter we have focused on first-order autonomous systems of differential equations such as the predator-prey systems described in Section 2.1. In particular, we have seen how such systems can be studied using vector fields and phase plane analysis and how solution curves in the phase plane relate to the $x(t)$- and $y(t)$-graphs of the solutions. In this lab project you will use these concepts and related numerical computations to study the behavior of the solutions to two different systems.

We have discussed predator-prey systems at length. These are systems in which one species benefits while the other species is harmed by the interaction of the two species. In this lab you will study two other types of systems—competitive and cooperative systems. A competitive system is one in which both species are harmed by interaction, for example, cars and pedestrians. A cooperative system is one in which both species benefit from interaction, for example, bees and flowers. Your overall goal is to understand what happens in both systems for all possible nonnegative initial conditions. Several pairs of cooperative and competitive systems are given at the end of this lab. (Your instructor will tell you which pair(s) of systems you should study.) The analytic techniques that are appropriate to analyze these systems have not been discussed so far, so you will employ mostly geometric/qualitative and numeric techniques to establish your conclusions. Since these are population models, you need consider only x and y in the first quadrant ($x \geq 0$ and $y \geq 0$).

Your report should include:

1. A brief discussion of all terms in each system. For example, what does the coefficient to the x term in equation for dx/dt represent? Which system is cooperative and which is competitive?

2. For each system, determine all relevant equilibrium points and analyze the behavior of solutions whose initial conditions satisfy either $x_0 = 0$ or $y_0 = 0$. Determine the curves in the phase plane along which the vector field is either horizontal or vertical. Which way does the vector field point along these curves?

3. For each system, describe all possible population evolution scenarios using the phase portrait as well as $x(t)$- and $y(t)$-graphs. Give special attention to the interpretation of the computer output in terms of the long-term behavior of the populations.

Your report: The text of your report should address the three items above, one at a time, in the form of a short essay. You should include a description of all "hand" computations that you did. You may include a limited number of pictures and graphs. (You should spend some time organizing the qualitative and numerical information since a few well-organized figures are much more useful than a long catalog.)

Systems:

Pair (1):

$$\text{A.}\quad \frac{dx}{dt} = -5x + 2xy$$
$$\frac{dy}{dt} = -4y + 3xy$$

$$\text{B.}\quad \frac{dx}{dt} = 6x - x^2 - 4xy$$
$$\frac{dy}{dt} = 5y - 2xy - 2y^2$$

Pair (2):

$$\text{A.}\quad \frac{dx}{dt} = -3x + 2xy$$
$$\frac{dy}{dt} = -5y + 3xy$$

$$\text{B.}\quad \frac{dx}{dt} = 5x - x^2 - 3xy$$
$$\frac{dy}{dt} = 8y - 3xy - 3y^2$$

Pair (3):

$$\text{A.}\quad \frac{dx}{dt} = -4x + 3xy$$
$$\frac{dy}{dt} = -3y + 2xy$$

$$\text{B.}\quad \frac{dx}{dt} = 5x - 2x^2 - 4xy$$
$$\frac{dy}{dt} = 7y - 4xy - 3y^2$$

Pair (4):

$$\text{A.}\quad \frac{dx}{dt} = -5x + 3xy$$
$$\frac{dy}{dt} = -3y + 2xy$$

$$\text{B.}\quad \frac{dx}{dt} = 9x - 2x^2 - 4xy$$
$$\frac{dy}{dt} = 8y - 5xy - 3y^2$$

3   LINEAR SYSTEMS

In Chapter 2 we focused on qualitative and numerical techniques for studying systems of differential equations. We did this because we can rarely find explicit formulas for solutions of a system with two or more dependent variables. The only exceptions to this are linear systems.

In this chapter we show how to use the algebraic and geometric form of the vector field to give the general solution of autonomous linear systems. Along the way, we find that understanding the qualitative behavior of a linear system is much easier than finding its general solution. The description of the qualitative behavior of linear systems leads to a classification scheme for these systems, which is very useful in applications. We also continue our study of models that yield linear systems, particularly the harmonic oscillator system.

3.1   PROPERTIES OF LINEAR SYSTEMS AND THE LINEARITY PRINCIPLE

In this chapter we investigate the behavior of the simplest types of systems of differential equations — autonomous linear systems. These systems are important both in their own right and as a tool in the study of nonlinear systems. We are able to classify the linear systems by their qualitative behavior and even give formulas for solutions.

Throughout this chapter we use two models repeatedly to illustrate the techniques we develop. One is the harmonic oscillator, the most important of all second-order equations. We derived this model in Sections 2.1 and 2.3. Now, using the techniques of this chapter, we can give a complete description of its solutions for all possible values of the parameters. The other model is an artificial one, which we present to illustrate all possibilities that can arise for planar linear systems. Study our analysis, but don't invest any money based on it.

The Harmonic Oscillator

The harmonic oscillator is a model for (among other things) the motion of a mass attached to a spring. The spring provides a restoring force that obeys Hooke's law, and the only other force considered is that due to damping. Let $y(t)$ be the position of the mass at time t, with $y = 0$ corresponding to the rest position of the spring. Newton's law of motion,

$$\text{force} = \text{mass} \times \text{acceleration},$$

when applied to a mass-spring system, yields the second-order differential equation

$$-ky - b\frac{dy}{dt} = m\frac{d^2y}{dt^2},$$

where m is the mass, k is the spring constant, and b is the damping coefficient. The $-ky$ term on the left-hand side comes from Hooke's law, and the $-b(dy/dt)$ term is the force from damping (see Section 2.3, page 179). This second-order equation is more commonly written as

$$m\frac{d^2y}{dt^2} + b\frac{dy}{dt} + ky = 0.$$

As we did in Sections 2.1 and 2.3, we can convert this equation into a linear system by letting $v = dy/dt$ be the velocity at time t. We obtain

$$\frac{dy}{dt} = v$$
$$\frac{dv}{dt} = -\frac{k}{m}y - \frac{b}{m}v.$$

Note that the derivatives dy/dt and dv/dt depend linearly on y and v. As we will see, the behavior of the solutions depends on the values of the parameters m, k, and b.

Two CD Stores

Harmonic oscillators do not display all possible behaviors we will encounter in this chapter, so we present the following apocryphal model from microeconomics.

After retiring from writing differential equations textbooks, Paul and Bob both decide to open small stores selling compact discs. "Paul's Rock and Roll CDs" and "Bob's Opera Only Discs" are located on the same block, and Paul and Bob soon become concerned about the effect each store has on the other. On the one hand, having two CD stores on the same block might help both stores by attracting customers to the neighborhood. On the other hand, the stores may compete with each other for limited supply of customers. Paul and Bob argue about this until they are so sick of arguing that they hire the famous mathematician Glen to settle the matter. Glen follows his own advice about keeping models as simple as possible and suggests the following system.

Let

$$x(t) = \text{daily profit of Paul's store at time } t; \text{ and}$$
$$y(t) = \text{daily profit of Bob's store at time } t.$$

That is, if $x(t) > 0$, then Paul's store is making money, but if $x(t) < 0$, then Paul's store is losing money. Since there is no hard information yet about how the profits of each store affect the change in profits of the other, Glen formulates the simplest possible model that allows each store to affect the other — a linear model. The system is

$$\frac{dx}{dt} = ax + by$$
$$\frac{dy}{dt} = cx + dy,$$

where a, b, c, and d are parameters. The rate of change of Paul's profits depends linearly on both Paul's profits and Bob's profits (and nothing else). The same assumptions apply to Bob's profits. In Chapter 5 we will see that using a model of this form is usually justified as long as both stores are operating near the break-even point.

We cannot yet use this model to predict future profits because we do not know the values of the parameters a, b, c, and d. However, we can develop a basic understanding of the significance of the signs and magnitudes of the parameters. Consider, for example, the parameter a. It measures the effect of Paul's profits on the rate of change dx/dt of that profit. Suppose, for instance, that a is positive. If Paul is making money, then $x > 0$ and so $ax > 0$. The ax term contributes positively to dx/dt, and so Paul makes more money in that case. In other words, Paul hopes that $a > 0$ if $x > 0$. On the other hand, being profitable ($x > 0$) could conceivably have a negative effect on Paul's profits. (For example, the store might be crowded, and customers might go elsewhere.) In this case profits would decrease, and under this assumption the parameter a should be negative in our model.

The parameter b measures the effect of Bob's profits on the rate of change of Paul's profits. If $b > 0$ and Bob makes money ($y > 0$), then Paul's profits also benefit because the by term contributes positively to dx/dt. On the other hand, if $b < 0$ then, when Bob makes money ($y > 0$), Paul's profits suffer. Conceivably we could interpret $b < 0$ as a measure of Bob's stealing customers from Paul.

Similarly, both Paul's profits and Bob's profits affect the rate of change of Bob's profits, and the parameters c and d have similar interpretations relative to dy/dt. This model assumes that only the profit of the two stores influences the change in those profits. These assumptions are clearly vast oversimplifications. However, this model does give us a simple situation for which we can interpret the solutions of various linear systems.

In Figure 3.1 we plot the phase portrait for the system

$$\frac{dx}{dt} = ax + by$$
$$\frac{dy}{dt} = cx + dy,$$

assuming that $a = d = 0$, $b = 1$, and $c = -1$, and we get circular solution curves. In Figure 3.2 we consider the case where $a = -1$, $b = 4$, $c = -3$, and $d = -1$. In this case the solution curves spiral toward the origin.

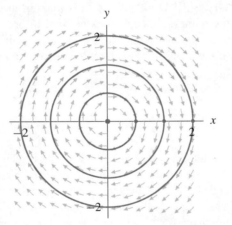

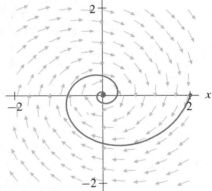

Figure 3.1
The direction field and three solution curves for the system

$$\frac{dx}{dt} = y$$
$$\frac{dy}{dt} = -x.$$

Note that all three curves are circles centered at the origin.

Figure 3.2
The direction field and a solution curve for the system

$$\frac{dx}{dt} = -x + 4y$$
$$\frac{dy}{dt} = -3x - y.$$

This solution curve spirals toward the origin as t increases.

In terms of the model, Figure 3.1 implies that both Paul's and Bob's profits periodically oscillate between making money and losing money. The solution curve in Figure 3.2 suggests that the profits oscillate while tending toward the point $(0, 0)$, which is the break-even point for both stores. The corresponding $x(t)$- and $y(t)$-graphs illustrate these behaviors (see Figures 3.3 and 3.4). As we will see, there are a number of other possible phase portraits for this model, depending on the values of the parameters a, b, c, and d. In this chapter we develop techniques to handle all possibilities.

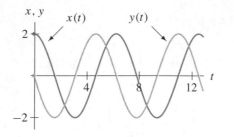

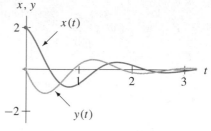

Figure 3.3
The $x(t)$- and $y(t)$-graphs corresponding to the solution curve in Figure 3.1, with initial condition $(x_0, y_0) = (2, 0)$.

Figure 3.4
The $x(t)$- and $y(t)$-graphs corresponding to the solution curve in Figure 3.2 with initial condition $(x_0, y_0) = (2, 0)$.

Linear Systems and Matrix Notation

In this chapter we mainly consider systems of differential equations of the form

$$\frac{dx}{dt} = ax + by$$
$$\frac{dy}{dt} = cx + dy,$$

where a, b, c, and d are constants (which may be 0). Such a system is said to be a **linear system with constant coefficients**. The constants a, b, c, and d are the **coefficients**. Both the harmonic oscillator model and the model of the CD stores are, up to changes in the names of the dependent variables and the coefficients, systems of this form.

The most important adjective — linear — refers to the fact that the equations for dx/dt and for dy/dt involve only first powers of the dependent variables. In other words they are linear functions of x and y. Since the coefficients a, b, c, and d are constants, this type of system is also *autonomous*, and therefore we know that distinct solution curves in the phase plane do not touch. These systems have two dependent variables, so we say that they are *planar* or *two-dimensional*. Since "two-dimensional, linear system with constant coefficients" is quite a mouthful, we usually just call these systems *planar linear systems* or even just *linear systems*.

We can use vector and matrix notation to write this system much more efficiently. Let **A** be the "2-by-2" square matrix (the 2×2 matrix)

$$\mathbf{A} = \begin{pmatrix} a & b \\ c & d \end{pmatrix},$$

and let

$$\mathbf{Y} = \begin{pmatrix} x \\ y \end{pmatrix}$$

denote the column vector of dependent variables. Then the **product** of a 2×2 matrix **A** and a column vector **Y** is the column vector **AY** given by

$$\mathbf{AY} = \begin{pmatrix} a & b \\ c & d \end{pmatrix} \begin{pmatrix} x \\ y \end{pmatrix} = \begin{pmatrix} ax + by \\ cx + dy \end{pmatrix}.$$

For example,

$$\begin{pmatrix} 5 & 2 \\ -1 & 3 \end{pmatrix} \begin{pmatrix} 3 \\ 4 \end{pmatrix} = \begin{pmatrix} 5 \cdot 3 + 2 \cdot 4 \\ -1 \cdot 3 + 3 \cdot 4 \end{pmatrix} = \begin{pmatrix} 23 \\ 9 \end{pmatrix}$$

and

$$\begin{pmatrix} (2-a) & \pi \\ e & y \end{pmatrix} \begin{pmatrix} y \\ 2v \end{pmatrix} = \begin{pmatrix} (2-a)y + 2\pi v \\ ey + 2yv \end{pmatrix}.$$

As in Chapter 2, if x and y are dependent variables, then we write

$$\mathbf{Y}(t) = \begin{pmatrix} x(t) \\ y(t) \end{pmatrix} \quad \text{and} \quad \frac{d\mathbf{Y}}{dt} = \begin{pmatrix} \dfrac{dx}{dt} \\ \dfrac{dy}{dt} \end{pmatrix}.$$

Using this matrix notation, we can write the two-dimensional linear system

$$\frac{dx}{dt} = ax + by$$

$$\frac{dy}{dt} = cx + dy$$

as

$$\begin{pmatrix} \dfrac{dx}{dt} \\ \dfrac{dy}{dt} \end{pmatrix} = \begin{pmatrix} ax + by \\ cx + dy \end{pmatrix} = \begin{pmatrix} a & b \\ c & d \end{pmatrix} \begin{pmatrix} x \\ y \end{pmatrix},$$

or more compactly as

$$\frac{d\mathbf{Y}}{dt} = \mathbf{A}\mathbf{Y},$$

where

$$\mathbf{A} = \begin{pmatrix} a & b \\ c & d \end{pmatrix} \quad \text{and} \quad \mathbf{Y} = \begin{pmatrix} x \\ y \end{pmatrix}.$$

The matrix $\mathbf{A}$ of coefficients of the system is called the **coefficient matrix**.

One advantage of the matrix notation is that it helps us see the similarities between first-order linear systems and first-order linear equations. Working with matrices also gives us some very useful algebraic tools, which we will exploit throughout this chapter.

Vector notation can be extended to include systems with any number n of dependent variables $y_1, y_2, \ldots, y_n$. The (constant coefficient) linear system with n dependent variables is

$$\frac{dy_1}{dt} = a_{11}y_1 + a_{12}y_2 + \cdots + a_{1n}y_n$$

$$\frac{dy_2}{dt} = a_{21}y_1 + a_{22}y_2 + \cdots + a_{2n}y_n$$

$$\vdots \qquad \qquad \vdots$$

$$\frac{dy_n}{dt} = a_{n1}y_1 + a_{n2}y_2 + \cdots + a_{nn}y_n.$$

In this case the coefficients of this system are $a_{11}, a_{12}, \ldots, a_{nn}$. Let

$$\mathbf{Y} = \begin{pmatrix} y_1 \\ y_2 \\ \vdots \\ y_n \end{pmatrix}, \quad \text{so} \quad \frac{d\mathbf{Y}}{dt} = \begin{pmatrix} \dfrac{dy_1}{dt} \\ \dfrac{dy_2}{dt} \\ \vdots \\ \dfrac{dy_n}{dt} \end{pmatrix}.$$

The coefficient matrix is the $n \times n$ matrix

$$\mathbf{A} = \begin{pmatrix} a_{11} & a_{12} & \cdots & a_{1n} \\ a_{21} & a_{22} & \cdots & a_{2n} \\ \vdots & \vdots & \ddots & \vdots \\ a_{n1} & a_{n2} & \cdots & a_{nn} \end{pmatrix},$$

and we have

$$\frac{d\mathbf{Y}}{dt} = \mathbf{AY} = \begin{pmatrix} a_{11} & a_{12} & \cdots & a_{1n} \\ a_{21} & a_{22} & \cdots & a_{2n} \\ \vdots & \vdots & \ddots & \vdots \\ a_{n1} & a_{n2} & \cdots & a_{nn} \end{pmatrix} \begin{pmatrix} y_1 \\ y_2 \\ \vdots \\ y_n \end{pmatrix}$$

$$= \begin{pmatrix} a_{11}y_1 + a_{12}y_2 + \ldots + a_{1n}y_n \\ a_{21}y_1 + a_{22}y_2 + \ldots + a_{2n}y_n \\ \vdots \\ a_{n1}y_1 + a_{n2}y_2 + \ldots + a_{nn}y_n \end{pmatrix}.$$

The number of dependent variables is called the **dimension** of the system, so this system is n-**dimensional**. For example, the three-dimensional system

$$\frac{dx}{dt} = \sqrt{2}\,x + y$$

$$\frac{dy}{dt} = z$$

$$\frac{dz}{dt} = -x - y + 2z$$

can be written as

$$\frac{d\mathbf{Y}}{dt} = \mathbf{AY},$$

where

$$\mathbf{Y} = \begin{pmatrix} x \\ y \\ z \end{pmatrix} \quad \text{and} \quad \mathbf{A} = \begin{pmatrix} \sqrt{2} & 1 & 0 \\ 0 & 0 & 1 \\ -1 & -1 & 2 \end{pmatrix}.$$

In this text we deal with primarily planar, or two-dimensional, systems. However, readers familiar with linear algebra will recognize that many of the concepts we discuss carry over to higher dimensional systems with little or no modification.

Linear systems are like other systems of differential equations, only simpler. All of the methods of Chapter 2 apply, and we use these methods to understand the associated vector fields, direction fields, and graphs of solutions. In addition, because linear systems are relatively simple algebraically, it is reasonable to hope that we can "read off" the behavior of solutions just from the coefficients. That is, if we are given the planar linear system

$$\frac{d\mathbf{Y}}{dt} = \mathbf{AY}, \quad \text{where } \mathbf{A} = \begin{pmatrix} a & b \\ c & d \end{pmatrix}$$

is the coefficient matrix, then we would like to understand the system completely if we know the four numbers a, b, c, and d. In fact we might even hope to come up with an explicit formula for the general solution. From the four numbers a, b, c, and d, we are able to give a geometric description of the behavior of the solutions in the xy-plane, describe the $x(t)$- and $y(t)$-graphs of solutions, and even give a formula for the general solution. Hence we are able to produce explicit formulas that solve any initial-value problem.

Equilibrium Points of Linear Systems and the Determinant

We start by looking for the simplest solutions — the equilibrium solutions. Recall that a point $\mathbf{Y}_0 = (x_0, y_0)$ is an equilibrium point of a system if and only if the vector field at $\mathbf{Y}_0$ is the zero vector. Since the vector field of a system at the point $\mathbf{Y}_0$ is given by the right-hand side of the differential equation evaluated at that point and since

$$\frac{d\mathbf{Y}}{dt} = \mathbf{AY}$$

for a linear system, we know that the vector field $\mathbf{F}(\mathbf{Y}_0)$ at $\mathbf{Y}_0$ for a linear system is given by

$$\mathbf{F}(\mathbf{Y}_0) = \mathbf{AY}_0.$$

In other words the vector at $\mathbf{Y}_0$ is computed by taking the product of the matrix $\mathbf{A}$ with the vector $\mathbf{Y}_0$. Consequently, the equilibrium points are the points $\mathbf{Y}_0$ such that

$$\mathbf{AY}_0 = \begin{pmatrix} 0 \\ 0 \end{pmatrix}.$$

That is,

$$\begin{pmatrix} a & b \\ c & d \end{pmatrix} \begin{pmatrix} x_0 \\ y_0 \end{pmatrix} = \begin{pmatrix} ax_0 + by_0 \\ cx_0 + dy_0 \end{pmatrix} = \begin{pmatrix} 0 \\ 0 \end{pmatrix}.$$

Written in scalar form, this vector equation is a pair of simultaneous linear equations

$$\begin{cases} ax_0 + by_0 = 0 \\ cx_0 + dy_0 = 0. \end{cases}$$

Clearly $(x_0, y_0) = (0, 0)$ is a solution to these equations. Therefore the point $\mathbf{Y}_0 = (0, 0)$ is an equilibrium point, and the constant function

$$\mathbf{Y}(t) = (0, 0) \quad \text{for all } t$$

is a solution to the linear system. This solution is often called the **trivial solution** of the system. (Note that this computation does not depend on the values of the coefficients a, b, c, and d. In other words *every* linear system has an equilibrium point at the origin.)

Any other equilibrium points (x_0, y_0) must also satisfy

$$\begin{cases} ax_0 + by_0 = 0 \\ cx_0 + dy_0 = 0. \end{cases}$$

To find them, assume for the moment that $a \neq 0$. Using the first equation, we get

$$x_0 = -\frac{b}{a} y_0.$$

The second equation then yields

$$c\left(-\frac{b}{a}\right) y_0 + dy_0 = 0,$$

which can be rewritten as

$$(ad - bc)y_0 = 0.$$

Hence either $y_0 = 0$ or $ad - bc = 0$. If $y_0 = 0$, then $x_0 = 0$, and once again we have the trivial solution. Therefore a linear system has nontrivial equilibrium points only if $ad - bc = 0$. This quantity, $ad - bc$, is a particularly important number associated with the 2×2 matrix $\mathbf{A}$.

DEFINITION   The **determinant** of a 2×2 matrix

$$\mathbf{A} = \begin{pmatrix} a & b \\ c & d \end{pmatrix}$$

is the number $ad - bc$. It is denoted $\det \mathbf{A}$.   ∎

With this definition, we are able to summarize the results of the above computation for equilibrium points of linear systems.

THEOREM   If $\mathbf{A}$ is a matrix with $\det \mathbf{A} \neq 0$, then the only equilibrium point for the linear system $d\mathbf{Y}/dt = \mathbf{AY}$ is the origin.   ∎

The argument above proves this theorem, provided the upper left-hand corner entry, a, of $\mathbf{A}$ is nonzero, but there is nothing special about this entry. By similar steps we can obtain the same result as long as at least one of the entries of A is nonzero (see Exercise 14). If all the entries of $\mathbf{A}$ are zero, then every point in the plane is an equilibrium point.

As an example, suppose

$$\mathbf{A} = \begin{pmatrix} 2 & 1 \\ -4 & 0.3 \end{pmatrix}.$$

Then $\det \mathbf{A} = (2)(0.3) - (1)(-4) = 4.6$. Since $\det \mathbf{A} \neq 0$, the only equilibrium point for linear system $d\mathbf{Y}/dt = \mathbf{AY}$ is the origin, $(0, 0)$.

An Important Property of the Determinant

The determinant is a quantity that pops up repeatedly throughout this chapter. For us its significance usually is whether or not it is zero. If we pick four numbers a, b, c, and d at random, then it is unlikely that the number $ad - bc$ is exactly zero. Thus matrices whose determinant is zero are often called **singular** or **degenerate**.

From the theorem we just discussed, we know that, if a linear system $d\mathbf{Y}/dt = \mathbf{AY}$ is **nondegenerate** ($\det \mathbf{A} \neq 0$), then it has exactly one equilibrium point, which is $(0, 0)$. In other words an initial condition of $(0, 0)$ corresponds to a solution curve that sits at $(0, 0)$ for all time. Any other initial condition yields a solution that changes with time.

In Section 3.2 we need to use the determinant again, so it is important to understand exactly what we verified when we justified this theorem. For linear systems, equilibria correspond to points $\mathbf{Y}_0$ for which

$$\mathbf{AY}_0 = \begin{pmatrix} 0 \\ 0 \end{pmatrix}.$$

Written in terms of scalars, this vector equation is identical to the the simultaneous system of linear equations

$$\begin{cases} ax_0 + by_0 = 0 \\ cx_0 + dy_0 = 0. \end{cases}$$

What we actually verified is that this system of equations has **nontrivial solutions** — solutions other than $(0, 0)$ — if and only if $\det \mathbf{A} = 0$.

The Linearity Principle

The solutions of linear systems have special properties that solutions of arbitrary systems do not have. These properties are so useful that we take advantage of them repeatedly. In fact they are exactly the reason that we will be so successful in our analysis of linear systems.

However, a note of caution is in order: It is important to make sure that the system under consideration actually *is* a linear system before you use any of these special properties. This is the equivalent of making sure that the car is in reverse before trying to back out of the garage. If the car is in drive instead of reverse when you start backing out of the garage, there can be dire consequences.

The most important property of linear systems is the **Linearity Principle**.

LINEARITY PRINCIPLE   Suppose $d\mathbf{Y}/dt = \mathbf{AY}$ is a linear system of differential equations.

1. If $\mathbf{Y}(t)$ is a solution of this system and k is any constant, then $k\mathbf{Y}(t)$ is also a solution.

2. If $\mathbf{Y}_1(t)$ and $\mathbf{Y}_2(t)$ are two solutions of this system, then $\mathbf{Y}_1(t) + \mathbf{Y}_2(t)$ is also a solution. ∎

Using the Linearity Principle (also called the **Principle of Superposition**), we can manufacture infinitely many new solutions from any given solution or pair of solutions. Taken together, the two parts of the Linearity Principle imply that, if $\mathbf{Y}_1(t)$ and

$Y_2(t)$ are solutions of the system and if k_1 and k_2 are any constants, then

$$k_1 Y_1(t) + k_2 Y_2(t)$$

is also a solution. A solution of the form $k_1 Y_1(t) + k_2 Y_2(t)$ is called a **linear combination** of the solutions $Y_1(t)$ and $Y_2(t)$. Given two solutions, we can produce infinitely many solutions by forming linear combinations of the original two.

For example, consider the partially decoupled linear system

$$\frac{d\mathbf{Y}}{dt} = \begin{pmatrix} 2 & 3 \\ 0 & -4 \end{pmatrix} \mathbf{Y}.$$

In Section 2.3 we found that

$$\mathbf{Y}_1(t) = \begin{pmatrix} e^{2t} \\ 0 \end{pmatrix} \quad \text{and} \quad \mathbf{Y}_2(t) = \begin{pmatrix} -e^{-4t} \\ 2e^{-4t} \end{pmatrix}$$

are solutions to this system (see page 177). We can double-check this by directly calculating both sides, $d\mathbf{Y}/dt$ and $\mathbf{AY}$, of the differential equation. For example, with $\mathbf{Y}_1(t)$ we have

$$\frac{d\mathbf{Y}_1}{dt} = \begin{pmatrix} 2e^{2t} \\ 0 \end{pmatrix} \quad \text{and} \quad \mathbf{AY}_1 = \begin{pmatrix} 2 & 3 \\ 0 & -4 \end{pmatrix}\begin{pmatrix} e^{2t} \\ 0 \end{pmatrix} = \begin{pmatrix} 2e^{2t} \\ 0 \end{pmatrix},$$

so $\mathbf{Y}_1(t)$ is a solution. (You should double-check that $\mathbf{Y}_2(t)$ is a solution. The verification is good practice with matrix arithmetic.)

The solution curves for $\mathbf{Y}_1(t)$ and $\mathbf{Y}_2(t)$ are shown in Figure 3.5. Note that each one is a line segment in the xy-plane. The solution curve for $\mathbf{Y}_1(t)$ approaches the equilibrium point at the origin as $t \to -\infty$, and the solution curve for $\mathbf{Y}_2(t)$ approaches the equilibrium point at the origin as $t \to \infty$. In the next section we exploit the geometry of solutions such as these to find them using only algebraic techniques.

The Linearity Principle tell us that any function of the form $k_1 \mathbf{Y}_1(t) + k_2 \mathbf{Y}_2(t)$ is also a solution to this system for any constants k_1 and k_2. To illustrate this fact, we

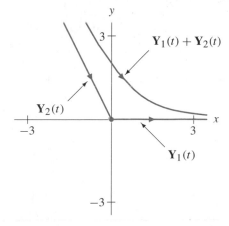

Figure 3.5
The Linearity Principle implies that the function $\mathbf{Y}_1(t) + \mathbf{Y}_2(t)$ is a solution of the system

$$\frac{d\mathbf{Y}}{dt} = \begin{pmatrix} 2 & 3 \\ 0 & -4 \end{pmatrix} \mathbf{Y}$$

because it is the sum of the two solutions $\mathbf{Y}_1(t)$ and $\mathbf{Y}_2(t)$.

check directly that $\mathbf{Y}_3(t) = -2\mathbf{Y}_1(t) + 5\mathbf{Y}_2(t)$ is a solution. Note that

$$\mathbf{Y}_3(t) = -2\mathbf{Y}_1(t) + 5\mathbf{Y}_2(t) = -2\begin{pmatrix} e^{2t} \\ 0 \end{pmatrix} + 5\begin{pmatrix} -e^{-4t} \\ 2e^{-4t} \end{pmatrix} = \begin{pmatrix} -2e^{2t} - 5e^{-4t} \\ 10e^{-4t} \end{pmatrix},$$

and therefore

$$\frac{d\mathbf{Y}_3}{dt} = \begin{pmatrix} -4e^{2t} + 20e^{-4t} \\ -40e^{-4t} \end{pmatrix}.$$

Also we compute

$$\mathbf{AY}_3 = \begin{pmatrix} 2 & 3 \\ 0 & -4 \end{pmatrix}\begin{pmatrix} -2e^{2t} - 5e^{-4t} \\ 10e^{-4t} \end{pmatrix}$$

$$= \begin{pmatrix} 2(-2e^{2t} - 5e^{-4t}) + 3(10e^{-4t}) \\ -4(10e^{-4t}) \end{pmatrix}$$

$$= \begin{pmatrix} -4e^{2t} + 20e^{-4t} \\ -40e^{-4t} \end{pmatrix}.$$

Since both computations yield the same function, this linear combination of the two solutions $\mathbf{Y}_1(t)$ and $\mathbf{Y}_2(t)$ is also a solution. (In the future we will not bother to double-check the consequences of the Linearity Principle.)

Again we emphasize that the solution curves for the solutions $\mathbf{Y}_1(t)$ and $\mathbf{Y}_2(t)$ possess a very special and useful geometric property. The fact that they form line segments is not typical of solution curves in general. In fact the typical solution curve of this system is not a straight line. For example, as we see in Figure 3.5, the solution curve of the solution $\mathbf{Y}_1(t) + \mathbf{Y}_2(t)$ is not straight.

Verification of the Linearity Principle

To show that the Linearity Principle holds in general, we first state the following two algebraic properties of matrix multiplication:

1. If $\mathbf{A}$ is a matrix and $\mathbf{Y}$ is a vector, then

$$\mathbf{A}(k\mathbf{Y}) = k\mathbf{AY}$$

for any constant k.

2. If $\mathbf{A}$ is a matrix and $\mathbf{Y}_1$ and $\mathbf{Y}_2$ are vectors, then

$$\mathbf{A}(\mathbf{Y}_1 + \mathbf{Y}_2) = \mathbf{AY}_1 + \mathbf{AY}_2.$$

We can verify these two facts for 2×2 matrices and 2-dimensional vectors by direct computation. For example, to verify property 2, let

$$\mathbf{A} = \begin{pmatrix} a & b \\ c & d \end{pmatrix}$$

be an arbitrary 2×2 matrix and let

$$\mathbf{Y}_1 = \begin{pmatrix} x_1 \\ y_1 \end{pmatrix} \quad \text{and} \quad \mathbf{Y}_2 = \begin{pmatrix} x_2 \\ y_2 \end{pmatrix}$$

be arbitrary vectors. Then

$$\mathbf{A}(\mathbf{Y}_1 + \mathbf{Y}_2) = \begin{pmatrix} a & b \\ c & d \end{pmatrix} \begin{pmatrix} x_1 + x_2 \\ y_1 + y_2 \end{pmatrix}$$

$$= \begin{pmatrix} a(x_1 + x_2) + b(y_1 + y_2) \\ c(x_1 + x_2) + d(y_1 + y_2) \end{pmatrix}$$

$$= \begin{pmatrix} ax_1 + ax_2 + by_1 + by_2 \\ cx_1 + cx_2 + dy_1 + dy_2 \end{pmatrix}$$

and

$$\mathbf{AY}_1 + \mathbf{AY}_2 = \begin{pmatrix} a & b \\ c & d \end{pmatrix} \begin{pmatrix} x_1 \\ y_1 \end{pmatrix} + \begin{pmatrix} a & b \\ c & d \end{pmatrix} \begin{pmatrix} x_2 \\ y_2 \end{pmatrix}$$

$$= \begin{pmatrix} ax_1 + by_1 \\ cx_1 + dy_1 \end{pmatrix} + \begin{pmatrix} ax_2 + by_2 \\ cx_2 + dy_2 \end{pmatrix}$$

$$= \begin{pmatrix} ax_1 + ax_2 + by_1 + by_2 \\ cx_1 + cx_2 + dy_1 + dy_2 \end{pmatrix}.$$

Thus property 2 holds. The verification of property 1 is left to the exercises (see Exercise 30).

Given these algebraic properties of matrix multiplication, we can verify the Linearity Principle using the standard rules of differentiation. Suppose $\mathbf{Y}_1(t)$ and $\mathbf{Y}_2(t)$ are solutions to $d\mathbf{Y}/dt = \mathbf{AY}$; that is, suppose

$$\frac{d\mathbf{Y}_1}{dt} = \mathbf{AY}_1 \quad \text{and} \quad \frac{d\mathbf{Y}_2}{dt} = \mathbf{AY}_2 \quad \text{for all } t.$$

For any constant k we have

$$\frac{d(k\mathbf{Y}_1)}{dt} = k\frac{d\mathbf{Y}_1}{dt} = k\mathbf{AY}_1 = \mathbf{A}(k\mathbf{Y}_1),$$

so $k\mathbf{Y}_1(t)$ is a solution to the system. Also

$$\frac{d(\mathbf{Y}_1 + \mathbf{Y}_2)}{dt} = \frac{d\mathbf{Y}_1}{dt} + \frac{d\mathbf{Y}_2}{dt} = \mathbf{AY}_1 + \mathbf{AY}_2 = \mathbf{A}(\mathbf{Y}_1 + \mathbf{Y}_2) \quad \text{for all } t.$$

As a result, $\mathbf{Y}_1(t) + \mathbf{Y}_2(t)$ is also a solution and the Linearity Principle is verified. We can see the advantage of the matrix and vector notation. To write out the above equations showing all the components would be a tedious exercise — in fact it is a tedious exercise at the end of this section (see Exercise 30).

Solving Initial-Value Problems

From the Linearity Principle we know that, given two solutions $\mathbf{Y}_1(t)$ and $\mathbf{Y}_2(t)$, we can make many more solutions of the form $k_1\mathbf{Y}_1(t) + k_2\mathbf{Y}_2(t)$ for any constants k_1 and k_2. This type of expression is called a *two-parameter family of solutions*, since we have two constants, k_1 and k_2, that we can adjust to obtain various solutions. It is reasonable to ask if these are all of the solutions or, put another way, if each solution is one of this form.

To see how the Linearity Principle is used to solve initial-value problems, we return to the differential equation

$$\frac{d\mathbf{Y}}{dt} = \begin{pmatrix} 2 & 3 \\ 0 & -4 \end{pmatrix} \mathbf{Y}$$

that was discussed earlier in this section. Suppose we want to find the solution $\mathbf{Y}(t)$ of this system with initial value $\mathbf{Y}(0) = (2, -3)$. We already know that

$$\mathbf{Y}_1(t) = \begin{pmatrix} e^{2t} \\ 0 \end{pmatrix} \quad \text{and} \quad \mathbf{Y}_2(t) = \begin{pmatrix} -e^{-4t} \\ 2e^{-4t} \end{pmatrix}$$

are solutions, and by direct evaluation we know that

$$\mathbf{Y}_1(0) = \begin{pmatrix} 1 \\ 0 \end{pmatrix} \quad \text{and} \quad \mathbf{Y}_2(0) = \begin{pmatrix} -1 \\ 2 \end{pmatrix},$$

so neither $\mathbf{Y}_1(t)$ nor $\mathbf{Y}_2(t)$ is the solution to the initial-value problem

$$\frac{d\mathbf{Y}}{dt} = \begin{pmatrix} 2 & 3 \\ 0 & -4 \end{pmatrix} \mathbf{Y}, \quad \mathbf{Y}(0) = \begin{pmatrix} 2 \\ -3 \end{pmatrix}.$$

But the Linearity Principle says that we can form any linear combination of $\mathbf{Y}_1(t)$ and $\mathbf{Y}_2(t)$ and still have a solution. Hence we seek k_1 and k_2 so that

$$k_1\begin{pmatrix} 1 \\ 0 \end{pmatrix} + k_2\begin{pmatrix} -1 \\ 2 \end{pmatrix} = \begin{pmatrix} 2 \\ -3 \end{pmatrix}.$$

This vector equation is equivalent to the simultaneous equations

$$\begin{cases} k_1 - k_2 = 2 \\ 2k_2 = -3. \end{cases}$$

The second equation yields $k_2 = -3/2$, and consequently, the first equation yields $k_1 = 1/2$. This computation implies

$$\frac{1}{2}\mathbf{Y}_1(0) - \frac{3}{2}\mathbf{Y}_2(0) = \begin{pmatrix} 2 \\ -3 \end{pmatrix},$$

so we consider the function

$$\mathbf{Y}(t) = \frac{1}{2}\mathbf{Y}_1(t) - \frac{3}{2}\mathbf{Y}_2(t)$$

$$= \frac{1}{2}\begin{pmatrix} e^{2t} \\ 0 \end{pmatrix} - \frac{3}{2}\begin{pmatrix} -e^{-4t} \\ 2e^{-4t} \end{pmatrix}$$

$$= \begin{pmatrix} \frac{1}{2}e^{2t} + \frac{3}{2}e^{-4t} \\ -3e^{-4t} \end{pmatrix}.$$

This function has the correct initial condition, and by the Linearity Principle we know that it must be a solution to the system. The Uniqueness Theorem tells us that this is the only function that solves the initial-value problem (see Section 2.4).

In this example we found the solution to the initial-value problem using the two solutions of the system that were already available plus a little arithmetic (but no calculus). By taking the appropriate linear combination of the two known solutions, we were able to find a solution with the desired initial conditions.

Maybe we were just lucky. Will we always be able to find the appropriate k_1 and k_2 no matter what initial condition we have? To check, suppose we consider the same differential equation with an arbitrary initial condition,

$$\frac{d\mathbf{Y}}{dt} = \begin{pmatrix} 2 & 3 \\ 0 & -4 \end{pmatrix}\mathbf{Y}, \quad \mathbf{Y}(0) = \begin{pmatrix} x_0 \\ y_0 \end{pmatrix},$$

and the two solutions $\mathbf{Y}_1(t)$ and $\mathbf{Y}_2(t)$ with which we started. To solve the initial-value problem, we need to find k_1 and k_2 so that

$$k_1\mathbf{Y}_1(0) + k_2\mathbf{Y}_2(0) = \mathbf{Y}(0) = \begin{pmatrix} x_0 \\ y_0 \end{pmatrix}.$$

In other words, given arbitrary x_0 and y_0, can we always find k_1 and k_2 such that

$$k_1\begin{pmatrix} 1 \\ 0 \end{pmatrix} + k_2\begin{pmatrix} -1 \\ 2 \end{pmatrix} = \begin{pmatrix} x_0 \\ y_0 \end{pmatrix}?$$

This vector equation is equivalent to the simultaneous system of equations

$$\begin{cases} k_1 - k_2 = x_0 \\ 2k_2 = y_0. \end{cases}$$

Since the second equation is so simple, we can always find k_1 and k_2 given x_0 and y_0. We use the second equation to find k_2 first, and then we find k_1 using this value of k_2 in the first equation.

Since we are able to solve every possible initial-value problem for the system

$$\frac{d\mathbf{Y}}{dt} = \begin{pmatrix} 2 & 3 \\ 0 & -4 \end{pmatrix}\mathbf{Y}$$

using a linear combination of

$$\mathbf{Y}_1(t) = \begin{pmatrix} e^{2t} \\ 0 \end{pmatrix} \quad \text{and} \quad \mathbf{Y}_2(t) = \begin{pmatrix} -e^{-4t} \\ 2e^{-4t} \end{pmatrix},$$

we have found the general solution to this system. It is the two-parameter family

$$\mathbf{Y}(t) = k_1 \mathbf{Y}_1(t) + k_2 \mathbf{Y}_2(t) = k_1 \begin{pmatrix} e^{2t} \\ 0 \end{pmatrix} + k_2 \begin{pmatrix} -e^{-4t} \\ 2e^{-4t} \end{pmatrix}.$$

Using vector addition, this two-parameter family can be written as

$$\mathbf{Y}(t) = \begin{pmatrix} k_1 e^{2t} - k_2 e^{-4t} \\ 2k_2 e^{-4t} \end{pmatrix}.$$

Linear Independence

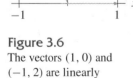

Figure 3.6
The vectors $(1, 0)$ and $(-1, 2)$ are linearly independent.

Note that in this example we used the Linearity Principle to produce infinitely many solutions starting with two given solutions $\mathbf{Y}_1(t)$ and $\mathbf{Y}_2(t)$. Then, because we were able to express an arbitrary initial condition as a linear combination of the initial conditions $\mathbf{Y}_1(0)$ and $\mathbf{Y}_2(0)$, we could use $\mathbf{Y}_1(t)$ and $\mathbf{Y}_2(t)$ to form the general solution.

Expressing arbitrary vectors as linear combinations of given vectors is a fundamental topic in linear algebra. In the two-dimensional case the key property that ensures that an arbitrary vector can be written as some linear combination of the given vectors (x_1, y_1) and (x_2, y_2) is that they do *not* lie on the same line through the origin. (Note that in the previous example the initial conditions $(1, 0)$ and $(-1, 2)$ do not lie on the same line through the origin — see Figure 3.6.) We say that the two vectors (x_1, y_1) and (x_2, y_2) are **linearly independent** if they do not lie on the same line through the origin or, equivalently, if neither one is a multiple of the other.

THEOREM   Suppose (x_1, y_1) and (x_2, y_2) are two linearly independent vectors in the plane. Then given any vector (x_0, y_0), there exist k_1 and k_2 so that

$$k_1 \begin{pmatrix} x_1 \\ y_1 \end{pmatrix} + k_2 \begin{pmatrix} x_2 \\ y_2 \end{pmatrix} = \begin{pmatrix} x_0 \\ y_0 \end{pmatrix}. \quad \blacksquare$$

Two linearly independent vectors can be combined via addition and scalar multiplication to form any other vector in the plane.

Note that the equation

$$k_1 \begin{pmatrix} x_1 \\ y_1 \end{pmatrix} + k_2 \begin{pmatrix} x_2 \\ y_2 \end{pmatrix} = \begin{pmatrix} x_0 \\ y_0 \end{pmatrix}$$

is really a simultaneous system of two linear equations

$$\begin{cases} x_1 k_1 + x_2 k_2 = x_0 \\ y_1 k_1 + y_2 k_2 = y_0 \end{cases}$$

in the two unknowns k_1 and k_2. We are given the x's and the y's, and we must solve for the k's. We can show that there are solutions by writing down formulas for k_1 and k_2

in terms of the x's and y's. As long as the denominators in these formulas are nonzero, the solutions exist. Solving systems of equations of this form involves the same sort of algebra as finding equilibrium points for linear systems. Hence it is not surprising that the determinant plays a role here too (see Exercises 31 and 32.)

The General Solution

The previous example and theorem illustrate the way in which we approach every linear system. Therefore it is worth summarizing the discussion.

THEOREM   Suppose $\mathbf{Y}_1(t)$ and $\mathbf{Y}_2(t)$ are solutions of the linear system $d\mathbf{Y}/dt = \mathbf{AY}$. If $\mathbf{Y}_1(0)$ and $\mathbf{Y}_2(0)$ are linearly independent, then for any initial condition $\mathbf{Y}(0) = (x_0, y_0)$ we can find constants k_1 and k_2 so that $k_1\mathbf{Y}_1(t) + k_2\mathbf{Y}_2(t)$ is the solution to the initial-value problem

$$\frac{d\mathbf{Y}}{dt} = \mathbf{AY}, \quad \mathbf{Y}(0) = \begin{pmatrix} x_0 \\ y_0 \end{pmatrix}. \quad \blacksquare$$

In this situation we say that the two-parameter family $k_1\mathbf{Y}_1(t) + k_2\mathbf{Y}_2(t)$, where k_1 and k_2 are arbitrary constants, is the **general solution** of the system.

By the Existence and Uniqueness Theorem for systems, we know that each initial-value problem for a linear system has exactly one solution. Given any two solutions $\mathbf{Y}_1(t)$ and $\mathbf{Y}_2(t)$ of a linear system with linearly independent initial conditions $\mathbf{Y}_1(0)$ and $\mathbf{Y}_2(0)$, we can form the general solution of the system by forming the two-parameter family $k_1\mathbf{Y}_1(t) + k_2\mathbf{Y}_2(t)$. By adjusting the constants k_1 and k_2, we can obtain the solution that satisfies any given initial condition.

This is really excellent progress. We now know that to find all the solutions to a linear system, we need to find only two particular solutions with linearly independent initial positions. Two solutions $\mathbf{Y}_1(t)$ and $\mathbf{Y}_2(t)$ of a linear system for which $\mathbf{Y}_1(0)$ and $\mathbf{Y}_2(0)$ are linearly independent are called **linearly independent solutions** of the linear system. (We will see in the exercises that if $\mathbf{Y}_1(t)$ and $\mathbf{Y}_2(t)$ are linearly independent solutions of a linear system, then the vectors $\mathbf{Y}_1(t)$ and $\mathbf{Y}_2(t)$ are linearly independent for all t; see Exercise 35.) The next step is to find a general way to come up with two linearly independent solutions $\mathbf{Y}_1(t)$ and $\mathbf{Y}_2(t)$. Much of the discussion in Sections 3.2, 3.4, and 3.5 involves techniques that do just that.

An Undamped Harmonic Oscillator

In Section 2.1 we studied the undamped harmonic oscillator given by the second-order differential equation

$$\frac{d^2y}{dt^2} = -y.$$

We guessed that $y_1(t) = \cos t$ is a solution and then checked our guess by verifying that

$$\frac{d^2y_1}{dt^2} + y_1 = \frac{d^2(\cos t)}{dt^2} + \cos t$$

$$= -\cos t + \cos t$$

$$= 0.$$

Similarly, we can check that $y_2(t) = \sin t$ is also a solution. Now that we have the Linearity Principle at our disposal, we can take this discussion one step further.

The second-order equation can be converted to the first-order system

$$\frac{dy}{dt} = v$$
$$\frac{dv}{dt} = -y,$$

which is a linear system. Using vector notation, we write

$$\mathbf{Y}(t) = \begin{pmatrix} y(t) \\ v(t) \end{pmatrix},$$

and the system can be represented as

$$\frac{d\mathbf{Y}}{dt} = \begin{pmatrix} 0 & 1 \\ -1 & 0 \end{pmatrix} \mathbf{Y}.$$

Recall that the second component of the vector-valued function $\mathbf{Y}(t)$ is $v = dy/dt$. We can use the solution $y_1(t)$ to form a vector-valued function

$$\mathbf{Y}_1(t) = \begin{pmatrix} y_1(t) \\ v_1(t) \end{pmatrix} = \begin{pmatrix} \cos t \\ -\sin t \end{pmatrix}.$$

Note that $\mathbf{Y}_1(t)$ is a solution to the system because

$$\frac{d\mathbf{Y}_1}{dt} = \begin{pmatrix} -\sin t \\ -\cos t \end{pmatrix}$$

and

$$\begin{pmatrix} 0 & 1 \\ -1 & 0 \end{pmatrix} \mathbf{Y}_1(t) = \begin{pmatrix} 0 & 1 \\ -1 & 0 \end{pmatrix} \begin{pmatrix} \cos t \\ -\sin t \end{pmatrix} = \begin{pmatrix} -\sin t \\ -\cos t \end{pmatrix}.$$

Similarly, the solution $y_2(t) = \sin t$ to the second-order equation yields

$$\mathbf{Y}_2(t) = \begin{pmatrix} \sin t \\ \cos t \end{pmatrix},$$

which is also a solution to the first-order system. (Double-checking this assertion is good practice with matrix notation.)

We have a first-order linear system with two dependent variables. Therefore we need two linearly independent solutions to obtain the general solution. At $t = 0$,

$$\mathbf{Y}_1(0) = \begin{pmatrix} 1 \\ 0 \end{pmatrix} \quad \text{and} \quad \mathbf{Y}_2(0) = \begin{pmatrix} 0 \\ 1 \end{pmatrix}.$$

That is, $\mathbf{Y}_1(0)$ lies on the y-axis and $\mathbf{Y}_2(t)$ lies on the v-axis. Consequently, these vectors are linearly independent, and the general solution to the first-order system is

$$\mathbf{Y}(t) = k_1 \begin{pmatrix} \cos t \\ -\sin t \end{pmatrix} + k_2 \begin{pmatrix} \sin t \\ \cos t \end{pmatrix}.$$

Recalling that solutions $\mathbf{Y}(t)$ to the first-order system are really functions of the form $\mathbf{Y}(t) = (y(t), v(t))$ where $y(t)$ is a solution to the original second-order equation, we obtain the general solution to the original second-order equation using the first component of $\mathbf{Y}(t)$. The result is

$$y(t) = k_1 \cos t + k_2 \sin t.$$

In Section 3.6 we will discuss a more immediate way to find the general solution of second-order equations such as this one, but it is important to realize that the Linearity Principle also applies to "linear" second-order equations such as the equation for a damped harmonic oscillator.

EXERCISES FOR SECTION 3.1

Recall the model

$$\frac{dx}{dt} = ax + by$$

$$\frac{dy}{dt} = cx + dy,$$

for Paul and Bob's CD stores, where $x(t)$ is Paul's daily profit, $y(t)$ is Bob's daily profit, and a, b, c, and d are parameters governing how the daily profit of each store affects the other. In Exercises 1–4, different choices of the parameters a, b, c, and d are specified. For each exercise write a brief paragraph describing the interaction between the stores, given the specified parameter values. [For example, suppose $a = 1$, $c = -1$, and $b = d = 0$. If Paul's store is making a profit ($x > 0$), then Paul's profit increases more quickly (because $ax > 0$). However, if Paul makes a profit, then Bob's profits suffer (because $cx < 0$). Since $b = d = 0$, Bob's current profits have no impact on his or Paul's future profits.]

1. $a = 1$, $b = -1$, $c = 1$, and $d = -1$ **2.** $a = 2$, $b = -1$, $c = 0$, and $d = 0$

3. $a = 1$, $b = 0$, $c = 2$, and $d = 1$ **4.** $a = -1$, $b = 2$, $c = 2$, and $d = -1$

In Exercises 5–7, rewrite the specified linear system in matrix form.

5.
$$\frac{dx}{dt} = 2x + y$$
$$\frac{dy}{dt} = x + y$$

6.
$$\frac{dx}{dt} = 3y$$
$$\frac{dy}{dt} = 3\pi y - 0.3x$$

7.
$$\frac{dp}{dt} = 3p - 2q - 7r$$
$$\frac{dq}{dt} = -2p + 6r$$
$$\frac{dr}{dt} = 7.3q + 2r$$

In Exercises 8–9, rewrite the specified linear system in component form.

8. $\left(\begin{array}{c} \dfrac{dx}{dt} \\ \dfrac{dy}{dt} \end{array} \right) = \left(\begin{array}{cc} -3 & 2\pi \\ 4 & -1 \end{array} \right) \left(\begin{array}{c} x \\ y \end{array} \right)$

9. $\left(\begin{array}{c} \dfrac{dx}{dt} \\ \dfrac{dy}{dt} \end{array} \right) = \left(\begin{array}{cc} 0 & \beta \\ \gamma & -1 \end{array} \right) \left(\begin{array}{c} x \\ y \end{array} \right)$

For the linear systems given in Exercises 10–13, sketch (using whatever technology you have available) the direction fields, several solutions, and the $x(t)$- and $y(t)$-graphs for the solution with initial condition $(x, y) = (1, 1)$.

10. $\dfrac{dx}{dt} = 2x + y$

$\dfrac{dy}{dt} = x + y$

11. $\dfrac{dx}{dt} = -2x + y$

$\dfrac{dy}{dt} = -x - 2y$

12. $\left(\begin{array}{c} \dfrac{dx}{dt} \\ \dfrac{dy}{dt} \end{array} \right) = \left(\begin{array}{cc} -3 & 2\pi \\ 4 & -1 \end{array} \right) \left(\begin{array}{c} x \\ y \end{array} \right)$

13. $\left(\begin{array}{c} \dfrac{dx}{dt} \\ \dfrac{dy}{dt} \end{array} \right) = \left(\begin{array}{cc} 5 & 1 \\ -1 & 6 \end{array} \right) \left(\begin{array}{c} x \\ y \end{array} \right)$

14. Let

$$A = \left(\begin{array}{cc} a & b \\ c & d \end{array} \right)$$

be a nonsingular matrix ($\det A \neq 0$).

   (a) Show that, if $a = 0$, then $b \neq 0$ and $c \neq 0$.

   (b) Suppose $a = 0$. Use the result of part (a) to show that the origin is the only equilibrium point.

Along with the verification given in the section, this result shows that, if $\det A \neq 0$, then the only equilibrium point for the system $dY/dt = AY$ is the origin.

15. Let

$$A = \left(\begin{array}{cc} a & b \\ c & d \end{array} \right)$$

be a nonzero matrix. That is, suppose that at least one of its entries is nonzero. Show that, if $\det A = 0$, then the system $dY/dt = AY$ has an entire line of equilibria. [*Hint:* First consider the case where $a \neq 0$. The x-component of the vector field at a point (x_0, y_0) is $ax_0 + by_0$. Consequently, in order for the vector field to vanish at a point (x_0, y_0), we must have $x_0 = (-b/a)y_0$. Use this observation and the fact that $\det A = 0$ to show that any point of the form $(-by_0/a, y_0)$ is an equilibrium point. What if we assume that entries of A other than a are nonzero?]

16. The general form of a linear, homogeneous, second-order equation with constant coefficients is

$$\frac{d^2 y}{dt^2} + p\frac{dy}{dt} + qy = 0.$$

    (a) Write the first-order system for this equation, and write this system in matrix form.

    (b) Show that if $q \neq 0$, then the origin is the only equilibrium point of the system.

    (c) Show that if $q \neq 0$, then the only solution of the second-order equation with y constant is $y(t) = 0$ for all t.

17. Consider the linear system corresponding to the second-order equation

$$\frac{d^2 y}{dt^2} + p\frac{dy}{dt} + qy = 0.$$

    (a) If $q = 0$ and $p \neq 0$, find all the equilibrium points.

    (b) If $q = p = 0$, find all the equilibrium points.

18. Convert the second-order equation

$$\frac{d^2 y}{dt^2} = 0$$

into a first-order system using $v = dy/dt$ as usual.

    (a) Find the general solution for the dv/dt equation.

    (b) Substitute this solution into the dy/dt equation, and find the general solution of the system.

    (c) Sketch the phase plane of the system.

19. Convert the third-order differential equation

$$\frac{d^3 y}{dt^3} + p\frac{d^2 y}{dt^2} + q\frac{dy}{dt} + ry = 0,$$

where p, q, and r are constants, to a three-dimensional linear system written in matrix form.

In Exercises 20–23, we consider the following model of the market for single-family housing in a community. Let $S(t)$ be the number of sellers at time t, and let $B(t)$ be the number of buyers at time t. We assume that there are natural equilibrium levels of buyers and sellers (made up of people who retire, change job locations, or wish to move for family reasons). The equilibrium level of sellers is S_0, and the equilibrium level of buyers is B_0.

    However, market forces can entice people to buy or sell under various conditions. For example, if the price of a house is very high, then house owners are tempted to sell their homes. If prices are very low, extra buyers enter the market looking for bargains. We let $b(t) = B(t) - B_0$ denote the deviation of the number of buyers from equilibrium at time t. So if $b(t) > 0$, then there are more buyers than usual, and we say it is a "seller's market." Presumably the competition of the extra buyers for the same number of houses for sale will force the prices up (the law of supply and demand).

    Similarly, we let $s(t) = S(t) - S_0$ denote the deviation of the number of sellers from the equilibrium level. If $s(t) > 0$, then there are more sellers on the market than usual; and if the number of buyers is low, there are too many houses on the market and prices decrease, which in turn affects decisions to buy or sell.

We can give a simple model of this situation as follows:

$$\frac{d\mathbf{Y}}{dt} = \mathbf{AY} = \begin{pmatrix} \alpha & \beta \\ \gamma & \delta \end{pmatrix} \begin{pmatrix} b \\ s \end{pmatrix}, \quad \text{where } \mathbf{Y} = \begin{pmatrix} b \\ s \end{pmatrix}.$$

The exact values of the parameters α, β, γ, and δ depend on the economy of a particular community. Nevertheless, if we assume that everybody wants to get a bargain when they are buying a house and to get top dollar when they are selling a house, then we can hope to predict whether the parameters are positive or negative even though we cannot predict their exact values.

Use the information given above to obtain information about the parameters α, β, γ, and δ. Be sure to justify your answers.

20. If there are more than the usual number of buyers competing for houses, we would expect the price of houses to rise, and this increase would make it less likely that new potential buyers will enter the market. What does this say about the parameter α?

21. If there are fewer than the usual number of buyers competing for the houses available for sale, then we would expect the price of houses to decrease. As a result, fewer potential sellers will place their houses on the market. What does this imply about the parameter γ?

22. Consider the effect on house prices if $s > 0$ and the subsequent effect on buyers and sellers. Then determine the sign of the parameter β.

23. Determine the most reasonable sign for the parameter δ.

24. Consider the linear system

$$\frac{d\mathbf{Y}}{dt} = \begin{pmatrix} 2 & 0 \\ 1 & 1 \end{pmatrix} \mathbf{Y}.$$

(a) Show that the two functions

$$\mathbf{Y}_1(t) = \begin{pmatrix} 0 \\ e^t \end{pmatrix} \quad \text{and} \quad \mathbf{Y}_2(t) = \begin{pmatrix} e^{2t} \\ e^{2t} \end{pmatrix}$$

are solutions to the differential equation.

(b) Solve the initial-value problem

$$\frac{d\mathbf{Y}}{dt} = \begin{pmatrix} 2 & 0 \\ 1 & 1 \end{pmatrix} \mathbf{Y}, \quad \mathbf{Y}(0) = \begin{pmatrix} -2 \\ -1 \end{pmatrix}.$$

25. Consider the linear system

$$\frac{d\mathbf{Y}}{dt} = \begin{pmatrix} 1 & -1 \\ 1 & 3 \end{pmatrix} \mathbf{Y}.$$

(a) Show that the function

$$\mathbf{Y}(t) = \begin{pmatrix} te^{2t} \\ -(t+1)e^{2t} \end{pmatrix}$$

is a solution to the differential equation.

(b) Solve the initial-value problem

$$\frac{d\mathbf{Y}}{dt} = \begin{pmatrix} 1 & -1 \\ 1 & 3 \end{pmatrix} \mathbf{Y}, \quad \mathbf{Y}(0) = \begin{pmatrix} 0 \\ 2 \end{pmatrix}.$$

In Exercises 26–29, a coefficient matrix for the linear system

$$\frac{d\mathbf{Y}}{dt} = \mathbf{A}\mathbf{Y}, \quad \text{where } \mathbf{Y}(t) = \begin{pmatrix} x(t) \\ y(t) \end{pmatrix}$$

is specified. Also two functions and an initial value are given. For each system:

(a) Check that the two functions are solutions of the system; if they are not solutions, then stop.

(b) Check that the two solutions are linearly independent; if they are not linearly independent, then stop.

(c) Find the solution to the linear system with the given initial value.

26.
$$\mathbf{A} = \begin{pmatrix} -2 & -1 \\ 2 & -5 \end{pmatrix}$$

Functions: $\mathbf{Y}_1(t) = (e^{-3t}, e^{-3t})$
$\mathbf{Y}_2(t) = (e^{-4t}, 2e^{-4t})$
Initial value: $\mathbf{Y}(0) = (2, 3)$

27.
$$\mathbf{A} = \begin{pmatrix} -2 & -1 \\ 2 & -5 \end{pmatrix}$$

Functions: $\mathbf{Y}_1(t) = (e^{-3t} - 2e^{-4t}, e^{-3t} - 4e^{-4t})$
$\mathbf{Y}_2(t) = (2e^{-3t} + e^{-4t}, 2e^{-3t} + 2e^{-4t})$
Initial value: $\mathbf{Y}(0) = (2, 3)$

28.
$$\mathbf{A} = \begin{pmatrix} -2 & -3 \\ 3 & -2 \end{pmatrix}$$

Functions: $\mathbf{Y}_1(t) = e^{-2t}(\cos 3t, \sin 3t)$
$\mathbf{Y}_2(t) = e^{-2t}(-\sin 3t, \cos 3t)$
Initial value: $\mathbf{Y}(0) = (2, 3)$

29.
$$\mathbf{A} = \begin{pmatrix} 2 & 3 \\ 1 & 0 \end{pmatrix}$$

Functions: $\mathbf{Y}_1(t) = (-e^{-t} + 12e^{3t}, e^{-t} + 4e^{3t})$
$\mathbf{Y}_2(t) = (-e^{-t}, 2e^{-t})$
Initial value: $\mathbf{Y}(0) = (2, 3)$

30. **(a)** Verify property 1, $\mathbf{A}k\mathbf{Y} = k\mathbf{A}\mathbf{Y}$, of matrix multiplication, where $\mathbf{Y}$ is a (two-dimensional) vector, $\mathbf{A}$ is a matrix, and k is a constant.

(b) Using scalar notation, write out and verify the Linearity Principle. (Aren't matrices nice?)

31. Show that the vectors (x_1, y_1) and (x_2, y_2) are linearly dependent—that is, not linearly independent—if any of the following conditions are satisfied.

(a) If $(x_1, y_1) = (0, 0)$.

(b) If $(x_1, y_1) = \lambda(x_2, y_2)$ for some constant λ.

(c) If $x_1 y_2 - x_2 y_1 = 0$. *Hint*: Assume x_1 is not zero; then $y_2 = x_2 y_1/x_1$. But $x_2 = x_2 x_1/x_1$, and we can use part b. The other cases are similar. Note that the quantity $x_1 y_2 - x_2 y_1$ is the determinant of the matrix

$$\begin{pmatrix} x_1 & y_1 \\ x_2 & y_2 \end{pmatrix}.$$

32. Given the vectors (x_1, y_1) and (x_2, y_2), show that they are linearly independent if the quantity $x_1 y_2 - x_2 y_1$ is nonzero (see part (c) of Exercise 31). [*Hint*: Suppose $x_2 \neq 0$. If (x_1, y_1) and (x_2, y_2) are on the same line through $(0, 0)$, then $(x_1, y_1) = \lambda(x_2, y_2)$ for some λ. But then $\lambda = x_1/x_2$ and $\lambda = y_1/y_2$. What does this say about x_1/x_2 and y_1/y_2? What if $x_2 = 0$?]

33. Suppose that $\mathbf{Y}_1(t) = (-e^{-t}, e^{-t})$ is a solution to some linear system $d\mathbf{Y}/dt = \mathbf{A}\mathbf{Y}$. For which of the following initial conditions can you give the explicit solution of the linear system?

(a) $\mathbf{Y}(0) = (-2, 2)$ **(b)** $\mathbf{Y}(0) = (3, 4)$ **(c)** $\mathbf{Y}(0) = (0, 0)$ **(d)** $\mathbf{Y}(0) = (3, -3)$

34. The Linearity Principle is a fundamental property of systems of the form $d\mathbf{Y}/dt = \mathbf{A}\mathbf{Y}$. However, you should not assume that it is true for systems that are not of this form, no matter how simple. For example, consider the system

$$\frac{dx}{dt} = 1$$
$$\frac{dy}{dt} = x.$$

The following computations show that the Linearity Principle does not hold for this system.

(a) Show that $\mathbf{Y}(t) = (t, t^2/2)$ is a solution to this system.

(b) Show that $2\mathbf{Y}(t)$ is *not* a solution.

An Extended Linearity Principle that applies to systems such as this one is discussed in Chapter 4.

35. Given solutions $\mathbf{Y}_1(t) = (x_1(t), y_1(t))$ and $\mathbf{Y}_2(t) = (x_2(t), y_2(t))$ to the system

$$\frac{d\mathbf{Y}}{dt} = \mathbf{A}\mathbf{Y}, \quad \text{where } \mathbf{A} = \begin{pmatrix} a & b \\ c & d \end{pmatrix},$$

we define the **Wronskian** of $Y_1(t)$ and $Y_2(t)$ to be the (scalar) function

$$W(t) = x_1(t)y_2(t) - x_2(t)y_1(t).$$

(a) Compute dW/dt.

(b) Use the fact that $Y_1(t)$ and $Y_2(t)$ are solutions of the linear system to show that

$$\frac{dW}{dt} = (a+d)W(t).$$

(c) Find the general solution of the differential equation $dW/dt = (a+d)W(t)$.

(d) Suppose that $Y_1(t)$ and $Y_2(t)$ are solutions to the system $dY/dt = AY$. Verify that if $Y_1(0)$ and $Y_2(0)$ are linearly independent, then $Y_1(t)$ and $Y_2(t)$ are also linearly independent for every t.

3.2 STRAIGHT-LINE SOLUTIONS

In Section 3.1 we discussed solutions of linear systems without worrying about how we came up with them (the rabbit-out-of-the-hat method). Often we used the time-honored method known as "guess and test." That is, we made a guess, then substituted the guess back into the equation and checked to see if it satisfied the system. However, the guess-and-test method is unsatisfying because it does not give us any understanding of where the formulas came from in the first place. In this section we use the geometry of the vector field to find special solutions of linear systems.

Geometry of Straight-Line Solutions

We begin by reconsidering an example from the previous section. The direction field for the linear system

$$\frac{dY}{dt} = AY, \quad \text{where } A = \begin{pmatrix} 2 & 3 \\ 0 & -4 \end{pmatrix},$$

is shown in Figure 3.7. Looking at the direction field, we see that there are two special lines through the origin. The first is the x-axis on which the vectors in the direction field

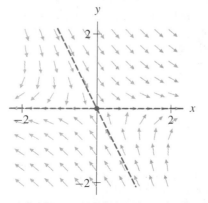

Figure 3.7

The direction field for the linear system

$$\frac{dY}{dt} = \begin{pmatrix} 2 & 3 \\ 0 & -4 \end{pmatrix} Y.$$

There are two special lines through the origin. On the x-axis, the vectors in the direction field all point directly away from the origin. On the distinguished line that runs from the second quadrant to the fourth quadrant, all vectors of the direction field point directly toward the origin.

all point directly away from the origin. The other special line runs from the second quadrant to the fourth quadrant. Along this line the vectors of the direction field all point directly toward the origin.

Because solution curves for the system are always tangent to the direction field, a solution that has its initial condition on the positive x-axis moves to the right, directly away from the origin. A solution with an initial condition on the negative x-axis moves to the left, directly away from the origin. Similarly, a solution with an initial condition in the second quadrant on the other special line moves directly toward the origin, and a solution with an initial condition in the fourth quadrant on this line moves directly toward the origin. Thus careful examination of the direction field suggests that there are solutions to this system that lie on straight lines through the origin in the phase plane.

In Section 3.1 we saw that

$$\mathbf{Y}_1(t) = \begin{pmatrix} e^{2t} \\ 0 \end{pmatrix} \quad \text{and} \quad \mathbf{Y}_2(t) = \begin{pmatrix} -e^{-4t} \\ 2e^{-4t} \end{pmatrix}$$

are two linearly independent solutions for the system $d\mathbf{Y}/dt = \mathbf{AY}$. Now let's consider the geometry of these solutions in the phase plane.

To plot the solution curve for $\mathbf{Y}_1(t)$, note that the x-coordinate of $\mathbf{Y}_1(t)$ is e^{2t} and the y-coordinate of $\mathbf{Y}_1(t)$ is always 0. Thus the solution curve lies on the positive x-axis. Moreover, $\mathbf{Y}_1(t) \to \infty$ as $t \to \infty$, and $\mathbf{Y}_1(t)$ tends to the origin as $t \to -\infty$. So $\mathbf{Y}_1(t)$ is a solution that tends directly away from the origin along the x-axis.

For $\mathbf{Y}_2(t)$, it is convenient to rewrite this solution in the form

$$\mathbf{Y}_2(t) = e^{-4t} \begin{pmatrix} -1 \\ 2 \end{pmatrix}.$$

This representation tells us that, as t varies, $\mathbf{Y}_2(t)$ is always a (positive) scalar multiple of the vector $(-1, 2)$. Since positive scalar multiples of a fixed vector always lie on the same ray from the origin, we see that $\mathbf{Y}_2(t)$ parameterizes the ray from $(0, 0)$ with slope -2 in the fourth quadrant (see Figure 3.7). As $t \to \infty$, $e^{-4t} \to 0$, so this solution tends toward the origin.

We see that the formulas for $\mathbf{Y}_1(t)$ and $\mathbf{Y}_2(t)$ confirm what we guessed by looking at the direction field. There are solutions of this system that lie on two distinguished straight lines in the phase plane.

Straight-line solutions are the simplest solutions (next to equilibrium points) for systems of differential equations. As these solutions move in the xy-plane along straight lines, it is important to remember that the speed at which they move depends on their position on the line. In this example, solutions go to $(0, 0)$ or escape to ∞ at an exponential rate, as can be seen in the $x(t)$- and $y(t)$-graphs for the solutions (see Figures 3.8 and 3.9).

From the geometry to the algebra of straight-line solutions

Assuming that the system has straight-line solutions (sadly, not all linear systems do), we turn our attention to finding formulas for them. The basic geometric observation is that, along a straight-line solution through the origin, the vector field must point either directly toward or directly away from $(0, 0)$ (see Figure 3.7). That is, if $\mathbf{V} = (x, y)$ is

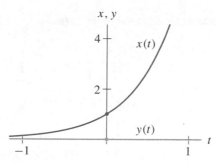

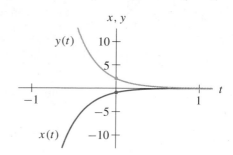

Figure 3.8
The $x(t)$- and $y(t)$-graphs of the straight-line solution

$$\mathbf{Y}_1(t) = \begin{pmatrix} e^{2t} \\ 0 \end{pmatrix}.$$

Figure 3.9
The $x(t)$- and $y(t)$-graphs of the straight-line solution

$$\mathbf{Y}_2(t) = \begin{pmatrix} -e^{-4t} \\ 2e^{-4t} \end{pmatrix}.$$

on a straight-line solution, then the vector field at (x, y) must point either in the same direction or in exactly the opposite direction as the vector from $(0,0)$ to (x, y).

We now turn this observation into an equation that we can solve to find straight-line solutions. For a linear system of the form $d\mathbf{Y}/dt = \mathbf{AY}$, the vector field at $\mathbf{V} = (x, y)$ is the product $\mathbf{AV}$ of $\mathbf{A}$ and $\mathbf{V}$, which in this example is

$$\begin{pmatrix} 2 & 3 \\ 0 & -4 \end{pmatrix} \begin{pmatrix} x \\ y \end{pmatrix}.$$

Hence we seek vectors $\mathbf{V} = (x, y)$ such that $\mathbf{AV}$ points in the same or in the opposite direction as the vector from $(0, 0)$ to (x, y) or, equivalently, for which there is some number λ such that

$$\mathbf{A} \begin{pmatrix} x \\ y \end{pmatrix} = \lambda \begin{pmatrix} x \\ y \end{pmatrix}.$$

If $\lambda > 0$, then the vector field points in the same direction as (x, y) — away from $(0, 0)$. If $\lambda < 0$, the vector field points in the opposite direction — toward $(0, 0)$.

Using vector notation, this equation can be written more economically as

$$\mathbf{AV} = \lambda \mathbf{V},$$

and it is important to remember that *this equation is the key equation for finding straight-line solutions of the linear system $d\mathbf{Y}/dt = \mathbf{AY}$.*

In our example we seek vectors $\mathbf{V} = (x, y)$ such that $\mathbf{AV} = \lambda \mathbf{V}$, which in coordinates is

$$\begin{pmatrix} 2 & 3 \\ 0 & -4 \end{pmatrix} \begin{pmatrix} x \\ y \end{pmatrix} = \lambda \begin{pmatrix} x \\ y \end{pmatrix}.$$

Multiplying we have

$$\begin{pmatrix} 2x + 3y \\ -4y \end{pmatrix} = \lambda \begin{pmatrix} x \\ y \end{pmatrix},$$

and we can rewrite this equation in the form

$$\begin{pmatrix} 2x + 3y \\ -4y \end{pmatrix} - \lambda \begin{pmatrix} x \\ y \end{pmatrix} = \begin{pmatrix} 0 \\ 0 \end{pmatrix},$$

which is equivalent to the system of simultaneous equations

$$\begin{cases} (2 - \lambda)x + 3y = 0 \\ (-4 - \lambda)y = 0. \end{cases}$$

There is one obvious solution to this system of equations, namely the trivial solution $(x, y) = (0, 0)$. But we already know that the origin is an equilibrium solution of this system, so this solution definitely does not give us a straight-line solution. What we need is a nonzero solution of this system of equations (one where at least one of x or y is nonzero).

To find a nonzero solution, it is important to notice that the simultaneous equations really have three unknowns, x, y, and λ, and in fact we need to determine λ before we can solve for x and y. If we write the simultaneous equations in matrix form, we have

$$\begin{pmatrix} 2 - \lambda & 3 \\ 0 & -4 - \lambda \end{pmatrix} \begin{pmatrix} x \\ y \end{pmatrix} = \begin{pmatrix} 0 \\ 0 \end{pmatrix},$$

and now we recall that we can use the determinant to see if such a system of equations has nontrivial solutions (see Section 3.1, page 220). These equations have nontrivial solutions if and only if

$$\det \begin{pmatrix} 2 - \lambda & 3 \\ 0 & -4 - \lambda \end{pmatrix} = 0.$$

Therefore by computing this determinant, we find that this system has nontrivial solutions if and only if

$$(2 - \lambda)(-4 - \lambda) - (3)(0) = 0.$$

This calculation tells us that we will have nontrivial solutions to our equations only if $\lambda = 2$ or if $\lambda = -4$. All other values of λ do not yield straight-line solutions. (Incidentally, recall that our two straight-line solutions $\mathbf{Y}_1(t)$ and $\mathbf{Y}_2(t)$ involve exponentials of the form e^{2t} and e^{-4t}. In a moment we will see that the appearance of $\lambda = 2$ and $\lambda = -4$ in the exponents is no accident.)

If $\lambda = -4$, then the simultaneous system of equations becomes simply

$$\begin{cases} 6x + 3y = 0 \\ 0 = 0. \end{cases}$$

The second equation always holds, so we need only choose x and y satisfying

$$6x + 3y = 0,$$

which simplifies to

$$y = -2x.$$

There is an entire line of vectors (x, y) that satisfy these equations, and one possible choice is $(x, y) = (-1, 2)$. Note that $(-1, 2)$ is exactly the initial condition for the straight-line solution

$$\mathbf{Y}_2(t) = e^{-4t} \begin{pmatrix} -1 \\ 2 \end{pmatrix}.$$

If $\lambda = 2$, the simultaneous equations become

$$\begin{cases} 3y = 0 \\ -2y = 0, \end{cases}$$

and both equations are satisfied if $y = 0$. Thus any vector of the form $(x, 0)$ with $x \neq 0$ gives a nontrivial solution. That is, anywhere along the x-axis, the vector field points directly away from $(0, 0)$ since $\lambda = 2 > 0$. One vector on this line is $(1, 0)$, which is the initial condition for the straight-line solution

$$\mathbf{Y}_1(t) = e^{2t} \begin{pmatrix} 1 \\ 0 \end{pmatrix}.$$

Eigenvalues and Eigenvectors

We return to this example in a moment, but first we generalize these computations so that we can apply them to any linear system. Consider the general linear system

$$\frac{d\mathbf{Y}}{dt} = \mathbf{AY}.$$

To find straight-line solutions through the origin, we must find nonzero vectors $\mathbf{V} = (x, y)$ such that the vector field at $\mathbf{V}$ points in the same direction as or directly opposite to $\mathbf{V} = (x, y)$. So we seek nonzero vectors $\mathbf{V} = (x, y)$ that satisfy

$$\mathbf{AV} = \lambda \mathbf{V}$$

for some scalar λ. This equation leads to the following definition.

DEFINITION   Given a matrix $\mathbf{A}$, a number λ is called an **eigenvalue** of $\mathbf{A}$ if there is a nonzero vector $\mathbf{V} = (x, y)$ for which

$$\mathbf{AV} = \mathbf{A} \begin{pmatrix} x \\ y \end{pmatrix} = \lambda \begin{pmatrix} x \\ y \end{pmatrix} = \lambda \mathbf{V}.$$

The vector $\mathbf{V}$ is called an **eigenvector** corresponding to the eigenvalue λ.   ∎

The word *eigen* is German for "own" or "self." An eigenvector is a vector where the vector field points in the same or opposite direction as the vector itself.

For example, consider the matrix

$$\mathbf{A} = \begin{pmatrix} 4 & 3 \\ -1 & 0 \end{pmatrix}.$$

The vector $(6, -2)$ is an eigenvector with the eigenvalue 3 because

$$\mathbf{A}\begin{pmatrix} 6 \\ -2 \end{pmatrix} = \begin{pmatrix} 4 & 3 \\ -1 & 0 \end{pmatrix}\begin{pmatrix} 6 \\ -2 \end{pmatrix} = \begin{pmatrix} 18 \\ -6 \end{pmatrix} = 3\begin{pmatrix} 6 \\ -2 \end{pmatrix}.$$

Also, the vector $(-1, 1)$ is an eigenvector with eigenvalue 1 because

$$\mathbf{A}\begin{pmatrix} -1 \\ 1 \end{pmatrix} = \begin{pmatrix} 4 & 3 \\ -1 & 0 \end{pmatrix}\begin{pmatrix} -1 \\ 1 \end{pmatrix} = \begin{pmatrix} -1 \\ 1 \end{pmatrix} = 1\begin{pmatrix} -1 \\ 1 \end{pmatrix}.$$

It is important to remember that being an eigenvector is a special property. For a typical matrix, most vectors are not eigenvectors. For example, $(2, 3)$ is not an eigenvector for **A** because

$$\mathbf{A}\begin{pmatrix} 2 \\ 3 \end{pmatrix} = \begin{pmatrix} 4 & 3 \\ -1 & 0 \end{pmatrix}\begin{pmatrix} 2 \\ 3 \end{pmatrix} = \begin{pmatrix} 17 \\ -2 \end{pmatrix},$$

and $(17, -2)$ is not a multiple of $(2, 3)$.

Lines of eigenvectors

Given a matrix **A**, if **V** is an eigenvector for eigenvalue λ, then any scalar multiple $k\mathbf{V}$ is also an eigenvector for λ. To verify this, we compute

$$\mathbf{A}(k\mathbf{V}) = k\mathbf{A}\mathbf{V} = k(\lambda\mathbf{V}) = \lambda(k\mathbf{V}),$$

where the first equality is a property of matrix multiplication and the second equality uses the fact that **V** is an eigenvector. Hence given an eigenvector **V** for the eigenvalue λ, the entire line of vectors through **V** and the origin are also eigenvectors for λ.

Computation of Eigenvalues

To find straight-line solutions of linear systems, we must find the eigenvalues and eigenvectors of the corresponding coefficient matrix. That is, we need to find the vectors $\mathbf{V} = (x, y)$ such that

$$\mathbf{A}\mathbf{V} = \mathbf{A}\begin{pmatrix} x \\ y \end{pmatrix} = \lambda\begin{pmatrix} x \\ y \end{pmatrix} = \lambda\mathbf{V}.$$

If

$$\mathbf{A} = \begin{pmatrix} a & b \\ c & d \end{pmatrix},$$

then we have

$$\begin{pmatrix} a & b \\ c & d \end{pmatrix}\begin{pmatrix} x \\ y \end{pmatrix} = \lambda\begin{pmatrix} x \\ y \end{pmatrix},$$

which is written in components as

$$\begin{cases} ax + by = \lambda x \\ cx + dy = \lambda y. \end{cases}$$

Thus we want nonzero solutions (x, y) to

$$\begin{cases} (a - \lambda)x + by = 0 \\ cx + (d - \lambda)y = 0. \end{cases}$$

From the determinant condition that we derived in Section 3.1 (page 220), we know that this system has nontrivial solutions if and only if

$$\det \begin{pmatrix} a - \lambda & b \\ c & d - \lambda \end{pmatrix} = 0.$$

We encounter this matrix each time we compute eigenvalues and eigenvectors, so we introduce some notation for it. The **identity matrix** is the 2×2 matrix

$$\mathbf{I} = \begin{pmatrix} 1 & 0 \\ 0 & 1 \end{pmatrix}.$$

This matrix is called the identity matrix because $\mathbf{IV} = \mathbf{V}$ for any vector $\mathbf{V}$. Also, $\lambda \mathbf{I}$ represents the matrix

$$\lambda \mathbf{I} = \begin{pmatrix} \lambda & 0 \\ 0 & \lambda \end{pmatrix}.$$

Computing the difference between the matrices $\mathbf{A}$ and $\lambda \mathbf{I}$ by subtracting corresponding entries yields

$$\mathbf{A} - \lambda \mathbf{I} = \begin{pmatrix} a - \lambda & b \\ c & d - \lambda \end{pmatrix}.$$

Thus our determinant condition for a nontrivial solution of the equation $\mathbf{AV} = \lambda \mathbf{V}$ may be written in the compact form

$$\det(\mathbf{A} - \lambda \mathbf{I}) = 0.$$

It is important to remember that the matrix $\mathbf{A} - \lambda \mathbf{I}$ is the matrix $\mathbf{A}$ with λ's subtracted from the upper-left and lower-right entries.

The Characteristic Polynomial

To find the eigenvalues of the matrix $\mathbf{A}$, we must find the values of λ for which

$$\det(\mathbf{A} - \lambda \mathbf{I}) = 0.$$

If we write this equation in terms of the entries of $\mathbf{A}$, we find

$$\det(\mathbf{A} - \lambda \mathbf{I}) = \det \begin{pmatrix} a - \lambda & b \\ c & d - \lambda \end{pmatrix} = (a - \lambda)(d - \lambda) - bc = 0,$$

which expands to the quadratic polynomial

$$\lambda^2 - (a + d)\lambda + (ad - bc) = 0.$$

This polynomial is called the **characteristic polynomial** of the system. Its roots are the eigenvalues of the matrix **A**.

A quadratic polynomial always has two roots, but these roots need not be real numbers, nor must they be distinct. If the roots of the characteristic polynomial are not real, we say that the matrix **A** has **complex eigenvalues**. We will study the behavior of solutions to systems with complex eigenvalues in Section 3.4, and the case of two equal roots is considered in Section 3.5.

Consider the matrix

$$\mathbf{A} = \begin{pmatrix} 2 & 3 \\ 0 & -4 \end{pmatrix}$$

that we discussed earlier in this section. This matrix has the characteristic polynomial

$$\det(\mathbf{A} - \lambda \mathbf{I}) = (2 - \lambda)(-4 - \lambda) - (3)(0) = \lambda^2 + 2\lambda - 8,$$

which has roots $\lambda_1 = 2$ and $\lambda_2 = -4$. As we saw earlier, these numbers are the eigenvalues of this matrix. (This example is somewhat unusual in that it is not necessary to expand the expression $\det(\mathbf{A} - \lambda \mathbf{I}) = (2 - \lambda)(-4 - \lambda) - (3)(0)$ into $\lambda^2 + 2\lambda - 8$ to determine the eigenvalues of **A**.)

Computation of Eigenvectors

The next step in the process of finding straight-line solutions of a system of differential equations is to find the eigenvectors associated to the eigenvalues. Suppose we are given a matrix

$$\mathbf{A} = \begin{pmatrix} a & b \\ c & d \end{pmatrix}$$

and we know that λ is an eigenvalue. To find a corresponding eigenvector, we must solve the equation $\mathbf{AV} = \lambda \mathbf{V}$ for the vector **V**. If we write

$$\mathbf{V} = \begin{pmatrix} x \\ y \end{pmatrix},$$

then $\mathbf{AV} = \lambda \mathbf{V}$ becomes a simultaneous system of linear equations in two unknowns, x and y. In fact the equations are

$$\begin{cases} ax + by = \lambda x \\ cx + dy = \lambda y. \end{cases}$$

Since λ is an eigenvalue, we know that there is at least an entire line of eigenvectors (x, y) that satisfy this system of equations. This infinite number of eigenvectors means that the equations are redundant. That is, either the two equations are equivalent, or one of the equations is always satisfied.

For example, suppose we are given the matrix

$$\mathbf{B} = \begin{pmatrix} 2 & 2 \\ 1 & 3 \end{pmatrix}.$$

We find the eigenvalues of $\mathbf{B}$ by finding the roots of the characteristic polynomial

$$\det(\mathbf{B} - \lambda\mathbf{I}) = (2 - \lambda)(3 - \lambda) - (2)(1) = 0,$$

which yields the quadratic equation

$$\lambda^2 - 5\lambda + 4 = 0.$$

The roots of this quadratic polynomial are $\lambda_1 = 4$ and $\lambda_2 = 1$, so 1 and 4 are the eigenvalues of $\mathbf{B}$.

To find an eigenvector $\mathbf{V}_1$ for $\lambda_1 = 4$, we must solve

$$\mathbf{B}\begin{pmatrix} x_1 \\ y_1 \end{pmatrix} = \begin{pmatrix} 2 & 2 \\ 1 & 3 \end{pmatrix}\begin{pmatrix} x_1 \\ y_1 \end{pmatrix} = 4\begin{pmatrix} x_1 \\ y_1 \end{pmatrix}.$$

Rewritten in terms of components, this equation is

$$\begin{cases} 2x_1 + 2y_1 = 4x_1 \\ x_1 + 3y_1 = 4y_1, \end{cases}$$

or, equivalently,

$$\begin{cases} -2x_1 + 2y_1 = 0 \\ x_1 - y_1 = 0. \end{cases}$$

Note that these equations are redundant (multiply both sides of the second by -2 to get the first). So any vector (x_1, y_1) that satisfies the second equation

$$x_1 - y_1 = 0$$

is an eigenvector. This equation specifies the line $y_1 = x_1$ in the plane. Any nonzero vector on this line is an eigenvector of $\mathbf{B}$ corresponding to the eigenvalue $\lambda_1 = 4$. For example, the vectors $(1, 1)$ and $(-\pi, -\pi)$ are two of the infinitely many eigenvectors for $\mathbf{B}$ corresponding to the eigenvalue $\lambda_1 = 4$.

For $\lambda_2 = 1$ we must solve

$$\mathbf{B}\begin{pmatrix} x_2 \\ y_2 \end{pmatrix} = \begin{pmatrix} 2 & 2 \\ 1 & 3 \end{pmatrix}\begin{pmatrix} x_2 \\ y_2 \end{pmatrix} = 1\begin{pmatrix} x_2 \\ y_2 \end{pmatrix}.$$

In terms of coordinates, this vector equation is the same as the system

$$\begin{cases} 2x_2 + 2y_2 = x_2 \\ x_2 + 3y_2 = y_2 \end{cases}$$

or, equivalently,

$$\begin{cases} x_2 + 2y_2 = 0 \\ x_2 + 2y_2 = 0. \end{cases}$$

Again these equations are redundant. So the eigenvectors corresponding to eigenvalue $\lambda_2 = 1$ are the nonzero vectors (x_2, y_2) that lie on the line $y_2 = -x_2/2$.

Straight-Line Solutions

After all of the algebra of the last few pages, it is time to return to the study of differential equations. To summarize what we have accomplished so far, suppose we are given a linear system of differential equations

$$\frac{d\mathbf{Y}}{dt} = \mathbf{AY}.$$

To find straight-line solutions, we first find the eigenvalues of $\mathbf{A}$ and then their associated eigenvectors. Once we have this information, we have determined the straight-line solutions.

To do this, suppose that λ is an eigenvalue with associated eigenvector $\mathbf{V} = (x, y)$. Then consider the function

$$\mathbf{Y}(t) = e^{\lambda t}\mathbf{V} = \begin{pmatrix} e^{\lambda t}x \\ e^{\lambda t}y \end{pmatrix}.$$

For each t, $\mathbf{Y}(t)$ is a scalar multiple of our eigenvector (x, y), so the curve given by $\mathbf{Y}(t)$ lies on the ray from the origin through (x, y). Moreover, $\mathbf{Y}(t)$ is a solution of the differential equation. We can check this assertion by substituting $\mathbf{Y}(t)$ in the differential equation. We compute

$$\frac{d\mathbf{Y}}{dt} = \frac{d}{dt}\begin{pmatrix} e^{\lambda t}x \\ e^{\lambda t}y \end{pmatrix} = \begin{pmatrix} \lambda e^{\lambda t}x \\ \lambda e^{\lambda t}y \end{pmatrix} = \lambda\mathbf{Y}(t).$$

On the other hand, we have

$$\mathbf{AY}(t) = \mathbf{A}e^{\lambda t}\mathbf{V} = e^{\lambda t}\mathbf{AV} = e^{\lambda t}\lambda\mathbf{V} = \lambda e^{\lambda t}\mathbf{V} = \lambda\mathbf{Y}(t)$$

since $\mathbf{V}$ is an eigenvector of $\mathbf{A}$. Comparing the results of these two computations, we see that

$$\frac{d\mathbf{Y}}{dt} = \mathbf{AY},$$

so $\mathbf{Y}(t)$ is indeed a solution.

This is an important observation: We obtain formulas for straight-line solutions using just the eigenvalues and eigenvectors of the matrix $\mathbf{A}$.

Sometimes we can do even better. Suppose we find two real, distinct eigenvalues λ_1 and λ_2 for the system with eigenvectors $\mathbf{V}_1$ and $\mathbf{V}_2$ respectively. Since $\mathbf{V}_1$ and $\mathbf{V}_2$ are eigenvectors for different eigenvalues, they must be linearly independent. That is, any scalar multiple of $\mathbf{V}_1$ is an eigenvector associated to λ_1. Consequently, $\mathbf{V}_2$ does not lie on the line through the origin determined by $\mathbf{V}_1$, and $\mathbf{V}_1$ and $\mathbf{V}_2$ are linearly independent. As a result, the two solutions

$$\mathbf{Y}_1(t) = e^{\lambda_1 t}\mathbf{V}_1 \quad \text{and} \quad \mathbf{Y}_2(t) = e^{\lambda_2 t}\mathbf{V}_2$$

are linearly independent. Therefore, using the Linearity Principle, the general solution of the system is $k_1\mathbf{Y}_1(t) + k_2\mathbf{Y}_2(t) = k_1 e^{\lambda_1 t}\mathbf{V}_1 + k_2 e^{\lambda_2 t}\mathbf{V}_2$.

THEOREM   Suppose the matrix $\mathbf{A}$ has a real eigenvalue λ with associated eigenvector $\mathbf{V}$. Then the linear system $d\mathbf{Y}/dt = \mathbf{A}\mathbf{Y}$ has the straight-line solution

$$\mathbf{Y}(t) = e^{\lambda t}\mathbf{V}.$$

Moreover, if λ_1 and λ_2 are distinct, real eigenvalues with eigenvectors $\mathbf{V}_1$ and $\mathbf{V}_2$ respectively, then the solutions $\mathbf{Y}_1(t) = e^{\lambda_1 t}\mathbf{V}_1$ and $\mathbf{Y}_2(t) = e^{\lambda_2 t}\mathbf{V}_2$ are linearly independent and

$$\mathbf{Y}(t) = k_1 e^{\lambda_1 t}\mathbf{V}_1 + k_2 e^{\lambda_2 t}\mathbf{V}_2$$

is the general solution of the system.   ∎

This is a powerful theorem. It lets us find solutions of linear systems of differential equations using only algebra. All we need to do is to find an eigenvalue and an associated eigenvector. There are no tedious or impossible integrations to perform. (One caveat here is that the eigenvalue must be real; we tackle the case of complex eigenvalues in Section 3.4.)

The theorem also explicitly provides the general solution of certain linear systems, namely those that have two distinct, real eigenvalues. We will treat the possibility that the eigenvalues of $\mathbf{A}$ are real but not distinct in Section 3.5.

Putting Everything Together

Now let's combine the geometry of the direction field with the algebra of this section to produce the general solution of a linear system of differential equations. Consider the linear system

$$\frac{d\mathbf{Y}}{dt} = \mathbf{B}\mathbf{Y} = \begin{pmatrix} 2 & 2 \\ 1 & 3 \end{pmatrix} \mathbf{Y}.$$

The direction field for this system is depicted in Figure 3.10. There appear to be two distinguished lines of eigenvectors, one cutting diagonally through the first and third quadrants, and another through the second and fourth quadrants. The associated eigenvalues are positive since the direction field points away from the origin.

To find formulas for corresponding straight-line solutions, we use the eigenvalues and eigenvectors of $\mathbf{B}$, which we computed earlier in the section. The eigenvalues of $\mathbf{B}$

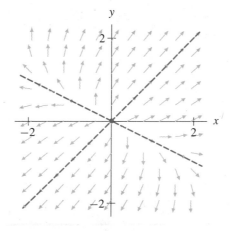

Figure 3.10

The direction field for the system

$$\frac{d\mathbf{Y}}{dt} = \mathbf{B}\mathbf{Y} = \begin{pmatrix} 2 & 2 \\ 1 & 3 \end{pmatrix} \mathbf{Y}.$$

Note the two distinguished lines of eigenvectors. The one in the first quadrant corresponds to the solution $\mathbf{Y}_1(t) = e^{4t}(1, 1)$ and the one in the second quadrant corresponds to the solution $\mathbf{Y}_2(t) = e^t(-2, 1)$.

are $\lambda_1 = 4$ and $\lambda_2 = 1$. The eigenvectors $V_1 = (x_1, y_1)$ associated to λ_1 satisfy the equation $y_1 = x_1$, and the eigenvectors $V_2 = (x_2, y_2)$ associated to λ_2 satisfy the equation $y_2 = -x_1/2$. In particular, we can use the vectors $V_1 = (1, 1)$ and $V_2 = (-2, 1)$ to produce two linearly independent straight-line solutions. The general solution is

$$\mathbf{Y}(t) = k_1 e^{4t} \begin{pmatrix} 1 \\ 1 \end{pmatrix} + k_2 e^t \begin{pmatrix} -2 \\ 1 \end{pmatrix}.$$

Note that there is nothing significant about our choice of $V_1 = (1, 1)$ and $V_2 = (-2, 1)$. For V_1 we can use any eigenvector associated to the eigenvalue $\lambda_1 = 4$, and for V_2 we can use any eigenvector associated to the eigenvalue $\lambda_2 = 1$.

A Harmonic Oscillator

Consider the harmonic oscillator with mass $m = 1$, spring constant $k = 10$, and damping coefficient $b = 7$. The second-order equation that models this oscillator is

$$\frac{d^2 y}{dt^2} + 7\frac{dy}{dt} + 10y = 0,$$

and the corresponding system is

$$\frac{d\mathbf{Y}}{dt} = \mathbf{CY}, \quad \text{where } \mathbf{C} = \begin{pmatrix} 0 & 1 \\ -10 & -7 \end{pmatrix} \quad \text{and} \quad \mathbf{Y} = \begin{pmatrix} y \\ v \end{pmatrix}.$$

The phase portrait is shown in Figure 3.11. Note that there appear to be straight-line solutions for this system.

The characteristic polynomial for the system is

$$(-\lambda)(-7 - \lambda) + 10 = \lambda^2 + 7\lambda + 10,$$

and the eigenvalues are $\lambda_1 = -5$ and $\lambda_2 = -2$. Note that both of these eigenvalues are negative.

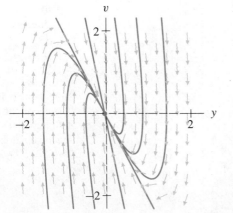

Figure 3.11

Phase portrait for

$$\frac{d\mathbf{Y}}{dt} = \begin{pmatrix} 0 & 1 \\ -10 & -7 \end{pmatrix} \mathbf{Y}.$$

This linear system is obtained from the harmonic oscillator

$$\frac{d^2 y}{dt^2} + 7\frac{dy}{dt} + 10y = 0,$$

where $\mathbf{Y} = (y, v)$ and $v = dy/dt$.

We compute the eigenvector for $\lambda_1 = -5$ by solving $\mathbf{CV}_1 = -5\mathbf{V}_1$. If $\mathbf{V}_1 = (y_1, v_1)$, we have

$$
\begin{cases}
v_1 = -5y_1 \\
-10y_1 - 7v_1 = -5v_1.
\end{cases}
$$

If we have done our arithmetic correctly, these two equations are redundant, and the desired eigenvectors must satisfy the equation $v_1 = -5y_1$. (It is a good idea to check the redundancy of these equations. If they are not redundant, then a mistake was made earlier in the computation.) Setting $y_1 = 1$, we obtain the eigenvector $\mathbf{V}_1 = (1, -5)$ corresponding to λ_1. Similarly, we can compute that an eigenvector for $\lambda_2 = -2$ is $\mathbf{V}_2 = (1, -2)$. (Do you notice anything special about these two eigenvectors?)

The general solution for this system is

$$
\mathbf{Y}(t) = k_1 e^{-5t} \begin{pmatrix} 1 \\ -5 \end{pmatrix} + k_2 e^{-2t} \begin{pmatrix} 1 \\ -2 \end{pmatrix}.
$$

Using this formula, we can find the exact position of the oscillator at any time. Moreover, we can also determine qualitative features of the model from these formulas. Each term in the expression for $\mathbf{Y}(t)$ contains an exponential of the form $e^{\lambda t}$ with $\lambda < 0$. Consequently each term tends to 0 as t increases. Note that this is consistent with the directions of the solution curves in the phase portrait (see Figure 3.11), but it is comforting to see everything fit together so nicely. Since $\mathbf{Y}(t) = (y(t), v(t))$, the general solution of the corresponding second-order equation is the first component of $\mathbf{Y}(t)$, that is,

$$
y(t) = k_1 e^{-5t} + k_2 e^{-2t}.
$$

One thing that we learned from the eigenvalues that we did not know from the phase portrait alone is the fact that every solution tends to zero at a rate that is at least comparable to the rate at which e^{-2t} tends to 0.

EXERCISES FOR SECTION 3.2

In Exercises 1–10,

(a) compute the eigenvalues;

(b) for each eigenvalue, compute the associated eigenvectors;

(c) using whatever technology is available, sketch the direction field for the system, and plot the straight-line solutions;

(d) for each eigenvalue, specify a corresponding straight-line solution and plot its $x(t)$- and $y(t)$-graphs; and

(e) if the system has two distinct eigenvalues, compute the general solution.

1. $\dfrac{d\mathbf{Y}}{dt} = \begin{pmatrix} 3 & 2 \\ 0 & -2 \end{pmatrix} \mathbf{Y}$

2. $\dfrac{d\mathbf{Y}}{dt} = \begin{pmatrix} -4 & -2 \\ -1 & -3 \end{pmatrix} \mathbf{Y}$

3.
$$\begin{pmatrix} \dfrac{dx}{dt} \\[2mm] \dfrac{dy}{dt} \end{pmatrix} = \begin{pmatrix} 4 & 2 \\ 1 & 3 \end{pmatrix} \begin{pmatrix} x \\ y \end{pmatrix}$$

4.
$$\begin{pmatrix} \dfrac{dx}{dt} \\[2mm] \dfrac{dy}{dt} \end{pmatrix} = \begin{pmatrix} 2 & 1 \\ -1 & 4 \end{pmatrix} \begin{pmatrix} x \\ y \end{pmatrix}$$

5.
$$\frac{dx}{dt} = -\frac{x}{2}$$
$$\frac{dy}{dt} = x - \frac{y}{2}$$

6.
$$\frac{dx}{dt} = 5x + 4y$$
$$\frac{dy}{dt} = 9x$$

7.
$$\begin{pmatrix} \dfrac{dx}{dt} \\[2mm] \dfrac{dy}{dt} \end{pmatrix} = \begin{pmatrix} 3 & 4 \\ 1 & 0 \end{pmatrix} \begin{pmatrix} x \\ y \end{pmatrix}$$

8.
$$\begin{pmatrix} \dfrac{dx}{dt} \\[2mm] \dfrac{dy}{dt} \end{pmatrix} = \begin{pmatrix} 2 & -1 \\ -1 & 1 \end{pmatrix} \begin{pmatrix} x \\ y \end{pmatrix}$$

9.
$$\frac{dx}{dt} = 2x + y$$
$$\frac{dy}{dt} = x + y$$

10.
$$\frac{dx}{dt} = -x - 2y$$
$$\frac{dy}{dt} = x - 4y$$

11. Solve the initial-value problem

$$\frac{dx}{dt} = -3x$$
$$\frac{dy}{dt} = -x + 2y,$$

where the initial condition $(x(0), y(0))$ is:

(a) $(1, 0)$ (b) $(0, 1)$ (c) $(-2, 1)$

12. Solve the initial-value problem

$$\frac{dx}{dt} = 3x$$
$$\frac{dy}{dt} = x - 2y,$$

where the initial condition $(x(0), y(0))$ is:

(a) $(1, 0)$ (b) $(0, 1)$ (c) $(2, 2)$

13. Solve the initial-value problem

$$\frac{d\mathbf{Y}}{dt} = \begin{pmatrix} -4 & 1 \\ 2 & -3 \end{pmatrix} \mathbf{Y}, \quad \mathbf{Y}(0) = \mathbf{Y}_0,$$

where the initial condition $\mathbf{Y}_0$ is:

(a) $\mathbf{Y}_0 = (1, 0)$ (b) $\mathbf{Y}_0 = (2, 1)$ (c) $\mathbf{Y}_0 = (-1, -2)$

14. Solve the initial-value problem

$$\frac{d\mathbf{Y}}{dt} = \begin{pmatrix} 4 & -2 \\ 1 & 1 \end{pmatrix} \mathbf{Y}, \quad \mathbf{Y}(0) = \mathbf{Y}_0,$$

where the initial condition $\mathbf{Y}_0$ is:

(a) $\mathbf{Y}_0 = (1, 0)$ (b) $\mathbf{Y}_0 = (2, 1)$ (c) $\mathbf{Y}_0 = (-1, -2)$

15. Show that a is the only eigenvalue and that every vector is an eigenvector for the matrix

$$\mathbf{A} = \begin{pmatrix} a & 0 \\ 0 & a \end{pmatrix}.$$

16. A matrix of the form

$$\mathbf{A} = \begin{pmatrix} a & b \\ 0 & d \end{pmatrix}$$

is called **upper triangular**. Suppose that $b \neq 0$ and $a \neq d$. Find the eigenvalues and eigenvectors of $\mathbf{A}$.

17. A matrix of the form

$$\mathbf{B} = \begin{pmatrix} a & b \\ b & d \end{pmatrix}$$

is called **symmetric**. Show that $\mathbf{B}$ has real eigenvalues and that, if $b \neq 0$, then $\mathbf{B}$ has two distinct eigenvalues.

18. Compute the eigenvalues of a matrix of the form

$$\mathbf{C} = \begin{pmatrix} a & b \\ c & 0 \end{pmatrix}.$$

Compare your results to those of Exercise 16.

19. Consider the second-order equation

$$\frac{d^2 y}{dt^2} + p\frac{dy}{dt} + qy = 0,$$

where p and q are positive.

(a) Convert this equation into a first-order, linear system.
(b) Compute the characteristic polynomial of the system.
(c) Find the eigenvalues.
(d) Under what conditions on p and q are the eigenvalues two distinct real numbers?
(e) Verify that the eigenvalues are negative if they are real numbers.

20. For the harmonic oscillator with mass $m = 1$, spring constant $k = 4$, and damping coefficient $b = 5$,

   (a) compute the eigenvalues and associated eigenvectors;

   (b) for each eigenvalue, pick an associated eigenvector $\mathbf{V}$, and determine the solution $\mathbf{Y}(t)$ with $\mathbf{Y}(0) = \mathbf{V}$;

   (c) for each solution derived in part (b), plot its solution curve in the yv-phase plane;

   (d) for each solution derived in part (b), plot its $y(t)$- and $v(t)$-graphs; and

   (e) for each solution derived in part (b), give a brief description of the behavior of the mass-spring system.

In Exercises 21–24, we return to Exercises 13–16 in Section 2.3. (For convenience, the equations are reproduced below.) For each second-order equation,

   (a) convert the equation to a first-order, linear system;

   (b) compute the eigenvalues and eigenvectors of the system;

   (c) for each eigenvalue, pick an associated eigenvector $\mathbf{V}$, and determine the solution $\mathbf{Y}(t)$ to the system; and

   (d) compare the results of your calculations in part (c) with the results that you obtained when you used the guess-and-test method of Section 2.3.

21. $\dfrac{d^2y}{dt^2} + 3\dfrac{dy}{dt} - 10y = 0$ **22.** $\dfrac{d^2y}{dt^2} + 3\dfrac{dy}{dt} + 2y = 0$

23. $\dfrac{d^2y}{dt^2} + 4\dfrac{dy}{dt} + y = 0$ **24.** $\dfrac{d^2y}{dt^2} + 5\dfrac{dy}{dt} + 6y = 0$

25. Verify that the linear system that models the harmonic oscillator with mass $m = 1$, spring constant $k = 4$, and damping coefficient $b = 1$ does not have real eigenvalues. Does this tell you anything about the phase portrait of this system?

3.3 PHASE PLANES FOR LINEAR SYSTEMS WITH REAL EIGENVALUES

In the preceding section we saw that straight-line solutions play a dominant role in finding the general solution of certain linear systems of differential equations. To solve such a system, we first use algebra to compute the eigenvalues and eigenvectors of the coefficient matrix. When we find a real eigenvalue and an associated eigenvector, we can write down the corresponding straight-line solution. Moreover, in the special case where we find two real, distinct eigenvalues, we can write down an explicit formula for the general solution of the system.

The sign of the eigenvalue plays an important role in determining the behavior of the corresponding straight-line solutions. If the eigenvalue is negative, the solution tends to the origin as $t \to \infty$. If the eigenvalue is positive, the solution tends away from the origin as $t \to \infty$. In this section we use the behavior of these straight-line solutions to determine the behavior of all solutions.

Saddles

One common type of linear system features both a positive and negative eigenvalue. For example, consider the linear system

$$\frac{d\mathbf{Y}}{dt} = \mathbf{AY}, \quad \text{where } \mathbf{A} = \begin{pmatrix} -3 & 0 \\ 0 & 2 \end{pmatrix}.$$

This is a particularly simple linear system, since it corresponds to the equations

$$\frac{dx}{dt} = -3x$$
$$\frac{dy}{dt} = 2y.$$

Note that dx/dt depends only on x and dy/dt depends only on y. That is, the system completely decouples. We can solve these two equations independently using methods from Chapter 1. However, in order to understand the geometry more fully, we use the methods of the previous two sections.

As usual, we first compute the eigenvalues of $\mathbf{A}$ by finding the roots of the characteristic polynomial

$$\det(\mathbf{A} - \lambda \mathbf{I}) = \det \begin{pmatrix} -3-\lambda & 0 \\ 0 & 2-\lambda \end{pmatrix} = (-3-\lambda)(2-\lambda) = 0.$$

Thus the eigenvalues of $\mathbf{A}$ are $\lambda_1 = -3$ and $\lambda_2 = 2$.

Next we compute the eigenvectors. For $\lambda_1 = -3$, we must solve the equation $\mathbf{AV} = -3\mathbf{V}$ for $\mathbf{V}$. If $\mathbf{V}_1 = (x_1, y_1)$, then we have

$$\begin{cases} -3x_1 = -3x_1 \\ 2y_1 = -3y_1. \end{cases}$$

So any nonzero vector $\mathbf{V}$ lying along the line $y = 0$ (the x-axis) in the plane is an eigenvector for $\lambda_1 = -3$. We choose $\mathbf{V}_1 = (1, 0)$. Therefore the solution

$$\mathbf{Y}_1(t) = e^{-3t}\mathbf{V}_1$$

is a straight-line solution whose solution curve is the positive x-axis. The solution tends to the origin as t increases.

In similar fashion we can check that any eigenvector corresponding to $\lambda_2 = 2$ lies along the y-axis. We choose $\mathbf{V}_2 = (0, 1)$ and obtain a second solution

$$\mathbf{Y}_2(t) = e^{2t}\mathbf{V}_2.$$

The general solution is therefore

$$\mathbf{Y}(t) = k_1 e^{-3t}\mathbf{V}_1 + k_2 e^{2t}\mathbf{V}_2 = \begin{pmatrix} k_1 e^{-3t} \\ k_2 e^{2t} \end{pmatrix}.$$

In Figure 3.12 we display the phase portrait for this system. The straight-line solutions lie on the axes, but all other solutions behave differently. In the figure we see that the other solutions seem to tend to infinity asymptotic to the y-axis and to come from infinity asymptotic to the x-axis. To see why, consider a solution $\mathbf{Y}(t)$ that is not a straight-line solution. Then

$$\mathbf{Y}(t) = k_1 e^{-3t}\mathbf{V}_1 + k_2 e^{2t}\mathbf{V}_2,$$

where both k_1 and k_2 are nonzero. When t is large and positive, the term e^{-3t} is very small. Therefore for large positive t, the vector $e^{-3t}\mathbf{V}_1$ in the general solution is negligible, and we have

$$\mathbf{Y}(t) \approx k_2 e^{2t}\mathbf{V}_2 = \begin{pmatrix} 0 \\ k_2 e^{2t} \end{pmatrix}.$$

That is, for large positive values of t, our solution behaves like a straight-line solution on the y-axis.

The opposite is true when we consider large negative values of t. In this case the term e^{2t} is very small, so we have

$$\mathbf{Y}(t) \approx k_1 e^{-3t}\mathbf{V}_1 = \begin{pmatrix} k_1 e^{-3t} \\ 0 \end{pmatrix},$$

which is a straight-line solution along the x-axis.

For example, the particular solution of this system that satisfies $\mathbf{Y}(0) = (1, 1)$ is

$$\mathbf{Y}(t) = \begin{pmatrix} e^{-3t} \\ e^{2t} \end{pmatrix}.$$

The x-coordinate of this solution tends to 0 as $t \to \infty$ and to infinity as $t \to -\infty$. The y-coordinate behaves in the opposite manner (see these $x(t)$- and $y(t)$-graphs in Figure 3.13).

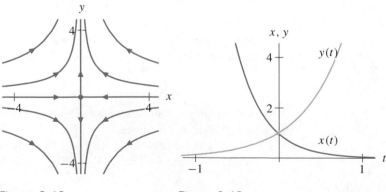

Figure 3.12
Phase portrait for the system

$$\frac{d\mathbf{Y}}{dt} = \mathbf{AY} = \begin{pmatrix} -3 & 0 \\ 0 & 2 \end{pmatrix}\mathbf{Y}.$$

Figure 3.13
The $x(t)$- and $y(t)$-graphs for the solution with initial position (1, 1).

Despite the fact that this example really consists of two one-dimensional differential equations, its phase portrait is entirely new. Along the axes we see the familiar phase lines for one-dimensional equations — a sink along the x-axis and a source on the y-axis. All other solutions tend to infinity as $t \to \pm\infty$. These solutions come from infinity in the direction of the eigenvectors corresponding to the one-dimensional sink, and they tend back to infinity in the direction of the one-dimensional source.

Any linear system for which we have one positive and one negative eigenvalue has similar behavior. An equilibrium point of this form is called a **saddle**. This name is supposed to remind you of a saddle for a horse. The path followed by a drop of water on a horse's saddle resembles the path of a solution of this type of linear system; it approaches the center of the seat in one direction and then veers off toward the ground in another.

Phase portraits for other saddles

The previous example is special in that the eigenvectors lie on the x- and y-axes. In general the eigenvectors for a saddle can lie on any two distinct lines through the origin. This makes the phase planes and the $x(t)$- and $y(t)$-graphs appear somewhat different in the general case.

For example, consider the system

$$\frac{d\mathbf{Y}}{dt} = \mathbf{B}\mathbf{Y}, \quad \text{where } \mathbf{B} = \begin{pmatrix} 8 & -11 \\ 6 & -9 \end{pmatrix}.$$

We first compute the eigenvalues of $\mathbf{B}$ by finding the roots of the characteristic polynomial

$$\det(\mathbf{B} - \lambda I) = \det \begin{pmatrix} 8 - \lambda & -11 \\ 6 & -9 - \lambda \end{pmatrix} = (8 - \lambda)(-9 - \lambda) + 66 = \lambda^2 + \lambda - 6 = 0.$$

The roots of this quadratic equation are $\lambda_1 = -3$ and $\lambda_2 = 2$, the eigenvalues of $\mathbf{B}$. These are exactly the same eigenvalues as in the previous example, so the origin is a saddle.

Next we compute the eigenvectors. For $\lambda_1 = -3$, the equations that give the eigenvectors (x_1, y_1) are

$$\begin{cases} 8x_1 - 11y_1 = -3x_1 \\ 6x_1 - 9y_1 = -3y_1. \end{cases}$$

So any nonzero vector that lies along the line $y = x$ in the plane serves as an eigenvector for $\lambda_1 = -3$. We choose $\mathbf{V}_1 = (1, 1)$. Therefore the solution

$$\mathbf{Y}_1(t) = e^{-3t}\mathbf{V}_1$$

is a straight-line solution lying on the line $y = x$. It tends to the origin as t increases.

Similar computations yield eigenvectors corresponding to $\lambda_2 = 2$ lying along the line $6x - 11y = 0$, for example $\mathbf{V}_2 = (11, 6)$. This in turn gives a straight-line solution of the form

$$\mathbf{Y}_2(t) = e^{2t}\mathbf{V}_2$$

that tends away from the origin as t increases. Thus the general solution is

$$\mathbf{Y}(t) = k_1 e^{-3t} \mathbf{V}_1 + k_2 e^{2t} \mathbf{V}_2.$$

As above we expect that, if k_1 and k_2 are nonzero, these solutions come from infinity in the direction of $\mathbf{V}_1$ and tend back to infinity in the direction of $\mathbf{V}_2$. In the phase plane we see the two straight-line solutions together with several other solutions (see Figure 3.14). The important point is that, once we have the eigenvalues and eigenvectors, we can immediately visualize the entire phase portrait.

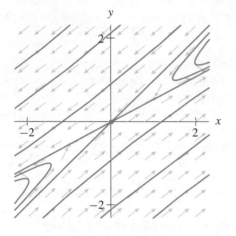

Figure 3.14

The direction field and phase portrait for the system

$$\frac{d\mathbf{Y}}{dt} = \mathbf{B}\mathbf{Y} = \begin{pmatrix} 8 & -11 \\ 6 & -9 \end{pmatrix} \mathbf{Y}.$$

The eigenvectors lie along the two distinguished lines that run through the first and third quadrants. Although some of the other solution curves look almost straight, they really curve slightly.

Sinks

Now consider the system of differential equations

$$\frac{d\mathbf{Y}}{dt} = \mathbf{C}\mathbf{Y}, \quad \text{where } \mathbf{C} = \begin{pmatrix} -1 & 0 \\ 0 & -4 \end{pmatrix}.$$

The matrix $\mathbf{C}$ has eigenvalues $\lambda_1 = -1$ and $\lambda_2 = -4$. Therefore we expect to have two straight-line solutions which tend to the origin as $t \to \infty$.

An eigenvector corresponding to $\lambda_1 = -1$ is $\mathbf{V}_1 = (1, 0)$, and an eigenvector for $\lambda_2 = -4$ is $\mathbf{V}_2 = (0, 1)$. Thus the general solution is

$$\mathbf{Y}(t) = k_1 e^{-t} \mathbf{V}_1 + k_2 e^{-4t} \mathbf{V}_2 = k_1 e^{-t} \begin{pmatrix} 1 \\ 0 \end{pmatrix} + k_2 e^{-4t} \begin{pmatrix} 0 \\ 1 \end{pmatrix} = \begin{pmatrix} k_1 e^{-t} \\ k_2 e^{-4t} \end{pmatrix}.$$

Since each term involves either e^{-t} or e^{-4t}, we know that every solution of this system tends to the origin. In Figure 3.15 we sketch the phase portrait for this system. In this picture we clearly see the straight-line solutions. As predicted, all other solutions tend to the origin. In fact whenever we have a linear system with two negative eigenvalues, all solutions tend to the origin. By analogy with autonomous, first-order equations, we call this type of equilibrium point a **sink**.

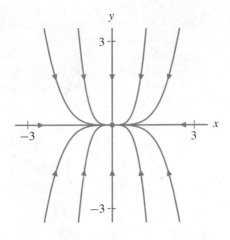

Figure 3.15
The phase portrait for the system

$$\frac{d\mathbf{Y}}{dt} = \mathbf{CY} = \begin{pmatrix} -1 & 0 \\ 0 & -4 \end{pmatrix}.$$

Note that all solution curves tend to the equilibrium point at the origin.

In Figure 3.15 it appears that every solution (with the exception of those on the y-axis) tends to the origin tangent to the x-axis. To see why, consider the general solution

$$\begin{pmatrix} x(t) \\ y(t) \end{pmatrix} = \begin{pmatrix} k_1 e^{-t} \\ k_2 e^{-4t} \end{pmatrix}.$$

If $k_1 \neq 0$, then we can solve for e^{-t} in $x(t) = k_1 e^{-t}$, and we obtain

$$e^{-t} = \frac{x(t)}{k_1}.$$

We then substitute this expression for e^{-t} into the formula for $y(t)$, and we obtain

$$
\begin{aligned}
y(t) &= k_2 e^{-4t} \\
&= k_2 (e^{-t})^4 \\
&= k_2 \left(\frac{x(t)}{k_1} \right)^4 \\
&= \frac{k_2}{k_1^4} (x(t))^4.
\end{aligned}
$$

In other words, each solution curve lies along a curve of the form

$$y = Kx^4$$

for some constant K if $k_1 \neq 0$. Since these curves are always tangent to the x-axis, we see why all solution curves whose initial conditions are not on the y-axis approach the equilibrium point at the origin along curves that are tangent to the x-axis.

More general sinks

In general, for any linear system with two distinct, negative eigenvalues, we have a similar phase portrait. For example, consider the system of differential equations

$$\frac{d\mathbf{Y}}{dt} = \mathbf{DY}, \quad \text{where } \mathbf{D} = \begin{pmatrix} -2 & -2 \\ -1 & -3 \end{pmatrix}.$$

The matrix $\mathbf{D}$ has eigenvalues $\lambda_1 = -4$ and $\lambda_2 = -1$. For $\lambda_1 = -4$, one eigenvector is $\mathbf{V}_1 = (1, 1)$, and for $\lambda_2 = -1$, one eigenvector is $\mathbf{V}_2 = (-2, 1)$. (Checking this is a good review of eigenvalues and eigenvectors. You should be able to check that these vectors are eigenvectors without recomputing from scratch.)

Thus this vector field has two linearly independent straight-line solutions that tend to the origin, and in fact the general solution is

$$\mathbf{Y}(t) = k_1 e^{-4t} \mathbf{V}_1 + k_2 e^{-t} \mathbf{V}_2$$

$$= k_1 e^{-4t} \begin{pmatrix} 1 \\ 1 \end{pmatrix} + k_2 e^{-t} \begin{pmatrix} -2 \\ 1 \end{pmatrix}$$

$$= \begin{pmatrix} k_1 e^{-4t} - 2k_2 e^{-t} \\ k_1 e^{-4t} + k_2 e^{-t} \end{pmatrix}.$$

Once we know that the eigenvalues for this system are -4 and -1, we know that every term in the general solution has a factor of e^{-4t} or e^{-t}. Hence every solution tends to the origin as $t \to \infty$, and the origin is a sink. The long-term behavior of solutions can be determined from the eigenvalues alone (without the eigenvectors or the formula for the general solution).

In Figure 3.16 we sketch the phase portrait for this system. In this picture we clearly see the straight-line solutions. As predicted, all other solutions tend to the origin as well. Again all solutions with the exception of the straight-line solutions associated to $\lambda_1 = -4$ seem to tend to the origin tangent to the line of eigenvectors for $\lambda_2 = -1$.

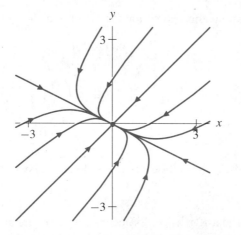

Figure 3.16
Phase portrait for the system

$$\frac{d\mathbf{Y}}{dt} = \mathbf{D}\mathbf{Y} = \begin{pmatrix} -2 & -2 \\ -1 & -3 \end{pmatrix} \mathbf{Y}.$$

All solutions tend to the equilibrium point at the origin, and all solutions with the exception of the straight-line solutions associated to $\lambda_1 = -4$ tend to the origin tangent to the line of eigenvectors for $\lambda_2 = -1$.

Direction of approach to the sink

To understand why solution curves approach the origin in the way that they do, we need to resort to some calculus. We compute the slope of the tangent line to any solution curve and then ask what happens to this slope as $t \to \infty$. Each solution curve is given by

$$\begin{pmatrix} x(t) \\ y(t) \end{pmatrix} = \begin{pmatrix} k_1 e^{-4t} - 2k_2 e^{-t} \\ k_1 e^{-4t} + k_2 e^{-t} \end{pmatrix}$$

for some choice of constants k_1 and k_2. From calculus we know that the slope of the tangent vector to a curve is given by dy/dx and

$$\frac{dy}{dx} = \frac{dy/dt}{dx/dt}.$$

Since $x(t) = k_1 e^{-4t} - 2k_2 e^{-t}$ and $y(t) = k_1 e^{-4t} + k_2 e^{-t}$, we have

$$\frac{dy/dt}{dx/dt} = \frac{-4k_1 e^{-4t} - k_2 e^{-t}}{-4k_1 e^{-4t} + 2k_2 e^{-t}}.$$

If we take the limit of this expression as $t \to \infty$, we end up with the indeterminate form $\frac{0}{0}$. It is tempting to use L'Hôpital's Rule, but this approach is destined to fail since the derivatives all involve exponential terms as well. The way to compute this limit is to multiply both numerator and denominator by e^t. Then the new expression is

$$\frac{dy/dt}{dx/dt} = \frac{-4k_1 e^{-3t} - k_2}{-4k_1 e^{-3t} + 2k_2}.$$

As $t \to \infty$, both exponential terms in this quotient tend to 0, and we see that the limit is $-k_2/(2k_2) = -1/2$ if $k_2 \neq 0$. That is, these solutions tend to the origin with slopes tending to $-1/2$ or, equivalently, tangent to the line of eigenvectors corresponding to the eigenvalue λ_2.

If $k_2 = 0$, our expression for dy/dx reduces to

$$\frac{dy/dt}{dx/dt} = \frac{-4k_1 e^{-4t}}{-4k_1 e^{-4t}} = 1,$$

which is exactly the slope for the straight-line solutions whose initial conditions lie along the line of eigenvectors associated to $\lambda_1 = -4$.

The discussion of the direction of approach to the equilibrium point may seem technical, but there really is a good qualitative reason that most solutions tend to $(0,0)$ tangent to the eigenvector corresponding to the eigenvalue -1. Recall that the vector field on the line of eigenvectors corresponding to the eigenvalue λ is simply the scalar product of λ and the position vector. Because $-4 < -1$, the vector field on the line of eigenvectors for the eigenvalue -4 at a given distance from the origin is much longer than those on the line of eigenvectors for the eigenvalue -1. So solutions on the line of eigenvectors for -4 tend to zero much more quickly than those for the eigenvalue -1. In particular, the solution $e^{-4t}\mathbf{V}_1$ tends to $(0,0)$ more quickly than $e^{-t}\mathbf{V}_2$.

In our general solution

$$\mathbf{Y}(t) = k_1 e^{-4t}\mathbf{V}_1 + k_2 e^{-t}\mathbf{V}_2,$$

if both k_1 and k_2 are nonzero, then the first term tends to the origin more quickly than the second. So when t is sufficiently large, the second term dominates, and we see that most solutions tend to zero along the direction of the eigenvectors for the eigenvalue closer to zero. The only exceptions are the solutions on the line of eigenvectors for the eigenvalue that is more negative. So, as in the previous example, provided that $k_2 \neq 0$, we can write $\mathbf{Y}(t) \approx k_2 e^{-t}\mathbf{V}_2$ as $t \to \infty$.

The case of an arbitrary sink with two eigenvalues $\lambda_1 < \lambda_2 < 0$ is entirely analogous. All solutions tend to the origin, and with the exception of those solutions with initial conditions that are eigenvectors corresponding to λ_1, all solutions tend to $(0,0)$ tangent to the line of eigenvectors for λ_2.

Sources

Consider the system

$$\frac{d\mathbf{Y}}{dt} = \mathbf{EY}, \quad \text{where } \mathbf{E} = \begin{pmatrix} 2 & 2 \\ 1 & 3 \end{pmatrix}.$$

In the previous section we computed that the eigenvalues of this matrix are $\lambda_1 = 4$ and $\lambda_2 = 1$. Also $\mathbf{V}_1 = (1, 1)$ is an eigenvector for the eigenvalue $\lambda_1 = 4$, and $\mathbf{V}_2 = (-2, 1)$ is an eigenvector for the eigenvalue 1. (Remember that you can check these assertions by computing $\mathbf{EV}_1$ and $\mathbf{EV}_2$.) Then $e^{4t}\mathbf{V}_1$ and $e^t\mathbf{V}_2$ are two linearly independent, straight-line solutions, and the general solution is

$$\mathbf{Y}(t) = k_1 e^{4t}\mathbf{V}_1 + k_2 e^t\mathbf{V}_2.$$

Since both eigenvalues of this system are positive, all nonzero solutions move away from the origin as $t \to \infty$.

The phase portrait for this system is shown in Figure 3.17. As in the previous example, we see two straight-line solutions, and all other solutions leave the origin in a direction tangent to the line of eigenvectors corresponding to the eigenvalue $\lambda_2 = 1$. The reason for this is essentially the same as the reason given for sinks earlier in the section. In fact the astute reader will note that $\mathbf{E} = -\mathbf{D}$ where $\mathbf{D}$ is the matrix specified in the previous example. Consequently, for the vector field, we have changed merely the direction of the arrows, and not the geometry of the solution curves. For this system instead of considering the behavior as $t \to \infty$, we consider the behavior as $t \to -\infty$. Now the eigenvalue 4 plays the role of the stronger eigenvalue. Solutions involving terms with e^{4t} tend to the origin much more quickly than those involving e^t as $t \to -\infty$.

In general once we know that both eigenvalues of a linear system are positive, we can conclude that all solutions tend away from the origin as t increases. We call the equilibrium point for a linear system with two positive eigenvalues a **source**. All solutions tend away from the equilibrium point as $t \to \infty$, and all except those on the line of eigenvectors corresponding to λ_1 leave the origin in a direction tangent to the line of eigenvectors corresponding to λ_2.

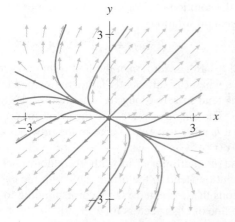

Figure 3.17
Phase portrait for the system

$$\frac{d\mathbf{Y}}{dt} = \mathbf{EY} = \begin{pmatrix} 2 & 2 \\ 1 & 3 \end{pmatrix}\mathbf{Y}.$$

Note that, since $\mathbf{E} = -\mathbf{D}$, we can obtain the phase portrait for this example from the phase portrait for $d\mathbf{Y}/dt = \mathbf{DY}$. The solution curves are identical, but solutions travel away from the origin as $t \to \infty$.

Stable and Unstable Equilibrium Points

Before considering one more example, we summarize the behavior of a linear system with two nonzero, real, distinct eigenvalues λ_1 and λ_2.

- If $\lambda_1 < 0 < \lambda_2$, then the origin is a saddle. There are two lines in the phase portrait that correspond to straight-line solutions. Solutions along one line tend toward $(0, 0)$ as t increases, and solutions on the other line tend away from $(0, 0)$. All other solutions come from and go to infinity.
- If $\lambda_1 < \lambda_2 < 0$, then the origin is a sink. All solutions tend to $(0, 0)$ as $t \to \infty$, and most tend to $(0, 0)$ in the direction of the λ_2-eigenvector.
- If $0 < \lambda_2 < \lambda_1$, then the origin is a source. All solutions tend away from $(0, 0)$ as $t \to \infty$, and most tend away in the direction of the λ_2-eigenvector.

A sink is said to be **stable** because nearby initial points yield solutions that tend back toward the equilibrium point as time increases. So if the initial condition is "bumped" a little bit away from the sink, the resulting solution does not stray far away from the initial point. Saddle and source equilibrium points are called **unstable** because there are initial conditions arbitrarily close to the equilibrium point whose solutions move away. Hence a small bump to an initial condition can have dramatic consequences. For a source, every initial condition near the equilibrium point corresponds to a solution that moves away. For a saddle, every initial condition except those on the straight-line solution tending to the equilibrium point (so almost every initial condition) corresponds to a solution that moves away.

If we run time backward, then a source looks like a sink with solutions tending toward it. Similarly, in backward time a sink looks like a source with solutions moving away from it. This is analogous to the situation for phase lines.

The saddle is a new type of equilibrium point that cannot occur in one-dimensional systems. Saddles need two dimensions in order to have one direction that is stable (corresponding to the negative eigenvalue) and another that is unstable (corresponding to the positive eigenvalue).

Paul's and Bob's CD Stores

Recall the model of Paul's and Bob's CD stores from Section 3.1. Suppose market research establishes that, if a store becomes popular, then it becomes too crowded and profits tend to decrease. Also all stores near a popular store suffer from the effect of overcrowding, and their profits also decrease. In other words, if Paul's profits become positive, profits of his store and of Bob's store tend to decrease, so parameters a and c should be negative. The same is true for Bob's store. As an example we let $a = -2$, $b = -3$, $c = -3$, and $d = -2$, so the linear system is

$$\frac{d\mathbf{Y}}{dt} = \begin{pmatrix} -2 & -3 \\ -3 & -2 \end{pmatrix} \mathbf{Y}.$$

All the coefficients are negative, so we might be tempted to say that this model predicts that profit for either store is impossible because, whenever one store starts to make money, it makes the rate of change of the profits of both stores smaller. However, we cannot always trust guesses. We use the tools that we have developed to study this system carefully.

To give an accurate sketch of the phase portrait, we first compute the eigenvalues and eigenvectors. The characteristic polynomial is

$$(-2 - \lambda)(-2 - \lambda) - 9 = \lambda^2 + 4\lambda - 5 = (\lambda - 1)(\lambda + 5),$$

and the eigenvalues are $\lambda_1 = -5$ and $\lambda_2 = 1$. Because one eigenvalue is positive and one is negative, the origin is a saddle (see Figure 3.18). We find an eigenvector for the eigenvalue $\lambda_1 = -5$ by solving

$$\begin{cases} -2x_1 - 3y_1 = -5x_1 \\ -3x_1 - 2y_1 = -5y_1, \end{cases}$$

and these equations have a line of solutions given by $x_1 = y_1$. So $(1, 1)$ is an eigenvector for the eigenvalue $\lambda_1 = -5$.

For the other eigenvalue $\lambda_2 = 1$, we must solve

$$\begin{cases} -2x_2 - 3y_2 = x_2 \\ -3x_2 - 2y_2 = y_2. \end{cases}$$

These equations have a line of solutions given by $x_2 = -y_2$. So $(-1, 1)$ is an eigenvector for the eigenvalue $\lambda_2 = 1$. We could now use this information to write down the general solution, but it is more useful to use it to sketch the phase portrait. We know that the diagonal $x_1 = y_1$ through the origin contains straight-line solutions and that these solutions tend toward the origin because the eigenvalue $\lambda_1 = -5$ is negative. The other diagonal line through the origin, $x_2 = -y_2$, contains straight-line solutions that move away from $(0, 0)$ as t increases. Every other solution is a linear combination of these two. So the only solutions that tend to $(0, 0)$ are those on the line $x = y$. As $t \to \infty$, all other solutions eventually move away from the origin in either the second or fourth quadrants (see Figure 3.18).

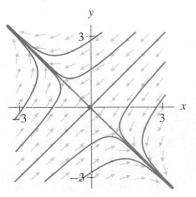

Figure 3.18
Phase portrait for the system

$$\frac{d\mathbf{Y}}{dt} = \begin{pmatrix} -2 & -3 \\ -3 & -2 \end{pmatrix} \mathbf{Y}.$$

The equilibrium point at the origin is a saddle, and most solutions tend to infinity asymptotic to the straight-line solutions whose solution curves lie in the second or fourth quadrants.

Analysis of the model

This model leads to some startling predictions for Paul's and Bob's profits. Suppose that at $t = 0$ both Paul and Bob are making a profit [$x(0) > 0$ and $y(0) > 0$]. If it happens that Paul and Bob are making *exactly* the same amount, then $x(0) = y(0)$ and the initial point is on the $x = y$ line. The solution with this initial condition tends to the origin as t increases; that is, Paul and Bob both make less and less profit as time increases, both of them tending toward the break-even point $(x, y) = (0, 0)$.

Next consider the case $x(0) > y(0)$ (even by just a tiny amount). Now the initial point is just below the diagonal $x = y$. The corresponding solution at first tends toward $(0, 0)$, but it eventually turns and follows the straight-line solution along the line $x = -y$ into the fourth quadrant. In this case $x(t) \to \infty$ but $y(t) \to -\infty$. In other words, Paul eventually makes a fortune, but Bob loses his shirt. The vector field is very small near $(0, 0)$, so the solution moves slowly when it is near the origin. But eventually it turns the corner and Paul gets rich and Bob loses out (see Figure 3.18).

On the other hand, suppose $y(0)$ is slightly larger than $x(0)$. Then the initial point is just above the line $x = y$. In this case the solution first tends toward $(0, 0)$, but it eventually "turns the corner" and tends toward infinity along the line $x = -y$ in the second quadrant. In this case $x(t) \to -\infty$ (Paul goes broke) and $y(t) \to \infty$ (Bob gets rich; see Figure 3.18).

In this example a tiny change in the initial condition causes a large change in the long-term behavior of the system. We emphasize that the difference in behavior of solutions takes a long time to appear because solutions move very slowly near the equilibrium point. This sensitive dependence on the choice of initial condition is caused by the straight-line solution through the origin. The solutions with $x(0) = y(0) + 0.01$ and $x(0) = y(0) + 0.02$ are both on the same side of the diagonal, so they both behave the same way in the long run. It is only when a small change pushes the initial condition to the other side of the straight-line solution along the diagonal that the big jump in the long-term behavior occurs (see Figure 3.18). For this reason, a straight-line solution of a saddle corresponding to the negative eigenvalue is sometimes called a **separatrix**, because it separates two different types of long-term behavior.

Common Sense versus Computation

The predictions of this model are not at all what we might have expected. The coefficient matrix

$$\begin{pmatrix} -2 & -3 \\ -3 & -2 \end{pmatrix}$$

consists of only negative numbers, so any increase in profits of either store has a negative effect on the rate of change of the profits. "Common sense" might suggest that neither store will ever show a profit. The behavior of the model is quite different.

One lesson to be learned for this simple-minded example is that, although it is always wise to compare the predictions of a model with common sense, common sense does not replace computation. Models are most valuable when they predict something unexpected.

EXERCISES FOR SECTION 3.3

In Exercises 1–8, we refer to linear systems from the exercises in Section 3.2. Sketch the phase portrait for the system specified.

1. The system in Exercise 1, Section 3.2   **2.** The system in Exercise 2, Section 3.2

3. The system in Exercise 3, Section 3.2   **4.** The system in Exercise 6, Section 3.2

5. The system in Exercise 7, Section 3.2   **6.** The system in Exercise 8, Section 3.2

7. The system in Exercise 9, Section 3.2   **8.** The system in Exercise 10, Section 3.2

In Exercises 9–12, we refer to initial-value problems from the exercises in Section 3.2. Sketch the solution curves in the phase plane and the $x(t)$- and $y(t)$-graphs for the solutions corresponding to the initial-value problems specified.

9. The initial-value problems in Exercise 11, Section 3.2

10. The initial-value problems in Exercise 12, Section 3.2

11. The initial-value problems in Exercise 13, Section 3.2

12. The initial-value problems in Exercise 14, Section 3.2

In Exercises 13–16, we refer to the second-order equations from the exercises in Section 3.2. Sketch the phase portrait for the second-order equations specified.

13. The second-order equation in Exercise 21, Section 3.2

14. The second-order equation in Exercise 22, Section 3.2

15. The second-order equation in Exercise 23, Section 3.2

16. The second-order equation in Exercise 24, Section 3.2

In Exercises 17–18, we consider the model of Paul's and Bob's CD stores from Section 3.1. Suppose Paul and Bob are both operating at the break-even point $(x, y) = (0, 0)$. For the models given below, state what happens if one of the stores starts to earn or lose just a little bit. That is, will the profits return to 0 for both stores? If not, does it matter which store starts to earn money first?

17.
$$
\begin{pmatrix} \dfrac{dx}{dt} \\ \dfrac{dy}{dt} \end{pmatrix} = \begin{pmatrix} 2 & 1 \\ 0 & -1 \end{pmatrix} \begin{pmatrix} x \\ y \end{pmatrix}
$$

18.
$$
\begin{pmatrix} \dfrac{dx}{dt} \\ \dfrac{dy}{dt} \end{pmatrix} = \begin{pmatrix} -2 & -1 \\ -1 & -1 \end{pmatrix} \begin{pmatrix} x \\ y \end{pmatrix}
$$

19. The slope field for the system

$$
\frac{dx}{dt} = -2x + \frac{1}{2}y
$$
$$
\frac{dy}{dt} = -y
$$

is shown to the right.

(a) Determine the type of the equilibrium point at the origin.

(b) Calculate all straight-line solutions.

(c) Plot the $x(t)$- and $y(t)$-graphs, $(t \geq 0)$, for the initial conditions $A = (2, 1)$, $B = (1, -2)$, $C = (-2, 2)$, and $D = (-2, 0)$.

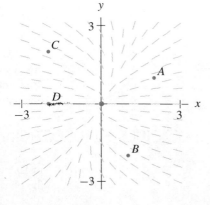

20. The slope field for the system

$$\frac{dx}{dt} = 2x + 6y$$

$$\frac{dy}{dt} = 2x - 2y$$

is shown to the right.

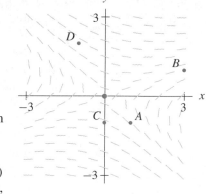

(a) Determine the type of the equilibrium point at the origin.

(b) Calculate all straight-line solutions.

(c) Plot the $x(t)$- and $y(t)$-graphs ($t \geq 0$) for the initial conditions $A = (1, -1)$, $B = (3, 1)$, $C = (0, -1)$, and $D = (-1, 2)$.

21. For the harmonic oscillator with mass $m = 1$, spring constant $k = 6$, and damping coefficient $b = 7$,

(a) write the second-order equation and the corresponding system,

(b) compute the characteristic polynomial,

(c) find the eigenvalues, and

(d) discuss the motion of the mass for the initial condition $(y(0), v(0)) = (2, 0)$. (How often does the mass cross the rest position $y = 0$? How quickly does the mass approach the equilibrium?)

22. Consider a harmonic oscillator with mass $m = 1$, spring constant $k = 1$, and damping coefficient $b = 4$. For the initial position $y(0) = 2$, find the initial velocity for which $y(t) > 0$ for all t and $y(t)$ reaches 0.1 most quickly. [*Hint*: It helps to look at the phase plane first.]

In Exercises 23–26, we consider a small pond inhabited by a species of fish. When left alone, the population of these fish settles into an equilibrium population. Suppose a few fish of another species are introduced to the pond. We would like to know if the new species survives and if the population of the native species changes much from its equilibrium population.

To determine the answers to these questions, we create a very simple model of the fish populations. Let $f(t)$ be the population of the native fish, and let f_0 denote the equilibrium population. We are interested in the change of the native fish population from its equilibrium level, so we let $x(t) = f(t) - f_0$; that is, $x(t)$ is the difference of the population of the native fish species from its equilibrium level. Let $y(t)$ denote the population of the introduced species. We note that, because $y(t)$ is an "absolute" population, it does not make sense to have $y(t) < 0$. So if $y(t)$ is ever equal to zero, we say that the introduced species has gone extinct. On the other hand, $x(t)$ can assume both positive and negative values because this variable measures the difference of the native fish population from its equilibrium level.

We are concerned with the behavior of these populations when both variables x and y are small, so the effects of terms involving x^2, y^2, xy, or higher powers are very,

very small. Consequently we ignore them in this model (see Section 5.1). Also we know that, if $x = y = 0$, then the native fish population is in equilibrium and none of the introduced species are there, so the population does not change; that is, $(x, y) = (0, 0)$ is an equilibrium point. Hence it is reasonable to use a linear system as a model.

For each model:

(a) Discuss briefly what sort of interaction between the species corresponds to the model; that is, do the introduced fish work to increase or decrease the native fish population, etc.

(b) Decide if the model agrees with the information above about the system. That is, will the population of the native species return to equilibrium if the introduced species is not present?

(c) Sketch the phase plane and describe the solutions of the linear system (using technology and information about eigenvalues and eigenvectors).

(d) State what predictions the model makes about what happens when a small number of the new species is introduced into the lake.

23. $\dfrac{d\mathbf{Y}}{dt} = \begin{pmatrix} -0.2 & -0.1 \\ 0.0 & -0.1 \end{pmatrix} \mathbf{Y}$

24. $\dfrac{d\mathbf{Y}}{dt} = \begin{pmatrix} -0.1 & 0.2 \\ 0.0 & 1.0 \end{pmatrix} \mathbf{Y}$

25. $\dfrac{d\mathbf{Y}}{dt} = \begin{pmatrix} -0.2 & 0.1 \\ 0.0 & -0.1 \end{pmatrix} \mathbf{Y}$

26. $\dfrac{d\mathbf{Y}}{dt} = \begin{pmatrix} 0.1 & 0.0 \\ -0.2 & 0.2 \end{pmatrix} \mathbf{Y}$

27. Consider the linear system

$$\frac{d\mathbf{Y}}{dt} = \begin{pmatrix} -2 & 1 \\ 0 & 2 \end{pmatrix} \mathbf{Y}.$$

(a) Show that $(0, 0)$ is a saddle.

(b) Find the eigenvalues and eigenvectors and sketch the phase plane.

(c) On the phase plane, sketch the solution curves with initial conditions $(1, 0.01)$ and $(1, -0.01)$.

(d) Estimate the time t at which the solutions with initial conditions $(1, 0.01)$ and $(1, -0.01)$ will be 1 unit apart.

3.4   COMPLEX EIGENVALUES

The techniques of the previous sections were based on the geometric observation that, for some linear systems, certain solution curves lie on straight lines in the phase plane. This geometric observation led to the algebraic notions of eigenvalues and eigenvectors. These, in turn, gave us the formulas for the general solution.

Unfortunately these ideas do not work for all linear systems. Geometrically we hit a road block when we encounter linear systems whose direction fields do not show any straight-line solutions (see Figure 3.19). In this case it is the algebra of eigenvalues and eigenvectors that leads to an understanding of the system. Even though the method is different, the goals are the same: Starting with the entries in the coefficient matrix, understand the geometry of the phase plane, the $x(t)$- and $y(t)$-graphs, and find the general solution.

Complex numbers

In this section and the rest of the book, we use complex numbers extensively. Complex numbers are numbers of the form $x + iy$, where x and y are real numbers and i is the "imaginary" number $\sqrt{-1}$. (There is a brief summary of the properties of complex numbers in the appendices.)

One word of caution: All mathematicians and almost everyone else denote the imaginary number $\sqrt{-1}$ by the letter i. Electrical engineers use the letter i for the current (because "current" starts with "c"), so they use the letter j for $\sqrt{-1}$.

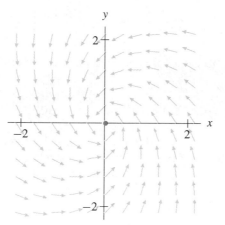

Figure 3.19
The direction field for

$$\frac{d\mathbf{Y}}{dt} = \begin{pmatrix} -2 & -3 \\ 3 & -2 \end{pmatrix} \mathbf{Y}.$$

Apparently there are no straight-line solutions.

A Linear System without Straight-Line Solutions

Consider the system

$$\frac{d\mathbf{Y}}{dt} = \mathbf{AY} = \begin{pmatrix} -2 & -3 \\ 3 & -2 \end{pmatrix} \mathbf{Y}.$$

From the direction field for this system (see Figure 3.19), we see that there are no solution curves that lie on straight lines. Instead, solutions spiral around the origin.

The characteristic polynomial of $\mathbf{A}$ is

$$\det(\mathbf{A} - \lambda \mathbf{I}) = (-2 - \lambda)(-2 - \lambda) + 9,$$

which simplifies to

$$\lambda^2 + 4\lambda + 13.$$

The eigenvalues are the roots of the characteristic polynomial, that is, the solutions of the equation

$$\lambda^2 + 4\lambda + 13 = 0.$$

Hence for this system the eigenvalues are the complex numbers $\lambda_1 = -2 + 3i$ and $\lambda_2 = -2 - 3i$. So how are we going to find solutions, and what information do complex eigenvalues give us?

General Solutions for Systems with Complex Eigenvalues

The most important thing to remember now is: Don't panic. Things are not nearly as complicated as they might seem. We cannot use the geometric ideas of Sections 3.2 and 3.3 to find solution curves that are straight lines because there aren't any straight-line solutions. However, the algebraic techniques we used in those sections work the same way for complex numbers as they do for real numbers. The rules of arithmetic for complex numbers are exactly the same as for real numbers, so all of the computations we did in the previous sections are still valid even if the eigenvalues are complex. Consequently our main observation about solutions of linear systems still holds even if the eigenvalues are complex. That is, given the linear system $d\mathbf{Y}/dt = \mathbf{AY}$, if λ is an eigenvalue for $\mathbf{A}$ and $\mathbf{Y}_0 = (x_0, y_0)$ is an eigenvector for the eigenvalue λ, then

$$\mathbf{Y}(t) = e^{\lambda t}\mathbf{Y}_0 = e^{\lambda t}\left(\begin{array}{c} x_0 \\ y_0 \end{array}\right) = \left(\begin{array}{c} e^{\lambda t}x_0 \\ e^{\lambda t}y_0 \end{array}\right)$$

is a solution. We can easily check this fact by differentiation, as we did before. We have

$$\frac{d\mathbf{Y}}{dt} = \frac{d}{dt}(e^{\lambda t}\mathbf{Y}_0) = \lambda e^{\lambda t}\mathbf{Y}_0 = e^{\lambda t}(\lambda\mathbf{Y}_0) = e^{\lambda t}\mathbf{AY}_0 = \mathbf{A}(e^{\lambda t}\mathbf{Y}_0) = \mathbf{AY}$$

because $\mathbf{Y}_0$ is an eigenvector with eigenvalue λ, so $\mathbf{Y}(t)$ is a solution. Of course, we need to make sense of the fact that the exponential is now a complex function and the fact that the eigenvector may contain complex entries, but this is no real problem. The important thing to notice here is that this computation is exactly the same if the numbers are real or complex.

Example revisited

For the system

$$\frac{d\mathbf{Y}}{dt} = \mathbf{AY}, \quad \text{where } \mathbf{A} = \left(\begin{array}{cc} -2 & -3 \\ 3 & -2 \end{array}\right),$$

we already know that the eigenvalues are $\lambda_1 = -2 + 3i$ and $\lambda_2 = -2 - 3i$. We now find the eigenvector for $\lambda_1 = -2 + 3i$ just as we would if λ_1 were real, by solving for $\mathbf{Y}_0$ in the system of equations

$$\mathbf{AY}_0 = \lambda_1\mathbf{Y}_0.$$

That is, we must find $\mathbf{Y}_0 = (x_0, y_0)$ such that

$$\begin{cases} -2x_0 - 3y_0 = (-2 + 3i)x_0 \\ 3x_0 - 2y_0 = (-2 + 3i)y_0, \end{cases}$$

which can be rewritten as

$$\begin{cases} -3ix_0 - 3y_0 = 0 \\ 3x_0 - 3iy_0 = 0. \end{cases}$$

Just as in the case of real eigenvalues, these equations are redundant. (Multiply both sides of the first equation by i to obtain the second equation and recall that $i^2 = -1$.)

Thus the solutions of these equations are all pairs of complex numbers (x_0, y_0) that satisfy $-3ix_0 - 3y_0 = 0$, or $x_0 = iy_0$. If we let $y_0 = 1$, then $x_0 = i$. In other words, the vector $(i, 1)$ is an eigenvector for eigenvalue $\lambda_1 = -2 + 3i$. We can double-check this by computing

$$\mathbf{A}\begin{pmatrix} i \\ 1 \end{pmatrix} = \begin{pmatrix} -2 & -3 \\ 3 & -2 \end{pmatrix}\begin{pmatrix} i \\ 1 \end{pmatrix} = \begin{pmatrix} -2i - 3 \\ 3i - 2 \end{pmatrix} = (-2 + 3i)\begin{pmatrix} i \\ 1 \end{pmatrix}.$$

Thus we know that

$$\mathbf{Y}(t) = e^{(-2+3i)t}\begin{pmatrix} i \\ 1 \end{pmatrix} = \begin{pmatrix} ie^{(-2+3i)t} \\ e^{(-2+3i)t} \end{pmatrix}$$

is a solution of the system.

Obtaining Real-Valued Solutions from Complex Solutions

So this is both good news and bad news. The good news is that we can find solutions to linear systems with complex eigenvalues. The bad news is that these solutions involve complex numbers. If this system is a model of populations, profits, or the position of a mechanical device, then only real numbers make sense. It is hard to imagine "$5i$" predators or a position of "$2 + 3i$" units from the rest position. In other words, the physical meaning of complex numbers is not readily apparent. We have to find a way of producing real solutions from complex solutions.

The key to getting real solutions from complex solutions is **Euler's formula**

$$e^{a+ib} = e^a e^{ib} = e^a(\cos b + i \sin b) = e^a \cos b + ie^a \sin b$$

for all real numbers a and b. (Using power series, one can verify that

$$e^{ib} = \cos b + i \sin b,$$

and Euler's formula follows using the laws of exponents. (See the appendix on complex numbers.) Euler's formula is how we exponentiate with complex exponents.

Euler's formula is one of the most remarkable identities in all of mathematics. It allows us to relate some of the most important functions and constants in a most intriguing way. For example, if we let $a = 0$ and $b = \pi$, then we obtain

$$e^{\pi i} = e^0 \cos \pi + ie^0 \sin \pi,$$

so

$$e^{\pi i} = -1.$$

That is, when we combine three of the most interesting numbers in mathematics, e, i and π, in the fashion $e^{i\pi}$, we obtain -1.

We use Euler's formula to define a complex-valued exponential function. We have

$$e^{(\alpha+i\beta)t} = e^{\alpha t}e^{i\beta t}$$

$$= e^{\alpha t}(\cos \beta t + i \sin \beta t)$$

$$= e^{\alpha t}\cos \beta t + ie^{\alpha t}\sin \beta t.$$

For the example above, this gives

$$e^{(-2+3i)t} = e^{-2t}\cos 3t + ie^{-2t}\sin 3t.$$

We can now rewrite the solution $\mathbf{Y}(t)$ as

$$\mathbf{Y}(t) = (e^{-2t}\cos 3t + ie^{-2t}\sin 3t)\begin{pmatrix} i \\ 1 \end{pmatrix}$$

$$= \begin{pmatrix} (e^{-2t}\cos 3t + ie^{-2t}\sin 3t)i \\ e^{-2t}\cos 3t + ie^{-2t}\sin 3t \end{pmatrix}$$

$$= \begin{pmatrix} ie^{-2t}\cos 3t - e^{-2t}\sin 3t \\ e^{-2t}\cos 3t + ie^{-2t}\sin 3t \end{pmatrix},$$

which in turn can be broken into

$$\mathbf{Y}(t) = \begin{pmatrix} -e^{-2t}\sin 3t \\ e^{-2t}\cos 3t \end{pmatrix} + i\begin{pmatrix} e^{-2t}\cos 3t \\ e^{-2t}\sin 3t \end{pmatrix}.$$

So far we have only rearranged the solution $\mathbf{Y}(t)$ to isolate the part that involves the number i. We now use the fact that we are dealing with a *linear* system to find the required real solutions.

THEOREM   Suppose $\mathbf{Y}(t)$ is a complex-valued solution to a linear system

$$\frac{d\mathbf{Y}}{dt} = \mathbf{AY} = \begin{pmatrix} a & b \\ c & d \end{pmatrix}\mathbf{Y},$$

where the coefficient matrix $\mathbf{A}$ has all real entries (a, b, c, and d are real numbers). Suppose

$$\mathbf{Y}(t) = \mathbf{Y}_{\text{re}}(t) + i\,\mathbf{Y}_{\text{im}}(t),$$

where $\mathbf{Y}_{\text{re}}(t)$ and $\mathbf{Y}_{\text{im}}(t)$ are real-valued functions of t. Then $\mathbf{Y}_{\text{re}}(t)$ and $\mathbf{Y}_{\text{im}}(t)$ are both solutions of the original system $d\mathbf{Y}/dt = \mathbf{AY}$.   ∎

It is important to note that there are no i's in the expression $\mathbf{Y}_{\text{im}}(t)$. We have factored out the i from this expression.

To verify this theorem, we use the fact that $\mathbf{Y}(t)$ is a solution. In other words,

$$\frac{d\mathbf{Y}}{dt} = \mathbf{AY} \quad \text{for all } t.$$

Now we replace $\mathbf{Y}(t)$ with $\mathbf{Y}_{\text{re}}(t) + i\,\mathbf{Y}_{\text{im}}(t)$ on both sides of the equation. On the left-hand side the usual rules of differentiation give

$$\frac{d\mathbf{Y}}{dt} = \frac{d(\mathbf{Y}_{\text{re}} + i\mathbf{Y}_{\text{im}})}{dt}$$

$$= \frac{d\mathbf{Y}_{\text{re}}}{dt} + i\frac{d\mathbf{Y}_{\text{im}}}{dt}.$$

On the right-hand side we use the fact that this is a linear system to obtain that

$$\mathbf{A}\mathbf{Y}(t) = \mathbf{A}(\mathbf{Y}_{\text{re}}(t) + i\ \mathbf{Y}_{\text{im}}(t))$$

$$= \mathbf{A}\ \mathbf{Y}_{\text{re}}(t) + i\mathbf{A}\ \mathbf{Y}_{\text{im}}(t).$$

So we have

$$\frac{d\mathbf{Y}_{\text{re}}}{dt} + i\frac{d\mathbf{Y}_{\text{im}}}{dt} = \mathbf{A}\mathbf{Y}_{\text{re}} + i\mathbf{A}\mathbf{Y}_{\text{im}}$$

for all t. Two complex numbers are equal only if both their real parts and their imaginary parts are equal. Hence the only way the above equation can hold is if

$$\frac{d\mathbf{Y}_{\text{re}}}{dt} = \mathbf{A}\mathbf{Y}_{\text{re}} \quad \text{and} \quad \frac{d\mathbf{Y}_{\text{im}}}{dt} = \mathbf{A}\mathbf{Y}_{\text{im}},$$

and this is exactly what it means to say $\mathbf{Y}_{\text{re}}(t)$ and $\mathbf{Y}_{\text{im}}(t)$ are solutions of $d\mathbf{Y}/dt = \mathbf{A}\mathbf{Y}$.

Completion of the first example

Recall that, for the system

$$\frac{d\mathbf{Y}}{dt} = \begin{pmatrix} -2 & -3 \\ 3 & -2 \end{pmatrix} \mathbf{Y},$$

we have shown that the complex vector-valued function

$$\mathbf{Y}(t) = \begin{pmatrix} -e^{-2t}\sin 3t \\ e^{-2t}\cos 3t \end{pmatrix} + i\begin{pmatrix} e^{-2t}\cos 3t \\ e^{-2t}\sin 3t \end{pmatrix}$$

is a solution. By taking real and imaginary parts, we know that both the real part

$$\mathbf{Y}_{\text{re}}(t) = \begin{pmatrix} -e^{-2t}\sin 3t \\ e^{-2t}\cos 3t \end{pmatrix}$$

and the imaginary part

$$\mathbf{Y}_{\text{im}}(t) = \begin{pmatrix} e^{-2t}\cos 3t \\ e^{-2t}\sin 3t \end{pmatrix}$$

are solutions of the original system. They are independent since their initial values $\mathbf{Y}_{\text{re}}(0) = (0, 1)$ and $\mathbf{Y}_{\text{im}}(0) = (1, 0)$ are independent. So the general solution of this system is

$$\mathbf{Y}(t) = k_1\begin{pmatrix} -e^{-2t}\sin 3t \\ e^{-2t}\cos 3t \end{pmatrix} + k_2\begin{pmatrix} e^{-2t}\cos 3t \\ e^{-2t}\sin 3t \end{pmatrix}$$

for constants k_1 and k_2. This can be rewritten in the form

$$\mathbf{Y}(t) = e^{-2t}\begin{pmatrix} -k_1\sin 3t + k_2\cos 3t \\ k_1\cos 3t + k_2\sin 3t \end{pmatrix}.$$

A little gift

Note that in the above example we only needed to compute the eigenvector corresponding to one of the two complex eigenvalues. By breaking the resulting complex-valued solution into its real and imaginary parts, we obtained a pair of independent solutions. So, in this case, using complex arithmetic means we only have to do half as much work. (If a matrix with real coefficients has complex eigenvalues, then the eigenvalues are related in a nice way. There is also a nice relationship among the eigenvectors. See Exercises 17–20.)

Qualitative analysis

The direction field for the system

$$\frac{d\mathbf{Y}}{dt} = \mathbf{AY}, \quad \text{where } \mathbf{A} = \begin{pmatrix} -2 & -3 \\ 3 & -2 \end{pmatrix}$$

indicates that solution curves spiral toward the origin (see Figure 3.20). The corresponding $x(t)$- and $y(t)$-graphs of solutions must alternate between positive and negative values with decreasing amplitude as the solution curve in the phase plane winds around the origin. The pictures of the solution curve and the $x(t)$- and $y(t)$-graphs confirm this, at least to some extent (see Figures 3.20 and 3.21). From these graphs it appears that the solution winds only once around the origin before reaching $(0, 0)$. Actually this solution spirals infinitely often, though these oscillations are difficult to detect. In Figure 3.22 we magnify a small portion of Figure 3.20. Indeed, the solution does continue to spiral.

The formula for the general solution of this system provides us with detailed behavior of this spiral. The oscillation in $x(t)$ and $y(t)$ are caused by the sine and cosine

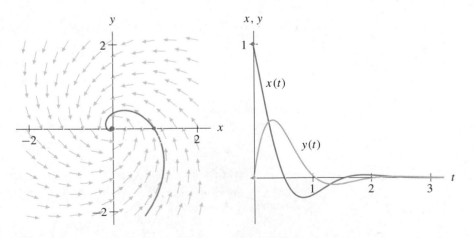

Figure 3.20

A solution curve in the phase plane for

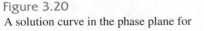

Figure 3.21

The $x(t)$- and $y(t)$-graphs of a solution to this differential equation.

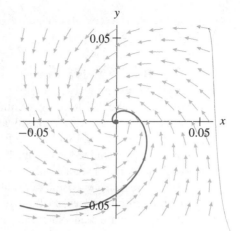

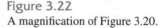

Figure 3.22
A magnification of Figure 3.20.

terms. These trigonometric expressions are all of the form $\sin 3t$ and $\cos 3t$, so when t increases through $2\pi/3$, these terms return to their original values. Hence the period of the oscillation around the origin is always $2\pi/3$, no matter how large t is or how close the solution comes to the origin. Meanwhile, solutions must approach the origin because of the exponential term e^{-2t}. This term shows that the amplitude of the oscillations of the $x(t)$- and $y(t)$-graphs decreases at this very fast rate. This also explains why it is difficult to see these oscillations near the origin.

Happily, this kind of description of solutions can be accomplished without resorting to computing the general solution of the system. In fact it can be obtained from the eigenvalues alone.

The Qualitative Behavior of Systems with Complex Eigenvalues

The discussion above can be generalized to any linear systems with complex eigenvalues. First, find a complex solution by finding the complex eigenvalues and eigenvectors. Then take the real and imaginary parts of this solution to obtain two independent solutions (see Exercise 19 for the verification that the real and imaginary parts of a solution are independent solutions). Finally, form the general solution in the usual way as a linear combination of the two independent particular solutions. This is sometimes a very tedious process, but it works. If the general solution is what we need, then we can find it.

Just as in the case of real eigenvalues, we can tell a tremendous amount about the system from the complex eigenvalues without doing all the detailed computations to obtain the general solution. Suppose

$$\frac{d\mathbf{Y}}{dt} = \mathbf{AY}$$

is a linear system with complex eigenvalues $\lambda_1 = \alpha + i\beta$ and $\lambda_2 = \alpha - i\beta$, $\beta \neq 0$. (Verifying that complex eigenvalues always come in pairs of this form is an interesting exercise — see Exercise 17). Then we know that the complex solutions have the form

$$\mathbf{Y}(t) = e^{(\alpha+i\beta)t}\mathbf{Y}_0,$$

where $\mathbf{Y}_0$ is a (complex) eigenvector of the matrix $\mathbf{A}$. We can rewrite this as

$$\mathbf{Y}(t) = e^{\alpha t}(\cos \beta t + i \sin \beta t)\mathbf{Y}_0.$$

Because $\mathbf{Y}_0$ is a constant, the real and imaginary parts of the solution $\mathbf{Y}(t)$ are a combination of two types of terms, the exponential and trigonometric terms. The effect of the exponential term on solutions depends on the sign of α. If $\alpha > 0$, then the $e^{\alpha t}$ term increases exponentially as $t \to \infty$, and the solution curve spirals off "toward infinity." If $\alpha < 0$, then the $e^{\alpha t}$ term tends to zero exponentially fast as t increases, so the solutions tend to the origin. If $\alpha = 0$, then the $e^{\alpha t}$ is identically 1 and the solutions oscillate with constant amplitude for all time. That is, the solutions are periodic.

The sine and cosine terms alternate from positive to negative and back again as t increases or decreases, so these terms make $x(t)$ and $y(t)$ oscillate. Hence the solutions in the xy-phase plane spiral around $(0, 0)$. The period of this oscillation is the time it takes to go around once (say from one crossing of the positive x-axis to the next). The period is determined by β. The functions $\sin \beta t$ and $\cos \beta t$ satisfy the equations

$$\sin \beta(t + 2\pi/\beta) = \sin \beta t$$
$$\cos \beta(t + 2\pi/\beta) = \cos \beta t,$$

so increasing t by $2\pi/\beta$ returns $\sin \beta t$ and $\cos \beta t$ to their original values (see Figure 3.23).

We can summarize these observations with the following classification. Given a linear system

$$\frac{d\mathbf{Y}}{dt} = \mathbf{A}\mathbf{Y}$$

that has complex eigenvalues $\lambda_1 = \alpha + i\beta$ and $\lambda_2 = \alpha - i\beta$, $\beta > 0$, the solutions spiral around the origin with a period of $2\pi/\beta$. Moreover:

- If $\alpha < 0$, then the solutions spiral toward the origin. In this case the origin is called a **spiral sink**.
- If $\alpha > 0$, then the solutions spiral away from the origin. In this case the origin is called a **spiral source**.
- If $\alpha = 0$, then the solutions are *periodic*. They return exactly to their initial conditions in the phase plane and repeat the same closed curve over and over. In this case the origin is called a **center**.

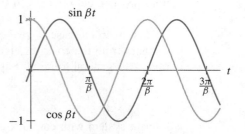

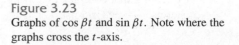

Figure 3.23
Graphs of $\cos \beta t$ and $\sin \beta t$. Note where the graphs cross the t-axis.

The question of which way the solutions spiral — clockwise or counterclockwise — can best be answered by looking at the direction field. Even one vector of the direction field is enough to tell which way a system with complex eigenvalues spirals. For example, if the direction field at $(1, 0)$ points down into the fourth quadrant, then the solutions must spiral in the clockwise direction. If the direction field at $(1, 0)$ points up into the first quadrant, then the solutions must spiral in the counterclockwise direction.

As we have seen, the $\sin \beta t$ and $\cos \beta t$ terms in the solutions make the solutions oscillate with a period of $2\pi/\beta$. This quantity is called the **natural period** of the system. Every solution takes this amount of time to wind once around the origin. The **natural frequency** is the reciprocal of the natural period, which is $\beta/2\pi$.

A word of caution concerning terminology: In some subjects the terms *natural frequency* and *natural period* are reserved for linear systems with eigenvalues having zero real part (that is, only for centers, not spiral sinks or sources — that convention is reasonable because solutions of the system are periodic only for centers). We use the terms *natural period* and *natural frequency* for any linear system with complex eigenvalues.

A Spiral Source

Consider the initial-value problem

$$\frac{d\mathbf{Y}}{dt} = \mathbf{BY}, \quad \mathbf{Y}(0) = \begin{pmatrix} 1 \\ 1 \end{pmatrix}, \quad \text{where } \mathbf{B} = \begin{pmatrix} 0 & 2 \\ -3 & 2 \end{pmatrix}.$$

The eigenvalues are the roots of the characteristic polynomial

$$\det(\mathbf{B} - \lambda\mathbf{I}) = (0 - \lambda)(2 - \lambda) + 6 = \lambda^2 - 2\lambda + 6,$$

so the eigenvalues are $\lambda = 1 \pm i\sqrt{5}$. Since the real parts of the eigenvalues are positive, the origin is a spiral source, and the natural period of the system is $2\pi/\sqrt{5}$. Thus the solution of the initial-value problem oscillates with increasing amplitude and constant period of $2\pi/\sqrt{5}$. The direction field (see Figure 3.24) shows that solutions spiral in the clockwise direction around the origin in the xy-phase plane.

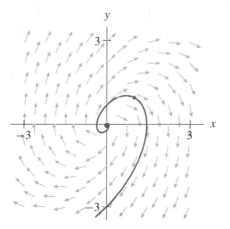

Figure 3.24
Direction field and the solution of the initial-value problem

$$\frac{d\mathbf{Y}}{dt} = \begin{pmatrix} 0 & 2 \\ -3 & 2 \end{pmatrix} \mathbf{Y} \quad \text{and} \quad \mathbf{Y}(0) = \begin{pmatrix} 1 \\ 1 \end{pmatrix}.$$

The solution spirals away from the origin.

To find the formula for the solution of the initial-value problem, we must find an eigenvector (x_0, y_0) for one of the eigenvalues, say $1 + i\sqrt{5}$. In other words, we solve

$$\mathbf{B}\begin{pmatrix} x_0 \\ y_0 \end{pmatrix} = (1 + i\sqrt{5})\begin{pmatrix} x_0 \\ y_0 \end{pmatrix},$$

which is equivalent to

$$\begin{cases} 2y_0 = (1 + i\sqrt{5})x_0 \\ -3x_0 + 2y_0 = (1 + i\sqrt{5})y_0. \end{cases}$$

Just as in the case of real eigenvalues, these two equations are redundant. (The second equation can be turned into the first by subtracting $2y_0$ from both sides, then multiplying both sides by $-(1 + i\sqrt{5})/3$.) We only need one eigenvector, so we choose any convenient value for x_0 and solve for y_0. If we set $x_0 = 2$, then we must have $y_0 = 1 + i\sqrt{5}$. Hence for the eigenvalue $\lambda = 1 + i\sqrt{5}$, the vector $(2, 1 + i\sqrt{5})$ is an eigenvector.

The corresponding complex solution is

$$\mathbf{Y}_1(t) = e^{(1+i\sqrt{5})t}\begin{pmatrix} 2 \\ 1 + i\sqrt{5} \end{pmatrix}.$$

Rewriting this using Euler's formula we obtain

$$\mathbf{Y}_1(t) = e^t\begin{pmatrix} 2\cos(\sqrt{5}\,t) \\ \cos(\sqrt{5}\,t) - \sqrt{5}\sin(\sqrt{5}\,t) \end{pmatrix} + ie^t\begin{pmatrix} 2\sin(\sqrt{5}\,t) \\ \sqrt{5}\cos(\sqrt{5}\,t) + \sin(\sqrt{5}\,t) \end{pmatrix}.$$

The general solution is

$$\mathbf{Y}(t) = k_1e^t\begin{pmatrix} 2\cos(\sqrt{5}\,t) \\ \cos(\sqrt{5}\,t) - \sqrt{5}\sin(\sqrt{5}\,t) \end{pmatrix} + k_2e^t\begin{pmatrix} 2\sin(\sqrt{5}\,t) \\ \sqrt{5}\cos(\sqrt{5}\,t) + \sin(\sqrt{5}\,t) \end{pmatrix},$$

where k_1 and k_2 are arbitrary constants. To solve the initial-value problem, we solve for k_1 and k_2 by setting the general solution at $t = 0$ equal to the initial condition $(1, 1)$, obtaining

$$k_1\begin{pmatrix} 2 \\ 1 \end{pmatrix} + k_2\begin{pmatrix} 0 \\ \sqrt{5} \end{pmatrix} = \begin{pmatrix} 1 \\ 1 \end{pmatrix}.$$

So $k_1 = 1/2$ and $k_2 = 1/(2\sqrt{5})$. The solution of the initial-value problem is

$$\mathbf{Y}(t) = \frac{1}{2}e^t\begin{pmatrix} 2\cos(\sqrt{5}\,t) \\ \cos(\sqrt{5}\,t) - \sqrt{5}\sin(\sqrt{5}\,t) \end{pmatrix} + \frac{1}{2\sqrt{5}}e^t\begin{pmatrix} 2\sin(\sqrt{5}\,t) \\ \sqrt{5}\cos(\sqrt{5}\,t) + \sin(\sqrt{5}\,t) \end{pmatrix}.$$

Centers

Consider an undamped harmonic oscillator with mass $m = 1$, spring constant $k = 2$, and with no damping ($b = 0$). The second-order equation is

$$\frac{d^2y}{dt^2} = -2y,$$

and the corresponding linear system is

$$\frac{d\mathbf{Y}}{dt} = \mathbf{C}\mathbf{Y}, \quad \text{where } \mathbf{C} = \begin{pmatrix} 0 & 1 \\ -2 & 0 \end{pmatrix} \quad \text{and} \quad \mathbf{Y} = \begin{pmatrix} y \\ v \end{pmatrix}.$$

The direction field for this system is given in Figure 3.25. We see from this picture that the solution curves encircle the origin. From this we predict that the eigenvalues for this system are complex. It is difficult to determine from the direction field picture if solution curves are periodic or if they spiral very slowly toward or away from the origin. Since these equations model a mechanical system for which we have assumed there is no damping, we might suspect that the solution curves are periodic. We can verify this by computing the eigenvalues for the system. As a bonus, the eigenvalues give us the period of the oscillations.

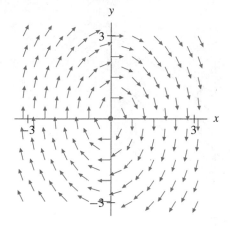

Figure 3.25

Direction field for the undamped harmonic oscillator system

$$\frac{d\mathbf{Y}}{dt} = \mathbf{C}\mathbf{Y}, \quad \text{where } \mathbf{C} = \begin{pmatrix} 0 & 1 \\ -2 & 0 \end{pmatrix}.$$

Although the direction field suggests that the eigenvalues of the system are complex, we cannot determine by looking at the direction field if the origin is a center, a spiral source, or a spiral sink.

The eigenvalues for the matrix $\mathbf{C}$ are the roots of its characteristic polynomial

$$\det(\mathbf{C} - \lambda\mathbf{I}) = (0 - \lambda)(0 - \lambda) + 2 = \lambda^2 + 2,$$

which are $\lambda = \pm i\sqrt{2}$. Hence the origin is a center and all solutions are periodic. The imaginary part of the eigenvalue is $\sqrt{2}$, so the natural period of the system is $(2\pi)/\sqrt{2}$. This means that every solution completes one oscillation in $\sqrt{2}\,\pi$ units of time regardless of its initial condition.

In fact all solution curves for this system lie on ellipses that encircle the origin. To see why, we compute the general solution of the system. Using methods of this section we first find that a complex eigenvector corresponding to the eigenvalue $i\sqrt{2}$ is $(1, i\sqrt{2})$, and we obtain the general solution

$$\mathbf{Y}(t) = k_1 \begin{pmatrix} \cos(\sqrt{2}\,t) \\ -\sqrt{2}\sin(\sqrt{2}\,t) \end{pmatrix} + k_2 \begin{pmatrix} \sin(\sqrt{2}\,t) \\ \sqrt{2}\cos(\sqrt{2}\,t) \end{pmatrix}.$$

Note that if $k_2 = 0$, we have

$$\begin{pmatrix} x(t) \\ y(t) \end{pmatrix} = \begin{pmatrix} k_1 \cos(\sqrt{2}\,t) \\ -k_1\sqrt{2}\sin(\sqrt{2}\,t) \end{pmatrix}.$$

Since we have

$$\frac{(x(t))^2}{k_1^2} + \frac{(y(t))^2}{2k_1^2} = 1,$$

this solution lies on an ellipse.

All solution curves for linear systems for which the origin is a center are ellipses (or circles). However, these ellipses need not have major and minor axes that lie along the x- and y-axes. For example, consider the linear system

$$\frac{d\mathbf{Y}}{dt} = \mathbf{DY}, \qquad \text{where } \mathbf{D} = \begin{pmatrix} -3 & 10 \\ -1 & 3 \end{pmatrix}.$$

The eigenvalues of this matrix are roots of $\lambda^2 + 1 = 0$, so $\lambda = \pm i$. The phase portrait consists entirely of ellipses, but they are not "centered" (see Figure 3.26 and Exercise 26).

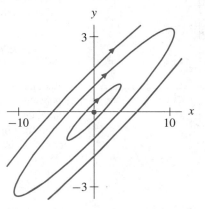

Figure 3.26
The phase portrait for the system

$$\frac{d\mathbf{Y}}{dt} = \begin{pmatrix} -3 & 10 \\ -1 & 3 \end{pmatrix} \mathbf{Y}.$$

Paul's and Bob's CD Stores Revisited

Recall the model for Paul's and Bob's compact disc stores from Section 3.1, using the linear system

$$\frac{d\mathbf{Y}}{dt} = \mathbf{AY}.$$

Now suppose that the coefficient matrix is

$$\mathbf{A} = \begin{pmatrix} 2 & 1 \\ -4 & -1 \end{pmatrix}.$$

We would like to predict the behavior of solutions to this system with as little computation as possible.

First we compute the eigenvalues from the characteristic polynomial

$$\det(\mathbf{A} - \lambda\mathbf{I}) = (2 - \lambda)(-1 - \lambda) + 4 = \lambda^2 - \lambda + 2 = 0.$$

The roots are $(1 \pm i\sqrt{7})/2$, so we know that solutions spiral around the equilibrium point at $(0, 0)$. Because the real part of the eigenvalues is $1/2$, the origin is a spiral source. This information tells us that every solution (except the equilibrium point $(0, 0)$) spirals away from $(0, 0)$ in bigger and bigger loops as t increases. We can determine the direction (clockwise or counterclockwise) and approximate shape of the solution curves by sketching the phase portrait (see Figure 3.27).

The $x(t)$- and $y(t)$-graphs of solutions oscillate with increasing amplitude. The period of these oscillations is $2\pi/(\sqrt{7}/2) = 4\pi/\sqrt{7} \approx 4.71$, and the amplitude increases like $e^{t/2}$. We sketch the qualitative behavior of the $x(t)$- and $y(t)$-graphs in Figure 3.28.

Either Paul and Bob will stay precisely at the break-even point $(x, y) = (0, 0)$, or the profits and losses of their stores will go up and down with increasing amplitude (a boom to bust to boom business cycle). Also the equilibrium point at the origin is unstable, so even a tiny profit or loss by either store eventually leads to large oscillations in the profits of both stores. It would be very difficult to predict this behavior from just looking at the linear system without any computations.

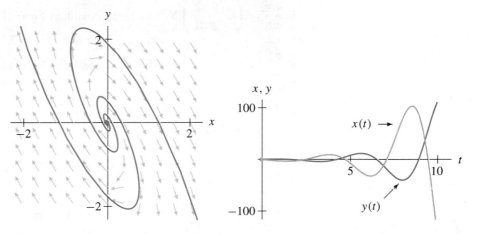

Figure 3.27
Phase portrait for
$$\frac{d\mathbf{Y}}{dt} = \begin{pmatrix} 2 & 1 \\ -4 & -1 \end{pmatrix} \mathbf{Y}.$$

Figure 3.28
$x(t)$- and $y(t)$- graphs of a solution for the system
$$\frac{d\mathbf{Y}}{dt} = \begin{pmatrix} 2 & 1 \\ -4 & -1 \end{pmatrix} \mathbf{Y}.$$

EXERCISES FOR SECTION 3.4

1. Rewrite the vector-valued function

$$\mathbf{Y}(t) = e^{(1+3i)t} \begin{pmatrix} 2+i \\ 1 \end{pmatrix}$$

in the form $\mathbf{Y}_{re}(t) + i\mathbf{Y}_{im}(t)$, where $\mathbf{Y}_{re}(t)$ and $\mathbf{Y}_{im}(t)$ are real-valued functions.

2. Rewrite the vector-valued function

$$\mathbf{Y}(t) = e^{(2+i)t} \begin{pmatrix} 1 \\ 4i \end{pmatrix}$$

in the form $\mathbf{Y}_{\text{re}}(t) + i\mathbf{Y}_{\text{im}}(t)$, where $\mathbf{Y}_{\text{re}}(t)$ and $\mathbf{Y}_{\text{im}}(t)$ are real-valued functions.

In Exercises 3–8, each linear system has complex eigenvalues. For each system,
  (a) find the eigenvalues;
  (b) determine if the origin is a spiral sink, a spiral source, or a center;
  (c) determine the natural period and natural frequency of the oscillations,
  (d) determine the direction of the oscillations in the phase plane (do the solutions go clockwise or counterclockwise around the origin?); and
  (e) sketch the xy-phase plane and the $x(t)$- and $y(t)$-graphs for the solutions with the indicated initial conditions.

3. $\dfrac{d\mathbf{Y}}{dt} = \begin{pmatrix} 0 & 2 \\ -2 & 0 \end{pmatrix} \mathbf{Y}$, with initial condition $\mathbf{Y}_0 = (1, 0)$

4. $\dfrac{d\mathbf{Y}}{dt} = \begin{pmatrix} 2 & 2 \\ -4 & 6 \end{pmatrix} \mathbf{Y}$, with initial condition $\mathbf{Y}_0 = (1, 1)$.

5. $\dfrac{d\mathbf{Y}}{dt} = \begin{pmatrix} -1 & 2 \\ -1 & -1 \end{pmatrix} \mathbf{Y}$, with initial condition $\mathbf{Y}_0 = (0, 1)$

6. $\dfrac{d\mathbf{Y}}{dt} = \begin{pmatrix} 0 & 2 \\ -2 & -1 \end{pmatrix} \mathbf{Y}$, with initial condition $\mathbf{Y}_0 = (-1, 1)$

7. $\dfrac{d\mathbf{Y}}{dt} = \begin{pmatrix} 2 & -6 \\ 2 & 1 \end{pmatrix} \mathbf{Y}$, with initial condition $\mathbf{Y}_0 = (2, 1)$

8. $\dfrac{d\mathbf{Y}}{dt} = \begin{pmatrix} 1 & 4 \\ -3 & 2 \end{pmatrix} \mathbf{Y}$, with initial condition $\mathbf{Y}_0 = (1, -1)$

In Exercises 9–14, the linear systems are the same as in Exercises 3–8. For each system,
  (a) find the general solution;
  (b) find the particular solution with the given initial value; and
  (c) sketch the $x(t)$- and $y(t)$-graphs of the particular solution. (Compare these sketches with the sketches you obtained in the corresponding problem from Exercises 3–8.)

9. $\dfrac{d\mathbf{Y}}{dt} = \begin{pmatrix} 0 & 2 \\ -2 & 0 \end{pmatrix} \mathbf{Y}$, with initial condition $\mathbf{Y}_0 = (1, 0)$

10. $\dfrac{d\mathbf{Y}}{dt} = \begin{pmatrix} 2 & 2 \\ -4 & 6 \end{pmatrix} \mathbf{Y}$, with initial condition $\mathbf{Y}_0 = (1, 1)$.

11. $\dfrac{d\mathbf{Y}}{dt} = \begin{pmatrix} -1 & 2 \\ -1 & -1 \end{pmatrix} \mathbf{Y}$, with initial condition $\mathbf{Y}_0 = (0, 1)$

12. $\dfrac{d\mathbf{Y}}{dt} = \begin{pmatrix} 0 & 2 \\ -2 & -1 \end{pmatrix} \mathbf{Y}$, with initial condition $\mathbf{Y}_0 = (-1, 1)$

13. $\dfrac{d\mathbf{Y}}{dt} = \begin{pmatrix} 2 & -6 \\ 2 & 1 \end{pmatrix} \mathbf{Y}$, with initial condition $\mathbf{Y}_0 = (2, 1)$

14. $\dfrac{d\mathbf{Y}}{dt} = \begin{pmatrix} 1 & 4 \\ -3 & 2 \end{pmatrix} \mathbf{Y}$, with initial condition $\mathbf{Y}_0 = (1, -1)$

15. The following six figures are graphs of functions $x(t)$. Two of them are $x(t)$-graphs of a solution of a linear system with complex eigenvalues, and the other four are not.

   (a) Identify which of the graphs are $x(t)$-graphs of a solution of a linear system.

   (b) For these two graphs, give the natural period of the system and classify the equilibrium point at the origin as a spiral sink, a spiral source, or a center.

   (c) For each of the other four graphs, describe how you can be sure that they are not the $x(t)$-graph of a linear system with complex eigenvalues.

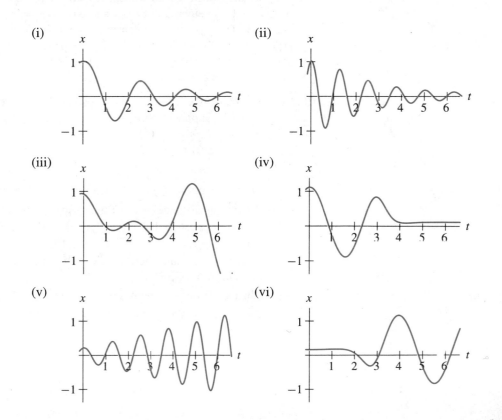

16. Show that a matrix of the form

$$A = \begin{pmatrix} a & b \\ -b & a \end{pmatrix}$$

with $b \neq 0$ must have complex eigenvalues.

17. Suppose that a and b are real numbers and that the polynomial $\lambda^2 + a\lambda + b$ has $\lambda_1 = \alpha + i\beta$ as a root with $\beta \neq 0$. Show that $\lambda_2 = \alpha - i\beta$, the complex conjugate of λ_1, must also be a root. [*Hint*: There are (at least) two ways to attack this problem. Either look at the form of the quadratic formula for the roots, or notice that

$$(\alpha + i\beta)^2 + a(\alpha + i\beta) + b = 0$$

and take the complex conjugate of both sides of this equation.]

18. Suppose that the matrix $\mathbf{A}$ with real entries has complex eigenvalues $\lambda = \alpha + i\beta$ and $\bar{\lambda} = \alpha - i\beta$ with $\beta \neq 0$. Show that the eigenvectors of $\mathbf{A}$ must be complex; that is, show that, if $\mathbf{Y}_0 = (x_0, y_0)$ is an eigenvector for $\mathbf{A}$, then either x_0 or y_0 or both have a nonzero imaginary part.

19. Suppose the matrix $\mathbf{A}$ with real entries has the complex eigenvalue $\lambda = \alpha + i\beta$, $\beta \neq 0$. Let $\mathbf{Y}_0$ be an eigenvector for λ and write $\mathbf{Y}_0 = \mathbf{Y}_1 + i\mathbf{Y}_2$, where $\mathbf{Y}_1 = (x_1, y_1)$ and $\mathbf{Y}_2 = (x_2, y_2)$ have real entries. Show that $\mathbf{Y}_1$ and $\mathbf{Y}_2$ are linearly independent. [*Hint*: Suppose they are not linearly independent. Then $(x_2, y_2) = k(x_1, y_1)$ for some constant k. Then $\mathbf{Y}_0 = (1 + ik)\mathbf{Y}_1$. Then use the fact that $\mathbf{Y}_0$ is an eigenvector of $\mathbf{A}$ and that $\mathbf{A}\mathbf{Y}_1$ contains no imaginary part.]

20. Suppose the matrix $\mathbf{A}$ with real entries has complex eigenvalues $\lambda = \alpha + i\beta$ and $\bar{\lambda} = \alpha - i\beta$. Suppose also that $\mathbf{Y}_0 = (x_1 + iy_1, x_2 + iy_2)$ is an eigenvector for the eigenvalue λ. Show that $\overline{\mathbf{Y}_0} = (x_1 - iy_1, x_2 - iy_2)$ is an eigenvector for the eigenvalue $\bar{\lambda}$. In other words, the complex conjugate of an eigenvector for λ is an eigenvector for $\bar{\lambda}$.

21. Consider the function $x(t) = e^{-\alpha t} \sin(\beta t)$ for $\alpha, \beta > 0$.

(a) What is the distance between successive zeros of this function? More precisely, if $t_1 < t_2$ are such that $x(t_1) = x(t_2) = 0$ and $x(t) \neq 0$ for $t_1 < t < t_2$, then what is $t_2 - t_1$?

(b) What is the distance between the first local maximum and the first local minimum of $x(t)$ for $t > 0$?

(c) What is the distance between the first two local maxima of $x(t)$ for $t > 0$?

(d) What is the distance between $t = 0$ and the first local maximum of $x(t)$ for $t > 0$?

22. Show that a function of the form

$$x(t) = k_1 \cos \beta t + k_2 \sin \beta t$$

can be written as

$$x(t) = K \cos(\beta t - \phi),$$

where $K = \sqrt{k_1^2 + k_2^2}$. (Sometimes a solution of a linear system with complex co-efficients is expressed in this form in order to clarify its behavior. The magnitude K gives the *amplitude* of the solution, and the angle ϕ is the *phase* of the solution.) [*Hint*: Pick ϕ such that $K \cos \phi = k_1$ and $K \sin \phi = k_2$.]

23. For the second-order equation

$$\frac{d^2 y}{dt^2} + p\frac{dy}{dt} + qy = 0 :$$

    (a) Write this equation as a first-order linear system.

    (b) What conditions on p and q guarantee that the eigenvalues of the corresponding linear system are complex?

    (c) What relationship between p and q guarantees that the origin is a spiral sink? What relationship guarantees that the origin is a spiral sink? What relationship guarantees that the origin is a center?

    (d) If the eigenvalues are complex, what conditions on p and q guarantee that solutions spiral around the origin in a clockwise direction?

24. The slope field for the system

$$\frac{dx}{dt} = -0.9x - 2y$$

$$\frac{dy}{dt} = x + 1.1y$$

is given to the right. Plot the $x(t)$- and $y(t)$-graphs for the initial conditions $A = (1, 1)$ and $B = (\ 2, 1)$. What do the graphs have in common?

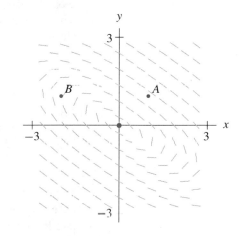

25. (Essay Question) We have seen that linear systems with real eigenvalues can be classified as sinks, sources, or saddles, depending on whether the eigenvalues are greater or less than zero. Linear systems with complex eigenvalues can be classified as spiral sources, spiral sinks, or centers, depending on the sign of the real part of the eigenvalue. Why is there not a type of linear system called a "spiral saddle"?

26. Consider the linear system

$$\frac{d\mathbf{Y}}{dt} = \begin{pmatrix} -3 & 10 \\ -1 & 3 \end{pmatrix} \mathbf{Y}.$$

Show that all solution curves in the phase portrait for this system are elliptical.

3.5 SPECIAL CASES: REPEATED AND ZERO EIGENVALUES

In the previous three sections we dealt with the linear systems

$$\frac{d\mathbf{Y}}{dt} = \mathbf{AY}$$

for which the 2×2 matrix $\mathbf{A}$ has either two distinct, nonzero real eigenvalues or complex eigenvalues. In these cases we were able to use the eigenvalues and eigenvectors to sketch the solutions in the xy-phase plane, to draw the $x(t)$- and $y(t)$-graphs, and to derive an explicit formula for the general solution. We have not yet dealt with the case where the characteristic polynomial of $\mathbf{A}$ has only one root (a double root), so there is only one eigenvalue. In previous sections we also classified the equilibrium point at the origin as a sink, source, saddle, spiral sink, spiral source, or center, depending on the sign of the eigenvalues (or the sign of their real parts). This classification scheme leaves out the cases where one or both eigenvalues are zero. In this section we will modify our methods to handle these remaining cases.

Most quadratic polynomials have two distinct, nonzero roots, so linear systems with only one eigenvalue or with an eigenvalue equal to zero are relatively rare. These systems are sometimes called *degenerate*. Nevertheless, they are still important. These special systems form the "boundaries" between the most common types of linear systems. Whenever we study linear systems that depend on a parameter and the system changes behavior or bifurcates as the parameter changes, these special systems play a crucial role (see Section 3.7).

A System with Repeated Eigenvalues

Consider the linear system

$$\frac{d\mathbf{Y}}{dt} = \mathbf{AY} = \begin{pmatrix} -2 & 1 \\ 0 & -2 \end{pmatrix} \mathbf{Y}.$$

The direction field for this system looks somewhat different from the vector fields we have considered thus far in that there appears to be only one straight line of solutions (see Figure 3.29).

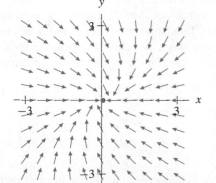

Figure 3.29
Direction field for the system

$$\frac{d\mathbf{Y}}{dt} = \begin{pmatrix} -2 & 1 \\ 0 & -2 \end{pmatrix} \mathbf{Y}.$$

The straight-line solutions all lie on the x-axis.

From an algebraic point of view, this matrix is also unusual. The eigenvalues of this system are the roots of the characteristic polynomial

$$\det(\mathbf{A} - \lambda \mathbf{I}) = (-2 - \lambda)(-2 - \lambda) - 0 = 0,$$

which has only one root at $\lambda = -2$. We say that $\mathbf{A}$ has a "repeated" eigenvalue $\lambda = -2$. We find the associated eigenvectors by solving $\mathbf{A}\mathbf{Y}_0 = -2\mathbf{Y}_0$ for $\mathbf{Y}_0 = (x_0, y_0)$. We have

$$\begin{cases} -2x_0 + y_0 = -2x_0 \\ \quad -2y_0 = -2y_0, \end{cases}$$

which yields $y_0 = 0$. Therefore all eigenvectors corresponding to the eigenvalue $\lambda = -2$ lie on the x-axis, so the only straight-line solutions for this system lie on this axis. The vector $(1, 0)$ is one eigenvector associated to $\lambda = -2$, so

$$\mathbf{Y}_1(t) = e^{-2t} \begin{pmatrix} 1 \\ 0 \end{pmatrix}$$

is a solution of this system. But this is only one solution and, as we know, we need two independent solutions to derive the general solution. Bummer.

On the other hand, this is not a complete catastrophe. Our goal is to understand the behavior of the solutions of this system. Writing a formula for the general solution certainly helps, but this is not the only option. We can always study the system using numerical and qualitative techniques.

To obtain a qualitative description of the solutions, we start with the one straight line of solutions. Because the eigenvalue is negative, we know that solutions tend to the origin along this line as t increases. Looking at the direction field (or using Euler's method), we can sketch other solutions (see Figure 3.30). Every solution tends to the origin as t increases, so $(0, 0)$ is a sink. For initial conditions that do not lie on the straight-line solution, it appears that the corresponding solutions make a turn and arrive at the origin in a direction that is tangent to the straight-line solution. These solutions look as if they are "trying to spiral" around the origin, but the line of solutions somehow "gets in the way." We see in the next section that linear systems with repeated eigenvalues form the "boundary" between linear systems that spiral and those with two independent lines of solutions.

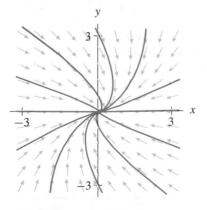

Figure 3.30
Phase portrait for the system

$$\frac{d\mathbf{Y}}{dt} = \begin{pmatrix} -2 & 1 \\ 0 & -2 \end{pmatrix} \mathbf{Y}.$$

Note that all solution curves approach the sink at the origin although there is only one line of eigenvectors.

The general solution for this example

In order to get some idea of how to compute the general solution of this system, we go back to the original equations of the system and rewrite it in components,

$$\frac{dx}{dt} = -2x + y$$
$$\frac{dy}{dt} = -2y.$$

In these equations we spot a lucky break. The second equation for dy/dt does not contain any x's, and the system partially decouples (see Section 2.3). We can consider the equation for dy/dt by itself,

$$\frac{dy}{dt} = -2y.$$

The general solution is $y(t) = k_2 e^{-2t}$ for some constant k_2. Now that we know $y(t)$, we can substitute it back into the first equation, which becomes

$$\frac{dx}{dt} = -2x + k_2 e^{-2t}.$$

We must be living a good life (or have chosen this example very carefully) because this equation is a first-order, linear equation for $x(t)$, which we can also solve (see Section 1.8). The general solution is

$$x(t) = k_2 t e^{-2t} + k_1 e^{-2t},$$

where k_1 is another constant. (The reader who does not remember how to compute the formulas for solutions of these equations should take this as a warning to review the analytic techniques developed in Chapter 1.) We can write these solutions in vector form as

$$\begin{pmatrix} x(t) \\ y(t) \end{pmatrix} = \begin{pmatrix} k_2 t e^{-2t} + k_1 e^{-2t} \\ k_2 e^{-2t} \end{pmatrix}.$$

Now we study this solution closely to see if there is any pattern that can be generalized. Hidden inside this solution is the straight-line solution we found above. By regrouping terms we find

$$\mathbf{Y}(t) = \begin{pmatrix} x(t) \\ y(t) \end{pmatrix} = k_1 e^{-2t} \begin{pmatrix} 1 \\ 0 \end{pmatrix} + k_2 e^{-2t} \begin{pmatrix} t \\ 1 \end{pmatrix}.$$

The first term consists of the straight-line solutions. The second term is new and different from any other solution to a linear system we have seen so far because of the t in the first component. By taking $k_1 = 1$ and $k_2 = 0$, we get the solution

$$\mathbf{Y}_1(t) = e^{-2t} \begin{pmatrix} 1 \\ 0 \end{pmatrix},$$

and by taking $k_1 = 0$ and $k_2 = 1$, we get

$$\mathbf{Y}_2(t) = e^{-2t} \begin{pmatrix} t \\ 1 \end{pmatrix} = t e^{-2t} \begin{pmatrix} 1 \\ 0 \end{pmatrix} + e^{-2t} \begin{pmatrix} 0 \\ 1 \end{pmatrix}.$$

These two solutions have linearly independent initial conditions, $\mathbf{Y}_1(0) = (1, 0)$ and $\mathbf{Y}_2(0) = (0, 1)$. So the solution $\mathbf{Y}(t)$ above is the general solution of the linear system.

The Form of the General Solution

The solution above provides some hints on how to find a second independent solution for systems with repeated eigenvalues but only one straight-line solution. The solution $\mathbf{Y}_2(t)$ consists of two terms. One term is $t\mathbf{Y}_1(t)$, the product of t with our known straight-line solution. The other term is the product of e^{-2t} and the vector $(0, 1)$. From the phase portrait, it is difficult to tell why $(0, 1)$ is significant. We will examine the situation algebraically.

Consider a system

$$\frac{d\mathbf{Y}}{dt} = \mathbf{AY} = \begin{pmatrix} a & b \\ c & d \end{pmatrix} \mathbf{Y},$$

where $\mathbf{A}$ has a repeated eigenvalue λ but only one line of eigenvectors. If $\mathbf{V}_1$ is an eigenvector, then

$$\mathbf{Y}_1(t) = e^{\lambda t} \mathbf{V}_1$$

is a straight-line solution. Following the pattern from the example above, we guess that there is another solution of the form

$$\mathbf{Y}_2(t) = e^{\lambda t}(t\mathbf{V}_1 + \mathbf{V}_2),$$

where $\mathbf{V}_2$ is some vector yet to be determined.

We know how to find the eigenvector $\mathbf{V}_1$. To find the vector $\mathbf{V}_2$, we must use the differential equation. If the curve $\mathbf{Y}_2(t)$ is a solution of the system, it must satisfy the differential equation

$$\frac{d\mathbf{Y}_2}{dt} = \mathbf{AY}_2 \quad \text{for all } t.$$

Using the formula for $\mathbf{Y}_2(t)$, we expand the left-hand side, obtaining

$$\frac{d\mathbf{Y}_2}{dt} = \frac{d(e^{\lambda t}(t\mathbf{V}_1 + \mathbf{V}_2))}{dt}$$

$$= \frac{d(e^{\lambda t}t\mathbf{V}_1)}{dt} + \frac{d(e^{\lambda t}\mathbf{V}_2)}{dt}$$

$$= \lambda e^{\lambda t} t\mathbf{V}_1 + e^{\lambda t}\mathbf{V}_1 + \lambda e^{\lambda t}\mathbf{V}_2.$$

Expanding the right-hand side, we obtain

$$\mathbf{AY}_2 = \mathbf{A}(e^{\lambda t}(t\mathbf{V}_1 + \mathbf{V}_2))$$

$$= e^{\lambda t}t\mathbf{AV}_1 + e^{\lambda t}\mathbf{AV}_2$$

$$= e^{\lambda t}t(\lambda\mathbf{V}_1) + e^{\lambda t}\mathbf{AV}_2$$

$$= \lambda e^{\lambda t}t\mathbf{V}_1 + e^{\lambda t}\mathbf{AV}_2,$$

because $\mathbf{V}_1$ is an eigenvector for $\mathbf{A}$ with eigenvalue λ.

If $\mathbf{Y}_2(t)$ is a solution, then the two sides of the differential equation must be equal, so

$$\lambda e^{\lambda t} t \mathbf{V}_1 + e^{\lambda t} \mathbf{V}_1 + \lambda e^{\lambda t} \mathbf{V}_2 = \lambda e^{\lambda t} t \mathbf{V}_1 + e^{\lambda t} \mathbf{A} \mathbf{V}_2$$

for all t. Simplifying, we obtain

$$e^{\lambda t}(\mathbf{V}_1 + \lambda \mathbf{V}_2) = e^{\lambda t}(\mathbf{A} \mathbf{V}_2)$$

or

$$\mathbf{V}_1 + \lambda \mathbf{V}_2 = \mathbf{A} \mathbf{V}_2,$$

where $\mathbf{V}_1$ is our known eigenvector for $\mathbf{A}$. This equation tells us how to determine which vector we should use for $\mathbf{V}_2$, namely, any solution of the equation

$$\mathbf{A} \mathbf{V}_2 - \lambda \mathbf{V}_2 = (\mathbf{A} - \lambda \mathbf{I}) \mathbf{V}_2 = \mathbf{V}_1.$$

Since we know $\mathbf{V}_1 = (x_1, y_1)$, the equation above can be solved for the unknown vector $\mathbf{V}_2 = (x_2, y_2)$. Writing it out in components, this equation is

$$\begin{cases} (a - \lambda)x_2 + by_2 = x_1 \\ cx_2 + (d - \lambda)y_2 = y_1. \end{cases}$$

We summarize this calculation in the following theorem.

THEOREM   Suppose $d\mathbf{Y}/dt = \mathbf{A}\mathbf{Y}$ is a linear system in which the 2×2 matrix $\mathbf{A}$ has a repeated real eigenvalue λ but only one line of eigenvectors. If $\mathbf{V}_1$ is an eigenvector, then the function $\mathbf{Y}_1(t) = e^{\lambda t} \mathbf{V}_1$ is a straight-line solution. There is another solution of the form

$$\mathbf{Y}_2(t) = e^{\lambda t}(t\mathbf{V}_1 + \mathbf{V}_2),$$

where $\mathbf{V}_2$ satisfies the equation $\mathbf{A}\mathbf{V}_2 - \lambda \mathbf{V}_2 = \mathbf{V}_1$. The solutions $\mathbf{Y}_1(t)$ and $\mathbf{Y}_2(t)$ are independent. Consequently the general solution has the form

$$\mathbf{Y}(t) = k_1 e^{\lambda t} \mathbf{V}_1 + k_2 e^{\lambda t}(t\mathbf{V}_1 + \mathbf{V}_2),$$

where k_1 and k_2 are arbitrary constants.   ∎

In fact there can be a lot of arithmetic involved in finding this general solution. Luckily finding the vector $\mathbf{V}_2$ is not so hard in most interesting cases.

Instant replay

In the example that we did "by hand" above, namely the system

$$\frac{d\mathbf{Y}}{dt} = \mathbf{A}\mathbf{Y} = \begin{pmatrix} -2 & 1 \\ 0 & -2 \end{pmatrix} \mathbf{Y},$$

we computed that $\lambda = -2$ is an eigenvalue, that $\mathbf{V}_1 = (1, 0)$ is an eigenvector, and that there is only one line of eigenvectors. So we have the straight-line solution

$$\mathbf{Y}_1(t) = e^{-2t} \begin{pmatrix} 1 \\ 0 \end{pmatrix}.$$

To find the vector $\mathbf{V}_2$, we must solve the system of equations

$$\begin{cases} (-2+2)x_2 + y_2 = 1 \\ (-2+2)y_2 = 0. \end{cases}$$

The second equation is $0 = 0$, but the first is $y_2 = 1$. Thus *any* vector whose second component is 1 can serve as $\mathbf{V}_2$. Usually we choose $\mathbf{V}_2$ as simple as possible, say $\mathbf{V}_2 = (0, 1)$, and we obtain the general solution

$$\mathbf{Y}(t) = k_1 e^{-2t} \begin{pmatrix} 1 \\ 0 \end{pmatrix} + k_2 e^{-2t} \left(t \begin{pmatrix} 1 \\ 0 \end{pmatrix} + \begin{pmatrix} 0 \\ 1 \end{pmatrix} \right) = \begin{pmatrix} k_1 e^{-2t} + k_2 t e^{-2t} \\ k_2 e^{-2t} \end{pmatrix},$$

which is exactly what we determined above.

Qualitative Analysis of Systems with Repeated Eigenvalues

Before discussing additional examples, we consider what the general solution tells us about the qualitative behavior of solutions. The form of the general solution is

$$\mathbf{Y}(t) = k_1 e^{\lambda t} \mathbf{V}_1 + k_2 e^{\lambda t}(t\mathbf{V}_1 + \mathbf{V}_2) = e^{\lambda t}(k_1 \mathbf{V}_1 + k_2 \mathbf{V}_2) + t e^{\lambda t} k_2 \mathbf{V}_1.$$

The dependence on t comes in two terms, the $e^{\lambda t}$ term and the $t e^{\lambda t}$ term.

If $\lambda < 0$, then both of these terms tend to zero as t increases, and therefore the equilibrium point at the origin is a sink. The $t e^{\lambda t}$ term is much larger than $e^{\lambda t}$ if t is large. Consequently $\mathbf{Y}(t) \approx t e^{\lambda t} k_2 \mathbf{V}_1$ when t is large. Thus the solution tends to $(0, 0)$ in a direction tangent to the line of eigenvectors (see Figure 3.31 for typical examples of phase portraits for these systems).

If $\lambda > 0$, then all solutions (except the equilibrium solution) tend to infinity as t increases, so $(0, 0)$ is a source. Again the $t e^{\lambda t}$ term dominates for large t if $k_2 \neq 0$.

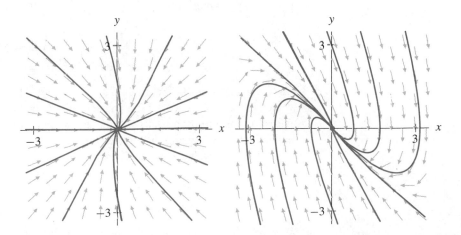

Figure 3.31
Typical phase portraits for systems with repeated eigenvalues.

If we use the formula for the general solution to draw the phase portrait, we see again that it looks as if the solutions (other than the one straight-line solution) are trying to spiral around $(0, 0)$. But they cannot spiral since the one straight-line solution gets in the way. The strange (and complicated) form of the general solution of systems with one straight-line of solutions is another symptom of the fact that these systems lie "between" systems with complex eigenvalues and those with real, distinct eigenvalues.

A Harmonic Oscillator with Repeated Eigenvalues

Consider the harmonic oscillator with mass $m = 1$, spring constant $k = 2$, and damping coefficient $b = 2\sqrt{2}$. The second-order equation that models the motion of the oscillator is

$$\frac{d^2y}{dt^2} + 2\sqrt{2}\frac{dy}{dt} + 2y = 0,$$

and its associated system is

$$\frac{d\mathbf{Y}}{dt} = \mathbf{BY} \quad \text{where } \mathbf{B} = \begin{pmatrix} 0 & 1 \\ -2 & -2\sqrt{2} \end{pmatrix}, \quad \mathbf{Y} = \begin{pmatrix} y \\ v \end{pmatrix},$$

and $v = dy/dt$. The eigenvalues of this system are the roots of the characteristic polynomial

$$\det(\mathbf{B} - \lambda\mathbf{I}) = (0 - \lambda)(-2\sqrt{2} - \lambda) + 2 = \lambda^2 + 2\sqrt{2}\lambda + 2.$$

Using the quadratic formula, we have

$$\lambda = \frac{-2\sqrt{2} \pm \sqrt{8 - 8}}{2},$$

so $-\sqrt{2}$ is a repeated eigenvalue.

Given the eigenvalue $\lambda = -\sqrt{2}$, we find an eigenvector $\mathbf{V}_1 = (y_1, v_1)$ by solving $\mathbf{BV}_1 = \lambda\mathbf{V}_1$, which is

$$\begin{pmatrix} 0 & 1 \\ -2 & -2\sqrt{2} \end{pmatrix} \begin{pmatrix} y_1 \\ v_1 \end{pmatrix} = -\sqrt{2} \begin{pmatrix} y_1 \\ v_1 \end{pmatrix}.$$

So the eigenvectors lie on the line $v_1 = -\sqrt{2}\,y_1$. For instance, one convenient eigenvector is $\mathbf{V}_1 = (1, -\sqrt{2})$. The phase portrait for this system is shown in Figure 3.32. All solutions approach $(0, 0)$ as t increases, and all solutions are tangent to the line of eigenvectors as they approach the origin.

The eigenvalue $\lambda = -\sqrt{2}$ and the eigenvector $\mathbf{V}_1$ give us the straight-line solution

$$\mathbf{Y}_1(t) = e^{-\sqrt{2}t} \begin{pmatrix} 1 \\ -\sqrt{2} \end{pmatrix}.$$

We find a second independent solution by first finding a vector $\mathbf{V}_2$ that satisfies

$$(\mathbf{B} - \lambda\mathbf{I})\mathbf{V}_2 = \mathbf{V}_1,$$

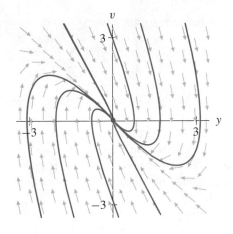

Figure 3.32
Direction field and solution curves for

$$\frac{d\mathbf{Y}}{dt} = \begin{pmatrix} 0 & 1 \\ -2 & -2\sqrt{2} \end{pmatrix} \mathbf{Y}.$$

Note that the solution curves approach the origin tangent to the straight-line solution.

which is

$$\begin{pmatrix} 0 & 1 \\ -2 & -2\sqrt{2} \end{pmatrix} \mathbf{V}_2 + \sqrt{2}\,\mathbf{V}_2 = \begin{pmatrix} 1 \\ -\sqrt{2} \end{pmatrix}.$$

One such vector is $\mathbf{V}_2 = (\sqrt{2}/2, 0)$. Hence a second solution of the system is given by

$$\mathbf{Y}_2(t) = e^{-\sqrt{2}\,t}\left(t\begin{pmatrix} 1 \\ -\sqrt{2} \end{pmatrix} + \begin{pmatrix} \sqrt{2}/2 \\ 0 \end{pmatrix} \right).$$

We can form the general solution

$$\mathbf{Y}(t) = k_1\mathbf{Y}_1(t) + k_2\mathbf{Y}_2(t) = k_1\begin{pmatrix} e^{-\sqrt{2}\,t} \\ -\sqrt{2}\,e^{-\sqrt{2}\,t} \end{pmatrix} + k_2\begin{pmatrix} (\sqrt{2}/2 + t)e^{-\sqrt{2}\,t} \\ -\sqrt{2}\,t e^{-\sqrt{2}\,t} \end{pmatrix}.$$

Paul's and Bob's CD Stores Revisited

Recall the model of Paul's and Bob's compact disc stores from Section 3.1. We suppose that the model has the form

$$\frac{d\mathbf{Y}}{dt} = \mathbf{AY} \quad \text{where } \mathbf{A} = \begin{pmatrix} -5 & 1 \\ -1 & -3 \end{pmatrix}\mathbf{Y}.$$

The coefficients imply that, if Bob is making money, then Paul's profits will increase. But if Paul is making money, then Bob's profits are hurt. In this model, rockers don't like opera.

The eigenvalues for this system are the roots of the characteristic polynomial

$$\det(\mathbf{A} - \lambda\mathbf{I}) = \lambda^2 + 8\lambda + 16.$$

There is only one eigenvalue, $\lambda = -4$. Hence the origin is a sink and all solutions tend to $(0, 0)$ as t increases. The profits of both Paul's and Bob's stores tend toward zero no matter what the initial condition.

To find the eigenvectors, we solve

$$\begin{cases} -5x_1 + y_1 = -4x_1 \\ -x_1 - 3y_1 = -4y_1, \end{cases}$$

which has the line $x_1 = y_1$ as its set of solutions. From this information we conclude that as t increases, every solution tends to $(0, 0)$ tangent to the line $x = y$. As the values of x and y tend to zero, they are almost equal.

By looking at the direction field, we can see which initial conditions lead to solutions that tend to zero along $x = y$ in the first quadrant and which initial conditions end up in the third quadrant, along with the corresponding $x(t)$- and $y(t)$-graphs (see Figures 3.33 and 3.34).

Because the solutions tend to the line $x = y$, the two stores have essentially the same profits or losses over the long term. This conclusion is a little surprising because the coefficients in the model imply that the profits of the two stores react in very different ways to the profits of the other store.

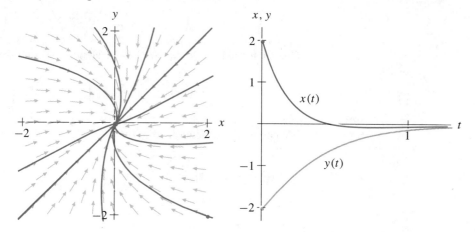

Figure 3.33
Phase portrait for the system

$$\frac{d\mathbf{Y}}{dt} = \begin{pmatrix} -5 & 1 \\ -1 & -3 \end{pmatrix} \mathbf{Y}.$$

Figure 3.34
The $x(t)$- and $y(t)$-graphs for the solution with the initial condition indicated in Figure 3.33.

Effect of a small change in the coefficients

We might wonder what would happen if the coefficients were changed just a little. Suppose Bob decides that he will help out Paul, so he puts up a tasteful sign saying

Open your mind to Rock and Roll

(But do it someplace else, like Paul's Rock and Roll CDs)

The more people who come to Bob's store (the larger y is), the more this sign helps Paul. So the parameter b increases from 1 to, say, 1.1. The new system is

$$\frac{d\mathbf{Y}}{dt} = \mathbf{BY} = \begin{pmatrix} -5 & 1.1 \\ -1 & -3 \end{pmatrix} \mathbf{Y}.$$

We can compute that the eigenvalues of this matrix are complex and have negative real part. Hence the origin is a spiral sink and all solutions still tend to $(0, 0)$. However, when the solutions are close to $(0, 0)$, they will oscillate instead of tending toward $(0, 0)$ along the line $y = x$. This small change in the system has made a distinct change in the qualitative behavior of the solutions. However, the oscillations are very subtle. In fact the phase portrait and $x(t)$- and $y(t)$-graphs for $d\mathbf{Y}/dt = \mathbf{BY}$ are indistinguishable from those of $d\mathbf{Y}/dt = \mathbf{AY}$ (compare Figures 3.33 and 3.34 to Figures 3.35 and 3.36).

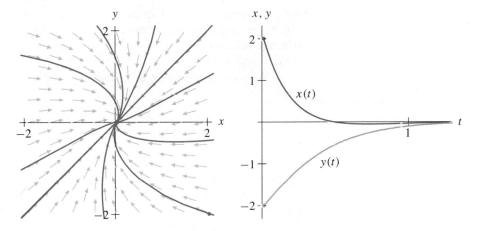

Figure 3.35
Phase portrait for

$$\frac{d\mathbf{Y}}{dt} = \begin{pmatrix} -5 & 1.1 \\ -1 & -3 \end{pmatrix} \mathbf{Y}.$$

Figure 3.36
The $x(t)$- and $y(t)$-graphs for the solution with the initial condition indicated in Figure 3.35.

Systems for Which Every Vector Is an Eigenvector

At this point we can deal with linear systems that have repeated real eigenvalues but only one line of eigenvectors. An example of a system with repeated eigenvalues and more than one line of eigenvectors is given by

$$\frac{d\mathbf{Y}}{dt} = \begin{pmatrix} a & 0 \\ 0 & a \end{pmatrix} \mathbf{Y}.$$

This system has a repeated eigenvalue $\lambda = a$, and *every* nonzero vector is an eigenvector for $\lambda = a$. In this case finding the general solution is equivalent to finding the general solutions of the two equations $dx/dt = ax$ and $dy/dt = ay$, that is, the system completely decouples (see Section 2.3). Because every vector is an eigenvector, every solution curve (except the equilibrium point at the origin) is a ray that approaches or leaves the origin as t increases. If $a > 0$, all the solutions tend to infinity as t increases (a source), whereas if $a < 0$, all solutions tend to zero as t increases (a sink). In the exercises we see that the only systems with one eigenvalue having more than one line of eigenvectors have coefficient matrix equal to $\lambda \mathbf{I}$, where λ is the eigenvalue (see Exercises 13 and 14). This case is very special.

Systems with Zero as an Eigenvalue

We are close to having a complete understanding of linear systems and their phase portraits. We can classify and sketch the behavior of solutions for the cases of real, complex, and repeated eigenvalues. The only case we have not yet explicitly considered is the case where one or both of the eigenvalues is zero. This case is important because it divides the linear systems with strictly positive eigenvalues (sources) and strictly negative eigenvalues (sinks) from those with one positive and one negative eigenvalue (saddles).

Suppose we have a linear system

$$\frac{d\mathbf{Y}}{dt} = \mathbf{AY},$$

and the matrix $\mathbf{A}$ has eigenvalues $\lambda_1 = 0$ and $\lambda_2 \neq 0$. Suppose $\mathbf{V}_1$ is an eigenvector for λ_1 and $\mathbf{V}_2$ is an eigenvector of λ_2.

We have two real, distinct eigenvalues, and all of the algebra we did in Section 3.3 applies. Hence the general solution is

$$\mathbf{Y}(t) = k_1 e^{\lambda_1 t} \mathbf{V}_1 + k_2 e^{\lambda_2 t} \mathbf{V}_2.$$

But $\lambda_1 = 0$, so $e^{\lambda_1 t} = e^{0t} = 1$ for all t. Hence

$$\mathbf{Y}(t) = k_1 \mathbf{V}_1 + k_2 e^{\lambda_2 t} \mathbf{V}_2,$$

and the general solution $\mathbf{Y}(t)$ depends on t only through the $\mathbf{V}_2$ term. If we take $k_2 = 0$, then we have $\mathbf{Y}(t) = k_1 \mathbf{V}_1$, which does not depend on t. So all the points $k_1 \mathbf{V}_1$, for any k_1, are equilibrium points. Every point on the line of eigenvectors for the eigenvalue $\lambda_1 = 0$ is an equilibrium point. If $\lambda_2 < 0$, then the second term in the general solution tends to zero as t increases, so the solution

$$\mathbf{Y}(t) = k_1 \mathbf{V}_1 + k_2 e^{\lambda_2 t} \mathbf{V}_2,$$

tends to the equilibrium point $k_1 \mathbf{V}_1$ along a line parallel to $\mathbf{V}_2$. If $\lambda_2 > 0$, then the solution above tends away from the line of equilibrium points as t increases. We have enough information to sketch the phase portraits.

An example with zero as an eigenvalue

Consider the system

$$\frac{d\mathbf{Y}}{dt} = \mathbf{AY}, \quad \text{where } \mathbf{A} = \begin{pmatrix} -3 & 1 \\ 3 & -1 \end{pmatrix}.$$

We compute the eigenvalues from the characteristic polynomial by solving

$$\det(\mathbf{A} - \lambda\mathbf{I}) = (-3 - \lambda)(-1 - \lambda) - 3 = 0,$$

which is

$$\lambda^2 + 4\lambda = 0.$$

The eigenvalues are $\lambda_1 = 0$ and $\lambda_2 = -4$.

The eigenvectors $\mathbf{V}_1 = (x_1, y_1)$ for $\lambda_1 = 0$ satisfy the equations

$$\begin{cases} -3x_1 + y_1 = 0x_1 \\ 3x_1 - y_1 = 0y_1. \end{cases}$$

They are on the line $y_1 = 3x_1$. For instance, $\mathbf{V}_1 = (1, 3)$ is an eigenvector for $\lambda_1 = 0$.

Similarly, the eigenvectors $\mathbf{V}_2 = (x_2, y_2)$ for $\lambda_2 = -4$ satisfy

$$\begin{cases} -3x_2 + y_2 = -4x_2 \\ 3x_2 - y_2 = 4y_2, \end{cases}$$

and these equations can be simplified to $x_2 + y_2 = 0$, so the solutions of these equations are on the line $x_2 = -y_2$. So $\mathbf{V}_2 = (-1, 1)$ is an eigenvector for $\lambda_2 = -4$.

From this information we can draw the phase portrait. There is a line of equilibrium points given by $y = 3x$, and every other solution approaches an equilibrium point on this line by following a line parallel to the line $y = -x$ (see Figure 3.37). The $x(t)$- and $y(t)$-graphs for the solution with initial condition $\mathbf{Y}(0) = (5, 2)$ and $\mathbf{Y}(0) = (1, 0)$ are given in Figure 3.38.

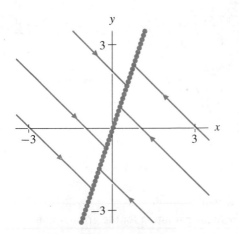

Figure 3.37
Phase portrait for the system

$$\frac{d\mathbf{Y}}{dt} = \begin{pmatrix} -3 & 1 \\ 3 & -1 \end{pmatrix} \mathbf{Y}.$$

Solutions tend toward the line of equilibrium points.

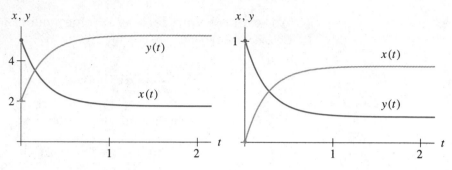

Figure 3.38
Graphs of $x(t)$ and $y(t)$ for solutions of

$$\frac{d\mathbf{Y}}{dt} = \begin{pmatrix} -3 & 1 \\ 3 & -1 \end{pmatrix} \mathbf{Y}$$

with initial positions $(5, 2)$ (left graphs) and $(1, 0)$ (right graphs).

EXERCISES FOR SECTION 3.5

In Exercises 1–4, each of the linear systems has one eigenvalue and one line of eigen-vectors. For each system,
(a) find the eigenvalue;
(b) find an eigenvector;
(c) sketch the direction field;
(d) sketch the phase plane, including the solution with the given initial condition; and
(e) sketch the $x(t)$- and $y(t)$-graphs of the solution with the given initial condition.

1. $\dfrac{d\mathbf{Y}}{dt} = \begin{pmatrix} -3 & 0 \\ 1 & -3 \end{pmatrix} \mathbf{Y}$, with initial condition $\mathbf{Y}_0 = (1, 0)$.

2. $\dfrac{d\mathbf{Y}}{dt} = \begin{pmatrix} 2 & 1 \\ -1 & 4 \end{pmatrix} \mathbf{Y}$, with initial condition $\mathbf{Y}_0 = (1, 0)$.

3. $\dfrac{d\mathbf{Y}}{dt} = \begin{pmatrix} -2 & -1 \\ 1 & -4 \end{pmatrix} \mathbf{Y}$, with initial condition $\mathbf{Y}_0 = (1, 0)$.

4. $\dfrac{d\mathbf{Y}}{dt} = \begin{pmatrix} 0 & 1 \\ -1 & -2 \end{pmatrix} \mathbf{Y}$, with initial condition $\mathbf{Y}_0 = (1, 0)$.

In Exercises 5–8, the linear systems are the same as those in Exercises 1–4. For each system,
(a) find the general solution;
(b) find the particular solution for the given initial condition; and

(c) sketch the $x(t)$- and $y(t)$-graphs of the solution. (Compare these sketches with the sketches you obtained in the corresponding problem from Exercises 1–4.)

5. $\dfrac{d\mathbf{Y}}{dt} = \begin{pmatrix} -3 & 0 \\ 1 & -3 \end{pmatrix} \mathbf{Y}$, with initial condition $\mathbf{Y}_0 = (1, 0)$.

6. $\dfrac{d\mathbf{Y}}{dt} = \begin{pmatrix} 2 & 1 \\ -1 & 4 \end{pmatrix} \mathbf{Y}$, with initial condition $\mathbf{Y}_0 = (1, 0)$.

7. $\dfrac{d\mathbf{Y}}{dt} = \begin{pmatrix} -2 & -1 \\ 1 & -4 \end{pmatrix} \mathbf{Y}$, with initial condition $\mathbf{Y}_0 = (1, 0)$.

8. $\dfrac{d\mathbf{Y}}{dt} = \begin{pmatrix} 0 & 1 \\ -1 & -2 \end{pmatrix} \mathbf{Y}$, with initial condition $\mathbf{Y}_0 = (1, 0)$.

9. Given a quadratic $\lambda^2 + \alpha\lambda + \beta$, what condition on α and β guarantees

(a) that the quadratic has a double root?

(b) that the quadratic has zero as a root?

10. Evaluate the limit of $te^{\lambda t}$ as $t \to \infty$ if

(a) $\lambda > 0$ (b) $\lambda < 0$

Be sure to justify your answer.

11. Consider the matrix

$$\mathbf{A} = \begin{pmatrix} 0 & 1 \\ -q & -p \end{pmatrix},$$

where p and q are positive. What condition on q and p guarantees:

(a) that A has two real eigenvalues?

(b) that A has complex eigenvalues?

(c) that A has only one eigenvalue and one line of eigenvectors?

12. Let

$$\mathbf{A} = \begin{pmatrix} a & b \\ c & d \end{pmatrix}.$$

Define the trace of $\mathbf{A}$ to be $\mathrm{tr}(\mathbf{A}) = a + d$. Show that $\mathbf{A}$ has only one eigenvalue if and only if $(\mathrm{tr}(\mathbf{A}))^2 - 4\det(\mathbf{A}) = 0$.

13. Suppose

$$\mathbf{A} = \begin{pmatrix} a & b \\ c & d \end{pmatrix}$$

is a matrix with eigenvalue λ such that every nonzero vector is an eigenvector with eigenvalue λ, that is, $\mathbf{AY} = \lambda\mathbf{Y}$ for every vector $\mathbf{Y}$. Show that $a = d = \lambda$ and $b = c = 0$. [*Hint*: Since $\mathbf{AY} = \lambda\mathbf{Y}$ for every $\mathbf{Y}$, try $\mathbf{Y} = (1, 0)$ and $\mathbf{Y} = (0, 1)$.]

14. Suppose λ is an eigenvector for the matrix

$$\mathbf{A} = \begin{pmatrix} a & b \\ c & d \end{pmatrix},$$

and suppose that there are two linearly independent eigenvectors $\mathbf{Y}_1$ and $\mathbf{Y}_2$ associated to λ. Show that every nonzero vector is an eigenvector with eigenvalue λ. What does this imply about a, b, c, and d?

15. Consider a harmonic oscillator with spring constant k, damping coefficient $b = 3$, and mass $m = 1$.

    (a) Find the value of k such that the resulting oscillator has repeated eigenvalues.

    (b) Find the eigenvalue and eigenvectors for this system.

    (c) Sketch the phase plane for this system.

    (d) Find the general solution for this system.

    (e) Find the solution with initial position $y(0) = 2$ and velocity $v(0) = 0$.

16. Consider a harmonic oscillator with spring constant $k = 2$, damping coefficient $b = 3$, and mass m.

    (a) Find the value of m such that the resulting oscillator has repeated eigenvalues.

    (b) Find the eigenvalue and eigenvectors for this system.

    (c) Sketch the phase plane for this system.

    (d) Find the general solution for this system.

    (e) Find the solution with initial position $y(0) = 0$ and velocity $v(0) = -2$.

In Exercises 17–19, each of the given linear systems has zero as an eigenvalue. For each system,

    (a) find the eigenvalues;

    (b) find the eigenvectors;

    (c) sketch the phase plane;

    (d) sketch the $x(t)$- and $y(t)$-graphs of the solution with initial condition $\mathbf{Y}_0 = (1, 0)$;

    (e) find the general solution; and

    (f) find the particular solution for the initial condition $\mathbf{Y}_0 = (1, 0)$ and compare it with your sketch from part (d).

17. $\dfrac{d\mathbf{Y}}{dt} = \begin{pmatrix} 0 & 2 \\ 0 & -1 \end{pmatrix} \mathbf{Y}$ **18.** $\dfrac{d\mathbf{Y}}{dt} = \begin{pmatrix} 2 & 4 \\ 3 & 6 \end{pmatrix} \mathbf{Y}$ **19.** $\dfrac{d\mathbf{Y}}{dt} = \begin{pmatrix} 4 & 2 \\ 2 & 1 \end{pmatrix} \mathbf{Y}$

20. Let $\mathbf{A} = \begin{pmatrix} a & b \\ c & d \end{pmatrix}$.

    (a) Show that, if one or both of the eigenvalues of $\mathbf{A}$ is zero, then the determinant of $\mathbf{A}$ is zero.

    (b) Show that, if $\det \mathbf{A} = 0$, then at least one of the eigenvalues of $\mathbf{A}$ is zero.

21. Find the eigenvalues and sketch the phase planes for the linear systems

(a) $\dfrac{d\mathbf{Y}}{dt} = \begin{pmatrix} 0 & 2 \\ 0 & 0 \end{pmatrix} \mathbf{Y}$ **(b)** $\dfrac{d\mathbf{Y}}{dt} = \begin{pmatrix} 0 & -2 \\ 0 & 0 \end{pmatrix} \mathbf{Y}$

22. Find the general solution for the linear systems

(a) $\dfrac{d\mathbf{Y}}{dt} = \begin{pmatrix} 0 & 2 \\ 0 & 0 \end{pmatrix} \mathbf{Y}$ **(b)** $\dfrac{d\mathbf{Y}}{dt} = \begin{pmatrix} 0 & -2 \\ 0 & 0 \end{pmatrix} \mathbf{Y}$

23. Consider the linear system

$$\frac{d\mathbf{Y}}{dt} = \begin{pmatrix} a & 0 \\ 0 & d \end{pmatrix} \mathbf{Y}.$$

(a) Find the eigenvalues.
(b) Find the eigenvectors.
(c) Suppose $a = d < 0$. Sketch the phase plane and compute the general solution. (What are the eigenvectors in this case?)
(d) Suppose $a = d > 0$. Sketch the phase plane and compute the general solution.

24. The slope field for the system

$$\frac{dx}{dt} = -3x - y$$
$$\frac{dy}{dt} = 4x + y$$

is shown to the right.

(a) Determine the type of the equilibrium point at the origin.
(b) Calculate all straight-line solutions.
(c) Plot the $x(t)$- and $y(t)$-graphs ($t \geq 0$) for the initial conditions $A = (-1, 2)$, $B = (-1, 1)$, $C = (-1, -2)$, and $D = (1, 0)$.

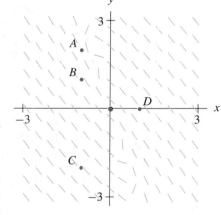

3.6 SECOND-ORDER LINEAR EQUATIONS

Throughout this chapter we have used the harmonic oscillator as an example. We have solved the second-order equation and its associated system of equations in a number of different cases. Now it is time to summarize all that we have learned about this important model.

Second-Order Equations versus First-Order Systems

As we know, the motion of a harmonic oscillator can be modeled by the second-order equation

$$m\frac{d^2y}{dt^2} + b\frac{dy}{dt} + ky = 0,$$

where $m > 0$ is the mass, $k > 0$ is the spring constant, and $b \geq 0$ is the damping coefficient. Since $m \neq 0$, we can also write this equation in the form

$$\frac{d^2y}{dt^2} + p\frac{dy}{dt} + qy = 0,$$

where $p = b/m$ and $q = k/m$ are nonnegative constants, and the corresponding linear system is

$$\frac{d\mathbf{Y}}{dt} = \begin{pmatrix} 0 & 1 \\ -q & -p \end{pmatrix}\mathbf{Y}.$$

As we will see in this section, any method to compute the general solution of the second-order equation also gives the general solution of the associated system, and vice versa. In particular we can use the Linearity Principle to produce new solutions from known ones by adding solutions and by multiplying solutions by constants. Therefore second-order equations of the form

$$a\frac{d^2y}{dt^2} + b\frac{dy}{dt} + cy = 0,$$

where a, b, and c are constants, are said to be **linear**. More precisely, these equations are **homogeneous, constant-coefficient**, linear, second-order equations. The constants a, b, and c are the coefficients, and the equation is homogeneous due to the fact that the right-hand side is 0. In Chapter 4 we will study the difference between homogeneous and nonhomogeneous differential equations in detail.

We can find the general solution of the linear system that models the harmonic oscillator by finding the eigenvalues and eigenvectors of the coefficient matrix. The arithmetic is not always pleasant, but the steps are clear. In this section we give a shortcut for finding the general solution of the corresponding second-order equation, and we relate this shortcut to the geometry and qualitative behavior of the solutions of both the second-order equation and the system.

A Free Gift from the Math Department

The shortcut method for finding the general solution of a second-order equation such as

$$\frac{d^2y}{dt^2} + 7\frac{dy}{dt} + 10y = 0,$$

for example, is to guess it. Given what we now know about solutions of the corresponding system, this is not as silly as it sounds. We know that the solutions of the system are made up of terms of the form $e^{\lambda t}\mathbf{V}$, where λ is an eigenvalue and $\mathbf{V}$ is an eigenvector. Hence if we are trying to guess the solution of the second-order equation, the most natural guess is

$$y(t) = e^{st},$$

where s is a constant to be determined. (From our point of view it makes more sense to use λ as the unknown constant. However, s is commonly used in applications, and for this discussion we will follow that practice.) Substituting the guess into the left-hand side of the second-order equation gives

$$\frac{d^2y}{dt^2} + 7\frac{dy}{dt} + 10y = \frac{d^2(e^{st})}{dt^2} + 7\frac{d(e^{st})}{dt} + 10e^{st}$$

$$= s^2e^{st} + 7se^{st} + 10e^{st}$$

$$= (s^2 + 7s + 10)e^{st}.$$

Since e^{st} is never zero, we must have

$$s^2 + 7s + 10 = 0$$

in order for $y(t) = e^{st}$ to be a solution. This quadratic equation has roots $s = -5$ and $s = -2$, so we know that $y_1(t) = e^{-5t}$ and $y_2(t) = e^{-2t}$ are solutions of the differential equation. (If this guess-and-test technique seems familiar, it should. We have already used this procedure when we studied analytic techniques for finding solutions to certain systems in Section 2.3.)

Applying the Linearity Principle, we see that any function of the form

$$y(t) = k_1e^{-5t} + k_2e^{-2t}$$

is also a solution for any choice of constants k_1 and k_2 (see Exercise 18 for a direct verification of this assertion). To see that this expression is in fact the general solution of the equation, we note that there is a one-to-one correspondence between solutions of

$$\frac{d^2y}{dt^2} + 7\frac{dy}{dt} + 10y = 0$$

and solutions of the associated system

$$\frac{dy}{dt} = v$$

$$\frac{dv}{dt} = -10y - 7v.$$

If we have a solution $\mathbf{Y}(t) = (y(t), v(t))$ to the system, then $y(t)$ is a solution to the second-order equation. If $y(t)$ is a solution to the equation, then we have

$$v(t) = -5k_1e^{-5t} - 2k_2e^{-2t},$$

since $v = dy/dt$. If we form the vector function

$$\mathbf{Y}(t) = \begin{pmatrix} y(t) \\ v(t) \end{pmatrix} = \begin{pmatrix} k_1e^{-5t} + k_2e^{-2t} \\ -5k_1e^{-5t} - 2k_2e^{-2t} \end{pmatrix},$$

we have a solution to the system that can be rewritten in the form

$$\mathbf{Y}(t) = k_1e^{-5t}\begin{pmatrix} 1 \\ -5 \end{pmatrix} + k_2e^{-2t}\begin{pmatrix} 1 \\ -2 \end{pmatrix}.$$

Solving the associated system

This form of the solution looks suspiciously familiar. If we write this second-order equation as a system in matrix notation, we obtain

$$\frac{d\mathbf{Y}}{dt} = \begin{pmatrix} 0 & 1 \\ -10 & -7 \end{pmatrix} \mathbf{Y},$$

which has $\lambda^2 + 7\lambda + 10$ as its characteristic polynomial. Note that this is exactly the same quadratic that we obtained earlier when we applied our guess-and-test technique to the the second-order equation (with s replaced by λ). The eigenvalues for this system are $\lambda_1 = -5$ and $\lambda_2 = -2$. Computing the associated eigenvectors, we find that one eigenvector corresponding to λ_1 is $(1, -5)$ and one eigenvector corresponding to λ_2 is $(1, -2)$. Thus using the eigenvalue/eigenvector methods of this chapter, we obtain exactly the same general solution,

$$\mathbf{Y}(t) = k_1 e^{-5t} \begin{pmatrix} 1 \\ -5 \end{pmatrix} + k_2 e^{-2t} \begin{pmatrix} 1 \\ -2 \end{pmatrix}.$$

If we had chosen different eigenvectors for the system, we would have obtained a slightly different form of the general solution. For example, $(-1, 5)$ is also an eigenvector corresponding to $\lambda_1 = -5$, and $(2, -4)$ is an eigenvector corresponding to $\lambda_1 = -2$. So our general solution may also be written as

$$\mathbf{Y}(t) = k_1 e^{-5t} \begin{pmatrix} -1 \\ 5 \end{pmatrix} + k_2 e^{-2t} \begin{pmatrix} 2 \\ -4 \end{pmatrix}.$$

But these solutions are precisely the same as those already obtained. (Replace k_1 with $-k_1$ and k_2 with $k_2/2$.) There really is no difference between this guessing method and the eigenvalue/eigenvector method.

Complex Eigenvalues

The method described above works in general for any second-order, linear equation, even those for which the characteristic polynomial has complex roots. For example, consider a harmonic oscillator with mass $m = 1$, damping coefficient $b = 2$, and spring constant $k = 2$. The second-order equation is

$$\frac{d^2 y}{dt^2} + 2\frac{dy}{dt} + 2y = 0.$$

As usual, we guess that $y(t) = e^{st}$ is a solution and obtain the characteristic equation

$$s^2 + 2s + 2 = 0.$$

Using the quadratic formula, we obtain the roots

$$s = \frac{-2 \pm \sqrt{4 - 8}}{2} = -1 \pm i.$$

Therefore we have a pair of complex solutions of this equation of the form $e^{(-1\pm i)t}$. As we did with systems, let's look more closely at one of these solutions. Consider $y(t) = e^{(-1+i)t}$. Using Euler's formula, we have

$$y(t) = e^{(-1+i)t} = e^{-t}e^{it} = e^{-t}(\cos t + i \sin t) = e^{-t}\cos t + ie^{-t}\sin t.$$

This solution is a complex-valued solution to a real differential equation, so, just as we argued in the case of systems, the real and imaginary parts of this function are themselves solutions of the original equation (see Exercise 19). That is, we have two real solutions given by $y_1(t) = e^{-t}\cos t$ and $y_2(t) = e^{-t}\sin t$. By the Linearity Principle, any solution of the form

$$y(t) = k_1 e^{-t}\cos t + k_2 e^{-t}\sin t$$

is also a solution. We can also obtain a vector solution to the associated system by differentiating $y(t)$ to obtain $v = dy/dt$. We have

$$\mathbf{Y}(t) = k_1 e^{-t} \begin{pmatrix} \cos t \\ -\cos t - \sin t \end{pmatrix} + k_2 e^{-t} \begin{pmatrix} \sin t \\ -\sin t + \cos t \end{pmatrix}.$$

This general solution to the system is exactly what we would have obtained had we used the eigenvalue/eigenvector methods.

The Method of the Lucky Guess

For a linear, second-order equation of the form

$$a\frac{d^2 y}{dt^2} + b\frac{dy}{dt} + cy = 0,$$

where a, b, and c are constants, we can compute the characteristic polynomial by guessing that $y(t) = e^{st}$ is a solution. We obtain

$$a\frac{d^2 y}{dt^2} + b\frac{dy}{dt} + cy = (as^2 + bs + c)e^{st},$$

and we see that the characteristic polynomial $as^2 + bs + c$ appears as the coefficient of e^{st}. Now that we have made this calculation once, we do not have to repeat it every time. We can just write down the characteristic polynomial immediately from the second-order equation.

In both the eigenvalue/eigenvector method for the system and the lucky guess method for the second-order equation, we must find the roots of the characteristic polynomial in order to compute the general solution. Whatever method we use, once we have the roots (that is, the eigenvalues), we can obtain the general solution. (We have already discussed examples with two distinct real eigenvalues and with complex eigenvalues. Later in this section we will see how to adapt this method in order to treat repeated eigenvalues.)

Finding the general solution via this lucky guess method is very efficient. We obtain the characteristic polynomial immediately from the second-order equation, and we can skip the work involved in finding the eigenvectors of the system. Consequently, we will use this method extensively in Chapter 4 where we will need to solve a number of second-order equations.

Indeed, this method is so efficient that one might be tempted to ask, "Do we really need systems, eigenvalues, eigenvectors, phase planes, and the rest of the ideas of this chapter?" The answer is "no," provided we care only about formulas and not about a qualitative understanding of solutions. It is also important to remember that this method does not generalize well to other linear systems.

A Classification of Harmonic Oscillators

We can now tell the full story about the solutions of the second-order equation

$$m\frac{d^2y}{dt^2} + b\frac{dy}{dt} + ky = 0$$

that models harmonic oscillators (among other things), and in doing so we will have occasion to use both the lucky guess method and the phase plane. Before starting our analysis, it is important to note that the mass m and the spring constant k are always positive but that the damping constant b can be either zero or positive. If $b = 0$, we have no damping and the oscillator is said to be undamped.

The undamped harmonic oscillator
The second-order equation for this case is simply

$$m\frac{d^2y}{dt^2} + ky = 0,$$

and the characteristic polynomial is

$$ms^2 + k = 0.$$

Since m and k are both positive, the eigenvalues are $\pm i\sqrt{k/m}$. This square root comes up so often that it is commonly written as $\omega = \sqrt{k/m}$.

We therefore have complex solutions of the form

$$e^{i\omega t} = \cos \omega t + i \sin \omega t.$$

Both the real and imaginary parts of this expression are solutions of the equation, so the general solution is

$$y(t) = k_1 \cos \omega t + k_2 \sin \omega t.$$

Each of these functions is a periodic function with period $2\pi/\omega = 2\pi\sqrt{m/k}$ (see Exercise 22 in Section 3.4).

Computing $v = dy/dt$, we obtain the vector form of the solution

$$\mathbf{Y}(t) = k_1 \begin{pmatrix} \cos \omega t \\ -\omega \sin \omega t \end{pmatrix} + k_2 \begin{pmatrix} \sin \omega t \\ \omega \cos \omega t \end{pmatrix}.$$

Each of these solutions generates an ellipse in the phase plane that begins at the point $(k_1, k_2\omega)$ and travels around the origin in the clockwise direction (see Exercise 20 in Section 2.1). Each solution returns to its initial position after $2\pi/\omega$ units of time. Therefore the quantity ω is called the **natural frequency** of the motion (see Section 3.4). The phase plane and the $y(t)$-graphs illustrate this periodicity (see Figure 3.39).

In terms of the actual undamped mass-spring system, these plots tell us that the mass either remains at rest forever or oscillates around its rest position without ever settling down. Without damping, the mass-spring system oscillates forever with the same amplitude and period. This regular behavior is why watches are often made with springs. Of course, physical systems have some damping, which explains why watches need winding every so often.

This type of motion is often called **simple harmonic motion**. One interesting observation about simple harmonic motion is that the period of the motion is determined solely by m and k. Therefore the period is independent of the initial condition (and consequently the amplitude of the motion.)

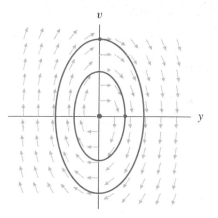

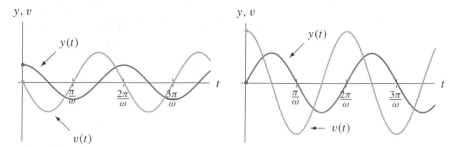

Figure 3.39
Solutions in the phase plane and the $y(t)$- and $v(t)$-graphs corresponding to an undamped harmonic oscillator with natural frequency ω.

Harmonic Oscillators with Damping

If damping is present, the mass-spring system behaves in several different ways, depending on the roots of the characteristic equation. For the harmonic oscillator equation

$$m\frac{d^2y}{dt^2} + b\frac{dy}{dt} + ky = 0,$$

the characteristic equation is

$$ms^2 + bs + k = 0$$

with roots given by the quadratic formula

$$\frac{-b \pm \sqrt{b^2 - 4mk}}{2m}.$$

Thus there are three possibilities for the roots of the characteristic equation.

- If b is small relative to $4km$ (or more precisely, if $b^2 - 4km < 0$), then we have complex roots. The real part of these roots is $-b/(2m)$, which is always negative. In this case the harmonic oscillator is said to be **underdamped**.
- If $b^2 - 4km > 0$, there are two distinct, real roots to this equation. In this case the oscillator is said to be **overdamped**.
- If $b^2 - 4km = 0$, we have repeated roots and the oscillator is said to be **critically damped**.

An underdamped oscillator

If b is relatively small but nonzero, the roots of the characteristic equation are complex with negative real parts. We expect spiraling in the phase plane for this system and corresponding oscillations for the $y(t)$-graphs.

For example, if $m = 1$, $b = 0.2$, and $k = 1.01$, the second-order equation for the motion of the oscillator is

$$\frac{d^2y}{dt^2} + 0.2\frac{dy}{dt} + 1.01y = 0,$$

and the roots of the characteristic polynomial $s^2 + 0.2s + 1.01$ are

$$\frac{-0.2 \pm \sqrt{0.04 - 4.04}}{2} = -0.1 \pm i.$$

Consequently the complex solution is

$$e^{(-0.1 \pm i)t} = e^{-0.1t}(\cos t + i \sin t)$$

and the general solution is

$$y(t) = k_1 e^{-0.1t} \cos t + k_2 e^{-0.1t} \sin t.$$

These solutions have a natural period of 2π, but the amplitude of the oscillations decays as time increases (see Figure 3.40). The corresponding motion of the spring is the familiar oscillation about the rest position, but the amplitude of successive oscillations decrease as t increases.

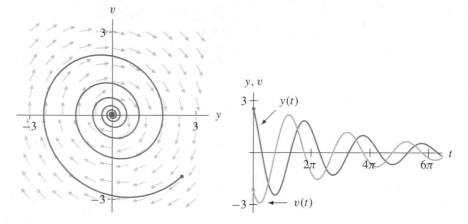

Figure 3.40
Solution in the phase plane and the $y(t)$- and $v(t)$-graphs for the underdamped harmonic oscillator

$$\frac{d^2y}{dt^2} + 0.2\frac{dy}{dt} + 1.01y = 0.$$

An overdamped oscillator

If the damping of the mass-spring system is relatively large, we expect somewhat different behavior for the motion of the mass. For example, if the system is submerged in a vat of peanut butter, we hardly expect the mass to oscillate about its rest position as in the underdamped case.

For example, the characteristic polynomial of the harmonic oscillator modeled by

$$\frac{d^2y}{dt^2} + 3\frac{dy}{dt} + y = 0$$

is $s^2 + 3s + 1 = 0$, and the eigenvalues are

$$s = \frac{-3 \pm \sqrt{5}}{2} \approx -1.5 \pm 1.12.$$

Both of these eigenvalues are real and negative. Hence all solutions of this equation tend to the rest position of the mass as time goes forward. But how do these solutions tend to this position? To answer this, we could write down the general solution of the second-order equation. However, since the answer we seek is a qualitative description of the motion of the oscillator, we can obtain it more directly using qualitative methods.

The system corresponding to the second-order equation above is

$$\frac{d\mathbf{Y}}{dt} = \begin{pmatrix} 0 & 1 \\ -1 & -3 \end{pmatrix} \mathbf{Y}.$$

Suppose $\mathbf{V}_1$ is an eigenvector corresponding to the eigenvalue $(-3 - \sqrt{5})/2$ and $\mathbf{V}_2$ is an eigenvector associated to the eigenvalue $(-3 + \sqrt{5})/2$.

We know that all solutions in the phase plane (except those on the line determined by $\mathbf{V}_1$) tend to the origin tangent to the $\mathbf{V}_2$ direction (see Figure 3.41). In particular, suppose we stretch or compress the spring and release the mass with no initial velocity ($v_0 = 0$). Our solution begins at a point on the y-axis, for example at $(3, 0)$. As t increases, such a solution tends directly to the origin without crossing the y- or v-axes (see Figure 3.41). The position $y(t)$ decreases to zero, and $v(t)$ is always negative (see Figure 3.42). In terms of the mass-spring system, the behavior of this solution means that the mass simply glides to its rest position without oscillating. The damping medium is so thick that the mass does not overshoot the rest position.

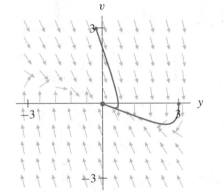

Figure 3.41
The direction field and two solution curves for

$$\frac{d\mathbf{Y}}{dt} = \begin{pmatrix} 0 & 1 \\ -1 & -3 \end{pmatrix} \mathbf{Y}.$$

One solution curve has initial condition $(y_0, v_0) = (3, 0)$, and the other solution curve has initial condition $(y_0, v_0) = (-0.25, 3)$.

However, for other initial conditions, it is possible for the mass to overshoot the rest position. For example, consider the solution to the system with initial condition $(-0.25, 3)$. According to our model, this initial condition corresponds to the situation where the spring is compressed and then released with a nonzero speed in the direction of the rest position. Note that the corresponding solution curve through this point crosses the y-axis and then turns and tends to the origin along the direction of $\mathbf{V}_2$ (see Figure 3.41).

The $y(t)$-graph for this initial condition $(-0.25, 3)$ is displayed in Figure 3.43. This graph shows that $y(t)$ initially increases and passes through $y = 0$ (the t-axis in Figure 3.43). Then $y(t)$ reaches a maximum and slowly decreases to 0 without touching $y = 0$ again.

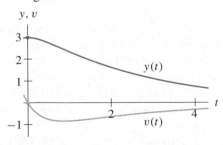

Figure 3.42
The $y(t)$- and $v(t)$-graphs for the solution of the harmonic oscillator system with initial condition $(3, 0)$.

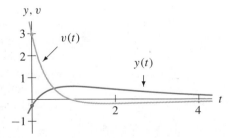

Figure 3.43
The $y(t)$- and $v(t)$-graphs for the solution of the harmonic oscillator system with initial condition $(-0.25, 3)$.

A critically damped oscillator

If the damping coefficient and the spring constant satisfy the equation

$$b^2 - 4km = 0,$$

then the characteristic equation has only one root, $s = -b/(2m)$.

As we know, this condition divides the phase portraits where solutions spiral toward the origin (spiral sinks) from the phase portraits that do not spiral. We call this oscillator "critically" damped because a small change in the damping constant changes the nature of the motion of the mass. If we decrease the damping just a tiny amount, the mass oscillates as it approaches its rest position. Increasing the damping puts us in the overdamped case, and there is no possibility of oscillation.

For example, suppose we consider a harmonic oscillator with mass $m = 1$ and spring constant $k = 2$, and we consider different values of the damping coefficient b. Then the second-order equation that models this oscillator is

$$\frac{d^2y}{dt^2} + b\frac{dy}{dt} + 2y = 0.$$

The roots of the characteristic equation $s^2 + bs + 2 = 0$ are

$$\frac{-b \pm \sqrt{b^2 - 8}}{2},$$

and consequently they are complex if $b < 2\sqrt{2}$ and real if $b > 2\sqrt{2}$. Repeated roots occur for $b = 2\sqrt{2}$.

Since we have already discussed the noncritical cases, we concentrate on the case where $b = 2\sqrt{2}$. In this case we know that the system has only one eigenvalue, $s = \sqrt{2}$, and we know that $y_1(t) = e^{-\sqrt{2}t}$ is one solution of this equation. In order to find the general solution, we need another solution that is not a multiple of $y_1(t)$, and therefore we turn to the method of the lucky guess.

But what should a second guess be? From the characteristic polynomial, we know that the natural guess, $y(t) = e^{st}$, will not be a solution unless $s = -\sqrt{2}$. To determine the desired $y_2(t)$, we think back to the case of repeated eigenvalues for systems (see Section 3.5). Recall that we found that the general solution contains terms of the form

$$te^{-\sqrt{2}t}\mathbf{V}$$

for some vector $\mathbf{V}$. Therefore we guess a solution of the form $y_2(t) = te^{-\sqrt{2}t}$. We can check this guess by substituting $y_2(t)$ into the left-hand side of the second-order equation, obtaining

$$\frac{d^2y_2}{dt^2} + 2\sqrt{2}\frac{dy_2}{dt} + 2y_2 = (-2\sqrt{2}\,e^{-\sqrt{2}t} + 2te^{-\sqrt{2}t}) +$$

$$2\sqrt{2}\,(e^{-\sqrt{2}t} - \sqrt{2}\,te^{-\sqrt{2}t}) + 2te^{-\sqrt{2}t}$$

$$= 0.$$

Thus the general solution is

$$y(t) = k_1e^{-\sqrt{2}t} + k_2te^{-\sqrt{2}t}.$$

Both $e^{-\sqrt{2}t}$ and $te^{-\sqrt{2}t}$ tend to 0 as t increases, so solutions tend to the rest position as we expect. Also these solutions do not involve sines or cosines, so the corresponding solutions do not oscillate about the rest position, again as expected.

We sketch the phase portrait for the corresponding system by first computing the eigenvectors associated to the eigenvalue $\lambda = -\sqrt{2}$. For this system, the eigenvectors satisfy $v = -\sqrt{2}\,y$. The phase portrait for this system is shown in Figure 3.44. All solutions approach $(0, 0)$ as t increases, and all solutions are tangent to the line $v = -\sqrt{2}\,y$ of eigenvectors as they approach the origin. This tangency prevents solutions from spiraling into the origin. (Having completed this analysis, you may find it informative to compare the discussion here with the discussion of this example found in Section 3.5.)

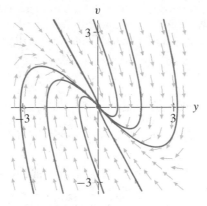

Figure 3.44

Direction field and solution curves for

$$\frac{d\mathbf{Y}}{dt} = \begin{pmatrix} 0 & 1 \\ -2 & -2\sqrt{2} \end{pmatrix} \mathbf{Y}.$$

Note that the solution curves approach the origin tangent to the straight-line solution.

Summary

We now have a complete picture of the behavior of harmonic oscillators modeled by the second-order, linear equation

$$m\frac{d^2y}{dt^2} + b\frac{dy}{dt} + ky = 0.$$

- If $b = 0$, the oscillator is undamped, and the equilibrium point at the origin in the phase plane is a center. All solutions are periodic, and the mass oscillates forever about its rest position. The (natural) period of the oscillations is $2\pi\sqrt{m/k}$.
- If $b > 0$ and $b^2 - 4km < 0$, the oscillator is underdamped. The origin in the phase plane is a spiral sink, and all other solutions spiral toward the origin. The mass oscillates back and forth as it tends to its rest position with period $4m\pi/\sqrt{4km - b^2}$.
- If $b > 0$ and $b^2 - 4km > 0$, the oscillator is overdamped. The origin in the phase plane is a real sink with two distinct eigenvalues. The mass tends to its rest position but does not oscillate.
- If $b > 0$ and $b^2 - 4km = 0$, the oscillator is critically damped. The system has exactly one eigenvalue, which is negative. All solutions tend to the origin tangent to the unique line of eigenvectors. As in the overdamped case, the mass tends to its rest position but does not oscillate.

The four cases just described completely classify the various long-term behaviors of all harmonic oscillators. In the next section we will derive a geometric way to classify these behaviors.

EXERCISES FOR SECTION 3.6

In Exercises 1–8, consider harmonic oscillators with mass m, spring constant k, and damping coefficient b. For the values specified,

   (a) write the second-order differential equation and the corresponding first-order system;

   (b) find the eigenvalues and eigenvectors of the linear system;

   (c) classify the oscillator (as underdamped, overdamped, critically damped, or undamped) and, when appropriate, give the natural period;

   (d) sketch the phase portrait of the associated linear system and include the solution curve for the given initial condition; and

   (e) sketch the $y(t)$- and $v(t)$-graphs of the solution with the given initial condition.

1. $m = 1, k = 7, b = 8$, with initial conditions $y(0) = -1, v(0) = 5$

2. $m = 1, k = 8, b = 6$, with initial conditions $y(0) = 1, v(0) = 0$

3. $m = 1, k = 5, b = 4$, with initial conditions $y(0) = 1, v(0) = 0$

4. $m = 1, k = 8, b = 0$, with initial conditions $y(0) = 1, v(0) = 4$

5. $m = 2, k = 1, b = 3$, with initial conditions $y(0) = 0, v(0) = 3$

6. $m = 9, k = 1, b = 6$, with initial conditions $y(0) = 1, v(0) = 1$

7. $m = 2, k = 3, b = 0$, with initial conditions $y(0) = 2, v(0) = -3$

8. $m = 2, k = 3, b = 1$, with initial conditions $y(0) = 0, v(0) = -3$

In Exercises 9–16, consider harmonic oscillators with mass m, spring constant k, and damping coefficient b. (The values of these parameters match up with those in Exercises 1–8). For the values specified,

   (a) find the general solution of the second-order equation that models the motion of the oscillator;

   (b) find the particular solution for the given initial condition; and

   (c) using the equations for the solution of the initial-value problem, sketch the $y(t)$- and $v(t)$-graphs. Compare these graphs to your sketches for the corresponding exercise from Exercises 1–8.

9. $m = 1, k = 7, b = 8$, with initial conditions $y(0) = -1, v(0) = 5$

10. $m = 1, k = 8, b = 6$, with initial conditions $y(0) = 1, v(0) = 0$

11. $m = 1, k = 5, b = 4$, with initial conditions $y(0) = 1, v(0) = 0$

12. $m = 1, k = 8, b = 0$, with initial conditions $y(0) = 1, v(0) = 4$

13. $m = 2, k = 1, b = 3$, with initial conditions $y(0) = 0, v(0) = 3$

14. $m = 9, k = 1, b = 6$, with initial conditions $y(0) = 1, v(0) = 1$

15. $m = 2, k = 3, b = 0$, with initial conditions $y(0) = 2, v(0) = -3$

16. $m = 2, k = 3, b = 1$, with initial conditions $y(0) = 0, v(0) = -3$

17. Construct a table of all the possible harmonic oscillator systems as follows:

    (a) The first column contains the type of oscillator.

    (b) The second column contains the eigenvalue condition that corresponds to this type of system.

    (c) The third column contains the condition on the parameters m, k, and b that is equivalent to the eigenvalue condition.

    (d) The fourth column contains the rate that solutions approach the origin and the natural period of the oscillator (if applicable).

    (e) The fifth column contains sample phase-plane diagrams.

    (f) The sixth column contains typical $y(t)$- and $v(t)$-graphs for solutions.

18. Suppose $y_1(t)$ and $y_2(t)$ are solutions of

$$\frac{d^2y}{dt^2} + p\frac{dy}{dt} + qy = 0.$$

Verify that $y(t) = k_1 y_1(t) + k_2 y_2(t)$ is also a solution for any choice of constants k_1 and k_2.

19. Suppose $y(t)$ is a complex-valued solution of

$$\frac{d^2y}{dt^2} + p\frac{dy}{dt} + qy = 0,$$

where p and q are real numbers. Show that, if $y(t) = y_{\text{re}}(t) + iy_{\text{im}}(t)$ where $y_{\text{re}}(t)$ and y_{im} are real valued, then both $y_{\text{re}}(t)$ and $y_{\text{im}}(t)$ are solutions of the second-order equation.

20. Consider a harmonic oscillator with mass $m = 1$ and spring constant $k = 3$, and let the damping coefficient b be a parameter. Then the motion of the oscillator is modeled by the equation

$$\frac{d^2y}{dt^2} + b\frac{dy}{dt} + 3y = 0.$$

For what value of b does the *typical* solution approach the equilibrium position most rapidly? (The equilibrium position is the point $(y, v) = (0, 0)$ where $v = dy/dt$.)

21. Consider a harmonic oscillator with mass $m = 1$, spring constant $k = 3$, and a (fixed) damping coefficient b. Then the motion of the oscillator is modeled by the equation

$$\frac{d^2y}{dt^2} + b\frac{dy}{dt} + 3y = 0.$$

What is the quickest rate at which a solution can approach the equilibrium state? (The equilibrium state is the point $(y, v) = (0, 0)$ where $v = dy/dt$. Your answer should depend on the value of b.)

22. An automobile's suspension system consists essentially of large springs with damping. When the car hits a bump, the springs are compressed. It is reasonable to use

a harmonic oscillator to model the up-and-down motion, where $y(t)$ measures the amount the springs are stretched or compressed and $v(t)$ is the vertical velocity of the bouncing car.

Suppose that you are working for a company that designs suspension systems for cars. One day your boss comes to you with the results of a market research survey indicating that most people want shock absorbers that "bounce twice" when compressed, then gradually return to their equilibrium position from above. That is, when the car hits a bump, the springs are compressed. Ideally they should expand, compress, and expand, then settle back to the rest position. After the initial bump, the spring would pass through its rest position three times and approach the rest position from the expanded state.

(a) Sketch a graph of the position of the spring after hitting a bump, where $y(t)$ denotes the state of the spring at time t, $y > 0$ corresponds to the spring being stretched, and $y < 0$ corresponds to the spring being compressed.

(b) Explain (politely) why the behavior pictured in the figure is impossible with standard suspension systems that are accurately modeled by the harmonic oscillator system.

(c) What is your suggestion for a choice of a harmonic oscillator system that most closely approximates the desired behavior? Justify your answer with an essay.

23. Suppose material scientists discover a new type of fluid called "magic-finger fluid." Magic-finger fluid has the property that, as an object moves through the fluid, it is accelerated in the direction that it travels ("anti-damping").

(a) Suppose the force F_{mf} that the magic-finger fluid applies to an object is proportional to the velocity of the object with proportionality constant b_{mf}. Formulate a linear, second-order differential equation for a mass-spring system moving in magic-finger fluid, assuming that the only forces involved are the natural restoring force F_s of the spring (given by Hooke's law) and the "anti-damping" force F_{mf}.

(b) Convert this mass-spring system to a first-order, linear system.

(c) Classify the possible behaviors of the linear system you constructed in part (b).

24. Consider a harmonic oscillator with $m = 1$, $k = 2$, and $b = 1$.

(a) What is the natural period?

(b) If m is increased slightly, does the natural period increase or decrease? How fast does it increase or decrease?

(c) If k is increased slightly, does the natural period increase or decrease? How fast does it increase or decrease?

(d) If b is increased slightly, does the natural period increase or decrease? How fast does it increase or decrease?

25. Suppose we wish to make a clock using a mass and a spring sliding on a table. We arrange for the clock to "tick" whenever the mass crosses $y = 0$. We use a spring with spring constant $k = 2$. If we assume there is no friction or damping ($b = 0$), then what mass m must be attached to the spring so that its natural period is one time unit?

26. As pointed out in the text, an undamped or underdamped harmonic oscillator can be used to make a clock. As in Exercise 25, if we arrange for the clock to tick whenever the mass passes the rest position, then the time between ticks is equal to one-half of the natural period of the oscillator.

(a) If dirt increases the coefficient of damping slightly for the harmonic oscillator, will the clock run fast or slow?

(b) Suppose the spring provides slightly less force for a given compression or extension as it ages. Will the clock run fast or slow?

(c) If grime collects on the harmonic oscillator and slightly increases the mass, will the clock run fast or slow?

(d) Suppose all of the above occur — the coefficient of damping increases slightly, the spring gets "tired," and the mass increases slightly — will the clock run fast or slow?

3.7 THE TRACE-DETERMINANT PLANE

In the previous sections we have encountered a number of different types of linear systems of differential equations. At this point it may seem that there are many different possibilities for these systems, each with its own characteristics. In order to put all of these examples in perspective, it is useful to pause and review the big picture.

One way to summarize everything that we have done so far is to make a table. As we have seen, the behavior of a linear system is governed by the eigenvalues and eigenvectors of the system, so our table should contain the following:

1. The name of the system (spiral sink, saddle, source, ...)

2. The eigenvalue conditions

3. One or two representative phase portraits

For example, we could begin to construct this table as in Table 3.1.

This list is by no means complete. In fact, one exercise at the end of this section is to compile a complete table (see Exercise 1). There are 8 other entries.

As is so often the case in mathematics, it is helpful to view information in several different ways. Since we are looking for "the big picture," why not try to summarize the different behaviors for linear systems in a picture rather than a table? One such picture is called the *trace-determinant plane*.

Trace and Determinant

Suppose we begin with the linear system $d\mathbf{Y}/dt = \mathbf{A}\mathbf{Y}$, where $\mathbf{A}$ is the matrix

$$\begin{pmatrix} a & b \\ c & d \end{pmatrix}.$$

The characteristic polynomial for $\mathbf{A}$ is

$$\det(\mathbf{A} - \lambda I) = (a - \lambda)(d - \lambda) - bc = \lambda^2 - (a + d)\lambda + ad - bc.$$

Table 3.1
Partial table of linear systems

Type	Eigenvalues	Phase Plane	Type	Eigenvalues	Phase Plane
Saddle	$\lambda_1 < 0 < \lambda_2$		Spiral Sink	$\lambda = a \pm ib$ $a < 0, b \neq 0$	
Sink	$\lambda_1 < \lambda_2 < 0$		Spiral Source	$\lambda = a \pm ib$ $a > 0, b \neq 0$	
Source	$0 < \lambda_1 < \lambda_2$		Center	$\lambda = \pm ib$ $b \neq 0$	

The quantity $a + d$ is called the **trace** of the matrix $\mathbf{A}$ and, as we know, the quantity $ad - bc$ is the determinant of $\mathbf{A}$. So the characteristic polynomial of $\mathbf{A}$ can be written more succinctly as

$$\lambda^2 - T\lambda + D,$$

where $T = a + d$ is the trace of $\mathbf{A}$ and $D = ad - bc$ is the determinant of $\mathbf{A}$. For example, if

$$\mathbf{A} = \begin{pmatrix} 1 & 2 \\ 3 & 4 \end{pmatrix},$$

then the characteristic polynomial is $\lambda^2 - 5\lambda - 2$, since $T = 5$ and $D = 4 - 6 = -2$. (Remember that the coefficient of the λ-term is $-T$. It is a common mistake to put this minus sign in the wrong place or even to forget it entirely.)

Since the characteristic polynomial of $\mathbf{A}$ depends only on T and D, it follows that the eigenvalues of $\mathbf{A}$ also depend only on T and D. If we solve the characteristic polynomial $\lambda^2 - T\lambda + D = 0$, we obtain the eigenvalues

$$\lambda = \frac{T \pm \sqrt{T^2 - 4D}}{2}.$$

From this formula we see immediately that the eigenvalues of $\mathbf{A}$ are complex if $T^2 - 4D < 0$, they are repeated if $T^2 - 4D = 0$, and they are real and distinct if $T^2 - 4D > 0$.

The Trace-Determinant Plane

We can now begin to paint the big picture for linear systems by examining the *trace-determinant* plane. We draw the T-axis horizontally and the D-axis vertically. Then the curve $T^2 - 4D = 0$, or equivalently $D = T^2/4$, is a parabola opening upward in this plane. We call it the *repeated-root* parabola. Above this parabola $T^2 - 4D < 0$, and below it $T^2 - 4D > 0$.

To use this picture, we first compute T and D for a given matrix and then locate the point (T, D) in this plane. Then we can immediately read off whether the eigenvalues are real, repeated, or complex, depending on the location of (T, D) relative to the repeated-root parabola (see Figure 3.45). For example, if

$$\mathbf{A} = \begin{pmatrix} 2 & 3 \\ 1 & 2 \end{pmatrix},$$

then $(T, D) = (4, 1)$, and the point $(4,1)$ lies below the curve $T^2 - 4D = 0$ (in this case, $T^2 - 4D = 12 > 0$), so the eigenvalues of A are real and distinct.

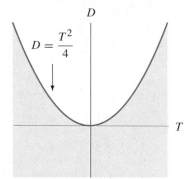

Figure 3.45
The shaded region corresponds to
$T^2 - 4D > 0$.

Refining the Big Picture

We can actually do much more with the trace-determinant plane. For example, if

$$T^2 - 4D < 0,$$

(the point (T, D) lies above the repeated-root parabola), then we know that the eigenvalues are complex and their real part is $T/2$. We have a spiral sink if $T < 0$, a spiral source if $T > 0$, and a center if $T = 0$. In the trace-determinant plane, the point (T, D) is located above the repeated-root parabola. If (T, D) lies to the left of the D-axis, the corresponding system has a spiral sink. If (T, D) lies to the right of the D-axis, the system has a spiral source. If (T, D) lies on the D-axis, then the system has a center. So our refined picture can be drawn this way (see Figure 3.46).

Real eigenvalues
We can also distinguish different regions in the trace-determinant plane where the linear system has real and distinct eigenvalues. In this case (T, D) lies below the repeated-

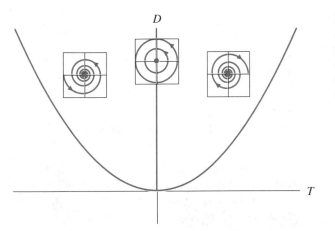

Figure 3.46
Above the repeated-root parabola we have centers along the D-axis, spiral sources to the right, and spiral sinks to the left.

root parabola. If $T^2 - 4D > 0$, the real eigenvalues are

$$\lambda = \frac{T \pm \sqrt{T^2 - 4D}}{2}.$$

If $T > 0$, the eigenvalue

$$\frac{T + \sqrt{T^2 - 4D}}{2}$$

is the sum of two positive terms and therefore is positive. Thus we only have to determine the sign of the other eigenvalue

$$\frac{T - \sqrt{T^2 - 4D}}{2}$$

to determine the type of the system.

If $D = 0$, then this eigenvalue is 0, so our matrix has one positive and one zero eigenvalue. If $D > 0$, then

$$T^2 - 4D < T^2.$$

Since we are considering the case where $T > 0$, we have

$$\sqrt{T^2 - 4D} < T$$

and

$$\frac{T - \sqrt{T^2 - 4D}}{2} > 0.$$

In this case both eigenvalues are positive, so the origin is a source.

On the other hand, if $T > 0$ but $D < 0$, then

$$T^2 - 4D > T^2,$$

so that

$$\sqrt{T^2 - 4D} > T$$

and

$$\frac{T - \sqrt{T^2 - 4D}}{2} < 0.$$

In this case the system has one positive and one negative eigenvalue, so the origin is a saddle.

In case $T < 0$ and $T^2 - 4D > 0$, we have:

- two negative eigenvalues if $D > 0$
- one negative and one positive eigenvalue if $D < 0$
- one negative eigenvalue and one zero eigenvalue if $D = 0$

Finally, along the repeated-root parabola we have repeated eigenvalues. If $T < 0$, both eigenvalues are negative; if $T > 0$, both are positive; and if $T = 0$, both are zero.

The full picture is displayed in Figure 3.47. Note that this picture gives us some of the same information that we compiled in our table earlier in this section.

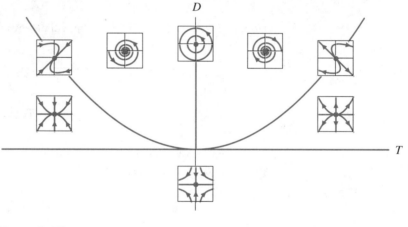

Figure 3.47
The big picture.

The Parameter Plane

The trace-determinant plane is an example of a *parameter plane*. The entries of the matrix **A** are parameters that we can adjust. When these entries change, the trace and determinant of the matrix also change, and our point (T, D) moves around in the parameter plane. As this point enters the various regions in the trace-determinant plane, we should envision the corresponding phase portraits changing accordingly. The trace-determinant plane is very much different from previous pictures we have drawn. It is a picture of a classification scheme of the behavior of all possible solutions to linear systems.

We must emphasize that the trace-determinant plane does not give complete information about the linear system at hand. For example, along the repeated-root parabola we have repeated eigenvalues, but we cannot determine whether we have one or many linearly independent eigenvectors. In order to make that distinction, we must actually calculate the eigenvectors. Similarly, we cannot determine the direction in which solutions wind about the origin if $T^2 - 4D < 0$. For example, both of the matrices

$$A = \begin{pmatrix} 0 & 1 \\ -1 & 0 \end{pmatrix} \quad \text{and} \quad B = \begin{pmatrix} 0 & -1 \\ 1 & 0 \end{pmatrix}$$

have trace 0 and determinant 1, but solutions of the system $d\mathbf{Y}/dt = \mathbf{AY}$ wind around the origin in the clockwise direction, whereas solutions of $d\mathbf{Y}/dt = \mathbf{BY}$ travel in the opposite direction.

The Harmonic Oscillator

We can also paint the same picture for the harmonic oscillator. Recall that this second-order equation is given by

$$m\frac{d^2y}{dt^2} + b\frac{dy}{dt} + ky = 0,$$

where $m > 0$ is the mass, $k > 0$ is the spring constant, and $b \geq 0$ is the damping coefficient. As a system we have

$$\frac{d\mathbf{Y}}{dt} = \begin{pmatrix} 0 & 1 \\ -k/m & -b/m \end{pmatrix},$$

so the trace $T = -b/m$ and the determinant $D = k/m$. We plot $T = -b/m$ on the horizontal axis and $D = k/m$ on the vertical axis as before.

Since m and k are positive and b is nonnegative, we are restricted to one-quarter of the picture for general linear systems, namely the second quadrant of the TD-plane. The picture is shown in Figure 3.48.

The repeated-root parabola in this case is $T^2 - 4D = b^2 - 4km = 0$. Above this parabola we have a spiral sink (if $b \neq 0$) or a center (if $b = 0$). Below the repeated-root parabola we have a sink with real distinct eigenvalues. On the parabola, we have repeated negative eigenvalues.

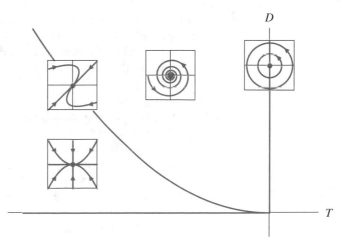

Figure 3.48
The trace-determinant plane for the harmonic oscillator.

Kathleen Alligood (1947–) received her Ph.D. in mathematics at the University of Maryland. She taught at the College of Charleston and Michigan State University before assuming her current position at George Mason University.

Alligood's research centers on the behavior of nonlinear systems but encompasses many of the topics described in this chapter. Nonlinear systems of differential equations may possess sinks just as linear systems do. However, it need not be the case that all solutions tend to the sink as in the linear case. Often the boundary of the set of solutions that tend to the sink is an extremely complicated mathematical object that contains infinitely many saddle points and their stable curves. Using techniques from topology, fractal geometry, and dynamical systems, Alligood and her coworkers were among the first to analyze the structure of these "fractal basin boundaries."

In the language of oscillators introduced in the previous section, if $(-b/m, k/m)$ lies above the repeated-root parabola and $b > 0$, we have an underdamped oscillator, or if $b = 0$, we have an undamped oscillator. If $(-b/m, k/m)$ lies on the repeated-root parabola, the oscillator is critically damped. Below the parabola, the oscillator is overdamped.

Navigating the Trace-Determinant Plane

One of the best uses of the trace-determinant plane is in the study of linear systems that depend on parameters. As the parameters change, so do the trace and determinant of the matrix. Consequently the phase portrait for the system also changes.

Usually, small changes in the parameters do not affect the qualitative behavior of the linear system very much. For example, a spiral sink remains a spiral sink and a saddle remains a saddle. Of course the eigenvalues and eigenvectors change as we vary the parameters, but the basic behavior of solutions remains more or less the same.

The critical loci

There are, however, certain exceptions to this scenario. For example, suppose that a change in parameters forces the point (T, D) to cross the positive D-axis from left to right. The corresponding linear system has changed from a spiral sink to a center and then immediately thereafter to a spiral source. Instead of all solutions tending to the equilibrium point at $(0, 0)$, suddenly we have a center, and then all of the nonequilibrium solutions tend to infinity. That is, the family of linear systems has encountered a bifurcation at the moment the point (T, D) crosses the D-axis.

The trace-determinant plane provides us with a chart of those locations where we can expect significant changes in the phase portrait. There are three such *critical loci*.

The first critical locus is the positive D-axis, as we saw above. A second critical line is the T- axis. If (T, D) crosses this line as our parameters vary, our system moves from a saddle to a sink, a source, or a center (or vice versa). The third critical locus is the repeated-root parabola where spirals turn into real sinks or sources.

There is one point in the trace-determinant plane where many different possibilities arise. If the trace and determinant are both zero, the chart shows that our system can change into any type of system whatsoever. All three of the critical loci meet at this point.

It is helpful to think of these three critical loci as fences. As long as we change parameters so that (T, D) does not pass over one of the fences, the linear system remains "unchanged" in the sense that the qualitative behavior of the solutions does not change. However, passing over a fence changes the behavior dramatically. The system undergoes a bifurcation.

A One-Parameter Family of Linear Systems

Consider the one-parameter family of linear systems $d\mathbf{Y}/dt = \mathbf{A}\mathbf{Y}$, where

$$\mathbf{A} = \begin{pmatrix} -2 & a \\ -2 & 0 \end{pmatrix},$$

which depends on the parameter a. As a varies, the determinant of this matrix is $2a$, but the trace is always -2. If we vary the parameter a from a large negative number to a large positive number, the corresponding point (T, D) in the trace-determinant plane moves vertically along the straight line $T = -2$ (see Figure 3.49). As a increases, we first travel from the saddle region into the region where we have a real sink. This change occurs when the system admits a zero eigenvalue, which in turn occurs at $a = 0$. As a continues to increase, we next move across the repeated-root parabola, and the system changes from having a sink with real eigenvalues to a spiral sink. This second bifurcation occurs when $T^2 - 4D = 0$, which for this example reduces to $D = 1$. Hence this bifurcation occurs at $a = 1/2$.

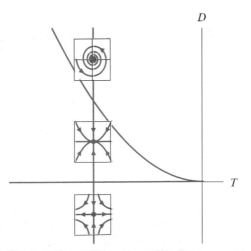

Figure 3.49
Motion in the trace-determinant plane corresponding to the one-parameter family of systems

$$\frac{d\mathbf{Y}}{dt} = \mathbf{A}\mathbf{Y}, \quad \text{where } \mathbf{A} = \begin{pmatrix} -2 & a \\ -2 & 0 \end{pmatrix}.$$

Bifurcation from sink to spiral sink

Let's investigate how the bifurcation from sink to spiral sink occurs in terms of the phase portraits of the corresponding systems. We need first to compute the eigenvalues and eigenvectors of the system. Of course these quantities depend on a. Since the characteristic polynomial is $\lambda^2 + 2\lambda + 2a = 0$, the eigenvalues are

$$\lambda = \frac{-2 \pm \sqrt{4 - 8a}}{2} = -1 \pm \sqrt{1 - 2a}.$$

As we deduced above, if $a > 1/2$, then $1 - 2a < 0$ and the eigenvalues are complex with negative real part. For $a < 1/2$, the eigenvalues

$$\lambda = -1 \pm \sqrt{1 - 2a}$$

are both real. In particular, if $0 < a < 1/2$, $\sqrt{1 - 2a} < 1$, so both eigenvalues are negative. Hence the origin is a sink with two straight lines of solutions (see Figure 3.50).

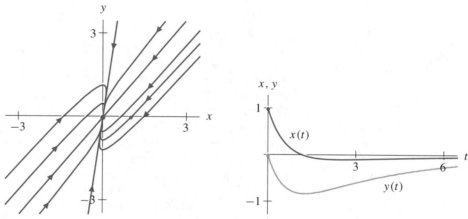

Figure 3.50
Phase portrait and the $x(t)$- and $y(t)$-graphs for the indicated solution for the one-parameter family with $a = 1/4$.

If we compute the eigenvectors for the eigenvalue $\lambda = -1 + \sqrt{1 - 2a}$, we find that they lie along the line

$$y = \left(\frac{1 + \sqrt{1 - 2a}}{a} \right) x.$$

Similarly, the eigenvectors corresponding to the eigenvalue $\lambda = -1 - \sqrt{1 - 2a}$ lie along the line

$$y = \left(\frac{1 - \sqrt{1 - 2a}}{a} \right) x.$$

Note that the slopes of both of these lines tends to 2 as a approaches $1/2$. That is, our two straight-line solutions merge to produce a single straight-line solution along the line $y = 2x$ as $a \to 1/2$ (see Figure 3.51).

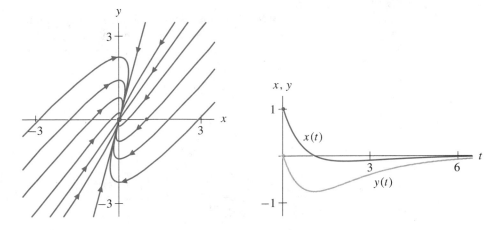

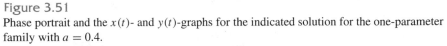

Figure 3.51
Phase portrait and the $x(t)$- and $y(t)$-graphs for the indicated solution for the one-parameter family with $a = 0.4$.

As a approaches $1/2$, the family of linear systems approaches a linear system with a repeated eigenvalue. At $a = 1/2$, the system is

$$\frac{d\mathbf{Y}}{dt} = \begin{pmatrix} -2 & 1/2 \\ -2 & 0 \end{pmatrix} \mathbf{Y},$$

whose characteristic polynomial is $\lambda^2 + 2\lambda + 1 = 0$. Hence the system has the repeated eigenvalue $\lambda = -1$. This system has a single line of eigenvectors that lie along the line $y = 2x$. The phase portrait and typical $x(t)$-graph are shown in Figure 3.52. Thus we see that the two independent eigenvectors come together to form the single line of eigenvectors as a approaches $1/2$.

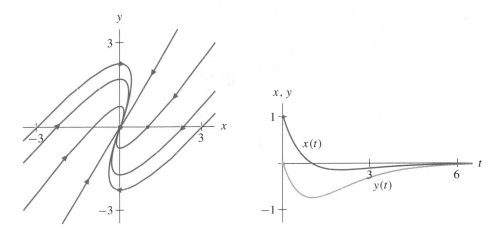

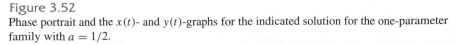

Figure 3.52
Phase portrait and the $x(t)$- and $y(t)$-graphs for the indicated solution for the one-parameter family with $a = 1/2$.

When the parameter crosses the repeated-root parabola, the origin becomes a spiral sink. The real part of the eigenvalue is -1, and the natural period is $2\pi/\sqrt{2a-1}$. For all values of a, solutions spiral toward the origin. If a is very large, solutions approach the origin at the exponential rate of e^{-t} with a very small period. The phase portrait and $x(t)$-graph for $a = 10$ are shown in Figure 3.53.

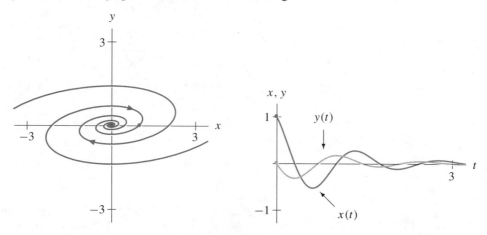

Figure 3.53
Phase portrait and $x(t)$-graph for the indicated solution for the one-parameter family with $a = 10$.

On the other hand, if a is just slightly larger than $1/2$, solutions still spiral toward the origin. However, the period of the oscillations, which is given by $2\pi/\sqrt{2a-1}$, is very large for a near $1/2$. To observe one oscillation, we must watch a solution for a long time. Since the solutions are tending to the origin at an exponential rate, these oscillations may be very hard to detect (see Figure 3.54, which is almost indistinguishable from Figure 3.52). In applications there may be very little practical difference between a very slowly oscillating solution decaying toward the origin and a solution that does not oscillate.

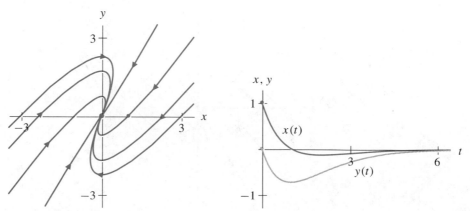

Figure 3.54
Phase portrait and $x(t)$-graph for the indicated solution for the one-parameter family with $a = 0.51$.

EXERCISES FOR SECTION 3.7

1. Construct a table of the possible linear systems as follows:

    (a) the first column contains the type of the system (sink, spiral sink, source, ...), if it has a name,

    (b) the second column contains the condition on the eigenvalues that corresponds to this case,

    (c) the third column contains a small picture of two or more possible phase portraits for this system, and

    (d) the fourth column contains $x(t)$- and $y(t)$-graphs of typical solutions indicated in your phase portraits.

[*Hint*: The most complete table contains 14 cases. Don't forget the double eigenvalue and zero eigenvalue cases.]

In Exercises 2–7, we consider the one-parameter families of linear systems depending on the parameter a. Each family therefore determines a curve in the trace-determinant plane. For each family,

  (a) sketch the corresponding curve in the trace-determinant plane;

  (b) in a brief essay, discuss different types of behaviors exhibited by the system as a increases along the real line (unless otherwise noted); and

  (c) identify the values of a where the type of the system changes. These are the bifurcation values of a.

2. $\dfrac{d\mathbf{Y}}{dt} = \begin{pmatrix} a & -1 \\ 2 & 0 \end{pmatrix} \mathbf{Y}$

3. $\dfrac{d\mathbf{Y}}{dt} = \begin{pmatrix} a & a^2 + a \\ 1 & a \end{pmatrix} \mathbf{Y}$

4. $\dfrac{d\mathbf{Y}}{dt} = \begin{pmatrix} a & a \\ 1 & 0 \end{pmatrix} \mathbf{Y}$

5. $\dfrac{d\mathbf{Y}}{dt} = \begin{pmatrix} a & \sqrt{1 - a^2} \\ 1 & 0 \end{pmatrix} \mathbf{Y}$

$$-1 \le a \le 1$$

6. $\dfrac{d\mathbf{Y}}{dt} = \begin{pmatrix} 2 & 0 \\ a & -3 \end{pmatrix} \mathbf{Y}$

7. $\dfrac{d\mathbf{Y}}{dt} = \begin{pmatrix} a & 1 \\ a & a \end{pmatrix} \mathbf{Y}$

8. Consider the two-parameter family of linear systems

$$\frac{d\mathbf{Y}}{dt} = \begin{pmatrix} a & 1 \\ b & 1 \end{pmatrix} \mathbf{Y}.$$

In the ab-plane, identify all regions where this system possesses a saddle, a sink, a spiral sink, and so on. [*Hint*: Draw a picture of the ab-plane and shade each point (a, b) of the plane a different color depending on the type of linear system for that choice (a, b) of parameters.]

9. Consider the two-parameter family of linear systems

$$\frac{d\mathbf{Y}}{dt} = \begin{pmatrix} a & b \\ b & a \end{pmatrix} \mathbf{Y}.$$

In the ab-plane, identify all regions where this system possesses a saddle, a sink, a spiral sink, and so on. [*Hint*: Draw a picture of the ab-plane and shade each point (a, b) of the plane a different color depending on the type of linear system for that choice (a, b) of the parameters.]

10. Consider the two-parameter family of linear systems

$$\frac{d\mathbf{Y}}{dt} = \begin{pmatrix} a & b \\ -b & a \end{pmatrix} \mathbf{Y}.$$

In the ab-plane, identify all regions where this system possesses a saddle, a sink, a spiral sink, and so on. [*Hint*: Draw a picture of the ab-plane and shade each point (a, b) of the plane a different color depending on the type of linear system for that choice (a, b) of parameters.]

In Exercises 11–13, we consider the equation

$$m\frac{d^2y}{dt^2} + b\frac{dy}{dt} + ky = 0$$

that models the motion of a harmonic oscillator with mass m, spring constant k, and damping coefficient b. In each exercise, we fix two values of these three parameters and obtain a one-parameter family of second-order equations. For each one-parameter family,

(a) rewrite the one-parameter family as a one-parameter family of linear systems,

(b) draw the curve in the trace-determinant plane obtained by varying the parameter, and

(c) in a brief essay, discuss the different types of behavior exhibited by this one-parameter family.

11. Consider

$$\frac{d^2y}{dt^2} + b\frac{dy}{dt} + 3y = 0.$$

That is, fix $m = 1$ and $k = 3$, and let $0 \le b < \infty$.

12. Consider

$$\frac{d^2y}{dt^2} + 2\frac{dy}{dt} + ky = 0.$$

That is, fix $m = 1$ and $b = 2$, and let $0 < k < \infty$.

13. Consider

$$m\frac{d^2y}{dt^2} + \frac{dy}{dt} + 2y = 0.$$

That is, fix $b = 1$ and $k = 2$, and let $0 < m < \infty$.

3.8 LINEAR SYSTEMS IN THREE DIMENSIONS

So far we have studied linear systems with two dependent variables. For these systems, the behavior of solutions and the nature of the phase plane can be determined by computing the eigenvalues and eigenvectors of the 2×2 coefficient matrix. Once we have found two solutions with linearly independent initial conditions, we can give the general solution.

In this section we show that the same is true for linear systems with three dependent variables. The eigenvalues and eigenvectors of the 3×3 coefficient matrix determine the behavior of solutions and the general solution. Three-dimensional linear systems have three eigenvalues, so the list of possible qualitatively distinct phase spaces is longer than for planar systems. Since we must deal with three scalar equations rather than two, the arithmetic can quickly become much more involved. The reader may wish to seek out software or a calculator capable of handling 3×3 matrices.

Linear Independence and the Linearity Principle

The general form of a linear system with three dependent variables is

$$\frac{dx}{dt} = a_{11}x + a_{12}y + a_{13}z$$

$$\frac{dy}{dt} = a_{21}x + a_{22}y + a_{23}z$$

$$\frac{dz}{dt} = a_{31}x + a_{32}y + a_{33}z,$$

where x, y, and z are the dependent variables and the coefficients a_{ij}, $(i, j = 1, 2, 3)$, are constants. We can write this system in matrix form as

$$\frac{d\mathbf{Y}}{dt} = \mathbf{AY},$$

where $\mathbf{A}$ is the coefficient matrix

$$\mathbf{A} = \begin{pmatrix} a_{11} & a_{12} & a_{13} \\ a_{21} & a_{22} & a_{23} \\ a_{31} & a_{32} & a_{33} \end{pmatrix}$$

and $\mathbf{Y}$ is the vector of dependent variables,

$$\mathbf{Y} = \begin{pmatrix} x \\ y \\ z \end{pmatrix}.$$

To specify an initial condition for such a system, we must give three numbers, x_0, y_0, and z_0.

The Linearity Principle holds for linear systems in all dimensions, so if $\mathbf{Y}_1(t)$ and $\mathbf{Y}_2(t)$ are solutions, then $k_1\mathbf{Y}_1(t) + k_2\mathbf{Y}_2(t)$ is also a solution for any constants k_1 and k_2.

Suppose $\mathbf{Y}_1(t)$, $\mathbf{Y}_2(t)$ and $\mathbf{Y}_3(t)$ are three solutions of the linear system

$$\frac{d\mathbf{Y}}{dt} = \mathbf{AY}.$$

If for any point (x_0, y_0, z_0) there exist constants k_1, k_2, and k_3 such that

$$k_1\mathbf{Y}_1(0) + k_2\mathbf{Y}_2(0) + k_3\mathbf{Y}_3(0) = (x_0, y_0, z_0),$$

then the general solution of the system is

$$\mathbf{Y}(t) = k_1\mathbf{Y}_1(t) + k_2\mathbf{Y}_2(t) + k_3\mathbf{Y}_3(t).$$

In order for three solutions $\mathbf{Y}_1(t)$, $\mathbf{Y}_2(t)$, and $\mathbf{Y}_3(t)$ to give the general solution, the three vectors $\mathbf{Y}_1(0)$, $\mathbf{Y}_2(0)$, and $\mathbf{Y}_3(0)$ must point in "different directions"; that is, no one of them can be in the plane through the origin and the other two. In this case the vectors $\mathbf{Y}_1(0)$, $\mathbf{Y}_2(0)$, and $\mathbf{Y}_3(0)$ (and the corresponding solutions) are said to be *linearly independent*. We present an algebraic technique for checking linear independence in the exercises (see Exercises 2 and 3).

An example
Consider the linear system

$$\frac{d\mathbf{Y}}{dt} = \mathbf{AY} = \begin{pmatrix} 0 & 0.1 & 0 \\ 0 & 0 & 0.2 \\ 0.4 & 0 & 0 \end{pmatrix} \begin{pmatrix} x \\ y \\ z \end{pmatrix}.$$

We can check that the functions

$$\mathbf{Y}_1(t) = e^{0.2t} \begin{pmatrix} 1 \\ 2 \\ 2 \end{pmatrix}$$

$$\mathbf{Y}_2(t) = e^{-0.1t} \begin{pmatrix} -\cos(\sqrt{0.03}\,t) - \sqrt{3}\sin(\sqrt{0.03}\,t) \\ -2\cos(\sqrt{0.03}\,t) + 2\sqrt{3}\sin(\sqrt{0.03}\,t) \\ 4\cos(\sqrt{0.03}\,t) \end{pmatrix}$$

$$\mathbf{Y}_3(t) = e^{-0.1t} \begin{pmatrix} -\sin(\sqrt{0.03}\,t) + \sqrt{3}\cos(\sqrt{0.03}\,t) \\ -2\sin(\sqrt{0.03}\,t) - 2\sqrt{3}\cos(\sqrt{0.03}\,t) \\ 4\sin(\sqrt{0.03}\,t) \end{pmatrix}$$

are solutions by substituting them into the differential equation. For example,

$$\frac{d\mathbf{Y}_1}{dt} = e^{0.2t} \begin{pmatrix} 0.2 \\ 0.4 \\ 0.4 \end{pmatrix}$$

and

$$\mathbf{AY}_1(t) = \begin{pmatrix} 0 & 0.1 & 0 \\ 0 & 0 & 0.2 \\ 0.4 & 0 & 0 \end{pmatrix} e^{0.2t} \begin{pmatrix} 1 \\ 2 \\ 2 \end{pmatrix} = e^{0.2t} \begin{pmatrix} 0.2 \\ 0.4 \\ 0.4 \end{pmatrix},$$

so $\mathbf{Y}_1(t)$ is a solution. The other two functions can be checked similarly (see Exercise 1). We can sketch the solution curves that correspond to these solutions in the three-dimensional phase space (see Figure 3.55).

The initial conditions of these three solutions are $\mathbf{Y}_1(0) = (1, 2, 2)$, $\mathbf{Y}_2(0) = (-1, -2, 4)$, and $\mathbf{Y}_3(0) = (\sqrt{3}, -2\sqrt{3}, 0)$. These vectors are shown in Figure 3.56, where we can see that none of them is in the plane determined by the other two; hence they are linearly independent.

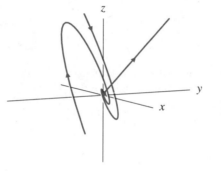

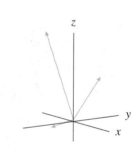

Figure 3.55
The solution curves of $\mathbf{Y}_1(t)$, $\mathbf{Y}_2(t)$, and $\mathbf{Y}_3(t)$.

Figure 3.56
Vectors $\mathbf{Y}_1(0) = (1, 2, 2)$,
$\mathbf{Y}_2(0) = (-1, -2, 4)$, and
$\mathbf{Y}_3(0) = (\sqrt{3}, -2\sqrt{3}, 0)$ in xyz-space.

For example, to find the solution $\mathbf{Y}(t)$ with initial position $\mathbf{Y}(0) = (2, 1, 3)$, we must solve

$$k_1 \mathbf{Y}_1(0) + k_2 \mathbf{Y}_2(0) + k_3 \mathbf{Y}_3(0) = (2, 1, 3),$$

which is equivalent to

$$\begin{cases} k_1 - k_2 + \sqrt{3}k_3 = 2 \\ 2k_1 - 2k_2 - 2\sqrt{3}k_3 = 1 \\ 2k_1 + 4k_2 = 3. \end{cases}$$

We obtain $k_1 = 4/3$, $k_2 = 1/12$ and $k_3 = \sqrt{3}/4$, and the solution is

$$\mathbf{Y}(t) = \frac{4}{3}\mathbf{Y}_1(t) + \frac{1}{12}\mathbf{Y}_2(t) + \frac{\sqrt{3}}{4}\mathbf{Y}_3(t).$$

Eigenvalues and Eigenvectors

The method for finding solutions of systems with three dependent variables is the same as that for systems with two variables. We begin by finding eigenvalues and eigenvectors. Suppose we are given a linear system $d\mathbf{Y}/dt = \mathbf{AY}$, where $\mathbf{A}$ is a 3×3 matrix of coefficients and $\mathbf{Y} = (x, y, z)$. An *eigenvector* for the matrix $\mathbf{A}$ is a nonzero vector $\mathbf{V}$ such that

$$\mathbf{AV} = \lambda \mathbf{V},$$

where λ is the *eigenvalue* for $\mathbf{V}$. If $\mathbf{V}$ is an eigenvector for $\mathbf{A}$ with eigenvalue λ, then

$$\mathbf{Y}(t) = e^{\lambda t} \mathbf{V}$$

is a solution of the linear system.

The method for finding eigenvalues and eigenvectors for a 3×3 matrix

$$\mathbf{A} = \begin{pmatrix} a_{11} & a_{12} & a_{13} \\ a_{21} & a_{22} & a_{23} \\ a_{31} & a_{32} & a_{33} \end{pmatrix}$$

is very similar to that for two-dimensional systems, only requiring more arithmetic. In particular we need the formula for the determinant of a 3×3 matrix.

DEFINITION   The *determinant* of the matrix $\mathbf{A}$ is

$$\det \mathbf{A} = a_{11}(a_{22}a_{33} - a_{23}a_{32}) - a_{12}(a_{21}a_{33} - a_{23}a_{31}) + a_{13}(a_{21}a_{32} - a_{22}a_{32}). \quad \blacksquare$$

Using the 3×3 identity matrix

$$\mathbf{I} = \begin{pmatrix} 1 & 0 & 0 \\ 0 & 1 & 0 \\ 0 & 0 & 1 \end{pmatrix},$$

we obtain the *characteristic polynomial* of $\mathbf{A}$ as

$$\det(\mathbf{A} - \lambda \mathbf{I}) = \det \begin{pmatrix} a_{11} - \lambda & a_{12} & a_{13} \\ a_{21} & a_{22} - \lambda & a_{23} \\ a_{31} & a_{32} & a_{33} - \lambda \end{pmatrix}.$$

As in the two-dimensional case, we have:

THEOREM   The eigenvalues of a 3×3 matrix $\mathbf{A}$ are the roots of its characteristic polynomial. $\blacksquare$

To find the eigenvalues of a 3×3 matrix, we must find the roots of a cubic polynomial. This is not as easy as finding the roots of a quadratic. Although there is a "cubic equation" analogous to the quadratic equation for finding the roots of a cubic, it is quite complicated. (It is used by computer algebra packages to give exact values of roots of cubics.) However, in cases where the cubic does not easily factor, we frequently turn to numerical techniques such as Newton's method for finding roots.

To find the corresponding eigenvectors, we must solve a system of three linear equations with three unknowns. Luckily there are many examples of systems that illustrate the possible behaviors in three dimensions and for which the arithmetic is manageable.

A diagonal matrix

The simplest type of 3×3 matrix is a *diagonal* matrix — the only nonzero terms lie on the diagonal. For example, consider the system

$$\frac{d\mathbf{Y}}{dt} = \mathbf{AY} = \begin{pmatrix} -3 & 0 & 0 \\ 0 & -1 & 0 \\ 0 & 0 & -2 \end{pmatrix} \begin{pmatrix} x \\ y \\ z \end{pmatrix}.$$

The characteristic polynomial of $\mathbf{A}$ is $(-3 - \lambda)(-1 - \lambda)(-2 - \lambda)$, which is simple because so many of the coefficients of $\mathbf{A}$ are zero. The eigenvalues are the roots of this polynomial, that is, the solutions of

$$(-3 - \lambda)(-1 - \lambda)(-2 - \lambda) = 0.$$

Thus the eigenvalues are $\lambda_1 = -3$, $\lambda_2 = -1$, and $\lambda_3 = -2$.

Finding the corresponding eigenvectors is also not too hard. For $\lambda_1 = -3$, we must solve

$$\mathbf{AV}_1 = -3\mathbf{V}_1$$

for $\mathbf{V}_1 = (x_1, y_1, z_1)$. The product $\mathbf{AV}_1$ is

$$\begin{pmatrix} -3 & 0 & 0 \\ 0 & -1 & 0 \\ 0 & 0 & -2 \end{pmatrix} \begin{pmatrix} x_1 \\ y_1 \\ z_1 \end{pmatrix} = \begin{pmatrix} -3x_1 \\ -y_1 \\ -2z_1 \end{pmatrix},$$

and therefore we want to solve

$$\begin{pmatrix} -3x_1 \\ -y_1 \\ -2z_1 \end{pmatrix} = -3 \begin{pmatrix} x_1 \\ y_1 \\ z_1 \end{pmatrix}$$

for x_1, y_1, and z_1. Solutions of this system of three equations with three unknowns are $y_1 = z_1 = 0$ and x_1 may have any (nonzero) value. So, in particular, $(1, 0, 0)$ is an eigenvector for $\lambda_1 = -3$. Similarly, we find that $(0, 1, 0)$ and $(0, 0, 1)$ are eigenvalues for $\lambda_2 = -1$ and $\lambda_3 = -2$, respectively. Note that $(1, 0, 0)$, $(0, 1, 0)$, and $(0, 0, 1)$ are linearly independent.

From these eigenvalues and eigenvectors we can construct solutions of the system

$$\mathbf{Y}_1(t) = e^{-3t} \begin{pmatrix} 1 \\ 0 \\ 0 \end{pmatrix} = \begin{pmatrix} e^{-3t} \\ 0 \\ 0 \end{pmatrix},$$

$$\mathbf{Y}_2(t) = e^{-t} \begin{pmatrix} 0 \\ 1 \\ 0 \end{pmatrix} = \begin{pmatrix} 0 \\ e^{-t} \\ 0 \end{pmatrix},$$

and

$$\mathbf{Y}_3(t) = e^{-2t} \begin{pmatrix} 0 \\ 0 \\ 1 \end{pmatrix} = \begin{pmatrix} 0 \\ 0 \\ e^{-2t} \end{pmatrix}.$$

Because this system is diagonal, we could have gotten this far "by inspection." If we write the system in components

$$\frac{dx}{dt} = -3x$$
$$\frac{dy}{dt} = -y$$
$$\frac{dz}{dt} = -2z,$$

we see that dx/dt depends only on x, dy/dt depends only on y, and dz/dt depends only on z. In other words, the system completely decouples, and each coordinate can be dealt with independently. The solutions of these differential equations are easy to find.

Now that we have three independent solutions, we can solve any initial-value problem for this system. For example, to find the solution $\mathbf{Y}(t)$ with $\mathbf{Y}(0) = (2, 1, 2)$, we must find constants k_1, k_2, and k_3 such that

$$(2, 1, 2) = k_1 \mathbf{Y}_1(0) + k_2 \mathbf{Y}_2(0) + k_3 \mathbf{Y}_3(0).$$

So $k_1 = 2$, $k_2 = 1$, and $k_3 = 2$, and $\mathbf{Y}(t) = (2e^{-3t}, e^{-t}, 2e^{-2t})$ is the required solution.

Figure 3.57 is a sketch of the phase space. Note that the coordinate axes are lines of eigenvectors, so they form straight-line solutions. Since all three of the eigenvalues are negative, solutions along all three of the axes tend toward the origin. Because every other solution can be made up as a linear combination of the solutions on the axes, all solutions must tend to the origin and it is natural to call the origin a sink.

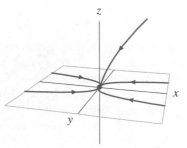

Figure 3.57
Phase space for $d\mathbf{Y}/dt = \mathbf{A}\mathbf{Y}$ for the diagonal matrix $\mathbf{A}$.

Three-dimensional behavior

Before giving a classification of linear systems in three dimensions, we give an example that has behavior very different from what is possible in two dimensions.

Consider the system

$$\frac{d\mathbf{Y}}{dt} = \mathbf{BY} = \begin{pmatrix} 0.1 & -1 & 0 \\ 1 & 0.1 & 0 \\ 0 & 0 & -0.2 \end{pmatrix} \begin{pmatrix} x \\ y \\ z \end{pmatrix}.$$

The characteristic polynomial of $\mathbf{B}$ is

$$((0.1 - \lambda)(0.1 - \lambda) + 1)(-0.2 - \lambda) = (\lambda^2 - 0.2\lambda + 1.01)(-0.2 - \lambda),$$

so the eigenvalues are $\lambda_1 = -0.2$, $\lambda_2 = 0.1 + i$, and $\lambda_3 = 0.1 - i$. Corresponding to the real negative eigenvalue λ_1, we expect to see a line of solutions that approach the origin in the phase space. By analogy to the two-dimensional case, the complex eigenvalues with positive real part correspond to solutions that spiral away from the origin. This is a "spiral saddle," which is not possible in two dimensions.

We could find the eigenvectors associated with each eigenvalue as above and find the general solution. The eigenvectors for the complex eigenvalues are complex, and to find the real solutions, we would have to take real and imaginary parts, just as in two dimensions. However, we are lucky again, and this system also decouples into

$$\frac{dx}{dt} = 0.1x - y$$

$$\frac{dy}{dt} = x + 0.1y$$

and

$$\frac{dz}{dt} = -0.2z.$$

In the xy-plane, the eigenvalues are $0.1 \pm i$, so the origin is a spiral source. Along the z-axis, all solutions tend toward zero as time increases (see Figure 3.58).

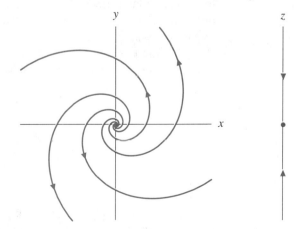

Figure 3.58
Phase plane for xy-system and phase line for z.

Combining these pictures we obtain a sketch of the three-dimensional phase space. Note that the z-coordinate of each solution decreases toward zero, while in the xy-plane solutions spiral away from the origin (see Figure 3.59).

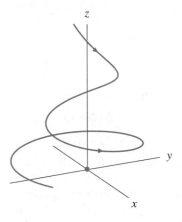

Figure 3.59
Phase space for $d\mathbf{Y}/dt = \mathbf{BY}$.

Classification of Three-Dimensional Linear Systems

Although there are more possible types of phase space pictures for three-dimensional linear systems than for two dimensions, the list is still finite. Just as for two dimensions, the nature of the system is determined by the eigenvalues. Real eigenvalues correspond to straight-line solutions that tend toward the origin if the eigenvalue is negative and away from the origin if the eigenvalue is positive. Complex eigenvalues correspond to spiraling. Negative real parts indicate spiraling toward the origin, whereas positive real parts indicate spiraling away from the origin.

Since the characteristic polynomial is a cubic, there are three eigenvalues (which might not all be distinct if there are repeated roots). It is always the case that at least one of the eigenvalues is real. The other two may be real or a complex conjugate pair (see exercises).

The most important types of three-dimensional linear systems can be divided into three categories: sinks, sources, and saddles. Examples of the other cases (which include systems with double eigenvalues and zero eigenvalues) are given in the exercises.

Sinks

We call the equilibrium point at the origin a sink if all solutions tend toward it as time increases. If all three eigenvalues are real and negative, then there are three straight lines of solutions, all of which tend toward the origin. Since every other solution is a linear combination of these solutions, all solutions tend to the origin as time increases (see Figure 3.57).

The other possibility for a sink is to have one real negative eigenvalue and two complex eigenvalues with negative real parts. This means that there is one straight line of solutions tending to the origin and a plane of solutions that spiral toward the origin. All other solutions exhibit both of these behaviors (see Figure 3.60).

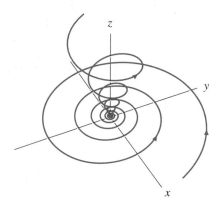

Figure 3.60
Example phase space for spiral sink.

Sources

There are two possibilities for sources as well. We can have either three real and positive eigenvalues or one real positive eigenvalue and a complex conjugate pair with positive real parts. An example of such a phase space is given in Figure 3.61. Note that this system looks just like the sink in Figure 3.60 except the directions of the arrows have been reversed, so solutions move away from the origin as time increases.

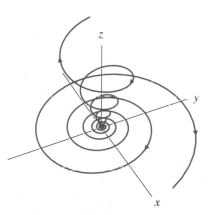

Figure 3.61
Example phase space for spiral source.

Saddles

The equilibrium point at the origin is a saddle if, as time increases to infinity, some solutions tend toward it while other solutions move away from it. This can occur in four different ways. If all the eigenvalues are real, then we could have one positive and two negative or two positive and one negative. In the first case, one positive and two negative, there is one straight line of solutions that tend away from the origin as time increases and a plane of solutions that tend toward the origin as time increases. In the other case, two positive and one negative, there is a plane of solutions that tend away from the origin as time increases and a line of solutions that tend toward the origin as time increases. In both cases, all other solutions will eventually move away from the origin as time increases or decreases (see Figure 3.62).

The other two cases occur if there is only one real eigenvalue and the other two are a complex conjugate pair. If the real eigenvalue is negative and the real parts of the complex eigenvalues are positive, then as time increases there is a straight line of solutions that tend toward the origin and a plane of solutions that tend away from it. All other solutions are a combination of these behaviors, so as time increases they spiral around the straight line of solutions in ever widening loops (see Figure 3.59). The other possibility is that the real eigenvalue is positive and the complex eigenvalues have negative real part. In this case there is a straight line of solutions that tend away from the origin as time increases and a plane of solutions that spiral toward the origin as time increases. Every other solution spirals around the straight line of solutions while moving away from the origin (see Figure 3.63).

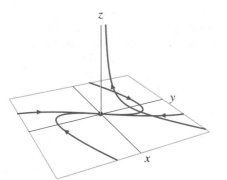

Figure 3.62
Example of a saddle with one positive and two negative eigenvalues.

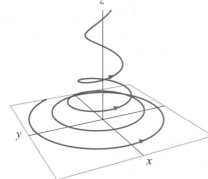

Figure 3.63
Example of a saddle with one real eigenvalue and a complex conjugate pair of eigenvalues.

An example revisited

We end this section by returning to the example that we used at the start of the section. All of the other examples in this section have been systems that decouple into systems of smaller dimension. Sadly, the general case is not so simple. This example doesn't look too complicated because the coefficient matrix has many zero entries. However, it does not immediately decouple into lower dimensional systems.

Consider the system

$$\frac{d\mathbf{Y}}{dt} = \mathbf{AY} = \begin{pmatrix} 0 & 0.1 & 0 \\ 0 & 0 & 0.2 \\ 0.4 & 0 & 0 \end{pmatrix} \begin{pmatrix} x \\ y \\ z \end{pmatrix}.$$

The characteristic polynomial for $\mathbf{A}$ is $-\lambda^3 + 0.008$, so the eigenvalues are the solutions of

$$-\lambda^3 + 0.008 = 0.$$

That is, the eigenvalues are the cube roots of 0.008. Every number has three cube roots if we consider complex as well as real roots. The cube roots of 0.008 are $\lambda_1 = 0.2$, $\lambda_2 = 0.2e^{i2\pi/3}$, and $\lambda_3 = 0.2e^{-i2\pi/3}$. The last two eigenvalues may be written as $\lambda_2 = -0.1 + i\sqrt{0.03}$ and $\lambda_3 = -0.1 - i\sqrt{0.03}$.

This system is a saddle with one positive real eigenvalue and a complex conjugate pair of eigenvalues with negative real parts. Solutions spiral tightly around the line of eigenvectors associated to the eigenvalue $\lambda_1 = 0.2$. In order to sketch the phase space, we must find the eigenvectors for these eigenvalues.

For $\lambda_1 = 0.2$, the eigenvalues are solutions of

$$\mathbf{A}\mathbf{V}_1 = 0.2\mathbf{V}_1,$$

which is written in coordinates as

$$\begin{cases} 0.1y_1 = 0.2x_1 \\ 0.2z_1 = 0.2y_1 \\ 0.4x_1 = 0.2z_1. \end{cases}$$

In particular $\mathbf{V}_1 = (1/2, 1, 1)$ is one such eigenvector. The vector $\mathbf{V}_1$ can be used to determine an entire line of eigenvectors in space.

To find the plane of solutions that spiral toward the origin, we must find the eigenvectors for $\lambda_2 = -0.1 + i\sqrt{0.03}$. That is, we must solve

$$\mathbf{A}\mathbf{V}_2 = (-0.1 + i\sqrt{0.03})\mathbf{V}_2$$

for $\mathbf{V}_2$. In other words,

$$y_2 = (-1 + i\sqrt{3})x_2$$
$$2z_2 = (-1 + i\sqrt{3})y_2$$
$$4x_2 = (-1 + i\sqrt{3})z_2.$$

One eigenvector associated to λ_2 is $\mathbf{V}_2 = (-1 + i\sqrt{3}, -2 - i2\sqrt{3}, 4)$. The corresponding solution to the system is

$$\mathbf{Y}_2(t) = e^{(-0.1 + i\sqrt{0.03})t}(-1 + i\sqrt{3}, -2 - i2\sqrt{3}, 4).$$

We can convert this into two real-valued solutions by taking real and imaginary parts. Since our goal is to find the plane on which solutions spiral, we need only look at the initial point $\mathbf{Y}_2(0) = (-1 + i\sqrt{3}, -2 - i2\sqrt{3}, 4)$. The initial points of the real and imaginary parts are $(-1, -2, 4)$ and $(\sqrt{3}, -2\sqrt{3}, 0)$, respectively. The plane on which solutions spiral toward the origin is the plane made up of all linear combinations of these two vectors. We can use this information to give a fairly accurate sketch of the phase space of this system (see Figure 3.55). We also sketch the graphs of the coordinate functions for one solution (see Figures 3.64 and 3.65). Note that for the example solution shown, all three coordinates tend to infinity as t increases because the eigenvector for the eigenvalue λ_1 has nonzero components for all three variables.

Three linearly independent solutions of this system are given in the first example of this section (see page 326). We can see from this example that linear systems in three dimensions can be quite complicated (even when many of the coefficients are zero). However, the qualitative behavior is still determined by the eigenvalues, so it is possible to classify these systems without completely solving them.

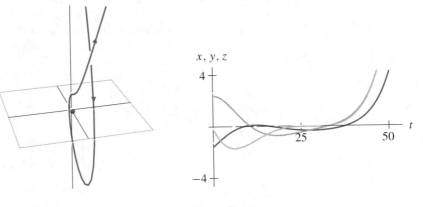

Figure 3.64
Phase space for system
$d\mathbf{Y}/dt = \mathbf{AY}$.

Figure 3.65
Graphs of $x(t)$, $y(t)$ and $z(t)$ for the indicated solution in Figure 3.64.

EXERCISES FOR SECTION 3.8

1. Consider the linear system

$$\frac{d\mathbf{Y}}{dt} = \mathbf{AY} = \begin{pmatrix} 0 & 0.1 & 0 \\ 0 & 0 & 0.2 \\ 0.4 & 0 & 0 \end{pmatrix} \begin{pmatrix} x \\ y \\ z \end{pmatrix}.$$

Check that the functions

$$\mathbf{Y}_2(t) = e^{-0.1t} \begin{pmatrix} -\cos(\sqrt{0.03}\,t) - \sqrt{3}\sin(\sqrt{0.03}\,t) \\ -2\cos(\sqrt{0.03}\,t) + 2\sqrt{3}\sin(\sqrt{0.03}\,t) \\ 4\cos(\sqrt{0.03}\,t) \end{pmatrix}$$

and

$$\mathbf{Y}_3(t) = e^{-0.1t} \begin{pmatrix} -\sin(\sqrt{0.03}\,t) + \sqrt{3}\cos(\sqrt{0.03}\,t) \\ -2\sin(\sqrt{0.03}\,t) - 2\sqrt{3}\cos(\sqrt{0.03}\,t) \\ 4\sin(\sqrt{0.03}\,t) \end{pmatrix}$$

are solutions to the system.

2. If a vector $\mathbf{Y}_3$ lies in the plane determined by the two vectors $\mathbf{Y}_1$ and $\mathbf{Y}_2$, then we can write $\mathbf{Y}_3$ as a linear combination of $\mathbf{Y}_1$ and $\mathbf{Y}_2$. That is,

$$\mathbf{Y}_3 = k_1\mathbf{Y}_1 + k_2\mathbf{Y}_2$$

for some constants k_1 and k_2. But then

$$k_1\mathbf{Y}_1 + k_2\mathbf{Y}_2 - \mathbf{Y}_3 = (0, 0, 0).$$

Show that, if

$$k_1\mathbf{Y}_1 + k_2\mathbf{Y}_2 + k_3\mathbf{Y}_3 = (0, 0, 0)$$

with not all of k_1, k_2, and $k_3 = 0$, then the vectors are not linearly independent. [*Hint*: Start by assuming that $k_3 \neq 0$ and show that $\mathbf{Y}_3$ is in the plane determined by $\mathbf{Y}_1$ and $\mathbf{Y}_2$. Then treat the other cases.] Note that this computation leads to the theorem that three vectors $\mathbf{Y}_1$, $\mathbf{Y}_2$ and $\mathbf{Y}_3$ are linearly independent if and only if the only solution of

$$k_1\mathbf{Y}_1 + k_2\mathbf{Y}_2 + k_3\mathbf{Y}_3 = (0, 0, 0)$$

is $k_1 = k_2 = k_3 = 0$.

3. Using the technique of Exercise 2, determine whether or not the following sets of three vectors are linearly independent.
   (a) $(1, 2, 1)$, $(1, 3, 1)$, $(1, 4, 1)$
   (b) $(2, 0, -1)$, $(3, 2, 2)$, $(1, -2, -3)$
   (c) $(1, 2, 0)$, $(0, 1, 2)$, $(2, 0, 1)$
   (d) $(-3, \pi, 1)$, $(0, 1, 0)$, $(-2, -2, -2)$

In Exercises 4–7, consider the linear system $d\mathbf{Y}/dt = \mathbf{AY}$ with the coefficient matrix $\mathbf{A}$ specified. Each of these systems decouples into a two-dimensional system and a one-dimensional system. For each exercise,

(a) compute the eigenvalues,

(b) determine how the system decouples,

(c) sketch the two-dimensional phase plane and one-dimensional phase line for the decoupled systems, and

(d) give a rough sketch of the phase portrait of the system.

4. $\mathbf{A} = \begin{pmatrix} 0 & 1 & 0 \\ -1 & 0 & 0 \\ 0 & 0 & 2 \end{pmatrix}$
$\qquad$
5. $\mathbf{A} = \begin{pmatrix} -2 & 3 & 0 \\ 3 & -2 & 0 \\ 0 & 0 & -1 \end{pmatrix}$

6. $\mathbf{A} = \begin{pmatrix} 1 & 0 & 3 \\ 0 & -1 & 0 \\ -3 & 0 & 1 \end{pmatrix}$
$\qquad$
7. $\mathbf{A} = \begin{pmatrix} 1 & 0 & 0 \\ 0 & 2 & -1 \\ 0 & -1 & 2 \end{pmatrix}$

Exercises 8–9 consider the properties of the cubic polynomial

$$p(\lambda) = \alpha\lambda^3 + \beta\lambda^2 + \gamma\lambda + \delta,$$

where α, β, γ, and δ are real numbers.

8. (a) Show that, if α is positive, then the limit of $p(\lambda)$ as $\lambda \to \infty$ is ∞ and the limit of $p(\lambda)$ as $\lambda \to -\infty$ is $-\infty$.

    (b) Show that, if α is negative, then the limit of $p(\lambda)$ as $\lambda \to \infty$ is $-\infty$ and the limit of $p(\lambda)$ as $\lambda \to -\infty$ is ∞.

    (c) Using the above, show that $p(\lambda)$ must have at least one real root (that is, at least one real number λ_0 such that $p(\lambda_0) = 0$). [*Hint*: Look at the graph of $p(\lambda)$.]

9. Suppose $a + ib$ is a root of $p(\lambda)$ (so $p(a + ib) = 0$). Show that $a - ib$ is also a root. [*Hint*: Remember that a complex number is zero if and only if both its real and imaginary parts are zero. Then compute $p(a + ib)$ and $p(a - ib)$.]

In Exercises 10–13, consider the linear system $d\mathbf{Y}/dt = \mathbf{BY}$ with the coefficient matrix $\mathbf{B}$ specified. These systems do not fit into the classification of the most common types of systems given in the text. However, the equations for dx/dt and dy/dt decouple from dz/dt. For each of these systems,

  (a) compute the eigenvalues,

  (b) sketch the xy-phase plane and the z-phase line, and

  (c) give a rough sketch of the phase portrait of the system.

10. $\mathbf{B} = \begin{pmatrix} -2 & 1 & 0 \\ 0 & -2 & 0 \\ 0 & 0 & -1 \end{pmatrix}$
    11. $\mathbf{B} = \begin{pmatrix} -2 & 1 & 0 \\ 0 & -2 & 0 \\ 0 & 0 & 1 \end{pmatrix}$

12. $\mathbf{B} = \begin{pmatrix} -1 & 2 & 0 \\ 2 & -4 & 0 \\ 0 & 0 & -1 \end{pmatrix}$
    13. $\mathbf{B} = \begin{pmatrix} -1 & 2 & 0 \\ 2 & -4 & 0 \\ 0 & 0 & 0 \end{pmatrix}$

In Exercises 14–15, consider the linear system $d\mathbf{Y}/dt = \mathbf{CY}$. These systems do not fit into the classification of the most common types of systems given in the text, and they do not decouple into lower dimensional systems. For each system,

  (a) compute the eigenvalues,

  (b) compute the eigenvectors, and

  (c) sketch (as best you can) the phase portrait of the system. [*Hint*: Use the eigenvalues and eigenvectors and also vectors in the vector field.]

14. $\mathbf{C} = \begin{pmatrix} -2 & 1 & 0 \\ 0 & -2 & 1 \\ 0 & 0 & -2 \end{pmatrix}$
    15. $\mathbf{C} = \begin{pmatrix} 0 & 1 & 0 \\ 0 & 0 & 1 \\ 0 & 0 & 0 \end{pmatrix}$

16. For the linear system

$$\frac{d\mathbf{Y}}{dt} = \mathbf{AY} = \begin{pmatrix} 2 & -1 & 0 \\ 0 & -2 & 3 \\ -1 & 3 & -1 \end{pmatrix} \begin{pmatrix} x \\ y \\ z \end{pmatrix}:$$

(a) Show that $\mathbf{V}_1 = (1, 1, 1)$ is an eigenvector of the coefficient matrix by computing $\mathbf{AV}_1$. What is the eigenvalue for this eigenvector?

(b) Find the other two eigenvalues for the matrix $\mathbf{A}$.

(c) Classify the system (source, sink, ...).

(d) Sketch (as best you can) the phase portrait. [*Hint*: Use the other eigenvalues and find the other eigenvectors.]

17. For the linear system

$$\frac{d\mathbf{Y}}{dt} = \mathbf{AY} = \begin{pmatrix} -4 & 3 & 0 \\ 0 & -1 & 1 \\ 5 & -5 & 0 \end{pmatrix} \begin{pmatrix} x \\ y \\ z \end{pmatrix}:$$

(a) Show that $\mathbf{V}_1 = (1, 1, 0)$ is an eigenvector of the coefficient matrix by computing $\mathbf{AV}_1$. What is the eigenvalue for this eigenvector?

(b) Find the other two eigenvalues for the matrix $\mathbf{A}$.

(c) Classify the system (source, sink, ...).

(d) Sketch (as best you can) the phase portrait. [*Hint*: Use the other eigenvalues and find the other eigenvectors.]

18. Consider the linear system

$$\frac{d\mathbf{Y}}{dt} = \mathbf{BY} = \begin{pmatrix} -10 & 10 & 0 \\ 28 & -1 & 0 \\ 0 & 0 & -8/3 \end{pmatrix} \begin{pmatrix} x \\ y \\ z \end{pmatrix}.$$

(This system is related to the Lorenz system studied in Section 2.5, and we will use the results obtained in this exercise when we return to the Lorenz equations in Section 5.5.)

(a) Find the characteristic polynomial and the eigenvalues.

(b) Find the eigenvectors.

(c) Sketch the phase portrait (as best you can).

(d) Comment on how the fact that the system "decouples" helps in the computations and in sketching the phase space.

Many years later, when Glen finally retires from writing math texts, he decides to join his friends and former collaborators Paul and Bob in the CD business. He opens a store specializing in New Age music located in between Paul's and Bob's stores. Let $z(t)$ be Glen's profits at time t (with $x(t)$ and $y(t)$ representing Paul's and Bob's profits at time t, respectively). Suppose the three stores affect each other in such a way that

$$\frac{dx}{dt} = -y + z$$
$$\frac{dy}{dt} = -x + z$$
$$\frac{dz}{dt} = z.$$

19. (a) If Glen makes a profit, does this help or hurt Paul's and Bob's profits?
     (b) If Paul and Bob are making profits, does this help or hurt Glen's profits?

20. Write this system in matrix form and find the eigenvalues. Use them to classify the system.

21. Suppose that at time $t = 0$ both Paul and Bob are making (equal) small profits, but Glen is just breaking even [$x(0) = y(0)$ are small and positive, but $z(0) = 0$].

     (a) Sketch the solution curve in the xyz-phase space.
     (b) Sketch the $x(t)$-, $y(t)$-, and $z(t)$-graphs of the solution.
     (c) Describe what happens to the profits of each store.

22. Suppose that at time $t = 0$ both Paul and Bob are just breaking even, but Glen is making a small profit [$x(0)$ and $y(0)$ are zero, but $z(0)$ is small and positive].

     (a) Sketch the solution curve in the xyz-phase space.
     (b) Sketch the $x(t)$-, $y(t)$-, and $z(t)$-graphs of the solution.
     (c) Describe what happens to the profits of each store.

LAB 3.1 Bifurcations in Linear Systems

In Chapter 3 we have studied techniques for solving linear systems. Given the coefficient matrix for the system, we can use these techniques to classify the system, describe the qualitative behavior of solutions, and give a formula for the general solution. In this lab we consider a two-parameter family of linear systems. The goal is to better understand how different linear systems are related to each other: In other words, what bifurcations occur in parameterized families of linear systems.

Consider the linear system

$$\frac{dx}{dt} = ax + by$$
$$\frac{dy}{dt} = -x - y,$$

where a and b are parameters that can take on any real value. In your report address the following items:

1. For each value of a and b, classify the linear system as source, sink, center, spiral sink, and so forth. Draw a picture of the ab-plane and indicate the values of a and b for which the system is of each type (that is, shade the values of a and b for which the system is a sink red, for which it is a source blue, and so forth). Be sure to describe all of the computations involved in creating this picture.

2. As the values of a and b are changed so that the point (a, b) moves from one region to another, the type of the linear system changes, that is, a bifurcation occurs. Which of these bifurcations is important for the long-term behavior of solutions? Which of these bifurcations corresponds to a dramatic change in the phase plane or the $x(t)$- and $y(t)$-graphs?

Your report: Address the items above in the form of a short essay. Include any computations necessary to produce the picture in part 1. You may include phase planes and/or graphs of solutions to illustrate your essay, but your answer should be complete and understandable without the pictures.

LAB 3.2 RLC Circuits

We have already seen examples of differential equations that serve as models of simple electrical circuits involving only a resistor, a capacitor, and a voltage source. In this lab we consider slightly more complicated circuits consisting of a resistor, a capacitor, an inductor, and a voltage source (see Figure 3.66). The behavior of the system can be described by specifying the current moving around the circuit and the changes in voltages across each component of the circuit. In this lab we take an axiomatic approach to the relationship between the current and the voltages. Readers interested in more information on the derivation of these laws are referred to texts in electric circuit theory.

In the tradition of electrical engineering, we let i denote the current moving around the circuit. We let v_T, v_C, and v_L denote the voltages across the voltage source, the

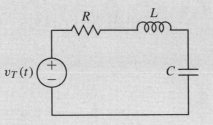

Figure 3.66
An RLC circuit.

capacitor, and the inductor, respectively. Also we let R denote the resistance, C the capacitance, and L the inductance of the associated components of the circuit. These are indicated in Figure 3.66. We think of v_T, R, C, and L as parameters set by the person building the circuit. The quantities i, v_C, and v_L depend on time.

We need the following basic relationships between the quantities above. First, Kirchhoff's voltage law states that the sum of the voltage changes around a closed loop must be zero. For our circuit this gives

$$v_T - Ri = v_C + v_L.$$

Next, we need the relationship between current and voltage in the capacitor and the inductor. In a capacitor the current is proportional to the rate of change of the voltage. The proportionality constant is the capacitance C. Hence we have

$$C\frac{dv_C}{dt} = i.$$

In an inductor the voltage is proportional to the rate of change of the current. The proportionality constant is the inductance L. Hence we have

$$L\frac{di}{dt} = v_L.$$

In this lab we consider the possible behavior of the circuit above for several different input voltages.

In your report address the following questions:

1. First, set the input voltage to zero, that is, assume $v_T = 0$. Using the three equations above, write a first-order system of differential equations with dependent variables i and v_C. [*Hint:* Use the first equation to eliminate v_L from the third equation. You should have R, C, and L as parameters in your system.]

2. Find the eigenvalues of the resulting system in terms of the parameters R, C, and L. What are the possible phase planes for your system given that R, C, and L are always nonnegative? Sketch the phase plane and the $v_C(t)$- and $i(t)$-graphs for each case.

3. Convert the first-order system of equations from part 1 into a second-order differential equation involving only v_C (and not i). (This is the form of the equation that you will typically find in electric circuit theory texts.)

4. Repeat part 1 assuming that v_T is nonzero. The resulting system will have $R, C, L,$ and v_T as parameters.

5. The units used in applications are volts and amps for voltages and currents, ohms for resistors, farads for capacitors, and henrys for inductors. A typical, off-the-shelf circuit might have parameter values $R = 2000$ ohms (or 2 kilo-ohms), $C = 2 \cdot 10^{-7}$ farads (or 0.2 micro farads), and $L = 1.5$ henrys. Assuming zero-input, $v_T = 0$, and the initial values of the current and voltage are $i(0) = 0$, $v_C(0) = 10$, describe the behavior of the current and voltage for this circuit.

6. Repeat part 5 using a voltage source of $v_T = 10$ volts.

Your report: In your report address the items above. Show all algebra and justify all steps. In parts 5 and 6 you may work either analytically or numerically. Give phase planes and graphs of solutions as appropriate.

LAB 3.3 Measuring Mass in Space

The effects on the human body of prolonged weightlessness during space flights is not completely understood. One important variable that must be monitored is the astronaut's "weight." However, weight refers to the force of gravity on a body. What actually must be measured is body mass. To perform this measurement, a mass-spring system is used, where the mass is the body of the astronaut. The astronaut sits in a special chair attached to springs. The frequency of the oscillation of the astronaut in the chair is measured and from this the mass is computed.*
In your report address the following items:.

1. Suppose the chair has a mass of 20kg. The system is initially calibrated by placing a known mass in the chair and measuring the period of oscillations. Suppose that a 25kg mass placed in the chair results in an oscillation with period of 1.3 seconds per oscillation. We assume that the coefficient of damping of the apparatus is very small (so as a first approximation we assume that there is no damping). What will be the period of oscillations of an astronaut with mass 60kg? What would your frequency of oscillation be?

2. Does it matter whether or not the calibration is done on the earth or in space? (It would be much better if it could be done on the earth since it is expensive to launch 25kg masses into space.)

3. Suppose an error is made during the calibration, and the actual frequency resulting when a 25kg mass is placed in the chair is 1.31 seconds instead of 1.3 seconds. How much error then results in the measurement of the mass of astronaut with mass 60kg? With mass 80kg?

4. Suppose a small amount of damping develops in the chair. How seriously does this affect the measurements? How could you determine if damping were present (that is, what measurements would you perform during the calibration phase)?

---

*This information comes from the "Q & A" column of the *New York Times*, Science Section, August 1, 1995, page C6.

Your report: In your report you should address each of the items above. Show all algebraic computations you perform and justify all assertions. While this lab does not require numerical approximations of solutions (since we can explicitly solve all the equations involved), you may include sketches of solutions and/or computer generated graphs *if appropriate*.

LAB 3.4   Find Your Own Harmonic Oscillator

In the text we claimed that the harmonic oscillator could be used as a simple model for many different situations. The key ingredients are a restoring force, which pushes the system back toward a rest position, and (perhaps) a damping force.

For this lab you are to find such a system. It may be mechanical, biological, psychological, political, financial, or whatever. You will have to do some analysis on your system, so you will need to be able to obtain some data on its "motion." You can either record the data yourself or use published data (with proper references).
In your report address the following items:

1. In a short essay, carefully describe the system you propose to model with a harmonic oscillator equation. State the dependent variable and describe the rest position, the restoring force, and the damping.

2. Collect data on the behavior of your system either by observing your system or from published sources (taking care to give proper references).

3. Use your data to estimate the parameter values appropriate for your system. Classify your system (as either overdamped, critically damped, or underdamped). Compare solutions of the harmonic oscillator equation with the data you have collected. Describe how well the solutions of the model fit the data and discuss any discrepancies.

Your report: In your report you should address the three items above in the form of a short essay. In part 2 describe how you obtained your data (either citing appropriate references or giving details of your experimental procedure). In part 3 be sure to address any problems that arose in using the harmonic oscillator as a model. (If your data differs significantly from your solutions, then you may want to adjust the parameter values that you are using or search for another situation that is better modeled by the harmonic oscillator equations.)

LAB 3.5   A Baby Bottle Harmonic Oscillator

If you performed the previous lab experiment, you undoubtedly saw that harmonic oscillators are everywhere. Here is a harmonic oscillator that one of the authors discovered late one night.

Consider a U-shaped tube or pipe partially filled with fluid. (For example, a common design of baby bottle has the shape shown in Figure 3.67.)

If held upright, the level of fluid will come to rest at an equal height on each side of the bottle. If we tip the bottle quickly to one side, then return it to level, there will be

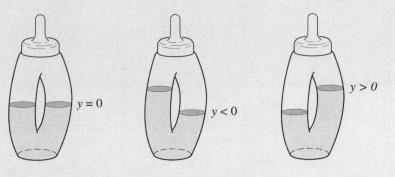

Figure 3.67
A U-shaped baby bottle.

more fluid in one side than in the other. Gravity pulling on this extra fluid will provide a restoring force, pushing the system back toward the rest position where both sides are at an equal level.

In your report address the following items:

1. Let $y(t)$ denote the height of the fluid in one side of bottle, with $y = 0$ corresponding to the rest position. Assume that the restoring force is proportional to the difference in the mass in the left- and right-hand sides of the bottle. Also assume, as a first approximation, that the damping force is proportional to the velocity. Use this to give a differential equation model of this system based on the harmonic oscillator equation.

2. Buy or borrow such a baby bottle and observe it. Estimate the natural period of the oscillations and the rate of decay of the amplitude of the oscillations. Use the observations to estimate the parameter values for the harmonic oscillator equation model.

3. Compare solutions of the harmonic oscillator equation in part 2 to your observations. In particular, if you increase the total mass in the bottle but leave the size of the initial displacement unchanged, does the model make the correct prediction of the change of behavior of the fluid in the bottle? For example, does the natural period increase or decrease in this situation?

Your report: Address each of the three questions above in the form of a short essay. In part 1 carefully describe the relationship between the forces and the differential equation model. In part 2 describe the experiments you performed to obtain data and the computations necessary for computing the coefficients in the harmonic oscillator equation from your data. In part 3 discuss any discrepancies between your data and the predictions of the model.

4

FORCING
AND RESONANCE

In this chapter we consider some particularly important nonautonomous, second-order equations and systems, along with the mathematical techniques used to study them. These examples are important because they occur in numerous applications and because the mathematical ideas developed to attack them are quite appealing.

We begin with a system called the forced harmonic oscillator. This system models (among other things) a mass attached to a spring sliding on a table. The only forces included in the model are the restoring force of the spring and the damping. Any other force affecting the motion of the mass is referred to as an external force. A harmonic oscillator with external force that depends only on time is a forced harmonic oscillator.

In the first four sections we give techniques for finding solutions of forced harmonic oscillator equations for typical forcing functions, and using these special equations as a guide, we discuss the phenomenon of resonance. In the final section we discuss recent models that explain the collapse of the Tacoma Narrows Bridge (see the cover of this book).

4.1   FORCED HARMONIC OSCILLATORS

In Sections 2.1 and 2.3 we introduced the harmonic oscillator equation as a model for a physical system of a mass attached to a spring that slides on a table. The mass is subject both to a restoring force provided by the spring and to damping. The restoring force is assumed to be proportional to the displacement, and the damping is assumed to be proportional to the velocity. The proportionality constants are parameters called the spring constant and the coefficient of damping. We obtain a linear, second-order, constant coefficient equation. In Section 3.6 we used the techniques from the study of linear systems to classify the possible behaviors of solutions of the harmonic oscillator equation. Moreover, we described an efficient method for generating the general solution of these equations.

In this chapter we consider the effect of "external forces," that is, forces other than the restoring force and the damping. Such forces may include shaking the table or pushing the mass.

Another device that can be very informally modeled using the harmonic oscillator system is the motion of a crystal glass after being gently tapped with a fork. The position variable y measures the amount the shape of the glass is deformed from its rest position. The restoring force pushes the glass back toward its original shape. After the tap by the fork, the ringing sound is caused by oscillations of the glass around the rest position, so this is an underdamped oscillator. The frequency of the sound is the natural frequency of the oscillations. As an example of an external force, we imagine an opera singer standing near the glass, singing a fixed note. The sound waves push against the glass and deform it slightly. The size and direction of this external force depend only on time.

In the first four sections of this chapter we consider how external forces affect the motion of a harmonic oscillator. In particular, how can an opera singer break the glass by singing a particular note?

An Equation for the Forced Harmonic Oscillator

Recall that the equations for the harmonic oscillator come from Newton's second law

$$\text{mass} \times \text{acceleration} = \text{force},$$

applied to the motion of a mass attached to a spring, sliding on a table. We let $y(t)$ denote the position of the mass m at time t, with $y = 0$ the rest position. The forces on the mass are the spring force, $-ky$, and the damping, $-b(dy/dt)$. Substituting into Newton's law gives

$$m\frac{d^2y}{dt^2} = -ky - b\frac{dy}{dt},$$

which is often written as

$$\frac{d^2y}{dt^2} + \frac{b}{m}\frac{dy}{dt} + \frac{k}{m}y = 0.$$

The parameters are $m > 0$, $k > 0$, and $b \geq 0$.

To include an external force, we must add another term to the right-hand side of this equation. This forcing term can be any function. We consider external forcing that depends only on time t, so the external force is given by a function $f(t)$. Typical examples of forcing functions include constant functions (for a force pushing at a constant strength against the mass) and $f(t) = e^{-at}$ (for a force whose strength decreases exponentially with time). A particularly important example is $f(t) = \sin \omega t$ or $\cos \omega t$, called sinusoidal forcing with period $2\pi/\omega$ (or frequency $\omega/(2\pi)$). This corresponds to a force that alternately pushes and pulls the mass back and forth in a periodic fashion (like sound waves pushing against a glass).

The new equation is

$$m\frac{d^2y}{dt^2} = -ky - b\frac{dy}{dt} + f(t),$$

which is frequently written

$$\frac{d^2y}{dt^2} + \frac{b}{m}\frac{dy}{dt} + \frac{k}{m}y = \frac{f(t)}{m}.$$

If we let $p = b/m$, $q = k/m$, and $g(t) = f(t)/m$, then we have

$$\frac{d^2y}{dt^2} + p\frac{dy}{dt} + qy = g(t).$$

This is a second-order, linear, constant-coefficient, **nonhomogeneous**, nonautonomous equation. The new adjective *nonhomogeneous* refers to the fact that the right-hand side of the equation is nonzero. We also refer to this as a **forced harmonic oscillator**, or a **forced equation**, for short. We often abuse terminology slightly and refer to $g(t)$ as the forcing function. The equation

$$\frac{d^2y}{dt^2} + p\frac{dy}{dt} + qy = 0$$

is said to be the corresponding **homogeneous** or **unforced** equation.

Our goal for the equation of the forced harmonic oscillator is the same as for the harmonic oscillator. Given the parameters p and q and the forcing function $g(t)$, describe the motion of the mass, produce graphs of the solutions and, if possible, give formulas for solutions. In this section we give examples of a method for finding explicit solutions of the forced harmonic oscillator system for special types of forcing functions.

The method used here is a "lucky-guess" method with the impressive name the *Method of Undetermined Coefficients*. An alternative, somewhat more systematic technique is the method of Laplace transforms given in Chapter 6. Again it must be emphasized that, as with nonlinear equations, there is no method that gives the general solution for arbitrary forcing functions. The equations we can solve explicitly are special cases. The goal of this section is to obtain the general solution for a few special cases and to study the behavior of solutions in these cases.

The Extended Linearity Principle

To find the general solution of an equation of the form

$$\frac{d^2y}{dt^2} + p\frac{dy}{dt} + qy = g(t),$$

we take full advantage of the fact that we already know how to find the general solution of the homogeneous equation

$$\frac{d^2y}{dt^2} + p\frac{dy}{dt} + qy = 0.$$

We could convert this second-order equation into a first-order system and find the general solution by computing eigenvalues and eigenvectors. However, in Section 3.6 we pointed out that the eigenvalues and the general solution can be obtained quickly by a guess-and-test method. We use this efficient method throughout this chapter.

The key to computing the general solution of a force harmonic oscillator equation is an extension of the Linearity Principle of Chapter 3.

EXTENDED LINEARITY PRINCIPLE   Consider a nonhomogeneous equation (a forced equation)

$$\frac{d^2y}{dt^2} + p\frac{dy}{dt} + qy = g(t)$$

and its corresponding homogeneous equation (the unforced equation)

$$\frac{d^2y}{dt^2} + p\frac{dy}{dt} + qy = 0.$$

1. Suppose $y_p(t)$ is a particular solution of the nonhomogeneous equation and $y_h(t)$ is a solution of the corresponding homogeneous equation. Then $y_h(t) + y_p(t)$ is also a solution of the nonhomogeneous equation.

2. Suppose $y_p(t)$ and $y_q(t)$ are two solutions of the nonhomogeneous equation. Then $y_p(t) - y_q(t)$ is a solution of the corresponding homogeneous equation.

Therefore, if $k_1y_1(t) + k_2y_2(t)$ is the general solution of the homogeneous equation, then

$$k_1y_1(t) + k_2y_2(t) + y_p(t)$$

is the general solution of the nonhomogeneous equation.   ∎

We verify this theorem by substituting the given functions into the nonhomogeneous equation. To verify the first assertion, we substitute $y_h(t) + y_p(t)$ into the nonhomogeneous equation and get

$$\frac{d^2}{dt}(y_h + y_p) + p\frac{d}{dt}(y_h + y_p) + q(y_h + y_p)$$

$$= \left(\frac{d^2y_h}{dt} + \frac{d^2y_p}{dt}\right) + p\left(\frac{dy_h}{dt} + \frac{dy_p}{dt}\right) + q\left(y_h + y_p\right)$$

$$= \left(\frac{d^2 y_h}{dt} + p\frac{dy_h}{dt} + qy_h \right) + \left(\frac{d^2 y_p}{dt} + p\frac{dy_p}{dt} + qy_p \right)$$

$$= 0 + g(t) = g(t),$$

so $y_h(t) + y_p(t)$ is a solution of the nonhomogeneous equation.

To verify the second assertion, we compute

$$\frac{d^2}{dt}(y_p - y_q) + p\frac{d}{dt}(y_p - y_q) + q(y_p - y_q)$$

$$= \left(\frac{d^2 y_p}{dt} - \frac{d^2 y_q}{dt} \right) + p\left(\frac{dy_p}{dt} - \frac{dy_q}{dt} \right) + q\left(y_p - y_q \right)$$

$$= \left(\frac{d^2 y_p}{dt} + p\frac{dy_p}{dt} + qy_p \right) - \left(\frac{d^2 y_q}{dt} + p\frac{dy_q}{dt} + qy_q \right)$$

$$= g(t) - g(t) = 0,$$

so $y_p(t) - y_q(t)$ is a solution of the homogeneous equation.

If $k_1 y_1(t) + k_2 y_2(t)$ is the general solution of the homogeneous equation, then the first half of the Extended Linearity Principle says that

$$k_1 y_1(t) + k_2 y_2(t) + y_p(t)$$

is a solution of the nonhomogeneous equation for *any* choice of the constants k_1 and k_2. The second half of the Extended Linearity Principle says that *any* solution $y_q(t)$ of the nonhomogeneous equation can be written as

$$k_1 y_1(t) + k_2 y_2(t) + y_p(t)$$

for some choice of k_1 and k_2. Therefore,

$$k_1 y_1(t) + k_2 y_2(t) + y_p(t)$$

is the general solution of the nonhomogeneous equation.

The Extended Linearity Principle gives us an algorithm for finding the general solution of equations for forced harmonic oscillators:

Step 1 Find the general solution of the corresponding unforced (homogeneous) second-order equation.

Step 2 Find *one* particular solution of the forced (nonhomogeneous) second-order equation.

Step 3 Add the results of Steps 1 and 2 to obtain the general solution of the forced equation.

Qualitative Implications of the Extended Linearity Principle

The Extended Linearity Principle gives us a powerful tool for understanding the qualitative behavior of the solutions of the *damped* forced harmonic oscillator equations. To see this, we first recall that the coefficients p and q in

$$\frac{d^2y}{dt^2} + p\frac{dy}{dt} + qy = 0$$

satisfy the inequalities $q > 0$ and $p \geq 0$. Whenever damping is present (as it is in any physical mechanical device), the damping coefficient p is positive, and the equilibrium point at the origin of the corresponding system is a sink (see Section 3.6). Consequently every solution of the unforced equation tends to zero.

Adding an external force $g(t)$, the Extended Linearity Principle implies that the equation

$$\frac{d^2y}{dt^2} + p\frac{dy}{dt} + qy = g(t)$$

has the general solution

$$k_1y_1(t) + k_2y_2(t) + y_p(t),$$

where $k_1y_1(t) + k_2y_2(t)$ is the general solution of the unforced equation and $y_p(t)$ is a solution of the forced equation. But we know that $k_1y_1(t) + k_2y_2(t)$ tends to zero. Hence for large t, we must have

$$k_1y_1(t) + k_2y_2(t) + y_p(t) \approx y_p(t).$$

That is, for large t, *all* solutions of the forced harmonic oscillator equation with nonzero damping ($p > 0$) are approximately the same. A schematic of this situation is shown in Figure 4.1. In other words, the initial conditions have no effect on the long-term behavior of the solutions.

Steady-state solution

In the above discussion any solution can play the role of $y_p(t)$. In engineering applications $y_p(t)$ is often called the **forced response**. The term **steady-state response** is used for the behavior of the forced response for large t. (The phrase "steady-state" is somewhat misleading since $y_p(t)$ need not be constant.)

A solution of the unforced harmonic oscillator (with any initial conditions) is called the **natural response**, or *free response*. The discussion above can be restated in the catch phrase "All solutions of a forced, damped harmonic oscillator approach the steady-state solution because the natural response dies out, leaving only the forced response."

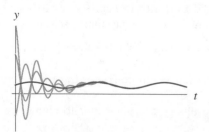

Figure 4.1
Typical graphs of solutions of a forced harmonic oscillator equation with nonzero damping ($p > 0$). The solution $y_p(t)$ is shown in dark blue.

An Example of the Method of Undetermined Coefficients

Consider the equation

$$\frac{d^2y}{dt^2} + 5\frac{dy}{dt} + 6y = e^{-t}.$$

To find the general solution, we follow the procedure suggested by the Extended Linearity Principle. The first step is to find the general solution of the unforced (homogeneous) equation

$$\frac{d^2y}{dt^2} + 5\frac{dy}{dt} + 6y = 0.$$

Using the methods of Section 3.6, the eigenvalues are -2 and -3. This is an overdamped system and the general solution is

$$y_h(t) = k_1e^{-2t} + k_2e^{-3t}.$$

The second step is to find one solution $y_p(t)$ of the forced (nonhomogeneous) equation

$$\frac{d^2y}{dt^2} + 5\frac{dy}{dt} + 6y = e^{-t}.$$

The method we use is a slightly sophisticated guess-and-test. Since we have the term e^{-t} on the right-hand side of the equation, a natural first guess is $y_p(t) = e^{-t}$. Substituting this guess into the left-hand side yields

$$\frac{d^2y_p}{dt^2} + 5\frac{dy_p}{dt} + 6y_p = e^{-t} - 5e^{-t} + 6e^{-t} = 2e^{-t},$$

which does not equal the right-hand side of the forced equation.

Our guess e^{-t} is therefore not a solution. However, it is off only by a constant. It is reasonable to hope that a constant multiple of e^{-t} is a solution, but the correct choice of constant is not clear. We avoid the issue for the moment and improve our guess to $y_p(t) = ke^{-t}$, where k is an unknown constant (or an "undetermined coefficient"). The hope is that by choosing the correct value for k we can make $y_p(t)$ a solution.

Substituting ke^{-t} into the left-hand side of the differential equation, we have

$$\frac{d^2y_p}{dt^2} + 5\frac{dy_p}{dt} + 6y_p = ke^{-t} - 5ke^{-t} + 6ke^{-t} = 2ke^{-t}.$$

In order for $y_p(t)$ to be a solution, we must choose $k - 1/2$. Hence $y_p(t) = e^{-t}/2$ is a solution of the forced equation.

Therefore the general solution of

$$\frac{d^2y}{dt^2} + 5\frac{dy}{dt} + 6y = e^{-t}$$

is the sum of the general solution of the unforced equation plus a particular solution of the forced equation. So the general solution is

$$y(t) = k_1e^{-2t} + k_2e^{-3t} + \frac{e^{-t}}{2}.$$

Several solutions with different initial conditions are shown in Figure 4.2. As predicted, these solutions all approach $y_p(t) = e^{-t}/2$ as t increases.

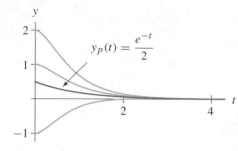

Figure 4.2
Several solutions of

$$\frac{d^2y}{dt^2} + 5\frac{dy}{dt} + 6y = e^{-t}.$$

Note that all of the graphs are asymptotic to $y_p(t) = e^{-t}/2$ as t increases.

The Method of Undetermined Coefficients

Deriving the solution to the unforced equation was discussed in Section 3.6. Thus what is new in the example just completed is the technique used to find one particular solution of the forced equation. We used a lucky-guess method. However, we did not leave everything to luck. The fact that the solution contains a term of the form e^{-t} is not so surprising since this expression forms the right-hand side of the equation. We therefore made an educated guess that the particular solution assumes the form $y_p(t) = ke^{-t}$ and then determined the specific value of k that made $y_p(t)$ a solution.

This "method" for finding particular solutions of forced harmonic oscillator equations is similar in spirit to our guess-and-test method for second-order homogeneous equations. It can be summarized in two steps. First, make a reasonable guess for a solution, but leave some coefficients undetermined. Then substitute the guess into the differential equation and solve for the values of the coefficients that make the guess a solution. This method is called the **Method of Undetermined Coefficients**.

Second Guessing

Whenever we use a guess-and-test method, we run the risk that our first guess (no matter how reasonable) is not a solution. Even leaving coefficients to be determined at the last moment may not be sufficient to obtain a solution. In this situation we just guess again. ("If at first, you don't succeed ....")

Consider the equation

$$\frac{d^2y}{dt^2} + 5\frac{dy}{dt} + 6y = e^{-2t}.$$

This equation differs from the previous example only in the forcing function. The general solution of the unforced equation is still

$$y_h(t) = k_1 e^{-2t} + k_2 e^{-3t}.$$

To find a particular solution of the forced equation, we make the reasonable guess $y_p(t) = ke^{-2t}$, with k as the undetermined coefficient. Substituting into the left-hand side of the forced equation gives

$$\frac{d^2y_p}{dt^2} + 5\frac{dy_p}{dt} + 6y_p = 4ke^{-2t} - 10ke^{-2t} + 6ke^{-2t} = 0.$$

This result is very upsetting. No matter what value we choose for k, substituting $y_p(t)$ into the left-hand side of the forced equation gives zero. Therefore there is no particular solution of the form $y_p(t) = ke^{-2t}$.

Choosing a second guess

The problem with our first guess is that for every k, $y_p(t) = ke^{-2t}$ is a solution of the unforced equation. When we substitute this $y_p(t)$ into the left-hand side of the differential equation, we are guaranteed to get zero. On the other hand, any guess for $y_p(t)$ should contain e^{-2t} as a factor for there to be any hope of obtaining e^{-2t} on the right-hand side.

We encountered this situation before when we discussed repeated roots of the characteristic equation (see Section 3.6). There we saw that a second, not-quite-so-obvious guess was kte^{-2t}. To see if a guess of this form makes sense, we first do some numerics. We use a numerical method to find an approximate solution $y(t)$ of the initial-value problem

$$\frac{d^2y}{dt^2} + 5\frac{dy}{dt} + 6y = e^{-2t}, \quad y(0) = 1, \ y'(0) = -2,$$

and we compare the result to e^{-2t}. (Note that we chose the same initial conditions for $y(t)$ as those of e^{-2t}. That is, at $t = 0$, $e^{-2t} = 1$ and $d(e^{-2t})/dt = -2$.) In Figure 4.3, we graph both the solution $y(t)$ and e^{-2t}. Note that the graphs of these two functions are difficult to distinguish. Close examination shows that the graph of the solution $y(t)$ is slightly but consistently above the graph of e^{-2t}. Since we expect that e^{-2t} appears as one factor in $y(t)$, it makes sense to graph the ratio $y(t)/e^{-2t}$ (see Figure 4.4). After an initial period of time, the ratio seems to follow a straight line. That is, for large t, the solution appears to be approximately kte^{-2t}.

Using this numerical experiment as motivation, we choose $y_p(t) = kte^{-2t}$ as our second guess where k is the undetermined coefficient. Substituting $y_p(t)$ into the left-hand side of the forced equation yields

$$\frac{d^2y_p}{dt^2} + 5\frac{dy_p}{dt} + 6y_p = \frac{d^2(kte^{-2t})}{dt} + 5\frac{d(kte^{-2t})}{dt} + 6kte^{-2t}$$

$$= 4kte^{2t} - 4ke^{2t} + 5(-2kte^{-2t} + ke^{-2t}) + 6kte^{-2t}$$

$$= ke^{-2t}$$

Hence if we choose $k = 1$, the function $y_p(t) = te^{-2t}$ is a solution of the forced

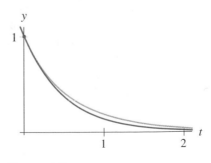

Figure 4.3
Solution of
$d^2y/dt^2 + 5dy/dt + 6y = e^{-2t}$ with
initial conditions $y(0) = 1$, $y'(0) = -2$
graphed with the function e^{-2t}.

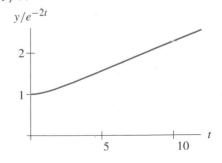

Figure 4.4
Ratio y/e^{-2t} of the two graphs in
Figure 4.3.

equation. The general solution of the forced equation is therefore

$$y(t) = k_1 e^{-2t} + k_2 e^{-3t} + te^{-2t}.$$

Figure 4.5 shows several solutions of the forced equation with different initial conditions. All solutions approach $y_p(t) = te^{-2t}$ (which approaches the t-axis) as predicted.

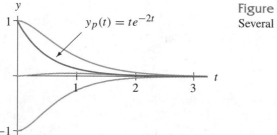

Figure 4.5
Several solutions of

$$\frac{d^2y}{dt^2} + 5\frac{dy}{dt} + 6y = e^{-2t}.$$

Rule of thumb for second guessing

The motivation described above is a good example of what is so unsatisfying about lucky-guess methods. While the numerical evidence for multiplying the first guess by t is compelling, it requires a careful eye to find it. In Chapter 6, we give other methods for coming up with this second guess.

It turns out that for linear equations of any order, the two examples we have discussed give a general rule of thumb for how to find particular solutions. The form of the guess depends on the right-hand side of the equation (the forcing function). If the first guess doesn't yield a particular solution (which occurs if the right-hand side is a solution of the homogeneous equation), we try a second guess by multiplying the first guess by t.

An Example from Electric Circuit Theory

We present an example of a nonhomogeneous second-order equation motivated by electric circuit theory (with realistic and rather messy numbers). Consider the resistor-inductor-capacitor (RLC) circuit pictured in Figure 4.6. Suppose that the voltage at the source is $V_S(t) = 2e^{-t/3}$ volts at time t; that is, the strength of the voltage source is decaying at an exponential rate.

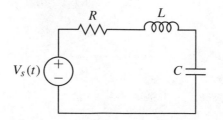

Figure 4.6
RLC circuit to be modeled.

Suppose that $R = 4k\Omega = 4000\Omega$, $C = 0.25\mu F = 0.25 \times 10^{-6}F$, and $L = 1.6H$ (where Ω, F, and H stand for ohms, farads, and henrys, standard units in electric circuit theory). The voltage $v(t)$ across the capacitor is a solution of the second-order equation

$$LC\frac{d^2v}{dt^2} + RC\frac{dv}{dt} + v = V_S(t).$$

Using the values given above, we have

$$0.4 \times 10^{-6}\frac{d^2v}{dt^2} + 10^{-3}\frac{dv}{dt} + v = 2e^{-t/3}.$$

We can put this equation into a more familiar form by multiplying through by 2.5×10^6, obtaining

$$\frac{d^2v}{dt^2} + 2500\frac{dv}{dt} + 2.5 \times 10^6 v = 5 \times 10^6 e^{-t/3}.$$

These numbers are imposing but are realistic for the standard units of measure in electric circuits. To find the general solution, we follow the procedure above. First, we find the general solution of the homogeneous equation. Then we find a particular solution of the nonhomogeneous equation.

We begin by computing the general solution of

$$\frac{d^2v}{dt^2} + 2500\frac{dv}{dt} + 2.5 \times 10^6 v = 0.$$

The eigenvalues are $(-1.25 \pm i\sqrt{0.9375}) \times 10^3$. Using the approximation 0.968 for $\sqrt{0.9375}$, the general solution is

$$v(t) = k_1 e^{-1250t} \sin(968t) + k_2 e^{-1250t} \cos(968t),$$

where k_1 and k_2 are constants. (Note that the solution decays incredibly quickly to zero, but oscillates very rapidly.)

Now to find a particular solution of the nonhomogeneous equation

$$\frac{d^2v}{dt^2} + 2500\frac{dv}{dt} + 2.5 \times 10^6 v = 5 \times 10^6 e^{-t/3},$$

we guess $v_p(t) = ke^{-t/3}$, where k is the undetermined coefficient. To find the appropriate value of k, we substitute the guess back into the differential equation. The left-hand side is

$$\frac{d^2v_p}{dt} + 2500\frac{dv_p}{dt} + 2.5 \times 10^6 v_p = \frac{d^2(ke^{-t/3})}{dt} + 2500\frac{d(ke^{-t/3})}{dt} + 2.5 \times 10^6 ke^{-t/3}$$

$$= \frac{k}{9}e^{-t/3} - 2500\frac{k}{3}e^{-t/3} + 2.5 \times 10^6 ke^{-t/3}$$

$$= \left(\frac{1}{9} - \frac{2500}{3} + 2.5 \times 10^6\right) ke^{-t/3}.$$

Next, we equate this left-hand side with the right-hand side of the nonhomogeneous equation, which is $5 \times 10^6 e^{-t/3}$. Cancelling the $e^{-t/3}$ term and solving for k, we see that $k \approx 2$. So our particular solution is (approximately) $v_p(t) = 2e^{-t/3}$.

We can now form the general solution

$$v(t) = k_1 e^{-1250t} \sin(968t) + k_2 e^{-1250t} \cos(968t) + 2e^{-t/3},$$

where k_1 and k_2 are constants that we adjust according to the initial conditions. The first two terms of the general solution oscillate very quickly and decrease very quickly to zero. Hence for large t, every solution is close to the particular solution. The graph of the solution that satisfies the initial condition $(v(0), v'(0)) = (1, 0)$ is shown in Figure 4.7. To see the oscillations, we must look very close to $t = 0$ (see Figure 4.8). In the language of circuit theory we say that the natural response (the solution of the homogeneous equation) decays quickly and every solution approaches the forced response, which is $2e^{-t/3}$.

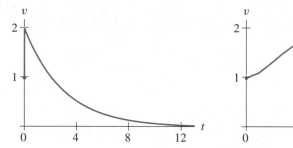

Figure 4.7
Solution of

$$\frac{d^2v}{dt^2} + 2500\frac{dv}{dt} + 2.5 \times 10^6 = 5 \times 10^6 e^{-t/3}$$

with initial condition $v(0) = 1$,
$v'(0) = 0$.

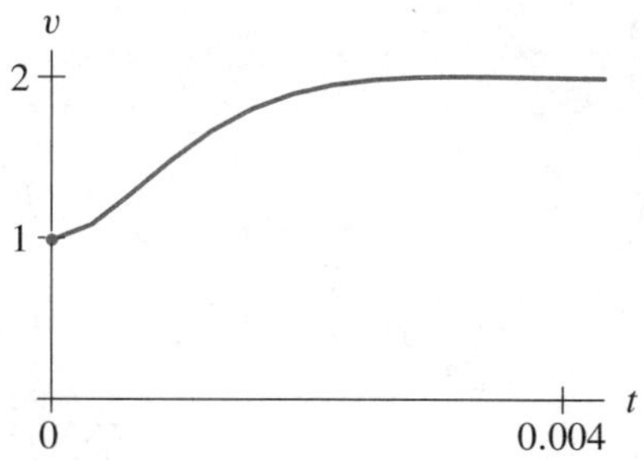

Figure 4.8
The same solution that is graphed in Figure 4.7. Here we show the graph only for values of t that are very close to 0 in order to illustrate the rapid increase from the initial value $v(0) = 1$.

EXERCISES FOR SECTION 4.1

In Exercises 1–6, find the general solution of the given differential equation.

1. $\dfrac{d^2y}{dt^2} + 6\dfrac{dy}{dt} + 8y = e^{-t}$

2. $\dfrac{d^2y}{dt^2} + 6\dfrac{dy}{dt} + 8y = 2e^{-3t}$

3. $\dfrac{d^2y}{dt^2} + 7\dfrac{dy}{dt} + 12y = 3e^{-2t}$

4. $\dfrac{d^2y}{dt^2} + 4\dfrac{dy}{dt} + 13y = e^{-t}$

5. $\dfrac{d^2y}{dt^2} + 4\dfrac{dy}{dt} + 13y = -3e^{-2t}$

6. $\dfrac{d^2y}{dt^2} + 7\dfrac{dy}{dt} + 10y = e^{-2t}$

In Exercises 7–10, find the solution of the given initial-value problem.

7. $\dfrac{d^2y}{dt^2} + 6\dfrac{dy}{dt} + 8y = e^{-t}$,

$y(0) = y'(0) = 0$

8. $\dfrac{d^2y}{dt^2} + 7\dfrac{dy}{dt} + 12y = 3e^{-t}$,

$y(0) = 2,\ y'(0) = 1$

9. $\dfrac{d^2y}{dt^2} + 4\dfrac{dy}{dt} + 13y = -3e^{-2t}$,

$y(0) = y'(0) = 0$

10. $\dfrac{d^2y}{dt^2} + 7\dfrac{dy}{dt} + 10y = e^{-2t}$,

$y(0) = y'(0) = 0$

In Exercises 11–16,

  (a) compute the general solution,

  (b) compute the solution with $y(0) = y'(0) = 0$, and

  (c) describe the long-term behavior of solutions in a brief paragraph.

11. $\dfrac{d^2y}{dt^2} + 4\dfrac{dy}{dt} + 3y = e^{-t/2}$

12. $\dfrac{d^2y}{dt^2} + 4\dfrac{dy}{dt} + 3y = e^{-2t}$

13. $\dfrac{d^2y}{dt^2} + 4\dfrac{dy}{dt} + 3y = e^{-4t}$

14. $\dfrac{d^2y}{dt^2} + 4\dfrac{dy}{dt} + 20y = e^{-t/2}$

15. $\dfrac{d^2y}{dt^2} + 4\dfrac{dy}{dt} + 20y = e^{-2t}$

16. $\dfrac{d^2y}{dt^2} + 4\dfrac{dy}{dt} + 20y = e^{-4t}$

17. Find the general solution of

$$\frac{d^2y}{dt^2} + 2\frac{dy}{dt} + y = e^{-t}.$$

  [*Hint*: Just keep guessing.]

18. One of the most common forcing functions is constant forcing. The equation for a harmonic oscillator with constant forcing is

$$\frac{d^2y}{dt^2} + p\frac{dy}{dt} + qy = c,$$

where p, q, and c are constants. Since we can find the general solution of the unforced equation by our usual methods, we can find the general solution of the forced equation if we can find one particular solution. Find one particular solution of the equation

$$\frac{d^2y}{dt^2} + p\frac{dy}{dt} + qy = c.$$

  [*Hint*: Guess a solution that looks like the forcing function.]

In Exercises 19–22, compute

(a) the general solution and

(b) the solution with $y(0) = y'(0) = 0$.

19. $\dfrac{d^2y}{dt^2} + 6\dfrac{dy}{dt} + 8y = 5$

20. $\dfrac{d^2y}{dt^2} + 5\dfrac{dy}{dt} + 6y = 2$

21. $\dfrac{d^2y}{dt^2} + 2\dfrac{dy}{dt} + 10y = 10$

22. $\dfrac{d^2y}{dt^2} + 4\dfrac{dy}{dt} + 6y = -8$

The discussion of the qualitative behavior of solutions of equations for damped forced harmonic oscillators relies on the fact that solutions of the unforced equation (natural response) tend to zero as t increases. If damping is not present, then the natural response does not tend to zero and the qualitative behavior of solutions is more complicated (see Section 4.3). However, the Extended Linearity Principle and the Method of Undetermined Coefficients apply to all nonhomogeneous, linear equations.

In Exercises 23–28,

(a) compute the general solution,

(b) compute the solution with $y(0) = y'(0) = 0$, and

(c) give a rough sketch and describe in a brief paragraph the long-term behavior of the solution in part (b).

23. $\dfrac{d^2y}{dt^2} + 9y = e^{-t}$

24. $\dfrac{d^2y}{dt^2} + 4y = 2e^{-2t}$

25. $\dfrac{d^2y}{dt^2} + 2y = -3$

26. $\dfrac{d^2y}{dt^2} + 4y = e^t$

27. $\dfrac{d^2y}{dt^2} + 9y = 6$

28. $\dfrac{d^2y}{dt^2} + 2y = -e^t$

29. In order to use the Method of Undetermined Coefficients to find a particular solution, we must be able to make a reasonable guess (up to multiplicative constants — the undetermined coefficients) of a particular solution. In this section we dealt with exponential forcing. If the forcing is some other type of function, then we must adjust our guess accordingly. For example, for an equation of the form

$$\frac{d^2y}{dt^2} + 4y = -3t^2 + 2t + 3,$$

the forcing function is the quadratic polynomial $g(t) = -3t^2 + 2t + 3$. It is reasonable to guess that a particular solution in this case is also a quadratic polynomial. Hence we guess $y_p(t) = at^2 + bt + c$. The constants a, b, and c are determined by substituting $y_p(t)$ into the equation.

(a) Find the general solution of the equation specified in this exercise.

(b) Find the particular solution with the initial condition $y(0) = 2$, $y'(0) = 0$.

In Exercises 30–33,

(a) compute the general solution,

(b) compute the solution with $y(0) = y'(0) = 0$, and

(c) give a rough sketch and describe in a brief paragraph the long-term behavior of the solution in part (b).

30. $\dfrac{d^2 y}{dt^2} + 2\dfrac{dy}{dt} = 3t + 2$

31. $\dfrac{d^2 y}{dt^2} + 4y = 3t + 2$

32. $\dfrac{d^2 y}{dt^2} + 3\dfrac{dy}{dt} + 2y = t^2$

33. $\dfrac{d^2 y}{dt^2} + 4y = t - \dfrac{t^2}{20}$

34. We can extend the Method of Undetermined Coefficients in order to solve equations whose forcing functions are sums of several types of functions. More precisely, suppose that $y_1(t)$ is a solution of the equation

$$\frac{d^2 y}{dt^2} + p\frac{dy}{dt} + qy = g(t)$$

and that $y_2(t)$ is a solution of the equation

$$\frac{d^2 y}{dt^2} + p\frac{dy}{dt} + qy = h(t).$$

Show that $y_1(t) + y_2(t)$ is a solution of the equation

$$\frac{d^2 y}{dt^2} + p\frac{dy}{dt} + qy = g(t) + h(t).$$

In Exercises 35–40,

(a) compute the general solution,

(b) compute the solution with $y(0) = y'(0) = 0$, and

(c) describe in a brief paragraph the long-term behavior of the solution in part (b).

35. $\dfrac{d^2 y}{dt^2} + 5\dfrac{dy}{dt} + 6y = e^{-t} + 4$

36. $\dfrac{d^2 y}{dt^2} + 3\dfrac{dy}{dt} + 2y = e^{-t} - 4$

37. $\dfrac{d^2 y}{dt^2} + 6\dfrac{dy}{dt} + 8y = 2t + e^{-t}$

38. $\dfrac{d^2 y}{dt^2} + 6\dfrac{dy}{dt} + 8y = 2t + e^{t}$

39. $\dfrac{d^2 y}{dt^2} + 4y = t + e^{-t}$

40. $\dfrac{d^2 y}{dt^2} + 4y = 6 + t^2 + e^{t}$

4.2 SINUSOIDAL FORCING

In this section we study the forced harmonic oscillator equation

$$\frac{d^2y}{dt^2} + p\frac{dy}{dt} + qy = g(t),$$

where $g(t)$ is a sine or cosine function. This type of external force occurs frequently in applications. Examples include the shaking of a building by an earthquake and the pressure waves of a sound striking a glass. These external forces share two qualitative features. First, they are periodic. That is, they repeat after a definite time T, called the period, so

$$g(t + T) = g(t)$$

for all t. Second, they have zero average. For each period, the push in one direction equals the pull in the other, which can be expressed analytically as

$$\int_0^T g(t)\,dt = 0.$$

The simplest and most familiar functions with this property are $\sin \omega t$ and $\cos \omega t$. The period of these functions is $\omega/(2\pi)$ and the frequency is $2\pi/\omega$.

In this section we apply the method of undetermined coefficients to study damped forced harmonic oscillators with sinusoidal forcing. In Section 4.3 we consider the undamped harmonic oscillator with sinusoidal forcing, and in Section 4.4 we study how the solutions of the damped harmonic oscillator with sinusoidal forcing depend on the forcing frequency and the damping parameter.

Sinusoidal Forcing of a Damped Harmonic Oscillator

To better understand the behavior of a harmonic oscillator with periodic forcing, we begin by finding the general solution of

$$\frac{d^2y}{dt^2} + 2\frac{dy}{dt} + 2y = \sin t$$

following the procedure given in Section 4.1.

The eigenvalues of the unforced equation

$$\frac{d^2y}{dt^2} + 2\frac{dy}{dt} + 2y = 0$$

are $-1 \pm i$. Hence this oscillator is underdamped, and the general solution is

$$y(t) = k_1 e^{-t} \cos t + k_2 e^{-t} \sin t.$$

To find one solution of the forced equation, we use the method of undetermined coefficients. At first glance it seems that $y_p(t) = k \sin t$ would be a reasonable guess for a solution of the forced equation. However, because of the dy/dt term, substituting this guess into the differential equation yields a term containing $\cos t$ on the left-hand side. No choice of constant k will eliminate this term, and there are no cosines on the right-hand side of the equation. We could make a more general guess of

$$y_p(t) = a \cos t + b \sin t$$

with two undetermined coefficients, but it is more efficient to take advantage of our knowledge of complex exponentials.

Complexification

First, consider the equation

$$\frac{d^2 y}{dt^2} + 2\frac{dy}{dt} + 2y = e^{it}.$$

A particular solution of this complex differential equation, which we call $y_c(t)$ (c for complex), has a real part and an imaginary part. That is,

$$y_c(t) = y_{re}(t) + i y_{im}(t).$$

Substituting $y_c(t)$ into the differential equation and using Euler's formula gives

$$\frac{d^2(y_{re} + i y_{im})}{dt^2} + 2\frac{d(y_{re} + i y_{im})}{dt} + 2(y_{re} + i y_{im}) = \cos t + i \sin t.$$

Taking real and imaginary parts on both sides of the equation, we have

$$\frac{d^2 y_{re}}{dt^2} + 2\frac{d y_{re}}{dt} + 2y_{re} = \cos t \quad \text{and} \quad \frac{d^2 y_{im}}{dt^2} + 2\frac{d y_{im}}{dt} + 2y_{im} = \sin t.$$

Hence the imaginary part, $y_{im}(t)$, is a particular solution of the original equation with forcing $\sin t$.

The advantage of using complex exponentials is that the appropriate guess for the particular solution is once again another exponential. For

$$\frac{d^2 y}{dt^2} + 2\frac{dy}{dt} + 2y = e^{it},$$

we guess $y_c(t) = ae^{it}$ and solve for the (complex) coefficient a that makes this guess a solution. Substituting into the left-hand side of the differential equation, we obtain

$$\frac{d^2 y_c}{dt^2} + 2\frac{d y_c}{dt} + 2y_c = -ae^{it} + 2aie^{it} + 2ae^{it}$$

$$= a(1 + 2i)e^{it}.$$

In order for $y_c(t)$ to be a solution, we must take

$$a = \frac{1}{1 + 2i} = \frac{1 - 2i}{5}$$

(see the appendices for a review of complex arithmetic). To determine the real and imaginary parts, we write

$$y_c(t) = \frac{1-2i}{5}(\cos t + i \sin t)$$

$$= \left(\frac{1}{5}\cos t + \frac{2}{5}\sin t\right) + i\left(-\frac{2}{5}\cos t + \frac{1}{5}\sin t\right).$$

Taking the imaginary part, we obtain a particular solution of the original equation with $\sin t$ forcing,

$$y_p(t) = -\frac{2}{5}\cos t + \frac{1}{5}\sin t.$$

Hence the general solution of

$$\frac{d^2y}{dt^2} + 2\frac{dy}{dt} + 2y = \sin t$$

is

$$y(t) = k_1 e^{-t}\sin t + k_2 e^{-t}\cos t + \left(-\frac{2}{5}\cos t + \frac{1}{5}\sin t\right).$$

Qualitative analysis

To determine the qualitative behavior of solutions of

$$\frac{d^2y}{dt^2} + 2\frac{dy}{dt} + 2y = \sin t,$$

we note that the unforced equation is an underdamped harmonic oscillator. As pointed out in Section 4.1, the portion of a solution arising from the solution of the unforced oscillator (the natural response) tends to zero, leaving only the particular solution of the forced oscillator (the forced response). From the general solution for this equation, we see that the natural response tends to zero like e^{-t} (see Figure 4.9.) Hence for large t, every solution is close to

$$y_p(t) = -\frac{2}{5}\cos t + \frac{1}{5}\sin t.$$

The behavior of $y_p(t)$ can be made much more evident by the use of trigonometric identities to combine the sine and cosine into one cosine function (see Exercise 22

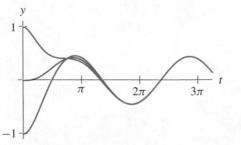

Figure 4.9
Several solutions of the equation

$$\frac{d^2y}{dt^2} + 2\frac{dy}{dt} + 2y = \sin t.$$

on page 280). However because we are using complex numbers to determine particular solutions, we use an alternative approach involving Euler's formula to combine these two terms.

Consider the solution $y_c(t) = ae^{it}$, where $a = (1 - 2i)/5$, to the complex differential equation. We write a in its polar form as

$$a = |a|e^{i\theta}$$

$$= \frac{1}{\sqrt{5}}e^{i\theta}$$

(see the appendix on complex numbers for a definition of polar form). In terms of the angle θ, we have

$$y_c(t) = ae^{it}$$

$$= \frac{1}{\sqrt{5}}e^{i\theta}e^{it}.$$

Since we add exponents when we multiply exponentials,

$$y_c(t) = \frac{1}{\sqrt{5}}e^{i(t+\theta)} = \frac{1}{\sqrt{5}}(\cos(t + \theta) + i\sin(t + \theta)).$$

Since our particular solution $y_p(t)$ is the imaginary part of $y_c(t)$, we see that

$$y_p(t) = \frac{1}{\sqrt{5}}\sin(t + \theta).$$

Since it is traditional to use cosine rather than sine in this context and to represent the angle θ in terms of degrees, we recall that

$$\sin \alpha = \cos(\alpha - 90°).$$

Thus

$$y_p(t) = \frac{1}{\sqrt{5}}\cos(t + \theta - 90°),$$

and we need only determine θ to complete the calculation. Since θ is the polar angle associated to the complex number a, we see that

$$\theta = \arctan\left(\frac{-2}{1}\right) \approx -63°.$$

Our particular solution can therefore be expressed as

$$y_p(t) = \frac{1}{\sqrt{5}}\cos(t - 153°).$$

The angle $153°$ is often denoted ϕ.

This form of the equation for $y_p(t)$ is particularly useful. We see without computation that $y_p(t)$ is a periodic function with amplitude $1/\sqrt{5}$ and period 2π. We also see that the graph of $y_p(t)$ can be obtained from the graph of cosine by shifting it to the right by $153°$ and "stretching" it by a factor of $1/\sqrt{5}$. The angle ϕ is called the **phase angle**.

It is useful to note that the complex number a, which is the undetermined coefficient in this computation, determines both the amplitude and the phase angle of the forced response. The amplitude is just the magnitude $|a|$ of a, and the phase angle can be calculated immediately from the polar angle of a. In Section 4.4, we study how the amplitude and phase shift depend on the parameters for a periodically forced harmonic oscillator.

Physical interpretation

Think of the harmonic oscillator as modeling a mass attached to a spring sliding on a table with damping and the external force arising from a periodic tilting of the table. The result above can be summarized by saying that the long-term behavior of the mass is to slide back and forth with the same frequency as the tilting of the table. Writing the particular solution of the forced equation in the form $A\cos(t + \phi)$, we can read off the amplitude A and the phase angle ϕ.

The behavior of the solutions in the above example is typical of the long-term behavior of solutions of equations for damped harmonic oscillators with sine or cosine forcing. All solutions approach a periodic solution (the forced response), which oscillates with the same period as the forcing. It is vital to remember that while the forcing function and the forced response look similar, they are not the same. The amplitude and phase of the forced response depend in a complicated way on the parameters of the harmonic oscillator and on the forcing frequency. We consider this dependence in detail in Section 4.4.

The Phase Plane of a Forced Harmonic Oscillator

So far, all of our pictures of solutions in this chapter have been of the solution as a function of time. We have used our knowledge of linear systems and linear, constant-coefficient, second-order equations to help find the general solution of forced harmonic oscillator equations, but we have not converted the forced equation into a first-order system.

The reason that we have not done this conversion is that a forced harmonic oscillator

$$\frac{d^2y}{dt^2} + p\frac{dy}{dt} + qy = g(t)$$

and its corresponding first-order system

$$\frac{dy}{dt} = v$$
$$\frac{dv}{dt} = -qy - pv + g(t)$$

are nonautonomous. The tangent vector of a solution curve in the phase plane depends on the position (y, v) *and* on the time t, so the vector field changes with time. Moreover, because the vector field changes with time, two solutions with the same (y, v)

values at different times can follow different paths. Thus solution curves can cross in the yv-plane without violating the Uniqueness Theorem.

Converting the second-order nonautonomous equation to a first-order system is still useful. There are no elementary numerical methods that apply to second-order equations, but Euler's method (as well as others, see Chapter 7) is essentially the same for autonomous and nonautonomous first-order systems. Also the solution curves in the yv-plane can help in understanding the behavior of solutions.

To demonstrate this and to gain more practice with the method of undetermined coefficients, we consider the equation

$$\frac{d^2y}{dt^2} + 2\frac{dy}{dt} + 10y = 4\cos 2t.$$

To find the general solution, we first note that the eigenvalues of the unforced equation

$$\frac{d^2y}{dt^2} + 2\frac{dy}{dt} + 10y = 0$$

are $-1 \pm 3i$. Hence the general solution of the unforced equation is

$$y(t) = k_1 e^{-t}\cos 3t + k_2 e^{-t}\sin 3t.$$

To find a particular solution of the forced equation, we complexify and obtain

$$\frac{d^2y}{dt^2} + 2\frac{dy}{dt} + 10y = 4e^{2it}.$$

The usual guess is $y_c(t) = ae^{2it}$. Substituting this guess into the left-hand side of the differential equation yields

$$\frac{d^2y_c}{dt^2} + 2\frac{dy_c}{dt} + 10y_c = -4ae^{2it} + 4aie^{2it} + 10ae^{2it}$$

$$= a(6 + 4i)e^{2it}$$

In order for ae^{2it} to be a solution, we must have

$$a(6 + 4i) = 4.$$

Therefore

$$a = \frac{4}{6 + 4i} = \frac{4(6 - 4i)}{52} = \frac{6 - 4i}{13},$$

so

$$y_c(t) = \frac{6 - 4i}{13}e^{2it}$$

$$= \frac{6 - 4i}{13}(\cos 2t + i\sin 2t)$$

$$= \left(\frac{6}{13}\cos 2t + \frac{4}{13}\sin 2t\right) + i\left(-\frac{4}{13}\cos 2t + \frac{6}{13}\sin 2t\right)$$

A particular solution $y_p(t)$ of the original equation with forcing $3\cos 2t$ is the real part of $y_c(t)$, which is

$$y_p(t) = \frac{6}{13}\cos 2t + \frac{4}{13}\sin 2t.$$

Hence the general solution of the forced equation

$$\frac{d^2y}{dt^2} + 2\frac{dy}{dt} + 10y = 4\cos 2t$$

is

$$y(t) = k_1 e^{-t}\cos 3t + k_2 e^{-t}\sin 3t + \frac{6}{13}\cos 2t + \frac{4}{13}\sin 2t,$$

which can be rewritten as

$$y(t) = k_1 e^{-t}\cos 3t + k_2 e^{-t}\sin 3t + \frac{2\sqrt{13}}{13}\cos(2t + \phi),$$

where the phase shift $\phi \approx -34°$.

Qualitative analysis: Graphs of solutions

We know that for any choice of initial conditions, the term

$$k_1 e^{-t}\cos 3t + k_2 e^{-t}\sin 3t$$

in the general solution quickly tends to zero. Hence every solution is approximately

$$\frac{2\sqrt{13}}{13}\cos(2t - 34°)$$

for large t. This function is periodic in t with period π and amplitude $2\sqrt{13}/13 \approx 0.55$.

If we rewrite the forced equation as a first-order system, we obtain

$$\frac{dy}{dt} = v$$

$$\frac{dv}{dt} = -10y - 2v + 4\cos 2t.$$

If $y(t)$ is a solution of the second-order equation, then $(y(t), v(t))$ is a solution of the system. We can graph solutions in both the ty- and yv-planes.

For example, to find the solution that satisfies $y(0) = y'(0) = 0$, we must solve

$$\begin{cases} k_1 + \dfrac{6}{13} = 0 \\[2mm] -k_1 + 3k_2 + \dfrac{8}{13} = 0. \end{cases}$$

Solving these equations, we find $k_1 = -6/13$ and $k_2 = -14/39$, and the solution with these initial conditions is

$$y(t) = -\frac{6}{13}e^{-t}\cos 3t - \frac{14}{39}e^{-t}\sin 3t + \frac{6}{13}\cos 2t + \frac{4}{13}\sin 2t.$$

We can graph this solution in the ty-plane and the corresponding solution of the system in the yv-plane (see Figure 4.10). Both these pictures show the solution oscillating around $y = 0$. After a short initial time period, the solution is essentially periodic with period π.

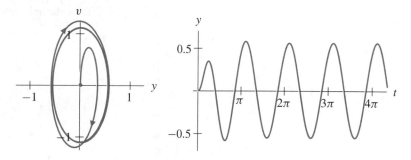

Figure 4.10
Graph of the solution of $d^2y/dt^2 + 2dy/dt + 10y = 4\cos 2t$ with initial conditions $y(0) = y'(0) = 0$ in the yv- and ty-planes.

As a second example, consider the solution with $y(0) = 2$ and $y'(0) = 0$. As above, we can solve for k_1 and k_2, obtaining $k_1 = 4/13$ and $k_2 = 20/13$. The graphs of this solution in the ty- and yv-plane are shown in Figure 4.11. Again, after a short time period, both representations of the solution settle into periodic behavior, oscillating around $y = 0$. In this case we can clearly see the solution in the yv-plane crossing over itself. As noted above, this is not a violation of the Uniqueness Theorem because the system is nonautonomous.

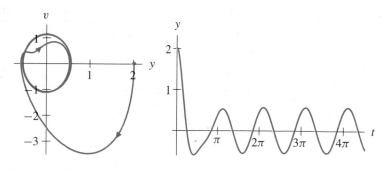

Figure 4.11
Graph of the solution of $d^2y/dt^2 + 2dy/dt + 10y = 4\cos 2t$ with initial conditions $y(0) = 2$ and $y'(0) = 0$ in the yv- and ty-planes.

Interpretation of the phase planes

Looking at the pictures of solutions in the yv-plane, we can make the following qualitative observations. The restoring force and the damping are proportional to y and $v = dy/dt$, respectively. For y and v large, the external force is much smaller than the restoring and damping forces of the unforced oscillator. Hence, far from the origin, solutions in the yv-plane look very similar to solutions of the corresponding unforced oscillator; that is, solutions tend toward the origin (see Figure 4.12). However, when y and v are close to zero, the external force is as large or larger than the restoring and damping forces. In this part of the yv-plane, the external force overcomes the damping and pushes the solution away from the origin (see Figure 4.13). Solutions tend toward the region of the yv-plane where the external forces and the damping "balance."

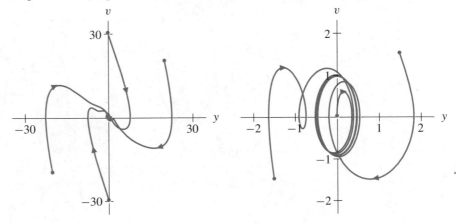

Figure 4.12
Several solutions of
$d^2y/dt^2 + 3dy/dt + 2y = 3\cos 2t$ with
initial conditions far from the origin.

Figure 4.13
Several solutions of
$d^2y/dt^2 + 3dy/dt + 2y = 3\cos 2t$ with
initial conditions close to the origin.

EXERCISES FOR SECTION 4.2

In Exercises 1–10, find the general solution of the given equation.

1. $\dfrac{d^2y}{dt^2} + 3\dfrac{dy}{dt} + 2y = \cos t$

2. $\dfrac{d^2y}{dt^2} + 3\dfrac{dy}{dt} + 2y = 5\cos t$

3. $\dfrac{d^2y}{dt^2} + 3\dfrac{dy}{dt} + 2y = \sin t$

4. $\dfrac{d^2y}{dt^2} + 3\dfrac{dy}{dt} + 2y = 2\sin t$

5. $\dfrac{d^2y}{dt^2} + 6\dfrac{dy}{dt} + 8y = \cos t$

6. $\dfrac{d^2y}{dt^2} + 6\dfrac{dy}{dt} + 8y = -4\cos 3t$

7. $\dfrac{d^2y}{dt^2} + 4\dfrac{dy}{dt} + 13y = 3\cos 2t$

8. $\dfrac{d^2y}{dt^2} + 4\dfrac{dy}{dt} + 20y = -\cos 5t$

9. $\dfrac{d^2y}{dt^2} + 4\dfrac{dy}{dt} + 20y = -3\sin 2t$

10. $\dfrac{d^2y}{dt^2} + 2\dfrac{dy}{dt} + y = \cos 3t$

In Exercises 11–14, find the solution of the given initial-value problem.

11. $\dfrac{d^2y}{dt^2} + 6\dfrac{dy}{dt} + 8y = \cos t$
$y(0) = y'(0) = 0$

12. $\dfrac{d^2y}{dt^2} + 6\dfrac{dy}{dt} + 8y = 2\cos 3t$
$y(0) = y'(0) = 0$

13. $\dfrac{d^2y}{dt^2} + 4\dfrac{dy}{dt} + 20y = -3\sin 2t$

$y(0) = y'(0) = 0$

14. $\dfrac{d^2y}{dt^2} + 2\dfrac{dy}{dt} + y = 2\cos 2t$

$y(0) = y'(0) = 0$

15. To find a particular solution of a forced equation with sine or cosine forcing, we solved the corresponding complexified equation. This method is particularly efficient, but there are other approaches. Find a particular solution via the Method of Undetermined Coefficients for the equation

$$\frac{d^2y}{dt^2} + 3\frac{dy}{dt} + y = \cos 3t$$

(a) by using the guess $y_p(t) = a\cos 3t + b\sin 3t$ (with a and b as undetermined coefficients) and

(b) by using the guess $y_p(t) = A\cos(3t + \phi)$ (with A and ϕ as undetermined coefficients).

Pictured below are the $y(t)$-graphs for several solutions of

$$\frac{d^2y}{dt^2} + p\frac{dy}{dt} + qy = \cos \omega t$$

for a choice of the parameters p, q, and ω.

In Exercises 16–19,

(a) match the figure to the choice of parameters, and

(b) write a short paragraph justifying your choice.

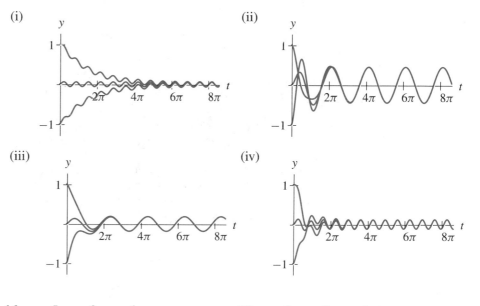

(i)   (ii)   (iii)   (iv)

16. $p = 5$, $q = 3$, $\omega = 1$

18. $p = 5$, $q = 1$, $\omega = 3$

17. $p = 1$, $q = 3$, $\omega = 1$

19. $p = 1$, $q = 1$, $\omega = 3$

20. Show that if $y_p(t)$ is a solution of

$$\frac{d^2y}{dt^2} + p\frac{dy}{dt} + qy = g(t),$$

then $ky_p(t)$ is a solution of

$$\frac{d^2y}{dt^2} + p\frac{dy}{dt} + qy = kg(t)$$

for any constant k.

21. Find the general solution of

$$\frac{d^2y}{dt^2} + 6\frac{dy}{dt} + 8y = 5\cos t.$$

[*Hint*: See Exercises 5 and 20.]

22. **(a)** Find the general solution of

$$\frac{d^2y}{dt^2} + 4\frac{dy}{dt} + 20y = 3 + 2\cos 2t.$$

[*Hint*: See Exercise 34 in Section 4.1.]

(b) Discuss the long-term behavior of solutions of this equation.

23. **(a)** Find the general solution of

$$\frac{d^2y}{dt^2} + 4\frac{dy}{dt} + 20y = e^{-t}\cos t.$$

[*Hint*: Consider the forcing function to be the real part of a complex exponential.]

(b) Discuss the long-term behavior of solutions of this equation.

24. **(a)** Find the general solution of

$$\frac{d^2y}{dt^2} + 4\frac{dy}{dt} + 20y = e^{-2t}\sin 4t.$$

[*Hint*: Keep guessing.]

(b) Discuss the long-term behavior of solutions of this equation.

In Exercises 25 and 26, we explore the different types of notation used to express solutions of harmonic oscillator equations with trigonometric forcing functions. Sadly, the differences in terminology sometimes make it quite difficult for mathematicians and engineers to communicate, even though we are talking about the same things.

25. Given a function

$$g(t) = a\cos\omega t + b\sin\omega t,$$

where a and b are constants, show that $g(t)$ is the real part of the complex function

$$ke^{i\phi}e^{i\omega t}.$$

[*Hint*: Use Euler's formula.]

Remark: The complex expression $ke^{i\phi}$ is called a **phasor**. If we know that $g(t)$ has the form $k\cos(\omega t + \phi)$, then we need know only the constants k and ϕ — the amplitude and the phase — to know the function g. Hence we can use the phasor $ke^{i\phi}$ as a notation for the function $g(t) = ke^{i\phi}e^{i\omega t}$.

26. Show that if $k_1 e^{i\phi_1}$ is the phasor representing $g_1(t)$ and $k_2 e^{i\phi_2}$ is the phasor representing $g_2(t)$ (both with frequency $\omega/(2\pi)$), then the phasor representing the sum $g_1(t) + g_2(t)$ is $k_1 e^{i\phi_1} + k_2 e^{i\phi_2}$. [*Hint*: Write the functions $g_1(t)$ and $g_2(t)$ as the real parts of the product of the phasors and $e^{i\omega t}$, add them, and simplify.]

Remark: Phasor terminology is widely used in electric circuit theory, in which an algebra of phasors is developed suited to dealing with currents and voltages in circuits with periodic voltage sources.

4.3 UNDAMPED FORCING AND RESONANCE

In Section 4.2 we considered the effect of periodic forcing on damped harmonic oscillators. In this section we turn our attention to periodically forced undamped oscillators. Despite the fact that physical systems always have some damping, there is a good reason to pay attention to the undamped case. If the physical system we are modeling has sufficiently small damping, then the undamped equation is a good approximation, at least for a certain time interval. As a consequence, the behavior of solutions of the undamped equation gives us considerable qualitative insight into the behavior of solutions if the damping is small.

The Extended Linearity Principle from Section 4.1 applies for any forced harmonic oscillator equation, damped or undamped. The general solution is the sum of the general solution of the unforced equation (natural response) and a particular solution of the forced equation (forced response). For the forced damped harmonic oscillator, we know that the damping term causes the natural response to tend to zero. Thus every solution approaches the forced response as time increases, regardless of the initial conditions. This is not the case for undamped equations. The detailed behavior of solutions depends on the initial conditions for all time. It is, however, still possible to give a qualitative description of solutions that applies for all initial conditions.

In this section we restrict attention to the case of sinusoidal forcing of undamped equations. Of particular interest is the way that the qualitative description of solutions depends on the frequency of the forcing term. A very dramatic change in the qualitative behavior of the solutions occurs as the frequency of the forcing function approaches the natural frequency of the equation. This phenomenon is called *resonance*.

An Undamped Harmonic Oscillator with Sinusoidal Forcing

We begin with a careful study of a particular undamped harmonic oscillator equation with sinusoidal forcing given by

$$\frac{d^2y}{dt^2} + 2y = \cos \omega t.$$

This equation contains a parameter ω that controls the frequency of the external forcing. The frequency of the forcing is $\omega/(2\pi)$. We first use the method of Section 4.1 to find the general solution of this equation.

The eigenvalues of the unforced equation

$$\frac{d^2y}{dt^2} + 2y = 0$$

are $\pm i\sqrt{2}$, so the general solution is

$$y_h(t) = k_1 \cos\sqrt{2}\,t + k_2 \sin\sqrt{2}\,t.$$

Since there is no damping, solutions oscillate for all time with the natural frequency $\sqrt{2}/(2\pi)$. The amplitude of the oscillation is determined by the values of k_1 and k_2, which in turn are determined by the initial conditions.

To find one solution of the forced equation

$$\frac{d^2y}{dt^2} + 2y = \cos\omega t$$

we use the Method of Undetermined Coefficients. We could use a combination of trigonometric functions as our guess of the particular solution, but as we will see, there are advantages to complexifying the equation as we did in Section 4.2. We consider

$$\frac{d^2y}{dt^2} + 2y = e^{i\omega t}.$$

The real part of a solution of this equation is a solution of the equation with forcing $\cos\omega t$. As usual, we guess $y_c(t) = ae^{i\omega t}$ and solve for a. Substituting $y_c(t)$ into the left-hand side of the equation, we find

$$\frac{d^2y_c}{dt^2} + 2y_c = -a\omega^2 e^{i\omega t} + 2ae^{i\omega t}$$

$$= a(2 - \omega^2)e^{i\omega t}.$$

In order for $y_c(t)$ to be a solution, we must have $a = 1/(2 - \omega^2)$, so

$$y_c(t) = \frac{1}{2 - \omega^2}\,e^{i\omega t} = \frac{1}{2 - \omega^2}\,(\cos\omega t + i\sin\omega t).$$

Hence a solution $y_p(t)$ of the equation

$$\frac{d^2y}{dt^2} + 2y = \cos\omega t$$

is the real part of $y_c(t)$, so

$$y_p(t) = \frac{1}{2 - \omega^2}\,\cos\omega t.$$

The general solution of the forced equation is therefore

$$y(t) = k_1 \cos\sqrt{2}\,t + k_2 \sin\sqrt{2}\,t + \frac{1}{2 - \omega^2}\,\cos\omega t.$$

Note that the solution depends on ω, which we think of as a parameter. We must omit the case $\omega = \sqrt{2}$ since this value makes the denominator $2 - \omega^2$ vanish. We deal with this case later in this section.

Qualitative Analysis

With the general solution in hand, we can consider the qualitative behavior of solutions of

$$\frac{d^2y}{dt^2} + 2y = \cos \omega t.$$

First note that, as predicted, the unforced response does not tend to zero. Also, for all t, the solution is a sum of sine and cosine terms of frequency $\sqrt{2}/(2\pi)$ and $\omega/(2\pi)$. For some values of t, both the forced and unforced responses are positive or both are negative, so they reinforce each other. For other values of t, the forced and unforced responses have opposite sign and cancel each other. We see these effects in the graphs of typical solutions (see Figures 4.14 and 4.15).

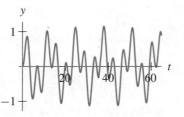

Figure 4.14
Solution of $d^2y/dt^2 + 2y = \cos \omega t$
for $\omega = 0.5$, $y(0) = y'(0) = 0$.

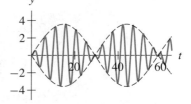

Figure 4.15
Solution of $d^2y/dt^2 + 2y = \cos \omega t$
for $\omega = 1.2$, $y(0) = y'(0) = 0$.

Beating

There is additional structure to the solutions that is more evident if $\omega = 1.2$ than if $\omega = 0.5$ (compare Figures 4.14 and 4.15). If $\omega = 1.2$, the amplitude of the oscillations increase and decrease in a very regular pattern. This phenomenon is called **beating**. Beating occurs when the natural response and forced response have approximately the same frequency. This is the case if $\omega = 1.2$, since the natural frequency, $\sqrt{2}/(2\pi) \approx 1.4/(2\pi)$, and the forcing frequency, $1.2/(2\pi)$, are close.

If these two frequencies are close, the two responses tend to reinforce or cancel each other over long time intervals. We can see this by looking at the natural and forced responses separately for a particular choice of initial conditions (see Figure 4.16). For

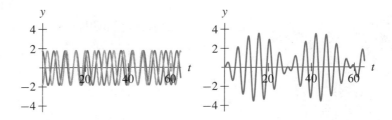

Figure 4.16
In the plot on the left, we graph the natural and forced responses for the initial-value problem $d^2y/dt^2 + 2y = \cos \omega t$ with $\omega = 1.2$, $y(0) = y'(0) = 0$. On the right, we graph the sum of these two functions. This sum is the desired solution.

certain time intervals these graphs oscillate in unison, and so their sum has large amplitude. Over other intervals the functions have opposite signs, and their sum is smaller. Incidentally, you can hear the phenomenon of beating when listening to a piano or a guitar that is slightly out of tune.

We can determine the period and frequency of the beats by using complex exponentials. We have already determined that the general solution of the differential equation is

$$y(t) = k_1 \cos \sqrt{2}\,t + k_2 \sin \sqrt{2}\,t + a \cos \omega t,$$

where $a = 1/(2 - \omega^2)$. Now we consider the particular solution $y(t)$ that satisfies the initial conditions $y(0) = 0$ and $y'(0) = 0$. These initial values imply that $k_1 = -a$ and $k_2 = 0$. Hence the solution to the initial-value problem is

$$y(t) = a \left(\cos \omega t - \cos \sqrt{2}\,t \right).$$

To simplify the notation involved in our computation with exponentials, we let

$$\alpha = \frac{\omega + \sqrt{2}}{2} \quad \text{and} \quad \beta = \frac{\omega - \sqrt{2}}{2}.$$

That is, α is the average of ω and $\sqrt{2}$, and β is one-half of the difference between ω and $\sqrt{2}$. Using Euler's formula, we can think of $y(t)$ as the real part of the complex function

$$y_c(t) = a \left(e^{i\omega t} - e^{i\sqrt{2}\,t} \right).$$

Note that $\omega t = (\alpha + \beta)t$ and $\sqrt{2}\,t = (\alpha - \beta)t$. Therefore,

$$y_c(t) = a \left(e^{i(\alpha+\beta)t} - e^{i(\alpha-\beta)t} \right),$$

and using the usual rules for manipulating exponents, we have

$$y_c(t) = a e^{i\alpha t} (e^{i\beta t} - e^{-i\beta t}).$$

Since $e^{i\beta t} = \cos \beta t + i \sin \beta t$ and $e^{-i\beta t} = \cos \beta t - i \sin \beta t$, we get

$$e^{i\beta t} - e^{-i\beta t} = 2i \sin \beta t,$$

and therefore

$$y_c(t) = a e^{i\alpha t} (2i \sin \beta t).$$

Recalling that the desired solution $y(t)$ is the real part of $y_c(t)$, we calculate

$$y_c(t) = a(\cos \alpha t + i \sin \alpha t)(2i \sin \beta t)$$
$$= a(-2 \sin \alpha t \, \sin \beta t - 2i \cos \alpha t \, \sin \beta t),$$

and obtain

$$y(t) = -2a \, [\sin \alpha t] \, [\sin \beta t]$$

$$= -2a \left[\sin \left(\frac{\omega + \sqrt{2}}{2} \right) t \right] \left[\sin \left(\frac{\omega - \sqrt{2}}{2} \right) t \right].$$

If $\omega \approx \sqrt{2}$, then the frequency of $\sin((\omega - \sqrt{2})t/2)$ is quite small. This is the frequency of the beats. Since the frequency is small and the period is the reciprocal of the frequency, the period of the function $\sin((\omega - \sqrt{2})t/2)$ is large. For example, if $\omega = 1.2$, the frequency of the beats is

$$\frac{\sqrt{2} - 1.2}{4\pi} \approx 0.017,$$

and the period is $4\pi/(\sqrt{2} - 1.2) \approx 58$. This slow oscillation is represented by the dashed outline in Figure 4.15.

On the other hand, the frequency of the term $\sin((\omega + \sqrt{2})t/2)$ is $(\sqrt{2} + \omega)/(4\pi)$, which is the average of the natural and forcing frequencies. If $\omega = 1.2$, this frequency is $(\sqrt{2} + 1.2)/(4\pi) \approx 0.21$, which is relatively large. The period is $4\pi/(\sqrt{2} + 1.2) \approx 4.76$. This period and frequency of the fast oscillations is visible in Figure 4.15.

Dependence of amplitude on forcing frequency

There is one more important feature of the general solution of

$$\frac{d^2y}{dt^2} + 2y = \cos \omega t$$

that we can obtain from the formula

$$y(t) = k_1 \cos \sqrt{2}\,t + k_2 \sin \sqrt{2}\,t + \frac{1}{2 - \omega^2} \cos \omega t.$$

The precise behavior of a solution depends on the values of k_1 and k_2, and these values are determined by the initial conditions. However, every solution contains the term

$$\frac{1}{2 - \omega^2} \cos \omega t.$$

The amplitude of the oscillations of this function is given by $1/|2 - \omega^2|$ (see Figure 4.17).

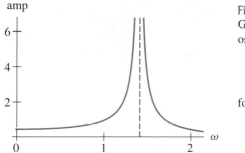

Figure 4.17
Graph of $1/|2 - \omega^2|$, the amplitude of the oscillations of solutions of

$$\frac{d^2y}{dt^2} + 2y = \cos \omega t$$

for large t.

From Figure 4.17 we can make some predictions about the behavior of unforced solutions. If ω is very large, then the coefficient $a = 1/(2 - \omega^2)$ is small. In this case solutions of the forced equation

$$\frac{d^2y}{dt^2} + 2y = \cos \omega t$$

are very close to solutions of the unforced equation. A large value of ω means the external force has very short period and large frequency. The above discussion predicts that such an external force with large frequency has only a small effect on solutions. The harmonic oscillator does not have enough time to respond to a push in one direction before the sign of the external force changes. A typical example is shown in Figure 4.18.

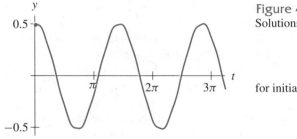

Figure 4.18
Solutions of

$$\frac{d^2y}{dt^2} + 2y = \cos 10t$$

for initial conditions $y(0) = 0.5$, $y'(0) = 0$.

On the other hand, if ω is close to $\sqrt{2}$, $1/|2-\omega^2|$ is very large since the denominator is small. In this case the external force has a dramatic effect on solutions. Since the forcing frequency and the natural frequency are nearly equal, the external force pushes and pulls the harmonic oscillator with almost the same frequency as the unforced oscillator. This is called *near resonant* forcing. In this case the period of a beat is very long, and the natural response and forced response reinforce each other for long stretches of time. Even solutions with relatively small initial conditions eventually oscillate with very large amplitude (see Figure 4.19).

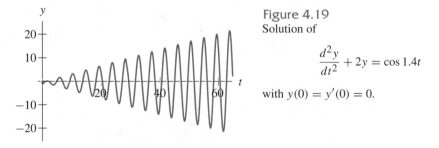

Figure 4.19
Solution of

$$\frac{d^2y}{dt^2} + 2y = \cos 1.4t$$

with $y(0) = y'(0) = 0$.

Resonance

In our study of solutions of

$$\frac{d^2y}{dt^2} + 2y = \cos \omega t,$$

we have avoided the case $\omega = \sqrt{2}$. That is, we have avoided the case where the frequency of the external force is exactly the same as the natural frequency. Forcing whose frequency is the same as the natural frequency of the oscillator is called **resonant forcing**, and the oscillator is said to be in **resonance**. Avoiding this case was necessary for

algebraic reasons, since we encountered $2 - \omega^2$ in the denominator in the general solution. From the earlier discussion for the case $\omega \approx \sqrt{2}$, we suspect that the solutions
have qualitatively interesting behavior in the case of resonant forcing.

Qualitative behavior
Setting $\omega = \sqrt{2}$, the differential equation becomes

$$\frac{d^2y}{dt^2} + 2y = \cos\sqrt{2}\,t.$$

We can predict the behavior of solutions of this equation in two ways. First, we can
study solutions of the forced equation

$$\frac{d^2y}{dt^2} + 2y = \cos\omega t$$

for ω very close to t. As pointed out above, the frequency of the beats decreases and
the amplitude of the forced response increases as ω approaches $\sqrt{2}$. A sequence of
solutions with ω approaching $\sqrt{2}$ is shown in Figure 4.20.

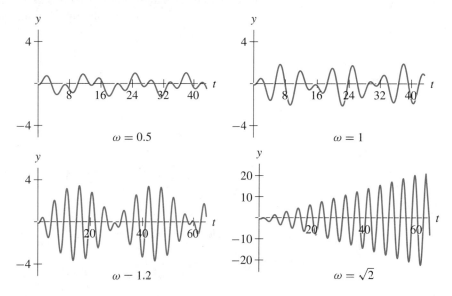

Figure 4.20
Solutions of $d^2y/dt^2 + 2y = \cos\omega t$ with $y(0) = y'(0) = 0$ for $\omega = 0.5$, $\omega = 1$,
$\omega = 1.2$, and $\omega = \sqrt{2}$.

Also in Figure 4.20, we see a numerical solution of

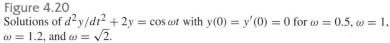

$$\frac{d^2y}{dt^2} + 2y = \cos\sqrt{2}\,t.$$

From these pictures we can make the reasonable prediction that the forced response of
the undamped harmonic oscillator with resonant forcing oscillates with an amplitude
that increases for all time.

The general solution in the resonant case

Next we compute the general solution for

$$\frac{d^2y}{dt^2} + 2y = \cos\sqrt{2}\,t.$$

The general solution of the unforced equation is still

$$k_1 \cos\sqrt{2}\,t + k_2 \sin\sqrt{2}\,t.$$

To find a particular solution of the forced equation, we proceed as before and consider

$$\frac{d^2y}{dt^2} + 2y = e^{i\sqrt{2}\,t}.$$

The natural first guess of $y_c(t) = ae^{i\sqrt{2}\,t}$ does not yield a solution for any value of a because it is a solution of the unforced equation. In Figure 4.20, we see that the solutions of the forced equation still oscillate, but the amplitude of the oscillation grows linearly with t. This motivates our second guess $y_c(t) = ate^{i\sqrt{2}\,t}$; that is, we multiply our first guess by t. (This guess also agrees with the rule-of-thumb for second-guessing in Section 4.1.) Using

$$\frac{dy_c}{dt} = a\left(e^{i\sqrt{2}\,t} + i\sqrt{2}\,te^{i\sqrt{2}\,t}\right)$$

and

$$\frac{d^2y_c}{dt^2} = 2a\left(i\sqrt{2}\,e^{i\sqrt{2}\,t} - te^{i\sqrt{2}\,t}\right),$$

we substitute $y_c(t)$ into the left-hand side of the differential equation and obtain

$$\frac{d^2y_c}{dt^2} + 2y_c = 2a\left(i\sqrt{2}\,e^{i\sqrt{2}\,t} - te^{i\sqrt{2}\,t}\right) + 2ate^{i\sqrt{2}\,t}$$

$$= 2ai\sqrt{2}\,e^{i\sqrt{2}\,t}.$$

Since this left-hand side must equal $e^{i\sqrt{2}\,t}$ if $y_c(t)$ is a solution, we have

$$a = \frac{1}{2i\sqrt{2}},$$

which can also be written as $a = -i/(2\sqrt{2})$ because $1/i = -i$. Thus

$$y_c(t) = -\frac{i}{2\sqrt{2}}\,t\,e^{i\sqrt{2}\,t}$$

$$= -\frac{i}{2\sqrt{2}}\,t\left(\cos\sqrt{2}\,t + i\sin\sqrt{2}\,t\right).$$

Since the particular solution $y_p(t)$ we desire is the real part of $y_c(t)$, we have

$$y_p(t) = \frac{1}{2\sqrt{2}}\,t\sin\sqrt{2}\,t,$$

and the general solution of

$$\frac{d^2y}{dt^2} + 2y = \cos\sqrt{2}\,t$$

is

$$k_1 \cos\sqrt{2}\,t + k_2 \sin\sqrt{2}\,t + \frac{1}{2\sqrt{2}}\,t\,\sin\sqrt{2}\,t.$$

From the general solution we can see that our prediction of the qualitative behavior of solutions is correct. Every solution oscillates with the natural frequency (which in this case is the same as the forcing frequency). For small times t, the solution resembles a solution of the unforced oscillator because the particular solution is close to zero for small t. For large t, the term

$$\frac{1}{2\sqrt{2}}\,t\,\sin\sqrt{2}\,t$$

dominates the other terms, so the amplitude of the oscillation grows linearly (see Figure 4.21).

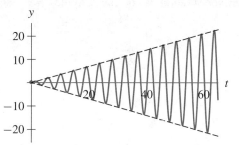

Figure 4.21
Solution of

$$\frac{d^2y}{dt^2} + 2y = \cos\sqrt{2}\,t$$

with $y(0) = y'(0) = 0$. The dashed lines are $y = \pm t/(2\sqrt{2})$.

A Heuristic Explanation of Resonance

There is nothing special about the particular equation we examined in this section. For any equation of the form

$$\frac{d^2y}{dt^2} + qy = \cos\omega t,$$

if the natural frequency, $\sqrt{q}/(2\pi)$, of the oscillator is equal to the frequency, $\omega/(2\pi)$, of the forcing, then every solution oscillates with unbounded amplitude. There is also nothing special about using cosine as the external force. For a typical equation of the form

$$\frac{d^2y}{dt^2} + qy = g(t),$$

where $g(t)$ is a periodic function with frequency $\sqrt{q}/(2\pi)$, every solution oscillates with unbounded amplitude.

Resonance is surprising for two reasons. First, the solutions become unbounded even though the external force is bounded. Second, the external force pushes exactly the same amount as it pulls. On average the external force is zero. Both of these properties might lead one to assume that the effect of periodic external forcing would not be large.

The explanation of resonance lies in the extra condition that the forcing frequency must be exactly equal to the natural frequency. If we think of the harmonic oscillator as a mass and spring on a table and the external force as the effect of tilting the table back and forth, then resonant forcing means the table is being tilted with the same period as the unforced motion of the mass. Suppose we start the mass at its rest position with zero velocity. The tilting of the table induces a motion of the mass. Since the periods of motion of the table and the mass are the same, the table is always tilted so as to increase the speed of the mass. With each period the mass moves farther from the rest position. Since this is a linear system, the period of the oscillations does not depend on the amplitude, and the effect described above keeps adding energy to the system until the spring breaks or the mass falls off the table. Using the same discussion, we can describe how it is possible for an opera singer to break a crystal glass by singing a particular note.*

Resonance can be either useful or dangerous. For example, resonance allows us to design devices that have a large response only in the presence of forcing of a particular frequency. On the other hand, if the unforced system corresponds to the swaying motion of a bridge and the external force comes from the shaking of an earthquake, a large amplitude solution can be catastrophic.

EXERCISES FOR SECTION 4.3

In Exercises 1–6, compute the general solution of the given equation.

1. $\dfrac{d^2y}{dt^2} + 9y = \cos t$

2. $\dfrac{d^2y}{dt^2} + 9y = 5\sin 2t$

3. $\dfrac{d^2y}{dt^2} + 4y = -\cos\dfrac{t}{2}$

4. $\dfrac{d^2y}{dt^2} + 4y = 3\cos 2t$

5. $\dfrac{d^2y}{dt^2} + 3y = \cos 3t$

6. $\dfrac{d^2y}{dt^2} + 5y = 5\sin 5t$

In Exercises 7–10, compute the solution of the given initial-value problem.

7. $\dfrac{d^2y}{dt^2} + 9y = \cos t$
   $y(0) = y'(0) = 0$

8. $\dfrac{d^2y}{dt^2} + 4y = 3\cos 2t$
   $y(0) = y'(0) = 0$

9. $\dfrac{d^2y}{dt^2} + 5y = 3\cos 2t$
   $y(0) = y'(0) = 0$

10. $\dfrac{d^2y}{dt^2} + 9y = \sin 3t$
    $y(0) = 1,\ y'(0) = -1$

---

*Experiments indicate that it requires a tone of approximately 140 dB held for 2 to 3 seconds to break a wine goblet of reasonably good quality. Producing such a tone is at the very limit of our ability to sing. See Haym Kruglak and René Pittet, "The Caruso Legend Lives On," *The Physics Teacher*, Vol. 17, p. 49, January 1979 and the references cited therein.

In Exercises 11–14, for the equation specified,

   (a) determine the frequency of the beats,

   (b) determine the frequency of the rapid oscillations, and

   (c) use the information from parts (a) and (b) to give a rough sketch of a typical solution. [*Hint*: You should be able to do this with no further calculation.]

11. $\dfrac{d^2y}{dt^2} + 4y = \cos\dfrac{9t}{4}$ **12.** $\dfrac{d^2y}{dt^2} + 10y = \sin 3t$

13. $\dfrac{d^2y}{dt^2} + 5y = 3\cos 2t$ **14.** $\dfrac{d^2y}{dt^2} + 6y = \cos 2t$

In Exercises 15–18, a $y(t)$-graph is given. This graph is the graph of the solution $y(t)$ to one of the following differential equations with initial conditions $y(0) = 0$ and $y'(0) = 0$. Determine which differential equation has $y(t)$ as a solution and provide a brief justification for your answer.

   (i) $\dfrac{d^2y}{dt^2} + 15y = \cos 4t$ (ii) $\dfrac{d^2y}{dt^2} + 16y = 5\cos 3t$

   (iii) $\dfrac{d^2y}{dt^2} + 16y = \dfrac{1}{2}\cos 4t$ (iv) $\dfrac{d^2y}{dt^2} + 16y = 10$

15.   **16.**

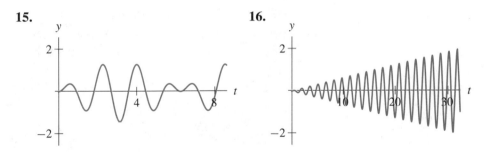

17.   **18.**

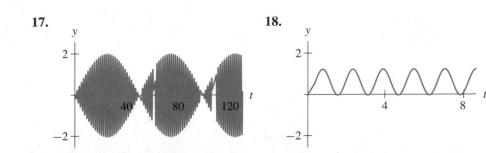

19. Consider the equation

$$\frac{d^2y}{dt^2} + 15y = \cos 4t + 2\sin t.$$

(a) Compute the general solution. [*Hint*: See Exercise 34 in Section 4.1.]

(b) Compute the solution with $y(0) = y'(0) = 0$.

(c) Sketch the graph of the solution of part (b).

(d) In a short essay, describe the behavior of the solution of part (b). How could you have predicted this behavior in advance?

20. Commercials for a maker of recording tape show a crystal glass being broken by the sound of a recording of an opera singer's voice. Is this a good test of the quality of the tape? Why or why not?

21. Legend has it that the marching band of a major university is forbidden from playing the song "Hey, Jude." The reason is that whenever the band played this song, the fans would stomp their feet in time to the music and after the second chorus, the stadium structure would begin to rock quite noticeably. The administration insisted that there was no danger that the stadium would collapse, but banned the song as a precaution. No other song has this effect on the stadium (although, presumably, the fans will stomp to other songs). Give a possible explanation of the stadium structure's violent reaction to the stomping accompanying the song "Hey, Jude" and to no other song.

22. Suppose the suspension system of the average car can be fairly well modeled by an underdamped harmonic oscillator with a natural period of 2 seconds. How far apart should speed bumps be placed so that a car traveling at 10 miles per hour over several bumps will bounce more and more violently with each bump?

23. Consider a unit mass sliding on a frictionless table attached to a spring, with spring constant $k = 16$. Suppose the mass is lightly tapped by a hammer every T seconds.

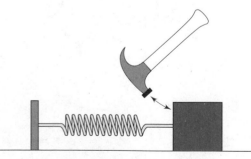

Mass tapped periodically with a hammer.

Suppose that the first tap occurs at time $t = 0$ and before that time the mass is at rest. Describe what happens to the motion of the mass for the following choices of the tapping period T:

(a) $T = 1$ (b) $T = 3/2$ (c) $T = 2$ (d) $T = 5/2$ (e) $T = 3$

4.4 AMPLITUDE AND PHASE OF THE STEADY STATE

In this section we return to the equation

$$\frac{d^2y}{dt^2} + p\frac{dy}{dt} + qy = \cos \omega t$$

for the periodically forced damped harmonic oscillator. Our goal is to establish a quantitative relationship between the behavior of the solutions and the parameters — particularly ω, which determines the forcing frequency, and p, which determines the damping. As we saw in Section 4.2, the long-term behavior of every solution is the same. It is determined by a particular solution $y_p(t)$ of the forced equation, which can be written in the form

$$y_p(t) = A \cos(\omega t + \phi).$$

Hence for large t, all solutions oscillate with the same frequency as the forcing. This is the *steady-state* behavior of solutions. The amplitude of the oscillation is A and the *phase angle* is ϕ. In this section we find expressions for A and ϕ as functions of ω, p, and q.

For small damping (p near zero), we expect the solutions to behave approximately the same as in the undamped case, at least for a bounded amount of time. The amplitude A of the forced response is large if the forcing frequency is close to the natural frequency of the unforced equation.

Behavior of the Steady State

To find a particular solution of the forced equation

$$\frac{d^2y}{dt^2} + p\frac{dy}{dt} + qy = \cos \omega t,$$

we follow the techniques introduced in Section 4.2 and use the Method of Undetermined Coefficients applied to the complexified equation

$$\frac{d^2y}{dt^2} + p\frac{dy}{dt} + qy = e^{i\omega t}.$$

Next, we guess $y_c(t) = ae^{i\omega t}$ and determine the value of a that makes $y_c(t)$ a solution. Substituting $y_c(t)$ into the left-hand side of the differential equation, we have

$$\frac{d^2y_c}{dt^2} + p\frac{dy_c}{dt} + qy_c = -a\omega^2 e^{i\omega t} + p(ia\omega e^{i\omega t}) + q(ae^{i\omega t})$$

$$= a\left((q - \omega^2) + ip\omega\right)e^{i\omega t}.$$

In order for $y_c(t)$ to be a solution, we must take

$$a = \frac{1}{b}$$

where b is the complex number $(q - \omega^2) + ip\omega$.

As we saw in Section 4.2, we can determine the amplitude A and phase angle ϕ of $y_p(t)$ from the polar form

$$a = |a|e^{i\phi} = Ae^{i\phi}$$

of the complex number a. (See the appendix on complex numbers for a review of the polar representation of a complex number.) In this case, since $a = 1/b$, we have

$$|a| = \frac{1}{|b|}$$

and the polar angle ϕ of a is the negative of the polar angle of b. Therefore the complex number

$$a = \frac{1}{b} = \frac{1}{(q - \omega^2) + i(p\omega)}$$

encapsulates the amplitude and phase angle of the forced response.

This is neat. Given any damped harmonic oscillator with forcing $\cos \omega t$, we can obtain the long-term behavior of the solution for any ω directly from the values of the parameters p, q, and ω.

Amplitude and Phase

Given $y_p(t) = A\cos(\omega t + \phi)$, it is convenient to have a formulas for A and ϕ. Since $A = |a| = 1/|b|$ and

$$|b| = \sqrt{(q - \omega^2)^2 + p^2\omega^2},$$

we have

$$A = \frac{1}{\sqrt{(q - \omega^2)^2 + p^2\omega^2}}.$$

Moreover we are assuming that both p and ω are positive, and thus the complex number b has a positive imaginary part. We can therefore consider its polar angle to be between 0 and π. Furthermore, since the polar angle of $a = 1/b$ is the negative of the polar angle of b, we consider ϕ to be the angle between $-\pi$ and 0 such that

$$\tan \phi = \frac{-p\omega}{q - \omega^2}.$$

Amplitude
To recap, for large t, we know that all solutions of

$$\frac{d^2y}{dt^2} + p\frac{dy}{dt} + qy = \cos \omega t$$

oscillate with the same amplitude A and with the same frequency, $\omega/(2\pi)$, as the forcing function. The formula for the amplitude of these oscillations is

$$A = \frac{1}{\sqrt{(q - \omega^2)^2 + p^2\omega^2}}.$$

In Figure 4.22 we show the graph of A as a function of ω for $q = 2$ and several values of p. The figures for other values of q are similar. If p is small, the damping is small.

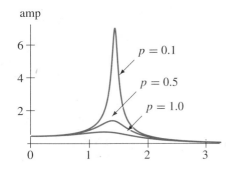

amp

Figure 4.22
Graphs of the amplitude

$$A = \frac{1}{\sqrt{(q - \omega^2)^2 + p^2\omega^2}}$$

with $q = 2$ and $p = 0.1$, $p = 0.5$, and $p = 1.0$.

It is not surprising to see that the amplitude A is large for $\omega \approx \sqrt{2}$, since if $p = 0$, this value of ω corresponds to resonant forcing. As p increases, two things happen: the maximum value of A decreases, and the value of ω where A attains its maximum moves to the left. (It is a good review of max-min problems to compute the maximum value of A and where it is attained. See Exercise 3.)

We can visualize the amplitude A in several ways. For example, if we fix q, then we can graph A as a function of the two variables p and ω (see Figure 4.23). Again, the most noticeable feature of this graph is the explosion of A near resonance for p near zero and ω near $\sqrt{q}$.

Another view of the amplitude is obtained by fixing the forcing frequency ω and graphing A as a function of p and q. Such a graph is shown in Figure 4.24. This graph shows that, by choosing p and q appropriately, we can obtain a large amplitude response. Ideally, we would like to see the graph of A as a function of all three parameters ω, p, and q, but this graph would live in four dimensions.

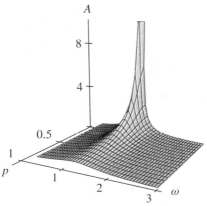

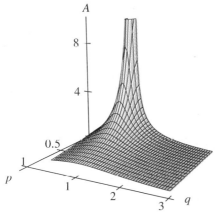

Figure 4.23
Graph of amplitude A as a function of p and ω for $q = 2$.

Figure 4.24
Graph of amplitude A as a function of p and q for $\omega = 1$.

Phase angle

A particular solution of the equation

$$\frac{d^2y}{dt^2} + p\frac{dy}{dt} + qy = \cos \omega t$$

can be written in the form

$$y_p(t) = A\cos(\omega t + \phi),$$

and we can think of the long-term behavior of solutions of this equation as translations and scalings of the external forcing $\cos \omega t$. The scaling is accomplished by multiplication by the amplitude A. The translation corresponds to the addition of the phase angle ϕ. The graph of

$$A\cos(\omega t + \phi) = A\cos\left(\omega\left(t + \frac{\phi}{\omega}\right)\right)$$

is obtained by sliding the graph of $A\cos\omega t$ by ϕ/ω to the right if $\phi/\omega < 0$ and to the left if $\phi/\omega > 0$.

By convention, we choose the phase angle ϕ between $-\pi$ and 0 so that it satisfies the condition

$$\tan \phi = \frac{-p\omega}{q - \omega^2}.$$

Ideally we would like to see the graph of ϕ as a function of all three parameters ω, p, and q, but this graph also requires four dimensions. In Figure 4.25 we fix $q = 2$ and sketch the graph of ϕ as a function of ω for several values of p. Similar graphs for other values of q are qualitatively the same. Note that when p is small, the phase angle ϕ stays close to zero for ω small. The graph of ϕ has a large slope when it passes through $-\pi/2 = 90°$ for ω near $\sqrt{q}$. For ω large, ϕ is asymptotic to $-\pi$. In other words, for an external force with small frequency, solutions are almost in phase with the forcing. For large forcing frequency, solutions are almost half a period out of phase with the forcing.

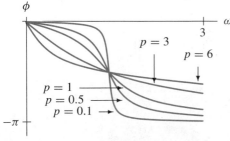

Figure 4.25
Phase angle ϕ for $q = 2$ with $p = 0.1$, $p = 0.5$, $p = 1.0$, $p = 3.0$, and $p = 6.0$. Note that when p is small, the phase angle ϕ stays close to zero for ω small. The graph of ϕ has a large slope when it passes through $-\pi/2 = 90°$ for ω near $\sqrt{q}$. For ω large, ϕ is asymptotic to $-\pi$.

More Parameters

There are two more parameters that we may wish to consider — the amplitude and phase of the forcing function. That is, we consider

$$\frac{d^2y}{dt^2} + p\frac{dy}{dt} + qy = F\cos(\omega t + \theta),$$

where F and θ are constants. Luckily these two new parameters add little complication to the discussion.

First, suppose $y_p(t)$ is a solution of the forced equation

$$\frac{d^2y}{dt^2} + p\frac{dy}{dt} + qy = g(t)$$

for some function $g(t)$. If F is a constant, then it follows from the linearity of the differential equation that the product $Fy_p(t)$ is a solution of

$$\frac{d^2y}{dt^2} + p\frac{dy}{dt} + qy = Fg(t)$$

(see Exercise 20 in Section 4.2). Multiplying the external force by a constant multiplies the forced response by the same constant.

Second, suppose again that $y_p(t)$ is a solution of the forced equation

$$\frac{d^2y}{dt^2} + p\frac{dy}{dt} + qy = g(t)$$

for some function $g(t)$. If θ is a constant, then $y_p(t+\theta)$ is a solution of the differential equation

$$\frac{d^2y}{dt^2} + p\frac{dy}{dt} + qy = g(t+\theta).$$

This assertion can be verified by defining a new independent variable $\tau = t + \theta$ and checking that $y_p(\tau)$ is a solution of

$$\frac{d^2y}{d\tau^2} + p\frac{dy_p}{d\tau} + qy_p = g(\tau)$$

(that is, we just shift the t-axis — see Exercise 5).

A Sobering Reminder

We have amassed a tremendous amount of information concerning the solutions of the equation

$$\frac{d^2y}{dt^2} + p\frac{dy}{dt} + qy = \cos\omega t,$$

especially if $p \geq 0$ and $q > 0$. It is important to remember that our calculations rely on the special form of this equation. Only the fact that the equation is linear allows us to use the Extended Linearity Principle to decompose the general solution into the general solution of the unforced equation plus a particular solution of the forced equation. Only the fact that the forcing function is a cosine function allows us to find a particular solution and manipulate it as we have done in the last three sections.

Nevertheless, the specific nature of the equations considered in this chapter does not diminish the importance of the calculations that we have done in the last four sections. We have learned a tremendous amount about particular examples. This is important for two reasons. First, as we have said many times, this particular differential equation is an excellent model for many different physical systems. Second, understanding this equation gives us a starting place for studying other types of equations — not all periodic forcing functions are sines or cosines and not all differential equations are linear.

EXERCISES FOR SECTION 4.4

1. The glass harmonica is a musical instrument invented by Ben Franklin. It consists of a collection of crystal glasses, one glass for each note. The musician rubs her slightly damp finger around the edge of the glass and the resulting vibrations of the glass make a very pure tone. The instrument can be tuned by pouring water into the glasses. The harder the musician pushes against the glass, the louder the note (but the frequency stays the same). Explain how the glass harmonica works.

2. Suppose an opera singer can break a glass by singing a particular note.

   (a) Will the singer have to sing a higher or a lower note to break an identical glass that is half full of water?

   (b) Suppose both notes are within the singer's range. Will it be harder or easier to break the glass when it is half full of water?

3. For large t, every solution of

$$\frac{d^2y}{dt^2} + p\frac{dy}{dt} + qy = \cos \omega t$$

oscillates with frequency $\omega/(2\pi)$ and amplitude A given by

$$A(\omega, p, q) = \frac{1}{\sqrt{(q - \omega^2)^2 + p^2\omega^2}}.$$

That is, the amplitude A is a function of the parameters ω, p, and q.

   (a) Compute $\partial A/\partial \omega$.

   (b) For fixed p and q, let $M(p, q)$ denote the maximum value of $A(\omega, p, q)$ as a function of ω. Compute an expression for $M(p, q)$. [*Hint*: This is a max-min problem from calculus.]

4. For the function $M(p, q)$ of Exercise 3,

   (a) for $q = 1$, plot $M(p, q)$ as a function of p,

   (b) explain why $M(p, q) = 1/\sqrt{p^2q - p^4/4}$ if the damping is small enough; and

   (c) explain why $M(p, q)$ is proportional to $1/p$ as $p \to 0$.

5. Given that $y_p(t)$ is a solution of

$$\frac{d^2y}{dt^2} + p\frac{dy}{dt} + qy = g(t),$$

show that $y_p(t + \theta)$ is a solution of

$$\frac{d^2y}{dt^2} + p\frac{dy}{dt} + qy = g(t + \theta).$$

6. In this section we computed a particular solution of the equation

$$\frac{d^2y}{dt^2} + p\frac{dy}{dt} + qy = \cos\omega t$$

of the form

$$y_p(t) = A\cos(\omega t + \phi)$$

where the phase angle ϕ satisfies the equation

$$\tan\phi = \frac{-p\omega}{q - \omega^2}$$

and $-180° < \phi < 0$. The angle ϕ is a function $\phi(\omega, p, q)$ of the parameters ω, p, and q.

   (a) Compute $\partial\phi/\partial\omega$.

   (b) Compute $\partial^2\phi/\partial\omega^2$.

   (c) For $q = 2$ and several values of p near zero, find the value of ω where ϕ changes most rapidly.

[*Hint*: This exercise is an excellent opportunity to test the power of symbolic differentiation packages.]

4.5 THE TACOMA NARROWS BRIDGE

On July 1, 1940, the $6 million Tacoma Narrows Bridge opened for traffic. On November 7, 1940, during a windstorm, the bridge broke apart and collapsed. During its short stand the bridge, a suspension bridge more than a mile long, was known as "Galloping Gertie" because the roadbed oscillated dramatically in the wind. The collapse of the bridge proved to be a scandal in more ways than one, including the fact that because the insurance premiums had been embezzled, the bridge was uninsured.*

    The roadbed of a suspension bridge hangs from vertical cables that are attached to cables strung between towers (see Figure 4.26 for a schematic picture). If we think of the vertical cables as long springs, then it is tempting to model the oscillations of

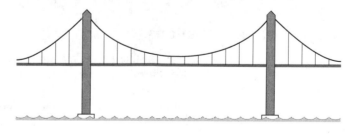

Figure 4.26
Schematic of a suspension bridge.

---

*The story of the bridge and its collapse can be found in Martin Braun, *Differential Equations and Their Applications*, Springer-Verlag, 1993, p. 173, and Matthys Levy and Mario Salvadori, *Why Buildings Fall Down: How Structures Fail*, W. W. Norton Co., 1992, p. 109.

the roadbed with a harmonic oscillator equation. We can think of the wind as somehow providing periodic forcing. It is very tempting to say, "Aha, the collapse must be due to resonance."

It turns out that things are not quite so simple. We know that to cause dramatic effects, the forcing frequency of a forced harmonic oscillator must be very close to its natural frequency. The wind seldom behaves in such a nice way for very long, and it would be very bad luck indeed if the oscillations caused by the wind happened to have a frequency almost exactly the same as the natural frequency of the bridge.

Recent research on the dynamics of suspension bridges (by two mathematicians, A. C. Lazer and P. J. McKenna[*]) indicates that the linear harmonic oscillator does not make an accurate model of the movement of a suspension bridge. The vertical cables do act like springs when they are stretched. That is, when the roadbed is below its rest position, the cables pull up. However, when the roadbed is significantly above its rest position, the cables are slack, so they do not push down. Hence the roadbed feels less force trying to pull it back into the rest position when it is pushed up than when it is pulled down (see Figure 4.27).

In this section we study a model for a system with these properties. The system of equations we study was developed by Lazer and McKenna from more complicated models of oscillations of suspension bridges. This system gives considerable insight into the possible behaviors of suspension bridges and even hints at how they can be made safer.

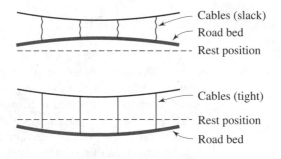

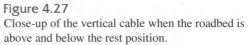

Figure 4.27
Close-up of the vertical cable when the roadbed is above and below the rest position.

Derivation of the Equations

The model we consider for the motion of the bridge uses only one variable to describe the position of the bridge. We assume that the bridge oscillates up and down as in Figure 4.27. We let $y(t)$ (measured in feet or meters) denote the vertical position of the center of the bridge, with $y = 0$ corresponding to the position where the cables are taut but not stretched. We let $y < 0$ correspond to the position in which the cables are stretched and $y > 0$ correspond to the position in which the cables are slack (see Figure 4.28). Of course, using one variable to study the motion of the bridge ignores many possible motions, and we comment on other models at the end of this section.

[*]See "Large-amplitude Periodic Oscillations in Suspension Bridges: Some New Connections with Non-linear Analysis" by A. C. Lazer and P. J. McKenna, in *SIAM Review*, Vol. 32, No. 4, 1990, pp. 537–578.

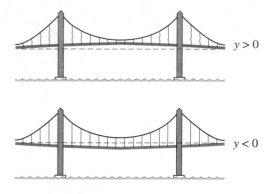

Figure 4.28
Positions of the bridge corresponding to $y > 0$ and
$y < 0$.

To develop a model for $y(t)$, we consider the forces that act on the center of the
bridge. Gravity provides a constant force in the negative direction of y. We also assume
that the cables provide a force that pulls the bridge up when $y < 0$ and that is proportional to y. On the other hand, when $y > 0$, the cable provides no force. When $y \neq 0$,
there is also a restoring force that pulls y back toward $y = 0$ due to the stretching of the
roadbed. Finally, there will also be some damping, which is assumed to be proportional
to dy/dt. We choose units so that the mass of the bridge is 1.

Based on these assumptions, the equation developed by Lazer and McKenna to
model a suspension bridge on a calm day (no wind) is

$$\frac{d^2y}{dt^2} + \alpha\frac{dy}{dt} + \beta y + c(y) = -g.$$

The first term is the vertical acceleration. The second term, $\alpha(dy/dt)$, arises from the
damping. Since suspension bridges are relatively flexible structures, we assume that
α is small. The term βy accounts for the force provided by the material of the bridge
pulling the bridge back toward $y = 0$. The function $c(y)$ accounts for the pull of the
cable when $y < 0$ (and the lack thereof when $y \geq 0$), and therefore it is given by

$$c(y) = \begin{cases} \gamma y, & \text{if } y < 0; \\ 0, & \text{if } y \geq 0. \end{cases}$$

The constant g represents the force due to gravity.

We can convert this to a system in the usual way, obtaining

$$\frac{dy}{dt} = v$$
$$\frac{dv}{dt} = -\beta y - c(y) - \alpha v - g.$$

This is an autonomous system. Simplifying the right-hand side of this system by combining the $-\beta y$ term with the terms in $c(y)$, we obtain

$$\frac{dy}{dt} = v$$

$$\frac{dv}{dt} = -h(y) - \alpha v - g,$$

where $h(y)$ is the piecewise-defined function

$$h(y) = \begin{cases} ay, & \text{if } y < 0; \\ \\ by, & \text{if } y \geq 0, \end{cases}$$

and $a = \beta + \gamma$ and $b = \beta$.

To study this example numerically, we choose particular values of the parameters. (These values are not motivated by any particular bridge.) Following Lazer and McKenna, we take $a = 17$, $b = 13$, $\alpha = 0.01$, and $g = 10$. So the system we study is

$$\frac{dy}{dt} = v$$

$$\frac{dv}{dt} = -h(y) - 0.01v - 10,$$

where

$$h(y) = \begin{cases} 17y, & \text{if } y < 0; \\ \\ 13y, & \text{if } y \geq 0. \end{cases}$$

We can easily compute that this system has only one equilibrium point, which is given by $(y, v) = (-10/17, 0)$. The y-coordinate of this equilibrium point is negative because gravity forces the bridge to sag a little, stretching the cables. Numerical results indicate that solutions spiral toward the equilibrium point very slowly. This is what we expect because there is a small amount of damping present. The direction field and solution curves are shown in Figures 4.29 and 4.30. The discontinuity of the direction field along the line $y = 0$ is difficult to see because the vector field is nearly vertical on both sides of the line.

The behavior of the solutions indicate that the bridge oscillates around the equilibrium position. The amplitude of the oscillation dies out slowly due to the small amount of damping. Because the forces controlling the motion change abruptly along $y = 0$, solutions also change direction when they cross from the left half-plane to the right half-plane.

The effect of wind

To add the effect of wind into the model, we add an extra term to the right-hand side of the equation. The effect of the wind is very difficult to quantify. Not only are there gusts of more or less random duration and strength, but also the way in which the wind interacts with the bridge can be very complicated. Even if we assume that the wind has constant speed and direction, the effect on the bridge need not be constant. As air moves past the bridge, swirls or vortices (like those at the end of an oar in water) form

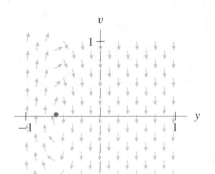

Figure 4.29
Direction field for the system

$$\frac{dy}{dt} = v$$

$$\frac{dv}{dt} = -h(y) - 0.01v - 10.$$

There is a spiral sink at $(y, v) = (-10/17, 0)$. In Figure 4.30, we see that solution curves for the system spiral slowly toward the equilibrium point.

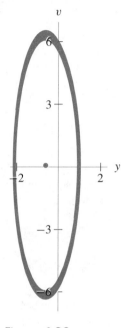

Figure 4.30
Solution curves spiral toward equilibrium very slowly.

above and below the roadbed. When they become large enough, these vortices "break off," causing the bridge to rebound. Hence even a constant wind can give a periodic push to the bridge.

Despite these complications, we assume, for simplicity, that the wind provides a forcing term of the form $\lambda \sin \mu t$. Since it is unlikely that turbulent winds will give a forcing term with constant amplitude λ or constant frequency $\mu/(2\pi)$, we will look for behavior of solutions that persist for a range of λ- and μ-values.

The system with forcing is given by

$$\frac{dy}{dt} = v$$

$$\frac{dv}{dt} = -h(y) - 0.01v - 10 + \lambda \sin \mu t.$$

This is a fairly simple model for the complicated behavior for a bridge moving in the wind, but we see below that even this simple model has solutions that behave in a surprising way.

Behavior of Solutions

We have seen that a linear system with damping and sinusoidal forcing has one periodic solution to which every other orbit tends as time increases, the steady-state solution. In other words, no matter what the initial conditions, the long-term behavior of the system will be the same. The amplitude and frequency of this periodic solution are determined by the amplitude and frequency of the forcing term (see Section 4.2).

The behavior of the system

$$\frac{dy}{dt} = v$$

$$\frac{dv}{dt} = -h(y) - 0.01v - 10 + \lambda \sin \mu t$$

is quite different. Below we describe results of a numerical study of these solutions of this system carried out by Glover, Lazer, and McKenna.*

*See "Existence and Stability of Large-scale Nonlinear Oscillations in Suspension Bridges" by J. Glover, A. C. Lazer, and P. J. McKenna, *ZAMP*, Vol. 40, 1989, pp. 171–200.

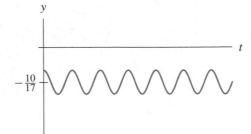

Figure 4.31
Solution of the system with small forcing.

If, for example, we choose $\mu = 4$ and λ very small ($\lambda < 0.05$), every solution tends toward a periodic solution with small magnitude near $y = -10/17$ (see Figure 4.31). For this periodic solution, $y(t)$ is negative for all t. Since this solution never crosses $y = 0$, it behaves just like the solution of the forced linear system

$$\frac{dy}{dt} = v$$

$$\frac{dv}{dt} = -17y - 0.01v - 10 + \lambda \sin 4t.$$

In terms of the behavior of the bridge, this means that in light winds, we expect the bridge to oscillate with small amplitude. Gravity keeps the bridge sagging downward and the cables are always stretched somewhat (see Figure 4.32). In this range, modeling the cables as linear springs is reasonable.

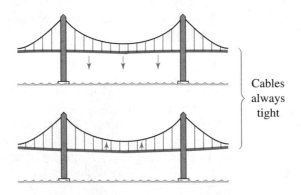

Cables always tight

Figure 4.32
Schematic of the bridge oscillating in light winds.

As λ increases, a new phenomenon is observed. Initial conditions near $(y, v) = (-10/17, 0)$ still yield solutions that oscillate with small amplitude (see Figure 4.33). However, if $y(0) = -10/17$ but $v(0)$ is large, solutions can behave differently. There is another periodic solution that oscillates around $y = -10/17$, but with (relatively) large amplitude (see Figure 4.34).

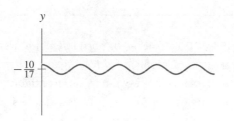

Figure 4.33
Solution of the forced system with larger forcing than in Figure 4.31 and initial conditions near the equilibrium.

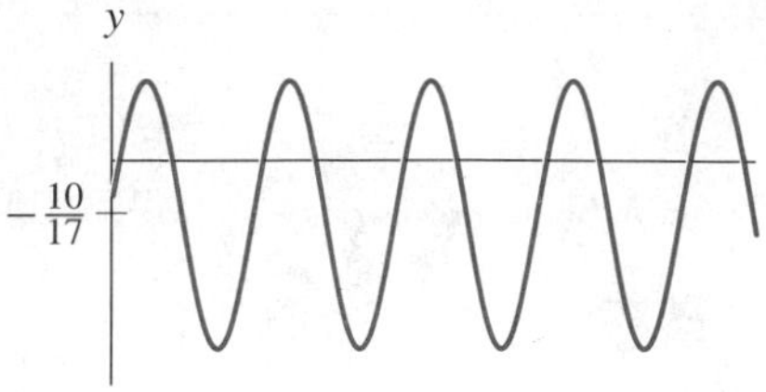

Figure 4.34
Solution of the forced system with the same large forcing as in Figure 4.33 but with initial conditions farther from the equilibrium.

This has dramatic implications for the behavior of the bridge. If the initial displacement is small, then we expect to see small oscillations as before. However, if a gust of wind gives the bridge a kick large enough to cause it to rise above $y = 0$, then the cables will go slack and the linear model will no longer be accurate (see Figure 4.35). In this situation the bridge can start oscillating with much larger amplitude, and these oscillations *do not die out*. So, in a moderate wind ($\lambda > 0.06$), a single strong gust could suddenly cause the bridge to begin oscillating with much larger amplitude, perhaps with devastating consequences.

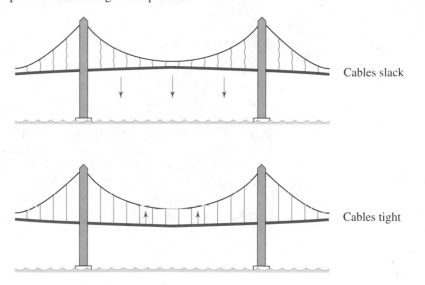

Figure 4.35
Schematic of large amplitude oscillations of the bridge.

Varying the parameters

As mentioned above, because the effects of the wind are not particularly regular, we should investigate the behavior of solutions as λ and μ are varied. It turns out that the large amplitude periodic solution persists for a fairly large range of λ and μ. This means that even in winds with uneven velocity and direction, the sudden jump in behavior to a persistent oscillation with large amplitude is possible.

Joseph McKenna (1948 –   ), born in Dublin, Ireland, received his Ph.D. in mathematics at the University of Michigan. For most of his professional life he has been involved in research in differential equations, applying them to diverse problems in soil physics, fluid dynamics, optics, biology, and most recently on flexing in bridges and ships.

His work with A. C. Lazer described in this section has received considerable attention in both the mathematics and engineering communities. Their research directly contradicts the long-standing view that resonance phenomena caused the collapse of the Tacoma Narrows Bridge. They have also suggested several alternative types of differential equations that govern the motion of such suspension bridges, including both the equation described in this chapter and a more complicated model involving nonlinear equations somewhat like those we discuss in the next chapter.

McKenna is currently Professor of Mathematics at the University of Connecticut.

Does This Explain the Tacoma Narrows Bridge Disaster?

As with any simple model of a complicated system, a note of caution is in order. To construct this model, we have made a number of simplifying assumptions. These include, but are not limited to, assuming that the bridge oscillates in one piece. The bridge can oscillate in two or more sections (see Figure 4.36). To include this in our model, we would have to include a new independent variable for the position along the bridge. The resulting model is a partial differential equation.

Another factor we have ignored is that the roadbed of the bridge has width as well as length. The final collapse of the Tacoma Narrows Bridge was preceded by violent twisting motions of the roadbed, alternately stretching and loosening the cables on either side of the road (see the cover of this book). Analysis of a model including

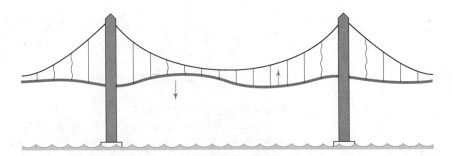

Figure 4.36
More complicated forms of oscillation of a suspension bridge.

the width gives considerable insight into the final moments before the bridge's collapse (see the paper by Lazer and McKenna cited on page 392).

This being said, the simple model discussed above still helps a great deal in understanding the behavior of the bridge. If this simple system can feature the surprising appearance of large amplitude periodic solutions, then it is not at all unreasonable to expect that more complicated and more accurate models will also exhibit this behavior. So this model does what it is supposed to do: It tells us what to look for when studying the behavior of a flexible suspension bridge.

EXERCISES FOR SECTION 4.5

For Exercises 1–3, recall that our simple model of a suspension bridge is

$$\frac{d^2y}{dt^2} + \alpha\frac{dy}{dt} + \beta y + c(y) = -g,$$

where α is the coefficient of damping, β is a parameter corresponding to the stiffness of the roadbed, the function $c(y)$ accounts for the pull of the cables, and g is the gravitational constant. For each of the following modifications of bridge design,

(a) discuss which parameters are changed, and

(b) discuss how you expect a change in the parameter values to affect the solutions. (For example, does the modification make the system look more or less like a linear system?)

1. The "stiffness" of the roadbed is increased, for example, by reinforcing the concrete or adding extra material that makes it harder for the roadbed to bend.

2. The coefficient of damping is increased.

3. The strength of the cables is increased.

4. The figure below is a schematic for an alternate suspension bridge design called the Lazer-McKenna light flexible long span suspension bridge. Why does this design avoid the problems of the standard suspension bridge design? Discuss this in a paragraph and give model equations similar to those in the text for this design.

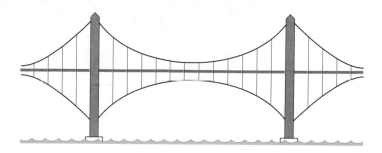

A schematic of the Lazer-McKenna light flexible long span suspension bridge.

In Exercises 5–8, we consider another application of the ideas in this section. Lazer and McKenna observe that the equations they study may also be used to model the up-and-down motion of an object floating in water, which can rise completely out of the water. This has serious implications for the behavior of a ship in heavy seas.

Suppose we have a cube made of a light substance floating in water. Gravity always pulls the cube downward. The cube floats at an equilibrium level at which the mass of the water displaced equals the mass of the cube. If the cube is higher or lower than the equilibrium level, then there is a restoring force proportional to the size of the displacement. We assume that the bottom and top of the cube stay parallel to the surface of the water at all times and that the system has a small amount of damping.

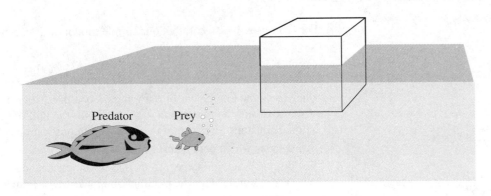

Cube floating in water.

5. Write a differential equation model for the up-and-down motion of the cube, assuming that it always stays in contact with the water and is never completely submerged.

6. Write a differential equation for the up-and-down motion of the cube, assuming that it always stays in contact with the water, but that it can be completely submerged.

7. Write a differential equation model for the up-and-down motion of the cube, assuming that it will never be completely submerged but can rise completely out of the water by some distance.

8. (a) Adjust each of the models in Exercises 5–7 to include the effect of waves on the motion of the cube (assuming the top and bottom remain parallel to the average water level).

   (b) Discuss the implications of the behavior of solutions of this system considered in the text for the motion of the cube.

LAB 4.1   A Periodically Forced RLC Circuit

In this lab we continue the study of simple RLC circuits that we began in a lab in Chapter 3 (see page 341). The circuit is shown in Figure 4.37. The parameters are the resistance R, the capacitance C, and the inductance L. The dependent variables we use are v_C, the voltage across the capacitor, and i, the current. (We follow the engineering convention of representing current as i. This variable does *not* represent the square root of -1.) In this lab we consider circuits in which the voltage source $v_T = v_T(t)$ is a time-dependent forcing term.

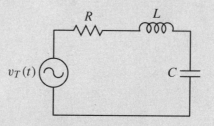

Figure 4.37
An RLC circuit.

From the lab in Chapter 3, we know that the voltage v_C and current i satisfy the system of differential equations

$$\frac{dv_C}{dt} = \frac{i}{C}$$

$$\frac{di}{dt} = -\frac{v_C}{L} - \frac{R}{L}i - \frac{v_T(t)}{L},$$

which is more commonly written as the second-order equation

$$LC\frac{d^2v_C}{dt^2} + RC\frac{dv_C}{dt} + v_C = v_T(t).$$

In this lab we consider the possible behavior of solutions to this equation (or the corresponding system) when $v_T(t) = a \sin \omega t$.

In your report, consider the following questions:

1. Assuming R, C, and L are all nonnegative, what types of long-term behavior are possible for solutions of the equation

$$LC\frac{d^2v_C}{dt^2} + RC\frac{dv_C}{dt} + v_C = a \sin \omega t?$$

Describe how the behavior of solutions depends on the parameters a and ω.

2. In a typical circuit R is on the order of 1000, C is on the order of 10^{-6}, and L is on the order of 1. Does this information help in limiting the possible behaviors of solutions?

3. Describe the solutions for various values of a and ω if $R = 2000$, $C = 2 \times 10^{-7}$, and $L = 1.5$.

Your report: In your report, address each of the items above, justifying all statements and showing all details. Give graphs of solutions as appropriate.

5

NONLINEAR SYSTEMS

In this chapter we study nonlinear autonomous systems. In Chapter 3 we saw that, using a combination of analytic and geometric techniques, we can understand linear systems completely. Here we combine these linear techniques with some additional qualitative methods to tackle nonlinear systems. While these techniques do not allow us to determine the phase portraits of all nonlinear systems, we see that we are able to handle some important nonlinear systems.

We first show how a nonlinear system can be approximated near an equilibrium point by a linear system. This process, known as "linearization," is one of the most frequently used techniques in applications. By studying the linear approximation, we can surmise the behavior of solutions of the nonlinear system, at least near the equilibrium point.

Next we give a qualitative method for extracting more information from direction fields. By looking at where one component of the direction field is zero (so the direction field is either vertical or horizontal), we obtain curves called "nullclines," which subdivide the phase space. When combined with linearization of equilibrium points, the nullclines can, in some cases, yield a complete description of the possible long-term behaviors of solutions.

In the remainder of the chapter, we study special types of models and the nonlinear systems associated with them. These special nonlinear systems are important both because they arise in applications and because the techniques involved in their analysis are delightful.

5.1   EQUILIBRIUM POINT ANALYSIS

From our work in Chapter 3, we are able to understand the solutions of linear systems both qualitatively and analytically. Unfortunately, nonlinear systems are in general much less amenable to the analytic and algebraic techniques that we have developed, but we can use the mathematics of linear systems to understand the behavior of solutions of nonlinear systems near their equilibrium points.

The Van der Pol Equation

To illustrate how to analyze the behavior of solutions near an equilibrium point, we begin with a simple but important nonlinear system — the Van der Pol system. Recall that the Van der Pol system is

$$\frac{dx}{dt} = y$$
$$\frac{dy}{dt} = -x + (1 - x^2)y.$$

We studied this system numerically in Section 2.4 (see page 187), and its direction field and phase plane are shown in Figure 5.1.

The only equilibrium point of this system is the origin, so let's examine how solutions near the origin behave. The direction field shows that the solutions circle around the origin, and if we plot numerical approximations of solutions near the origin, we see a picture that is reminiscent of a spiral source (see Figure 5.2).

We can understand why solutions spiral away from the origin by approximating the Van der Pol system with another system that is much easier for us to analyze. Although the system is nonlinear, there is only one nonlinear term, the x^2y term in the equation for dy/dt. If x and y are small, then this term is much smaller than any of the other terms in the equation. For example, if both x and y are 0.1, then the x^2y term

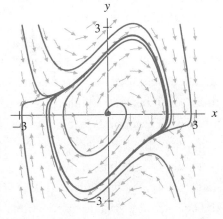

Figure 5.1
Direction field and phase plane for the Van der Pol system.

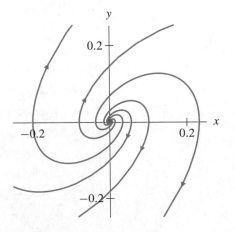

Figure 5.2
Phase plane for the Van der Pol system near the origin.

is 0.001, which is significantly smaller than either x or y. If both x and y are 0.01, then $x^2 y = 10^{-6}$, which is again much smaller than either x or y. Perhaps we can approximate the Van der Pol system by one in which we simply neglect the $x^2 y$ term, at least if both x and y are close to 0. If we drop this term from the system, then we are left with

$$\frac{dx}{dt} = y$$
$$\frac{dy}{dt} = -x + y,$$

which is a linear system. Consequently, the techniques of Chapter 3 apply. The eigenvalues of the linear system are $(1 \pm \sqrt{3}\,i)/2$, and since they are complex with a real part that is positive, we know that the solutions of the linear system spiral away from the origin.

    The linear system and the Van der Pol system have vector fields that are very close to each other near the equilibrium point at the origin. Since solutions of the linear system spiral away from the origin, it is not surprising that solutions of the Van der Pol system that start near the origin also spiral away.

    The technique we applied above is called **linearization**. Near the equilibrium point, we approximate the nonlinear system by an appropriate linear system. For initial conditions near the equilibrium point, the solutions of the nonlinear system and the linear approximation remain close at least for some interval of time.

A Competing Species Model

Let x and y denote the populations of two species that compete for resources. An increase in either species has an adverse effect on the growth rate of the other species. An example of a model of such a system is

$$\frac{dx}{dt} = 2x \left(1 - \frac{x}{2}\right) - xy$$
$$\frac{dy}{dt} = 3y \left(1 - \frac{y}{3}\right) - 2xy.$$

Although the terms involved in these equations are based on reasonable assumptions, we choose coefficients to simplify our discussion rather than to model any particular species.

    Note that, for a given value of x, if y increases then the $-xy$ term causes dx/dt to decrease. Similarly, for a given value of y, if x increases then the $-2xy$ term causes dy/dt to decrease. An increase in the population of either species causes a decrease in the rate of growth of the other species.

Qualitative analysis

We begin our analysis of this system by noting that if $y = 0$, we have $dy/dt = 0$. In other words, if the y's are extinct, they stay extinct. If $y = 0$,

$$\frac{dx}{dt} = 2x \left(1 - \frac{x}{2}\right),$$

which is a logistic population model with a carrying capacity of 2. The phase line of this equation agrees with the x-axis of the phase plane. In particular, $(0, 0)$ and $(2, 0)$ are equilibrium points of the system. Similarly, if $x = 0$, we have $dx/dt = 0$, so the phase line of

$$\frac{dy}{dt} = 3y\left(1 - \frac{y}{3}\right)$$

agrees with the y-axis of the phase plane of the system, and $(0,3)$ is another equilibrium point.

By the Uniqueness Theorem, solutions with initial conditions in the first quadrant must remain in the first quadrant for all time. That is, the axes coincide with solution curves, and the Uniqueness Theorem guarantees that solutions cannot cross. Because this model refers to populations and negative populations do not make much sense, we limit our attention to solutions that are contained in the first quadrant only.

We find the equilibrium points by solving for x and y in the system of equations

$$\begin{cases} 2x\left(1 - \frac{x}{2}\right) - xy = 0 \\ 3y\left(1 - \frac{y}{3}\right) - 2xy = 0, \end{cases}$$

which can be rewritten in the form

$$\begin{cases} x(2 - x - y) = 0 \\ y(3 - y - 2x) = 0. \end{cases}$$

The first equation is satisfied if $x = 0$ or if $2 - x - y = 0$, and the second equation is satisfied if either $y = 0$ or if $3 - y - 2x = 0$.

Suppose first that $x = 0$. Then the equation $y = 0$ yields an equilibrium point at the origin, and the equation $3 - y - 2x = 0$ yields an equilibrium point at $(0, 3)$.

Now suppose that $2 - x - y = 0$. Then the equation $y = 0$ yields an equilibrium point at $(2, 0)$, and the equation $3 - y - 2x = 0$ yields an equilibrium point at $(1, 1)$. (Solve the equations $2 - x - y = 0$ and $3 - y - 2x = 0$ simultaneously.) Hence the equilibrium points are $(0, 0)$, $(0, 3)$, $(2, 0)$, and $(1, 1)$.

Linearization of this system about $(1, 1)$

The equilibrium point $(1, 1)$ is of particular interest. Its existence indicates that it is possible for these two species to coexist in equilibrium. If we numerically compute the phase plane for this system (see Figure 5.3), solutions seem to have only three different types of long-term behaviors. Some solutions tend to $(2, 0)$, some tend to $(0, 3)$, and others tend to $(1, 1)$.

Two important questions remain. First, what solutions tend to the equilibrium point $(1, 1)$? In particular, is the set of these solutions large enough that we could hope to find an example of such a solution in nature? Second, what separates the initial conditions that yield solutions for which x tends to zero from those solutions for which y tends to zero? To answer these questions, we study the system near the equilibrium point $(1, 1)$ using linearization.

Linear systems always have an equilibrium point at the origin. Hence the first step in comparing the nonlinear system near the equilibrium point $(1, 1)$ to a linear system is to move the equilibrium point to the origin via a change of variables. Once the

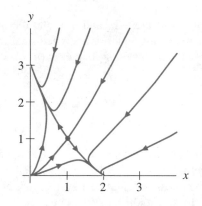

Figure 5.3
Phase plane for the system

$$\frac{dx}{dt} = 2x\left(1 - \frac{x}{2}\right) - xy$$

$$\frac{dy}{dt} = 3y\left(1 - \frac{y}{3}\right) - 2xy.$$

equilibrium point is at the origin, we can use the same ideas as we did in the Van der Pol example to identify the linear approximation.

To move the equilibrium point to the origin, we introduce two new variables, u and v, by the formulas $u = x - 1$ and $v = y - 1$. Note that u and v are both close to 0 when (x, y) is close to $(1, 1)$. To obtain the system in the new variables, we first compute

$$\frac{du}{dt} = \frac{d(x-1)}{dt} = \frac{dx}{dt}$$

$$\frac{dv}{dt} = \frac{d(y-1)}{dt} = \frac{dy}{dt}.$$

The right-hand sides of the system in the new variables are given by

$$\frac{dx}{dt} = 2x\left(1 - \frac{x}{2}\right) - xy$$

$$= 2(u+1)\left(1 - \frac{u+1}{2}\right) - (u+1)(v+1)$$

$$= -u - v - u^2 - uv,$$

and

$$\frac{dy}{dt} = 3y\left(1 - \frac{y}{3}\right) - 2xy$$

$$= 3(v+1)\left(1 - \frac{v+1}{3}\right) - 2(u+1)(v+1)$$

$$= -2u - v - 2uv - v^2.$$

In terms of the new variables, we have

$$\frac{du}{dt} = -u - v - u^2 - uv$$

$$\frac{dv}{dt} = -2u - v - 2uv - v^2.$$

As we expect, the origin is an equilibrium point for this system. The expression for du/dt involves the linear terms $-u$ and $-v$ and the nonlinear terms $-u^2$ and $-uv$.

For dv/dt the linear terms are $-2u$ and $-v$, and the nonlinear terms are $-2uv$ and $-v^2$. Near the origin the nonlinear terms are much smaller than the linear terms, and we therefore approximate the nonlinear system near $(u, v) = (0, 0)$ with the linear system

$$\frac{du}{dt} = -u - v$$

$$\frac{dv}{dt} = -2u - v.$$

The eigenvalues of this system are $-1 \pm \sqrt{2}$. One of these numbers is positive and the other is negative, so $(u, v) = (0, 0)$ is a saddle point for the linear system (see Figure 5.4). Since the linear and nonlinear systems are approximately the same, we expect the phase plane for the nonlinear system near the equilibrium point $(x, y) = (1, 1)$ to look like the uv-phase plane of the linear system.

We conclude that there are only two curves of solutions in the xy-plane that tend toward the equilibrium point $(1, 1)$ as t increases. Hence the solutions that tend to $(1, 1)$ form a very small set. Even if initial conditions are chosen exactly on this curve, arbitrarily small perturbations can push the initial conditions to one side or the other. Since our model is only a very simplified version of the dynamics of the populations, leaving out innumerable sources of such small perturbations, we do not expect to see solutions leading to the $(1, 1)$ equilibrium point in nature.

On the other hand, the curve of solutions that tend to $(1, 1)$ do play an important role in the study of this system. Looking at the phase plane, we note that this curve divides the first quadrant into two regions. In one region one species survives, and in the other region the other species survives (see Figure 5.3).

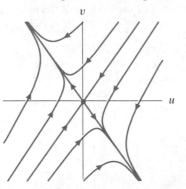

Figure 5.4
Phase plane for

$$\frac{du}{dt} = -u - v$$

$$\frac{dv}{dt} = -2u - v,$$

the linear approximation of the competitive system near $(x, y) = (1, 1)$, which is the same point as $(u, v) = (0, 0)$.

A Nonpolynomial Example

The two examples above have vector fields that are polynomials. When the vector field is a polynomial and the equilibrium point under consideration is the origin, it is very easy to identify which terms are linear and which are nonlinear. However, not all vector fields are polynomials.

Consider the system

$$\frac{dx}{dt} = y$$

$$\frac{dy}{dt} = -y - \sin x.$$

This system is a model for the motion of a damped pendulum, and we will study it extensively in Sections 5.3 and 5.4. The equilibria of this system occur at the points $(x, y) = (0, 0)$, $(\pm\pi, 0)$, $(\pm 2\pi, 0)$, ... Suppose we want to study the solutions that are close to the equilibrium point at $(0, 0)$. Since dy/dt includes a $\sin x$ term, it is not immediately clear what the linear terms of this system are. However, from calculus we know that the power series expansion of $\sin x$ is

$$\sin x = x - \frac{x^3}{3!} + \frac{x^5}{5!} - \dots .$$

Therefore we can write

$$\frac{dx}{dt} = y$$

$$\frac{dy}{dt} = -y - \left(x - \frac{x^3}{3!} + \frac{x^5}{5!} \dots \right).$$

Near the origin we drop the nonlinear terms, and we are left with the linear system

$$\frac{dx}{dt} = y$$

$$\frac{dy}{dt} = -y - x.$$

The eigenvalues of this system are $(-1 \pm \sqrt{3}\,i)/2$. Since these eigenvalues are complex with negative real parts, we expect the corresponding equilibrium point for the nonlinear system to be a spiral sink (see Figures 5.5 and 5.6).

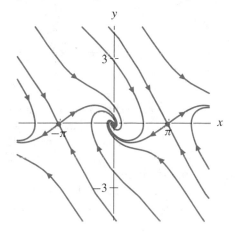

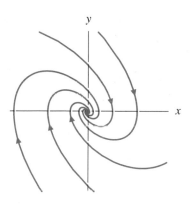

Figure 5.5
Phase plane for the system

$$\frac{dx}{dt} = y$$

$$\frac{dy}{dt} = -y - \sin x.$$

Figure 5.6
Phase plane for the system

$$\frac{dx}{dt} = y$$

$$\frac{dy}{dt} = -y - x.$$

Linearization

The next step is to make the process of linearization more orderly. In the preceding examples we used a change of variables to move the equilibrium point to the origin. Then we identified the linear and nonlinear terms using calculus (if necessary).

Consider the general form of a nonlinear system

$$\frac{dx}{dt} = f(x, y)$$
$$\frac{dy}{dt} = g(x, y).$$

Suppose that (x_0, y_0) is an equilibrium point for this system. We wish to understand what happens to solutions near (x_0, y_0) — that is, to linearize the system near (x_0, y_0). We introduce new variables

$$u = x - x_0$$
$$v = y - y_0$$

that move the equilibrium point to the origin. If x and y are close to the equilibrium point (x_0, y_0), then both u and v are close to 0.

Since $x = u + x_0$ and $y = v + y_0$ and the numbers x_0 and y_0 are constants, the system written in terms of u and v is

$$\frac{du}{dt} = \frac{d(x - x_0)}{dt} = \frac{dx}{dt} = f(x, y) = f(x_0 + u, y_0 + v)$$
$$\frac{dv}{dt} = \frac{d(y - y_0)}{dt} = \frac{dy}{dt} = g(x, y) = g(x_0 + u, y_0 + v).$$

Therefore we have

$$\frac{du}{dt} = f(x_0 + u, y_0 + v)$$
$$\frac{dv}{dt} = g(x_0 + u, y_0 + v).$$

If $u = v = 0$, the right-hand side of this system vanishes, so we have moved the equilibrium point to the origin in the uv-plane.

Now we would like to be able to eliminate the "higher-order" or nonlinear terms in the expressions for du/dt and dv/dt. Since these expressions may include exponentials, logarithms, and trigonometric functions, it is not always clear what the linear terms are. In this case it is necessary to study f and g more closely.

The basic idea of differential calculus is that it is possible to study a function by studying the "best linear approximation" of that function. For functions of two variables, the best linear approximation at a particular point is given by the tangent plane. Hence we have

$$f(x_0 + u, y_0 + v) \approx f(x_0, y_0) + \left[\frac{\partial f}{\partial x}(x_0, y_0)\right] u + \left[\frac{\partial f}{\partial y}(x_0, y_0)\right] v,$$

where the right-hand side of this equation is the equation for the tangent plane to the graph of f at (x_0, y_0). (This expression is also the first-degree Taylor polynomial approximation of f.)

Thus we can rewrite the system for du/dt and dv/dt as

$$\frac{du}{dt} = f(x_0, y_0) + \left[\frac{\partial f}{\partial x}(x_0, y_0)\right] u + \left[\frac{\partial f}{\partial y}(x_0, y_0)\right] v + \dots$$

and

$$\frac{dv}{dt} = g(x_0, y_0) + \left[\frac{\partial g}{\partial x}(x_0, y_0)\right] u + \left[\frac{\partial g}{\partial y}(x_0, y_0)\right] v + \dots,$$

where "$\dots$" stands for the terms that make up the difference between the tangent plane and the function. These are precisely the terms we wish to ignore when forming the linear approximation of the system. Since $f(x_0, y_0) = 0$ and $g(x_0, y_0) = 0$, we can use matrix notation to write the system more succinctly as

$$\begin{pmatrix} \dfrac{du}{dt} \\ \dfrac{dv}{dt} \end{pmatrix} = \begin{pmatrix} \dfrac{\partial f}{\partial x}(x_0, y_0) & \dfrac{\partial f}{\partial y}(x_0, y_0) \\ \dfrac{\partial g}{\partial x}(x_0, y_0) & \dfrac{\partial g}{\partial y}(x_0, y_0) \end{pmatrix} \begin{pmatrix} u \\ v \end{pmatrix} + \dots.$$

The 2×2 matrix of partial derivatives in this expression is called the **Jacobian matrix** of the system at (x_0, y_0).

Hence the **linearized system** at the equilibrium point (x_0, y_0) is

$$\begin{pmatrix} \dfrac{du}{dt} \\ \dfrac{dv}{dt} \end{pmatrix} = \begin{pmatrix} \dfrac{\partial f}{\partial x}(x_0, y_0) & \dfrac{\partial f}{\partial y}(x_0, y_0) \\ \dfrac{\partial g}{\partial x}(x_0, y_0) & \dfrac{\partial g}{\partial y}(x_0, y_0) \end{pmatrix} \begin{pmatrix} u \\ v \end{pmatrix}.$$

We use this "linearized" system to study the behavior of solutions of the nonlinear system near the equilibrium point (x_0, y_0). Note that we need only know the partial derivatives of the components of the vector field at the equilibrium point to create the linearized system. We do not need to compute the change of variables moving the equilibrium point to the origin.

As always, the derivative of a nonlinear function provides only a local approximation. Hence the solutions of the linearized system are close to solutions of the nonlinear system only near the equilibrium point. How close to the equilibrium point we must be for the linear approximation to be any good depends on the size of the nonlinear terms.

More Examples of Linearization

Consider the nonlinear system

$$\frac{dx}{dt} = -2x + 2x^2$$

$$\frac{dy}{dt} = -3x + y + 3x^2.$$

There are two equilibrium points for this system — $(0, 0)$ and $(1, 0)$. To understand solutions that start near these points, we first compute the Jacobian matrix

$$\begin{pmatrix} \dfrac{\partial f}{\partial x}(x, y) & \dfrac{\partial f}{\partial y}(x, y) \\ \dfrac{\partial g}{\partial x}(x, y) & \dfrac{\partial g}{\partial y}(x, y) \end{pmatrix} = \begin{pmatrix} -2 + 4x & 0 \\ -3 + 6x & 1 \end{pmatrix}$$

since $f(x, y) = -2x + 2x^2$ and $g(x, y) = -3x + y + 3x^2$. At the two equilibrium points, we have

$$
\begin{pmatrix}
\dfrac{\partial f}{\partial x}(0, 0) & \dfrac{\partial f}{\partial y}(0, 0) \\[2mm]
\dfrac{\partial g}{\partial x}(0, 0) & \dfrac{\partial g}{\partial y}(0, 0)
\end{pmatrix}
=
\begin{pmatrix}
-2 & 0 \\
-3 & 1
\end{pmatrix}
$$

and

$$
\begin{pmatrix}
\dfrac{\partial f}{\partial x}(1, 0) & \dfrac{\partial f}{\partial y}(1, 0) \\[2mm]
\dfrac{\partial g}{\partial x}(1, 0) & \dfrac{\partial g}{\partial y}(1, 0)
\end{pmatrix}
=
\begin{pmatrix}
2 & 0 \\
3 & 1
\end{pmatrix}.
$$

Near $(0, 0)$ the phase plane for the nonlinear system should resemble that of the linearized system

$$
\frac{d\mathbf{Y}}{dt} = \begin{pmatrix} -2 & 0 \\ -3 & 1 \end{pmatrix} \mathbf{Y}.
$$

The eigenvalues of this linear system are -2 and 1, so the origin is a saddle. We can compute that $(0, 1)$ is an eigenvector for the eigenvalue 1 and that $(1, 1)$ is an eigenvector for the eigenvalue -2. Using this information we can sketch the phase plane for the linear system (see Figure 5.7).

At the other equilibrium point, $(1, 0)$, the linearized system is

$$
\frac{d\mathbf{Y}}{dt} = \begin{pmatrix} 2 & 0 \\ 3 & 1 \end{pmatrix} \mathbf{Y}.
$$

Here the eigenvalues are 2 and 1, so the origin is a source for this system. Using the fact that $(0, 1)$ is an eigenvector for the eigenvalue 1 and $(1, 3)$ is an eigenvector for the eigenvalue 2, we can sketch the phase plane (see Figure 5.8).

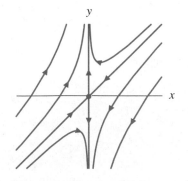

Figure 5.7
Phase plane for the system

$$
\frac{d\mathbf{Y}}{dt} = \begin{pmatrix} -2 & 0 \\ -3 & 1 \end{pmatrix} \mathbf{Y}.
$$

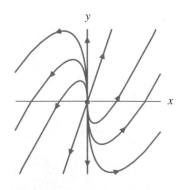

Figure 5.8
Phase plane for the system

$$
\frac{d\mathbf{Y}}{dt} = \begin{pmatrix} 2 & 0 \\ 3 & 1 \end{pmatrix} \mathbf{Y}.
$$

Near the two equilibrium points, the phase plane for the nonlinear system resembles that of the linearized systems. Solution curves (numerically approximated by the computer) are shown in Figure 5.9. If we magnify the phase plane near $(0, 0)$ and $(1, 0)$, we see that the solution curves do indeed look like those of the corresponding linearized systems (see Figure 5.9).

Classification of Equilibrium Points

The fundamental idea underlying the technique of linearization is to use a linear system to approximate the behavior of solutions of a nonlinear system near an equilibrium point. The solutions of the nonlinear system near the equilibrium point are close to solutions of the approximating linear system, at least for a short time interval. For most systems, the information gained by studying the linearization is enough to determine the long-term behavior of solutions of the nonlinear system near the equilibrium point.

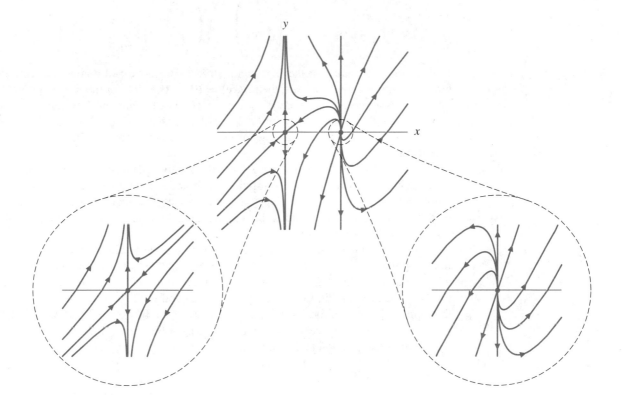

Figure 5.9
Solution curves and magnifications near the fixed points for the system

$$\frac{dx}{dt} = -2x + 2x^2$$

$$\frac{dy}{dt} = -3x + y + 3x^2.$$

For example, consider the system

$$\frac{dx}{dt} = f(x, y)$$
$$\frac{dy}{dt} = g(x, y).$$

Suppose (x_0, y_0) is an equilibrium point and let

$$\mathbf{J} = \begin{pmatrix} \dfrac{\partial f}{\partial x}(x_0, y_0) & \dfrac{\partial f}{\partial y}(x_0, y_0) \\ \dfrac{\partial g}{\partial x}(x_0, y_0) & \dfrac{\partial g}{\partial y}(x_0, y_0) \end{pmatrix}$$

be the Jacobian matrix at (x_0, y_0). The linearized system is

$$\begin{pmatrix} \dfrac{du}{dt} \\ \dfrac{dv}{dt} \end{pmatrix} = \mathbf{J} \begin{pmatrix} u \\ v \end{pmatrix}.$$

If all of the eigenvalues of the Jacobian matrix are negative real numbers or complex numbers with negative real part, then $(u, v) = (0, 0)$ is a sink for the linear system and all solutions approach $(u, v) = (0, 0)$ as $t \to \infty$. For the nonlinear system, solutions that start near the equilibrium point $(x, y) = (x_0, y_0)$ approach it as $t \to \infty$. Hence we say that (x_0, y_0) is a **sink**. If the eigenvalues are complex, then (x_0, y_0) is said to be a **spiral sink**.

Similarly, if the Jacobian matrix has only positive eigenvalues or complex eigenvalues with positive real part, then solutions with initial conditions near the equilibrium point (x_0, y_0) move away from (x_0, y_0) as t increases. The equilibrium point (x_0, y_0) of the nonlinear system is said to be a **source**. If the eigenvalues are complex, then (x_0, y_0) is called a **spiral source**.

If the Jacobian matrix has one positive and one negative eigenvalue, then the equilibrium point (x_0, y_0) is called a **saddle** for the nonlinear system. As for a linear system with a saddle equilibrium point at the origin, there are exactly two curves of solutions that approach the equilibrium point as t increases and exactly two curves of solutions that approach the equilibrium point as t decreases (see Figures 5.7 and 5.9). For the nonlinear system, these curves of solutions need not be straight lines. All other solutions with initial position near (x_0, y_0) move away as t increases and as t decreases.

A Reminder

It is important to remember that this classification of equilibrium points for nonlinear systems says nothing about the behavior of solutions of the nonlinear system with initial positions far from (x_0, y_0).

Separatrices

The four special solution curves that tend toward a saddle equilibrium point as $t \to \infty$ or as $t \to -\infty$ are called **separatrices**. (One of these curves by itself is called a *separatrix*.) Separatrices are of special importance because they separate solutions with different behaviors. The two separatrices on which solutions tend toward the saddle as $t \to \infty$ are called **stable separatrices**, while those on which solutions tend toward the saddle as $t \to -\infty$ are called **unstable separatrices**.

In the system

$$\frac{dx}{dt} = -2x + 2x^2$$

$$\frac{dy}{dt} = -3x + y + 3x^2$$

studied above, the origin is a saddle (see Figures 5.7 and 5.9). The stable separatrix that tends toward the origin separates the strip of the phase plane bounded by the lines $x = 0$ and $x = 1$ into two parts. Initial conditions in this strip that are above the separatrix yield solutions where $y(t) \to \infty$ as t increases, while initial conditions in this strip that are below the separatrix yield solutions where $y(t) \to -\infty$ as t increases (see Figure 5.10).

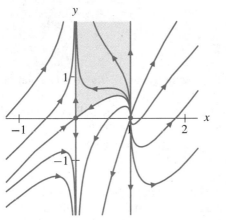

Figure 5.10
Separatrices of $(0, 0)$ for the system

$$\frac{dx}{dt} = -2x + 2x^2$$

$$\frac{dy}{dt} = -3x + y + 3x^2,$$

and regions of the strip between $x = 0$ and $x = 1$ with different long-term behaviors.

When Linearization Fails

Unfortunately, for some equilibria of some systems, the information given by the linearized system is not enough to determine the complete behavior of solutions of the nonlinear system near the equilibrium point.

For example, consider the system

$$\frac{dx}{dt} = y - (x^2 + y^2)x$$

$$\frac{dy}{dt} = -x - (x^2 + y^2)y.$$

The origin is an equilibrium point for this system, and its linearized system is

$$\begin{pmatrix} \dfrac{du}{dt} \\ \dfrac{dv}{dt} \end{pmatrix} = \begin{pmatrix} 0 & 1 \\ -1 & 0 \end{pmatrix} \begin{pmatrix} u \\ v \end{pmatrix}.$$

The eigenvalues of this linear system are $\pm i$, and hence it is a center. All nonzero solutions of the linearized system are periodic. In fact each solution curve is a circle centered at the origin.

However, there are no periodic solutions for the nonlinear system. To see why, consider the vector field as a sum of two vector fields, the linear vector field $\mathbf{V}_1(x, y) = (y, -x)$ and the nonlinear vector field $\mathbf{V}_2(x, y) = (-(x^2 + y^2)x, -(x^2 + y^2)y)$. The linear vector field $\mathbf{V}_1$ corresponds to the linearized system. It is always tangent to circles centered at the origin. On the other hand, the vector field $\mathbf{V}_2(x, y) = (-(x^2 + y^2)x, -(x^2 + y^2)y)$ always points directly toward $(0, 0)$ since it is a scalar multiple of the field $(-x, -y)$. (The scalar is the positive number $x^2 + y^2$.) The result of adding $\mathbf{V}_1(x, y)$ and $\mathbf{V}_2(x, y)$ is a vector field that always has a negative radial component. Thus, solutions to the nonlinear system spiral toward $(0, 0)$ (see Figure 5.11).

Note that if we merely change the signs of the higher-order terms in the above system, then the resulting system

$$\frac{dx}{dt} = y + (x^2 + y^2)x$$

$$\frac{dy}{dt} = -x + (x^2 + y^2)y$$

has the same linearization near $(0, 0)$, but now solutions spiral away from the origin.

In this example the solutions of the nonlinear system near the origin and the solutions of the linearized system are still approximately the same, at least for a short amount of time. However, since the linearized system is a center, any small perturbation can change the long-term behavior of the solutions. Even the very small perturbation caused by the inclusion of the nonlinear terms can turn the center into a spiral sink or a spiral source.

Fortunately there are only two situations in which the long-term behavior of solutions near an equilibrium point of the nonlinear system and its linearization can differ. One is when the linearized system is a center. The other is when the linearized system has zero as an eigenvalue (see the exercises). In every other case, the long-term behavior of solutions of the nonlinear system near an equilibrium point is the same as the solutions of its linearization.

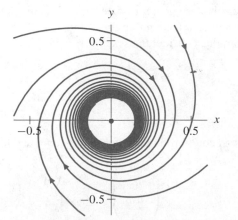

Figure 5.11
The solution curves for the system

$$\frac{dx}{dt} = y - (x^2 + y^2)x$$

$$\frac{dy}{dt} = -x - (x^2 + y^2)y$$

spiral slowly toward the equilibrium point at the origin even though the linearization at the origin is a center.

EXERCISES FOR SECTION 5.1

1. Consider the three systems

(i)
$$\frac{dx}{dt} = 2x + y$$
$$\frac{dy}{dt} = -y + x^2$$

(ii)
$$\frac{dx}{dt} = 2x + y$$
$$\frac{dy}{dt} = y + x^2$$

(iii)
$$\frac{dx}{dt} = 2x + y$$
$$\frac{dy}{dt} = -y - x^2.$$

All three have an equilibrium point at $(0, 0)$. Which two systems have phase planes with the same "local picture" near $(0, 0)$? Justify your answer. [*Hint*: Very little computation is required for this exercise, but be sure to give a complete justification.]

2. Consider the following three systems:

(i)
$$\frac{dx}{dt} = 3 \sin x + y$$
$$\frac{dy}{dt} = 4x + \cos y - 1$$

(ii)
$$\frac{dx}{dt} = -3 \sin x + y$$
$$\frac{dy}{dt} = 4x + \cos y - 1$$

(iii)
$$\frac{dx}{dt} = -3 \sin x + y$$
$$\frac{dy}{dt} = 4x + 3 \cos y - 3.$$

All three have an equilibrium point at $(0, 0)$. Which two systems have phase planes with the same "local picture" near $(0, 0)$? Justify your answer. [*Hint*: Very little computation is required for this exercise, but be sure to give a complete justification.]

3. Consider the system

$$\frac{dx}{dt} = -2x + y$$
$$\frac{dy}{dt} = -y + x^2.$$

(a) Find the linearized system for the equilibrium point $(0, 0)$.
(b) Classify $(0, 0)$ (as either a source, sink, center, ...).
(c) Sketch the phase plane for the nonlinear system near $(0, 0)$.
(d) Repeat parts (a)–(c) for the equilibrium point at $(2, 4)$.

4. Consider the system

$$\frac{dx}{dt} = -x$$
$$\frac{dy}{dt} = -4x^3 + y.$$

(a) Show that the origin is the only equilibrium point.
(b) Find the linearized system at the origin.
(c) Classify the linearized system and sketch its phase plane.

5. Consider the system in Exercise 4.

   (a) Find the general solution of the equation $dx/dt = -x$. [*Hint*: This is as easy as it looks.]

   (b) Using the solution to part (a) in place of x, find the general solution of the equation

$$\frac{dy}{dt} = -4x^3 + y.$$

   [*Hint*: This provides some review of linear equations in Section 1.8.]

   (c) Use the results from parts (a) and (b) to form the general solution of the system.

   (d) Find the solution curves of the system that tend toward the origin as $t \to \infty$.

   (e) Find the solution curves of the system that tend toward the origin as $t \to -\infty$.

   (f) Sketch the solution curves in the phase plane corresponding to these solutions. These are the separatrices.

   (g) Compare the sketch of the linearized system that you obtained in Exercise 4 with a sketch of the separatrix solutions for the equilibrium point at the origin for this system. In what ways are the two pictures the same? How do they differ?

6. For the competing species population model

$$\frac{dx}{dt} = 2x \left(1 - \frac{x}{2}\right) - xy$$

$$\frac{dy}{dt} = 3y \left(1 - \frac{y}{3}\right) - 2xy$$

   studied in this section, we showed that the equilibrium point $(1, 1)$ is a saddle.

   (a) Find the linearized system near each of the other equilibrium points.

   (b) Classify each equilibrium point (as either a source, a sink, a saddle, ...).

   (c) Sketch the phase plane of each linearized system.

   (d) Give a brief description of the phase plane near each equilibrium point of the nonlinear system.

In Exercises 7–16, we restrict attention to the first quadrant ($x, y \geq 0$). For each system,

   (a) find and classify all equilibria (in the first quadrant),

   (b) sketch the phase plane of the system near each equilibrium point, and

   (c) describe the behaviors of solutions near each equilibrium point.

7.
$$\frac{dx}{dt} = x(-x - 3y + 150)$$
$$\frac{dy}{dt} = y(-2x - y + 100)$$

8.
$$\frac{dx}{dt} = x(10 - x - y)$$
$$\frac{dy}{dt} = y(30 - 2x - y)$$

9.
$$\frac{dx}{dt} = x(100 - x - 2y)$$
$$\frac{dy}{dt} = y(150 - x - 6y)$$

10.
$$\frac{dx}{dt} = x(-x - y + 100)$$
$$\frac{dy}{dt} = y(-x^2 - y^2 + 2500)$$

11.
$$\frac{dx}{dt} = x(-x - y + 40)$$
$$\frac{dy}{dt} = y(-x^2 - y^2 + 2500)$$

12.
$$\frac{dx}{dt} = x(-4x - y + 160)$$
$$\frac{dy}{dt} = y(-x^2 - y^2 + 2500)$$

13.
$$\frac{dx}{dt} = x(-8x - 6y + 480)$$
$$\frac{dy}{dt} = y(-x^2 - y^2 + 2500)$$

14.
$$\frac{dx}{dt} = x(2 - x - y)$$
$$\frac{dy}{dt} = y(y - x^2)$$

15.
$$\frac{dx}{dt} = x(2 - x - y)$$
$$\frac{dy}{dt} = y(y - x)$$

16.
$$\frac{dx}{dt} = x(x - 1)$$
$$\frac{dy}{dt} = y(x^2 - y)$$

17. Consider the system

$$\frac{dx}{dt} = -x^3$$
$$\frac{dy}{dt} = -y + y^2.$$

It has equilibrium points at $(0, 0)$ and $(0, 1)$.

   (a) Find the linearized system at $(0, 0)$.

   (b) Find the eigenvalues and eigenvectors and sketch the phase plane of the linearized system at $(0, 0)$.

   (c) Find the linearized system at $(0, 1)$.

   (d) Find the eigenvalues and eigenvectors and sketch the phase plane of the linearized system at $(0, 1)$.

   (e) Sketch the phase plane of the nonlinear system. [*Hint*: The system decouples, so first draw a phase line for each of the individual equations.]

   (f) Why do the phase planes for the linearized systems and the phase plane for the nonlinear system near the equilibrium points look so different?

18. If a nonlinear system depends on a parameter, then the equilibrium points can change as the parameter varies. In other words, as the parameter changes, a bifurcation can occur. Consider the one-parameter system family of systems

$$\frac{dx}{dt} = x^2 - a$$
$$\frac{dy}{dt} = -y(x^2 + 1),$$

where a is the parameter.

   (a) Show that the system has no equilibrium points if $a < 0$.

   (b) Show that the system has two equilibrium points if $a > 0$.

   (c) Show that the system has exactly one equilibrium point if $a = 0$.

   (d) Find the linearization of the equilibrium point for $a = 0$ and compute the eigenvalues of this linear system.

Remark: The system changes from having no equilibrium points to having two equilibrium points as the parameter a is increased through $a = 0$. We say that the system has a bifurcation at $a = 0$ and that $a = 0$ is a bifurcation value of the parameter.

19. Continuing the study of the nonlinear system given in Exercise 18,

    (a) use the direction field to sketch the phase plane for the system if $a = -1$,

    (b) use the direction field and the linearization at the equilibrium point to sketch the phase plane for $a = 0$, and

    (c) use the direction field and the linearization at the equilibrium points to sketch the phase plane for $a = 1$.

Remark: The transition from a system with no equilibrium points to a system with one saddle and one sink via a system with one equilibrium point with zero as an eigenvalue is typical of bifurcations that create equilibria.

In Exercises 20–25, each system depends on the parameter a. In each exercise,

  (a) find all equilibrium points,

  (b) determine all values of a at which a bifurcation occurs, and

  (c) in a short paragraph complete with pictures, describe the phase plane at, before, and after each bifurcation value.

20.
$$\frac{dx}{dt} = y - x^2$$
$$\frac{dy}{dt} = y - a$$

21.
$$\frac{dx}{dt} = y - x^2$$
$$\frac{dy}{dt} = a$$

22.
$$\frac{dx}{dt} = y - x^2$$
$$\frac{dy}{dt} = y - x - a$$

23.
$$\frac{dx}{dt} = y - ax^3$$
$$\frac{dy}{dt} = y - x$$

24.
$$\frac{dx}{dt} = y - x^2 + a$$
$$\frac{dy}{dt} = y + x^2 - a$$

25.
$$\frac{dx}{dt} = y - x^2 + a$$
$$\frac{dy}{dt} = y + x^2$$

26. The system

$$\frac{dx}{dt} = x(-x - y + 70)$$
$$\frac{dy}{dt} = y(-2x - y + a)$$

is a model for a pair of competing species for which dy/dt depends on the parameter a. Find the two bifurcation values of a. Describe the fate of the x and y populations before and after each bifurcation.

27. Suppose two species X and Y are to be introduced to an island. It is known that the two species compete, but the precise nature of their interaction is unknown. We assume that the populations $x(t)$ and $y(t)$ of X and Y, respectively, are modeled by a system

$$\frac{dx}{dt} = f(x, y)$$
$$\frac{dy}{dt} = g(x, y).$$

(a) Suppose $f(0, 0) = g(0, 0) = 0$; that is, $(0, 0)$ is an equilibrium point. What does this say about the ability of X and Y to migrate to the island?

(b) Suppose that a small population consisting solely of one species reproduces rapidly. What can you conclude about the values of $\partial f/\partial x$ and $\partial g/\partial y$ at $(0, 0)$?

(c) Since X and Y compete for resources, the presence of either of the species will decrease the rate of growth of the population of the other. What does this say about $\partial f/\partial y$ and $\partial g/\partial x$ at $(0, 0)$?

(d) Using the assumptions from parts (a)–(c), what type(s) of equilibrium point could $(0, 0)$ possibly be? [There may be more than one possibility. If so, specify all of them.]

(e) For each of the possibilities listed in part (d), sketch a possible phase plane near $(0, 0)$.

Be sure to justify all answers.

28. For the two species X and Y of Exercise 27, suppose that both X and Y reproduce very slowly. Also suppose that competition between these two species is very intense.

(a) What can you conclude about $\partial f/\partial x$ and $\partial g/\partial y$ at $(0, 0)$?

(b) What can you conclude about $\partial f/\partial y$ and $\partial g/\partial x$ at $(0, 0)$?

(c) What type(s) of equilibrium point can $(0, 0)$ possibly be? [There may be more than one possibility. If so, specify all of them.]

(d) For each of the possibilities listed in part (c), sketch a possible phase plane near $(0, 0)$.

Remember to justify all answers.

29. For the species X and Y in Exercises 27 and 28, suppose that X reproduces very quickly if it is on the island with no Y's present and that species Y reproduces slowly if there are no X's present. Also suppose that the growth rate of species X is decreased a relatively large amount by the presence of Y but that species Y is indifferent to X's population.

(a) What can you say about $\partial f/\partial x$ and $\partial g/\partial y$ at $(0, 0)$?

(b) What can you say about $\partial f/\partial y$ and $\partial g/\partial x$ at $(0, 0)$?

(c) What are the possible type(s) for the equilibrium point at $(0, 0)$? [There may be more than one possibility. If so, specify all of them.]

(d) For each of the types listed above, sketch the phase plane near $(0, 0)$.

Be sure to justify all answers.

30. Suppose two similar countries Y and Z are engaged in an arms race. Let $y(t)$ and $z(t)$ denote the size of the stockpiles of arms of Y and Z, respectively. We model this situation with the system of differential equations

$$\frac{dy}{dt} = h(y, z)$$

$$\frac{dz}{dt} = k(y, z).$$

Suppose that all we know about the functions h and k are the two assumptions:

(i) If country Z's stockpile of arms is not changing, then any increase in size of Y's stockpile of arms results in a decrease in the rate of arms building in country Y. The same is true for country Z.

(ii) If either country increases its stockpile, the other responds by increasing its rate of arms production.

(a) What do the assumptions imply about $\partial h / \partial y$ and $\partial k / \partial z$?

(b) What do the assumptions imply about $\partial h / \partial z$ and $\partial k / \partial y$?

(c) What types of equilibrium points are possible for this system? Justify your answer. [*Hint*: Suppose you have an equilibrium point. What do your results in parts (a) and (b) imply about the Jacobian matrix at that equilibrium point?]

5.2 QUALITATIVE ANALYSIS

The process of linearization discussed in Section 5.1 gives us a powerful technique for understanding the behavior of solutions of a nonlinear system near an equilibrium point. Unfortunately it provides "local" information only — information that can be used only near equilibrium points. (Making predictions based on linearizations far away from equilibria can have drastic consequences — see Section 4.5.)

So far our only general techniques for the study of the behavior of nonlinear systems away from equilibrium points are numerical. Indeed, a careful numerical study of a system can give considerable insight into the behavior of its solutions. Unfortunately it is difficult to know if enough initial conditions have been tested to observe all of the possible behaviors of solutions. In this section we develop qualitative techniques that can be used in combination with linearization and numerics.

Competing Species

Recall the system

$$\frac{dx}{dt} = 2x \left(1 - \frac{x}{2}\right) - xy$$

$$\frac{dy}{dt} = 3y \left(1 - \frac{y}{3}\right) - 2xy,$$

where x and y are populations of two species that compete for resources (see Section 5.1).

In Section 5.1 we determined that the equilibrium points are $(0, 0)$, $(0, 3)$, $(2, 0)$, and $(1, 1)$. By linearizing, we found that the point $(1, 1)$ is a saddle. There is one curve of solutions that approach $(1, 1)$ as $t \to \infty$, and this curve separates the phase plane into two regions. The use of numerical methods suggests that solutions that do not tend to (1, 1) tend either to (0, 3) or (2, 0) as t increases (see Figure 5.12). To verify this observation and to better understand the behavior of solutions, we look more closely at the direction field.

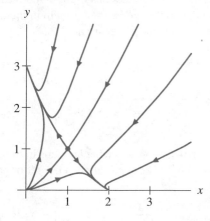

Figure 5.12

Phase portrait for the competing species system

$$\frac{dx}{dt} = 2x\left(1 - \frac{x}{2}\right) - xy$$

$$\frac{dy}{dt} = 3y\left(1 - \frac{y}{3}\right) - 2xy.$$

This computer-generated phase portrait suggests that solutions that do not tend to (1, 1) tend either to (0, 3) or to (2, 0).

Nullclines

The direction field for the competing species system is sketched in Figure 5.13. We can use this picture to provide a rough sketch of solution curves, but with a bit more quali-tative analysis, we can give a much more complete picture of the behavior of solutions. One tool for this analysis is the **nullcline**.

DEFINITION   For the system

$$\frac{dx}{dt} = f(x, y)$$

$$\frac{dy}{dt} = g(x, y),$$

the *x-nullcline* is the set of points (x, y) where $f(x, y)$ is zero — that is, the level curve where $f(x, y)$ is zero. The *y-nullcline* is the set of points where $g(x, y)$ is zero.

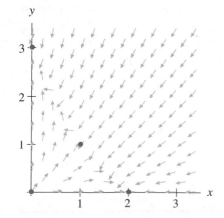

Figure 5.13

Direction field for the competing species system

$$\frac{dx}{dt} = 2x\left(1 - \frac{x}{2}\right) - xy$$

$$\frac{dy}{dt} = 3y\left(1 - \frac{y}{3}\right) - 2xy.$$

It is difficult to judge the long-term behavior of solutions solely from the direction field.

Along the x-nullcline the x-component of the vector field is zero, and consequently the vector field is vertical. It points either straight up or straight down. Similarly, on the y-nullcline the y-component of the vector field is zero, so the vector field is horizontal. It points either left or right. Because both $f(x, y)$ and $g(x, y)$ must be zero at an equilibrium point, the intersections of the nullclines are the equilibrium points.

To show how nullclines can be used to help in the qualitative analysis of systems, we consider our competing species example

$$\frac{dx}{dt} = 2x\left(1 - \frac{x}{2}\right) - xy$$
$$\frac{dy}{dt} = 3y\left(1 - \frac{y}{3}\right) - 2xy.$$

Recall that we are interested only in what happens in the first quadrant $(x, y \geq 0)$, since this is a population model. The x-nullcline is the set of points (x, y) that satisfy

$$\frac{dx}{dt} = 2x\left(1 - \frac{x}{2}\right) - xy = 0.$$

Since this equation is equivalent to $x(2 - x - y) = 0$, the x-nullcline consists of two lines, $x = 0$ and $y = -x+2$. On these lines the x-component of the vector field is zero. Thus the vector field is vertical along these lines. In the remainder of the phase plane the x-component of the vector field is either positive or negative. If $dx/dt > 0$, then solutions move toward the right. If $dx/dt < 0$, then solutions move toward the left. In Figure 5.14 we mark part of the x-nullcline with vertical lines as a reminder that the vector field is vertical along the nullcline. We can label the regions off the x-nullcline as either "right" or "left" depending on whether dx/dt is positive or negative.

Similarly, the y-nullcline is the set of points where $dy/dt = 0$. This is the set of points that satisfy

$$3y\left(1 - \frac{y}{3}\right) - 2xy = y(3 - y - 2x) = 0.$$

This set also consists of two lines, $y = 0$ and $y = -2x + 3$. On these lines the y-component of the vector field is zero, so the vector field is horizontal. On the rest of the phase plane, either $dy/dt > 0$ and solutions move upward, or $dy/dt < 0$ and solutions move downward. In Figure 5.15 we mark part of the y-nullcline with horizontal segments as a reminder that the vector field is horizontal along the nullcline. We label the regions off the y-nullcline as either "up" or "down" depending on the sign of dy/dt.

Analysis using nullclines

We combine the x- and y-nullclines in Figure 5.16. The equilibrium points occur at the intersections of the x- and y-nullclines. The nullclines divide the first quadrant into four regions labeled A, B, C, and D. We can use this picture to give a detailed analysis of the behavior of solutions of this system.

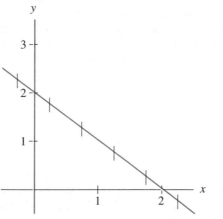

Figure 5.14
The x-nullclines for the system

$$\frac{dx}{dt} = 2x\left(1 - \frac{x}{2}\right) - xy$$

$$\frac{dy}{dt} = 3y\left(1 - \frac{y}{3}\right) - 2xy.$$

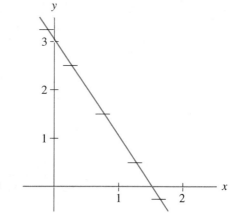

Figure 5.15
The y-nullclines for the system

$$\frac{dx}{dt} = 2x\left(1 - \frac{x}{2}\right) - xy$$

$$\frac{dy}{dt} = 3y\left(1 - \frac{y}{3}\right) - 2xy.$$

First consider the triangular region A (see Figure 5.16). The segment $0 < y < 3$ on the y-axis is a solution curve, and the vector field on the other two sides of A point into A. Hence a solution that begins in region A at time zero will remain in A for all positive time. There is no way the solution can leave. Region A lies in the "left" ($dx/dt < 0$) and the "up" ($dy/dt > 0$) regions in the phase plane, so we can label it "left-up." That is, as t increases, solutions in region A must move toward the upper-left corner of the region — toward the equilibrium point (0, 3).

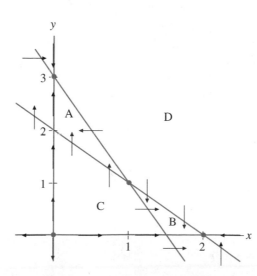

Figure 5.16
The x- and y-nullclines for

$$\frac{dx}{dt} = 2x\left(1 - \frac{x}{2}\right) - xy$$

$$\frac{dy}{dt} = 3y\left(1 - \frac{y}{3}\right) - 2xy.$$

The nullclines divide the first quadrant into four regions marked A, B, C, and D.

Similarly, solutions cannot leave region B, and they move "right-down" as t increases. Hence all solutions in region B tend to the equilibrium point $(2, 0)$ as t increases. In Figure 5.17, we display two solutions of the competing species model together with their $x(t)$- and $y(t)$-graphs.

Solutions in region C move "right-up." There are three possibilities for what happens to these solutions. They may leave region C and enter A, they may leave C and enter B, or they may approach the equilibrium point $(1, 1)$. Since the point $(1, 1)$ is a saddle, the solutions that enter A are divided from those that enter B by the stable separatrix of $(1, 1)$.

Similarly, solutions in region D move left-down, tending toward regions A, B, or the equilibrium point $(1, 1)$. Again the stable separatrix of $(1, 1)$ separates the solutions that enter A (and tend to $(0, 3)$) from those that enter B (and tend to $(2, 0)$).

This analysis gives us a fairly complete qualitative picture of the behavior of solutions of the competing species model. We know that most solutions tend to an equilibrium population with one species extinct and the other at its carrying capacity (see Figure 5.17). The stable separatrix of the saddle $(1, 1)$ divides the two long-term behaviors.

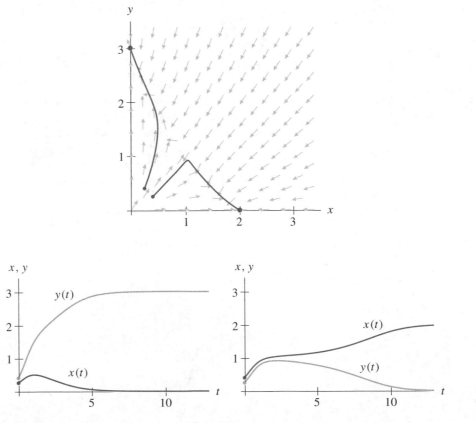

Figure 5.17
Two solutions in the phase plane for the competing species system above with their corresponding $x(t)$- and $y(t)$-graphs.

Important observations

In this example both nullclines consist of straight lines. However, in general a nullcline can be any type of curve, and we will soon study examples with such nullclines.

Also note that there are two very different kinds of nullclines in this example. For the nullclines along the x- and y-axes, the vector field was (coincidentally) tangent to these lines. Consequently, solutions that start on these lines stay on them forever. We can use techniques from Section 1.6 to analyze these solutions completely, since these nullclines really are phase lines for a one-dimensional equation.

The vector field is not tangent to the other two nullclines in this example. It points across these lines, and as a result these nullclines give us information only about the direction of solutions as they cross the nullcline.

In particular it is important to note the difference between a straight-line solution (as discussed in Chapter 3) and a nullcline. A straight-line solution is a solution curve that corresponds to a real eigenvector for a linear system. A nullcline is a curve along which the vector field is either purely horizontal or purely vertical. It is possible for a nullcline and a straight-line solution to coincide for some linear systems, but in general nullclines and straight-line solutions are different.

Nullclines That Are Not Lines

As a somewhat more complicated example of how nullclines are used, consider

$$\frac{dx}{dt} = 2x\left(1 - \frac{x}{2}\right) - xy$$

$$\frac{dy}{dt} = y\left(\frac{9}{4} - y^2\right) - x^2 y.$$

We can still interpret this as a competing species model because the growth rate of each species decreases when the population of the other increases. In this model the additional complications in the dy/dt equation are for the sake of illustration of our techniques. As is usual with population models, we consider only solutions in the first quadrant.

The x-nullcline satisfies the equation

$$x(2 - x - y) = 0,$$

which consists of the two lines $x = 0$ and $y = -x + 2$. The y-nullcline is the set of points (x, y) that satisfy

$$y\left(-x^2 - y^2 + \frac{9}{4}\right) = 0.$$

These points lie either on the line $y = 0$ or on the circle $x^2 + y^2 = 9/4 = (3/2)^2$. The intersection points of the x- and y-nullclines give the equilibrium points $(0, 0)$, $(0, 3/2)$, $(1 + \sqrt{2}/4, 1 - \sqrt{2}/4) \approx (1.35, 0.65)$, $(1 - \sqrt{2}/4, 1 + \sqrt{2}/4) \approx (0.65, 1.35)$, and $(2, 0)$ (see Figure 5.18). (The point $(0, -3/2)$ is also an equilibrium point, but since it lies outside the first quadrant, it is not important to our analysis.)

As in the previous example, the x- and y-axes consist of solution curves. If $y = 0$, then $dy/dt = 0$ and we have

$$\frac{dx}{dt} = 2x\left(1 - \frac{x}{2}\right),$$

which is the same logistic equation as in the previous example. If $x = 0$, we have $dx/dt = 0$ and

$$\frac{dy}{dt} = y\left(-y^2 + \frac{9}{4}\right).$$

The phase line for $dy/dt = y(-y^2 + 9/4)$ has equilibrium points at $y = 0$, $y = 3/2$ (and $y = -3/2$ but we are not considering $y < 0$). On the phase line, $y = 0$ is a source and $y = 3/2$ is a sink (see Figure 5.19).

The x- and y-nullclines divide the first quadrant into 5 regions labeled A, B, C, D, and E in Figure 5.18. A solution that enters in regions A, B, or C remains in that region as t increases since the vector field on the boundaries of these regions never points out of these regions. By labeling the regions with the direction of the vector field (such as "right-up" in region D), we see that solutions in A and B tend toward the equilibrium point $(0.65, 1.35)$, whereas those in region C tend toward $(2, 0)$. In Figure 5.19 we sketch the phase plane and $x(t)$- and $y(t)$-graphs for two of these solutions. Solutions in regions D and E either enter one of the regions A, B, or C or else they tend to one of the equilibrium points $(0.65, 1.35)$ or $(1.35, 0.65)$. Again we must have separatrix solutions that divide the solutions that tend to $(0.65, 1.35)$ from those that tend to $(2, 0)$. By linearizing at $(1.35, 0.65)$, we can confirm that it is a saddle.

Using qualitative analysis involving nullclines, we can conclude that all solutions in this model tend toward equilibrium points as t increases, as in the previous model. The choice of which equilibrium point the solution tends toward depends on the location of the initial condition. Unlike the previous example, there is a large set of solutions that tend to the equilibrium point $(0.65, 1.35)$, where the populations of both species are positive (mutual coexistence). The only solutions that approach the equilibrium $(0, 3/2)$ are those on the y-axis. On the other hand, there is a large set of initial

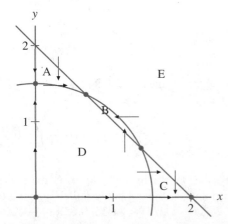

Figure 5.18
Nullclines for the system

$$\frac{dx}{dt} = 2x\left(1 - \frac{x}{2}\right) - xy$$
$$\frac{dy}{dt} = y\left(\frac{9}{4} - y^2\right) - x^2 y.$$

The nullclines separate the first-quadrant into five regions.

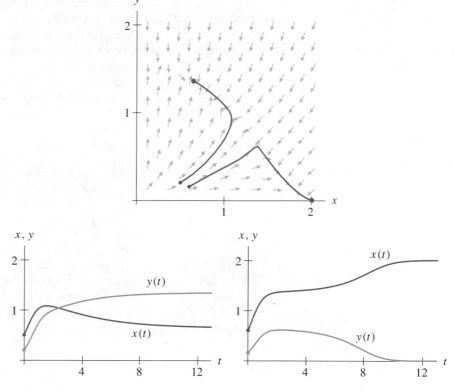

Figure 5.19
Phase plane for modified competing species system with $x(t)$- and $y(t)$-graphs for indicated solutions.

conditions that give solutions approaching the equilibrium point $(2, 0)$, where the y-species is extinct and the x-species is at its carrying capacity.

Using All Our Tools

Consider the system

$$\frac{dx}{dt} = x + y - x^3$$

$$\frac{dy}{dt} = -0.5x.$$

Systems of this form arise in the study of nerve cells. Roughly speaking, the variable $x(t)$ represents voltage across the boundary of a nerve cell at time t, and $y(t)$ represents the permeability of the cell wall at time t. A rapid change in x corresponds to the nerve cell "firing."

Information from the nullclines

The x-nullcline for this system is the curve $y = -x + x^3$. Above this curve the x-component of the vector field is positive, and below it the x-component is negative.

Hence, above the x-nullcline solutions move to the right, and below the x-nullcline solutions move to the left. The y-nullcline is the line $x = 0$, the y-axis. To the right of the y-nullcline we have $dy/dt < 0$, and to the left of the y-nullcline we have $dy/dt > 0$. Hence, in the right half of the phase plane solutions move down, and in the left half plane solutions move up (see Figure 5.20). The x- and y-nullclines divide the phase plane into four regions. Using the above qualitative analysis, we can conclude that all solutions must circulate clockwise around the origin — the only equilibrium point of the system (see Figure 5.21).

Information from linearization

The linearized system at the origin is

$$\frac{dx}{dt} = x + y$$
$$\frac{dy}{dt} = -0.5x,$$

which has eigenvalues $(1 \pm i)/2$. Hence the origin is a spiral source.

This analysis is applicable only near the origin. Since the term we dropped to obtain the linearization is $-x^3$, the linear approximation is no longer valid once this term has significant size. Once x^3 is comparable in size to x, it is not safe to use the linearization to study the nonlinear system.

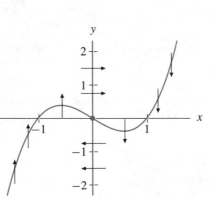

Figure 5.20
The x- and y-nullclines for the system

$$\frac{dx}{dt} = x + y - x^3$$
$$\frac{dy}{dt} = -0.5x.$$

The nullclines for this system separate the plane into four regions.

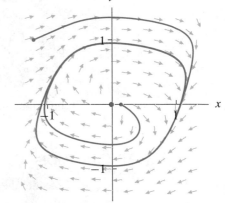

Figure 5.21
Solutions of the system

$$\frac{dx}{dt} = x + y - x^3$$
$$\frac{dy}{dt} = -0.5x.$$

Note that solutions with initial conditions close to the origin spiral outward, while those with initial conditions far from the origin spiral inward.

Information from numerical approximations of solutions

To get a more detailed idea of the behavior of solutions of this system, we use numerical methods to compute some approximate solutions. From the linearization at the origin, we know that initial conditions near the origin yield solutions that spiral outward. If we take an initial condition far from the origin, the numerics show that the resulting solution spirals inward.

From the Uniqueness Theorem we know that solution curves never cross. Hence the solutions that start near the origin must eventually stop spiraling outward, otherwise they would cross the inward spiraling solutions. Between the outward and inward spiraling solutions, there must be at least one solution that spirals neither outward nor inward. This solution is periodic. Numerical evidence indicates that there is only one periodic solution and that all other solutions (except the equilibrium solution at the origin) spiral toward this periodic solution. The $x(t)$- and $y(t)$-graphs of two solutions are shown in Figure 5.22. From this picture we see that both of these solutions converge toward the same type of periodic behavior.

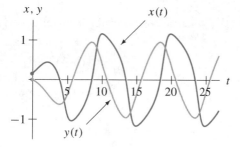

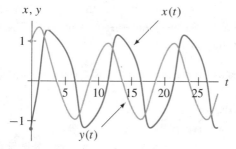

Figure 5.22
The $x(t)$- and $y(t)$-graphs for two solutions of the system

$$\frac{dx}{dt} = x + y - x^3$$
$$\frac{dy}{dt} = -0.5x.$$

The initial condition for the first solution is close to the origin, whereas the initial condition for the second solution lies outside the solution curve associated to the periodic solution. Note that both graphs indicate that these two solutions behave the same over the long term.

Mathematical Toolbox

In this section we have analyzed three first-order systems of differential equations, employing analytical, qualitative, and numerical techniques. As we have seen, the analysis of systems of differential equations is much more difficult than the analysis of differential equations with only one dependent variable. We need many different types of techniques to handle systems, and we must be willing to use whichever technique is appropriate. As with any craft, the ability to choose the appropriate tool for the problem at hand is a crucial skill.

EXERCISES FOR SECTION 5.2

In Exercises 1–3, sketch the x- and y-nullclines of the system specified. Then find all equilibrium points. Using the direction of the vector field between the nullclines, describe the possible fate of the solution curves corresponding to the initial conditions (a), (b), and (c).

1. $\dfrac{dx}{dt} = 2 - x - y$

$\dfrac{dy}{dt} = y - x^2$

(a) $x_0 = 2$, $y_0 = 1$

(b) $x_0 = 0$, $y_0 = -1$

(c) $x_0 = 0$, $y_0 = 0$

2. $\dfrac{dx}{dt} = 2 - x - y$

$\dfrac{dy}{dt} = y - |x|$

(a) $x_0 = -1$, $y_0 = 1$

(b) $x_0 = 2$, $y_0 = 1$

(c) $x_0 = 2$, $y_0 = 2$

3. $\dfrac{dx}{dt} = x(x - 1)$

$\dfrac{dy}{dt} = x^2 - y$

(a) $x_0 = -1$, $y_0 = 0$

(b) $x_0 = 0.8$, $y_0 = 0$

(c) $x_0 = 1$, $y_0 = 3$

4. Consider the Volterra-Lotka system, a model of two competitive species,

$$\frac{dx}{dt} = x(-Ax - By + C)$$

$$\frac{dy}{dt} = y(-Dx - Ey + F)$$

where x, $y \geq 0$ and where the parameters A–F are all positive.

(a) Can there ever be more than one equilibrium point for which both $x > 0$ and $y > 0$ (that is, where the species "coexist in equilibrium")? If so, give an example of such values of A through F. If not, why not?

(b) What condition on parameters A through F guarantees that there is at least one equilibrium point with $x > 0$ and $y > 0$?

In Exercises 5–14, we restrict attention to the first quadrant (x, $y \geq 0$). For each system,

(a) sketch the nullclines,

(b) sketch the phase plane, and

(c) write a brief paragraph describing the possible behaviors of solutions. [*Hint*: You may wish to use information obtained in Exercises 7–16 of Section 5.1.]

5. $\dfrac{dx}{dt} = x(-x - 3y + 150)$

$\dfrac{dy}{dt} = y(-2x - y + 100)$

6. $\dfrac{dx}{dt} = x(10 - x - y)$

$\dfrac{dy}{dt} = y(30 - 2x - y)$

7. $\dfrac{dx}{dt} = x(100 - x - 2y)$

$\dfrac{dy}{dt} = y(150 - x - 6y)$

8. $\dfrac{dx}{dt} = x(-x - y + 100)$

$\dfrac{dy}{dt} = y(-x^2 - y^2 + 2500)$

9.
$$\frac{dx}{dt} = x(-x - y + 40)$$
$$\frac{dy}{dt} = y(-x^2 - y^2 + 2500)$$

10.
$$\frac{dx}{dt} = x(-4x - y + 160)$$
$$\frac{dy}{dt} = y(-x^2 - y^2 + 2500)$$

11.
$$\frac{dx}{dt} = x(-8x - 6y + 480)$$
$$\frac{dy}{dt} = y(-x^2 - y^2 + 2500)$$

12.
$$\frac{dx}{dt} = x(2 - x - y)$$
$$\frac{dy}{dt} = y(y - x^2)$$

13.
$$\frac{dx}{dt} = x(2 - x - y)$$
$$\frac{dy}{dt} = y(y - x)$$

14.
$$\frac{dx}{dt} = x(x - 1)$$
$$\frac{dy}{dt} = y(x^2 - y)$$

15. The **Volterra-Lotka system** of differential equations for competitive species is

$$\frac{dx}{dt} = x(-Ax - By + C)$$
$$\frac{dy}{dt} = y(-Dx - Ey + F),$$

where $x, y \geq 0$ and the parameters A–F are all nonnegative.

Some species live in a "cooperative" manner — each species helping the other to survive and prosper (for example, flowers and honey bees).

(a) How would you alter the Volterra-Lotka system described above to give a general form for a set of systems modeling cooperative species?

(b) What do the nullclines look like for your cooperative system? Are there equilibrium points? Are there conditions on the parameters that guarantee that there are equilibrium points with both x and y positive?

Exercises 16–20 refer to the models of chemical reactions created in Exercises 25–30 of Section 2.1. Here $a(t)$ is the amount of substance A in a solution, and $b(t)$ is the amount of substance B in a solution at time t. We only need to consider nonnegative $a(t)$ and $b(t)$.

For each system,

(a) sketch the nullclines and draw in the vector field along the nullcline,

(b) label the regions in the ab-phase plane created by the nullclines and determine which general direction the vector field points in each region (that is, increasing or decreasing a, increasing or decreasing b), and

(c) identify the regions that solutions cannot leave and determine the fate of solutions in these regions as time increases.

16.
$$\frac{da}{dt} = -\frac{ab}{2}$$
$$\frac{db}{dt} = -\frac{ab}{2}$$

17.
$$\frac{da}{dt} = 2 - \frac{ab}{2}$$
$$\frac{db}{dt} = \frac{3}{2} - \frac{ab}{2}$$

18.
$$\frac{da}{dt} = 2 - \frac{ab}{2} - \frac{a^2}{3}$$
$$\frac{db}{dt} = \frac{3}{2} - \frac{ab}{2}$$

19.
$$\frac{da}{dt} = 2 - \frac{ab}{2} + \frac{b^2}{3}$$
$$\frac{db}{dt} = \frac{3}{2} - \frac{ab}{2} - \frac{b^2}{3}$$

20.
$$\frac{da}{dt} = 2 - \frac{ab}{2} - \frac{2ab^2}{3}$$
$$\frac{db}{dt} = \frac{3}{2} - \frac{ab}{2} - \frac{ab^2}{3}$$

Exercises 21–23 refer to the system

$$\frac{dx}{dt} = y$$
$$\frac{dy}{dt} = x - x^2.$$

21. Sketch the nullclines and find the direction of the vector field along the nullclines.

22. Show that there is at least one solution in each of the second and fourth quadrants that tends to the origin as $t \to \infty$. Similarly, show that there is at least one solution in each of the first and third quadrants that tends to the origin as $t \to -\infty$.

23. Find the linearized system near the equilibrium points $(0, 0)$ and $(1, 0)$. Use information obtained from the linearized systems and Exercises 21 and 22 to describe the phase plane. What is it that you still do not know about the phase plane?

5.3 HAMILTONIAN SYSTEMS

As we have emphasized many times, nonlinear systems of differential equations are almost impossible to solve explicitly. We have also seen that solution curves of systems may behave in many different ways and that there are no qualitative techniques that are guaranteed to work in all cases. Fortunately there are certain special types of nonlinear systems that arise often in practice and for which there are special techniques that enable us to gain some understanding of the phase plane. In this and the next section we will discuss two of these special types of nonlinear systems. But first, we pause for a story.

How This Book Came to Be

Paul and Glen have been writing a differential equations textbook for the past ten years. They want their book to be filled with brilliant and witty new ideas about differential equations, but they are finding that new ideas are hard to come by.

More troubling to them is the following observation that they have both made over the years. Whenever Paul comes to the office in the morning with a witty new idea, they both work feverishly on his idea, and at first their creative juices flow. They work diligently, but eventually the idea doesn't pan out. So their enthusiasm wanes and with it, their creativity. On the other hand, whenever Glen comes in the morning with a new idea, no matter how trivial, something different happens. They again both work feverishly. Sometimes Paul's enthusiasm wanes, but Glen is always excited. There is good give and take; they never give up. At least something comes of the idea and the book progresses. Now this bothered Paul and Glen. Why should their creative energy depend so critically on who has the first idea? They decided to model their plight with a system of differential equations.

They let $x(t)$ denote Glen's and $y(t)$ Paul's level of enthusiasm at time t. Now $x(t)$ and $y(t)$ are difficult to measure because of a lack of standardized units of enthusiasm. However, it is clear that $y(t) > 0$ means that Paul is enthusiastic, whereas $y(t) < 0$ means that Paul is glum. When $y(t) = 0$, Paul just sits there, neither happy nor sad, and similarly for Glen.

Both Glen and Paul have observed that Glen's enthusiasm changes at a rate directly proportional to Paul's level of enthusiasm. When Paul is excited, Glen gets more enthusiastic, but when Paul is glum, Glen loses his verve. A simple equation catching this behavior is

$$\frac{dx}{dt} = y.$$

Paul is a bit more difficult to categorize. When Glen is mildly enthusiastic, Paul's enthusiasm goes up. But when Glen gets wildly excited, Paul starts to lose enthusiasm. Apparently, when the torrent of ideas spilling out of Glen becomes too great, Paul gets a headache and tunes Glen out. On the other hand, when Glen is down, Paul is really glum. For the rate of change of Paul's enthusiasm then, use

$$\frac{dy}{dt} = x - x^2.$$

The graph of $dy/dt = x - x^2$ as a function of x captures exactly Paul's enthusiasm level (see Figure 5.23). We can see that $dy/dt > 0$ if $0 < x < 1$, but $dy/dt < 0$ otherwise.

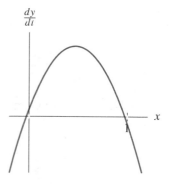

Figure 5.23
The graph of $dy/dt = x - x^2$
as a function of x.

The system of differential equations they settle on is

$$\frac{dx}{dt} = y$$

$$\frac{dy}{dt} = x - x^2.$$

The direction field for this system is shown in Figure 5.24. There are two equilibrium points, one at the origin and one at $x = 1$, $y = 0$. The technique of linearization from Section 5.1 can be used to study the solutions near the equilibrium points. At the origin the Jacobian matrix is

$$\begin{pmatrix} 0 & 1 \\ 1 & 0 \end{pmatrix},$$

which has eigenvalues ± 1. Hence the origin is a saddle.

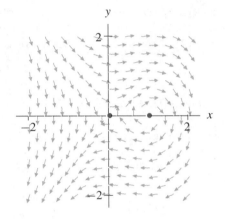

Figure 5.24
The direction field for

$$\frac{dx}{dt} = y$$
$$\frac{dy}{dt} = x - x^2.$$

The system has two equilibrium points, and by linearizing we know that the equilibrium point at the origin is a saddle.

The equilibrium point at $(x, y) = (1, 0)$ has Jacobian matrix

$$\begin{pmatrix} 0 & 1 \\ -1 & 0 \end{pmatrix},$$

which has eigenvalues $\pm i$. The linearized system is a center. As we saw in Section 5.1, this is one of the cases when the long-term behavior of solutions of the nonlinear system near the equilibrium point is not completely determined by the linearization. The behavior of the nonlinear system near the equilibrium point $(1, 0)$ could be that of a spiral sink, a spiral source, or a center.

Numerical approximations of solutions give the phase plane shown in Figure 5.25, which implies that solutions behave in a very regular way. The solutions with initial conditions near $(1, 0)$ seem to form closed loops corresponding to periodic solutions. Also, the unstable and stable separatrices emerging from the saddle at the origin into the first and fourth quadrants seem to form a single loop.

The qualitative techniques we have studied so far do not allow us to predict that this system will have such regular solution curves. Also, because the numerical approximations of solutions are only approximations, we should be cautious. Distinguishing solution curves that form closed loops from those that spiral very slowly can be very difficult. We would like to have techniques that can be used to verify the special behavior of this system.

Conserved quantities

Paul and Glen were so mystified by the behavior of their system that they decided to show it to their friend Bob. Bob exclaimed that he had seen this system before and that it had a conserved quantity. Noting the confused looks he got from Paul and Glen, he explained:

DEFINITION   A real-valued function $H(x, y)$ of the two variables x and y is a **conserved quantity** for a system of differential equations if H is constant along all solution curves of the system. That is, if $(x(t), y(t))$ is a solution of the system, then $H(x(t), y(t))$ is constant. In other words,

$$\frac{d}{dt} H(x(t), y(t)) = 0. \quad \blacksquare$$

Bob remembered that

$$H(x, y) = \frac{1}{2} y^2 - \frac{1}{2} x^2 + \frac{1}{3} x^3$$

is a conserved quantity for the system in question. To check this, suppose $(x(t), y(t))$ is a solution of the system. Then we compute

$$\frac{d}{dt} H(x(t), y(t)) = \frac{\partial H}{\partial x} \cdot \frac{dx}{dt} + \frac{\partial H}{\partial y} \cdot \frac{dy}{dt}$$
$$= (-x + x^2) \cdot y + y \cdot (x - x^2)$$
$$= 0,$$

where the first equality follows from the Chain Rule, and the second equality uses the fact that $(x(t), y(t))$ is a solution to replace dx/dt with y and dy/dt with $x - x^2$.

This means that the solution curves always lie along the level curves of H. The level curves of H are shown in Figure 5.26. We see in this picture that around the equilibrium point $(1, 0)$ the level curves of H form closed circles, and the branches of

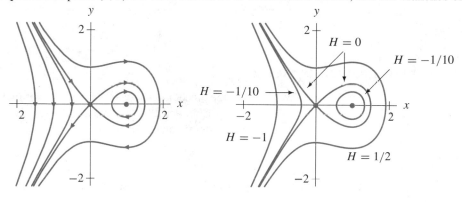

Figure 5.25
The phase plane of the system

$$\frac{dx}{dt} = y$$
$$\frac{dy}{dt} = x - x^2.$$

Figure 5.26
The level curves of H. Compare the solution curves shown in Figure 5.25 to these level curves.

the level curve emanating from the origin toward the right connect into a loop. This agrees with our analysis of the phase plane for the system (see Figure 5.25).

The fact that the right-hand branches of the stable and unstable separatrices of the origin form a single loop is very special. This type of solution curve is a **saddle connection**. Inside the saddle connection, all solution curves are periodic, whereas outside the saddle connection, all solutions have the property that both $x(t)$ and $y(t)$ tend to $-\infty$ as $t \to \infty$. This explains, to Paul and Glen's great relief, why their daily productivity depends so crucially on who has the first idea. If $y(0) > 0$ but $x(0) = 0$, the solution curve eventually tends to $x = y = -\infty$. But if $x(0) > 0$, $y(0) = 0$ [with $x(0)$ not too large], then both $x(t)$ and $y(t)$ are periodic, with $x(t) > 0$ for all t. We sketch graphs of $x(t)$ and $y(t)$ for initial conditions near $(0, 0)$ in Figure 5.27.

This was a source of great satisfaction to both Paul and Glen. So much was their happiness that they invited Bob to become a coauthor of their book. Bob reluctantly agreed, but only after Glen and Paul promised to listen to *The Marriage of Figaro* in its entirety.

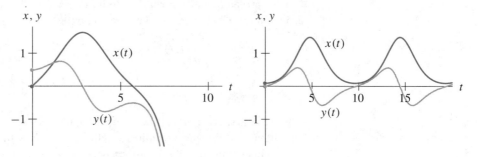

Figure 5.27
Graphs of $x(t)$ and $y(t)$ for solutions with initial conditions $y(0) > 0$, $x(0) = 0$ and $y(0) = 0$, $0 < x(0) < 1$.

Hamiltonian Systems

What makes the preceding analysis work is the fact that the system of differential equations modeling Glen and Paul's enthusiasm levels is a special kind of system called a Hamiltonian system (named for William Rowan Hamilton [1805–1865], an Irish mathematician).

DEFINITION   A system of differential equations is called a **Hamiltonian system** if there exists a real-valued function $H(x, y)$ such that

$$\frac{dx}{dt} = \frac{\partial H}{\partial y}$$
$$\frac{dy}{dt} = -\frac{\partial H}{\partial x}$$

for all x and y. The function H is called the *Hamiltonian function* for the system.   ∎

Note that H is always a conserved quantity for such a system. We can verify this by letting $(x(t), y(t))$ be any solution of the system. Then

$$\frac{d}{dt} H(x(t), y(t)) = \frac{\partial H}{\partial x} \cdot \frac{dx}{dt} + \frac{\partial H}{\partial y} \cdot \frac{dy}{dt}$$

$$= \frac{\partial H}{\partial x} \left(\frac{\partial H}{\partial y} \right) + \frac{\partial H}{\partial y} \left(-\frac{\partial H}{\partial x} \right)$$

$$= 0.$$

The first equality is the Chain Rule, and the second equality uses the fact that the system is Hamiltonian and that $(x(t), y(t))$ is a solution to replace dx/dt with $\partial H/\partial y$ and dy/dt with $-\partial H/\partial x$.

So, as above, solution curves of the system lie along the level curves of H. Sketching the phase plane for a Hamiltonian system is the same as sketching the level sets of the Hamiltonian function.

Examples of Hamiltonian Systems: The Harmonic Oscillator

Recall that the undamped harmonic oscillator system is

$$\frac{dy}{dt} = v$$

$$\frac{dv}{dt} = -qy,$$

where q is a positive constant. If we let

$$H(y, v) = \frac{1}{2} v^2 + \frac{q}{2} y^2,$$

then

$$\frac{dy}{dt} = v = \frac{\partial H}{\partial v}$$

and

$$\frac{dv}{dt} = -qy = -\frac{\partial H}{\partial y}.$$

Hence the undamped harmonic oscillator system is a Hamiltonian system. The level sets of the function H are ellipses in the yv-plane that correspond to the phase-plane picture of the solutions of the undamped harmonic oscillator (see Section 3.4). This Hamiltonian function is sometimes called the **energy function** for the oscillator.

The Nonlinear Pendulum

Consider a pendulum made of a light rigid rod of length l with a ball at one end of mass m. The ball is called the *bob*, and the rigid rod is called the *arm* of the pendulum. We assume that all the mass of the pendulum is in the bob, neglecting the mass of the rod. The other end of the rigid rod is attached to the wall in such a way that it can turn through an entire circle in a plane perpendicular to the ground. The position of the bob

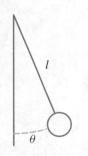

Figure 5.28
A pendulum with
rod length l and
angle θ.

at time t is given by an angle $\theta(t)$, which we choose to measure in the counterclockwise direction with $\theta = 0$ corresponding to the downward vertical axis (see Figure 5.28).

We assume that there are only two forces acting on the pendulum: gravity and friction. The constant gravitational force equal to mg is in the downward direction, where g is the acceleration of gravity near earth ($g \approx 9.8$ m/s^2). Only the component of this force tangent to the circle of motion affects the motion of the pendulum. This component is $-mg\sin\theta$ (see Figure 5.29). There is also a force due to friction, which we assume to be proportional to the velocity of the bob.

The position of the bob at time t is given by the point $(l\sin\theta(t), -l\cos\theta(t))$ on the circle of radius l (remember that $\theta = 0$ corresponds to straight down). The speed of the bob is the length of the velocity vector, which is $l\,d\theta/dt$. The component of the acceleration that points along the direction of the motion of the bob has length $l\,d^2\theta/dt^2$. We take the force due to friction to be proportional to the velocity, so this force is $-b(l\,d\theta/dt)$, where $b > 0$ is a parameter that corresponds to the *coefficient of damping*.

Putting all this together with Newton's law "force equals mass times acceleration," we find the equations of motion,

$$ml\frac{d^2\theta}{dt^2} = -bl\frac{d\theta}{dt} - mg\sin\theta$$

or

$$\frac{d^2\theta}{dt^2} + \frac{b}{m}\frac{d\theta}{dt} + \frac{g}{l}\sin\theta = 0.$$

We can write this as a nonlinear system in the usual manner by letting $v = d\theta/dt$ be the velocity. Then the equations become

$$\frac{d\theta}{dt} = v$$

$$\frac{dv}{dt} = -\frac{b}{m}v - \frac{g}{l}\sin\theta.$$

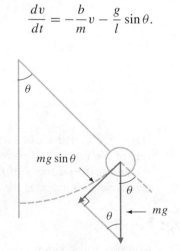

Figure 5.29
Decomposition of the force of gravity into components
along the pendulum arm and tangent to the circle of
motion of the pendulum bob.

Remember that θ is an angular variable, so that $\theta = 0$, $\theta = 2\pi$, $\theta = 4\pi$, and so forth all measure the same position: the pendulum passing through its lowest point. The values $\theta = \pi$, 3π, 5π, ... occur when the pendulum is at the highest point of its circular motion.

The ideal pendulum

In practice there is always a force due to friction acting on a pendulum. For the moment, however, suppose there is no friction in this model. This is an "ideal" case that does not occur in the real world. However, it is not an unreasonable model for a very well-built and well-lubricated pendulum.

When no friction is present, the coefficient b vanishes. For convenience we suppose that the pendulum arm has unit length; that is, we suppose $l = 1$. (We consider the effect of adjusting the length of the arm in the exercises.) The equations of motion of the ideal pendulum with unit length arm are

$$\frac{d\theta}{dt} = v$$
$$\frac{dv}{dt} = -g \sin \theta.$$

The first step is to find the equilibrium points. We must have $v = 0$ from the first equation and $\sin \theta = 0$ from the second, so that $v = 0$, $\theta = n\pi$, where n is any integer, yield all the equilibrium points. When θ is an even multiple of π and $v = 0$, the bob hangs motionless in a downward position, an obvious equilibrium position for the pendulum. When θ is an odd multiple of π and $v = 0$, there is also a rest position corresponding to the bob balanced perfectly motionless in an upright position. This kind of equilibrium is hard to see in practice: Just when you manage to balance the pendulum perfectly, someone twitches, and the resulting current of air moves the pendulum out of equilibrium and into motion in one direction or the other.

We give a method below for determining whether a given system is a Hamiltonian system and, if so, how to compute the Hamiltonian. For the moment we use the rabbit-out-of-the-hat method and claim that the differential equations for the ideal pendulum form a Hamiltonian system, with the Hamiltonian function given by

$$H(\theta, v) = \frac{1}{2}v^2 - g \cos \theta.$$

This follows from the computation of partial derivatives of H,

$$\frac{\partial H}{\partial \theta} = g \sin \theta = -\frac{dv}{dt}, \quad \frac{\partial H}{\partial v} = v = \frac{d\theta}{dt}.$$

We can describe the graph of the function H as follows: For each fixed θ, the $v^2/2$ term implies that the graph of H is a parabola in the v direction. The critical points occur at $v = 0$, $g \sin \theta = 0$ or $\theta = 0, \pm\pi, \pm2\pi, \ldots$. The points $v = 0$, $\theta = 0$, $\pm2\pi, \pm4\pi, \ldots$ are local minima, and the points $v = 0$, $\theta = \pm\pi, \pm3\pi, \ldots$ are saddle points of the graph.

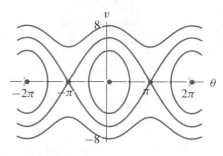

Figure 5.30
Level curves for the ideal pendulum.

The level curves of H are plotted in Figure 5.30. We know that the vector field is always tangent to these curves. The θ component of the vector field equals v, so the vector field points to the right when $v > 0$ and to the left when $v < 0$. Using this fact we can assign directions along these curves and thereby determine the phase portrait. This is shown in Figure 5.31.

There are three different types of solution curves present in the phase-plane picture. These three types are shown schematically in Figure 5.32 and labeled A, B, and C. Around the equilibrium points at $v = 0$, $\theta = 0$, $\pm 2\pi$, $\pm 4\pi$, ..., we find periodic solutions traversed in the clockwise direction. These are the type A solutions in Figure 5.32. Since the θ-value along these solution curves never reaches π, 3π, ..., it follows that the pendulum never passes through the upright position. Consequently, the pendulum simply oscillates back and forth periodically with the maximum and minimum θ-values determined by where the solution curve crosses the θ-axis. This is the usual swinging motion we associate with a pendulum. The graph of $\theta(t)$ for such a solution is shown in Figure 5.33.

On the other hand, a solution curve such as B in Figure 5.32 corresponds to the pendulum rotating forever in a counterclockwise direction. Note that $v \neq 0$ along such a solution curve, so the pendulum never reaches a point where its velocity is 0. The graph of $\theta(t)$ for such a solution is increasing for all t if $v > 0$ and decreasing for all t if $v < 0$ (see Figure 5.33).

The intermediate types of solutions (see C in Figure 5.32) are separatrices of sad-

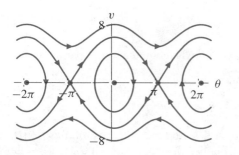

Figure 5.31
Phase plane for the ideal pendulum.

Figure 5.32
Special solutions for the pendulum equations.

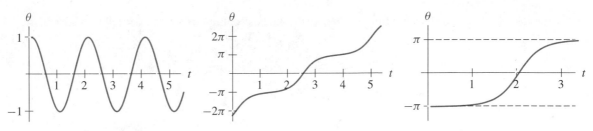

Figure 5.33
Graphs of $\theta(t)$ for the three different types of solutions represented in Figure 5.32.

dle equilibrium points forming saddle connections. These solutions tend toward and come from the upright equilibrium position as time tends to $\pm\infty$. The $\theta(t)$-graph of such a solution is shown in Figure 5.33. In order to be on such a solution, we must choose the initial angle and velocity perfectly. With slightly too high an initial velocity, the pendulum swings past the vertical position; with slightly to low an initial velocity, the pendulum falls back. The separatrices of the saddle equilibrium points separate the "oscillating" solutions, which oscillate back and forth, from the "rotating" solutions, which swing around and around.

The value of the Hamiltonian function for the ideal pendulum at a particular point (y, v) is called the **energy**. The physical principle of conservation of energy applies to the ideal pendulum in the same way that it does to the undamped harmonic oscillator. This is one way we could predict that the undamped harmonic oscillator and the ideal pendulum are Hamiltonian systems. A more mathematical approach is given next.

Finding Hamiltonian Systems

Hamiltonian systems are special kinds of systems of differential equations in several senses. As we have seen, we can "solve" a Hamiltonian system in the plane in a qualitative sense once we know the Hamiltonian function. All we need to do is plot the level curves of H and sketch in the directions to determine the phase portrait. Unfortunately Hamiltonian systems are fairly rare. Given a system, we would like to be able to determine whether or not it is a Hamiltonian system and, if it is, to determine the Hamiltonian function.

Suppose we have a system of equations

$$\frac{dx}{dt} = f(x, y)$$
$$\frac{dy}{dt} = g(x, y)$$

and wish to check whether it is Hamiltonian. We ask whether there exists a function $H(x, y)$ such that for all (x, y),

$$f(x, y) = \frac{\partial H}{\partial y}(x, y)$$

and

$$g(x, y) = -\frac{\partial H}{\partial x}(x, y).$$

If such a function H exists (and has continuous second partial derivatives), then

$$\frac{\partial^2 H}{\partial x \partial y} = \frac{\partial^2 H}{\partial y \partial x}.$$

Therefore if the system is Hamiltonian, then

$$\frac{\partial f}{\partial x} = \frac{\partial^2 H}{\partial x \partial y} = \frac{\partial^2 H}{\partial y \partial x} = -\frac{\partial g}{\partial y}.$$

That is, to check whether a system may be Hamiltonian, we compute $\partial f/\partial x$ and $\partial g/\partial y$ and check if $\partial f/\partial x = -\partial g/\partial y$. If this equation does not hold for all (x, y), then the system is not Hamiltonian.

For example, consider the nonlinear system

$$\frac{dx}{dt} = f(x, y) = x + y^2$$
$$\frac{dy}{dt} = g(x, y) = y^2 - x.$$

We compute

$$\frac{\partial f}{\partial x} = 1 \neq -2y = -\frac{\partial g}{\partial y}.$$

Therefore this system is not a Hamiltonian system.

Constructing Hamiltonian functions

For the system

$$\frac{dx}{dt} = f(x, y)$$
$$\frac{dy}{dt} = g(x, y),$$

if

$$\frac{\partial f}{\partial x} = -\frac{\partial g}{\partial y},$$

then the system is Hamiltonian. We can verify this by actually constructing the Hamiltonian function as follows: If we are to have

$$f(x, y) = \frac{\partial H}{\partial y}(x, y),$$

then integrating both sides of this equation with respect to y, we must have

$$H(x, y) = \int f(x, y) \, dy$$

up to a "constant of integration" that may depend on x. That is, we can write

$$H(x, y) = \int f(x, y) \, dy + \phi(x),$$

where ϕ is some function of x alone that is to be determined. To find ϕ, we simply differentiate $H(x, y)$ with respect to x and equate the result with $-g(x, y)$. We find

$$\frac{\partial H}{\partial x} = \frac{\partial}{\partial x} \int f(x, y)\, dy + \phi'(x) = -g(x, y).$$

That is,

$$\phi'(x) = -g(x, y) - \frac{\partial}{\partial x} \int f(x, y)\, dy.$$

Integration of the right-hand side with respect to x then determines ϕ and consequently the Hamiltonian H.

Two examples

Recall the system of differential equations governing Glen and Paul's book-writing dilemma:

$$\frac{dx}{dt} = f(x, y) = y$$
$$\frac{dy}{dt} = g(x, y) = x - x^2.$$

Since

$$\frac{\partial f}{\partial x} = 0 = -\frac{\partial g}{\partial y},$$

we know this system is Hamiltonian. To find H we first integrate f with respect to y, finding

$$H(x, y) = \int y\, dy = \frac{1}{2}y^2 + \phi(x).$$

Next we must have

$$x - x^2 = -\frac{\partial H}{\partial x}$$

$$= -\frac{\partial}{\partial x}\left(\frac{1}{2}y^2 + \phi(x)\right)$$

$$= -\phi'(x).$$

Integrating $\phi'(x) = -x + x^2$ with respect to the variable x, we find

$$\phi(x) = -\frac{1}{2}x^2 + \frac{1}{3}x^3,$$

so

$$H(x, y) = \frac{1}{2}y^2 - \frac{1}{2}x^2 + \frac{1}{3}x^3.$$

This is exactly the function we found earlier. Of course, we could add an arbitrary constant to H, but this would not change the shape of the level curves of H.

As another example, consider the system

$$\frac{dx}{dt} = f(x, y) = -x \sin y + 2y$$

$$\frac{dy}{dt} = g(x, y) = -\cos y.$$

This system is Hamiltonian since

$$\frac{\partial f}{\partial x} = -\sin y = -\frac{\partial g}{\partial y}$$

for all (x, y). So we first integrate f with respect to y, finding

$$H(x, y) = x \cos y + y^2 + \phi(x).$$

Then we must have

$$-\cos y = -\frac{\partial H}{\partial x}$$

$$= -\frac{\partial}{\partial x} \left(x \cos y + y^2 + \phi(x) \right)$$

$$= -\cos y - \phi'(x).$$

Therefore we can choose ϕ to be any constant, say 0, and our Hamiltonian function is

$$H(x, y) = x \cos y + y^2.$$

We study this system more closely in the exercises

Equilibrium Points of Hamiltonian Systems

Hamiltonian systems have a number of special properties not shared by general systems of differential equations. For example, suppose (x_0, y_0) is our equilibrium point for the Hamiltonian system

$$\frac{dx}{dt} = \frac{\partial H}{\partial y}(x, y)$$

$$\frac{dy}{dt} = -\frac{\partial H}{\partial x}(x, y).$$

The Jacobian matrix at this equilibrium point is given by

$$\begin{pmatrix} \dfrac{\partial^2 H}{\partial x \partial y} & \dfrac{\partial^2 H}{\partial y^2} \\ -\dfrac{\partial^2 H}{\partial x^2} & -\dfrac{\partial^2 H}{\partial y \partial x} \end{pmatrix},$$

where each of these partial derivatives is evaluated at (x_0, y_0). Since

$$\frac{\partial^2 H}{\partial x \partial y} = \frac{\partial^2 H}{\partial y \partial x},$$

Zhihong Xia (1962 –   ) received his bachelor's degree in astronomy from Nanjing University, China, and his doctorate in mathematics from Northwestern University. In his doctoral thesis, he combined his mathematical skills and astronomical background to settle a 100-year-old question in celestial mechanics. Using a variety of analytic and qualitative techniques involving Hamiltonian systems, he showed that certain solutions of the nonlinear system of differential equations known as the n-body problem could escape to infinity in finite time.

In 1993, Xia received the prestigious Blumenthal Award for the research in his doctoral thesis. He has taught at Harvard University and Georgia Institute of Technology. He is currently Professor of Mathematics at Northwestern University.

the Jacobian matrix assumes the form

$$\begin{pmatrix} \alpha & \beta \\ \gamma & -\alpha \end{pmatrix},$$

where $\alpha = \partial^2 H/\partial x\,\partial y$, $\beta = \partial^2 H/\partial y^2$, and $\gamma = -\partial^2 H/\partial x^2$.

The characteristic polynomial of this matrix is

$$(\alpha - \lambda)(-\alpha - \lambda) - \beta\gamma = \lambda^2 - \alpha^2 - \beta\gamma.$$

This means that the eigenvalues are given by the roots of

$$\lambda^2 = \alpha^2 + \beta\gamma,$$

which are

$$\lambda = \pm\sqrt{\alpha^2 + \beta\gamma}.$$

Thus we see that there are only three possibilities for the eigenvalues:

1. If $\alpha^2 + \beta\gamma > 0$, both eigenvalues are real and have opposite signs.

2. If $\alpha^2 + \beta\gamma < 0$, both eigenvalues are imaginary with real part equal to zero.

3. If $\alpha^2 + \beta\gamma = 0$, then 0 is the only eigenvalue.

In particular it is not possible to have eigenvalues that are complex and have nonzero real parts for a Hamiltonian system in the plane. So solutions cannot spiral into or away from an equilibrium point. Similarly, we cannot have two positive or two negative eigenvalues. Thus Hamiltonian systems cannot have equilibrium points that are sinks or sources. This gives us another preliminary indication whether a given system is Hamiltonian: If the direction field indicates the presence of sinks or sources, it is not Hamiltonian. Two phase planes are pictured in Figure 5.34. We can say for sure that the phase plane on the left does not come from a Hamiltonian system because it has an equilibrium point that is a sink. The phase plane on the right might be that of a Hamiltonian system. The only way to determine whether it is Hamiltonian is to study the formulas for the system as we have done above.

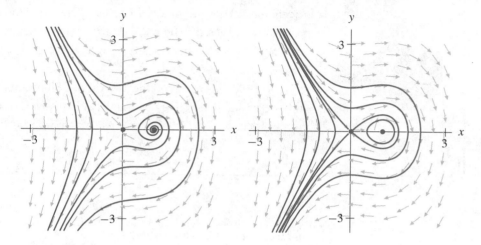

Figure 5.34
The phase plane on the left cannot be a Hamiltonian system, whereas the phase plane on the right might be Hamiltonian.

EXERCISES FOR SECTION 5.3

1. For the system

$$
\frac{dx}{dt} = y
$$
$$
\frac{dy}{dt} = x^3 - x,
$$

(a) show that the system is Hamiltonian with Hamiltonian function

$$
H(x, y) = \frac{y^2}{2} + \frac{x^2}{2} - \frac{x^4}{4},
$$

(b) sketch the level sets of H, and

(c) sketch the phase plane for the system. Include a description of all equilibrium points and any saddle connections.

2. For the system

$$
\frac{dx}{dt} = x \cos(xy)
$$
$$
\frac{dy}{dt} = -y \cos(xy),
$$

(a) show that the system is a Hamiltonian system with Hamiltonian function

$$H(x, y) = \sin(xy),$$

(b) sketch the level sets of H, and

(c) sketch the phase plane of the system. Include a description of all equilibrium points and any saddle connections.

3. For the system

$$\frac{dx}{dt} = -x \sin y + 2y$$

$$\frac{dy}{dt} = -\cos y,$$

(a) check that the system is a Hamiltonian system with Hamiltonian function

$$H(x, y) = x \cos y + y^2,$$

(b) sketch the level sets of H, and

(c) sketch the phase plane of the system. Include a description of all equilibrium points and any saddle connections.

[*Hint*: Take advantage of whatever technology you have available for drawing level sets and phase planes. Then interpret the pictures you obtain.]

In Exercises 4–8, we continue the study of the ideal pendulum system with bob mass m and arm length l given by

$$\frac{d\theta}{dt} = v$$

$$\frac{dv}{dt} = -\frac{g}{l} \sin \theta.$$

4. (a) What is the linearization of the ideal pendulum system above at the equilibrium point $(0, 0)$?

(b) Using $g = 9.8$ meters/sec^2, how should l and m be chosen so that small swings of the pendulum have period 1 second?

5. For the linearization of the ideal pendulum above at $(0, 0)$, the period of the oscillation is independent of the amplitude. Does the same statement hold for the ideal pendulum itself? Is the period of oscillation the same no matter how high the ideal pendulum swings? If not, will the period be shorter or longer for high swings?

6. An ideal pendulum clock — a clock containing an ideal pendulum that "ticks" once for each swing of the pendulum arm — keeps perfect time when the pendulum makes very high swings. Will the clock run slow or fast if the amplitude of the swings is very small?

7. (a) If the arm length of the ideal pendulum is doubled from l to $2l$, what is the effect on the period of small amplitude swinging solutions?

(b) What is the rate of change of the period of small amplitude swings as l is varied?

8. Will an ideal pendulum clock that keeps perfect time on earth run slow or fast on the moon?

In Exercises 9–12, determine whether the given system is Hamiltonian. If so, find a Hamiltonian function.

9. $\dfrac{dx}{dt} = x - 3y^2$

$\dfrac{dy}{dt} = -y$

10. $\dfrac{dx}{dt} = \sin x \cos y$

$\dfrac{dy}{dt} = -\cos x \sin y$

11. $\dfrac{dx}{dt} = x \cos y$

$\dfrac{dy}{dt} = -y \cos x$

12. $\dfrac{dx}{dt} = 1$

$\dfrac{dy}{dt} = y$

13. Which of the following phase planes could possibly correspond to that of a Hamiltonian system? If not, explain why.

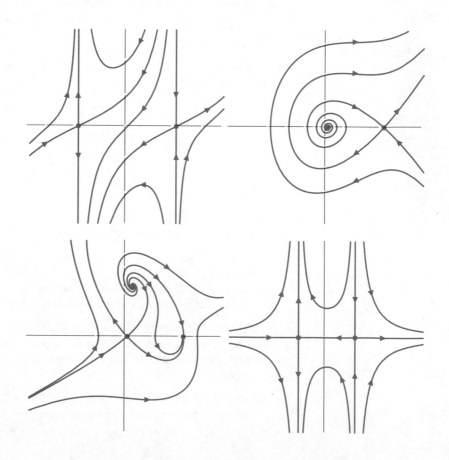

14. Consider the system of differential equations of the form

$$\frac{dx}{dt} = F(y)$$

$$\frac{dy}{dt} = G(x).$$

That is, dx/dt depends only on y, and dy/dt depends only on x. Show that this system is Hamiltonian. What is the Hamiltonian function?

15. Consider the nonlinear system

$$\frac{dx}{dt} = 1 - y^2$$

$$\frac{dy}{dt} = x(1 + y^2).$$

(a) Is this a Hamiltonian system?

(b) Use a computer to plot the phase portrait of this system. Do you think that there is conserved quantity for this system? Can you find one?

(c) Suppose we multiply this vector field by the function

$$f(x, y) = \frac{1}{1 + y^2}.$$

The new vector field is

$$\frac{dx}{dt} = \frac{1 - y^2}{1 + y^2}$$

$$\frac{dy}{dt} = x.$$

Is this system Hamiltonian?

(d) Can you now find a conserved quantity for the original system?

(e) Why does this method work?

16. Consider the nonlinear system

$$\frac{dx}{dt} = -yx^2$$

$$\frac{dy}{dt} = x + 1.$$

(a) Is this a Hamiltonian system?

(b) Can you find a conserved quantity? [*Hint*: Using the technique of the previous exercise, multiply this system by a function that puts the new system in the form

$$\frac{dx}{dt} = F(y)$$

$$\frac{dy}{dt} = G(x).]$$

17. Consider the system of differential equations

$$\frac{dx}{dt} = 1 - y^2$$
$$\frac{dy}{dt} = x(2 - y).$$

Find a conserved quantity for this system.

In Exercises 18–20, we study bifurcations of parameterized families of Hamiltonian systems. Because the equilibrium points of Hamiltonian systems assume a special form, their bifurcations do also.

18. Consider the system

$$\frac{dx}{dt} = y$$
$$\frac{dy}{dt} = x^2 - a,$$

where a is a parameter.

(a) Verify that this system is a Hamiltonian system and that

$$H(x, y) = \frac{y^2}{2} - \frac{x^3}{3} + ax$$

is a Hamiltonian function.

(b) Show that the system has equilibrium points at $(x, y) = (\pm\sqrt{a}, 0)$ if $a \geq 0$ and no equilibrium points if $a < 0$. Thus, $a = 0$ is a bifurcation value of the parameter.

(c) Linearize the system at each of the equilibrium points and determine the behavior of solutions near the equilibrium points.

(d) Sketch the level curves of H (and hence the phase plane of the system) for $a = -1$, $a = -0.5$, $a = 0$, $a = 0.5$, and $a = 1$. [*Hint*: Think about the effect of changing a on the graph of H. It is worthwhile to employ whatever graphing technology you have available.]

(e) In a brief paragraph, describe the bifurcation that takes place at $a = 0$.

19. Consider the system of differential equations

$$\frac{dx}{dt} = x^2 + 2xy$$
$$\frac{dy}{dt} = a - 2xy - y^2,$$

which depend on the parameter a. Describe in detail the bifurcation that occurs at $a = 0$.

20. Suppose Glen and Paul decide to modify the model of their enthusiasm levels because Paul notes that his enthusiasm level also goes down at a constant rate over time. Therefore the new equations are

$$\frac{dx}{dt} = y$$

$$\frac{dy}{dt} = x - x^2 - a.$$

(a) Describe in detail any major bifurcations that occur in this family.

(b) Suppose $a = 1$. What happens to Glen and Paul's book no matter how enthusiastic both authors start out?

5.4 DISSIPATIVE SYSTEMS

The Hamiltonian systems discussed in the previous chapter are obviously idealized systems. In the real world, pendulums do not swing forever in a periodic motion. Rather they eventually wind down; energy is dissipated. In this section we will discuss in detail these types of dissipative systems. We begin by modifying the ideal pendulum by considering a small damping term.

The Nonlinear Pendulum with Friction

Recall that the second-order equation governing the motion of the pendulum is (see page 440)

$$\frac{d^2\theta}{dt^2} + \frac{b}{m}\frac{d\theta}{dt} + \frac{g}{l}\sin\theta = 0,$$

where b is the coefficient of damping, m is the mass of the pendulum bob, l is the length of the pendulum arm, and g is the acceleration of gravity ($g \approx 9.8\text{m/s}^2$). All the parameters are positive. The term

$$\frac{b}{m}\frac{d\theta}{dt}$$

results from damping due, for example, to air resistance or friction at the pivot point of the pendulum arm. We assumed that $b = 0$ in our idealized system. In this section, we assume that $b \neq 0$.

Introducing the velocity $v = d\theta/dt$, this second-order differential equation can be written as a nonlinear system in the usual manner,

$$\frac{d\theta}{dt} = v$$

$$\frac{dv}{dt} = -\frac{b}{m}v - \frac{g}{l}\sin\theta.$$

This system is no longer Hamiltonian, since our test for existence of a Hamiltonian function fails because (see page 444):

$$\frac{\partial}{\partial\theta}(v) = 0$$

but

$$-\frac{\partial}{\partial v}\left(-\frac{b}{m}v - \frac{g}{l}\sin\theta\right) = \frac{b}{m} \neq 0.$$

The equilibrium points and nullclines

We begin the study of the damped pendulum system

$$\frac{d\theta}{dt} = v$$

$$\frac{dv}{dt} = -\frac{b}{m}v - \frac{g}{l}\sin\theta$$

by finding the equilibrium points. As for the ideal pendulum, the equilibrium points occur at $(\theta, v) = (0, 0), (\pm\pi, 0), (\pm 2\pi, 0), \ldots$.

The Jacobian matrix of the vector field at (θ, v) is

$$\begin{pmatrix} 0 & 1 \\ -\frac{g}{l}\cos\theta & -\frac{b}{m} \end{pmatrix}.$$

If $(\theta, v) = (0, 0), (\pm 2\pi, 0), (\pm 4\pi, 0), \ldots$ (the equilibrium points where the pendulum is hanging downward), this matrix is

$$\begin{pmatrix} 0 & 1 \\ -\frac{g}{l} & -\frac{b}{m} \end{pmatrix}.$$

The linearized system at these equilibrium points has this matrix as its coefficient matrix. The eigenvalues of this matrix are roots of the characteristic equation

$$-\lambda\left(-\frac{b}{m} - \lambda\right) + \frac{g}{l} = 0,$$

or

$$\lambda^2 + \frac{b}{m}\lambda + \frac{g}{l} = 0.$$

Hence the eigenvalues are

$$-\frac{b}{2m} \pm \sqrt{\left(\frac{b}{2m}\right)^2 - \frac{g}{l}}.$$

There are three different cases, depending on the sign of $(b/(2m))^2 - g/l$. If this quantity is negative, then the eigenvalues of the Jacobian matrix are complex. The real part is $-b/(2m)$, which is negative, so in this case the equilibrium is a spiral sink. If $(b/(2m))^2 - g/l > 0$, we find two real distinct eigenvalues. Since

$$0 < \left(\frac{b}{2m}\right)^2 - \frac{g}{l} < \left(\frac{b}{2m}\right)^2,$$

we have that

$$\frac{b}{2m} > \sqrt{\left(\frac{b}{2m}\right)^2 - \frac{g}{l}},$$

and both eigenvalues are negative. Finally, if $(b/(2m))^2 - g/l = 0$, we have a repeated eigenvalue $-b/(2m)$, which is again negative.

Thus in each case we find that the equilibrium points at $(\theta, v) = (0, 0)$, $(\pm 2\pi, 0)$, $(\pm 4\pi, 0)$, ... are sinks. If we focus attention on the case of a pendulum with only a small amount of friction, then b is very small. For b sufficiently small, we have that $(b/(2m))^2 - g/l$ is negative and the points $(\theta, v) = (0, 0)$, $(\pm 2\pi, 0)$, $(\pm 4\pi, 0)$, ... are spiral sinks. We deal only with this case here (see the exercises for the other cases).

At the other equilibrium points $(\theta, v) = (\pm \pi, 0)$, $(\pm 3\pi, 0)$, ..., which correspond to the pendulum standing upright, the Jacobian matrix is

$$\begin{pmatrix} 0 & 1 \\ \dfrac{g}{l} & -\dfrac{b}{m} \end{pmatrix},$$

and this is the coefficient matrix for the linearized system at these points. The eigenvalues are

$$-\frac{b}{2m} \pm \sqrt{\left(\frac{b}{2m}\right)^2 + \frac{g}{l}}.$$

These eigenvalues are always real because b, m, l, and g are all positive. The eigenvalue

$$-\frac{b}{2m} + \sqrt{\left(\frac{b}{2m}\right)^2 + \frac{g}{l}}$$

is positive, whereas the other eigenvalue,

$$-\frac{b}{2m} - \sqrt{\left(\frac{b}{2m}\right)^2 + \frac{g}{l}},$$

is negative. Hence the equilibrium points $(\theta, v) = (\pm \pi, 0)$, $(\pm 3\pi, 0)$, ... are saddles.

To find the θ- and v-nullclines, note that $d\theta/dt = 0$ if $v = 0$, so the vector field points up or down along the θ-axis. The v-nullcline is where $dv/dt = 0$, which occurs on the curve

$$v = -\frac{mg}{bl} \sin \theta;$$

so the vector field points in a horizontal direction along this curve. We can determine the direction of the vector field in a region between two nullclines by checking the direction at a single point (see Figure 5.35).

Coupled with our analysis at the equilibrium points, we can now describe the behavior of solution curves near the equilibrium points as shown in Figure 5.36. Note that the solution curves circulate about the equilibrium points at $(\theta, v) = (0, 0)$, $(\pm 2\pi, 0)$, $(\pm 4\pi, 0)$, ... in a clockwise direction. We can verify that the stable and unstable separatrices at $(\theta, v) = (\pm \pi, 0)$, $(\pm 3\pi, 0)$, ... point in the directions indicated in Figure 5.36 by finding the linearization at these points or by using the direction field (see Figure 5.35).

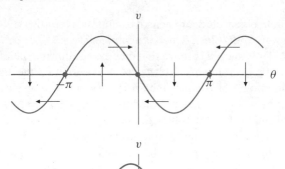

Figure 5.35
Nullclines for the system

$$\frac{d\theta}{dt} = v$$

$$\frac{dv}{dt} = -\frac{b}{m}v - \frac{g}{l}\sin\theta.$$

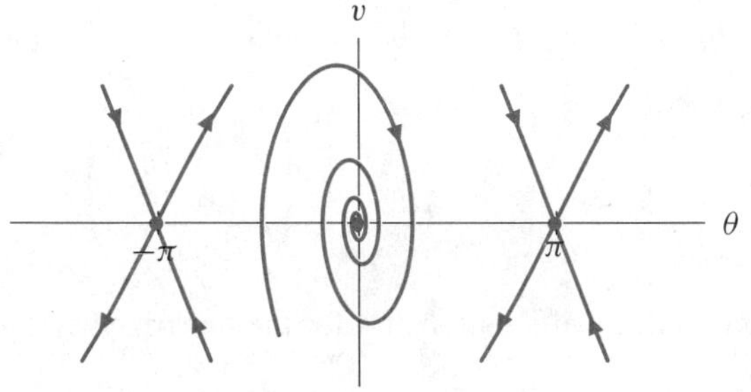

Figure 5.36
Phase plane near the equilibrium points for the system

$$\frac{d\theta}{dt} = v$$

$$\frac{dv}{dt} = -\frac{b}{m}v - \frac{g}{l}\sin\theta.$$

The Effects of Dissipation

At this point Figure 5.36 contains all the information about the phase plane of the pendulum with small damping available by the qualitative techniques that we have at our disposal. We do not have the benefit of a conserved quantity as we did in the frictionless case, so we cannot determine the entire phase plane. However, using the Hamiltonian function

$$H(\theta, v) = \frac{1}{2}v^2 - \frac{g}{l}\cos\theta$$

that we developed in the last section for the ideal pendulum, we can complete the picture. If $(\theta(t), v(t))$ is a solution of the ideal pendulum system (with $b = 0$), then

$$\frac{d}{dt}H(\theta(t), v(t)) = 0.$$

If $b \neq 0$, this is no longer true. However, if $(\theta(t), v(t))$ is a solution of the pendulum system with $b \neq 0$, we have

$$\frac{dH}{dt} = \frac{\partial H}{\partial \theta}\frac{d\theta}{dt} + \frac{\partial H}{\partial v}\frac{dv}{dt}$$

$$= \left(\frac{g}{l}\sin\theta\right)v + v\left(-\frac{b}{m}v - \frac{g}{l}\sin\theta\right)$$

$$= -\frac{b}{m}v^2$$

$$\leq 0.$$

That is, the function H is no longer constant along solution curves of the system, but rather $H(\theta(t), v(t))$ is decreasing whenever $v(t) \neq 0$. Hence solution curves in the θv-phase plane cross level sets of H moving from larger to smaller H values.

This provides us with a strategy for determining the phase plane. First, we draw the level curves of H as we did before (see Figure 5.37). If the value of b/m is small and the velocity v is small, the value of H decreases slowly along solutions. We can sketch the phase plane by sketching curves that flow in the same general direction as the solution curves of the ideal pendulum but that move from level curves of higher H value toward lower H value (see Figure 5.38).

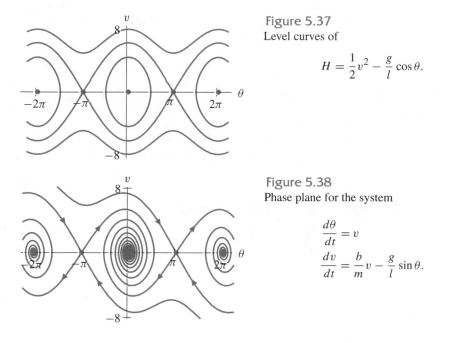

Figure 5.37
Level curves of

$$H = \frac{1}{2}v^2 - \frac{g}{l}\cos\theta.$$

Figure 5.38
Phase plane for the system

$$\frac{d\theta}{dt} = v$$

$$\frac{dv}{dt} = \frac{b}{m}v - \frac{g}{l}\sin\theta.$$

The fact that H is decreasing along solution curves allows us to determine the fate of stable and unstable separatrices emanating from saddle points: The unstable separatrices must fall into adjacent sinks, whereas the stable separatrices come from "infinity" (see Figure 5.38). Also, the equilibrium point analysis we performed earlier agrees with what we have just done. The whole picture begins to fit together in a neat package, with the local analysis near equilibrium points fully complementing the global structure provided by all the level curves of H. Finally, if we graph $\theta(t)$ for a typical solution curve, we see that the pendulum eventually tends to oscillate about the downward-pointing rest position, as it should (see Figure 5.39).

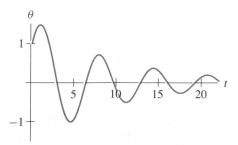

Figure 5.39
A $\theta(t)$-graph for a typical solution to the system

$$\frac{d\theta}{dt} = v$$

$$\frac{dv}{dt} = \frac{b}{m}v - \frac{g}{l}\sin\theta.$$

Lyapunov Functions

The function H above plays a very different role for the damped pendulum than it does for the ideal pendulum. Solutions of the damped pendulum system move across level sets of H from higher to lower values. The function H is called a **Lyapunov function**, after the Russian mathematician Aleksandr Mikhailovich Lyapunov (1857–1918). This idea can be generalized as:

DEFINITION   A function $L(x, y)$ is called a **Lyapunov function** for a system of differential equations if, for every solution $(x(t), y(t))$ that is not an equilibrium solution of the system,

$$\frac{d}{dt} L(x(t), y(t)) \leq 0$$

for all t with strict inequality except for a discrete set of t's.   ∎

So the value of a Lyapunov function never increases and usually decreases along a nonequilibrium solution. A Lyapunov function can be a great help in drawing the phase plane for a system. Solution curves must cross level sets of the Lyapunov function from higher to lower values.

The Damped Harmonic Oscillator

Consider the damped harmonic oscillator system

$$\frac{dy}{dt} = v$$
$$\frac{dv}{dt} = -qy - pv,$$

where q and p are positive constants. If $p = 0$, we saw in Section 5.3 (page 439) that the system is Hamiltonian, with Hamiltonian function

$$H(y, v) = \frac{1}{2}v^2 + \frac{q}{2}y^2.$$

If $p > 0$ and $(y(t), v(t))$ is a solution of the system, then we can compute the rate of change of H along the solution by

$$\frac{d}{dt} H(y(t), v(t)) = \frac{\partial H}{\partial y}\frac{dy}{dt} + \frac{\partial H}{\partial v}\frac{dv}{dt}$$

$$= qy \cdot v + v(-qy - pv)$$

$$= -pv^2$$

$$\leq 0.$$

Hence $H(y(t), v(t))$ decreases at a nonzero rate (except when $v = 0$). This implies that H is a Lyapunov function for the damped harmonic oscillator system. The level sets of H are ellipses in the yv-plane, so solutions of the damped harmonic oscillator system (with $p > 0$) cross these level sets from larger to smaller values of H. Since H has one global minimum at the origin and no other critical points, all solutions must tend toward the origin as time increases (see Figures 5.40 and 5.41). This agrees with our analysis of the damped harmonic oscillator from the point of view of linear systems and will be useful in Section 6.2.

As we mentioned before (see page 439), the value of H at a point (y, v) is called the energy of the harmonic oscillator in position y with velocity v. Saying that H decreases along solutions of the damped harmonic oscillator system is a precise way of saying that the damping dissipates energy. The same holds for the pendulum with friction.

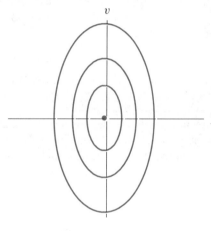

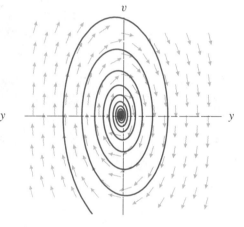

Figure 5.40
Level sets for

$$H(y, v) - \frac{1}{2}v^2 + \frac{q}{2}y^2.$$

With the exception of the point at the origin, each level set is an ellipse.

Figure 5.41
Phase plane for the damped harmonic oscillator

$$\frac{dy}{dt} = v$$

$$\frac{dv}{dt} = -qy - pv$$

with p small.

Gradient Systems

Not all systems have Lyapunov functions. Finding a Lyapunov function for a system, even if we somehow know that the system has a Lyapunov function, can be extremely challenging. However, for some modeling problems the physical system being modeled provides a motivation for a "lucky guess" of a Lyapunov function. This is the case for the damped pendulum and the damped harmonic oscillator. The Hamiltonian function of the undamped system becomes a Lyapunov function if damping is added.

In some cases the Lyapunov function is an integral part of the construction of the system. Such systems are called **gradient systems**. We begin with an example.

Navigation by Smell

Many animal species use smell to navigate through their environment. A lobster, for example, can use its antennae to detect very small concentrations of chemicals in the water around it (that is, it "smells" the water). With this ability the lobster can determine which direction a smell is coming from and hence find food in the murky deep. Although it is not clear how lobsters accomplish this, one possible method is to determine local variations in the concentration and move in the direction in which the concentration increases fastest.*

To describe how we can design a mechanical lobster that navigates by smell, we begin by assuming our lobster can only move on a flat two-dimensional plane. Let $S(x, y)$ equal the concentration at (x, y) of the chemicals making up the smell of a dead fish. We know from vector calculus that at (x, y) the direction in which S increases the fastest is given by the gradient vector

$$\nabla S(x, y) = \left(\frac{\partial S}{\partial x}, \frac{\partial S}{\partial y} \right).$$

This defines a vector field on the xy-plane. We assume that our model lobster always moves in the direction that increases the smell the fastest. Hence the velocity of our lobster's motion points in the same direction as the gradient of S. If we let $(x(t), y(t))$ denote the location of the lobster at time t, then the velocity vector is given by

$$\left(\frac{dx}{dt}, \frac{dy}{dt} \right) = \nabla S(x(t), y(t)).$$

Written in terms of scalar equations, we have

$$\frac{dx}{dt} = \frac{\partial S}{\partial x}$$
$$\frac{dy}{dt} = \frac{\partial S}{\partial y}.$$

We use this system of equations to determine our lobster's motion. If the initial position is (x_0, y_0) at time $t = 0$, then the solution curve in the phase plane satisfying this initial condition is our prediction of a lobster's path.

Two dead fish

Suppose $S(x, y)$ is defined for $-2 \leq x \leq 2$ and $-2 \leq y \leq 2$ and is given by the formula

$$S(x, y) = \frac{x^2}{2} - \frac{x^4}{4} - \frac{y^2}{2} + 8.$$

The graph of S is shown in Figure 5.42. We restrict attention to the region of the xy-plane with x and y between 2 and -2 because this function S takes on negative values when x or y is large, and negative concentrations are not physically meaningful. More

---

*The ability of lobsters to do this "computation" in turbulent fluids is only beginning to be duplicated by technology. See Lipkin, R., "Tracking Undersea Scent," *Science News*, *147*, 5 (1994): 78, for a discussion of a "robo-lobster."

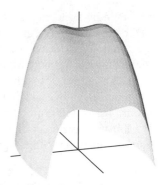

Figure 5.42
Graph of the function
$S(x, y) = x^2/2 - x^4/4 - y^2/2 + 8.$

realistic, but more algebraically complicated choices for S are considered in the exercises.

We can see (or compute by the techniques of vector calculus) that S has a local maximum at each of the points $(-1, 0)$ and $(1, 0)$ and a saddle point at $(0, 0)$. A function of this sort might arise if there were a smelly dead fish at each of the points $(-1, 0)$ and $(1, 0)$.

The gradient of S gives rise to a vector field and, as above, this specifies a system of differential equations

$$\frac{dx}{dt} = \frac{\partial S}{\partial x} = x - x^3$$
$$\frac{dy}{dt} = \frac{\partial S}{\partial y} = -y.$$

By construction, the velocity vector of a solution of this system is equal to the gradient of S and hence points in the direction for which S increases the fastest. If $(x(t), y(t))$ is a solution, we compute the rate of change of S along the solution by

$$\frac{d}{dt} S(x(t), y(t)) = \frac{\partial S}{\partial x} \frac{dx}{dt} + \frac{\partial S}{\partial y} \frac{dy}{dt}$$

$$= (x - x^3)(x - x^3) + (-y)(-y)$$

$$= (x - x^3)^2 + y^2$$

$$\geq 0.$$

Hence S increases at every point where $\nabla S \neq 0$. These are precisely the points where the vector field of the system is zero — that is, the equilibrium points of the system.

To sketch the phase plane of the system

$$\frac{dx}{dt} = \frac{\partial S}{\partial x} = x - x^3$$
$$\frac{dy}{dt} = \frac{\partial S}{\partial y} = -y,$$

we begin by sketching the level sets of S (see Figure 5.43). The direction field points in the direction of the gradient of S, which is perpendicular to the level sets. To sketch the solution curves, we draw curves always perpendicular to the level sets of S moving in the direction of increasing S (see Figure 5.44).

We can verify by linearization that the equilibrium points $(\pm 1, 0)$ are sinks, while the point $(0, 0)$ is a saddle (see the exercises). Lobsters that start with $x(0) < 0$ tend toward the equilibrium point $(-1, 0)$, whereas those with $x(0) > 0$ tend toward $(1, 0)$. If a lobster starts precisely on the y-axis $(x(0) = 0)$, then it will tend toward the saddle equilibrium point at the origin, unable to decide between the two fish.

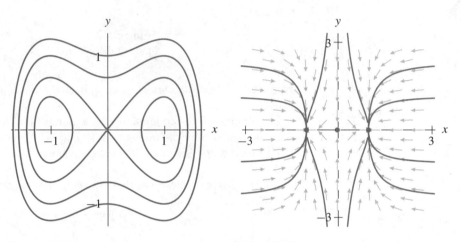

Figure 5.43
Level sets of

$$S(x, y) = \frac{x^2}{2} - \frac{x^4}{4} - \frac{y^2}{2} + 8.$$

Figure 5.44
Phase plane for the system

$$\frac{dx}{dt} = \frac{\partial S}{\partial x} = x - x^3$$
$$\frac{dy}{dt} = \frac{\partial S}{\partial y} = -y.$$

General Form of Gradient Systems

The vector field for the system in the preceding example is the gradient of the function S, hence it is not unreasonable to call it a gradient system.

DEFINITION   A system of differential equations is called a **gradient system** if there is a function G such that

$$\frac{dx}{dt} = \frac{\partial G}{\partial x}$$
$$\frac{dy}{dt} = \frac{\partial G}{\partial y}$$

for all (x, y).   ∎

As in the example, if $(x(t), y(t))$ is a solution for a gradient system with vector field given by the gradient of $G(x, y)$, then

$$\frac{d}{dt}G(x(t), y(t)) = \frac{\partial G}{\partial x}\frac{dx}{dt} + \frac{\partial G}{\partial y}\frac{dy}{dt}$$

$$= \left(\frac{\partial G}{\partial x}\right)^2 + \left(\frac{\partial G}{\partial y}\right)^2$$

$$\geq 0.$$

That is, the value of G increases along every orbit except at critical points of G. The function $-G$ decreases along orbits and hence is a Lyapunov function for this system.

Properties of Gradient Systems

Just as in the case of Hamiltonian systems, gradient systems enjoy several special properties that make their phase planes relatively simple. For one thing, gradient systems cannot possess periodic solutions. Since the gradient function must increase along all nonequilibrium solutions, a solution curve can never return to its starting place.

There are also restrictions on the types of equilibrium points that occur in a gradient system. Suppose

$$\frac{dx}{dt} = \frac{\partial G}{\partial x}$$

$$\frac{dy}{dt} = \frac{\partial G}{\partial y}$$

is a gradient system. Then the Jacobian matrix associated with this vector field is

$$\begin{pmatrix} \dfrac{\partial^2 G}{\partial x^2} & \dfrac{\partial^2 G}{\partial y \partial x} \\[2ex] \dfrac{\partial^2 G}{\partial x \partial y} & \dfrac{\partial^2 G}{\partial y^2} \end{pmatrix}.$$

Because the mixed partial derivatives of G are equal, this matrix assumes the form

$$\begin{pmatrix} \alpha & \beta \\ \beta & \gamma \end{pmatrix},$$

where $\alpha = \partial^2 G/\partial x^2$, $\beta = \partial^2 G/\partial y \partial x$, and $\gamma = \partial^2 G/\partial y^2$. The eigenvalues are roots of the equation

$$\lambda^2 - (\alpha + \gamma)\lambda + \alpha\gamma - \beta^2 = 0,$$

which are

$$\frac{\alpha + \gamma \pm \sqrt{(\alpha + \gamma)^2 - 4\alpha\gamma + 4\beta^2}}{2}.$$

Simplifying, we see that the eigenvalues equal

$$\frac{1}{2}(\alpha + \gamma) \pm \frac{1}{2}\sqrt{(\alpha - \gamma)^2 + 4\beta^2}.$$

Since the term $(\alpha - \gamma)^2 + 4\beta^2 \geq 0$, it follows that these eigenvalues never have an imaginary part; that is, the eigenvalues are always real. Gradient systems do not have spiral sinks, spiral sources, or centers.

It is important to note that not all systems of differential equations that possess Lyapunov functions are gradient systems. For example, the linear system

$$\frac{dx}{dt} = -x + y$$

$$\frac{dy}{dt} = -x - y$$

has a Lyapunov function given by $L(x, y) = x^2 + y^2$. To verify this assertion, we compute the derivative of L along a solution $(x(t), y(t))$ by

$$\frac{d}{dt} L(x(t), y(t)) = 2x \frac{dx}{dt} + 2y \frac{dy}{dt}$$

$$= 2x(-x + y) + 2y(-x - y)$$

$$= -2(x^2 + y^2)$$

$$\leq 0.$$

Since $dL/dt = 0$ only at the origin, this function decreases along all nonequilibrium solution curves. But L cannot be a gradient system, since the eigenvalues at (0,0) are complex $(-1 \pm i)$; so the origin is a spiral sink, which cannot occur in gradient systems.

EXERCISES FOR SECTION 5.4

1. Consider the system

$$\frac{dx}{dt} = -x^3$$

$$\frac{dy}{dt} = -y^3.$$

(a) Verify that

$$L(x, y) = \frac{x^2}{2} + \frac{y^2}{2}$$

is a Lyapunov function for the system.

(b) Sketch the level sets of L.

(c) What can you conclude about the phase plane of the system from the information in parts (a) and (b) above? (Sketch the phase plane and write a short essay describing what you know about the phase plane and how you know it.)

2. Consider the system

$$\frac{dx}{dt} = y$$

$$\frac{dy}{dt} = -x - \frac{y}{4} + x^2.$$

(a) Verify that the function

$$L(x, y) = \frac{y^2}{2} + \frac{x^2}{2} - \frac{x^3}{3}$$

is a Lyapunov function for the system.

(b) Sketch the level sets of L.

(c) What can you conclude about the phase plane of the system from the information in parts (a) and (b)? (Sketch the phase plane and write a short essay describing what you know about the phase plane and how you know it.)

3. Consider the system

$$\frac{dx}{dt} = y$$
$$\frac{dy}{dt} = -4x - 0.1y.$$

(a) Verify that all solutions tend toward the origin as t increases, and sketch the phase plane. [*Hint*: The system is linear.]

(b) Verify that

$$L(x, y) = x^2 + y^2$$

is *not* a Lyapunov function for the system.

(c) Verify that

$$K(x, y) = 2x^2 + \frac{y^2}{2}$$

is a Lyapunov function for the system.

In Exercises 4–11, we consider the damped pendulum system

$$\frac{d\theta}{dt} = v$$
$$\frac{dv}{dt} = -\frac{g}{l}\sin\theta - \frac{b}{m}v,$$

where b is the damping coefficient, m is the mass of the bob, l is the length of the arm, and g is the acceleration of gravity ($g \approx 9.8\text{m/s}^2$).

4. What relationship must hold between the parameters b, m, and l for the period of a small swing back and forth of the damped pendulum to be one second?

5. Suppose we have a pendulum clock that uses a slightly damped pendulum to keep time (that is, b is positive but $b \approx 0$). The clock "ticks" each time the pendulum arm crosses $\theta = 0$. If the mass of the pendulum bob is increased, does the clock run fast or slow?

6. Suppose we have a pendulum clock that uses only a slightly damped pendulum to keep time. The clock "ticks" each time the pendulum arm crosses $\theta = 0$.

(a) As the clock "winds down" (so the amplitude of the swings decreases), does the clock run slower or faster?

(b) If the initial push of the pendulum is large so that the pendulum swings very close to the vertical, will the clock run too fast or too slow?

7. For fixed values of b and l, for what values of the mass m will the pendulum be usable as a clock?

8. Suppose we take $l = 9.8$m (so $g/l = 1$), $m = 1$, and b large, say $b = 4$. For the damped pendulum system above and with this choice of parameter values, do the following.

   (a) Find the eigenvalues and eigenvectors of the linearized system at the equilibrium point $(0, 0)$.

   (b) Find the eigenvalues and eigenvectors of the linearized system at the equilibrium point $(\pi, 0)$.

   (c) Sketch the phase plane near the equilibrium points.

   (d) Sketch the entire phase plane. [*Hint*: Begin by sketching the level sets of H as in the text.]

9. Suppose we have a pendulum clock that uses a slightly damped oscillator to keep time. Suppose that the clock "ticks" each time the pendulum arm crosses $\theta = 0$, but the arm must reach a height of $\theta = \pm 0.1$ to record the swing (that is, if the entire swing takes place with $-0.1 < \theta < 0.1$, then the clock doesn't tick). Suppose one tick is one second. In terms of the parameters b, m, and l, give a rough estimate of how long the clock can keep accurate time. Comment on why pendulum clocks must be wound.

10. **(a)** For the slightly damped pendulum ($b > 0$ but b close to zero) find the set of all initial conditions $(\theta(0), v(0))$ that execute exactly two complete revolutions for $t > 0$ (that is, pass the vertical position exactly twice) before settling into a back and forth swinging motion. Sketch the phase plane for the slightly damped pendulum and shade these initial conditions.

   (b) Repeat part (a) for solutions that execute exactly five complete revolutions for $t > 0$ before settling into back and forth swinging motion.

11. Suppose that rather than adding damping to the ideal pendulum, we add a small amount of "antidamping"; that is, we take b slightly negative in the damped pendulum system. Physically this would mean that whenever the velocity is nonzero, the pendulum is accelerated in the direction of motion.

   (a) Linearize and classify the equilibrium points in this situation.

   (b) Sketch the phase plane for this system.

   (c) Describe in a brief paragraph the behavior of a solution with initial condition near $\theta = v = 0$.

12. Let $G(x, y) = x^3 - 3xy^2$.

   (a) What is the gradient system with vector field given by the gradient of G?

   (b) Sketch the graph of G and the level sets of G.

   (c) Sketch the phase plane of the gradient system in part (a).

13. Let $G(x, y) = x^2 - y^2$.

(a) What is the gradient system with vector field given by the gradient of G?

(b) Classify the equilibrium point at the origin. [*Hint*: This system is linear.]

(c) Sketch the graph of G and the level sets of G.

(d) Sketch the phase plane of the gradient system in part (a).

Remark: This is why saddle equilibrium points have the name *saddle*.

14. Let $G(x, y) = x^2 + y^2$.

(a) What is the gradient system with vector field given by the gradient of G?

(b) Classify the equilibrium point at the origin. [*Hint*: The system is linear.]

(c) Sketch the graph of G and the level sets of G.

(d) Sketch the phase plane of the gradient system in part (a).

15. For the two dead fish example given by the system

$$\frac{dx}{dt} = x - x^3$$
$$\frac{dy}{dt} = -y,$$

(a) find the linearized system for the equilibrium point at the origin and verify that the origin is a saddle,

(b) find the linearized system for the equilibrium point at $(1, 0)$ and verify that this point is a sink,

(c) from the eigenvalues and eigenvectors of the system in part (b), determine from which direction the model lobster will approach the equilibrium point $(1, 0)$, and

(d) check that the linearized system at the equilibrium point $(-1, 0)$ is the same as that at $(1, 0)$.

16. The system for the two dead fish example

$$\frac{dx}{dt} = x - x^3$$
$$\frac{dy}{dt} = y,$$

has the special property that the equations "decouple"; that is, the equation for dx/dt depends only on x, and the equation for dy/dt depends only on y.

(a) Sketch the phase lines for the dx/dt and dy/dt equations.

(b) Using these phase lines, sketch the phase plane of the system.

17. Suppose the smell of a bunch of dead fish in the region $-2 \leq x \leq 2, -2 \leq y \leq 2$ is given by the function

$$S(x, y) = x^2 + y^2 - \frac{x^4 + y^4}{4} - 3x^2 y^2 + 100.$$

(a) What is the gradient system whose vector field is the gradient of S?

(b) Using a numerical solver, sketch the phase plane for this system.

(c) How many dead fish are there, and where are they?

(d) Using the results from part (b), sketch the level sets of S.

(e) Why is the model not realistic for large values of x or y?

18. A reasonable model for the smell at (x, y) of a dead fish located at (x_1, y_1) is given by

$$S_1(x, y) = \frac{1}{(x - x_1)^2 + (y - y_1)^2 + 1}.$$

That is, S_1 is given by 1 over the distance to the dead fish squared plus 1.

(a) Form the function S giving the total smell from three dead fish located at $(1, 0)$, $(-1, 0)$, and $(0, 2)$.

(b) Sketch the level sets of S.

(c) Sketch the phase plane of the gradient system

$$\frac{dx}{dt} = \frac{\partial S}{\partial x}$$
$$\frac{dy}{dt} = \frac{\partial S}{\partial y}.$$

(d) Write out explicitly the formulas for the right-hand sides of the equations in part (c).

(e) Why did we use distance squared plus 1 instead of just distance squared in the definition of S_1? Why did we use distance squared plus 1 instead of just distance plus 1 in the definition of S_1?

19. Suppose

$$\frac{dx}{dt} = f(x, y)$$
$$\frac{dy}{dt} = g(x, y)$$

is a gradient system. That is, there exists a function $G(x, y)$ such that $f = \partial G/\partial x$ and $g = \partial G/\partial y$.

(a) Verify that if f and g have continuous partial derivatives, then

$$\frac{\partial f}{\partial y} = \frac{\partial g}{\partial x}$$

for all (x, y).

(b) Use this to show that the system

$$\frac{dx}{dt} = x^2 + 3xy$$
$$\frac{dy}{dt} = 2x + y^3$$

is *not* a gradient system.

20. The following phase portrait cannot occur for a gradient system. Explain why.

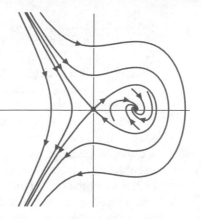

21. Let

$$H(y, v) = \frac{1}{2}v^2 + V(y)$$

for some function V and consider the associated Hamiltonian system

$$\frac{dy}{dt} = \frac{\partial H}{\partial v} = v$$

$$\frac{dv}{dt} = -\frac{\partial H}{\partial y} = -\frac{dV}{dy}.$$

Let k be a positive constant. Give a physical interpretation of the relationship between the Hamiltonian system and the system

$$\frac{dy}{dt} = v$$

$$\frac{dv}{dt} = -\frac{dV}{dy} - kv.$$

Show that H is a Lyapunov function for this system.

22. Consider the Hamiltonian system

$$\frac{dx}{dt} = \frac{\partial H}{\partial y}$$

$$\frac{dy}{dt} = -\frac{\partial H}{\partial x}$$

and the gradient system

$$\frac{dx}{dt} = \frac{\partial H}{\partial x}$$

$$\frac{dy}{dt} = \frac{\partial H}{\partial y},$$

where H is the same function in each case. What can you say about the relationship between the two phase portraits of these two systems?

5.5   NONLINEAR SYSTEMS IN THREE DIMENSIONS

We saw in Section 2.5 that solutions of differential equations with three dependent variables are curves in three-dimensional space. These curves can loop around each other in very complicated ways. In Section 3.8 we studied the behavior of linear systems with three dependent variables. The behavior of linear systems can be determined by the eigenvalues and eigenvectors. However, the list of possible behaviors is much longer than for planar systems.

In this section we consider two examples of nonlinear systems in three dimensions. The first is a population model with three species forming a food chain. For the second example, we return to the Lorenz equations studied in Section 2.5. We can use the technique of linearization of equilibrium points and numerical approximation (Euler's method) to obtain a little bit of information about these systems. As we have emphasized in the previous sections, there are very few all-purpose tools for differential equations in three and higher dimensions. In fact, these systems are an area of active research in mathematics.

A Food Chain Model

We have studied the behavior of populations living in isolation and of pairs of species interacting in predator-prey, cooperative, and competitive systems. These systems only scratch the surface of the types of complicated interactions found in nature. One possibility is the formation of a food chain of three or more species. A single species can be both predator and prey. An example that has been studied recently involves the Balsam fir tree, the moose, and the wolf. The fir trees are eaten by the moose, and the moose (particularly young and infirm individuals) are eaten by wolves. A natural question is whether or not changes in the wolf population affect the tree population. Recent studies of the tree/moose/wolf population in Isle Royal National Park* indicate that changes in the wolf population may affect the trees.

We form a model of a system with three species, each one eaten by the next. For convenience, we call these species trees, moose, and wolves in analogy to the example above. Let

$$x(t) = \text{population of trees at time } t,$$
$$y(t) = \text{population of moose at time } t, \text{ and}$$
$$z(t) = \text{population of wolves at time } t.$$

We assume that each of the populations in isolation can be modeled with a logistic equation and that the effect of interaction between species is proportional to the product of the populations. The behavior of solutions will depend on the parameters chosen for the growth rate, carrying capacity, and effect of interaction. In order to get an idea of how solutions of systems of this form behave, we begin by taking all of the parameters to be 1. (This certainly is not the case for trees, moose, and wolves, but it makes the

---

*See McLaren, B. E. and Peterson, R. O., "Wolves, Moose and Tree Rings on Isle Royale," *Science, 266* (1994): 1555.

arithmetic that follows much more pleasant.) Our model is

$$\frac{dx}{dt} = x(1-x) - xy$$

$$\frac{dy}{dt} = y(1-y) + xy - yz$$

$$\frac{dz}{dt} = z(1-z) + yz.$$

Note that the growth of the tree population is decreased by the presence of moose (the $-xy$ term), the growth of the moose population is increased by the trees (the $+xy$ term) but decreased by the wolves (the $-yz$ term), and the growth of the wolf population is increased by the moose (the $+yz$ term).

We can find the equilibrium points of this system by setting the right-hand sides of the equations to zero and solving for x, y, and z. The equilibrium points are $(0,0,0)$, $(1,0,0)$, $(0,1,0)$, $(0,0,1)$, $(1,0,1)$, and $(2/3, 1/3, 4/3)$. Of these, only the point $(2/3, 1/3, 4/3)$ has all three coordinates nonzero, so the three species can coexist in equilibrium at these populations. The Jacobian matrix for this system at (x,y,z) is

$$\begin{pmatrix} 1-2x-y & -x & 0 \\ y & 1-2y+x-z & -y \\ 0 & z & 1-2z+y \end{pmatrix}.$$

So at the equilibrium point $(2/3, 1/3, 4/3)$, the Jacobian is

$$\begin{pmatrix} -2/3 & -2/3 & 0 \\ 1/3 & -1/3 & -1/3 \\ 0 & 4/3 & -4/3 \end{pmatrix}.$$

The characteristic polynomial of the matrix is

$$-\left(\lambda^3 + \frac{7}{3}\lambda^2 + \frac{20}{9}\lambda + \frac{8}{9}\right),$$

and the eigenvalues are

$$\lambda_1 = -1 \quad \text{and} \quad \lambda_2, \lambda_3 = -\frac{2}{3} \pm \frac{2}{3}i.$$

Since λ_1 is negative and λ_2 and λ_3 have negative real part, this equilibrium point is a sink, and all solutions with initial conditions sufficiently close to this point will tend toward it as t increases. In fact numerical simulations show that every solution with initial conditions in the first octant (where x, y, and z are all positive) tend to this equilibrium point. Hence we expect that over the long run the populations will settle at these values and the species will coexist in equilibrium.

This model does not include many factors in the environment that can have a dramatic effect on the populations. For example, in 1981 a disease decreased the wolf population dramatically. More recently, the particularly harsh winter of 1996–97 caused a crash in the moose population.* Constructing a model that includes such factors is at least as difficult as long-range prediction of the weather. However, we see below that we can study the effect of such events by adjusting parameters in the simple model.

Suppose that a disease affects the population of wolves, the predator at the top of the food chain. Suppose this disease kills a certain small fraction of the population each year. We can adjust our model to include this element by adding the term $-\gamma z$ to the dz/dt equation, where γ is a parameter indicating the fraction of the population killed by the disease per unit time. The new model is

$$\frac{dx}{dt} = x(1-x) - xy$$

$$\frac{dy}{dt} = y(1-y) + xy - yz$$

$$\frac{dz}{dt} = z(1-z) - \gamma z + yz.$$

We can again compute the equilibrium points for this system, which will now depend on the value of γ. The only equilibrium point with all three coordinates nonzero is $((2-\gamma)/3, (1+\gamma)/3, (4-2\gamma)/3))$. Assuming that γ is small, this point will still be a sink. We can ask what effect adjusting γ has on the populations. To do this we compute the derivative with respect to γ of each of the coordinates of the equilibrium point. For $z = (4-2\gamma)/3$, the derivative with respect to γ is

$$\frac{d}{d\gamma}\left(\frac{4-2\gamma}{3}\right) = -\frac{2}{3},$$

which is negative. This is as we expect, since an increase in the effect of the disease should decrease the number of wolves. For $y = (1+\gamma)/3$, the derivative with respect to γ is

$$\frac{d}{d\gamma}\left(\frac{1+\gamma}{3}\right) = \frac{1}{3},$$

which is positive. So an increase in γ is good for the moose population, again as we would expect. Finally, for $x = (2-\gamma)/3$, the derivative with respect to γ is

$$\frac{d}{d\gamma}\left(\frac{2-\gamma}{3}\right) = -\frac{1}{3},$$

which is negative. An increase in γ decreases the equilibrium population of trees.

There are two important remarks to be made. First, although it is easy to make simple models of systems representing long food chains, it is not so easy to find such systems in nature. Most predators have a variety of prey species to choose from. When one prey species is in short supply, the predator changes its diet.

Second, care must be taken when modifying a system to account for changing situations. Although the discussion above might seem obvious from common sense, if

---

*See "Winter Devastates Island's Moose," by Les Line in the *New York Times* Science Section, April 1, 1997.

we had chosen to modify the equations to take into account the wolf disease by reducing the growth-rate parameter of the wolf population, then we would obtain different results (see the exercises).

Lorenz Equations

The Lorenz equations are

$$\frac{dx}{dt} = \sigma(y - x)$$

$$\frac{dy}{dt} = \rho x - y - xz$$

$$\frac{dz}{dt} = -\beta z + xy,$$

where σ, ρ, and β are parameters. In Section 2.5 we fixed the parameter values to $\sigma = 10$, $\beta = 8/3$, and $\rho = 28$ to obtain the system studied by Lorenz:

$$\frac{dx}{dt} = 10(y - x)$$

$$\frac{dy}{dt} = 28x - y - xz$$

$$\frac{dz}{dt} = -\frac{8}{3}z + xy.$$

Numerical approximations that we studied in that section showed that solutions looped around the phase space in a very complicated way (see Figure 5.45). With the aid of the technique of linearization near the equilibrium points, we can gain a little more insight into the behavior of these orbits. (However, the full story will have to wait for Chapter 8.)

The equilibrium points are $(0, 0, 0)$, $(6\sqrt{2}, 6\sqrt{2}, 27)$, and $(-6\sqrt{2}, -6\sqrt{2}, 27)$. We will linearize around each of these points to determine the local phase-space picture.

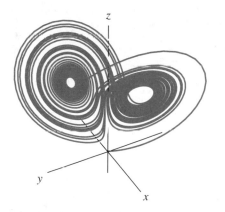

Figure 5.45
One solution curve of the Lorenz system in xyz-phase space.

The Jacobian matrix at (x, y, z) is

$$\begin{pmatrix} -10 & 10 & 0 \\ 28 - z & -1 & -x \\ y & x & -8/3 \end{pmatrix}.$$

At the origin the Jacobian is

$$\begin{pmatrix} -10 & 10 & 0 \\ 28 & -1 & 0 \\ 0 & 0 & -8/3 \end{pmatrix},$$

so the linear system that approximates the Lorenz system near the origin is

$$\frac{dx}{dt} = -10x + 10y$$

$$\frac{dy}{dt} = 28x - y$$

$$\frac{dz}{dt} = -\frac{8}{3}z.$$

This system decouples, since the equations for dx/dt and dy/dt do not depend on z, and the equation for dz/dt does not depend on x or y. We have already studied this system in Section 3.8 and found that the origin is a sink in the z direction and a saddle in the xy-plane. In the full three-dimensional picture, the origin is a saddle with a plane of initial conditions tending toward the origin as t increases and a line of initial conditions tending toward the origin as t decreases. The picture of the phase space is reproduced in Figure 5.46.

Next we consider the equilibrium point $(6\sqrt{2}, 6\sqrt{2}, 27)$. The Jacobian at this point is

$$\begin{pmatrix} -10 & 10 & 0 \\ 1 & -1 & -6\sqrt{2} \\ 6\sqrt{2} & 6\sqrt{2} & -8/3 \end{pmatrix}.$$

The eigenvalues are $\lambda_1 \approx -13.8$, $\lambda_2 \approx 0.094 + 10.2i$, and $\lambda_3 \approx 0.094 - 10.2i$, so this point is a spiral saddle that has a line of solutions that tend toward the equilibrium point as t increases and a plane of solutions that spiral toward the equilibrium point as t decreases. Solutions approach the equilibrium point quickly along the direction of the eigenvector of the negative eigenvalue, then spiral slowly away along the plane corresponding to the complex eigenvalues. By computing the eigenvectors, we can obtain the orientation of the straight line of solutions and the plane of spiraling solutions. This yields a good representation of the phase space near the equilibrium point, which we sketch in Figure 5.46.

We can compute the linearization for the equilibrium point $(-6\sqrt{2}, -6\sqrt{2}, 27)$ in the same manner, and we find that it is also a spiral saddle. These computations give us the "local" picture of the phase space near the equilibrium points (see Figure 5.46). Placing this next to a picture of a solution of the full nonlinear Lorenz system, we start to see some order in the way the solution behaves (see Figure 5.47). The solution approaches one of the equilibrium points $(\pm 6\sqrt{2}, \pm 6\sqrt{2}, 27)$ along the straight line corresponding to the negative eigenvalue. When it gets close to the equilibrium point, it begins to spiral away. When the spiral is large enough, the solution becomes involved with the saddle at the origin and either returns to repeat its previous pattern or goes to the equilibrium point "on the other side."

Reality Check

The study of the local behavior near the equilibrium points has given us a bit more insight into the complicated behavior of solutions of the Lorenz system. However, the local pictures do not tell the whole story. This is what makes the study of this and other three-dimensional systems so hard. Calculus and linearization techniques tell us a great deal about small parts of the phase space, but the behavior of solutions also depends on what is going on "globally." The development of the tools needed to study these systems as well as understand individual models is an active area of mathematical research.

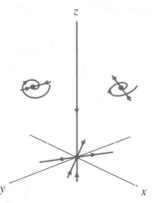

Figure 5.46
Local phase-space pictures around the three equilibrium points.

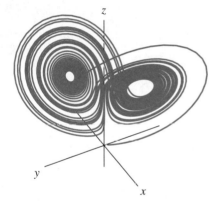

Figure 5.47
A solution curve of the nonlinear Lorenz system.

EXERCISES FOR SECTION 5.5

Exercises 1–4 refer to the three-species food chain model

$$\frac{dx}{dt} = x(1 - x) - xy$$
$$\frac{dy}{dt} = y(1 - y) + xy - yz$$
$$\frac{dz}{dt} = z(1 - z) + yz.$$

from the text. In the section we studied the linearization for the equilibrium point where all three species coexisted. The other equilibrium points are listed below. For each point,

(a) find the Jacobian at the point,

(b) find the eigenvalues and eigenvectors,

(c) classify the equilibrium point,

(d) sketch the phase space of the linearized system, and

(e) discuss in a few sentences what happens to a solution with initial value near the equilibrium point. There will be several cases, depending on which populations are nonzero. You need not consider negative populations, but consider all combinations of zero and positive populations that are near the equilibrium point.

1. $(1, 0, 0)$ **2.** $(0, 1, 0)$ **3.** $(0, 0, 1)$ **4.** $(1, 0, 1)$

5. Discuss why there is an equilibrium point at $(1, 0, 1)$ with trees and wolves coexisting at their carrying capacities, but there are not equilibrium points at $(0, 1, 1)$ and $(1, 1, 0)$.

    (a) First give the "mathematical" reason by studying the differential equations.

    (b) Relate this to the model to see if your conclusions are "physically reasonable."

6. Suppose that a moose disease enters the region of study and that βy moose per time unit are killed by the disease (where β is small).

    (a) Modify the model to include the moose disease.

    (b) What are the equilibrium points for the new model (recall β is small)?

    (c) How does an increase in β affect the equilibrium populations of trees, moose, and wolves where all three species coexist?

7. In the section we modified the original tree-moose-wolf model to include the effect of a wolf disease by subtracting a term ζy from the wolf population. It is tempting to claim that the effect of a disease among the wolves changes the growth-rate parameter of the wolf population. That is, to change the dz/dt equation to

$$\frac{dz}{dt} = \zeta z(1 - z) + yz,$$

where ζ is a constant near 1. The equations for dx/dt and dy/dt remain unchanged.

    (a) Find the equilibrium points of this new system [*Hint*: The equilibrium points will depend on the parameter ζ. Remember, ζ is close to 1.]

    (b) What effect does a decrease in ζ have on the equilibrium population where all three species coexist?

    (c) Is the change in the equilibrium point what you expected? Why or why not?

8. Besides bad weather during the winter of 1996-97, the moose population on Isle Royale was also attacked by winter wood ticks.* We can think of the ticks as another predator on moose.

    (a) Suppose the ticks eat only moose blood. Incorporate the effect of the ticks into the model. [*Hint*: Introduce a new dependent variable.]

    (b) Discuss the equilibria of your new model.

---

*See "Winter Devastates Island's Moose," by Les Line, *New York Times* Science Section, April 1, 1997.

5.6 PERIODIC FORCING OF NONLINEAR SYSTEMS AND CHAOS

Thus far in this chapter we have discussed only autonomous nonlinear systems. On the other hand, in Chapter 4 we encountered many important second-order equations that were nonautonomous. In particular, periodically forced harmonic oscillator equations depended explicitly on time and exhibited quite interesting behavior. For example, in Section 4.2, we found that periodic forcing affects the amplitude of the oscillating solutions and could even cause the system to "blow up" in the case of resonant forcing. Also, in Section 4.5 we studied a particular nonlinear system with periodic forcing (the Tacoma Narrows bridge) and saw that it had surprising behavior as well. In this section we continue the study of periodically forced nonlinear systems.

As in the Lorenz system, these systems are examples of systems that often exhibit *chaotic* behavior. In this context, the word chaotic means that solutions exhibit infinitely many different qualitative behaviors. Moreover, these different solutions are packed closely together, so that any change in the initial conditions has a radical effect on the long-term behavior of the solution. For these reasons, it is impossible to find analytic solutions of a chaotic system since there are so many different types of solutions. Also a complete qualitative analysis of solutions is impossible for the same reason. Instead we offer a few glimpses of the startling behavior of these systems so that we can recognize "chaos" when we see it in other systems. We note that these systems are the subject of active research in mathematics. If our presentation seems incomplete, it is because the complete story has yet to be discovered.

A Periodically Forced Duffing's Equation

We begin with an example of a nonlinear system that we "completely understand," that is, we can draw the phase plane and describe the behavior of every solution. We then add a periodic forcing term to this system and develop techniques for gaining insight for the resulting nonautonomous system.

The nonlinear system we consider is a Duffing's equation of the form

$$\frac{dy}{dt} = v$$
$$\frac{dv}{dt} = y - y^3.$$

This is a Hamiltonian system with energy $H(y, v) = v^2/2 - y^2/2 + y^4/4$. Hence we can draw the phase plane for this system by drawing the level curves of H (see Figure 5.48).

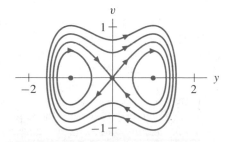

Figure 5.48
Phase plane for the system

$$\frac{dy}{dt} = v$$
$$\frac{dv}{dt} = y - y^3.$$

The system has two center equilibrium points at $(\pm 1, 0)$ and a saddle equilibrium point at $(0, 0)$. The solutions that tend to the saddle as $t \to -\infty$ also tend to it as $t \to +\infty$. All other solutions are periodic, either looping around one of the centers or looping around all three equilibrium points.

We modify this system by adding a sinusoidal forcing term to obtain the system

$$\frac{dy}{dt} = v$$
$$\frac{dv}{dt} = y - y^3 + \epsilon \sin t.$$

Note that the amplitude of the forcing is given by ϵ and the period is 2π (see the exercises for the interpretation of this system as a second-order equation).

The Return Map

As we pointed out in Section 4.1, the usual tools for dealing with systems — the vector field, direction field, and phase plane — are not relevant to nonautonomous systems because the vector field changes with time. We really need a three-dimensional picture with y-, v-, and t-axes. We would then think of solutions as moving along curves in the three-dimensional yvt-space. An attempt to render such a drawing is given in Figure 5.49. This figure contains only three solutions and yet it is fairly difficult to visualize. Also, to understand the long-term behavior of solutions, we would have to extend the t-axis quite far. We need to find a better way to depict solutions.

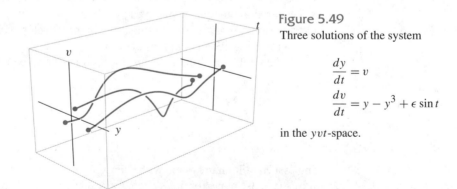

Figure 5.49
Three solutions of the system

$$\frac{dy}{dt} = v$$
$$\frac{dv}{dt} = y - y^3 + \epsilon \sin t$$

in the yvt-space.

To represent the solutions of this system, we use an idea that is due to the person who first saw and understood the ramifications of chaotic systems, the mathematician Henri Poincaré. The goal is to replace the three-dimensional picture in Figure 5.49 by a two-dimensional picture. Fix an initial point $y(0) = y_0$, $v(0) = v_0$. In Figure 5.50 this point is represented by a point in the plane $t = 0$. If we draw the plane $t = T$ for some later time T in the yvt-space, we can ask where the solution that starts at (y_0, v_0) at time $t = 0$ meets this plane. By following the solution curve until time $t = T$, we can determine where this meeting point is (see Figure 5.50).

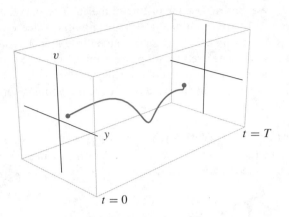

Figure 5.50
Initial point on the plane $t = 0$ and the intersection of the solution on the plane $t = T$.

Now we take advantage of the fact that the forcing term is periodic. Take $T = 2\pi$. The solution starting at $t = 0$ intersects the plane $t = 2\pi$ at some point $(y_1, v_1) = (y(2\pi), v(2\pi))$. One way to follow this solution further is to extend the t-axis. However, because the forcing is periodic, the differential equations are periodic with period 2π. Hence the curve that starts at (y_1, v_1) when $t = 2\pi$ is a translation by 2π time units of the curve that starts at (y_1, v_1) on the plane $t = 0$. So we can follow the solution for another 2π time units without adding to the t-axis. After 2π more time units, the solution hits the plane $t = 2\pi$ again, this time at a point (y_2, v_2), which is equal to $(y(4\pi), v(4\pi))$. We can repeat the process by moving the point (y_2, v_2) to the plane $t = 0$, and then following the solution for another 2π time units (see Figure 5.51).

This picture is still very complicated when t is large since there are many strands of the solution running from $t = 0$ to $t = 2\pi$. However, the behavior of the solution can be recovered from a simpler picture. Suppose we turn this picture so that the t-axis goes directly into the paper. Then we see only the points where the solution pierces the plane $t = 0$, that is, the points $(y_0, v_0) = (y(0), v(0))$, $(y_1, v_1) = (y(2\pi), v(2\pi))$, $(y_2, v_2) = (y(4\pi), v(4\pi))$, ... This sequence of points is enough to give a good idea of the long-term behavior of the solution.

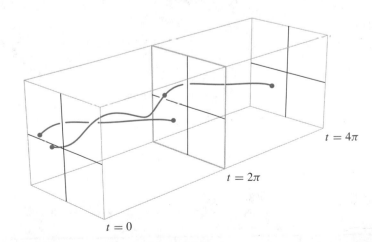

Figure 5.51
The phase space from $0 \leq t \leq 4\pi$. Note that the section from 2π to 4π is just a translation of the section from 0 to 2π.

We thus replace the picture of the solution curves in the three-dimensional phase space with a function that takes a given initial value (y_0, v_0) in the plane $t = 0$ to its "first return" point (y_1, v_1), also in the plane $t = 0$. This function is called the **return map** or **Poincaré return map**; we think of points in the plane $t = 0$ as being taken or "mapped" to the next point of intersection with this plane by the return map. By repeatedly applying this function, we see the points where the solution pierces the plane in succession.

Return map for the unforced system

To illustrate this procedure, we first consider a simple case. Let $\epsilon = 0$, so that we have the unforced system

$$\frac{dy}{dt} = v$$
$$\frac{dv}{dt} = y - y^3.$$

This is an autonomous system, but we use the ideas above to give a picture of the return map in this special case. If we start with an initial condition near the equilibrium point at $(1, 0)$, then the solution in yvt-space spirals around this equilibrium point (see Figure 5.52). The return map for this solution picks out points on this solution curve at 2π time intervals and thus yields a sequence of points on a loop around the $(1, 0)$. If we take a different initial condition farther from the $(1, 0)$, the solution spirals at a different rate around the $(1, 0)$, and the resulting return map yields a sequence of points that progress more slowly around the $(1, 0)$. The equilibrium points of the system give solutions that always return to the same point on the plane. Such points are called **fixed points** for the Poincaré return map. Finally, the solutions that tend from the saddle equilibrium point back to itself give a sequence of points that progress from the origin and back to the origin via the Poincaré return map (see Figure 5.53).

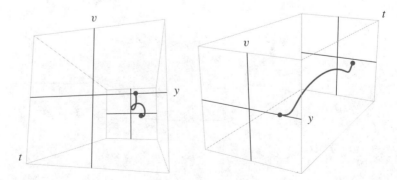

Figure 5.52
In the yvt-space, two views of a solution with initial point near $(1, 0)$ for the unforced system.

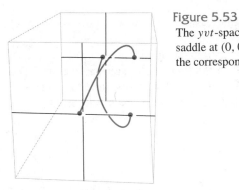

Figure 5.53
The yvt-space picture of a solution that tends from the saddle at $(0, 0)$ as $t \to -\infty$ back to $(0, 0)$ as $t \to +\infty$ and the corresponding Poincaré return map.

In Figure 5.54, we depict the results of applying the return map to several different initial conditions in the unforced system. In this case the Poincaré return map picture looks very much like the phase plane. Initial conditions near the $(\pm 1, 0)$ give solutions that stay on closed loops around $(\pm 1, 0)$, so in the Poincaré map picture these solutions give sequences of points on loops around the $(\pm 1, 0)$. The distribution of points around these loops is different at different radii because the rate at which the solutions spiral around the origin decreases as the radius increases. The orbits connecting the saddle fixed point to itself give sequences running from the origin back to the origin.

Just as with the phase space, it is very instructive to watch these pictures as they are computed. The order in which the sequence of points appears for the return map gives much more information than the static picture. However, we can still use the completed Poincaré map picture to draw conclusions about solutions.

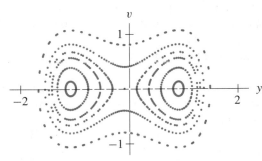

Figure 5.54
The Poincaré return map for many different solutions of the system

$$\frac{dy}{dt} = v$$

$$\frac{dv}{dt} = y - y^3.$$

Return Map for the Forced Nonlinear System

We now return to the forced nonlinear system

$$\frac{dy}{dt} = v$$

$$\frac{dv}{dt} = y - y^3 + \epsilon \sin t.$$

To compute solutions numerically we need to choose a value of ϵ. For the sake of illustration we take $\epsilon = 0.06$; other values of ϵ are considered in the exercises. Our goal is to use the Poincaré return map pictures to interpret the behavior of solutions of this system.

We fix an initial condition near the center $(1, 0)$, compute the solution in the yvt-space and then the Poincaré return map picture. The resulting image is given in Figure 5.55. The $y(t)$-graph of this solution is given in Figure 5.56. We can use this picture to predict the behavior of the solution: At least for times that are multiples of 2π, the position of the solution stays relatively near $(1, 0)$.

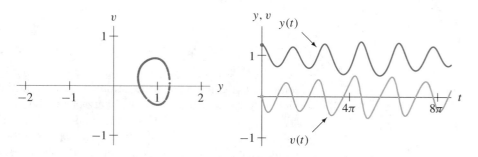

Figure 5.55
Poincaré return map for a solution with initial point near $(1, 0)$ for the forced Duffing system with $\epsilon = 0.06$.

Figure 5.56
The $y(t)$- and $v(t)$-graphs of the solutions in Figure 5.55.

The solutions that start near the saddle at the origin are much more interesting. The Poincaré return map for such a solution is given in Figure 5.57. Instead of a simple curve connecting the origin to itself all on one side of the phase plane, we see a cloud of points on both sides of the origin.

Figure 5.57 allows us to predict that a solution that starts near the origin behaves in a wild way. The y-coordinate takes on both positive and negative values. Moreover, because the points appear to follow no particular pattern, we can predict that the $y(t)$-graph of this solution oscillates without any particular pattern. This indeed is what we observe in Figure 5.58.

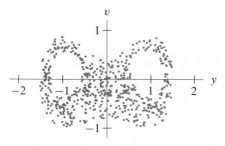
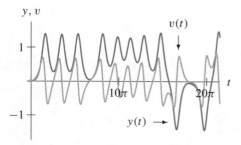
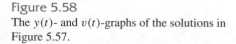

Figure 5.57
Poincaré return map for a solution with initial point near the origin for the forced nonlinear system with $\epsilon = 0.06$.

Figure 5.58
The $y(t)$- and $v(t)$-graphs of the solutions in Figure 5.57.

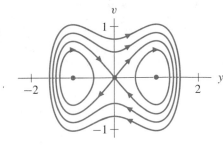

Figure 5.59

Phase plane for the unforced system

$$\frac{dy}{dt} = v$$

$$\frac{dv}{dt} = y - y^3.$$

We can get some idea of why this solution behaves the way it does by looking again at the phase plane of the unforced system (see Figure 5.59). A solution that starts near the origin, say just to the right of $(0, 0)$ near the separatrix leaving the origin, moves away from the origin for a while. Then it loops back and returns close to $(0, 0)$. The forcing term is very small, so it has little effect on the solution when it is far from $(0, 0)$, since the vector field has relatively large magnitude in this region. So we expect that the solution will return to this vicinity near the incoming separatrix. When the solution is close to the origin, the vector field of the unforced system is small, so the forcing becomes more important. When the solution is close to $(0, 0)$, if the forcing term is pushing the solution up (that is, $0.06 \sin t > 0$), then the solution moves toward the region $v > 0$ and hence follows the outgoing separatrix back into the right half-plane. If, on the other hand, the forcing term is negative when the solution is close to the origin, then it is possible that there is enough "push" to move the solution below the incoming separatrix. In this situation the solution proceeds into the left half-plane. Once in the left half-plane, the solution makes a loop around the left equilibrium point.

So, with each loop the solution of the forced system must "make a choice" when it returns to a neighborhood of the origin. The solution can go into either the right or left half-plane. Which direction the solution goes depends on the position of the solution relative to the origin and the sign of the forcing term when it arrives near the origin. So the choice depends on timing.

As a consequence we cannot predict what happens when the initial condition is changed slightly. The solution moves slowly when it is near $(0, 0)$, so a slight change of initial condition that pushes the solution closer to the origin makes a significant difference in the amount of time the solution spends near $(0, 0)$. This in turn affects the time at which the solution returns to a neighborhood of the origin, and hence it can affect which side of the yv-plane it next enters. Hence, a slight difference in the initial condition can make a radical difference in the long-term behavior of the solution. This is demonstrated in Figure 5.60 where the $y(t)$-graphs of two solutions with initial

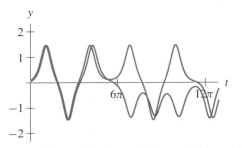

Figure 5.60

The $y(t)$-graphs of two solutions whose initial points are very close together.

conditions that are very close together are shown. The solutions stay close for a while, but at some later time they are far enough apart so that they make different decisions about which way to turn when they are near the origin. After this time the solutions are radically different.

Reality check

This sort of behavior is rather unnerving. We know that the system is "deterministic." The behavior of solutions is completely determined by the right-hand side of the system of differential equations. However, when we look at solutions, they behave in a way that does not seem to have any particular pattern. We have even anthropomorphized the solutions, saying things like "the solution decides which way to go." Solutions don't think. They don't have to think because their behavior is determined by the right-hand side of the differential equation. What is going on is that a slight change in the initial conditions can cause a huge change in the long-term behavior of the system.

While this sort of behavior seems unusual compared with the differential equations we have studied in previous sections, it is not at all unusual in nature. Physical systems like the flow of water in a turbulent stream, the weather patterns on the earth, and even the flipping of a coin, all behave this way. These systems are deterministic in that they follow strict laws of physics. This does not mean that they are predictable. A small change in initial conditions can make a radical difference in their behavior.

It is even dangerous to trust numerical simulations of these systems. We know that every numerical method gives only approximations of solutions. There are small errors in each step of the simulation. For a system like the one above, a small error in the numerical approximation gives us an approximate solution that is slightly different from the intended solution. But nearby solutions can have radically different long-term behavior. So a numerical simulation can give results that are very far from the desired solution.

This is one reason that long-range (beyond five days) weather prediction is usually not very accurate. Incomplete knowledge of weather systems and errors in numerical simulations yield predictions that can be far from correct.

The Periodically Forced Pendulum

As a second example of a periodically forced nonlinear system we return to the system modeling the motion of a pendulum. We can think of a pendulum sitting on a table that is being shaken periodically. It turns out that much of the behavior observed above also occurs for this system.

The equations

The equations of the periodically forced pendulum with mass 1 and arm length 1 are

$$\frac{d\theta}{dt} = v$$

$$\frac{dv}{dt} = -g \sin \theta + \epsilon \sin t,$$

where g is the gravitational constant. The forcing term $\epsilon \sin t$ models an external force that periodically pushes the pendulum clockwise and counterclockwise with amplitude ϵ and period 2π. For convenience we assume that units of time and distance have been chosen so that $g = 1$ and our system is

$$\frac{d\theta}{dt} = v$$

$$\frac{dv}{dt} = -\sin\theta + \epsilon\sin t.$$

The return map

We construct the return map for the forced pendulum system exactly as above. The period of the forcing term is again 2π, so we follow solutions in $\theta v t$-space starting on the plane $t = 0$ and marking where they cross the plane $t = 2\pi$.

For the examples below, we fix $\epsilon = 0.01$; other values of ϵ are considered in the exercises. In Figure 5.61 we show the Poincaré return map for a solution with initial condition near $(0, 0)$. The resulting thick loop corresponds to a solution that oscillates with varying amplitude. The $\theta(t)$-graph of the same solution in Figure 5.62 shows this oscillation.

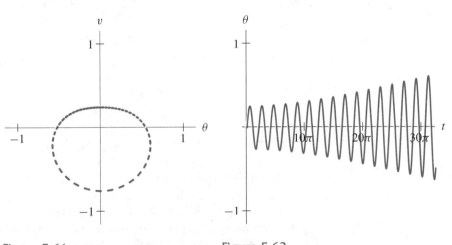

Figure 5.61
Poincaré return map for the periodically forced pendulum system with $\epsilon = 0.01$ for a solution with initial condition near $(0, 0)$.

Figure 5.62
Graph in the $t\theta$-plane of the solution in Figure 5.61. (The amplitude of the oscillation remains bounded for all t.)

Initial conditions near $(0, 0)$ for the forced pendulum system correspond to starting the pendulum at a small angle and with small velocity. In this situation the unforced (undamped) pendulum oscillates forever with small constant amplitude. The addition of the forcing term means that, just as for the forced harmonic oscillator, the forcing sometimes pushes with the direction of motion, making the pendulum swing higher, and sometimes pushes against the direction of motion, making the pendulum swing less. Unlike the harmonic oscillator, the period of the pendulum swing depends on the amplitude. Hence forcing adds and subtracts energy from the system in a way that, over the long term, is quite complicated. (This is not so evident from the picture because the forcing is small.)

Solutions near the saddle equilibrium points

Figure 5.63 shows the Poincaré return map for a solution of the periodically forced pendulum equation ($\epsilon = 0.01$) with initial condition near $(-\pi, 0)$. These points form a "cloud" with no particular structure. Also, the θ coordinate becomes quite large. This means that the pendulum arm has completely rotated many times in one direction.

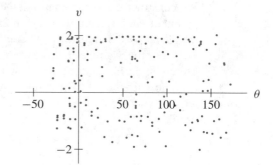
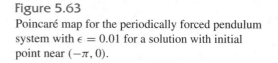

Figure 5.63
Poincaré map for the periodically forced pendulum system with $\epsilon = 0.01$ for a solution with initial point near $(-\pi, 0)$.

Just as in the forced Duffing system, whenever the solution approaches a saddle equilibrium point it must "decide" which way to turn. If it stays in the upper half-plane, then the θ-coordinate of the solution increases by a multiple of 2π before returning to the vicinity of another saddle. If it chooses to go into the lower half-plane, then the θ-coordinate decreases by 2π before another choice is made. If we graph the θ-coordinate of the solution above on the $t\theta$-plane, we see that it moves in a very wild way. In particular, it is possible that the θ-coordinate of the solution can become very large positive (or negative) if the pendulum repeatedly "decides" to rotate the same direction (see Exercises 5–8).

The decision of which way the solution turns when it is close to a saddle depends on the sign of the forcing term. At that time the behavior of the solution depends very delicately on the initial condition. If we start two solutions near $(-\pi, 0)$ with almost the same initial condition, then eventually they split apart and become radically different (see Figure 5.64).

From the pictures above, we can deduce the existence of some interesting behavior in the forced pendulum system. A solution of the forced pendulum system with initial condition near $(-\pi, 0)$ corresponds to an initial placement of the pendulum that is almost vertical but with very little velocity. Of course the pendulum swings down. During the swing the forcing term has very little effect on the motion of the pendulum.

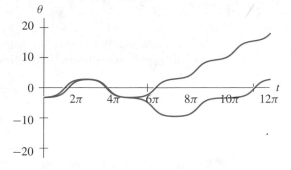
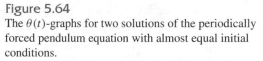

Figure 5.64
The $\theta(t)$-graphs for two solutions of the periodically
forced pendulum equation with almost equal initial
conditions.

When the pendulum swings up to the almost vertical position, it slows down
again and the effect of the forcing term is more pronounced. When the pendulum is
near the top of the swing, the forcing term either "pushes it over the top" so that it
makes another turn in the same direction, or "pulls it back" so that it swings back the
way it came. Which way the pendulum goes depends on the sign of the forcing term,
which in turn depends on the time the pendulum arrives at the top of its swing. Since
the pendulum moves very slowly near the top of the swing, a small change in initial
conditions can make a big change in the timing and hence cause a change in direction
of the pendulum.

We emphasize that the sort of physical argument given above is not meant to replace the Poincaré return map pictures and the analysis of the solutions. Although the
physical argument makes sense, it does not say whether a forcing term with $\epsilon = 0.01$ is
large enough to cause this sort of behavior.

Moral

The moral of this section is that even systems like the pendulum that we "understand"
can become very complicated when additional terms like periodic forcing are added.
The periodically forced pendulum system does not appear at first glance to be all that
complicated, but we see from the Poincaré return map that its solutions behave quite
unpredictably. A small change in the initial position frequently has a radical effect on
the behavior of the solution.

If this sort of behavior, which we now call **chaos**, can be observed in a system
as simple as the periodically forced pendulum, it should not be surprising that it can
also be found in nature. This is not a new discovery. Henri Poincaré first considered
the possibility of the existence of "chaos" in nature in the 1880s. He was studying the
motion of a small planet (an asteroid) under the gravitational influence of a star (the
sun) and a large planet (Jupiter). Poincaré developed the return map to investigate the
behavior of this system. What is remarkable is that Poincaré did not have the advantage
of looking at numerical simulations of solutions as we do today. Nevertheless, he could
see that the return map would behave in a very complicated fashion, and he wisely said
he would not attempt to draw it.

EXERCISES FOR SECTION 5.6

In Exercises 1–4, Poincaré return map pictures are given for four different orbits with four different values of ϵ and initial conditions for the periodically forced Duffing equation

$$\frac{dy}{dt} = v$$

$$\frac{dv}{dt} = y - y^3 + \epsilon \sin t.$$

described in the text. Also, four $y(t)$-graphs for solutions of this system are given.

(a) Match the Poincaré return map picture with the $y(t)$-graph.

(b) Describe in a brief essay how you made the match and describe the qualitative behavior of the solution.

(i)

(ii)

(iii)

(iv)

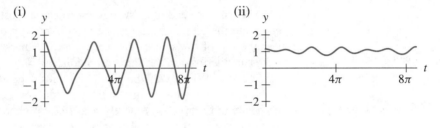

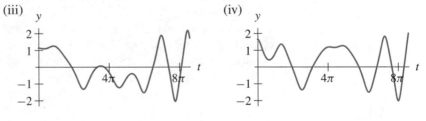

1. For $\epsilon = 0.1$, $y(0) = 1.1$, $v(0) = 0$.

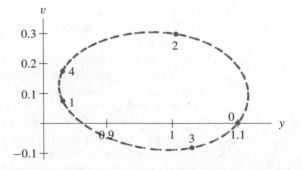

Poincaré return map with 500 iterates. The first four returns are indicated.

2. For $\epsilon = 0.4$, $y(0) = 1.1$, $v(0) = 0$.

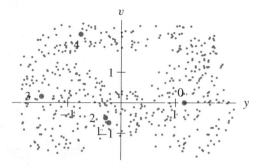

Poincaré return map with 500 iterates. The first
four returns are indicated.

3. For $\epsilon = 0.1$, $y(0) = 1.6$, $v(0) = 0$.

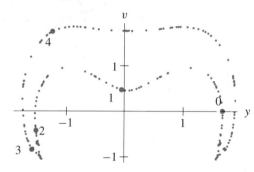

Poincaré return map with 200 iterates. The first
four returns are indicated.

4. For $\epsilon = 0.5$, $y(0) = 1.6$, $v(0) = 0$.

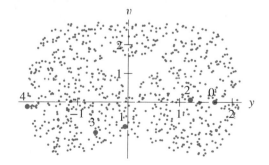

Poincaré return map with 800 iterates. The first
four returns are indicated.

In Exercises 5–8, the Poincaré return map pictures are given for four different orbits with four different values of ϵ and initial conditions for the periodically forced pendulum system

$$\frac{d\theta}{dt} = v$$

$$\frac{dv}{dt} = -\sin\theta + \epsilon \sin t$$

described in the text. Also, four $\theta(t)$-graphs for solutions of this system are given.

(a) Match the Poincaré return map picture with the $\theta(t)$-graph.

(b) Describe in a brief essay how you made the match and describe the qualitative behavior of the solution.

(c) Describe the behavior of the pendulum arm when it follows the indicated solution.

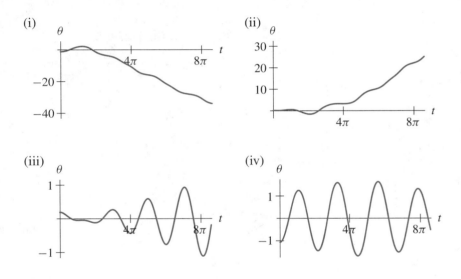

5. For $\epsilon = 0.1$, $\theta(0) = .2$, $v(0) = 0$.

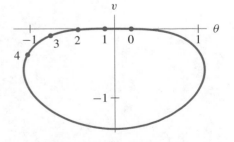

Poincaré return map with 1,000 iterates. The first four returns are indicated.

6. For $\epsilon = 0.5$, $\theta(0) = .2$, $v(0) = 0$.

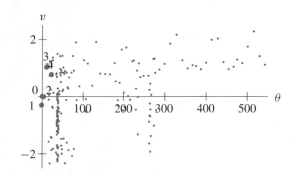

Poincaré return map with 250 iterates. The first four returns are indicated.

7. For $\epsilon = 0.1$, $\theta(0) = -1.06$, $v(0) = 0$.

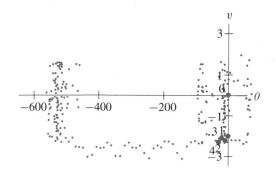

Poincaré return map with 400 iterates. The first four returns are indicated.

8. For $\epsilon = 0.5$, $\theta(0) = -1.06$, $v(0) = 0$.

Poincaré return map with 250 iterates. The first four returns are indicated.

9. The Poincaré return map for a solution of the periodically forced harmonic oscillator

$$\frac{dy}{dt} = v$$

$$\frac{dv}{dt} = -3y + 0.2\sin t$$

is given below.

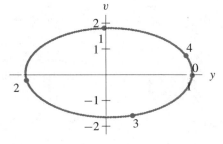

Poincaré return map for the forced harmonic oscillator system with initial conditions $y(0) = 1$, $v(0) = 0$.

(a) Find the solution with the initial condition $y(0) = 1$, $v(0) = 0$.

(b) Discuss why the Poincaré return map for this solution looks the way it does.

(c) What would the Poincaré return map for the solution with $y(0) = 4$, $v(0) = 0$ look like?

(d) Discuss why the Poincaré return map for the forced harmonic oscillator system is different from the Poincaré map of the forced pendulum.

10. Describe and sketch the Poincaré return map for the solution of the forced harmonic oscillator system

$$\frac{dy}{dt} = v$$

$$\frac{dv}{dt} = -4y + 0.1\sin 2t.$$

[*Hint:* You should be able to do this problem qualitatively, with only a small amount of computation. What is the natural frequency of the system?]

LAB 5.1   Hard and Soft Springs

In this lab we continue our study of second-order equations by considering "nonlinear springs." In Section 3.6 we developed the model of a spring based on Hooke's law. Hooke's law asserts that the restoring force of a spring is proportional to its displacement, and this assumption leads to the second-order equation

$$m\frac{d^2y}{dt^2} + k_s y = 0.$$

Since the resulting differential equation is linear, we say that the spring is linear. In this case the restoring force is $-k_s y$. In addition we assume that the friction or damping force is proportional to the velocity. The resulting second-order equation is

$$m\frac{d^2y}{dt^2} + k_d\frac{dy}{dt} + k_s y = 0.$$

Hooke's law is an idealized model that works well for small oscillations. In fact the restoring force of a spring is roughly linear if the displacement of the spring from its equilibrium position is small, but it is generally more accurate to model the restoring force by a cubic of the form $-ky + by^3$, where b is small relative to k. If b is negative, the spring is said to be hard, and if b is positive, the spring is soft. In this lab we consider the behavior of hard and soft springs for particular values of the parameters. (Your instructor will tell you which parameter value(s) from Table 5.1 to use.)

In your report, you should analyze the phase planes and $y(t)$- and $v(t)$-graphs to describe the long-term behavior of the solutions to the equations:

1. (Hard spring with no damping) The first equation that you should study is the hard spring with no damping; that is, $k_d = 0$ and $b = b_1$. Examine solutions using both their graphs and the phase plane. Consider the periods of the periodic solutions that have the initial condition $v(0) = 0$. Sketch the graph of the period as a function of the initial condition $y(0)$. Is there a minimum period? A maximum period? If so, how do you interpret these extrema?

2. (Hard spring with damping) Now use the given value of k_d and $b = b_1$ to introduce damping into the discussion. What happens to the long-term behavior of solutions in this case? Determine the value of the damping parameter that separates the underdamped case from the overdamped case.

3. (Soft spring with no damping) Consider the soft spring that corresponds to the positive value b_2 of b. Over what range of y-values is this model reasonable? Consider the periods of the periodic solutions that have the initial condition $v(0) = 0$. Sketch the graph of the period as a function of the initial condition $y(0)$. Is there a minimum period? A maximum period? Use the phase portrait to help justify your answer.

4. (Soft spring with damping) Using the given values of k_d and $b = b_2$, what happens to the long-term behavior of solutions in this case? Determine the value of the damping parameter that separates the underdamped case from the overdamped case.

5. From a physical point of view, what's the difference between a hard spring and a soft spring?

493

Your report: Address each of the five items in the form of a short essay. You may illustrate your essay with pictures of phase planes and graphs of solutions. However, your essay should be complete and understandable without the pictures. Make sure you relate the behavior of the solutions to the motion of the associated mass and spring systems.

Table 5.1

Choices for the parameter values. Assume the mass $m = 1$ unless you are told otherwise by your instructor.

Choice	k_s	k_d	b_1	b_2
1	0.1	0.15	−0.005	0.005
2	0.2	0.20	−0.008	0.008
3	0.3	0.20	−0.009	0.009
4	0.2	0.20	−0.005	0.005
5	0.1	0.10	−0.005	0.005
6	0.3	0.20	−0.007	0.007
7	0.3	0.15	−0.007	0.007
8	0.1	0.15	−0.004	0.004
9	0.2	0.15	−0.005	0.005
10	0.3	0.20	−0.008	0.008

LAB 5.2   Higher Order Approximations of the Pendulum

In previous chapters we studied the behavior of second-order, homogeneous linear equations (like the harmonic oscillator) by reducing them to first-order linear systems. This "reduction" technique can be applied to nonlinear equations as well, and in this lab we study the ideal pendulum and approximations to the pendulum using this technique.

In the text we modeled the ideal pendulum by the second-order, nonlinear equation

$$\frac{d^2\theta}{dt^2} + \frac{g}{l}\sin\theta = 0,$$

where θ is the angle from the vertical, g is the gravitational constant ($g = 32$ ft/s^2), and l is the length of the rod of the pendulum, that is, the radius of the circle on which the mass travels. In this lab we compare the results of numerical simulation of this model with the results obtained from two approximations to this model. The first approximation is a linear approximation given by

$$\frac{d^2\theta}{dt^2} + \frac{g}{l}\theta = 0.$$

The second approximation is a cubic approximation

$$\frac{d^2\theta}{dt^2} + \frac{g}{l}\left(\theta - \frac{\theta^3}{6}\right) = 0.$$

Recall from calculus that the expression $\theta - \theta^3/6$ represents the first two terms in the power series expansion of $\sin\theta$ about $\theta = 0$. We are especially interested in how close the solutions of the approximations of the ideal pendulum equation are to the original ideal pendulum equation. In particular we are interested in how closely the periods of the periodic orbits of the approximations of the pendulum equation relate to the periods of the periodic orbits of original equation. Your instructor will tell you what value of the parameter l (the length of the pendulum arm) you should use. Your report should include:

1. A phase portrait analysis for all three equations. Compare and contrast these phase portraits from the point of view of how well the linear and cubic equations approximate the ideal pendulum.

2. In order to study how the periods of the periodic orbits are related, consider the one-parameter family of initial conditions parameterized by θ_0, where $\theta(0) = \theta_0$ and $\theta'(0) = 0$ (no initial velocity). In other words you should study the various solutions that begin at a given angle with zero velocity. For what intervals of initial conditions do the periods of the periodic orbits of

$$\frac{d^2\theta}{dt^2} + \frac{g}{l}\theta = 0$$

and

$$\frac{d^2\theta}{dt^2} + \frac{g}{l}\left(\theta - \frac{\theta^3}{6}\right) = 0$$

closely approximate the periods of the periodic orbits of the ideal pendulum? (The computation of the periods in the linear approximation can be done exactly using the techniques of Chapter 3. Analytic techniques exist for computing the periods of the periodic orbits of the other two equations, but in this lab you should work numerically.) You should plot graphs of the period as a function of θ_0 using a relatively small table (5, 10, or 15 entries) of periods obtained using direct numerical simulation of the model.

3. Another family of initial conditions is $\theta(0) = 0$ and $\theta'(0) = v_0$. In this family the initial velocity is the parameter. Initially the pendulum points straight down with a given velocity v_0. What changes from your results in part 2 above?

4. Suppose you are a clock maker who makes clocks based on the motion of a pendulum. For each of the three equations, what would you do to double the period of the oscillation?

Your report: Address each of the items above in the form of a short essay. Be as systematic as possible when collecting data, and present this data in a concise and clear format. You may illustrate your essay with pictures of phase planes and graphs of solutions or of the data that you collect. However, your essay should be complete and understandable without the pictures.

LAB 5.3   A Family of Predator-Prey Equations

In this laboratory exercise, you will study a one-parameter family of nonlinear, first-order systems consisting of predator-prey equations. The family is

$$\frac{dx}{dt} = 9x - \alpha x^2 - 3xy$$
$$\frac{dy}{dt} = -2y + xy,$$

where $\alpha \geq 0$ is a parameter. In other words, for different values of α we have different systems. The variable x is the population (in some scaled units) of prey, and y is the population of predators. For a given value of α, we want to understand what happens to these populations as $t \to \infty$.

You should investigate the phase portraits of these equations for various values of α in the interval $0 \leq \alpha \leq 5$. To get started, you might want to try $\alpha = 0, 1, 2, 3, 4$, and 5. Think about what the phase portrait means in terms of the evolution of the x and y populations. Where are the equilibrium points? What types are they? What happens to a typical solution curve? Also, consider the behavior of the special solutions where either $x = 0$ or $y = 0$ (solution curves lying on the x- or y-axes).

Determine the bifurcation values of α. That is, determine the values of α where nearby α's lead to "different" behaviors in the phase portrait. For example, $\alpha = 0$ is a bifurcation value because for $\alpha = 0$ the long-term behavior of the populations is dramatically different than the long-term behavior of the populations if α is slightly positive.

Your report: After you have determined all of the bifurcation values for α in the interval $0 \leq \alpha \leq 5$, study enough specific values of α to be able to discuss all of the various population evolution scenarios for these systems. In your report, you should describe these scenarios using the phase portraits and $x(t)$- and $y(t)$-graphs. Your report should include:

1. A brief discussion of the significance of the various terms in the system. For example, what does the $9x$ represent? What does the $3xy$ term represent?

2. A discussion of all bifurcations including the bifurcation at $\alpha = 0$. For example, a bifurcation occurs between $\alpha = 3$ and $\alpha = 5$. What does this bifurcation mean for the predator population?

Address the questions above in the form of a short essay, and support your assertions with selected illustrations. (Please remember that, although one good illustration may be worth 1000 words, 1000 illustrations are usually worth nothing.)

6 | LAPLACE TRANSFORMS

In this chapter we study an analytic technique for finding formulas for solutions of certain differential equations using an operation called the Laplace transform. This technique is particularly effective on linear, constant-coefficient equations and is quite different from our previous methods. The Laplace transform lets us replace the operations of integration and differentiation with algebraic computations. As a technique for solving initial-value problems, the Laplace transform is sometimes more efficient and sometimes less efficient than our previous techniques.

The importance of the Laplace transform is that it is useful in several different types of applications. For example, the Laplace transform lets us deal efficiently with linear, constant-coefficient differential equations that have discontinuous forcing functions. These discontinuities include simple jumps that model the action of a switch. Using Laplace transforms, we can also make a meaningful mathematical model of the impulse force provided by a hammer blow or an explosion. Laplace transforms also find solutions for a forced harmonic oscillator using solutions generated by a completely different forcing term. This technique is useful when dealing with a system that is modeled by an unknown linear, constant-coefficient differential equation for which we provide the forcing function (see Section 6.5). Finally, Laplace transforms give us a way of extending our qualitative analysis of homogeneous linear equations using eigenvalues to nonhomogeneous linear equations (see Section 6.6).

6.1 LAPLACE TRANSFORMS

Integral Transforms

In this chapter we study a tool, the **Laplace transform**, for solving differential equations. The Laplace transform is one of many different types of *integral transforms*. In general, integral transforms address the question: How much is a given function $y(t)$ "like" a particular standard function?

For example, if $y(t)$ represents a radio signal, we might want to compare it to the function $\sin \omega t$, a sine wave with frequency $\omega/(2\pi)$. By adjusting the parameter ω, we could test how well $y(t)$ fits sine waves with different frequencies. Ideally, for each value of ω, we would like a number that indicates how much $y(t)$ is like $\sin \omega t$. One way to accomplish this comparison is to compute the integral

$$\int_{-N}^{N} y(t) \sin \omega t \, dt$$

for large N. If $y(t)$ is oscillating with frequency $\omega/(2\pi)$ and is positive when $\sin \omega t$ is positive, then this integral is very large. If $y(t)$ has some other frequency, then the signs of $y(t)$ and $\sin \omega t$ differ at some times t, so there is cancellation in the integral and its value is smaller.

To use this idea in differential equations, it is natural to compare $y(t)$ to the function that comes up most often, the exponential function. In fact we could use the complex exponential and compute

$$\int_{-\infty}^{\infty} y(t) e^{-zt} \, dt,$$

where $z = s + i\omega$ is a complex parameter. The larger the value of this integral for a particular z, the more $y(t)$ is "like" e^{zt}. In particular, if $y(t) = e^{zt}$, then the integrand is the constant function 1, and the integral is infinite.

In practice it turns out to be easier to write z in terms of its real and imaginary parts as $z = s + i\omega$ and to compute the two transforms

$$\int_{-\infty}^{\infty} y(t) e^{-st} \, dt \quad \text{and} \quad \int_{-\infty}^{\infty} y(t) e^{-i\omega t} \, dt$$

separately. The integral

$$\int_{-\infty}^{\infty} y(t) e^{-i\omega t} \, dt$$

is called the *Fourier transform* of the function $y(t)$, and its value at a particular value of ω is a measure of the extent to which $y(t)$ is oscillating with frequency $\omega/(2\pi)$. The imaginary part of this quantity is the comparison of $y(t)$ to $\sin \omega t$ discussed above, and the real part is the comparison of $y(t)$ to $\cos \omega t$. Although the Fourier transform has many important applications in differential equations, it is not discussed in this chapter. Rather, we focus on the integral $\int y(t) e^{-st} \, dt$.

Laplace Transforms

The Laplace transform of a (given) function $y(t)$ uses integration to compare $y(t)$ to the exponential functions e^{st}.

DEFINITION   The **Laplace transform** function Y of the function y is defined by

$$Y(s) = \int_0^\infty y(t)\, e^{-st}\, dt$$

for all numbers s for which this improper integral converges. ■

For example, if $y(t) = e^{2t}$, then its Laplace transform $Y(s)$ is determined by evaluating the improper integral

$$Y(s) = \int_0^\infty e^{2t}\, e^{-st}\, dt$$

$$= \int_0^\infty e^{(2-s)t}\, dt$$

$$= \lim_{b \to \infty} \int_0^b e^{(2-s)t}\, dt$$

$$= \lim_{b \to \infty} \left[\frac{1}{2-s} e^{(2-s)t} \Big|_0^b \right]$$

$$= \frac{1}{2-s} \lim_{b \to \infty} \left[e^{(2-s)b} - e^0 \right].$$

Since

$$\frac{1}{2-s} \lim_{b \to \infty} e^{(2-s)b} = \begin{cases} \infty, & \text{if } s < 2; \\ 0, & \text{if } s > 2, \end{cases}$$

we see that the improper integral for $Y(s)$ does not converge if $s \le 2$ and that

$$Y(s) = \frac{1}{s-2} \quad \text{if} \quad s > 2.$$

In other words, the Laplace transform function $Y(s)$ for the function $y(t) = e^{2t}$ is

$$Y(s) = \begin{cases} \dfrac{1}{s-2}, & \text{if } s > 2; \\[2mm] \text{undefined}, & \text{if } s \le 2. \end{cases}$$

This computation involves a number of technical details with improper integrals that are important, but don't let them cause you to lose touch with the underlying idea involved in the definition. The Laplace transform $Y(s)$ associated to $y(t)$ is a function that measures how close $y(t)$ is to the exponential functions e^{st} for all values of s. In this example we found that the Laplace transform of e^{2t} is a function $Y(s)$ that is very small if s is much larger than 2, but as s approaches 2, the values of $Y(s)$ increase until they become unbounded at $s = 2$.

Surprisingly we can use this idea to solve differential equations. As we will soon see, we can use the Laplace transform to convert a differential equation into an algebraic equation, which is often easier to solve. This conversion is similar to the translation of an English sentence into a sentence in Chinese. Both sentences have the same meaning, but the words and grammar are very different. The function $y(t)$ represents a phenomenon in terms of the independent variable t "in the time domain," and the function $Y(s)$ represents the same phenomenon in the s-domain.

More formally, we say that the Laplace transform defines a mathematical transformation that takes a function $y(t)$ into its transformed function $Y(s)$. From a strict mathematical point of view, this transform is just one (particularly nice) function that converts a function $y(t)$ into a new function $Y(s)$, and we use the script letter $\mathcal{L}$ to represent this function. In other words, we say that $Y = \mathcal{L}[y]$. In particular our computation of the Laplace transform of e^{2t} is often written as

$$\mathcal{L}[e^{2t}] = \frac{1}{s - 2} \quad \text{for } s > 2.$$

This way of writing the Laplace transform is somewhat sloppy in that it assumes that the independent variable of the transformed function is the variable s, but this lack of precision hardly ever causes confusion.

In order to be certain that the Laplace transform exists (that is, the improper integral converges) for at least some values of s, we must restrict attention to continuous or piecewise continuous functions $y(t)$ for which there are positive constants K and M (that depend on y) such that $|y(t)| < e^{Mt}$ for $t \geq K$. Such functions are said to have no more than "exponential growth." As a practical matter, most functions encountered in applications have this property.

Also note that the Laplace transform is defined as an integral over the interval $0 \leq t < \infty$ (rather than over the entire real line $-\infty < t < \infty$). Use of this interval is necessary since, for $s > 0$, $e^{-st} \to \infty$ as $t \to \infty$. If the Laplace transform were to be evaluated over the entire real line, then the restrictions on the function $y(t)$ in order to ensure convergence of the integral would be much more stringent. In most applications we are interested primarily in the future ($t \geq 0$) anyway.

Computation of Laplace Transforms of Exponential Functions

For a function $y(t)$ and a given number s, the value of the Laplace transform at s measures the degree to which the function $y(t)$ resembles the function e^{st} on the interval $t \geq 0$. To check this assertion, we compute the Laplace transform of an arbitrary exponential function. If $y(t) = e^{at}$, then

$$\mathcal{L}[e^{at}] = \int_0^\infty e^{at}\, e^{-st}\, dt$$

$$= \int_0^\infty e^{(a-s)t}\, dt.$$

Recall that this improper integral is actually a limit of integrals as the upper limit of integration goes to infinity. Hence

$$\mathcal{L}[e^{at}] = \lim_{b \to \infty} \int_0^b e^{(a-s)t} \, dt$$

$$= \lim_{b \to \infty} \left[\frac{1}{a-s} e^{(a-s)t} \Big|_0^b \right]$$

$$= \frac{1}{a-s} \lim_{b \to \infty} \left[e^{(a-s)b} - e^0 \right]$$

$$= \frac{1}{s-a} \quad \text{if } s > a.$$

If $s \leq a$, then the improper integral diverges. Strictly speaking, the Laplace transform of e^{at} is the rational function $1/(s-a)$ restricted to the interval $s > a$. Note that our earlier calculation of $\mathcal{L}[e^{2t}]$ was simply a special case of this calculation.

There is another special case of this computation that is worth mentioning at this point. Note that if $a = 0$, then $e^{at} = 1$ for all t. Consequently we have also computed the Laplace transform of the function that is constantly 1. Since $a = 0$, we have

$$\mathcal{L}[1] = \frac{1}{s} \quad \text{for } s > 0.$$

We often need the results of these computations as well as the transforms of other frequently encountered functions. Therefore, we provide a table of Laplace transforms as well as important properties of this transformation on page 557.

Properties of the Laplace Transform

There are many transforms that convert one function into another, but the Laplace transform has one very special property that is the basis for its success in solving differential equations.

LAPLACE TRANSFORM OF DERIVATIVES Given a function $y(t)$ with Laplace transform $\mathcal{L}[y]$, the Laplace transform of dy/dt is

$$\mathcal{L}\left[\frac{dy}{dt}\right] = s\mathcal{L}[y] - y(0). \qquad \blacksquare$$

To verify this theorem we use the definition of $\mathcal{L}[dy/dt]$ and compute

$$\mathcal{L}\left[\frac{dy}{dt}\right] = \int_0^\infty \frac{dy}{dt} e^{-st} \, dt.$$

Using integration by parts with the choices $u = e^{-st}$ and $dv = (dy/dt) \, dt$, we have $du = -s \, e^{-st} \, dt$ and $v = y(t)$, and therefore

$$\mathcal{L}\left[\frac{dy}{dt}\right] = \lim_{b \to \infty} \left[y(t) e^{-st} \Big|_0^b \right] + \int_0^\infty y(t) s e^{-st} \, dt.$$

Our earlier assumption that $y(t)$ has at most exponential growth is important here. Since $|y(t)| < e^{Mt}$ for some constant M,

$$\lim_{t \to \infty} y(t)\, e^{-st} = 0$$

for sufficiently large s. Hence,

$$\mathcal{L}\left[\frac{dy}{dt}\right] = -y(0) + \int_0^\infty y(t)\, s\, e^{-st}\, dt$$

$$= -y(0) + s\mathcal{L}[y].$$

This formula is the fundamental property of the Laplace transform that lets us essentially replace the calculus operation of differentiation in the t-domain with the algebraic operation of multiplication by s in the s-domain. Of course, this description is not quite right since we also have to remember to subtract $y(0)$, but in any case we think of the Laplace transform as an operation that turns a problem in calculus into a problem in algebra.

Turning a problem in differential equations into a problem in algebra would not be very useful if the Laplace transform did not have reasonable algebraic properties. However, due to its definition in terms of an integral, the Laplace transform has nice linearity properties.

LINEARITY OF THE LAPLACE TRANSFORM   Given functions f and g and a constant c,

$$\mathcal{L}[f + g] = \mathcal{L}[f] + \mathcal{L}[g]$$

$$\mathcal{L}[cf] = c\mathcal{L}[f].$$

In other words, the transform "operator" $\mathcal{L}$ is a linear operator.   ∎

To verify these properties, we use the linearity properties of integration. That is,

$$\mathcal{L}[f + g] = \int_0^\infty (f(t) + g(t))\, e^{-st}\, dt$$

$$= \int_0^\infty f(t)\, e^{-st}\, dt + \int_0^\infty g(t)\, e^{-st}\, dt$$

$$= \mathcal{L}[f] + \mathcal{L}[g],$$

and

$$\mathcal{L}[cf] = \int_0^\infty cf(t)\, e^{-st}\, dt = c\int_0^\infty f(t)\, e^{-st}\, dt = c\mathcal{L}[f].$$

With the derivative formula and the linearity of $\mathcal{L}$ established, we can now use the Laplace transform to solve differential equations.

Solving Differential Equations Using the Laplace Transform

Consider the initial-value problem

$$\frac{dy}{dt} = y - 4e^{-t}, \quad y(0) = 1.$$

We first study this problem using qualitative techniques so that we know what to expect for the analytical solution.

Qualitative analysis

The slope field for

$$\frac{dy}{dt} = y - 4e^{-t}$$

is given in Figure 6.1. It indicates that the solution with $y(0) = 1$ decreases for t near zero. Once the solution crosses below $y = 0$, both terms on the right-hand side of the differential equation are negative, so the solution continues to decrease. For large t, the equation is close to the equation

$$\frac{dy}{dt} = y,$$

and hence we expect the solution $y(t) \to -\infty$ as $t \to \infty$.

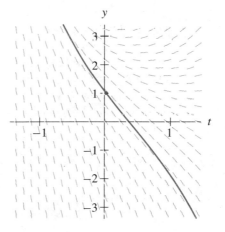

Figure 6.1
Slope field for $dy/dt = y - 4e^{-t}$ and solution with $y(0) = 1$.

Solution using Laplace transforms

This equation is linear, so we could use the techniques of Section 1.8. However, the Laplace transform provides an alternate method of solution.

Starting with the initial-value problem

$$\frac{dy}{dt} = y - 4e^{-t}, \quad y(0) = 1,$$

the first step is to take the Laplace transform of both sides

$$\mathcal{L}\left[\frac{dy}{dt}\right] = \mathcal{L}[y - 4e^{-t}].$$

Then using the previous formula for the Laplace transform of a derivative to simplify the left-hand side and the linearity of Laplace transform on the right-hand side, we obtain

$$s\mathcal{L}[y] - y(0) = \mathcal{L}[y] - 4\mathcal{L}[e^{-t}],$$

and we substitute the initial condition $y(0) = 1$ to get

$$s\mathcal{L}[y] - 1 = \mathcal{L}[y] - 4\mathcal{L}[e^{-t}].$$

Earlier we computed that $\mathcal{L}[e^{at}] = 1/(s - a)$, so we can apply this formula with $a = -1$ to obtain

$$s\mathcal{L}[y] - 1 = \mathcal{L}[y] - \frac{4}{s+1}.$$

The unknown of the original differential equation is the function y, so we solve this equation for the Laplace transform of y, obtaining

$$\mathcal{L}[y] = \frac{1}{s-1} - \frac{4}{(s-1)(s+1)}.$$

These calculations yield the Laplace transform of the solution of the initial-value problem. Note that we used only arithmetic to obtain $\mathcal{L}[y]$. (The calculus is hidden in the computation of the Laplace transforms of exponentials.) In a certain sense we have now solved the initial-value problem.

Unfortunately we are not looking for the Laplace transform $\mathcal{L}[y]$ of the solution; we really want the actual solution $y(t)$. Somehow we must "undo" or take an "inverse" Laplace transform. That is, we must figure out what function $y(t)$ has the function

$$\frac{1}{s-1} - \frac{4}{(s-1)(s+1)}$$

as its Laplace transform.

Inverse Laplace Transform

To use Laplace transforms to solve a differential equation, we first compute the Laplace transform of both sides of the equation. Next we solve for the Laplace transform of the solution. Finally, to find the solution, we find a function with the given Laplace transform. The last step is called taking the **inverse Laplace transform**. The notation for this inverse transform is $\mathcal{L}^{-1}$, that is,

$$\mathcal{L}^{-1}[F] = f \quad \text{if and only if} \quad \mathcal{L}[f] = F.$$

There is a uniqueness property for inverse Laplace transforms which states that, if f is a continuous function with $\mathcal{L}[f] = F$, then f is the only continuous function whose Laplace transform is F. This uniqueness property allows us to say "the" inverse Laplace transform of F instead of "an" inverse Laplace transform of F.

Because the Laplace transform is a linear operator, that is,

$$\mathcal{L}[f + g] = \mathcal{L}[f] + \mathcal{L}[g] \quad \text{and} \quad \mathcal{L}[cf] = c\mathcal{L}[f]$$

for any functions $f(t)$ and $g(t)$ and any constant c, it follows that the inverse Laplace transform is also a linear operator,

$$\mathcal{L}^{-1}[f + g] = \mathcal{L}^{-1}[f] + \mathcal{L}^{-1}[g] \quad \text{and} \quad \mathcal{L}^{-1}[cf] = c\mathcal{L}^{-1}[f].$$

Linearity is important because it allows us to compute the inverse Laplace transform of a complicated sum by computing the inverse Laplace transform of each summand.

Inverse Laplace transforms are generally computed in much the same way as antiderivatives in calculus are computed. We work with a small list of "known" transforms, and to compute the inverse Laplace transform of a complicated function $F(s)$, we first break it into a sum of functions whose inverse transforms we already know.

Examples of inverse Laplace transforms

When we applied the Laplace transform to the initial-value problem earlier in this section, we arrived at the expression

$$\mathcal{L}[y] = \frac{1}{s - 1} - \frac{4}{(s - 1)(s + 1)}$$

for the Laplace transform of the solution y of the given initial-value problem. Hence, we find y by computing the inverse Laplace transform

$$y = \mathcal{L}^{-1}\left[\frac{1}{s - 1} - \frac{4}{(s - 1)(s + 1)}\right].$$

By linearity we have

$$y = \mathcal{L}^{-1}\left[\frac{1}{s - 1}\right] - \mathcal{L}^{-1}\left[\frac{4}{(s - 1)(s + 1)}\right],$$

and from the formula for $\mathcal{L}[e^{at}]$ we know that

$$\mathcal{L}^{-1}\left[\frac{1}{s - 1}\right] = e^t.$$

To compute the inverse Laplace transform

$$\mathcal{L}^{-1}\left[\frac{4}{(s - 1)(s + 1)}\right]$$

we must do a little algebra. The idea is to rewrite the term

$$\frac{4}{(s - 1)(s + 1)}$$

as a combination of functions that are known Laplace transforms. In this case (and quite often when using Laplace transforms) we use the technique of partial fractions. That is, we write

$$\frac{4}{(s-1)(s+1)} = \frac{A}{s-1} + \frac{B}{s+1}$$

and solve for the constants A and B, obtaining

$$\frac{4}{(s-1)(s+1)} = \frac{2}{s-1} - \frac{2}{s+1}.$$

Each of the terms on the right-hand side is recognizable as the Laplace transform of an exponential function. Hence

$$\mathcal{L}^{-1}\left[\frac{4}{(s-1)(s+1)}\right] = \mathcal{L}^{-1}\left[\frac{2}{s-1}\right] - \mathcal{L}^{-1}\left[\frac{2}{s+1}\right]$$

$$= 2\mathcal{L}^{-1}\left[\frac{1}{s-1}\right] - 2\mathcal{L}^{-1}\left[\frac{1}{s+1}\right]$$

$$= 2e^t - 2e^{-t}.$$

Completion of the initial-value problem

We showed above that the Laplace transform of the solution y of the initial-value problem

$$\frac{dy}{dt} = y - 4e^{-t}, \quad y(0) = 1$$

is

$$\mathcal{L}[y] = \frac{1}{s-1} - \frac{4}{(s-1)(s+1)}.$$

Hence

$$y = \mathcal{L}^{-1}\left[\frac{1}{s-1} - \frac{4}{(s-1)(s+1)}\right]$$

$$= \mathcal{L}^{-1}\left[\frac{1}{s-1}\right] - \mathcal{L}^{-1}\left[\frac{4}{(s-1)(s+1)}\right]$$

$$= e^t - \left(2e^t - 2e^{-t}\right)$$

$$= -e^t + 2e^{-t}.$$

is the solution. We see that $y(t) \to -\infty$ as $t \to \infty$, as we predicted earlier.

An RC Circuit Example

Consider the RC circuit in Figure 6.2. We let $v_c(t)$ denote the voltage across the capacitor, R the resistance, C the capacitance, and $V(t)$ the voltage supplied by the voltage

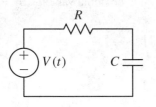

Figure 6.2
RC circuit.

source. From electric circuit theory we know that the voltage $v_c(t)$ satisfies the differential equation

$$RC\frac{dv_c}{dt} + v_c = V(t).$$

Suppose the voltage source $V(t)$ has constant value 3 and the initial voltage across the capacitor is $v_c(0) = 4$. With the quantities $R = 2$ and $C = 1$ as in Figure 6.2 (rather unrealistic quantities if we use the usual units of ohms for R, farads for C, and volts for V), the initial-value problem is

$$2\frac{dv_c}{dt} + v_c = 3, \quad v_c(0) = 4.$$

Qualitative analysis

Rewriting this initial-value problem in the standard form

$$\frac{dv_c}{dt} = -\frac{v_c}{2} + \frac{3}{2}, \quad v_c(0) = 4,$$

we see that this equation is autonomous with a sink at $v_c = 3$ and no other equilibrium points. Hence, the solution of the initial-value problem tends to $v_c = 3$ as $t \to \infty$, which is confirmed by the slope field and phase line (see Figure 6.3).

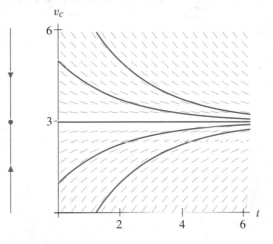

Figure 6.3
Phase line and slope field for $dv_c/dt = (-v_c + 3)/2$.

Solution using Laplace transforms

The differential equation

$$\frac{dv_c}{dt} = -\frac{v_c}{2} + \frac{3}{2}$$

is both separable and linear, but in this section we choose to use the method of Laplace transforms to find a formula for the solution. First, taking the Laplace transform of both sides of the equation gives

$$\mathcal{L}\left[\frac{dv_c}{dt}\right] = -\frac{1}{2}\mathcal{L}[v_c] + \mathcal{L}\left[\frac{3}{2}\right].$$

Using the formula for $\mathcal{L}[dv_c/dt]$, simplifying, and substituting the initial value $v_c(0) = 4$ yields

$$s\mathcal{L}[v_c] - 4 = -\frac{1}{2}\mathcal{L}[v_c] + \frac{3}{2}\mathcal{L}[1].$$

We have seen earlier that $\mathcal{L}[1] = 1/s$. Therefore we have

$$s\mathcal{L}[v_c] - 4 = -\frac{1}{2}\mathcal{L}[v_c] + \frac{3}{2s}.$$

Solving for $\mathcal{L}[v_c]$ yields

$$\left(s + \frac{1}{2}\right)\mathcal{L}[v_c] = 4 + \frac{3}{2s},$$

which implies that

$$\mathcal{L}[v_c] = \frac{4}{s + 1/2} + \frac{3}{2s(s + 1/2)}$$

$$= \frac{4}{s + 1/2} + \frac{3}{2}\frac{1}{s(s + 1/2)}.$$

To compute the inverse Laplace transform, we use the partial fractions decomposition

$$\frac{1}{s(s + 1/2)} = \frac{2}{s} - \frac{2}{s + 1/2}.$$

We obtain

$$\mathcal{L}[v_c] = \frac{4}{s + 1/2} + \frac{3}{2}\left(\frac{2}{s} - \frac{2}{s + 1/2}\right)$$

$$= \frac{1}{s + 1/2} + \frac{3}{s}.$$

Hence,

$$v_c(t) = \mathcal{L}^{-1}\left[\frac{1}{s + 1/2}\right] + \mathcal{L}^{-1}\left[\frac{3}{s}\right]$$

$$= e^{-t/2} + 3,$$

Note that this solution is consistent with the phase line and slope field for this equation.

EXERCISES FOR SECTION 6.1

In Exercises 1–4, compute the Laplace transform of the given function from the definition.

1. $f(t) = 3$ (the constant function)   **2.** $g(t) = t$

3. $h(t) = -5t^2$ **4.** $k(t) = t^5$

5. Verify that

$$\mathcal{L}[t^n] = \frac{n!}{s^{n+1}} \quad (s > 0).$$

[*Hint*: To do this carefully requires mathematical induction.]

6. Using

$$\mathcal{L}[t^n] = \frac{n!}{s^{n+1}} \quad (s > 0),$$

give a formula for the Laplace transform of an arbitrary nth degree polynomial

$$p(t) = a_0 + a_1 t + a_2 t^2 + \ldots + a_n t^n,$$

where the a_i's are constants.

In Exercises 7–14, find the inverse Laplace transform of the given function.

7. $\dfrac{1}{s - 3}$ **8.** $\dfrac{5}{(s - 1)(s - 2)}$ **9.** $\dfrac{2}{3s + 5}$

10. $\dfrac{14}{(3s + 2)(s - 4)}$ **11.** $\dfrac{5}{3s}$ **12.** $\dfrac{4}{s(s + 3)}$

13. $\dfrac{2s + 1}{(s - 1)(s - 2)}$ **14.** $\dfrac{2s^2 + 3s - 2}{s(s + 1)(s - 2)}$

In Exercises 15–24,

  (a) compute the Laplace transform of both sides of the equation;
  (b) substitute the initial conditions and solve for the Laplace transform of the solution;
  (c) find a function whose Laplace transform is the same as the solution; and
  (d) check that you have found the solution of the initial-value problem.

15. $\dfrac{dy}{dt} = -y + e^{-2t}, \quad y(0) = 2$ **16.** $\dfrac{dy}{dt} + 5y = e^{-t}, \quad y(0) = 2$

17. $\dfrac{dy}{dt} + 7y = 1, \quad y(0) = 3$ **18.** $\dfrac{dy}{dt} + 4y = 6, \quad y(0) = 0$

19. $\dfrac{dy}{dt} + 9y = 2, \quad y(0) = -2$ **20.** $\dfrac{dy}{dt} = -y + 2, \quad y(0) = 4$

21. $\dfrac{dy}{dt} = -y + e^{-2t}, \quad y(0) = 1$ **22.** $\dfrac{dy}{dt} = 2y + t, \quad y(0) = 0$

23. $\dfrac{dy}{dt} = -y + t^2, \quad y(0) = 1$ **24.** $\dfrac{dy}{dt} + 4y = 2 + 3t, \quad y(0) = 1$

25. Find the general solution of the equation

$$\frac{dy}{dt} = 2y + 2e^{-3t}.$$

(This equation is linear, but please use the method of Laplace transforms.)

26. Suppose $g(t) = \int f(t)\,dt$; that is, $g(t)$ is the antiderivative of $f(t)$. Express the Laplace transform of $g(t)$ in terms of the Laplace transform of $f(t)$.

27. All of the examples in this section and all the differential equations in this exercise set are linear. Trying to use the Laplace transform on even the simplest nonlinear equation leads quickly to headaches. Try using the Laplace transform to find the solution of the initial-value problem

$$\frac{dy}{dt} = y^2, \quad y(0) = 1.$$

Give a short explanation outlining where you got stuck and why.

6.2 DISCONTINUOUS FUNCTIONS

In Section 6.1 we saw that using Laplace transforms to find the solutions of a linear differential equation involves entirely different ideas than our previous methods. The operations of differentiation and integration are replaced with algebra. However, Laplace transforms are not a panacea. They apply only to linear equations and even though they replace calculus with algebra, the algebra can be very complicated.

Given these limitations, it is important to ask why we need another method for solving linear equations. The remainder of this chapter is devoted to applications of the Laplace transform that let us study new types of equations and that give us new insights into familiar equations.

In applications discontinuous functions arise naturally. For example, the sudden introduction of a new species or disease affecting a population or the turning on or off of a light switch are discontinuous phenomena. Differential equations containing discontinuous functions are difficult to handle analytically using our previous methods, but Laplace transforms can sometimes tame these discontinuities, as the following examples show.

Laplace transform of a Heaviside function

For $a \geq 0$, let $u_a(t)$ denote the function

$$u_a(t) = \begin{cases} 0, & \text{if } t < a; \\ 1, & \text{if } t \geq a. \end{cases}$$

Thus $u_a(t)$ has a discontinuity at $t = a$ where it jumps from 0 to 1. For example, the function $u_2(t)$ has a discontinuity at $t = 2$, where it jumps from 0 to 1 (see Figure 6.4). A function of this form is called a step function, or **Heaviside function** (named for the engineer Oliver Heaviside). It is useful when modeling discontinuous processes such as turning on a light switch.

The Laplace transform of $u_a(t)$ is

$$\mathcal{L}[u_a] = \int_0^\infty u_a(t) e^{-st}\,dt.$$

To compute this integral, we use the definition of u_a and break the computation into two pieces,

$$\mathcal{L}[u_a] = \int_0^a u_a(t) e^{-st}\,dt + \int_a^\infty u_a(t) e^{-st}\,dt.$$

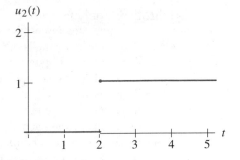

Figure 6.4
Graph of the Heaviside function $u_2(t)$.

The first integral is zero because $u_a(t)$ is zero for $t < a$. We can simplify the second integral because $u_a(t) = 1$ for $t \geq a$, hence

$$\mathcal{L}[u_a] = \int_a^\infty e^{-st}\,dt$$

$$= \lim_{b\to\infty} \int_a^b e^{-st}\,dt$$

$$= \lim_{b\to\infty} \left. \frac{e^{-st}}{-s}\right|_a^b$$

$$= \lim_{b\to\infty} \frac{e^{-sb}}{-s} - \frac{e^{-as}}{-s}$$

$$= 0 + \frac{e^{-as}}{s}.$$

We have established the formula

$$\mathcal{L}[u_a] = \frac{e^{-as}}{s}.$$

Even though the original function u_a has a discontinuity at $t = a$, the Laplace transform is continuous for all $s > 0$. This *smoothing property* is a very useful aspect of Laplace transforms.

A Differential Equation with a Discontinuity

Consider the initial-value problem

$$\frac{dy}{dt} = -y + u_3(t), \quad y(0) = 2.$$

We can rewrite this differential equation in the form

$$\frac{dy}{dt} = \begin{cases} -y, & \text{if } t < 3; \\ -y + 1, & \text{if } t \geq 3. \end{cases}$$

Qualitative analysis

Since this equation really consists of a pair of autonomous first-order equations, we can use the qualitative methods of Chapter 1 to analyze the behavior of solutions. The slope field is shown in Figure 6.5. For $t < 3$ the equation $dy/dt = -y$ has a sink at $y = 0$, and all solutions approach this equilibrium point. For $t \geq 3$ the equation $dy/dt = -y + 1$ has a sink at $y = 1$, which attracts all other solutions. Hence the solution of our initial-value problem initially decreases toward $y = 0$. Then, at $t = 3$, the solution begins to approach $y = 1$. If at time $t = 3$ we have $y(3) < 1$, then the solution ceases to decrease and begins to increase toward 1. If at time $t = 3$ we have $y(3) > 1$, the solution continues to decrease toward $y = 1$.

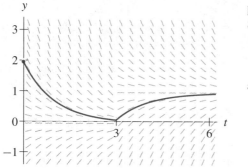

Figure 6.5
Slope field for

$$\frac{dy}{dt} = -y + u_3(t)$$

and solution with $y(0) = 2$.

In order to decide if the solution of the initial-value problem increases or decreases toward 1 as $t \to \infty$, we must compute the value of the solution at time $t = 3$. The initial-value problem

$$\frac{dy}{dt} = -y, \quad y(0) = 2$$

has solution $y(t) = 2e^{-t}$. Hence $y(3) = 2e^{-3}$ and we see that $y(3) < 1$. Consequently the solution increases for $t > 3$ toward $y = 1$.

Note that to obtain even a qualitative description of the solution we had to find the analytic form of the solution up to time $t = 3$, when the term $u_3(t)$ "turns on."

Solution using Laplace transforms

As in Section 6.1, to solve the initial-value problem

$$\frac{dy}{dt} = -y + u_3(t), \quad y(0) = 2$$

using Laplace transforms, we first take the Laplace transform of both sides of the differential equation (and use the linearity of the Laplace transform) to obtain

$$\mathcal{L}\left[\frac{dy}{dt}\right] = -\mathcal{L}[y] + \mathcal{L}[u_3].$$

Using the rule for the Laplace transform of a derivative and our computation of $\mathcal{L}[u_3] = e^{-3s}/s$, we have

$$s\mathcal{L}[y] - y(0) = -\mathcal{L}[y] + \frac{e^{-3s}}{s}.$$

Substituting $y(0) = 2$ and solving for $\mathcal{L}[y]$ gives

$$\mathcal{L}[y] = \frac{2}{s+1} + \frac{e^{-3s}}{s(s+1)}.$$

Hence the solution is

$$y = \mathcal{L}^{-1}\left[\frac{2}{s+1}\right] + \mathcal{L}^{-1}\left[\frac{e^{-3s}}{s(s+1)}\right].$$

Now

$$\mathcal{L}^{-1}\left[\frac{2}{s+1}\right] = 2e^t,$$

but to compute

$$\mathcal{L}^{-1}\left[\frac{e^{-3s}}{s(s+1)}\right]$$

we need another rule for computing Laplace transforms.

Shifting the Origin on the t-Axis

Given a function $f(t)$ suppose that we wish to consider a function $g(t)$, which is the same as the function $f(t)$ but shifted so that time $t = 0$ for f corresponds to some later time, say $t = a$, for g. For time $t \leq a$ we let $g(t) = 0$. (So g is the same as f except that g is "turned on" at time $t = a$.) An efficient way to write $g(t)$ is

$$g(t) = u_a(t)f(t - a).$$

Note that

$$g(a) = u_a(a)f(a - a) = f(0)$$

and, if $b > 0$,

$$g(a + b) = u_a(a + b)f(b) = f(b)$$

as desired. For example, if $f(t) = e^{-t}$ and $a = 3$, then $g(t) = u_3(t)e^{-(t-3)}$ (see Figures 6.6 and 6.7).

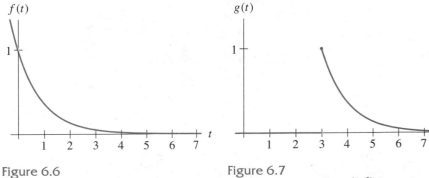

Figure 6.6
Graph of $f(t) = e^{-t}$.

Figure 6.7
Graph of $g(t) = u_3(t)e^{-(t-3)}$.

To compute the Laplace transform of $g(t)$, we return to the definition

$$\mathcal{L}[g] = \int_0^\infty g(t)e^{-st}dt.$$

Using the fact that $g(t) = 0$ for $t < a$ and that $g(t) = f(t - a)$ for $t \geq a$, we obtain

$$\mathcal{L}[g] = \int_a^\infty f(t - a)e^{-st}dt.$$

The u-substitution $u = t - a$ in the integral gives

$$\mathcal{L}[g] = \int_0^\infty f(u)e^{-s(u+a)}du$$

$$= e^{-sa}\int_0^\infty f(u)e^{-su}du$$

$$= e^{-sa}\mathcal{L}[f] = e^{-sa}F(s).$$

Hence we can express the Laplace transform of $g(t) = u_a(t)f(t - a)$ in terms of the Laplace transform of $f(t)$ by the following rule:

If $\mathcal{L}[f] = F(s)$, then $\mathcal{L}[u_a(t)f(t - a)] = e^{-as}F(s)$.

(This rule, along with the other rules we develop for Laplace transforms, can be found in the table on page 557).

For the example just given, if $g(t) = u_3(t)e^{-(t-3)}$, then the Laplace transform of $g(t)$ is

$$\mathcal{L}[g] = e^{-3s}\mathcal{L}[e^{-t}] = \frac{e^{-3s}}{s+1}.$$

Completion of the initial-value problem
For the initial-value problem

$$\frac{dy}{dt} = -y + u_3(t), \quad y(0) = 2,$$

we showed that

$$y = \mathcal{L}^{-1}\left[\frac{2}{s+1}\right] + \mathcal{L}^{-1}\left[\frac{e^{-3s}}{s(s+1)}\right].$$

Using partial fractions, we can write

$$\frac{1}{s(s+1)} = \frac{1}{s} - \frac{1}{s+1}$$

so

$$\mathcal{L}^{-1}\left[\frac{e^{-3s}}{s(s+1)}\right] = \mathcal{L}^{-1}\left[\frac{e^{-3s}}{s}\right] - \mathcal{L}^{-1}\left[\frac{e^{-3s}}{s+1}\right]$$

$$= u_3(t) - u_3(t)e^{-(t-3)}.$$

Hence
$$y(t) = 2e^{-t} + u_3(t)(1 - e^{-(t-3)}).$$

Note that the second term is nonzero only for $t \geq 3$. That is, this term "turns on" precisely when the $u_3(t)$-term in the differential equation "turns on."

An RC Circuit with an Exponentially Decaying Voltage Source

Consider the RC circuit in Figure 6.8 with voltage source
$$V(t) = 2u_4(t)e^{-(t-4)}.$$

The voltage source is turned on to a voltage 2 at time $t = 4$, then it decays exponentially as t increases (see Figure 6.9). The differential equation for the voltage v_c across the capacitor is
$$RC\frac{dv_c}{dt} + v_c = V(t).$$

Suppose the initial voltage across the capacitor is $v_c(0) = 5$. Taking the (unrealistic) values $R = 1$ and $C = 1/3$ and using $V(t)$ given above, we have the initial-value problem
$$\frac{1}{3}\frac{dv_c}{dt} + v_c = 2u_4(t)e^{-(t-4)}, \qquad v_c(0) = 5.$$

Multiplying both sides by 3 and moving the v_c term to the right-hand side, we obtain
$$\frac{dv_c}{dt} = -3v_c + 6u_4(t)e^{-(t-4)}, \qquad v_c(0) = 5.$$

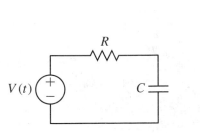

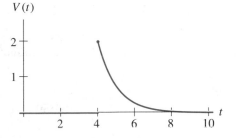

Figure 6.8
RC circuit.

Figure 6.9
Voltage source $V(t) = 2u_4(t)e^{-(t-4)}$.

Qualitative analysis

Again we can rewrite this differential equation as
$$\frac{dv_c}{dt} = \begin{cases} -3v_c, & \text{if } t < 4; \\ -3v_c + 6e^{-(t-4)}, & \text{if } t \geq 4. \end{cases}$$

The solution with initial condition $v_c(0) = 5$ for $0 \leq t < 4$ is $v_c(t) = 5e^{-3t}$. This solution decreases toward $v_c = 0$ quickly as t increases. For $t \geq 4$ the positive term $6e^{-(t-4)}$ causes solutions near $v_c = 0$ to increase. However, because $6e^{-(t-4)}$ tends to zero as t increases, solutions eventually tend to zero. Hence we expect the solution of the initial-value problem to decrease quickly toward $v_c = 0$ for $0 \leq t < 4$, then increase for t slightly larger than 4, then decrease again toward zero as t increases. This behavior is confirmed by the slope field and solution in Figure 6.10.

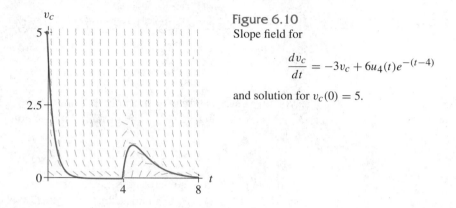

v_c

5

2.5

0

4   8 t

Figure 6.10
Slope field for

$$\frac{dv_c}{dt} = -3v_c + 6u_4(t)e^{-(t-4)}$$

and solution for $v_c(0) = 5$.

Solution using Laplace transforms

Starting with the initial-value problem

$$\frac{dv_c}{dt} = -3v_c + 6u_4(t)e^{-(t-4)}, \quad v_c(0) = 5,$$

we take Laplace transforms of both sides of the differential equation to obtain

$$\mathcal{L}\left[\frac{dv_c}{dt}\right] = -3\mathcal{L}[v_c] + 6\mathcal{L}[u_4(t)e^{-(t-4)}],$$

which simplifies to

$$s\mathcal{L}[v_c] - v_c(0) = -3\mathcal{L}[v_c] + \frac{6e^{-4s}}{s+1}.$$

Substituting $v_c(0) = 5$ and solving for $\mathcal{L}[v_c]$ gives

$$\mathcal{L}[v_c] = \frac{5}{s+3} + \frac{6e^{-4s}}{(s+3)(s+1)}.$$

Thus

$$v_c = \mathcal{L}^{-1}\left[\frac{5}{s+3}\right] + \mathcal{L}^{-1}\left[\frac{6e^{-4s}}{(s+3)(s+1)}\right].$$

Using the partial fractions decomposition

$$\frac{6}{(s+3)(s+1)} = \frac{-3}{s+3} + \frac{3}{s+1},$$

we have

$$v_c = \mathcal{L}^{-1}\left[\frac{5}{s+3}\right] - 3\mathcal{L}^{-1}\left[\frac{e^{-4s}}{s+3}\right] + 3\mathcal{L}^{-1}\left[\frac{e^{-4s}}{s+1}\right].$$

Hence the solution of the initial-value problem is

$$v_c(t) = 5e^{-3t} - 3u_4(t)e^{-3(t-4)} + 3u_4(t)e^{-(t-4)}.$$

For $0 \leq t < 4$ the solution is $v_c(t) = 5e^{-3t}$, which decreases quickly toward zero. For $t \geq 4$ the solution is

$$v_c(t) = 5e^{-3t} - 3e^{-3(t-4)} + 3e^{-(t-4)}.$$

Each term in this solution tends to 0 as t increases, as predicted.

EXERCISES FOR SECTION 6.2

1. For $a \geq 0$, let $g_a(t)$ denote the function

$$g_a(t) = \begin{cases} 1, & \text{if } t < a; \\ 0, & \text{if } t \geq a. \end{cases}$$

(a) Give an expression for $g_a(t)$ using the Heaviside function $u_a(t)$.

(b) Compute $\mathcal{L}[g_a]$.

2. (a) Suppose $a \geq 0$. Compute the Laplace transform of the function

$$r_a(t) = \begin{cases} 0, & \text{if } t < a; \\ k(t - a) & \text{if } t \geq a. \end{cases}$$

[*Hint*: Your solution should contain a and k as parameters.]

(b) Sketch the graph of $r_a(t)$ and comment on why it is called a *ramp function*.

3. Suppose $a \geq 0$. Compute the Laplace transform of the function

$$g_a(t) = \begin{cases} t/a, & \text{if } t < a; \\ 1, & \text{if } t \geq a \end{cases}$$

[*Hint*: Your solution should contain a as a parameter.]

In Exercises 4–7, find the inverse Laplace transform of the given function.

4. $\dfrac{e^{-2s}}{s - 3}$

5. $\dfrac{e^{-3s}}{(s - 1)(s - 2)}$

6. $\dfrac{4e^{-2s}}{s(s + 3)}$

7. $\dfrac{14e^{-s}}{(3s + 2)(s - 4)}$

In Exercises 8–13, solve the given initial-value problem.

8. $\dfrac{dy}{dt} + 7y = u_2(t), \quad y(0) = 3$

9. $\dfrac{dy}{dt} + 9y = u_5(t), \quad y(0) = -2$

10. $\dfrac{dy}{dt} = -y + 2u_3(t), \quad y(0) = 4$

11. $\dfrac{dy}{dt} = -y + u_2(t)e^{-2(t-2)}, \quad y(0) = 1$

12. $\dfrac{dy}{dt} = u_2(t), \quad y(0) = 3$

13. $\dfrac{dy}{dt} = -y + u_1(t)(t - 1), \quad y(0) = 2$

14. Solve the initial-value problem

$$\frac{dy}{dt} = -2y + u_2(t)e^{-t}, \quad y(0) = 3.$$

[*Warning*: This is harder than it looks.]

15. Suppose $a \geq 0$. Find the general solution of

$$\frac{dy}{dt} = -y + u_a(t).$$

[*Hint*: Your solution should contain a as a parameter.]

16. Suppose $f(t)$ is a periodic function with period T; that is,

$$f(t + T) = f(t) \quad \text{for all } t.$$

Show that

$$\mathcal{L}[f] = \frac{1}{1 - e^{-Ts}} \int_0^T f(t)e^{-st}\, dt.$$

17. Compute the Laplace transform of the square wave

$$w(t) = \begin{cases} 1, & \text{if } 2n \leq t < 2n + 1 \text{ for some integer } n; \\ -1, & \text{if } 2n + 1 \leq t < 2n + 2 \text{ for some integer } n. \end{cases}$$

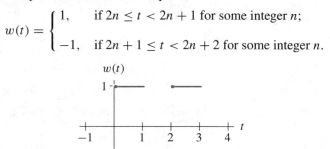

Periodic square wave with period 2.

18. Compute the Laplace transform of the sawtooth function

$$z(t) = t - [t],$$

where $[t]$ denotes the greatest integer less than or equal to t. [*Hint*: Use the formula developed in Exercise 16.]

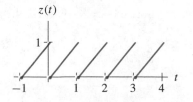

A sawtooth wave with period 1.

19. Consider the initial-value problem

$$\frac{dy}{dt} = -y + w(t), \quad y(0) = 0,$$

where $w(t)$ is the square wave in Exercise 17.

(a) Compute the Laplace transform of the solution.

(b) Describe the qualitative behavior of this solution. [*Hint*: Knowing the Laplace transform of the solution may not help with this part, but see Section 6.6.]

20. Consider the initial-value problem

$$\frac{dy}{dt} = -y + z(t), \quad y(0) = 0,$$

where $z(t)$ is the sawtooth wave in Exercise 18.

(a) Compute the Laplace transform of the solution.

(b) Describe the qualitative behavior of this solution. [*Hint*: Knowing the Laplace transform of the solution may not help with this part, but see Section 6.6.]

6.3 SECOND-ORDER EQUATIONS

In this section we extend the method of Laplace transforms to second-order, constant-coefficient, forced linear equations, that is, equations of the form

$$\frac{d^2y}{dt^2} + p\frac{dy}{dt} + qy = f(t),$$

where p and q are constants. We have dealt with equations of this form for particular forcing functions $f(t)$ in Chapter 4. Laplace transforms allow us to study a larger class of forcing functions, including some with discontinuities.

To begin, we recall that the functions $\sin \omega t$, $\cos \omega t$, $e^{at} \sin \omega t$, and $e^{at} \cos \omega t$ appear very often in the study of second-order equations. Hence our first task is to study the Laplace transform of these functions.

Laplace Transform of Sine and Cosine

To compute the Laplace transform of $\sin \omega t$, we could use the definition and evaluate the integral

$$\mathcal{L}[\sin \omega t] = \int_0^\infty \sin \omega t \, e^{-st} \, dt.$$

Another approach is to use complex exponentials. These alternatives lead to good exercises in integration by parts and partial fractions with complex numbers, respectively (see Exercises 1–4 and 19–21).

The method we use to compute $\mathcal{L}[\sin \omega t]$ is to take advantage of the fact that we already know that $y(t) = \sin \omega t$ is the solution of the initial-value problem

$$\frac{d^2y}{dt^2} + \omega^2 y = 0, \quad y(0) = 0, \quad y'(0) = \omega.$$

Hence we can compute $\mathcal{L}[\sin \omega t]$ by computing the Laplace transform of the solution of this initial-value problem.

First note that two applications of the rule for the Laplace transform of a derivative yields

$$\mathcal{L}\left[\frac{d^2y}{dt^2}\right] = s\mathcal{L}\left[\frac{dy}{dt}\right] - y'(0)$$

$$= s(s\mathcal{L}[y] - y(0)) - y'(0)$$

$$= s^2\mathcal{L}[y] - sy(0) - y'(0).$$

To solve the initial-value problem

$$\frac{d^2y}{dt^2} + \omega^2 y = 0, \quad y(0) = 0, \quad y'(0) = \omega$$

using Laplace transforms, we first take the Laplace transform of both sides of the differential equation, which yields

$$\mathcal{L}\left[\frac{d^2y}{dt^2}\right] + \omega^2\mathcal{L}[y] = \mathcal{L}[0].$$

Simplifying and using the fact that $\mathcal{L}[0] = 0$, we have

$$s^2\mathcal{L}[y] - sy(0) - y'(0) + \omega^2\mathcal{L}[y] = 0.$$

Substituting the initial conditions $y(0) = 0$ and $y'(0) = \omega$ gives

$$s^2\mathcal{L}[y] - \omega + \omega^2\mathcal{L}[y] = 0$$
$$\left(s^2 + \omega^2\right)\mathcal{L}[y] = \omega,$$

and therefore

$$\mathcal{L}[y] = \frac{\omega}{s^2 + \omega^2}.$$

But we already know that the solution of the initial-value problem is $y(t) = \sin \omega t$, so we have

$$\mathcal{L}[\sin \omega t] = \frac{\omega}{s^2 + \omega^2}.$$

To compute $\mathcal{L}[\cos \omega t]$ we could repeat this computation using the initial-value problem

$$\frac{d^2y}{dt^2} + \omega^2 y = 0, \quad y(0) = 1, \quad y'(0) = 0$$

whose solution is $y(t) = \cos \omega t$ (see Exercise 5). Alternately, we can use the fact that

$$\cos \omega t = \frac{1}{\omega}\frac{d(\sin \omega t)}{dt}.$$

Applying Laplace transforms to this both sides of this equation gives

$$\mathcal{L}[\cos \omega t] = \frac{1}{\omega} \mathcal{L}\left[\frac{d(\sin \omega t)}{dt}\right]$$

$$= \frac{1}{\omega}(s\mathcal{L}[\sin \omega t] - \sin 0)$$

$$= \frac{s}{\omega}\mathcal{L}[\sin \omega t]$$

$$= \frac{s}{\omega}\frac{\omega}{s^2 + \omega^2},$$

so

$$\mathcal{L}[\cos \omega t] = \frac{s}{s^2 + \omega^2}.$$

(These formulas are included in the table of Laplace transforms on page 557.)

Shifting the Origin on the s-Axis

Next we consider the Laplace transforms of the functions $e^{at}\sin \omega t$ and $e^{at}\cos \omega t$. In general, the Laplace transform of a product is quite complicated. However, if one of the factors is an exponential, that factor combines nicely with the factor e^{-st} in the definition of the transform.

Suppose we are given a function $f(t)$ and we know that its Laplace transform is $F(s)$. To compute the Laplace transform of $e^{at}f(t)$, we recall the definition and write

$$\mathcal{L}[e^{at}f(t)] = \int_0^\infty e^{at}f(t)e^{-st}\,dt$$

$$= \int_0^\infty f(t)e^{-(s-a)t}\,dt$$

$$= F(s - a)$$

That is,

If $\mathcal{L}[f] = F(s)$, then $\mathcal{L}[e^{at}f(t)] = F(s - a)$.

In other words, multiplying $f(t)$ by e^{at} corresponds to shifting the argument of the Laplace transform F of f by a.

For example,

$$\mathcal{L}[\cos 2t] = \frac{s}{s^2 + 4},$$

so to compute the Laplace transform of $e^{-3t}\cos 2t$, we replace s by $s + 3$. We obtain

$$\mathcal{L}[e^{-3t}\cos 2t] = \frac{s + 3}{(s + 3)^2 + 4} = \frac{s + 3}{s^2 + 6s + 13}.$$

We can use this rule to compute inverse Laplace transforms as well. For example, to compute

$$\mathcal{L}^{-1}\left[\frac{1}{s^2 + 2s + 5}\right]$$

we first note that the roots of

$$s^2 + 2s + 5$$

are $-1 \pm 2i$. If the denominator had real roots, we could factor it and use partial fractions to compute the inverse transform. Since the roots in this case are complex, we write

$$s^2 + 2s + 5 = (s + 1)^2 + 4.$$

(Recall that this algebraic computation is called "completing the square." For a review of this technique, see Exercises 11–14.) Next, we note that

$$\frac{1}{(s + 1)^2 + 4} = F(s + 1),$$

where

$$F(s) = \frac{1}{s^2 + 4}.$$

We know that

$$\mathcal{L}^{-1}\left[\frac{1}{s^2 + 4}\right] = \frac{1}{2}\mathcal{L}^{-1}\left[\frac{2}{s^2 + 4}\right] = \frac{1}{2}\sin 2t.$$

So, using the rule above, we have

$$\mathcal{L}^{-1}\left[\frac{1}{(s + 1)^2 + 4}\right] = \frac{1}{2}e^{-t}\sin 2t.$$

A Forced Harmonic Oscillator

In Chapter 4 the forced harmonic oscillator equation was introduced as a model of a mass attached to a spring sliding back and forth on a table with an external force caused by, for example, tilting the table. As a first example of the Laplace transform method for such an equation, consider the initial-value problem

$$\frac{d^2 y}{dt^2} + 4y = 3\cos t, \quad y(0) = y'(0) = 0.$$

This equation models a unit mass sliding on a frictionless table with spring constant 4 and a periodic external force $3\cos t$.

Qualitative analysis

The characteristic polynomial for the unforced harmonic oscillator

$$\frac{d^2 y}{dt^2} + 4y = 0$$

is

$$s^2 + 4 = 0.$$

Hence the eigenvalues are $s = \pm 2i$, and the natural period is $2\pi/2 = \pi$. Since the forcing has period 2π, the forced equation

$$\frac{d^2 y}{dt^2} + 4y = 3\cos t$$

is not a resonant system, and every solution remains bounded for all time. In fact solutions are combinations of trigonometric functions with period π and 2π.

Solution using Laplace transforms

To use the Laplace transform on the initial-value problem

$$\frac{d^2 y}{dt^2} + 4y = 3\cos t, \quad y(0) = y'(0) = 0,$$

we take the Laplace transform of both sides of the differential equation, obtaining

$$\mathcal{L}\left[\frac{d^2 y}{dt^2}\right] + 4\mathcal{L}[y] = 3\mathcal{L}[\cos t].$$

Simplifying we have

$$s^2 \mathcal{L}[y] - s y(0) - y'(0) + 4\mathcal{L}[y] = \frac{3s}{s^2 + 1}.$$

Substituting the initial conditions yields

$$(s^2 + 4)\mathcal{L}[y] = \frac{3s}{s^2 + 1},$$

which gives

$$\mathcal{L}[y] = \frac{3s}{(s^2 + 4)(s^2 + 1)}.$$

Therefore, the desired solution is

$$y(t) = \mathcal{L}^{-1}\left[\frac{3s}{(s^2 + 4)(s^2 + 1)}\right].$$

Using partial fractions, we have

$$y(t) = \mathcal{L}^{-1}\left[\frac{-s}{s^2 + 4}\right] + \mathcal{L}^{-1}\left[\frac{s}{s^2 + 1}\right],$$

which gives

$$y(t) = -\cos 2t + \cos t.$$

This solution is bounded and behaves as predicted (see Figure 6.11).

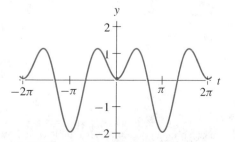

Figure 6.11
Solution of

$$\frac{d^2 y}{dt^2} + 4y = 3\cos t$$

with $y(0) = y'(0) = 0$.

An Oscillator with Discontinuous Forcing

Next we consider the initial-value problem

$$\frac{d^2y}{dt^2} + 2\frac{dy}{dt} + 5y = h(t), \quad y(0) = y'(0) = 0,$$

where $h(t) = 1 - u_7(t)$ is given by

$$h(t) = \begin{cases} 1, & \text{if } t < 7; \\ 0, & \text{if } t \geq 7; \end{cases}$$

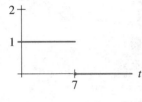

$h(t)$

Figure 6.12
Graph of the function $h(t)$.

(see Figure 6.12).

This is a model for a unit mass attached to a spring with spring constant 5 sliding on a table with damping coefficient 2 (see Chapter 4). We can interpret the initial conditions and forcing term as follows: Suppose at time $t = 0$ the mass is held at rest at $y = 0$. When $t < 7$ the table is tilted so gravity provides a unit force stretching the spring. At time $t = 7$ the table is suddenly returned to the level position (see Figure 6.13).

Qualitative analysis
The characteristic polynomial of the equation

$$\frac{d^2y}{dt^2} + 2\frac{dy}{dt} + 5y = 0$$

is

$$s^2 + 2s + 5 = 0.$$

The eigenvalues are $s = -1 \pm 2i$, so this oscillator is underdamped. Hence when $0 < t < 7$ the mass slides down the table, then oscillates while approaching an equilibrium position for which the spring is slightly stretched. For $t > 7$, after the table is returned to level, the mass slides back toward the original rest position, then oscillates with exponentially decreasing amplitude.

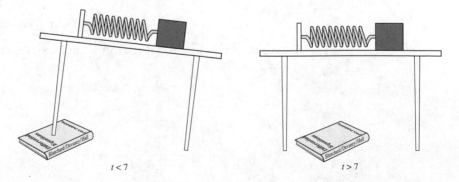

$t < 7$ $t > 7$

Figure 6.13
Schematic of a mass-spring system on a tilted table ($t < 7$) and a level table ($t > 7$).

Solution using Laplace transforms

We begin as before by taking the Laplace transform of both sides of the differential equation. Using linearity, we obtain

$$\mathcal{L}\left[\frac{d^2y}{dt^2}\right] + 2\mathcal{L}\left[\frac{dy}{dt}\right] + 5\mathcal{L}[y] = \mathcal{L}[h(t)].$$

Hence

$$(s^2\mathcal{L}[y] - sy(0) - y'(0)) + 2(s\mathcal{L}[y] - y(0)) + 5\mathcal{L}[y] = \mathcal{L}[h(t)].$$

Collecting terms, substituting the initial conditions $y(0) = 0$ and $y'(0) = 0$, and evaluating $\mathcal{L}[h(t)] = \mathcal{L}[1 - u_7(t)]$, we have

$$(s^2 + 2s + 5)\mathcal{L}[y] = \frac{1}{s} - \frac{e^{-7s}}{s},$$

which yields

$$\mathcal{L}[y] = \frac{1}{s(s^2 + 2s + 5)} - \frac{e^{-7s}}{s(s^2 + 2s + 5)}.$$

That is,

$$y = \mathcal{L}^{-1}\left[\frac{1}{s(s^2 + 2s + 5)} - \frac{e^{-7s}}{s(s^2 + 2s + 5)}\right].$$

To compute the right-hand side, we first use the partial fractions decomposition

$$\frac{1}{s(s^2 + 2s + 5)} = \frac{1}{5s} - \frac{s + 2}{5(s^2 + 2s + 5)}.$$

We know that

$$\mathcal{L}^{-1}\left[\frac{1}{5s}\right] = \frac{1}{5},$$

so we need to work only with the second term. The quadratic denominator in this term has complex roots, so we first complete the square, giving

$$s^2 + 2s + 5 = (s + 1)^2 + 4.$$

We then write this expression in the more convenient form

$$\frac{s + 2}{5(s^2 + 2s + 5)} = \frac{1}{5}\left(\frac{s + 1}{(s + 1)^2 + 4} + \frac{1}{2}\frac{2}{(s + 1)^2 + 4}\right).$$

Thus

$$\mathcal{L}^{-1}\left[\frac{1}{5s} - \frac{s + 2}{5(s^2 + 2s + 5)}\right] = \frac{1}{5} - \frac{1}{5}e^{-t}\left(\cos 2t + \frac{1}{2}\sin 2t\right).$$

The inverse Laplace transform of

$$\frac{e^{-7s}}{s(s^2 + 2s + 5)}$$

involves the Heaviside function $u_7(t)$ and the same partial fraction decomposition as before. Our final result is

$$y(t) = \frac{1}{5} - \frac{e^{-t}}{5}\left(\cos 2t + \frac{1}{2}\sin 2t\right)$$

$$- \frac{u_7(t)}{5} + \frac{u_7(t)e^{-(t-7)}}{5}\left(\cos(2(t-7)) + \frac{1}{2}\sin(2(t-7))\right).$$

The solution for $0 \le t < 7$ oscillates toward the equilibrium point $(y, v) = (1/5, 0)$ (where $v = dy/dt$), which is an equilibrium when the table is tilted. For $t > 7$ the terms $1/5$ and $-u_7(t)/5$ cancel. The remaining terms all tend to oscillate with decreasing amplitude to the equilibrium point $(y, v) = (0, 0)$ as predicted (see Figure 6.14).

Note that the graph of $y(t)$ is differentiable at $t = 7$, but the graph of $v(t) = dy/dt$ has a corner at $t = 7$. Since the forcing "turns off" at $t = 7$, the second derivative of y has a jump discontinuity, so the first derivative has a corner.

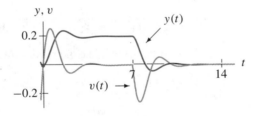

Figure 6.14
The $y(t)$- and $v(t)$-graphs of the solution of the initial-value problem

$$\frac{d^2y}{dt^2} + 2\frac{dy}{dt} + 5y = h(t),$$

with $y(0) = 0$ and $y'(0) = 0$.

Sinusoidal Forcing

We can imagine many different types of forcing that can suddenly appear or disappear. For example, suppose that we again have a mass attached to a spring on a table. For $t \le 5$ we suppose the table is horizontal so there is no external force on the mass. At time $t = 5$ an earthquake begins to shake the table, rocking it back and forth, providing a periodic forcing term.

As an example, let the mass $m = 1$ and the spring constant $k_s = 4$. We assume the oscillator is undamped, so $k_d = 0$. For the forcing term we take the function $3u_5(t)\sin(t - 5)$, which has amplitude 3 and period 2π and begins at time $t = 5$. The differential equation is

$$\frac{d^2y}{dt^2} + 4y = 3u_5(t)\sin(t - 5).$$

We take initial conditions $y(0) = 1$ and $y'(0) = 0$.

Qualitative analysis

For $t < 5$ this equation represents an unforced, undamped harmonic oscillator with natural period π, so the solution with initial conditions $y(0) = 1$ and $y'(0) = 0$ oscillates with constant amplitude. At time $t = 5$ the forcing turns on. The period of the forcing is 2π, so this is not a resonant system and all solutions are bounded. In fact all solutions are combinations of trigonometric functions with the natural and forcing periods.

Solution using Laplace transforms

To apply the Laplace transform method to the initial-value problem

$$\frac{d^2y}{dt^2} + 4y = 3u_5(t)\sin(t - 5), \quad y(0) = 1, \quad y'(0) = 0,$$

we apply the Laplace transform to both sides of the differential equation, obtaining

$$\mathcal{L}\left[\frac{d^2y}{dt^2}\right] + 4\mathcal{L}[y] = \mathcal{L}[3u_5(t)\sin(t - 5)].$$

Thus

$$s^2\mathcal{L}[y] - sy(0) - y'(0) + 4\mathcal{L}[y] = \frac{3e^{-5s}}{s^2 + 1}.$$

Substituting the initial conditions and simplifying, we obtain

$$(s^2 + 4)\mathcal{L}[y] - s = \frac{3e^{-5s}}{s^2 + 1}.$$

So

$$y = \mathcal{L}^{-1}\left[\frac{s}{s^2 + 4}\right] + \mathcal{L}^{-1}\left[\frac{3e^{-5s}}{(s^2 + 4)(s^2 + 1)}\right].$$

Using partial fractions, we can write the second term on the right as

$$\frac{3e^{-5s}}{(s^2 + 4)(s^2 + 1)} = -\frac{e^{-5s}}{s^2 + 4} + \frac{e^{-5s}}{s^2 + 1},$$

which gives

$$y = \mathcal{L}^{-1}\left[\frac{s}{s^2 + 4}\right] - \mathcal{L}^{-1}\left[\frac{e^{-5s}}{s^2 + 4}\right] + \mathcal{L}^{-1}\left[\frac{e^{-5s}}{s^2 + 1}\right].$$

Using

$$\mathcal{L}^{-1}\left[\frac{s}{s^2 + 4}\right] = \cos 2t, \quad \mathcal{L}^{-1}\left[\frac{1}{s^2 + 1}\right] = \sin t,$$

and

$$\mathcal{L}^{-1}\left[\frac{2}{s^2 + 4}\right] = \sin 2t,$$

we obtain

$$y(t) = \cos 2t + u_5(t)(\sin(t - 5) - \tfrac{1}{2}\sin(2(t - 5))).$$

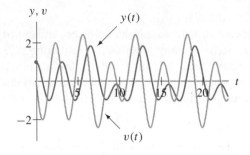

Figure 6.15
The $y(t)$- and $v(t)$-graphs for the solution of the initial-value problem

$$\frac{d^2y}{dt^2} + 4y = 3u_5(t)\sin(t-5),$$

where $y(0) = 1$ and $y'(0) = 0$.

The $y(t)$- and $v(t)$-graphs for this solution are shown in Figure 6.15. Note that the solution behaves as predicted. It is $y(t) = \cos 2t$ until time $t = 5$, at which time the forcing is turned on. There is no discontinuity in the second derivative because the value of $3u_5(t)\sin(t-5)$ is zero at $t = 5$. The third derivative d^3y/dt^3 is discontinuous at $t = 5$, but this is very difficult to see in the graphs.

Resonant Forcing

Finally, we consider the case of an undamped harmonic oscillator with resonant forcing. In Chapter 4 we saw that the behavior of resonant oscillators is much different from that of nonresonant cases. Solutions oscillate with an amplitude that grows linearly with time, and hence they are unbounded (see Figure 6.16). Finding formulas for solutions for a resonant oscillator by the methods of Chapter 4 involved "second-guessing" and could be quite tedious. With the Laplace transform method, finding the solution of a resonant equation is almost as easy as the nonresonant case once we add one additional Laplace transform to our list of known transforms.

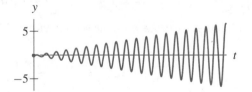

Figure 6.16
Solution of the undamped oscillator with resonant forcing

$$\frac{d^2y}{dt^2} + 9y = \cos 3t$$

with $y(0) = y'(0) = 0$.

As an example, we consider

$$\frac{d^2y}{dt^2} + 9y = \cos 3t, \quad y(0) = y'(0) = 0.$$

Both the natural period and the forcing period of this undamped oscillator are $2\pi/3$, so this is a resonant oscillator.

We begin, as always, by taking the Laplace transform of both sides of the differential equation,

$$\mathcal{L}\left[\frac{d^2y}{dt^2}\right] + 9\mathcal{L}[y] = \mathcal{L}[\cos 3t].$$

Simplifying and substituting the initial conditions yields

$$s^2 \mathcal{L}[y] + 9\mathcal{L}[y] = \frac{s}{s^2 + 9},$$

which reduces to

$$\mathcal{L}[y] = \frac{s}{(s^2 + 9)^2}.$$

Hence

$$y(t) = \mathcal{L}^{-1}\left[\frac{s}{(s^2 + 9)^2}\right].$$

This is the first time that we have encountered a Laplace transform of the form

$$\frac{s}{(s^2 + w^2)^2},$$

and the method of partial fractions does not help because this fraction is already in its partial fractions form. In this case we resort to our table of transforms (see page 557), and we obtain

$$\mathcal{L}[t \sin \omega t] = \frac{2\omega s}{(s^2 + \omega^2)^2}.$$

In our case we write

$$y(t) = \mathcal{L}^{-1}\left[\frac{s}{(s^2 + 9)^2}\right] = \frac{1}{6}\mathcal{L}\left[\frac{6s}{(s^2 + 9)^2}\right],$$

so

$$y(t) = \frac{1}{6}t \sin 3t$$

(see Figure 6.16). The only difference between this computation and the nonresonant examples above is that we used a different entry from the table when finding the inverse Laplace transform. We must remember that, even though the computation is similar, the long-term behavior of the solution of a resonant equation is vastly different from a nonresonant equation.

Analysis of the Laplace Transform Method

If it were a technique only for finding closed-form solutions for constant-coefficient, linear equations, the Laplace transform would quickly go the way of the dinosaur and the slide rule. Laplace transforms do replace calculus computations with algebraic computations, but computer algebra systems replace calculus computations with typing.

The real power of Laplace transforms comes in their unique approach to differential equations. We have already seen that Laplace transforms help us solve differential equations involving discontinuities. In the next section Laplace transforms are used in an even more fundamental way in order to model a forcing term that corresponds to a sudden push like a tap of a hammer or a jolt from an explosion. Laplace transforms are essential to the construction of a mathematically meaningful model of such a forcing term.

In Section 6.5 we use the ability of Laplace transforms to convert calculus into algebra to obtain a method of comparing the solutions arising from a harmonic oscillator with different forcing functions. In Section 6.6 we use Laplace transforms to obtain a generalization of the notion of eigenvalue of a constant-coefficient linear equation that applies to forced (nonautonomous) equations. It is these theoretical uses of the Laplace transform that make this technique worth learning.

EXERCISES FOR SECTION 6.3

In Exercises 1–4, compute the Laplace transform of the given function by computing the integral. (These exercises are mainly a review of integration by parts.)

1. $\sin \omega t$

2. $\cos \omega t$

3. $e^{at} \sin \omega t$

4. $e^{at} \cos \omega t$

5. Use the fact that $y(t) = \cos \omega t$ is the solution of the initial-value problem

$$\frac{d^2 y}{dt^2} + \omega^2 y = 0, \quad y(0) = 1, \quad y'(0) = 0$$

to verify the formula for $\mathcal{L}[\cos \omega t]$.

In Exercises 6–10, compute the Laplace transform of the given function. [*Hint*: One way to do these exercises is to use integration by parts. Alternately, in Exercise 6, differentiate both sides of the formula for $\mathcal{L}[\cos \omega t]$ with respect to ω. The others can be handled similarly.]

6. $f(t) = t \sin \omega t$

7. $f(t) = t \cos \omega t$

8. $f(t) = t e^{at}$

9. $f(t) = t^2 e^{at}$

10. $f(t) = t^n e^{at}$

In Exercises 11–14, write the given quadratic in the form $(s+\alpha)^2 + \beta^2$ (that is, complete the square).

11. $s^2 + 2s + 10$

12. $s^2 - 4s + 5$

13. $s^2 + s + 1$

14. $s^2 + 6s + 10$

In Exercises 15–18, compute the inverse Laplace transform of the given functions (compare with Exercises 11–14).

15. $\dfrac{1}{s^2 + 2s + 10}$

16. $\dfrac{s}{s^2 - 4s + 5}$

17. $\dfrac{2s + 3}{s^2 + s + 1}$

18. $\dfrac{s + 1}{s^2 + 6s + 10}$

19. Compute (from the definition) the Laplace transform

$$\mathcal{L}\left[e^{(a+ib)t}\right],$$

where a and b are real constants.

20. Suppose $y(t)$ is complex-valued and write $y(t) = y_{re}(t) + iy_{im}(t)$ where $y_{re}(t)$ and $y_{im}(t)$ are the real and imaginary parts of $y(t)$ respectively. Show that $\mathcal{L}[y_{re}]$ is the real part of $\mathcal{L}[y]$ and $\mathcal{L}[y_{im}]$ is the imaginary part of $\mathcal{L}[y]$.

21. Use Exercises 19 and 20 to compute $\mathcal{L}\left[e^{at}\cos\omega t\right]$ and $\mathcal{L}\left[e^{at}\sin\omega t\right]$.

22. Compute

$$\mathcal{L}^{-1}\left[\frac{1}{s^2 + 2s + 5}\right]$$

as follows:

   (a) Verify that $s^2 + 2s + 5 = (s + 1 + 2i)(s + 1 - 2i)$.

   (b) Use partial fractions to write

$$\frac{1}{s^2 + 2s + 5} = \frac{A}{s + 1 + 2i} + \frac{B}{s + 1 - 2i},$$

    where A and B are (complex) constants.

   (c) Use Exercise 19 to take the inverse Laplace transform of each term computed in part (b).

   (d) Simplify the sum of the inverse Laplace transforms obtained in part (c) to obtain the desired inverse Laplace transform.

In Exercises 23–26, use partial fractions and complex exponentials to compute the inverse Laplace transform of the given functions.

23. $\dfrac{1}{s^2 + 2s + 10}$

24. $\dfrac{s}{s^2 - 4s + 5}$

25. $\dfrac{2s + 3}{s^2 + s + 1}$

26. $\dfrac{s + 1}{s^2 + 6s + 10}$

In Exercises 27–32,

   (a) compute the Laplace transform of both sides of the differential equation,

   (b) substitute in the initial conditions and simplify to obtain the Laplace transform of the solution, and

   (c) find the solutions by taking the inverse Laplace transform.

27. $\dfrac{d^2y}{dt^2} + 4y = \cos 2t, \quad y(0) = -2, \quad y'(0) = 0$

28. $\dfrac{d^2y}{dt^2} + 2\dfrac{dy}{dt} + 5y = \sin 3t, \quad y(0) = 2, \quad y'(0) = 0$

29. $\dfrac{d^2y}{dt^2} + 3y = u_4(t)\cos(5(t-4))$, $y(0) = 0$, $y'(0) = -2$

30. $\dfrac{d^2y}{dt^2} + 9y = u_5(t)\sin(3(t-5))$, $y(0) = 2$, $y'(0) = 0$

31. $\dfrac{d^2y}{dt^2} + 3y = w(t)$, $y(0) = 2$, $y'(0) = 0$,   where

$$w(t) = \begin{cases} t, & \text{if } 0 \le t < 1; \\ 1, & \text{if } t \ge 1. \end{cases}$$

32. $\dfrac{d^2y}{dt^2} + \dfrac{dy}{dt} + 3y = w(t)$, $y(0) = 2$, $y'(0) = -2$,

where $w(t)$ is the function defined in Exercise 31.

33. **(a)** Find the solution of the initial-value problem

$$\dfrac{d^2y}{dt^2} + 9y = 0.1\cos 3t, \quad y(0) = 2, \quad y'(0) = 0.$$

[*Hint*: It is not necessary to use Laplace transforms, but doing so will help you in the next exercise.]

(b) In a short paragraph, describe the behavior of the solution in part (a).

34. **(a)** Find the solution of the initial-value problem

$$\dfrac{d^2y}{dt^2} + 9y = 0.1\left((1 - u_5(t))\cos 3t + u_{12}(t)\cos 3t\right), \; y(0) = 2, \; y'(0) = 0.$$

(b) In a short paragraph, describe the behavior of the solution in part (a).

35. If $w(t)$ is the square wave of Exercise 17 in Section 6.2:

(a) Compute the Laplace transform of the solution of the initial-value problem

$$\dfrac{d^2y}{dt^2} + 20\dfrac{dy}{dt} + 200y = w(t), \quad y(0) = 1, \quad y'(0) = 0.$$

(b) Describe the behavior of the solution. [*Hint*: Knowing the Laplace transform of the solution may not help with this part.]

36. If $z(t)$ is the sawtooth wave of Exercise 18 in Section 6.2:

(a) Compute the Laplace transform of the solution of the initial-value problem

$$\dfrac{d^2y}{dt^2} + 20\dfrac{dy}{dt} + 200y = z(t), \quad y(0) = 1, \quad y'(0) = 0.$$

(b) Describe the behavior of the solution. [*Hint*: Knowing the Laplace transform of the solution may not help with this part.]

6.4 DELTA FUNCTIONS AND IMPULSE FORCING

We have seen that Laplace transforms make it possible for us to work with equations that have discontinuous forcing terms. In this section we consider another type of discontinuous forcing function known as an impulse function.

Impulse Forcing

Impulse forcing is the term used to describe a very quick push or pull on a system, such as a blow with a hammer or the effect of an explosion. For example, suppose we have an unforced oscillator that satisfies the equation

$$\frac{d^2y}{dt^2} + 2\frac{dy}{dt} + 26y = 0.$$

This equation models a unit mass attached to a spring with spring constant 26 sliding on a table with damping coefficient 2 (see Chapter 4).

Now suppose we strike the mass with a hammer once at time $t = 4$ (see Figure 6.17). We can write the forced equation as

$$\frac{d^2y}{dt^2} + 2\frac{dy}{dt} + 26y = g(t),$$

where

$$g(t) = \begin{cases} \text{very big,} & \text{when } t = 4; \\ 0, & \text{when } t \neq 4. \end{cases}$$

To make this precise enough to have any hope of obtaining a solution, we need some sort of formula for $g(t)$. As a first attempt to derive such a formula, we might assume that the hammer delivers a large constant force over a short time interval. More precisely, we assume that the hammer is in contact with the mass during the time interval from $4 - \Delta t$ to $4 + \Delta t$ for some small time $\Delta t > 0$, and that the force on the mass

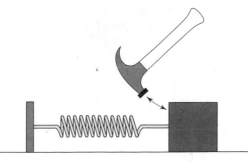

Figure 6.17
Harmonic oscillator struck by hammer.

during this interval is the constant k. Then the forcing function becomes

$$g_{\Delta t}(t) = \begin{cases} k, & \text{if } 4 - \Delta t \leq t \leq 4 + \Delta t; \\ 0, & \text{otherwise.} \end{cases}$$

Here we think of both Δt and k as parameters. The parameters k and Δt are closely related. The external force is only nonzero for a time interval of length $2\Delta t$. Hence the smaller we take Δt, the larger we should take k in order to deliver the same total push.

We choose $k = 1/(2\Delta t)$ so that k becomes large as $\Delta t \to 0$. Therefore

$$g_{\Delta t}(t) = \begin{cases} \dfrac{1}{2\Delta t}, & \text{if } 4 - \Delta t \leq t \leq 4 + \Delta t; \\ 0, & \text{otherwise.} \end{cases}$$

With this choice of k, the area under the graph of $g_{\Delta t}(t)$ is the same as the area of a rectangle with base $2\Delta t$ and height $1/(2\Delta t)$. That is, the area is 1 no matter what we choose for Δt (see Figure 6.18).

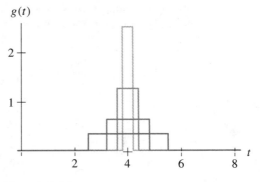

Figure 6.18
Graph of the function $g_{\Delta t}$ for several values of Δt.

Since we are modeling a hammer strike that is delivered very quickly, we would like to make the time interval Δt as small as possible. So we consider what happens as $\Delta t \to 0$. Informally, as $\Delta t \to 0$, we are compressing the same amount of force into a shorter and shorter time interval.

An "instantaneous" force would be represented by the limit as $\Delta t \to 0$. But

$$\lim_{\Delta t \to 0} g_{\Delta t}(t)$$

is a confusing object. For any $t \neq 4$, the limit of $g_{\Delta t}(t)$ as $\Delta t \to 0$ is 0 because for small Δt, the time t is outside the interval on which $g_{\Delta t}(t)$ is positive. On the other hand,

$$\lim_{\Delta t \to 0} g_{\Delta t}(4)$$

does not exist because $g_{\Delta t}(4) \to \infty$ as $\Delta t \to 0$. The "function"

$$\delta_4(t) = \lim_{\Delta t \to 0} g_{\Delta t}(t)$$

is called the **Dirac delta function**. This "function" is zero for all values of t except $t = 4$, where it is infinite.

To a mathematician, this "definition" is worse than useless. According to this definition there is no way to tell the difference between $\delta_4(t)$ and $2\delta_4(t)$ because both of these "functions" are zero for $t \neq 4$ and infinite for $t = 4$. On the other hand, the definite integral of $\delta_4(t)$ seems to make sense. Indeed, the total area under the graph of $g_{\Delta t}$ is always 1, no matter how small we take Δt. So it looks as though we may use Laplace transforms to rescue the Dirac delta function, turning it from nonsense into a useful mathematical object.

The Laplace Transform of a Dirac Delta Function

To make sense of this mess, we turn to the Laplace transform. For any $\Delta t > 0$ we can compute

$$\mathcal{L}[g_{\Delta t}] = \int_0^\infty g_{\Delta t}(t)e^{-st}\,dt$$

$$= \int_{4-\Delta t}^{4+\Delta t} \frac{1}{2\Delta t} e^{-st}\,dt$$

$$= \left(\frac{1}{2\Delta t}\right)\left(\frac{e^{-st}}{-s}\right)\Bigg|_{t=4-\Delta t}^{t=4+\Delta t}$$

$$= \left(\frac{1}{2\Delta t}\right)\left(-\frac{e^{-s(4+\Delta t)} - e^{-s(4-\Delta t)}}{s}\right)$$

$$= \left(\frac{e^{-4s}}{s}\right)\left(\frac{e^{s\Delta t} - e^{-s\Delta t}}{2\Delta t}\right).$$

Taking the limit of this quantity as $\Delta t \to 0$, we have

$$\lim_{\Delta t \to 0} \mathcal{L}[g_{\Delta t}] = \lim_{\Delta t \to 0} \left(\frac{e^{-4s}}{s}\right)\left(\frac{e^{s\Delta t} - e^{-s\Delta t}}{2\Delta t}\right)$$

$$= \left(\frac{e^{-4s}}{s}\right)\lim_{\Delta t \to 0} e^{-s\Delta t}\left(\frac{e^{2s\Delta t} - 1}{2\Delta t}\right)$$

$$= e^{-4s}$$

(see Exercise 1).

The limit of the Laplace transform of $g_{\Delta t}$ as $\Delta t \to 0$ is a perfectly normal function. As long as we take Laplace transform *before* taking the limit, everything stays nice. Hence we define the Laplace transform of $\delta_4(t)$ as

$$\mathcal{L}[\delta_4] = e^{-4s}.$$

We could just as easily put the impulse force at time $t = a$, obtaining a Dirac delta function $\delta_a(t)$ that "is zero for $t \neq a$ and ∞ for $t = a$." We have the following definition:

DEFINITION   The Laplace transform of $\delta_a(t)$ is $\mathcal{L}[\delta_a] = e^{-as}$.   ∎

We use $\delta_a(t)$ to model the force delivered by a tap of a hammer at time $t = a$. As we observed above, as a function, $2\delta_a(t)$ is impossible to distinguish from $\delta_a(t)$. However, taking the Laplace transforms, we use linearity to see that

$$\mathcal{L}[2\delta_a] = 2\mathcal{L}[\delta_a] = 2e^{-as}.$$

Using this, we can effectively model a hammer tap of twice the original strength. In fact we can treat the Laplace transform of the $\delta_a(t)$ in the same way we treat any other Laplace transform.

Impulse Forcing of a Damped Harmonic Oscillator

We now return to the harmonic oscillator equation with mass 1, spring constant 26, and coefficient of damping 2, which is struck by a hammer at $t = 4$. For this example we choose initial conditions $y(0) = 1$ and $y'(0) = 0$. That is, the spring is stretched one unit from its rest position and then released with velocity 0. Using the Dirac delta function, we can write the initial-value problem

$$\frac{d^2y}{dt^2} + 2\frac{dy}{dt} + 26y = \delta_4(t), \quad y(0) = 1, \quad y'(0) = 0.$$

Qualitative analysis

For $0 \leq t < 4$ we have an unforced, underdamped harmonic oscillator, since the characteristic equation has roots $-1 \pm 5i$. The graph of the solution with initial condition $y(0) = 1$ and $y'(0) = 0$ oscillates with decreasing amplitude (see Figure 6.19). At $t = 4$ the impulse force is, speaking informally, infinite. Because the force is the product of mass and acceleration, d^2y/dt^2 is infinite at $t = 4$. Hence the velocity $v = dy/dt$ has a discontinuity at $t = 4$ and the solution $y(t)$ abruptly changes in amplitude. After $t = 4$ we again have an underdamped oscillator.

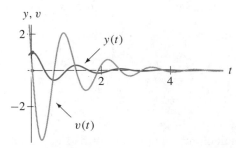

Figure 6.19
The $y(t)$- and $v(t)$-graphs for the solution of unforced oscillator

$$\frac{d^2y}{dt^2} + 2\frac{dy}{dt} + 26y = 0$$

with initial conditions $y(0) = 1$ and $y'(0) = 0$.

Unfortunately this is only a rough qualitative description of the behavior of the solution using mathematically suspicious phrases like "the acceleration is infinite." To see what is really going on, we use the Laplace transform.

Solution using Laplace transform

Taking the Laplace transform of both sides of the equation, we obtain

$$\mathcal{L}\left[\frac{d^2y}{dt^2}\right] + 2\mathcal{L}\left[\frac{dy}{dt}\right] + 26\mathcal{L}[y] = \mathcal{L}[\delta_4].$$

Using the formulas for the Laplace transform of a derivative and second derivative and solving for $\mathcal{L}[y]$, we find

$$s^2\mathcal{L}[y] - sy(0) - y'(0) + 2s\mathcal{L}[y] - 2y(0) + 26\mathcal{L}[y] = \mathcal{L}[\delta_4].$$

Substituting the initial conditions and evaluating $\mathcal{L}[\delta_4]$, we have

$$s^2\mathcal{L}[y] - s + 2s\mathcal{L}[y] - 2 + 26\mathcal{L}[y] = e^{-4s}.$$

Solving for $\mathcal{L}[y]$ yields

$$\mathcal{L}[y] = \frac{s+2}{s^2 + 2s + 26} + \frac{e^{-4s}}{s^2 + 2s + 26}.$$

Hence

$$y(t) = \mathcal{L}^{-1}\left[\frac{s+2}{s^2 + 2s + 26}\right] + \mathcal{L}^{-1}\left[\frac{e^{-4s}}{s^2 + 2s + 26}\right]$$

$$= \mathcal{L}^{-1}\left[\frac{s+2}{(s+1)^2 + 25}\right] + \frac{1}{5}\mathcal{L}^{-1}\left[\frac{5e^{-4s}}{(s+1)^2 + 25}\right],$$

and the solution is

$$y(t) = e^{-t}\cos 5t + \tfrac{1}{5}e^{-t}\sin 5t + \tfrac{1}{5}u_4(t)e^{-(t-4)}\sin(5(t-4)).$$

The solution has the predicted form (see Figure 6.20). Note that the $v(t)$-graph has a jump discontinuity at $t = 4$, just as the impulse force is applied. Thus the slope of the tangent line to the graph of $y(t)$ changes abruptly at $t = 4$. Note that there is a "corner" in the graph of $y(t)$ at $t = 4$ (see Figure 6.21).

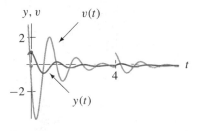

Figure 6.20
The $y(t)$- and $v(t)$-graphs for
$d^2y/dt^2 + 2dy/dt + 26y = \delta_4(t)$ with
$y(0) = 1$ and $y'(0) = 0$.

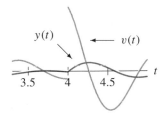

Figure 6.21
Magnification of the corner and
jump discontinuity of the $y(t)$-
and $v(t)$-graphs at $t = 4$.

An Undamped Oscillator with Impulse Forcing

In applications it is common to suppose that the impulse force is applied at $t = 0$. For example, consider the undamped oscillator

$$\frac{d^2y}{dt^2} + 4y = \delta_0(t).$$

This is a model of a unit mass attached to a spring with spring constant 4, sliding on a frictionless table. We assume that, for all $t < 0$, the mass is at the equilibrium point, at rest at $y = 0$. When $t = 0$, the mass is given a sharp push in the positive y-direction. So at $t = 0$ we have $y(0) = 0$, but $y'(0)$ is undefined since the acceleration at this time is infinite. To indicate that prior to $t = 0$ the mass was at rest, we write the initial condition $y'(0) = 0^-$. Thus we consider the initial-value problem

$$\frac{d^2y}{dt^2} + 4y = \delta_0(t), \quad y(0) = 0, \quad y'(0) = 0^-.$$

Qualitative analysis

Since there is no external force for $t > 0$, once the mass is set in motion at time $t = 0$ it oscillates forever with constant period and amplitude. The period of oscillation is the natural period (which is π). To determine the amplitude of the oscillations, we compute the solution.

Solution using Laplace transforms

Taking the Laplace transform of both sides of the differential equation gives

$$\mathcal{L}\left[\frac{d^2y}{dt^2}\right] + 4\mathcal{L}[y] = \mathcal{L}[\delta_0].$$

Simplifying and using the fact that $\mathcal{L}[\delta_0] = e^{-0s} = 1$, we find

$$s^2\mathcal{L}[y] - sy(0) - y'(0) + 4\mathcal{L}[y] = 1.$$

Substituting the initial conditions $y(0) = y'(0) = 0$ (just before the push of the impulse force) and solving for $\mathcal{L}[y]$ yields

$$\mathcal{L}[y] = \frac{1}{s^2 + 4},$$

so

$$y(t) = \mathcal{L}^{-1}\left[\frac{1}{s^2 + 4}\right].$$

Hence the solution is

$$y(t) = \tfrac{1}{2}\sin 2t.$$

That is, the solution oscillates with amplitude $1/2$ for $t > 0$ (see Figure 6.22).

We should be careful to note that the formula that we just obtained is the solution for $t > 0$. The solution for all t is

$$y(t) = \begin{cases} \frac{1}{2}\sin 2t, & \text{if } t > 0; \\ 0, & \text{if } t \leq 0. \end{cases}$$

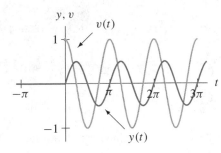

Figure 6.22
The $y(t)$- and $v(t)$-graphs of the solution of the
initial-value problem

$$\frac{d^2y}{dt^2} + 4y = \delta_0(t),$$

with $y(0) = y'(0) = 0^-$.

Note that the derivative of this solution does not exist at $t = 0$ (see Figure 6.22). For $t < 0$ the velocity is identically 0, but for t slightly positive the velocity is close to 1. Thus $v(t)$ has a jump discontinuity at $t = 0$. This observation agrees with our earlier qualitative analysis since the acceleration is infinite at $t = 0$.

Periodic Impulse Forcing

The qualitative approach can be useful, even when complicated forcing terms with discontinuities are involved. As an example, consider an undamped oscillator that we tap with a hammer every a time units, that is, at times $t = a, t = 2a, t = 3a, \ldots$. This leads to an initial-value problem of the form

$$\frac{d^2y}{dt^2} + y = \sum_{n=1}^{\infty} \delta_{na}(t), \quad y(0) = 0, \quad y'(0) = 0.$$

The long-term effect of this striking will depend on what the mass is doing when the strike arrives. If the mass is moving away from the hammer, then the hammer strike will speed up the mass; but if the mass is moving toward the hammer, the blow will slow it down or, perhaps, reverse its direction.

This situation is similar to that of sinusoidal forcing—if the period of the forcing is the same as the natural period of the oscillator, then the impact of the hammer will always catch the mass in the same state. Hence a resonance phenomenon will occur and the amplitudes of the oscillations will increase without bound. If the forcing period and natural period are different, then the strikes of the hammer will sometimes slow the mass and sometimes speed it up.

EXERCISES FOR SECTION 6.4

1. Compute the limit

$$\lim_{\Delta t \to 0} e^{-s\Delta t} \left(\frac{e^{2s\Delta t} - 1}{2\Delta t} \right).$$

In Exercises 2–5, solve the given initial-value problem.

2. $\dfrac{d^2y}{dt^2} + 3y = 5\delta_2(t), \quad y(0) = 0, \quad y'(0) = 0$

3. $\dfrac{d^2y}{dt^2} + 2\dfrac{dy}{dt} + 5y = \delta_3(t), \quad y(0) = 1, \quad y'(0) = 1$

4. $\dfrac{d^2y}{dt^2} + 2\dfrac{dy}{dt} + 2y = -2\delta_2(t), \quad y(0) = 2, \quad y'(0) = 0$

5. $\dfrac{d^2y}{dt^2} + 2\dfrac{dy}{dt} + 3y = \delta_1(t) - 3\delta_4(t), \quad y(0) = 0, \quad y'(0) = 0$

6. **(a)** Discuss the qualitative behavior of the solution of the initial-value problem

$$\frac{d^2y}{dt^2} + 2\frac{dy}{dt} + 3y = \delta_4(t), \quad y(0) = 1, \quad y'(0) = 0.$$

(b) Compute the solution of this initial-value problem.

(c) Graph this solution. Write a paragraph comparing your description of the qualitative behavior with the graph.

7. There is a relationship between the Heaviside function $u_a(t)$ and the Dirac delta function $\delta_a(t)$ that can be observed from their Laplace transforms.

(a) Show that, for $a > 0$, $\mathcal{L}[\delta_a] = s\mathcal{L}[u_a] - u_a(0)$.

(b) What relationship does this suggest between $u_a(t)$ and $\delta_a(t)$?

(c) Is there any way we could understand this relationship in terms of everyday calculus (without Laplace transforms)?

8. Calculate the Laplace transform of

$$g(t) = \sum_{n=1}^{\infty} \delta_{na}(t),$$

where $a > 0$. [*Hint*: See Exercise 16 in Section 6.2.]

9. **(a)** Find the Laplace transform of the solution of the initial-value problem

$$\frac{d^2y}{dt^2} + 2y = \sum_{n=1}^{\infty} \delta_n(t), \quad y(0) = 0, \quad y'(0) = 0.$$

(b) Find the solution of the initial-value problem in part (a) (that is, take the inverse Laplace transform of your answer in part (a)).

(c) What is the long-term qualitative behavior of this solution?

10. **(a)** Find the Laplace transform of the solution of the initial-value problem

$$\frac{d^2y}{dt^2} + y = \sum_{n=1}^{\infty} \delta_{2n\pi}(t), \quad y(0) = 0, \quad y'(0) = 0.$$

(b) Find the solution of the initial-value problem in part (a) (that is, take the inverse Laplace transform of your answer to part (a)).

(c) What is the long-term qualitative behavior of this solution?

6.5 CONVOLUTIONS

When using Laplace transforms to find the solution of an initial-value problem, the hardest step is usually the last—computing the inverse Laplace transform. The complication arises when we have to break a complicated product into the sum of two or more simpler terms using partial fractions. While the arithmetic involved is elementary, it can be very complicated. The probability of making a careless error is high (at least it is for these authors). It would be nice if there were a product rule for inverse Laplace transforms, that is, some way to compute the inverse Laplace transform of a product from the inverse Laplace transform of each of its factors.

In this section we derive such a product rule. Unfortunately, using this rule turns out almost always to be more complicated than doing the partial fractions decomposition. However, the product rule has other surprising applications. In particular, it lets us compute solutions of a forced harmonic oscillator equation using the solutions of the same harmonic oscillator with a different forcing function.

The Product Rule for Inverse Laplace Transforms

Suppose we wish to compute the inverse Laplace transform of a product, that is

$$\mathcal{L}^{-1}[F(s)G(s)]$$

where we know $\mathcal{L}^{-1}[F] = f(t)$ and $\mathcal{L}^{-1}[G] = g(t)$. From the definition of Laplace transform, we have

$$F(s) = \int_0^\infty f(\tau)e^{-s\tau}\, d\tau \quad \text{and} \quad G(s) = \int_0^\infty g(u)e^{-su}\, du.$$

(Note that we are using τ and u as the variables of integration. This is just notation, but it will be important in the computation that follows.) Hence

$$F(s)G(s) = \left(\int_0^\infty f(\tau)e^{-s\tau}\, d\tau \right) \left(\int_0^\infty g(u)e^{-su}\, du \right).$$

The first integral contains no u's, so it may be moved inside the second integral, giving

$$F(s)G(s) = \int_0^\infty \left(\int_0^\infty f(\tau)e^{-s\tau}\, d\tau \right) g(u)e^{-su}\, du.$$

Combining this into a double integral and simplifying gives

$$F(s)G(s) = \int_0^\infty \int_0^\infty f(\tau)g(u)e^{-s(\tau+u)}\, d\tau\, du.$$

Next, we make a change of variables, replacing the variable τ with the new variable $t = \tau + u$. Note that $\tau = t - u$, $dt/du = 1$, and $0 < \tau < \infty$ implies $u < t < \infty$, so we have

$$F(s)G(s) = \int_0^\infty \int_u^\infty f(t-u)g(u)e^{-st}\, dt\, du.$$

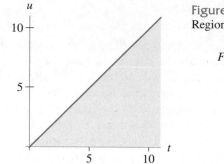

Figure 6.23
Region of integration for the double integral

$$F(s)G(s) = \int_0^\infty \int_u^\infty f(t-u)g(t)e^{-st}\, dt\, du.$$

The region of integration is shown in Figure 6.23. Changing the order of integration in the double integral gives

$$F(s)G(s) = \int_0^\infty \int_0^t f(t-u)g(u)e^{-st}\, du\, dt.$$

We can segregate the terms that contain u, yielding

$$F(s)G(s) = \int_0^\infty \left(\int_0^t f(t-u)g(u)\, du \right) e^{-st}\, dt.$$

Stepping back to look at this, we get a surprise. If we let

$$h(t) = \int_0^t f(t-u)g(u)\, du,$$

then

$$F(s)G(s) = \int_0^\infty h(t)e^{-st}\, dt,$$

or in other words,

$$F(s)G(s) = \mathcal{L}[h].$$

That is, we have expressed the product of $F(s)$ and $G(s)$ as a Laplace transform. Taking inverse Laplace transforms of both sides of this equation gives

$$\mathcal{L}^{-1}[F(s)G(s)] = h(t) = \int_0^t f(t-u)g(u)\, du.$$

This computation motivates the definition of convolution.

DEFINITION   The **convolution** $f * g$ of two functions $f(t)$ and $g(t)$ is the function defined by

$$(f * g)(t) = \int_0^t f(t-u)g(u)\, du. \quad \blacksquare$$

With this notation we can rewrite the results of our computation as

$$\mathcal{L}^{-1}[F(s)G(s)] = f * g,$$

which is equivalent to

$$\mathcal{L}^{-1}[F(s)G(s)] = \mathcal{L}^{-1}[F] * \mathcal{L}^{-1}[G].$$

In words, the rule is "The inverse Laplace transform of a product is the convolution of the inverse Laplace transforms."

At first glance, it appears from the formula that $f * g$ is not the same as $g * f$ since the integrals look different. However, we know that

$$\mathcal{L}^{-1}[F(s)G(s)] = \mathcal{L}^{-1}[G(s)F(s)],$$

so this cannot be the case (see Exercise 6 to rectify this difficulty).

A disappointing example

As a first (disappointing) attempt to use this rule, we consider

$$\mathcal{L}^{-1}\left[\frac{3s}{(s^2+1)(s^2+4)}\right] = \mathcal{L}^{-1}\left[\left(\frac{3}{s^2+1}\right)\left(\frac{s}{s^2+4}\right)\right].$$

Now

$$\mathcal{L}^{-1}\left[\frac{3}{s^2+1}\right] = 3\mathcal{L}^{-1}\left[\frac{1}{s^2+1}\right] = 3\sin t$$

and

$$\mathcal{L}^{-1}\left[\frac{s}{s^2+4}\right] = \cos 2t.$$

Hence, applying the product rule, we have

$$\mathcal{L}^{-1}\left[\left(\frac{3}{s^2+1}\right)\left(\frac{s}{s^2+4}\right)\right] = (3\sin t) * (\cos 2t)$$

$$= \int_0^t 3\sin(t-u)\cos 2u \, du.$$

The integral we end up with is computable, but rather unpleasant (see Exercise 5).

On the other hand, computing this inverse Laplace transform via partial fractions gives

$$\mathcal{L}^{-1}\left[\frac{3s}{(s^2+1)(s^2+4)}\right] = \mathcal{L}^{-1}\left[\frac{s}{s^2+1}\right] \quad \mathcal{L}^{-1}\left[\frac{s}{s^2+4}\right],$$

so

$$\mathcal{L}^{-1}\left[\frac{3s}{(s^2+1)(s^2+4)}\right] = \cos t - \cos 2t.$$

This example is typical. While doing the partial fractions decomposition is tedious, it is less tedious than computing the convolution. However, for what it's worth, we do get the rather surprising formula

$$\cos t - \cos 2t = 3\int_0^t \sin(t-u)\cos 2t \, du$$

since both sides have the same Laplace transform.

It appears that the product rule above is not very useful for computing inverse Laplace transforms. However, as we show next, convolution does have applications that are much more interesting and important.

Delta Forcing and Convolution

To illustrate one of the uses of convolutions, we consider the equation

$$\frac{d^2y}{dt^2} + p\frac{dy}{dt} + qy = f(t),$$

where p and q are constants.

Before we deal with the general forcing term $f(t)$, we consider the special case where $f(t) = \delta_0(t)$, the Dirac delta function at $t = 0$. We consider the initial-value problem

$$\frac{d^2y}{dt^2} + p\frac{dy}{dt} + qy = \delta_0(t), \quad y(0) = 0, \quad y'(0) = 0^-,$$

where the superscript "$-$" as usual indicates that $y'(t) = 0$ for $t < 0$. To keep the notation straight later on, we let $\zeta(t)$ denote the solution of this initial-value problem. (In electric circuit theory, $\zeta(t)$ is called the *impulse response*.)

To obtain an expression for $\zeta(t)$ we take the Laplace transform of both sides of the differential equation, giving

$$\mathcal{L}\left[\frac{d^2\zeta}{dt^2}\right] + p\mathcal{L}\left[\frac{d\zeta}{dt}\right] + q\mathcal{L}[\zeta] = \mathcal{L}[\delta_0].$$

Using the rules for the Laplace transform of a derivative and the fact that $\mathcal{L}[\delta_0] = 1$, we obtain

$$s^2\mathcal{L}[\zeta] - s\zeta(0) - \zeta'(0) + ps\mathcal{L}[\zeta] - p\zeta(0) + q\mathcal{L}[\zeta] = 1.$$

Substituting the initial conditions and solving for $\mathcal{L}[\zeta]$ yields

$$\mathcal{L}[\zeta] = \frac{1}{s^2 + ps + q}.$$

Now let's return to the general case

$$\frac{d^2y}{dt^2} + p\frac{dy}{dt} + qy = f(t), \quad y(0) = y'(0) = 0.$$

Using the Laplace transform exactly as above, we compute that the Laplace transform of the solution $y(t)$ is

$$\mathcal{L}[y] = \frac{\mathcal{L}[f]}{s^2 + ps + q}.$$

Now we can take advantage of the product rule for inverse Laplace transforms by writing

$$\mathcal{L}[y] = \frac{1}{s^2 + ps + q}\mathcal{L}[f] = \mathcal{L}[\zeta]\mathcal{L}[f].$$

So the solution of the initial-value problem with $y(0) = y'(0) = 0$ and forcing function $f(t)$ is

$$y(t) = \mathcal{L}^{-1}[\mathcal{L}[\zeta]\mathcal{L}[f]] = (\zeta * f)(t).$$

We can compute the solution $y(t)$ knowing only the forcing function $f(t)$ and the solution $\zeta(t)$ for the differential equation with the same left-hand side but with the forcing function $\delta_0(t)$. If we know $\zeta(t)$ and $f(t)$ we do not even need to know the coefficients p and q, that is, *we do not need to know the differential equation!*

A computation using convolutions

Consider the initial-value problem

$$\frac{d^2y}{dt^2} + 4y = \delta_0(t), \quad y(0) = y'(0) = 0.$$

Letting $\zeta(t)$ denote the solution, we can compute, by the usual steps, that

$$\mathcal{L}[\zeta] = \frac{1}{s^2 + 4}$$

so that $\zeta(t) = \frac{1}{2}\sin 2t$.

We can now use the observation just made to see that the solution $y(t)$ of

$$\frac{d^2y}{dt^2} + 4y = f(t), \quad y(0) = y'(0) = 0$$

with new forcing function $f(t)$ has as its Laplace transform

$$\mathcal{L}[y] = \mathcal{L}[\zeta]\mathcal{L}[f] = \frac{1}{s^2 + 4}\mathcal{L}[f].$$

So

$$y(t) = \left(\frac{1}{2}\sin 2t\right) * f = \int_0^u \frac{1}{2}\sin(2(t - u))f(u)\, du.$$

For example, if $f(t) = e^{-t}$, then the solution of the initial-value problem

$$\frac{d^2y}{dt^2} + 4y = e^{-t}, \quad y(0) = y'(0) = 0$$

is

$$y(t) = \left(\frac{1}{2}\sin 2t\right) * e^{-t} = \int_0^t \frac{1}{2}\sin(2(t - u))e^{-u}\, du.$$

This is an unpleasant integral to do by hand, but a computer algebra system gives

$$y(t) = -\frac{4}{5}e^{-t} - \frac{1}{5}\cos 2t - \frac{1}{10}\sin 2t.$$

Again, we have replaced the usual method for computing solutions with a convolution integral. As we have already noted, this may not decrease the difficulty of the computation. However, this technique differs fundamentally from our previous methods. We are able to compute the solution of

$$\frac{d^2y}{dt^2} + 4y = f(t), \quad y(0) = y'(0) = 0$$

using only the solution of

$$\frac{d^2y}{dt^2} + 4y = \delta_0(t), \quad y(0) = y'(0) = 0.$$

That is, we can compute new solutions (with new forcing functions) from old solutions without referring back to the differential equation.

Computing Solutions from Experimental Data

Suppose we have a "black box," perhaps an RLC circuit or a mechanical device, which we know can be modeled by an equation of the form

$$\frac{d^2y}{dt^2} + p\frac{dy}{dt} + qy = 0,$$

but we do not know the coefficients p and q. Suppose the box has an input where we can apply any forcing function $f(t)$ we choose. Suppose also that the box has an output where we can record the solution of the initial-value problem

$$\frac{d^2y}{dt^2} + p\frac{dy}{dt} + qy = f(t), \quad y(0) = y'(0) = 0$$

(see Figure 6.24).

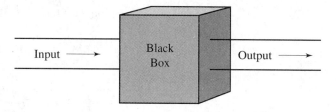

Figure 6.24
Black box with input and output.

The previous discussion gives us a method for predicting the output $y(t)$ for a given input $f(t)$. First, we do an experiment. We input the delta function $\delta_0(t)$ and carefully record the output $\zeta(t)$ that satisfies

$$\frac{d^2\zeta}{dt^2} + p\frac{d\zeta}{dt} + q\zeta = \delta_0(t), \quad \zeta(0) = 0, \quad \zeta'(0) = 0^-.$$

Then we numerically compute the Laplace transform $\mathcal{L}[\zeta]$.

For the input forcing function $f(t)$, let $y(t)$ denote the output. That is, $y(t)$ satisfies the initial-value problem

$$\frac{d^2y}{dt^2} + p\frac{dy}{dt} + qy = f(t), \quad y(0) = y'(0) = 0.$$

Thus we obtain $\mathcal{L}[y] = \mathcal{L}[\zeta]\mathcal{L}[f]$, and therefore

$$y(t) = (\zeta * f)(t) = \int_0^t \zeta(t - u)f(u)\,du.$$

Since we know both ζ and f, at least numerically, we can numerically compute approximations for $\mathcal{L}[y]$ and y. That is, we can predict the output for a given input from data from a completely different input. This corresponds to solving an initial-value problem from another solution obtained experimentally. We can find a solution of an initial-value problem without ever knowing the differential equation.

Sobering Reminder

From the point of view of differential equations, the above discussion can be a bit distressing. We have found a way to compute solutions from data, completely avoiding the differential equation (we do not even need to know what the differential equation is). However, it must be remembered that this technique requires a system (black box) that can be modeled by a special type of differential equation—a linear, constant-coefficient equation. Also, all of the discussion above requires that the initial conditions be zero (the rest position with zero input). This corresponds to a device that quickly returns to its rest position after the forcing is removed. Finally, it was only by understanding the differential equation and the Laplace transform that we could derive this method.

While these methods allow us to compute formulas or numerical approximations of solutions, it gives us little insight into the qualitative behavior of solutions. Nevertheless, Laplace transforms can be used in the qualitative study of linear, constant-coefficient equations. We turn to this topic in the next section.

EXERCISES FOR SECTION 6.5

In Exercises 1–4, compute the convolution $f * g$ for the given functions f and g.

1. $f(t) = 1$ and $g(t) = e^{-t}$

2. $f(t) = e^{-at}$ and $g(t) = e^{-bt}$

3. $f(t) = \cos t$ and $g(t) = u_2(t)$

4. $f(t) = u_2(t)$ and $g(t) = u_3(t)$

5. Compute $(3 \sin t) * \cos 2t$ by computing the integral. [*Hint*: Use trigonometric identities first.]

6. Show that convolution is a commutative operation. In other words, show that $f * g = g * f$ for every two functions f and g.

7. Suppose the solution $\zeta(t)$ of the initial-value problem

$$\frac{d^2y}{dt^2} + p\frac{dy}{dt} + qy = \delta_0(t), \quad y(0) = y'(0) = 0$$

has a Laplace transform $\mathcal{L}[\zeta]$ whose value at $s = 0$ is $1/5$ and whose value at $s - 2$ is $1/17$. Find p and q.

8. Verify that the solution $\eta(t)$ of the initial-value problem

$$\frac{dy}{dt} + ay = f(t), \quad y(0) = 0$$

has Laplace transform $\mathcal{L}[\eta] = \mathcal{L}[\zeta]\mathcal{L}[f]$, where $\zeta(t)$ is the solution of the initial-value problem

$$\frac{dy}{dt} + ay = \delta_0(t), \quad y(0) = 0^-.$$

9. Let $\zeta(t)$ be the solution of the initial-value problem

$$\frac{d^2y}{dt^2} + p\frac{dy}{dt} + qy = \delta_0(t), \quad y(0) = y'(0) = 0^-.$$

Let a and b be constants and let $f(t)$ be an arbitrary function.

(a) Find an expression for the Laplace transform of the solution of the initial-value problem

$$\frac{d^2y}{dt^2} + p\frac{dy}{dt} + qy = 0, \quad y(0) = a, \quad y'(0) = 0$$

in terms of a, p, q, and $\mathcal{L}[\zeta]$.

(b) Find an expression for the Laplace transform of the solution of the initial-value problem

$$\frac{d^2y}{dt^2} + p\frac{dy}{dt} + qy = 0, \quad y(0) = 0, \quad y'(0) = b$$

in terms of b, p, q, and $\mathcal{L}[\zeta]$.

(c) Find an expression for the Laplace transform of the solution of the initial-value problem

$$\frac{d^2y}{dt^2} + p\frac{dy}{dt} + qy = f(t), \quad y(0) = a, \quad y'(0) = b$$

in terms of a, b, p, q, $\mathcal{L}[f]$, and $\mathcal{L}[\zeta]$.

10. Suppose we know $\eta(t)$, the solution of the initial-value problem

$$\frac{d^2y}{dt^2} + p\frac{dy}{dt} + qy = u_0(t), \quad y(0) = y'(0) = 0.$$

Find a formula for the solution $y(t)$ of the initial-value problem

$$\frac{d^2y}{dt^2} + p\frac{dy}{dt} + qy = f(t), \quad y(0) = y'(0) = 0$$

in terms of η and f.

11. Suppose $y_1(t)$ is the solution of the initial-value problem

$$\frac{d^2y}{dt^2} + p\frac{dy}{dt} + qy = f_1(t), \quad y(0) = y'(0) = 0.$$

(a) Compute an expression for $\mathcal{L}[y_1]$.

(b) Suppose $y_2(t)$ is the solution of the initial-value problem

$$\frac{d^2y}{dt^2} + p\frac{dy}{dt} + qy = f_2(t), \quad y(0) = y'(0) = 0.$$

for a different forcing function $f_2(t)$. Show that

$$\frac{\mathcal{L}[f_2]}{\mathcal{L}[y_2]} = \frac{\mathcal{L}[f_1]}{\mathcal{L}[y_1]}.$$

(c) Show that

$$\mathcal{L}[y_2] = \mathcal{L}[f_2]\frac{\mathcal{L}[y_1]}{\mathcal{L}[f_1]}.$$

(This implies that we could use the the solution with any forcing function and zero initial conditions to compute solutions for other forcing functions.)

6.6 THE QUALITATIVE THEORY OF LAPLACE TRANSFORMS

As we have stressed many times, finding the solution of a differential equation means considerably more than finding a formula. Understanding the qualitative behavior of the solution is frequently much more important than the formula. So far we have used the Laplace transform mainly to find formulas for solutions. In this section we use the Laplace transform to study the qualitative behavior of solutions.

We have used Laplace transforms on constant-coefficient, linear equations. In Chapter 3 we saw that we could solve autonomous, linear equations using eigenvalues and eigenvectors. More important, using the eigenvalues, we could give a qualitative description of solutions without nearly as much arithmetic as was involved in finding a formula for the general solution.

The concept of eigenvalues, as we have presented it, does not naturally extend to nonhomogeneous equations such as the forced harmonic oscillator. On the other hand, the Laplace transform works equally well with homogeneous and nonhomogeneous equations. There is an extensive theory that uses the Laplace transform to extend the idea of eigenvalues to nonhomogeneous equations in order to study the qualitative nature of solutions. We study a small part of this qualitative theory in this section.

Homogeneous Second-Order Equations

We begin our investigation of the use of Laplace transforms in the qualitative theory of differential equations by studying a familiar friend, the second-order homogeneous equation

$$\frac{d^2y}{dt^2} + p\frac{dy}{dt} + qy = 0.$$

We have several methods available that solve this equation and that are easier to use than Laplace transforms, but our goal now is to learn more about Laplace transforms.

Taking the Laplace transform of both sides of this equation, we obtain

$$\mathcal{L}\left[\frac{d^2y}{dt^2}\right] + p\mathcal{L}\left[\frac{dy}{dt}\right] + q\mathcal{L}[y] = 0.$$

Using the rules concerning Laplace transforms of derivatives, this equation becomes

$$\left(s^2\mathcal{L}[y] - sy(0) - y'(0)\right) + p\left(s\mathcal{L}[y] - y(0)\right) + q\mathcal{L}[y] = 0,$$

which can be simplified to

$$(s^2 + ps + q)\mathcal{L}[y] - (s + p)y(0) - y'(0) = 0.$$

Thus the Laplace transform of the solution with initial conditions $y(0)$ and $y'(0)$ is

$$\mathcal{L}[y] = \frac{(s+p)y(0)}{s^2 + ps + q} + \frac{y'(0)}{s^2 + ps + q}.$$

The important point to notice here is that the denominator of the Laplace transform of the solution is the quadratic polynomial

$$s^2 + ps + q.$$

We have seen this quadratic in two other contexts already. It is the characteristic polynomial obtained from the differential equation by guessing a solution of the form $y(t) = e^{st}$. It is also the characteristic polynomial of the linear system

$$\frac{dy}{dt} = v$$

$$\frac{dv}{dt} = -qy - pv$$

that corresponds to this second-order equation. This is not a coincidence.

If the roots of the characteristic polynomial (which are the eigenvalues of the system) are λ and μ, then we can write

$$s^2 + ps + q = (s - \lambda)(s - \mu).$$

We first treat the case in which λ and μ are real. The solution y is given by

$$y = \mathcal{L}^{-1}\left[\frac{(s+p)y(0)}{(s-\lambda)(s-\mu)}\right] + \mathcal{L}^{-1}\left[\frac{y'(0)}{(s-\lambda)(s-\mu)}\right].$$

Using partial fractions, we can break this into a sum of fractions with denominators $(s - \lambda)$ and $(s - \mu)$. Every term of the solution will have a factor involving either $e^{\lambda t}$ or $e^{\mu t}$. Qualitatively we know that if λ and μ are both positive, then the origin is a source; if they are both negative, then the origin is a sink; and if one is positive and the other is negative, then the origin is a saddle.

If $\lambda = \alpha + i\beta$ and $\mu = \alpha - i\beta$, then

$$(s - \lambda)(s - \mu) = (s - (\alpha + i\beta))(s - (\alpha - i\beta))$$

$$= ((s - \alpha) - i\beta)((s - \alpha) + i\beta)$$

$$= (s - \alpha)^2 + \beta^2.$$

In this case we can write y as the inverse Laplace transform of functions with the quadratic $(s - \alpha)^2 + \beta^2$ in the denominator and with either a constant or a constant multiple of s in the numerator. The inverse Laplace transforms of these terms are multiples of either $e^{\alpha t} \sin \beta t$ or $e^{\alpha t} \cos \beta t$. Again this is exactly what we expect. In this case the solutions oscillate and the long-range behavior is determined by the sign of α. The amplitude increases if α is positive and decreases if α is negative. The frequency of oscillation is determined by β and is given by $\beta/(2\pi)$.

We can summarize this discussion as follows: For homogeneous, second-order, constant-coefficient equations, the qualitative behavior of solutions is determined by the values of s for which the denominator of the Laplace transform of the solution is zero. (These values of s are precisely the same as the eigenvalues of the corresponding linear system.) This motivates some terminology.

DEFINITION   Suppose $F(s)$ is a rational function; that is

$$F(s) = \frac{G(s)}{H(s)},$$

where $G(s)$ and $H(s)$ are polynomials with no common factors and the degree of G is less than the degree of H. The **poles** of F are the values of s for which $H(s) = 0$. If $F(s)$ is the sum of rational functions,

$$F(s) = \frac{G_1(s)}{H_1(s)} + \frac{G_1(s)}{H_1(s)} + \cdots + \frac{G_n(s)}{H_n(s)},$$

the poles of F are found by first rewriting F as a single fraction, canceling any common terms in the numerator and denominator, and then finding the poles of the resulting rational function.   ∎

We can summarize this computation by saying that, for a homogeneous, second-order, constant-coefficient, linear equation, the poles of the Laplace transform of the general solution are the same as the eigenvalues of the corresponding linear system.

Nonhomogeneous Second-Order Equations

When we consider a nonhomogeneous linear differential equation, the qualitative techniques that we used for homogeneous linear equations and systems no longer apply. A nonautonomous system is very different from an autonomous system because the vector field for the corresponding system changes with time.

However, when using the Laplace transform, there is not much difference between a homogeneous and a nonhomogeneous equation. The arithmetic for nonhomogeneous equations is slightly more complicated, but the basic method is the same. Hence we are led to consider the poles of the Laplace transform of solutions of nonhomogeneous equations in hopes of obtaining the same sort of qualitative information we can get from the eigenvalues of a homogeneous equation.

For example, consider the initial-value problem

$$\frac{d^2 y}{dt^2} + 2\frac{dy}{dt} + 2y = e^{-t/10}, \quad y(0) = 4, \quad y'(0) = 1.$$

Taking the Laplace transform of both sides of the equation and solving for $\mathcal{L}[y]$, we obtain

$$\mathcal{L}[y] = \frac{(s+2)y(0)}{s^2 + 2s + 2} + \frac{y'(0)}{s^2 + 2s + 2} + \frac{1}{(s^2 + 2s + 2)(s + \frac{1}{10})},$$

and substituting the initial conditions gives

$$\mathcal{L}[y] = \frac{4(s+2)}{s^2 + 2s + 2} + \frac{1}{s^2 + 2s + 2} + \frac{1}{(s^2 + 2s + 2)(s + \frac{1}{10})}.$$

To find the poles, we note that the least common denominator of the sum for $\mathcal{L}[y]$ is

$$(s^2 + 2s + 2)(s + \tfrac{1}{10}).$$

The roots of $s^2 + 2s + 2$ are $-1 \pm i$, and so the poles are $s = -1 \pm i$ and $s = -\tfrac{1}{10}$. Using partial fractions, we can rewrite the Laplace transform of the solution as a sum of terms with denominators

$$s + \tfrac{1}{10} \quad \text{and} \quad s^2 + 2s + 2 = (s + 1)^2 + 1$$

From the table of Laplace transforms, we see that the only terms that can appear in the solution are $e^{-t/10}$, $e^{-t} \sin t$, and $e^{-t} \cos t$.

Just as for the homogeneous case, the poles of the Laplace transform, $s = -\tfrac{1}{10}$ and $s = -1 \pm i$, tell us the qualitative behavior of the solution. The solution decreases toward zero because all of the poles are negative or have negative real parts. The solution is not monotonic but oscillates with a natural frequency of 2π because of the complex poles. Finally, the rate at which the solution approaches zero is determined by the exponential term with the exponent whose real part is closest to zero. Hence, this solution approaches zero at the same rate as $e^{-t/10}$.

This qualitative information agrees with the graph of the solution given in Figure 6.25, for the oscillations contributed by the $\sin t$ and $\cos t$ terms are hardly visible. What is remarkable is that the description we obtain using the poles of the Laplace transform is very similar to the description of solutions of linear, homogeneous equations that we derive using the eigenvalues. We informally think of the poles of the Laplace transform of solutions as an extension of the idea of eigenvalues to nonhomogeneous equations.

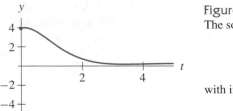

Figure 6.25
The solution of

$$\frac{d^2 y}{dt^2} + 2\frac{dy}{dt} + 2y = e^{-t/10}$$

with initial conditions $y(0) = 4$ and $y'(0) = 1$.

Classification Using Poles

Collecting these ideas, we can state some rules concerning the relation between the poles of the Laplace transform of a solution and the qualitative behavior of that solution.

If all of the poles are either negative or have negative real parts, then the solution tends toward the origin as t increases. The rate of approach is exponential with exponent equal to the real part of the pole closest to zero. If one or more of the poles is positive or has positive real part, then the solution is unbounded. It tends to infinity exponentially if the pole with largest real part is real. If the pole with the largest real part is complex, then the solution oscillates and the amplitude of the oscillation grows exponentially. We emphasize that so far we have talked only about solutions that tend to infinity or to zero at an exponential rate. It is traditional in this area to call differential equations for which all solutions tend to zero as t increases *stable*. Differential equations for which one or more solutions tend to infinity as t increases are called *unstable*.

The part of the complex plane to the left of the imaginary axis is called the left half-plane. The left half-plane contains the negative real numbers and complex numbers that have negative real part. The positive reals and the complex numbers with positive real part make up the right half-plane. The condition for stability is efficiently summarized by:

All the poles in the left half-plane implies stability.

and

One or more poles in the right half-plane implies instability.

We summarize this information qualitatively in Figure 6.26. In the special case in

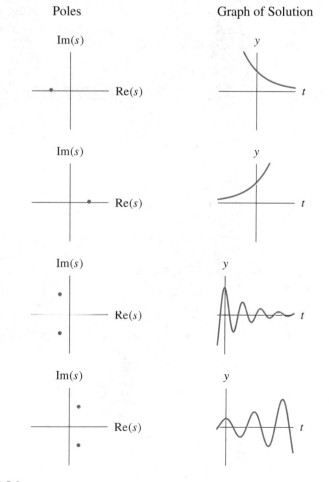

Figure 6.26

Schematic representation of the information contained in the poles of the Laplace transform. Each picture on the left gives the location of the poles in the complex s-plane while the picture on the right is the $y(t)$-graph of the corresponding solution.

which all of the poles lie on the imaginary axis, the situation is more complicated. The solution may oscillate or we may encounter resonance. We will not deal further with this special but important case (however, see Exercises 5–8).

Another Example with a Moral

Consider the differential equation

$$\frac{d^2y}{dt^2} + \frac{dy}{dt} + 3y = u_2(t)e^{-(t-2)/10}\sin(t-2).$$

This is an underdamped harmonic oscillator with a sinusoidal forcing term that is turned on at time $t = 2$ and that decreases exponentially as t increases. Taking the Laplace transform, we obtain

$$s^2\mathcal{L}[y] - sy(0) - y'(0) + s\mathcal{L}[y] - y(0) + 3\mathcal{L}[y] = \mathcal{L}[u_2(t)e^{-(t-2)/10}\sin(t-2)].$$

Solving for $\mathcal{L}[y]$ gives

$$\mathcal{L}[y] = \frac{(s+1)y(0)}{s^2+s+3} + \frac{y'(0)}{s^2+s+3} + \frac{e^{-2s}}{((s+\frac{1}{10})^2+1)(s^2+s+3)}.$$

The poles are $(-1 \pm i\sqrt{11})/2$ and $-\frac{1}{10} \pm i$. The poles $(-1 \pm i\sqrt{11})/2$ are the eigenvalues of the unforced equation and represent the natural response of the system. The poles $-\frac{1}{10} \pm i$ represent the forced response of the system. From this we see that the natural response decays exponentially to zero (like $e^{-t/2}$), and the solution approaches a steady-state oscillation that has period 2π and amplitude decreasing more slowly (like $e^{-t/10}$—see Figure 6.27).

However, we must be careful. If we change the equation to

$$\frac{d^2y}{dt^2} + \frac{dy}{dt} + 3y = (1 - u_2(t))e^{-(t-2)/10}\sin(t-2),$$

then we can compute that the poles of the Laplace transform are exactly the same as before. However, the second equation has a forcing term that turns off at time $t = 2$, so the long-term behavior of solutions is the same as for the unforced, underdamped harmonic oscillator; that is, it tends to zero relatively quickly (like $e^{-t/2}$—see Figure 6.28).

The moral is that one should always expect the unexpected. Relying blindly on a technique, without examining the equation or the underlying question (or physical system, if there is one), invites disaster.

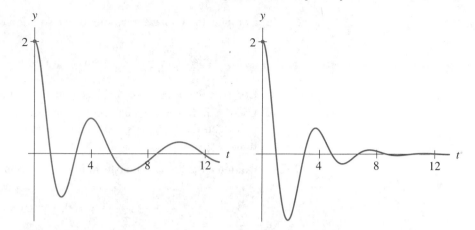

Figure 6.27
Solution of $d^2y/dt^2 + dy/dt + 3y = u_2(t)e^{-(t-2)/10}\sin(t - 2)$ with initial conditions $y(0) = 2$ and $y'(0) = 0$.

Figure 6.28
Solution of $d^2y/dt^2 + dy/dt + 3y = (1 - u_2(t))e^{-(t-2)/10}\sin(t - 2)$ with initial conditions $y(0) = 2$ and $y'(0) = 0$.

EXERCISES FOR SECTION 6.6

In Exercises 1–4,
   (a) compute the Laplace transform of the solution,
   (b) find the poles of the Laplace transform of the solution, and
   (c) discuss the behavior of the solution.

1. $\dfrac{d^2y}{dt^2} + 2\dfrac{dy}{dt} + 2y = e^{-2t}\sin 4t, \quad y(0) = 2, \quad y'(0) = -2$

2. $\dfrac{d^2y}{dt^2} + \dfrac{dy}{dt} + 5y = u_2(t)\sin(4(t - 2)), \quad y(0) = -2, \quad y'(0) = 0$

3. $\dfrac{d^2y}{dt^2} + \dfrac{dy}{dt} + 8y = (1 - u_4(t))\cos(t - 4), \quad y(0) = 0, \quad y'(0) = 0$

   [*Hint*: Recall that $\cos(t - 4) = (\cos t)(\cos 4) + (\sin t)(\sin 4)$.]

4. $\dfrac{d^2y}{dt^2} + \dfrac{dy}{dt} + 3y = (1 - u_2(t))e^{-(t-2)/10}\sin(t - 2), \quad y(0) = 1, \quad y'(0) = 2$

   [*Hint*: Recall that $\sin(t - 2) = (\cos 2)(\sin t) - (\sin 2)(\cos t)$.]

5. **(a)** Compute the Laplace transform of the solution of the initial-value problem

$$\frac{d^2y}{dt^2} + 16y = 0, \quad y(0) = 1, \quad y'(0) = 1.$$

   (b) Compute the poles of the Laplace transform of the solution.

   (c) Use this to formulate a conjecture on what having poles on the imaginary axis for the Laplace transform implies about the qualitative behavior of the solution.

6. **(a)** Compute the Laplace transform of the solution of the initial-value problem

$$\frac{d^2y}{dt^2} + 4y = \sin 2t, \quad y(0) = 0, \quad y'(0) = 0.$$

(b) Compute the poles of the Laplace transform of the solution.

(c) Use this to formulate a conjecture concerning what the occurrence of "double poles" on the imaginary axis implies about the qualitative behavior of solutions.

7. **(a)** Compute the Laplace transform for the solution of the initial-value problem

$$\frac{d^2y}{dt^2} + 2\frac{dy}{dt} + y = 0, \quad y(0) = 1, \quad y'(0) = 2.$$

(b) What are the poles of the Laplace transform?

(c) Use this to formulate a conjecture concerning what the occurrence of a "double pole" in the Laplace transform implies about the qualitative behavior of solutions.

8. **(a)** Compute the Laplace transform of the solution of

$$\frac{d^2y}{dt^2} + 16y = t, \quad y(0) = 1, \quad y'(0) = 1.$$

(b) Compute the poles of the Laplace transform of the solution.

(c) Use this to formulate a conjecture on what having a "double pole" at zero for the Laplace transform implies about the qualitative behavior of the solution.

9. In Exercise 17 of Section 6.2, we considered the equation

$$\frac{d^2y}{dt^2} + 20\frac{dy}{dt} + 200y = w(t), \quad y(0) = 1, \quad y'(0) = 0,$$

where $w(t)$ is the square wave.

(a) Compute the Laplace transform of the solution of this initial-value problem.

(b) What are the poles of the solution?

(c) Describe the long-term behavior of the solution.

10. In Exercise 18 of Section 6.2, we considered the equation

$$\frac{d^2y}{dt^2} + 20\frac{dy}{dt} + 200y = z(t) \quad y(0) = 1, \quad y'(0) = 0$$

where $z(t)$ is the sawtooth wave.

(a) Compute the Laplace transform of the solution of this initial-value problem.

(b) What are the poles of the solution?

(c) Describe the long-term behavior of the solution.

Table 6.1
Frequently Encountered Laplace Transforms

$y(t) = \mathcal{L}^{-1}[Y]$	$Y(s) = \mathcal{L}[y]$	$y(t) = \mathcal{L}^{-1}[Y]$	$Y(s) = \mathcal{L}[y]$
$y(t) = e^{at}$	$Y(s) = \dfrac{1}{s-a}$ $(s > a)$	$y(t) = t^n$	$Y(s) = \dfrac{n!}{s^{n+1}}$ $(s > 0)$
$y(t) = \sin \omega t$	$Y(s) = \dfrac{\omega}{s^2 + \omega^2}$	$y(t) = \cos \omega t$	$Y(s) = \dfrac{s}{s^2 + \omega^2}$
$y(t) = e^{at} \sin \omega t$	$Y(s) = \dfrac{\omega}{(s-a)^2 + \omega^2}$	$y(t) = e^{at} \cos \omega t$	$Y(s) = \dfrac{s-a}{(s-a)^2 + \omega^2}$
$y(t) = t \sin \omega t$	$Y(s) = \dfrac{2\omega s}{(s^2 + \omega^2)^2}$	$y(t) = t \cos \omega t$	$Y(s) = \dfrac{s^2 - \omega^2}{(s^2 + \omega^2)^2}$
$y(t) = u_a(t)$	$Y(s) = \dfrac{e^{-as}}{s}$ $(s > 0)$	$y(t) = \delta_a(t)$	$Y(s) = e^{-as}$

Table 6.2
Rules for Laplace Transforms:
Given functions $y(t)$ and $w(t)$ with $\mathcal{L}[y] = Y(s)$ and $\mathcal{L}[w] = W(s)$ and constants α and a

Rule for Laplace Transform	Rule for Inverse Laplace Transform
$\mathcal{L}\left[\dfrac{dy}{dt}\right] = s\mathcal{L}[y] - y(0) = sY(s) - y(0)$	
$\mathcal{L}[y + w] = \mathcal{L}[y] + \mathcal{L}[w] = Y(s) + W(s)$	$\mathcal{L}^{-1}[Y + W] = \mathcal{L}^{-1}[Y] + \mathcal{L}^{-1}[W] = y(t) + w(t)$
$\mathcal{L}[\alpha y] = \alpha \mathcal{L}[y] = \alpha Y(s)$	$\mathcal{L}^{-1}[\alpha Y] = \alpha \mathcal{L}^{-1}[Y] = \alpha y(t)$
$\mathcal{L}[u_a(t)y(t - a)] = e^{-as}\mathcal{L}[y] = e^{-as}Y(s)$	$\mathcal{L}^{-1}[e^{-as}Y] = u_a(t)y(t - a)$
$\mathcal{L}[e^{at}y(t)] = Y(s - a)$	$\mathcal{L}^{-1}[Y(s - a)] = e^{at}\mathcal{L}^{-1}[Y] = e^{at}y(t)$

LAB 6.1   Convolutions

In Section 6.5 we saw how it is possible to compute the solution of a nonhomogeneous, constant-coefficient, linear differential equation

$$\frac{d^2y}{dt^2} + p\frac{dy}{dt} + qy = f(t), \quad y(0) = y'(0) = 0$$

for the forcing function $f(t)$ from the solution of

$$\frac{d^2y}{dt^2} + p\frac{dy}{dt} + qy = \delta_0(t), \quad y(0) = y'(0) = 0^-.$$

In this process we did not need to know the constants p and q. This technique applies to linear, constant-coefficient equations of any order.

In this lab we ask you to carry out these computations. Table 6.3 contains solutions of initial-value problem

Table 6.3
Solutions for initial conditions zero and forcing $\delta_0(t)$

Time	$y_1(t)$	$y_2(t)$	$y_3(t)$	$y_4(t)$	Time	$y_1(t)$	$y_2(t)$	$y_3(t)$	$y_4(t)$
0.1	0.0854	0.6065	0.0042	0.0905	2.1	−0.0195	0.0000	0.0074	0.2572
0.2	0.1425	0.3679	0.0136	0.1637	2.2	−0.0169	0.0000	0.0070	0.2438
0.3	0.1779	0.2231	0.0243	0.2222	2.3	−0.0139	0.0000	0.0064	0.2306
0.4	0.1927	0.1353	0.0336	0.2681	2.4	−0.0109	0.0000	0.0058	0.2177
0.5	0.1921	0.0821	0.0398	0.3033	2.5	−0.0080	0.0000	0.0052	0.2052
0.6	0.1801	0.0498	0.0424	0.3293	2.6	−0.0054	0.0000	0.0046	0.1931
0.7	0.1604	0.0302	0.0417	0.3476	2.7	−0.0031	0.0000	0.0041	0.1814
0.8	0.1361	0.0183	0.0384	0.3595	2.8	−0.0012	0.0000	0.0036	0.1703
0.9	0.1100	0.0111	0.0337	0.3659	2.9	0.0002	0.0000	0.0032	0.1596
1.0	0.0840	0.0067	0.0283	0.3679	3.0	0.0013	0.0000	0.0028	0.1494
1.1	0.0597	0.0041	0.0231	0.3662	3.1	0.0020	0.0000	0.0025	0.1396
1.2	0.0381	0.0025	0.0186	0.3614	3.2	0.0024	0.0000	0.0023	0.1304
1.3	0.0198	0.0015	0.0149	0.3543	3.3	0.0026	0.0000	0.0021	0.1217
1.4	0.0050	0.0009	0.0123	0.3452	3.4	0.0025	0.0000	0.0019	0.1135
1.5	−0.0062	0.0006	0.0105	0.3347	3.5	0.0023	0.0000	0.0018	0.1057
1.6	−0.0141	0.0003	0.0094	0.3230	3.6	0.0021	0.0000	0.0016	0.0984
1.7	−0.0191	0.0002	0.0088	0.3106	3.7	0.0017	0.0000	0.0015	0.0915
1.8	−0.0217	0.0001	0.0084	0.2975	3.8	0.0014	0.0000	0.0013	0.0850
1.9	−0.0223	0.0001	0.0081	0.2842	3.9	0.0010	0.0000	0.0012	0.0789
2.0	−0.0214	0.0000	0.0078	0.2707	4.0	0.0007	0.0000	0.0011	0.0733

$$\frac{d^2y}{dt^2} + qy = \delta_0(t), \quad y(0) = 0^-,$$

$$\frac{d^2y}{dt^2} + p\frac{dy}{dt} + qy = \delta_0(t), \quad y(0) = y'(0) = 0^-,$$

or

$$\frac{d^3y}{dt^3} + r\frac{d^2y}{dt^2} + p\frac{dy}{dt} + qy = \delta_0(t), \quad y(0) = y'(0) = y''(0) = 0^-$$

for various choices of constants r, p, and q. From this data, you can compute solutions for different forcing functions by computing convolutions. The arithmetic involved in computing the convolutions numerically can be oppressive, so it is worthwhile to take advantage of whatever technology you have available.

In your report, address the following questions:

1. For each of the given solutions, compute the approximate solution if the delta function on the right-hand side is replaced with $f(t) = \cos 4t$. Initial conditions remain zero.

2. For each of the given solutions, compute the approximate solution if the delta function on the right-hand side is replaced with $g(t) = 3u_1(t) + u_2(t) - 4u_3(t)$. Initial conditions remain zero.

3. For each of the given solutions, compute the approximate solution if the delta function on the right-hand side is replaced with $h(t) = e^{-t}$. Initial conditions remain zero.

4. For each of the given solutions, determine the differential equation. That is, determine the order of the equation and approximate values of the coefficients.

Your report: In your report you should summarize your technique for computing the convolutions and present the solutions both as graphs and tables. Since these are numerical calculations, you should also discuss the accuracy of your answers.

LAB 6.2 Poles

In Section 6.6 we discussed the relationship between the poles of the Laplace transform of a solution and the qualitative behavior of the solution. The general rule of thumb is that the real part of the poles gives the exponential growth and decay rate of the solution, whereas the imaginary part of the poles gives the period of oscillation. This makes it easy to pick out exponential growth of solutions (poles with positive real part) or to determine the exponential rate of decay (pole with negative real part closest to zero).

While exponential growth and decay are the most dramatic aspects of solutions, there can also be linear growth. This occurs, for example, with resonant forcing of undamped harmonic oscillator equations. This behavior can be identified in the poles of the solution. The goal of this lab is to make those identifications.

Your report should address the following items:

1. Find the poles of the solutions of each of the following initial-value problems:

(a) $\dfrac{dy}{dt} + y = e^{-t}, \quad y(0) = 0$

(b) $\dfrac{dy}{dt} + y = t, \quad y(0) = 0$

(c) $\dfrac{dy}{dt} + y = t^2, \quad y(0) = 0$

(d) $\dfrac{d^2y}{dt^2} + 9y = \sin 3t, \quad y(0) = y'(0) = 0$

(e) $\dfrac{d^2y}{dt^2} + 9y = t \quad y(0) = y'(0) = 0$

(f) $\dfrac{d^2y}{dt^2} + 9y = t^2 \quad y(0) = y'(0) = 0$

(g) $\dfrac{d^2y}{dt^2} + 3\dfrac{dy}{dt} + 2y = e^{-t} \quad y(0) = y'(0) = 0$

(h) $\dfrac{d^2y}{dt^2} + 2\dfrac{dy}{dt} + y = e^{-t} \quad y(0) = y'(0) = 0$

2. For each initial-value problem in part 1, find the solution (either numerically or analytically). Determine the rate of growth or decay of the solution. [*Hint*: Determining the exact rate of growth can be subtle if you are working numerically. For equations (a), (g), and (h), you should compare the solution to e^{-t}.]

3. Based on your results in parts 1 and 2, form a conjecture regarding the relationship between multiple poles and growth or decay rates of solutions.

4. Compute a formula for

$$\mathcal{L}^{-1}\left[\frac{d}{ds}\mathcal{L}[f]\right]$$

in terms of $f(t)$. Write a short essay relating this computation to your work in parts 1, 2, and 3.

Your report: In part 3 state as clear and concise a conjecture as possible that is consistent with the results of parts 1 and 2. Then make sure your conjecture is consistent with your formula in part 4.

7 | NUMERICAL METHODS

Errors in numerical approximations of solutions to initial-value problems are inevitable. Usually these errors can be controlled by reducing the step size, but reducing the step size increases the arithmetic. Consequently, there is a trade-off between accuracy and computation time.

We begin this chapter with an analysis of the errors involved with Euler's method to obtain an understanding of the relationship between error and step size. We see what is meant by the statement that Euler's method is a first-order approximation scheme.

In the second and third sections, we give methods that achieve a given accuracy with fewer steps and less arithmetic. The methods we consider are improved Euler's method and Runge-Kutta. These methods generally produce much more accurate results for a given step size.

Computers and calculators can use only a finite amount of memory to store each number in a computation. However, most numbers have infinite decimal expansions. Hence computers and calculators make small errors in each calculation due to the representation of numbers by the machines involved. In the final section we consider the effect of these errors on our approximations.

7.1 NUMERICAL ERROR IN EULER'S METHOD

Of the three approaches to the study of differential equations—the analytic, the qualitative, and the numeric—the most commonly used technique in science and engineering is numerical approximation. This comes as no surprise since there are many efficient software packages available that automate a great deal of the process of obtaining numerical solutions. Even when relatively simple analytic or qualitative techniques exist, there is a temptation to "just let the computer do it."

There is nothing wrong with this mode of operation, as long as two important points are kept in mind. First, the computer gives numbers and graphs, not interpretation and insight. Understanding the qualitative behavior of an approximate solution and its implications for the physical system under consideration is still the job of the human.

Second, numerical methods give *approximate* solutions. There is always a certain degree of error, and if the time interval over which the solution is appropriated is lengthened, the error will usually increase. While the error may remain insignificant for some differential equations, we have seen "chaotic" examples where small errors amplify very quickly (see Sections 2.5, 5.6, 8.4, and 8.5).

So far we have avoided any discussion of the errors that arise with numerical procedures such as Euler's method. However, if we are going to rely on computers, we should be aware of these errors, and we should know how to determine whether our appropriate solutions have errors that are within acceptable ranges.

When we use a computer to approximate the solution of a differential equation, there are two very different ways in which we introduce error into the computation. One is the error that is inherent in the approximation scheme that we use, and the other is the error that results from the use of finite arithmetic during the computations. In this chapter we primarily discuss the errors that arise from the approximation procedure. However, in Section 7.4 we discuss some of the interesting and practical issues that stem from the arithmetic of the computer.

Step Size and Euler's Method

Recall from Section 1.4 that Euler's method is an iterative scheme that approximates the solution to an initial-value problem of the form

$$\frac{dy}{dt} = f(t, y), \quad y(t_0) = y_0.$$

The approximate solution is given by the iteration

$$y_{k+1} = y_k + f(t_k, y_k)\,\Delta t,$$

where Δt is the step size and y_0 is the initial value $y(t_0)$. There is a simple relationship between the step size Δt, the length of the interval over which we would like to approximate the solution, and the number of steps involved in the computations. In particular, if we want to approximate the solution over the interval from t_0 to t_n (using n steps), we divide the total length of the interval into n equal-length subintervals. In other words,

$$\Delta t = \frac{t_n - t_0}{n}.$$

The rule-of-thumb discussed in Section 1.4 is that the smaller the step size, the more accurate the approximate solution. However, the smaller the step size, the larger the number of steps, so the longer it takes the computer to approximate the solution. It is useful to be able to predict roughly how large a step size to take in order to achieve a particular accuracy.

We begin by studying how the accuracy of Euler's method changes with step size for a particular example. When we introduced Euler's method for first-order equations in Section 1.4, we considered the initial-value problem

$$\frac{dy}{dt} = -2ty^2, \quad y(0) = 1.$$

This equation is separable and the solution is

$$y(t) = \frac{1}{1+t^2}.$$

Since we have a formula for the solution, we can compare the results of Euler's method for different step sizes to the exact values of the solution.

In Section 1.4 we applied Euler's method to this initial-value problem over the interval $0 \le t \le 2$ using $\Delta t = 0.1$ (20 steps). For this differential equation the Euler's method iteration scheme is

$$y_{k+1} = y_k - 2t_k y_k^2 \Delta t,$$

and we obtained the approximation $y_{20} = 0.1933\ldots$, which is a good approximation of the exact value $y(2) = 1/(1 + 2^2) = 0.2$. Increasing the number of steps to 2000 ($\Delta t = 0.001$) yields the more accurate approximation $y_{2000} = 0.199937\ldots$, which is an even better approximation of $y(2) = 0.2$. In Section 1.4 we simply observed that increasing the number of steps in the overall computation seems to improve the accuracy. Now we would like to be more precise about exactly how the accuracy is improved as the number of steps is increased.

First, we take a closer look at the result of the computation that used 2000 steps over the interval $0 \le t \le 2$. Denote the error in the computation by e_{2000}. We have

$$e_{2000} = |0.2 - y_{2000}| = |0.2 - 0.199937| = 0.000063.$$

Now suppose we double the number of steps. After starting at the same initial condition and taking 4000 steps with $\Delta t = 0.0005$, we obtain the approximation $y_{4000} = 0.199969$. In this case we have the error

$$e_{4000} = |0.2 - y_{4000}| = |0.2 - 0.199969| = 0.000031.$$

In this example we see that the ratio of the errors is

$$\frac{e_{4000}}{e_{2000}} = \frac{0.000031}{0.000063} \approx 0.492.$$

When we doubled the number of steps in this example, we halved the error.

This straightforward computation leads us to wonder how the accuracy of the approximation depends on the step size and the number of steps. Since Euler's method is so easy to implement, we can play around with various values to see what happens. Using the computer we first compute the value of y_n for various choices of the number n of steps. Then we calculate the error e_n for these values. In Figure 7.1 we plot the results for $n = 100, 200, 300, \ldots, 6000$. This graph is a plot of the accuracy for Euler's method applied to the initial-value problem

$$\frac{dy}{dt} = -2ty^2, \quad y(0) = 1,$$

as a function of step size. If we ignore the scales on the axes for a moment, we note that this graph looks especially familiar. In fact if we let n denote the total number of steps, this graph looks remarkably like the graph of $1/n$. Indeed, in Figure 7.2 we plot a graph of a function of the form K/n. (Actually $K = 0.126731$, but it is not important how we came up with this number.) Note the agreement between the data points and the graph of this function. The remainder of this section is devoted to an explanation of why these results are typical.

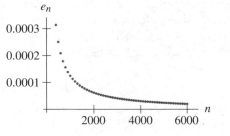

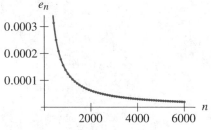

Figure 7.1
The error in Euler's method as a function of the number n of steps used in the approximation. Euler's method is applied to $dy/dt = -2ty^2$, $y(0) = 1$, over the interval $0 \le t \le 2$.

Figure 7.2
The error in Euler's method from Figure 7.1 fit to the function $0.126731/n$. Note that it is almost impossible to distinguish the graph of $0.126731/n$ from the data points e_n.

The Error in the First Step of the Method

Given an initial-value problem of the form

$$\frac{dy}{dt} = f(t, y), \quad y(t_0) = y_0,$$

and a step size Δt, we know that the value of y_1 in the first Euler step is determined by assuming that the graph of the solution is a line segment that starts at (t_0, y_0) and has slope $f(t_0, y_0)$. First, we need to determine how to estimate the error made in this initial step. The tool that lets us accomplish this estimate is Taylor's Theorem with remainder.

Taylor's Theorem says that

$$y(t_1) = y(t_0) + y'(t_0)\, \Delta t + y''(\xi_1)\frac{(\Delta t)^2}{2}$$

for some number ξ_1 between t_0 and t_1. Since we know that $y(t)$ is a solution to the initial-value problem, we have $y'(t_0) = f(t_0, y_0)$ and $y(t_0) = y_0$. In other words,

$$y(t_1) = y_0 + f(t_0, y_0)\, \Delta t + y''(\xi_1)\frac{(\Delta t)^2}{2}.$$

Euler's method corresponds to dropping the $(\Delta t)^2$ term, that is, taking the linear approximation of the solution from t_0 to t_1. Thus the difference $|y(t_1) - y_1|$ between the first Euler approximation y_1 and the actual value $y(t_1)$ of the solution is

$$\left| y''(\xi_1)\frac{(\Delta t)^2}{2} \right|.$$

It is convenient to have some notation for this term, so we denote it τ_1 to represent the **truncation** error. It is called the truncation error because it results from a "truncation" of the Taylor series for $y(t_1)$. Euler's method is based on the Taylor approximation of $y(t_1)$, but it throws out all nonlinear terms (terms of degree 2 or higher).

For the first step of Euler's method, the truncation error is the error, but that won't be true for subsequent steps. So we introduce notation for the **total error** after k steps. By definition, the total error after k steps is the difference between the kth Euler approximation and the actual value $y(t_k)$ at time t_k, and we denote it by e_k. More precisely,

$$e_k = |y(t_k) - y_k|.$$

Note that for the first step ($k = 1$), the truncation error and the total error are identical. That is,

$$e_1 = \tau_1.$$

Usually the truncation error and the total error agree only at the first step.

At this point it is natural to wonder whether the formula

$$\tau_1 = \left| y''(\xi_1)\frac{(\Delta t)^2}{2} \right|$$

is something that we can really compute if we don't already know the solution $y(t)$. The key point to keep in mind is the fact that $y(t)$ is a solution to the differential equation. Since we know that $dy/dt = f(t, y)$, we can compute the second derivative of $y(t)$ using the multivariable Chain Rule. Since

$$\frac{dy}{dt} = f(t, y(t)),$$

we have

$$\frac{d^2y}{dt^2} = \frac{\partial f}{\partial t}\frac{dt}{dt} + \frac{\partial f}{\partial y}\frac{dy}{dt}$$

$$= \frac{\partial f}{\partial t} + \frac{\partial f}{\partial y} f(t, y).$$

Therefore we can bound the error e_1 using an upper bound M_1 for the quantity

$$\frac{\partial f}{\partial t} + \frac{\partial f}{\partial y} \, f(t, y)$$

along the graph of $y(t)$ between $y(t_0)$ and $y(t_1)$. Since we are given $f(t, y)$, we can actually compute the quantities in this expression.

Suppose M_1 is a constant that is large enough so that

$$\left| \frac{\partial f}{\partial t} + \frac{\partial f}{\partial y} \, f(t, y) \right| < M_1$$

along the graph of $y(t)$ (or at least the part of the graph we are considering). If we are able to find such a bound M_1, then the error in the first step is bounded according to the inequality

$$e_1 = \tau_1 \leq M_1 \frac{(\Delta t)^2}{2}$$

(see Figure 7.3).

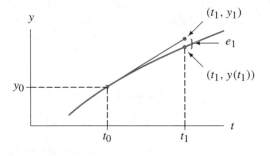

Figure 7.3
The error e_1 in the first step of Euler's method.

The Error in the Second Step

Unfortunately, once we have taken the first step, we are no longer on the graph of $y(t)$, and consequently our error analysis is more complicated. Basically there are two sources of error in the second step. If we knew the value of $y(t_1)$ exactly, then the error e_2 in the second step would again be the truncation error τ_2. In other words, from Taylor's Theorem again we have

$$y(t_2) = y(t_1) + y'(t_1) \, \Delta t + y''(\xi_2) \frac{(\Delta t)^2}{2}$$

for some number ξ_2 between t_1 and t_2, and using the differential equation to replace $y'(t_1)$ with $f(t_1, y(t_1))$, we would have

$$y(t_2) = y(t_1) + f(t_1, y(t_1)) \, \Delta t + y''(\xi_2) \frac{(\Delta t)^2}{2}.$$

The truncation error at this step is therefore

$$\tau_2 = \left| y''(\xi_2) \frac{(\Delta t)^2}{2} \right|.$$

Comparing the formula

$$y(t_2) = y(t_1) + f(t_1, y(t_1)) \, \Delta t + y''(\xi_2) \frac{(\Delta t)^2}{2}$$

from Taylor's Theorem with the formula

$$y_2 = y_1 + f(t_1, y_1) \, \Delta t$$

for the second approximation y_2 from Euler's method, we see that, not only are we missing the quadratic term, but Euler's method uses the slope $f(t_1, y_1)$ at (t_1, y_1) rather than the slope $f(t_1, y(t_1))$ at $(t_1, y(t_1))$ (see Figure 7.4). Therefore when we estimate the error e_2, we have

$$e_2 = |y(t_2) - y_2|$$

$$\leq |y(t_1) - y_1| + |f(t_1, y(t_1)) - f(t_1, y_1)| \, \Delta t + |y''(\xi_2)| \frac{(\Delta t)^2}{2}$$

$$\leq e_1 + |f(t_1, y(t_1)) - f(t_1, y_1)| \, \Delta t + M_1 \frac{(\Delta t)^2}{2}$$

if we can find a bound M_1 on the second derivative as we did when we estimated τ_1. In other words, the total error after the second step is bounded by the total error after the first step, the truncation error τ_2 associated to the second step, and a third term, $|f(t_1, y(t_1)) - f(t_1, y_1)| \, \Delta t$, which accounts for the fact that we are using the slope $f(t_1, y_1)$ at the point (t_1, y_1) rather than the slope $f(t_1, y(t_1))$ at the point $(t_1, y(t_1))$ on the actual solution curve. Figure 7.4 illustrates the contribution that each of these three terms makes to the overall error.

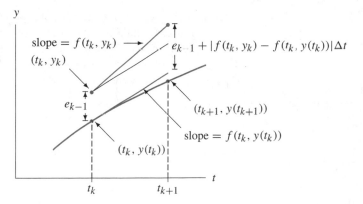

Figure 7.4
The relationship between the error e_{k-1} and the error e_k.

To obtain an estimate for the error e_2 in a relatively convenient form, we need to bound the difference between $f(t_1, y(t_1))$ and $f(t_1, y_1)$. Since these two values of f come from points along the line $t = t_1$, we have the estimate

$$|f(t_1, y(t_1)) - f(t_1, y_1)| \leq M_2 \cdot e_1,$$

where M_2 is the maximum of the partial derivative $\partial f / \partial y$ along the line segment between $(t_1, y(t_1))$ and (t_1, y_1). We therefore obtain the estimate

$$e_2 \leq e_1 + M_2 \, e_1 \, \Delta t + M_1 \frac{(\Delta t)^2}{2} = (1 + M_2 \, \Delta t) \, e_1 + M_1 \frac{(\Delta t)^2}{2}.$$

The analysis of the error e_k at the kth step is essentially the same as the analysis that we just completed for the second step. If M_1 is a bound for the second derivative along the graph of the solution and if M_2 is a bound for the partial derivative $\partial f / \partial y$, then the error e_k is bounded by

$$e_k \leq (1 + M_2 \, \Delta t) e_{k-1} + M_1 \frac{(\Delta t)^2}{2}.$$

Note that this formula is recursive in that it expresses e_k in terms of e_{k-1} (it is really a discrete dynamical system). The term

$$M_1 \frac{(\Delta t)^2}{2}$$

accounts for the truncation error τ_k at the kth step, and the term

$$(1 + M_2 \, \Delta t) e_{k-1}$$

represents the magnification of the error at the $k - 1$st step by a factor due to the fact that we are computing the right-hand side of the differential equation at the Euler approximation (t_k, y_k) rather than at the actual point $(t_k, y(t_k))$ on the solution.

The Error After n Steps

If we are going to approximate the solution $y(t)$ using n steps, then we would definitely like a compact formula for the error e_n in terms of quantities that are easy to compute directly from the differential equation, that is, directly from the right-hand side $f(t, y)$. The recursive formula that we just derived does not really fit these criteria. However, in Exercise 11 we show that, if we find maximums for the expressions

$$M_1 = \left| \frac{\partial f}{\partial t} + \frac{\partial f}{\partial y} f(t, y) \right| \quad \text{and} \quad M_2 = \left| \frac{\partial f}{\partial y} \right|$$

that come up in the estimates, then we can use this recursive formula to bound the total error in Euler's method. Over the interval $t_0 \leq t \leq t_n$, we can derive a bound of the form

$$e_n \leq C \cdot \Delta t,$$

where C is a constant that is determined by the values of M_1, M_2, and the total length of the t-interval over which the solution is approximated. Note that the value of C *does not depend* on the number of steps used in Euler's method. Moreover, since $\Delta t = (t_n - t_0)/n$, we can rewrite this estimate as

$$e_n \leq \frac{K}{n},$$

where K is the product of the constant C and the length of the total interval $t_0 \leq t \leq t_n$. From our point of view, K is just another constant that does not depend on the number of steps used in Euler's method.

What's most important is that we have theoretical bounds that can be expressed either as

$$e_n \leq C \cdot \Delta t \quad \text{or as} \quad e_n \leq \frac{K}{n}.$$

Roughly speaking, if we halve Δt, then we halve the error in Euler's method. Expressed in terms of the number of steps, this estimate says that if we double the number of steps, then we double the accuracy of the approximation.

The Initial-Value Problem $dy/dt = -2ty^2$, $y(0) = 1$

At the beginning of this section, we compared the results of Euler's method with the analytic solution to the initial-value problem

$$\frac{dy}{dt} = -2ty^2, \quad y(0) = 1.$$

We saw that the error was roughly proportional to the step size Δt. This is consistent with the theoretical estimates that we just discussed. However, using the theoretical estimates mentioned above (and developed in more detail in Exercise 11) to choose the appropriate step size to guarantee a desired error is usually overkill.

To see why, let's consider Euler's method applied to the initial-value problem

$$\frac{dy}{dt} = -2ty^2, \quad y(0) = 1,$$

more carefully. We have been estimating the solution over the interval $0 \leq t \leq 2$. Without even trying to derive an analytic solution to the problem, we know that the solution is decreasing (why?) and that it is never zero, because $y(t) = 0$ for all t is an equilibrium solution. Therefore we know that the graph of the actual solution stays in the rectangle

$$\{(t, y) \mid 0 \leq t \leq 2, \ 0 \leq y \leq 1\}$$

in the ty-plane. Using ideas similar to those discussed above, we can determine that the approximate values $y_1, y_2, y_3, \ldots, y_n$ produced by Euler's method must also lie in the same rectangle. In theory we can now estimate the expressions M_1 and M_2 involving the relevant partial derivatives over this rectangle and come up with a bound on the total error in the approximation. In fact, we do exactly this in Exercise 12. In the end we obtain the bound

$$e_n \leq (16,661,456) \, \Delta t.$$

In other words, we need to take roughly 34 million steps to guarantee an accuracy of 1 over this interval of length 2. Looking at the data from our experiments at the beginning of the section, we see that this is an awful estimate. With ten steps we get an error that is less than 0.015.

Why is the theoretical estimate so bad? The problem comes from trying to estimate the total error e_n using upper bounds on M_1 and M_2 that hold throughout the rectangle even though it turns out that the actual solution and the Euler approximates remain close, and their graphs avoid the places where the values of M_1 and M_2 are large. In general, the problem is that we don't have a very good idea of where the graph of the solution is going to be until we make an accurate approximation.

In practice the theoretical bounds for the total error are replaced by a running total of estimates of the error that may be introduced at each step. Although this analysis is not 100% trustworthy, it tends to give a better indication of the accuracy of our approximation.

To illustrate the point, we return to the initial-value problem

$$\frac{dy}{dt} = -2ty^2, \quad y(0) = 1,$$

and we estimate the solution over the interval $0 \leq t \leq 2$ with 100 steps. Since we have the analytic solution in this case, we can compute the error $e_k = |y(t_k) - y_k|$ at each step. The results of this calculation are shown in Figure 7.5. In addition, we keep a running estimate of the total error by appealing to the recursive formula

$$e_k \leq (1 + M_2 \Delta t) e_{k-1} + M_1 \frac{(\Delta t)^2}{2}$$

derived above. However, at each step we estimate

$$M_1 = \left| \frac{\partial f}{\partial t} + \frac{\partial f}{\partial y} f(t, y) \right| \quad \text{and} \quad M_2 = \left| \frac{\partial f}{\partial y} \right|$$

based on the current value of y_k. That is, after we compute y_k from (t_{k-1}, y_{k-1}), we then calculate M_1 at the point $(t_{k-1} + \Delta t/2, (y_{k-1} + y_k)/2)$ and M_2 at the point $(t_{k-1}, (y_{k-1} + y_k)/2)$. (These points are "halfway" points among the set of all possibilities.) Then given the resulting values for M_1 and M_2, we use the approximation

$$e_k \approx (1 + M_2 \Delta t) e_{k-1} + M_1 \frac{(\Delta t)^2}{2}.$$

This gives us a running estimate for the total error e_k, and the estimate is based on the same principles that justify our theoretical results. A plot of these estimates is shown

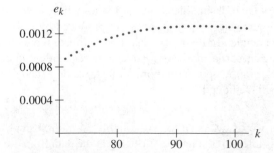

Figure 7.5
The cumulative error e_k after the kth step in the Euler approximation for the initial-value problem

$$\frac{dy}{dt} = -2ty^2, \quad y(0) = 1,$$

over the interval $0 \leq t \leq 2$ with $n = 100$ steps.

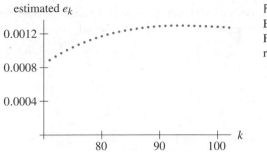

estimated e_k

Figure 7.6
Estimates of the errors e_k as shown in
Figure 7.5 based on the evaluation of
relevant partial derivatives.

in Figure 7.6. Note that this graph is almost identical to the graph of the actual error as
shown in Figure 7.5.

In general this type of local estimation of the errors involved in the calculation
is much easier to achieve. In fact the calculation can be left to the computer, and the
estimates of the errors e_k can be one of the results of the computation.

The Order of Euler's Method

The most important lesson of this section is that the size of the error in Euler's method
for an initial-value problem over a fixed interval of time is proportional to the step size,
or in other words, it is inversely proportional to the number of steps that we use over a
fixed interval. We say that Euler's method is a **first-order** method because the error is,
at worst, proportional to the first power of the step size—note the 1 in the exponent of
the bound

$$e_n \leq C \cdot (\Delta t)^1.$$

In the next two sections we introduce methods that have higher orders. Higher-order
methods are more typical of professionally written numerical solvers.

EXERCISES FOR SECTION 7.1

1. Reproduce Figure 7.1 by performing the following calculations:
   (a) Solve the initial-value problem $dy/dt = -2ty^2$, $y(0) = 1$.
   (b) Given the interval $0 \leq t \leq 2$ and $n = 20$, calculate the Euler approximation y_{20} to $y(2)$. (Double-check your calculations by comparing your results to
   those given at the beginning of the section.)
   (c) Calculate the total error e_{20} involved in your approximation.
   (d) Repeat the calculations in parts (b) and (c) using $n = 1000, 2000, \ldots, 6000$.
   (e) If your computer is up to the task, compute e_n for additional intermediate values of n. (Figure 7.1 was done with n ranging from 100 to 6000 by steps of
   100.)
   (f) Some computer math systems can determine the best fit for this data to a function of the form K/n. If your computer can do this, determine the best value
   of K for this initial-value problem. Plot the data and the graph of K/n on the
   same axes.

In Exercises 2–5, repeat the steps in Exercise 1 to arrive at plots like Figures 7.1 and 7.2 for the initial-value problem and interval specified.

2. $\dfrac{dy}{dt} = 1 - y, \quad y(0) = 0, \quad 0 \le t \le 1$

3. $\dfrac{dy}{dt} = ty, \quad y(0) = 1, \quad 0 \le t \le \sqrt{2}$

4. $\dfrac{dy}{dt} = -y^2, \quad y(0) = 1/2, \quad 0 \le t \le 2$

5. $\dfrac{dy}{dt} = 1 - t + 3y, \quad y(0) = 1, \quad 0 \le t \le 1$

6. In this section we state that the error e_{2000} is 0.000063 for the Euler approximation of the initial-value problem $dy/dt = -2ty^2$, $y(0) = 1$ over the interval $0 \le t \le 2$. Estimate how many steps n are needed to obtain an approximation y_n that has an error e_n no larger than 0.000001.

7. Reproduce Figure 7.6 by performing the following calculations:

    (a) Calculate the expressions

$$M_1 = \frac{\partial f}{\partial t} + \frac{\partial f}{\partial y} f(t, y) \quad \text{and} \quad M_2 = \frac{\partial f}{\partial y}$$

    for $f(t, y) = -2ty^2$.

    (b) Given the initial-value problem

$$\frac{dy}{dt} = -2ty^2, \quad y(0) = 1,$$

    the interval $0 \le t \le 2$, and $n = 100$ steps, calculate the point (t_1, y_1) that corresponds to the first step of Euler's method. Evaluate the quantity M_1 at the point $(t_0 + \Delta t/2, (y_0 + y_1)/2)$, and use this value to estimate the error e_1. How does your estimate compare with the true error e_1?

    (c) Calculate the point (t_2, y_2) that corresponds to the second step of Euler's method. Evaluate M_1 at the point $(t_1 + \Delta t/2, (y_1 + y_2)/2)$ and M_2 at the point $(t_1, (y_1 + y_2)/2)$. Use these values to estimate the error e_2. How does your estimate compare with the true error e_2?

    (d) Repeat part (c) for the remaining 98 steps.

    (e) Make two plots—one that plots the true errors e_k versus k and the other that plots the estimated error versus k.

In Exercises 8–9, repeat parts (b)–(d) of Exercise 7 to estimate the errors e_k for the initial-value problem and interval specified. Why don't we ask for a comparison of this estimate of e_k with the true value of e_k?

8. $\dfrac{dy}{dt} = t - y^3, \quad y(0) = 1, \quad 0 \le t \le 1$

9. $\dfrac{dy}{dt} = \sin ty, \quad y(0) = 3, \quad 0 \le t \le 3$

10. Using what you learned in calculus, show that $1 + \alpha < e^\alpha$ if $\alpha > 0$.

11. In this exercise we derive the theoretical bounds mentioned in the section. Let R be a rectangle $\{(t, y) \mid a \le t \le b, c \le y \le d\}$ in the ty-plane and suppose that $f(t, y)$ is continuously differentiable on R. Given an initial-value problem

$$\frac{dy}{dt} = f(t, y), \quad y(t_0) = y_0,$$

and an interval $t_0 \le t \le t_n$ such that the point (t_0, y_0) is in R and $t_n \le b$, then we can bound the error e_n involved in the Euler approximation assuming the Euler approximate values $y_1, y_2, \dots, y_n$ all lie between c and d. To do so, let

$$M_1 = \max \left| \frac{\partial f}{\partial t} + \frac{\partial f}{\partial y} f \right| \text{ on } R$$

and let

$$M_2 = \max \left| \frac{\partial f}{\partial y} \right| \text{ on } R.$$

(a) Show that

$$e_1 \le M_1 \frac{(\Delta t)^2}{2}.$$

(b) Show that

$$e_2 \le e_1 + M_2 e_1 \Delta t + M_1 \frac{(\Delta t)^2}{2}$$

and explain the significance of each of these three terms.

(c) Explain why

$$e_{k+1} \le (1 + M_2 \Delta t) e_k + M_1 \frac{(\Delta t)^2}{2}.$$

(d) Let $K_1 = 1 + M_2 \Delta t$ and $K_2 = M_1 (\Delta t)^2 / 2$. Show that

$$e_3 \le (K_1^2 + K_1 + 1) K_2.$$

(e) Show that

$$e_n \le \left(K_1^{n-1} + K_1^{n-2} + \cdots + K_1 + 1 \right) K_2.$$

(f) Explain why

$$K_1^{n-1} + K_1^{n-2} + \cdots + K_1 + 1 = \frac{K_1^n - 1}{K_1 - 1},$$

and thus

$$e_n \le \left(\frac{K_1^n - 1}{K_1 - 1} \right) K_2.$$

(g) Using the definitions of K_1 and K_2, show that

$$e_n \le \frac{M_1}{2M_2} \left((1 + M_2 \Delta t)^n - 1 \right) \Delta t.$$

(h) Use the result of Exercise 10 to conclude

$$e_n \leq \frac{M_1}{2M_2} \left(e^{(M_2 \Delta t)n} - 1 \right) \Delta t.$$

(i) Finally, conclude that

$$e_n \leq \frac{M_1}{2M_2} \left(e^{M_2(t_n - t_0)} - 1 \right) \Delta t.$$

(j) Explain why this justifies the inequality $e_n \leq C \cdot \Delta t$ given in this section.

(k) In what way is this bound for the total error different from the "estimates" shown in Figure 7.6 and computed in Exercises 7–9?

12. Given the initial-value problem

$$\frac{dy}{dt} = -2ty^2, \quad y(0) = 1,$$

and the interval $0 \leq t \leq 2$, carry out the theoretical bound on e_n given by the results of Exercise 11 as follows:

(a) Let R be the rectangle $\{(t, y) \mid 0 \leq t \leq 2, 0 \leq y \leq 1\}$. Determine the maximum values for M_1 and M_2 on R.

(b) Using the results of Exercise 11, derive the constant C for which we are certain that the inequality $e_n \leq C \cdot \Delta t$ holds.

(c) Using the value of C determined in part (b), find K such that $e_n \leq K/n$.

(d) Explain why you think these two estimates are so conservative compared to what we know from computations like those given at the beginning of the section.

7.2 IMPROVING EULER'S METHOD

Euler's method is a convenient numerical algorithm in many ways. It is easy to understand and to implement. However, for numerical work where a high degree of accuracy is essential, Euler's method is not the algorithm of choice. There are algorithms that usually are more accurate and that require fewer calculations to attain that accuracy. In this section and the subsequent one, we present two of these algorithms in order to illustrate how we can derive and implement more accurate algorithms.

Higher-Order Algorithms

As we mentioned in the last section, we can interpret Euler's method in terms of Taylor approximation. Given a solution to the initial-value problem

$$\frac{dy}{dt} = f(t, y), \quad y(t_0) = y_0,$$

then its Taylor series

$$y(t_1) = y_0 + y'(t_0)\Delta t + \frac{y''(t_0)}{2}(\Delta t)^2 + \ldots$$

can be rewritten as

$$y(t_1) = y_0 + f(t_0, y_0)\Delta t + \frac{y''(t_0)}{2}(\Delta t)^2 + \ldots.$$

Euler's method can be interpreted as the approximation that results from truncating this series at the linear term. That is,

$$y(t_1) \approx y_1 = y_0 + f(t_0, y_0)\Delta t.$$

One way to improve the accuracy of Euler's method is to use more of the Taylor series. Rather than truncating at the linear term, we include the quadratic term and obtain a more accurate approximation. In other words, we can approximate $y(t_1)$ by a new y_1 where

$$y(t_1) \approx y_1 = y_0 + f(t_0, y_0)\Delta t + \frac{y''(t_0)}{2}(\Delta t)^2.$$

To do this we need to know $y''(t_0)$, the second derivative of the solution. The only information we have about the solution y is that it satisfies the differential equation $dy/dt = f(t, y)$. We can differentiate both sides to obtain an expression for $y''(t_0)$, that is

$$\frac{d^2y}{dt^2} = \frac{\partial f}{\partial t}\frac{dt}{dt} + \frac{\partial f}{\partial y}\frac{dy}{dt}$$

$$= \frac{\partial f}{\partial t} + \frac{\partial f}{\partial y} f(t, y).$$

In principle there is no problem performing this calculation and using the results to implement a numerical algorithm that is more accurate than Euler's method. In practice this is seldom done because the calculation requires knowledge of the partial derivatives of f, and if we want to program a "black box" in a traditional computer language such as Fortran or C, we would have to provide these partial derivatives as well as the original function f from the differential equation. These days there exist computer languages that can calculate these derivatives, so this restriction is no longer a problem. Nevertheless, some very clever algorithms have been developed that provide approximation schemes with equivalent accuracy without using the partial derivatives of f, and these algorithms are commonly used. They will be the ones that we focus on in this chapter.

Numerical Approximation and Numerical Integration

To understand how these methods were developed, it is useful to think about Euler's method in terms of the numerical integration techniques that are used to approximate the definite integral. By the Fundamental Theorem of Calculus, we know that the solution of the initial-value problem

$$\frac{dy}{dt} = f(t, y), \quad y(t_0) = y_0$$

satisfies the equation

$$y(t_1) = y(t_0) + \int_{t_0}^{t_1} f(\tau, y(\tau))\, d\tau$$

because $dy/dt = f(t, y)$. In other words,

$$y(t_1) - y(t_0) = \int_{t_0}^{t_1} f(\tau, y(\tau))\, d\tau.$$

This is the "integral equation" equivalent of the original differential equation. In Euler's method we approximate this difference by the step

$$y_1 - y_0 = f(t_0, y_0)\, \Delta t.$$

Geometrically this approximation is really the approximation of the integral—the "area" under the graph of $f(t, y(t))$—by the area of a rectangle with height $f(t_0, y_0)$ and width Δt (see Figure 7.7). We use the value of the derivative $y'(t_0) = f(t_0, y_0)$ at the left-hand endpoint of the interval $t_0 \leq t \leq t_1$ to approximate the value of $y(t_1)$. Thus Euler's method is analogous to approximating a Riemann integral by left-hand Riemann sums.

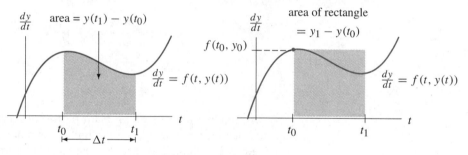

Figure 7.7
The first step of Euler's method interpreted as a Riemann approximation to an area.

From your younger days when you took calculus, you may recall that the left-hand Riemann sums converge to the value of the integral as the number of subdivisions increases but that there are better ways to approximate this integral. A simple variation on left-hand approximation is trapezoidal approximation. In this case we try to approximate the area under the graph of $f(t, y(t))$ by trapezoids. The width of each trapezoid is Δt, and the heights are given by the values $f(t_0, y(t_0))$, $f(t_1, y(t_1))$, $f(t_2, y(t_2))$, ... (see Figure 7.8). The area of the kth trapezoid is

$$\left(\frac{f(t_{k-1}, y(t_{k-1})) + f(t_k, y(t_k))}{2} \right) \Delta t.$$

It follows that

$$y(t_k) \approx y(t_{k-1}) + \left(\frac{f(t_{k-1}, y(t_{k-1})) + f(t_k, y(t_k))}{2} \right) \Delta t.$$

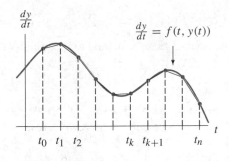

Figure 7.8
Approximating the area under the graph of $dy/dt = f(t, y(t))$ using trapezoidal approximation.

This formula for $y(t_k)$ suggests the approximation scheme

$$y_k = y_{k-1} + \left(\frac{f(t_{k-1}, y_{k-1}) + f(t_k, y_k)}{2} \right) \Delta t.$$

There is only one thing wrong with this scheme: The number y_k appears on both sides of the equation. In other words, you need to know y_k to compute y_k. (But if you already knew it, you wouldn't need the formula.) We have to get rid of the y_k on the right-hand side of the approximation scheme.

One way to turn this equation into a useful approximation scheme is to replace the y_k on the right-hand side with some other reasonable value. At this point the only way we know to approximate y_k is to use Euler's method, so that is what we do. In other words, we replace y_k on the right-hand side by the Euler approximation to y_k. This yields the approximation scheme

$$y_k = y_{k-1} + \left(\frac{f(t_{k-1}, y_{k-1}) + f(t_k, y_{k-1} + f(t_{k-1}, y_{k-1})\Delta t)}{2} \right) \Delta t.$$

This complicated formula can be more clearly thought of as a sequence of steps.

Improved Euler's Method

Given the initial condition $y(t_0) = y_0$ and the step size Δt, compute the point (t_{k+1}, y_{k+1}) from (t_k, y_k) as follows:

1. Use the differential equation to compute the slope $m_k = f(t_k, y_k)$.

2. Calculate the value $\tilde{y}_{k+1}$ that results from one application of Euler's method. That is,

$$\tilde{y}_{k+1} = y_k + f(t_k, y_k)\Delta t.$$

3. Calculate $t_{k+1} = t_k + \Delta t$ and use the differential equation to compute the slope $n_k = f(t_{k+1}, \tilde{y}_{k+1})$ at the point $(t_{k+1}, \tilde{y}_{k+1})$.

4. Compute y_{k+1} by

$$y_{k+1} = y_k + \left(\frac{m_k + n_k}{2} \right) \Delta t.$$

Improved Euler's Method for $f(t, y) = -2ty^2$, $y(0) = 1$

In Section 1.4 we applied Euler's method to the initial-value problem

$$\frac{dy}{dt} = -2ty^2, \quad y(0) = 1,$$

and in the last section we used this example to illustrate the typical errors involved in Euler's method. Now let's compare our previous results with the results from improved Euler's method.

We begin with $\Delta t = 0.1$. In this example the function $f(t, y) = -2ty^2$, and its value at the initial condition is $f(0, 1) = 0$. Therefore the initial slope is $m_0 = 0$. One step of Euler's method yields the point $(t_1, \tilde{y}_1) = (0.1, 1.0)$. Improved Euler's method uses the value of $f(t, y)$ at this new point to help determine the value of y_1. In other words, we compute $n_0 = f(0.1, 1.0) = -0.2$, and average this slope with m_0 to obtain the slope that we use to compute y_1. In this case the average is

$$\frac{m_0 + n_0}{2} = \frac{0.0 - 0.2}{2} = -0.1.$$

Hence we use this slope to calculate y_1. We obtain

$$y_1 = y_0 + (-0.1)\Delta t$$

$$= 1.0 + (-0.1)(0.1) = 0.99.$$

Before computing another step by hand, we compare the result of this calculation to the result that Euler's method provides. At the initial condition the differential equation yields the value 0 for the slope. Therefore Euler's method assumes that the slope over the subinterval $0 \leq t \leq 0.1$ vanishes, and therefore $y_1 = y_0 = 1.0$. On the other hand, improved Euler's method considers the slope at the point $(0.1, 1.0)$ as well as the slope at $(0.0, 1.0)$. When these two slopes are averaged, we obtain a (small) negative value for the slope to use over the subinterval $0 \leq t \leq 0.1$. Using this average slope makes sense because we can see from the differential equation that $f(t, y) = -2ty^2$ is negative for all nonzero t throughout the subinterval. Of course, there is no reason to believe that this average of the two slopes is going to be a better choice than the slope from Euler's method, but it is a good idea to try this average.

The process is basically the same for the next step. The first step yielded the point $(t_1, y_1) = (0.1, 0.99)$. Consequently, we use the differential equation to produce the slope that Euler's methods uses. We have

$$m_1 = f(t_1, y_1) = f(0.1, 0.99) = -0.19602.$$

So, starting at $(0.1, 0.99)$, we take one step of Euler's method via the computation

$$\tilde{y}_2 = y_1 + m_1 \Delta t$$

$$= 0.99 + (-0.19602)(0.1)$$

$$= 0.970398.$$

Thus for our other slope n_1, we compute $n_1 = f(t_2, \tilde{y}_2) = -0.37669$. We average these two slopes and obtain

$$\frac{m_1 + n_1}{2} = \frac{(-0.19602) + (-0.37669)}{2} = -0.286344.$$

Using this average we can now compute the second step

$$y_2 = y_1 + (-0.286344)\,\Delta t$$
$$= 0.99 + (-0.286344)(0.1)$$
$$= 0.961366.$$

That's enough by hand. Turning to the computer we obtain the results shown in Table 7.1.

Table 7.1
Improved Euler's method for $dy/dt = -2ty^2$ with $\Delta t = 0.1$

k	t_k	y_k
0	0	1
1	0.1	0.99
2	0.2	0.961366
3	0.3	0.917246
$\vdots$	$\vdots$	$\vdots$
19	1.9	0.217670
20	2.0	0.200695

Recall that we know the exact solution $y(t) = 1/(1 + t^2)$ for this initial-value problem, and its value at $t = 2$ is $y(2) = 0.2$. Thus the error in this computation is

$$|0.2 - 0.200695| = 0.000695.$$

When we used Euler's method with the same step size, we obtained the approximation $y(2) \approx 0.193342$ (see page 563). The error in that approximation was

$$|0.2 - 0.193342| = 0.00658.$$

Note that improved Euler's method is roughly ten times more accurate than Euler's method in this example.

Cost

For a given step size, using improved Euler's method instead of Euler's method yields a good improvement in accuracy. The down side is that each step of improved Euler's method involves two steps of Euler's method. The first step is used to get the value of $\tilde{y}_{k+1}$, and the second computes y_{k+1}. Each step of improved Euler's method takes roughly twice the amount of arithmetic as one step of Euler's method.

In the example above, we gained ten times the accuracy by using the improved Euler's method over the Euler's method at a cost of doing twice the arithmetic. This is a good return (particularly if a computer is doing the arithmetic).

The Order of Improved Euler

In the previous section we saw that Euler's method has order 1. In other words, the error is roughly proportional to the reciprocal of the number of subdivisions employed in the approximation. In order to compare the two methods, we need to know how the accuracy of improved Euler's method behaves as a function of the step size. As in the previous section, we consider the initial-value problem

$$\frac{dy}{dt} = -2ty^2, \quad y(0) = 1,$$

which has the exact solution $y(t) = 1/(1 + t^2)$. Using the computer we approximate the value $y(2) = 0.2$ twice. First we compute the approximation with 1000 steps. The error e_{1000} in this approximation is

$$e_{1000} = 2.59 \times 10^{-7}.$$

Then we repeat the computation using twice as many steps. In this case we get a more accurate answer with an error of

$$e_{2000} = 6.48 \times 10^{-8}.$$

These errors are definitely small, but to get a sense of how this method depends on the step size, we compute the ratio

$$\frac{e_{1000}}{e_{2000}} \approx 4.003.$$

We see that, if we double the number of steps (that is, halve the step size), we decrease the error by a factor of 4 $(= 2^2)$. Consequently we say that improved Euler's method is a numerical method with order 2.

Using estimates like the ones that we discussed in the last section, it can be shown that, given appropriate bounds on f and its derivatives, the error for improved Euler's method behaves like a quadratic function in its step size. That is, the error in a given approximation using improved Euler's method is proportional to $(\Delta t)^2$. Since Δt is simply the length of the t-interval in question divided by the number of steps n, we can also express the error as

$$e_n = \frac{K}{n^2}.$$

for some constant K. In Figure 7.9, we graph the error as a function of the number of steps for Euler's method and improved Euler's method. This plot illustrates the advantage of a second-order method over a first-order method.

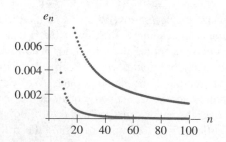

Figure 7.9

The errors related to Euler's method and improved Euler's method for the initial-value problem $dy/dt = -2ty^2$, $y(0) = 1$, over the interval $0 \le t \le 2$.

EXERCISES FOR SECTION 7.2

In Exercises 1–8, use improved Euler's method with the given number n of steps to approximate the solution to the initial-value problem specified. Your answer should include a table of approximate values of the dependent variable. It should also include a sketch of the graph of the approximate solution. These initial-value problems are the same as those in Exercises 1–8 of Section 1.4. Compare the graphs that you get from improved Euler's method with those that you get from Euler's method. If your computer system has a built-in routine for the numerical solution of differential equations, compare these graphs with those that come from the built-in numerical solver.

1. $\dfrac{dy}{dt} = 2y + 1, \quad y(0) = 3, \quad 0 \leq t \leq 2, \quad n = 4$

2. $\dfrac{dy}{dt} = t - y^2, \quad y(0) = 1, \quad 0 \leq t \leq 1, \quad n = 4$

3. $\dfrac{dy}{dt} = y^2 - 2y + 1, \quad y(0) = 2, \quad 0 \leq t \leq 2, \quad n = 4$

4. $\dfrac{dy}{dt} = \sin y, \quad y(0) = 1, \quad 0 \leq t \leq 3, \quad n = 6$

5. $\dfrac{dw}{dt} = (3 - w)(w + 1), \quad w(0) = 4, \quad 0 \leq t \leq 5, \quad n = 5$

6. $\dfrac{dw}{dt} = (3 - w)(w + 1), \quad w(0) = 0, \quad 0 \leq t \leq 5, \quad n = 10$

7. $\dfrac{dy}{dt} = e^{2/y}, \quad y(0) = 2, \quad 0 \leq t \leq 2, \quad n = 4$

8. $\dfrac{dy}{dt} = e^{2/y}, \quad y(1) = 2, \quad 1 \leq t \leq 3, \quad n = 4$

In Exercises 9–12, use improved Euler's method to approximate the solution to the given initial-value problem over the interval specified. For each initial-value problem,

(a) determine an analytic solution (see Exercises 2–5 in Section 7.1); and

(b) using improved Euler's method with $n = 4$, determine an approximate value y_4 and the total error e_4.

(c) How many steps of improved Euler's method would you expect to use to obtain an approximation for which $e_n \leq 0.0001$?

9. $\dfrac{dy}{dt} = 1 - y, \quad y(0) = 0, \quad 0 \leq t \leq 1$

10. $\dfrac{dy}{dt} = ty, \quad y(0) = 1, \quad 0 \leq t \leq \sqrt{2}$

11. $\dfrac{dy}{dt} = -y^2, \quad y(0) = 1/2, \quad 0 \leq t \leq 2$

12. $\dfrac{dy}{dt} = 1 - t + 3y, \quad y(0) = 1, \quad 0 \leq t \leq 1$

13. In this section we showed that the total error e_{20} in the approximation of the initial-value problem $dy/dt = -2ty^2$, $y(0) = 1$, over the interval $0 \leq t \leq 2$ is

$$e_{20} = 0.000695.$$

(a) How many steps should you use to get an approximate value y_n that has an associated error e_n no greater than 0.0001?

(b) Calculate that approximate value y_n.

(c) Calculate the total error in that case.

In Exercises 14–17, produce a plot like Figure 7.9 for the given initial-value problem over the interval specified. See the directions to Exercise 1 in Section 7.1 for more details. However, restrict the values of n to numbers in the interval $10 \leq n \leq 100$.

14. $\dfrac{dy}{dt} = 1 - y$, $\quad y(0) = 0$, $\quad 0 \leq t \leq 1$

15. $\dfrac{dy}{dt} = ty$, $\quad y(0) = 1$, $\quad 0 \leq t \leq \sqrt{2}$

16. $\dfrac{dy}{dt} = -y^2$, $\quad y(0) = 1/2$, $\quad 0 \leq t \leq 2$

17. $\dfrac{dy}{dt} = 1 - t + 3y$, $\quad y(0) = 1$, $\quad 0 \leq t \leq 1$

7.3 THE RUNGE-KUTTA METHOD

Improved Euler's method shows us that if we intelligently modify Euler's method and do a little bit more work in each step, we can get a significantly more accurate approximation to our solution. This observation leads us to wonder if there is a "standard" method for the numerical approximation of solutions of differential equations. We want an algorithm that minimizes the number of computations that have to be performed while it maximizes the accuracy.

In fact for most practical purposes, there is such an algorithm. It is called the (fourth-order) Runge-Kutta method. It is named after the two German mathematicians who developed the method approximately 100 years ago. There are more sophisticated algorithms that are required for specialized problems, but the Runge-Kutta method is often the numerical method used for the numerical approximation of solutions.

The Runge-Kutta Method

In improved Euler's method applied to the differential equation $dy/dt = f(t, y)$, we used an average of two slopes to compute each value. In other words, to compute y_k from y_{k-1}, we used the average of values of the right-hand side $f(t, y)$ at $t = t_{k-1}$ and $t = t_k$. By analogy with numerical integration, improved Euler's method is similar to the trapezoidal rule.

For numerical integration, there are algorithms that are usually more efficient than the trapezoidal rule. Simpson's rule approximates the area under the graph using parabolic interpolation and generally gives a better estimate for the integral. In fact

Simpson's rule can be interpreted as a weighted average of values where twice as much weight is given to the values of the function at the midpoints of the subintervals than is given to the endpoints of the subintervals. The Runge-Kutta method is similar to Simpson's rule in the way that it considers a weighted average.

We begin with a description of the algorithm and then illustrate its implementation and accuracy using the example from previous sections. Refer to Figure 7.10 as we describe how the various slopes used in the method are derived.

To calculate the value y_{k+1} from y_k, we use four slopes given by the function $f(t, y)$ that defines the differential equation. These slopes are denoted m_k, n_k, q_k, and p_k. We also need several intermediate variables. Thus the bookkeeping in the Runge-Kutta method is more formidable than improved Euler's method, so these variables wear many different hats.

The four slopes are determined successively as follows:

1. The first slope m_k is calculated just as in Euler's method; that is, $m_k = f(t_k, y_k)$.

2. In the improved Euler's method we used m_k to produce a second slope corresponding to a point with $t = t_{k+1} = t_k + \Delta t$. The Runge-Kutta method does almost the same thing except that it goes only halfway along the t-axis to $\tilde{t} = t_k + \Delta t/2$. That

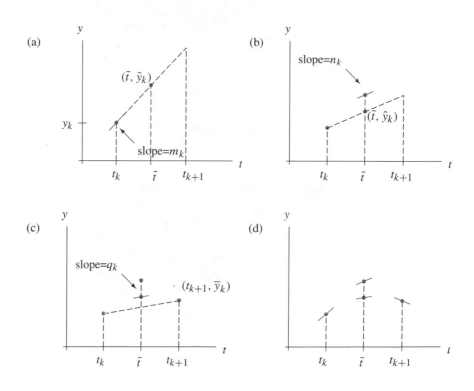

Figure 7.10
Part (d) illustrates the four slopes that are used in the Runge-Kutta method. Parts (a)–(c) illustrate how the points at which the slopes are evaluated are determined.

is, we use m_k to produce a point $(\tilde{t}, \tilde{y}_k)$ where

$$\tilde{y}_k = y_k + m_k \frac{\Delta t}{2}$$

(see Figure 7.10(a)). Once we have determined this point, we use the function $f(t, y)$ to determine the second slope n_k by

$$n_k = f(\tilde{t}, \tilde{y}_k).$$

3. Now we repeat the previous step of the algorithm where we use the slope n_k in place of the slope m_k. In other words, we go from (t_k, y_k) to the line $t = \tilde{t}$ along a line of slope n_k. Thus we obtain a new number $\hat{y}_k$ where

$$\hat{y}_k = y_k + n_k \frac{\Delta t}{2}$$

(see Figure 7.10(b)). Given this point on the line $t = \tilde{t}$, we compute the third slope q_k by

$$q_k = f(\tilde{t}, \hat{y}_k).$$

4. Finally, we obtain our fourth slope by using q_k to produce a point on the line $t = t_{k+1}$. We get

$$\overline{y}_k = y_k + q_k \Delta t$$

(see Figure 7.10(c)). Once we have this fourth point we compute the slope at that point by

$$p_k = f(t_{k+1}, \overline{y}_k).$$

5. Now that we have collected the four slopes (see Figure 7.10(d)), we take a weighted average and compute the next step using this average slope. We count each of the slopes that come from the points with $t = \tilde{t}$ twice as heavily as we count the other two slopes. In other words, our weighted average is

$$\frac{m_k + 2n_k + 2q_k + p_k}{6}.$$

The actual step that we take is therefore

$$y_{k+1} = y_k + \left(\frac{m_k + 2n_k + 2q_k + p_k}{6} \right) \Delta t.$$

This seems complicated, but it works well. Basically the slopes are chosen so that we are approximating the solution up to terms of order 4. As we shall see, this gives us an algorithm that is very efficient.

The Initial-Value Problem $dy/dt = -2ty^2$, $y(0) = 1$

Let's illustrate a couple of steps of the algorithm using our old friend

$$\frac{dy}{dt} = -2ty^2, \quad y(0) = 1.$$

In the previous section, we approximated the solution to this initial-value problem using improved Euler's method with $\Delta t = 0.1$. Now we use the same Δt with the Runge-Kutta method.

We begin with the initial slope

$$m_0 = f(t_0, y_0) = f(0, 1) = 0.$$

This slope determines

$$\tilde{y}_0 = y_0 + m_0 \frac{\Delta t}{2} = 1 + 0 \left(\frac{0.1}{2} \right) = 1.$$

Given $\tilde{y}_0$ and $\tilde{t} = t_0 + \Delta t/2 = 0 + 0.05 = 0.05$, we obtain the second slope we need by

$$n_0 = f(\tilde{t}, \tilde{y}_0) = f(0.05, 1.0) = -0.1.$$

In turn this slope determines the y-value $\hat{y}_0$ via

$$\begin{aligned}
\hat{y}_0 &= y_0 + n_0 \frac{\Delta t}{2} \\
&= 1.0 + (-0.1)(0.05) \\
&= 0.995.
\end{aligned}$$

Applying $f(t, y)$ to the point $(\tilde{t}, \hat{y}_0)$ determines a third slope

$$q_0 = f(\tilde{t}, \hat{y}_0) = -0.0990025,$$

and using this slope, we obtain the fourth point with y-value $\overline{y}_0$ by

$$\overline{y}_0 = y_0 + q_0 \Delta t = 1.0 + (-0.0990025)(0.1) = 0.9901.$$

Given this fourth point we compute the slope at that point by

$$p_0 = f(t_{k+1}, \overline{y}_0) = -0.19606.$$

Now that we have these four slopes, we are able to compute the weighted average that we use to compute the slope that determines the first step. We have

$$\begin{aligned}
\frac{m_0 + 2n_0 + 2q_0 + p_0}{6} &= \left(\frac{0 + 2(-0.1) + 2(-0.0990025) + -0.19606}{6} \right) \\
&- -0.0990108.
\end{aligned}$$

This weighted average yields the first step

$$y_1 = y_0 - 0.0990108 \, \Delta t = 1.0 + (-0.0990108)(0.1) = 0.990099.$$

(If you try to reproduce these calculations on your calculator and you get a slightly different answer, don't be concerned. Your calculator may carry a different number of significant digits in the calculation, and this difference may affect your computation slightly.)

Doing these calculations by hand is definitely not the way to go now that we have easy access to programmable calculators, spreadsheets, and general systems for doing mathematics by computers such as Maple, *Mathematica*, and MATLAB. Later in this section we provide codes that can be used in various computing environments to implement Runge-Kutta. They are not too complicated, and using them definitely beats doing these computations by hand.

Computing 20 steps via the Runge-Kutta method yields the results shown in Table 7.2. Note that the approximation of $y(2) = 0.2$ is very good, especially considering the relatively large step size that we used. Compare these results to those found in Table 7.1. The error obtained in improved Euler's method was 0.000658, whereas the error here is 0.000001. Switching to the Runge-Kutta method in this example improves the accuracy by a factor of 658. (Actually the improvement is more like a factor of 1000, but that is not clear from Table 7.2 because we show the results only to six significant digits.)

Table 7.2
Runge-Kutta method for $dy/dt = -2ty^2$ with $\Delta t = 0.1$

k	t_k	y_k
0	0	1
1	0.1	0.990099
2	0.2	0.961538
3	0.3	0.917431
⋮	⋮	⋮
19	1.9	0.216920
20	2.0	0.200001

The Order of the Runge-Kutta Method

Euler's method has order 1 and improved Euler's method has order 2. Now let's see if we are right about Runge-Kutta's order. To find the order of the Runge Kutta method, we recall that the solution of the initial-value problem

$$\frac{dy}{dt} = -2ty^2, \quad y(0) = 1$$

is $y(t) = 1/(1 + t^2)$. If we approximate this solution over the interval $0 \leq t \leq 2$ using the Runge-Kutta method with ten steps, we find that the error in this approximation is

$$e_{10} = 1.095 \times 10^{-5},$$

and if we approximate the solution using 20 steps, we get an error of

$$e_{20} = 6.541 \times 10^{-7}.$$

The ratio

$$\frac{e_{10}}{e_{20}} = 16.75$$

indicates that if we double the number of steps, the error improves by roughly a factor of 16 ($= 2^4$). This is what we expect from a numerical procedure that has order 4.

Programming Runge-Kutta

Although the Runge-Kutta method seems somewhat complicated at first, it really isn't especially difficult to program. Here we give three implementations — one for a graphing calculator, one for *Mathematica*, and one in the C programming language. Some of the exercises at the end of this section require that you implement the method using whatever technology is available to you. These three implementations should give you some guidance as to how to proceed.

The TI graphing calculator

This code assumes that the initial t-value is stored in A, the end value is stored in B, the number of steps is stored in N, the initial y-value is stored in Y, and $Y_1 = f(X,Y)$. The code uses X to represent t and Y to represent y in our notation.

```
:(B - A)/N -> H
:A -> X
:0 -> I
:Lbl 1
:Disp Y
:Pause
:Y -> Z
:Y1 -> C
:X + H/2 -> X
:Z + C*H/2 -> Y
:Y1 -> D
:Z + D*H/2 -> Y
:Y1 -> E
:X + H/2 -> X
:Z + E*H -> Y
:Y1 -> F
:Z + (C + 2*D + 2*E + F)*H/6 -> Y
:IS>(I,N)
:Goto 1
```

Mathematica code

The *Mathematica* code is more compact, but using it effectively requires some knowledge of commands such as `ListPlot` and `NestList`. We begin by defining a function that takes one step of the Runge-Kutta method.

```
RungeKuttaStep[f_, vars_, {tk_, yk }, deltat_]:=
Module[{k1, k2, k3, k4, tmiddle, tend},
    k1 = f /. {vars[[1]] -> tk, vars[[2]] -> yk};
    tmiddle = tk + deltat/2;
    k2 = f /. {vars[[1]] -> tmiddle,
               vars[[2]] -> yk + k1 deltat/2};
    k3 = f /. {vars[[1]] -> tmiddle,
               vars[[2]] -> yk + k2 deltat/2};
    tend = tk + deltat;
    k4 = f /. {vars[[1]] -> tend,
               vars[[2]] -> yk + k3 deltat};
    {tend, yk + (k1 + 2 k2 + 2 k3 + k4) deltat/6}
    ]
```

For example, the command

```
RungeKuttaStep[-2 t y^2, {t, y}, {0,1}, 0.1]
```

returns

```
{0.1, 0.990099}
```

We then approximate the solution to the initial-value problem

$$\frac{dy}{dt} = -2ty^2, \quad y(0) = 1$$

over the interval $0 \le t \le 2$ using 20 steps by the command

```
RKresults =
    NestList[RungeKuttaStep[-2 t y^2, {t, y}, #, 0.1]&,
        {0, 1}, 20]
```

This command stores the resulting points in the variable RKresults, and hence we are able to plot the approximate solution using ListPlot. For example,

```
ListPlot[RKresults,
        PlotJoined -> True,
        AxesOrigin -> {0,0}]
```

C code for Runge-Kutta

The following is a C program that applies the Runge-Kutta method to the initial-value problem

$$\frac{dy}{dt} = -2ty^2, \quad y(0) = 1.$$

It does the approximation over the interval $0 \le t \le 2$ with 20 steps. You should change the values of T0, Tn, NUMSTEPS, and Y0 at the top and the definition of $f(t, y)$ in the return statement to apply the code to a different initial-value problem.

```
/*

            Runge-Kutta approximation of
                 dy/dt = f(t,y)
*/


/* need to set these values and the function f(t,y) below
    before compilation */

#define T0 0.0
#define Tn 2.0
#define NUMSTEPS 20
#define Y0 1.0


#include <stdio.h>
#include <math.h>
```

```
/*
    this subroutine contains the formula for the
    first-order differential equation,
    that is, dy/dt = f(t,y)
*/

double f(t,y)
    double t,y;
{
  return(-2.0 * t * y * y);
}



main()
{
  double tk = T0, tmid, yk = Y0,
         k1, k2, k3, k4,
          deltat = (double)(Tn - T0)/NUMSTEPS,
          f();

  printf("%f %f\n", tk, yk);
  while (tk < Tn){
    k1 = f(tk, yk);
    tmid = tk + deltat/2.0;
    k2 = f(tmid, yk + k1 * deltat/2.0);
    k3 = f(tmid, yk + k2 * deltat/2.0);
    tk = tk + deltat;
    k4 = f(tk, yk + k3 * deltat);
    yk = yk + (k1 + 2.0 * k2 + 2.0 * k3 + k4) * deltat/6.0;
    printf("%f %f\n", tk, yk);
  }
}
```

The Runge-Kutta Method for Systems

The generalization of the Runge-Kutta method to first-order systems is straightforward if we take advantage of our vector notation for systems. Moreover, once we implement Runge-Kutta for systems, we also have an efficient method for the numerical study of higher-order differential equations, since higher-order equations can be rewritten as first-order systems. In fact this is the only way to study second- and higher-order differential equations numerically

Consider the first-order system

$$\frac{d\mathbf{Y}}{dt} = \mathbf{F}(t, \mathbf{Y}), \quad \mathbf{Y}(t_0) = \mathbf{Y}_0.$$

To apply the Runge-Kutta method to this system, we need to compute four vectors $\mathbf{K}_1$, $\mathbf{K}_2$, $\mathbf{K}_3$, and $\mathbf{K}_4$ at each step of the algorithm. These vectors are analogous to the four slopes that we computed for first-order equations, and they are defined in essentially the same way. Each of these vectors also depends on the index k, but we will not indicate this dependence to keep the formulas more readable. That is, suppose we have the point $(t_k, \mathbf{Y}_k)$. Then the vectors are

$$\mathbf{K}_1 = \mathbf{F}(t_k, \mathbf{Y}_k)$$
$$\mathbf{K}_2 = \mathbf{F}(t_k + \Delta t/2, \mathbf{Y}_k + (\Delta t/2)\mathbf{K}_1)$$
$$\mathbf{K}_3 = \mathbf{F}(t_k + \Delta t/2, \mathbf{Y}_k + (\Delta t/2)\mathbf{K}_2)$$
$$\mathbf{K}_4 = \mathbf{F}(t_k + \Delta t, \mathbf{Y}_k + \Delta t\, \mathbf{K}_3).$$

Once we have these vectors, then we can form a weighted average just as in the case of first-order equations. We get the vector

$$\mathbf{K} = \left(\frac{\mathbf{K}_1 + 2\mathbf{K}_2 + 2\mathbf{K}_3 + \mathbf{K}_4}{6} \right),$$

and we use this vector $\mathbf{K}$ to make the step from $(t_k, \mathbf{Y}_k)$ to

$$(t_{k+1}, \mathbf{Y}_{k+1}) = (t_k + \Delta t, \mathbf{Y}_k + \Delta t\, \mathbf{K}).$$

If we are plotting solution curves in the phase plane, then we simply plot the points $\mathbf{Y}_k$.

Figures 7.11 and 7.12 illustrate the results of Euler's method versus the Runge-Kutta method on the predator-prey system we discussed at length in Section 2.1. Note that although Euler's method is not accurate enough to catch the closed orbits in the phase plane, the Runge-Kutta method does a nice job of capturing the periodic nature of this system.

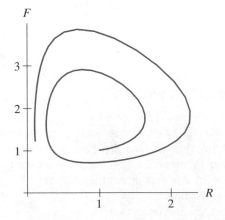

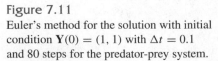

Figure 7.11
Euler's method for the solution with initial condition $\mathbf{Y}(0) = (1, 1)$ with $\Delta t = 0.1$ and 80 steps for the predator-prey system.

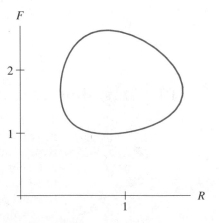

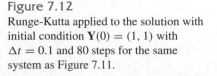

Figure 7.12
Runge-Kutta applied to the solution with initial condition $\mathbf{Y}(0) = (1, 1)$ with $\Delta t = 0.1$ and 80 steps for the same system as Figure 7.11.

EXERCISES FOR SECTION 7.3

In Exercises 1–5, use the Runge-Kutta method with the given number n of steps to approximate the solution to the initial-value problem specified. Your answer should include a table of approximate values of the dependent variable. It should also include a sketch of the graph of the approximate solution. These initial-value problems also appear in Section 7.2. Compare the graphs that you get from the Runge-Kutta method to those that come from Euler's method and improved Euler's method. If your computer has a built-in routine for the numerical solution of differential equations, compare these graphs with those that come from the built-in numerical solver.

1. $\dfrac{dy}{dt} = 2y + 1$, $\quad y(0) = 3$, $\quad 0 \le t \le 2$, $\quad n = 4$

2. $\dfrac{dy}{dt} = t - y^2$, $\quad y(0) = 1$, $\quad 0 \le t \le 1$, $\quad n = 4$

3. $\dfrac{dy}{dt} = (3 - y)(y + 1)$, $\quad y(0) = 0$, $\quad 0 \le t \le 5$, $\quad n = 10$

4. $\dfrac{dy}{dt} = e^{2/y}$, $\quad y(0) = 2$, $\quad 0 \le t \le 2$, $\quad n = 4$

5. $\dfrac{dy}{dt} = e^{2/y}$, $\quad y(1) = 2$, $\quad 1 \le t \le 3$, $\quad n = 4$

6. Consider the initial-value problem

$$\frac{dy}{dt} = -2ty^2, \quad y(0) = 1$$

over the interval $0 \le t \le 2$.

(a) Calculate the Runge-Kutta approximation to the solution with $n - 4$ steps.

(b) Calculate the total error e_4 associated with this approximation.

(c) How many steps are necessary to approximate the solution with an error of less than 0.0001?

7. Consider the predator-prey system

$$\frac{dR}{dt} = 2R - 1.2RF$$

$$\frac{dF}{dt} = -F + 1.2RF$$

and the initial condition $(R_0, F_0) = (1.0, 1.0)$.

(a) Calculate the Runge-Kutta approximation to this system for five steps with $\Delta t = 1.0$. Plot the results in the phase plane.

(b) With the aid of a calculator or a computer, repeat the calculation in part (a) with $\Delta t = 0.1$ and $n = 80$ steps. Plot the result in the phase plane.

(c) Using the results of part (b), plot the $R(t)$- and $F(t)$-graphs for this solution.

8. Use Runge-Kutta to approximate the solution to the second-order equation

$$\frac{d^2y}{dt^2} + (y^2 - 3)\frac{dy}{dt} + 5y = 0$$

that satisfies the initial condition $(y(0), y'(0)) = (1, 0)$. Use $\Delta t = 0.1$ and $n = 100$. Plot your results both in the yv-phase plane (where $v = y'$) and as the graph of $y(t)$.

7.4 THE EFFECTS OF FINITE ARITHMETIC

In theory the errors involved in the use of numerical approximation procedures decrease as the step sizes decrease, and our preference for the Runge-Kutta method stems from the fact that this method is a fourth-order method that isn't too complicated to implement. It is tempting to assume that the computer or calculator is doing exactly what it was told to do. Unfortunately this is not the case. Floating-point arithmetic has only a finite amount of accuracy, and this limitation introduces another source of error in each step of the numerical method.

In this section we briefly take a look at a couple of informative examples so that you are aware of some of the pitfalls that are associated with computer arithmetic. As always, we start with our favorite example so that we know the exact solution, and then we discuss what should be done when we do not know the solution. Of course, the latter case is the situation in which we typically employ these algorithms.

The Initial-Value Problem $dy/dt = -2ty^2$, $y(0) = 1$ Yet Again

In the previous section we found that the Runge-Kutta method gives us a very accurate approximation to the solution $y(t) = 1/(1 + t^2)$ over the interval $0 \leq t \leq 2$ using only the step size $\Delta t = 0.1$. This is a fairly large step size. In theory we can make the error as small as we like by decreasing the step size. To see if this works in practice,

Table 7.3

The error in the value of $y(2)$ associated with the Runge-Kutta method for $dy/dt = -2ty^2$, $y(0) = 1$ for 2^n subdivisions of the interval $0 \leq t \leq 2$

n	2^n	e_{2^n}	n	2^n	e_{2^n}
1	2	0.01741182739635213639	10	1024	0.0000000000000008953949
2	4	0.00040567218499903968	11	2048	0.0000000000000000552336
3	8	0.00002714430679354174	12	4096	0.0000000000000000080491
4	16	0.00000161801719431032	13	8192	0.0000000000000000024980
5	32	0.00000009769105269175	14	16384	0.0000000000000000094369
6	64	0.00000000598843688526	15	32768	0.0000000000000000072164
7	128	0.00000000037049122104	16	65536	0.0000000000000000077716
8	256	0.00000000002303576774	17	131072	0.0000000000000000144329
9	512	0.00000000000143582368	18	262144	0.0000000000000000355271

we try an experiment. (Such experiments are an excellent use of home computers, a much better use of electricity than the drawing of flying toasters.)

We start by subdividing the interval $0 \leq t \leq 2$ into two subintervals (so there are two steps and $\Delta t = 1.0$), and we calculate the error e_2. The Runge-Kutta method with $\Delta t = 1.0$ yields the approximation $y_2 = 0.182588$ of the exact value $y(2) = 0.2$, and therefore the error is

$$e_2 = |0.2 - 0.182588| = 0.0174118.$$

It's remarkable that Runge-Kutta gives such an accurate approximation with such a large Δt.

The theory says that, if we double the number of steps in the Runge-Kutta method, the error should decrease by a factor of 16. Therefore we double the number of steps (and halve the step size) and recompute the approximate solution. Repeating this procedure, doubling the number of steps each time, we obtain the data in Table 7.3.

Note that the error improves as expected for the first 10 doublings. Note that, for $n = 13$, we have an approximation that agrees with the exact value to 15 decimal places. Then, at $n = 14$, the error starts to increase. Calculating more steps does not seem to help (see Table 7.4).

Table 7.4
The error associated with the Runge-Kutta method for $dy/dt = -2ty^2$ for 2^n subdivisions

n	2^n	e_{2^n}	n	2^n	e_{2^n}
12	4096	0.00000000000000080491	20	1048576	0.00000000000000224820
13	8192	0.00000000000000024980	21	2097152	0.00000000000000621725
14	16384	0.00000000000000094369	22	4194304	0.00000000000001271205
15	32768	0.00000000000000072164	23	8388608	0.00000000000001187939
16	65536	0.00000000000000077716	24	16777216	0.00000000000001057487
17	131072	0.00000000000000144329	25	33554432	0.00000000000002037259
18	262144	0.00000000000000355271	26	67108864	0.00000000000000949241
19	524288	0.00000000000000130451			

The Effects of Finite Arithmetic

At a certain point the efficiency of the numerical method is defeated by the effects of round-off errors in the arithmetic of the computer. Unless we are lucky for some reason, we cannot expect to get an answer that is any more accurate than the arithmetic allows. This is like the concept of significant digits in the sciences.

There are various ways to go about modeling the accuracy of the computer arithmetic, and like our models they lead to somewhat different predictions about the effects of finite arithmetic. For our purposes we'll adopt one of the standard models that says the total error, denoted by $e(\Delta t)$, as a function of step size Δt for the Runge-Kutta method is roughly

$$e(\Delta t) = c_1 (\Delta t)^4 + \frac{c_2}{\Delta t},$$

where c_1 and c_2 are constants determined by the differential equation, its initial condition, and interval of which the solution is being approximated. The first term, $c_1(\Delta t)^4$, is the term that says that Runge-Kutta is a fourth-order method. The second term, $c_2/(\Delta t)$, is the term that measures the round-off errors for small values of Δt. Presumably, the constant c_2 is very small, and thus Δt must be very small before the effects of this term are noticeable.

We can observe the effects of round-off if we repeat the calculations done above with single-precision arithmetic (see Table 7.5). Note that, in this case, we are better off with 128 subdivisions than we are with millions of subdivisions. In fact by the time we get to 16 million subdivisions, our approximation is about as accurate as it was with 2 subdivisions.

Table 7.5

The error associated to the Runge-Kutta method implemented in single precision for $dy/dt = -2ty^2$ for 2^n subdivisions

n	2^n	e_{2^n}	n	2^n	e_{2^n}
1	2	0.01741183	13	8192	0.00000031
2	4	0.00040567	14	16384	0.00000002
3	8	0.00002714	15	32768	0.00000001
4	16	0.00000161	16	65536	0.00000035
5	32	0.00000012	17	131072	0.00000324
6	64	0.00000002	18	262144	0.00000173
7	128	0.00000001	19	524288	0.00002292
8	256	0.00000007	20	1048576	0.00000132
9	512	0.00000001	21	2097152	0.00062698
10	1024	0.00000005	22	4194304	0.00052327
11	2048	0.00000012	23	8388608	0.00506006
12	4096	0.00000004	24	16777216	0.01590731

What to Do in Practice

In practice we want to use these numerical procedures when we do not know analytic representations for the solutions of a differential equation. Since it is usually impossible to carry out meaningful estimates from the theory, how do we decide what is a reasonable strategy to follow? If we accept the assertion that the error is roughly determined by an expression of the form

$$e(\Delta t) = c_1(\Delta t)^4 + \frac{c_2}{\Delta t},$$

then we want to use a value of Δt that minimizes the first term as much as possible before the second term starts to have an effect. However, since we don't know the actual value of the solution, how can we predict where this minimum is going to occur?

We compute multiple approximations where we double the number of steps, and we look for our approximations to "converge" to a certain value. For example, if we apply Runge-Kutta to the initial-value problem

$$\frac{dy}{dt} = y^2 - t, \quad y(1) = 1,$$

over the interval $1 \le t \le 3$ with various numbers of subdivisions, we get the results given in Table 7.6.

Note that we have the approximation $y(3) \approx -1.49389744$ with both 64 and 128 subdivisions. Given the repetition, we compute one more approximation with 256 subdivisions, and we are fairly confident that the actual value of the solution to five decimal places is -1.49389. It's also a relatively safe bet that the next digit in the decimal expansion is 7, but if we need a more precise answer, we repeat the calculations using more precision.

Table 7.6
Runge-Kutta applied to the initial-value
problem $dy/dt = y^2 - t, y(1) = 1$

n	$y_n \approx y(3)$
2	-1.39687991
4	-1.47455096
8	-1.49300539
16	-1.49384940
32	-1.49389482
64	-1.49389744
128	-1.49389744
256	-1.49389791
512	-1.49389756

EXERCISES FOR SECTION 7.4

In Exercises 1–3, use Runge-Kutta to produce a table similar to Table 7.6 to approximate the value of the solution at the right-hand endpoint of the interval specified. What number of steps corresponds to the optimal approximation?

1. $\dfrac{dy}{dt} = t - y^3, \quad y(0) = 2, \quad 0 \le t \le 1$

2. $\dfrac{dy}{dt} = \sin ty, \quad y(0) = 3, \quad 0 \le t \le 3$

3. $\dfrac{dy}{dt} = (\cos y)(t^2 - y), \quad y(0) = 0, \quad 0 \le t \le 2$

LAB 7.1 | Errors of Numerical Approximations

In this chapter we studied three different methods for finding numerical approximations of solutions of differential equations: Euler, improved Euler, and Runge-Kutta. The accuracy of these methods depends on the step size Δt. In this lab we obtain additional insight into the relationship between the error in numerical approximations and the step size by studying some special equations and solutions.

In your report address the following items:

1. Use Euler's method to approximate the solution of the initial-value problem

$$\frac{dy}{dt} = y^2 - 4ty + 4t^2y - 4y + 8t - 3, \quad y(0) = -1.$$

How does the approximation change when the step size is changed? Interpret your results. [*Hint*: Look at the slope field for this equation. Also, look at Section 1.9.]

2. Use Euler's method, improved Euler's method, and Runge-Kutta to approximate the solution of the initial-value problem

$$\frac{dy}{dt} = 2\sqrt{y}, \quad y(0) = 1.$$

Find the formula for the solution and evaluate the error as a function of the step size for each of the methods. Are any of the methods better or worse than you would have expected? [*Hint*: The geometry of the slope field will not be as helpful here, but the special form of the formula of the solution is helpful.]

3. Repeat part 2 for the initial-value problem

$$\frac{dy}{dt} = y + t^3 - 6, \quad y(0) = 0.$$

Your report: In your report, you should address each of the items above. You should *not* include any tables of numerical solutions. You may include a limited number of tables and/or graphs of how the error in your approximate solutions depends on the step size. The goal of this lab is to interpret why the error depends on the step size the way it does.

LAB 7.2 | Lost in Space

The motion of planets and satellites in space is governed by the Newton's laws of motion and gravitational attraction. Newton's law of motion states that the total force is the product of the mass and the acceleration, and Newton's law of gravitation asserts that the force due to gravity between two bodies is inversely proportional to the distance between the bodies. This force is an attracting force, pulling each body directly toward the other. As in the case of the harmonic oscillator in Chapter 2 and the pendulum in

Chapter 5, these laws give rise to second-order differential equations for the positions of the bodies as a function of time.

It is frequently said that the differential equations governing the motion of the moon and the planets were the first differential equations. The ability to compute the position of heavenly bodies (like the moon) was of great financial importance during the time of Newton. Knowing the precise location of the moon and planets at any given time was essential in navigation of the open sea. These equations are still of great theoretical importance in the development of techniques for the study of differential equations. They are also of great practical importance for space navigation. The type of numerical approximation techniques we have discussed are used for computing orbits of satellites to the moon and planets (and back — see the movie *Apollo 13*).

In this lab suppose you are on a mission in deep space, and after a misunderstanding, your roommate leaves you adrift in your space suit with no fuel. Not being completely without a heart, your roommate also leaves behind an escape capsule. However, you and the escape capsule are 100 meters apart and you are both stationary. "Not to worry," you think, "two objects attract each other via the force of gravity, so the escape capsule and I will come together eventually." Suppose your mass (in the space suit) is 150 kg and the mass of the escape capsule is 10^6 kg. Suppose that you must be within 1 meter of the capsule in order to grab it. How long will it take you to reach the capsule?

First we need the differential equations. Since both you and the capsule start at rest and the only force is gravity, which acts along the line connecting you with the capsule, both you and the capsule will remain on that line. Consequently we only need one number to denote the relative positions of you and the capsule. Let $x_1(t)$ be the position of the capsule at time t and $x_2(t)$ denote your position at time t. We assume that coordinates have been chosen so that $x_2 > x_1$. We are most interested in the distance between you and the capsule, so let $x(t) = x_2(t) - x_1(t)$. By Newton's laws we know that the product of your mass and your acceleration is equal to the force of gravity between you and the capsule and that this force is proportional to the product of your masses divided by the square of the distance between you. That is,

$$150 \frac{d^2 x_2}{dt^2} = -G \frac{150 \times 10^6}{|x_2 - x_1|^2},$$

where the minus sign indicates that the force is pulling in the direction of decreasing x_2 and G is the "gravitation constant." This is the same as

$$\frac{d^2 x_2}{dt^2} = -G \frac{10^6}{|x_2 - x_1|^2}.$$

Similarly, the acceleration of the capsule is given by

$$\frac{d^2 x_1}{dt^2} = G \frac{150}{|x_2 - x_1|^2}.$$

Subtracting these equations and using $x = x_2 - x_1$, we obtain the differential equation

$$\frac{d^2 x}{dt^2} = -G \frac{10^6 + 150}{x^2}$$

for $x(t)$. This is a nonlinear, second-order differential equation. The gravitation constant G has been carefully measured by experiment and is given by $G \approx 6.67 \times 10^{-17}$. (The units are kg m^3/sec^2.)

In your report, address the following items:

1. Convert the second-order differential equation

$$\frac{d^2 x}{dt^2} = -G \frac{10^6 + 150}{x^2}$$

into a first-order system.

2. Using numerical techniques, approximate the time necessary for you to come to within 1 meter of the escape capsule, given that you started 100 meters away and both you and the capsule have initial velocity 0. At what velocity are you and the capsule coming together when you are 1 meter apart? (You may use whatever numerical method you like.)

3. Verify that your computation of the time until you are within 1 meter of the escape capsule is accurate to within 1 second.

Your report: In your report you should address these three items. Be sure to explain your methods, particularly for part 3. You may use graphs and/or charts to illustrate your methods, but do not submit long lists of numbers.

8 | DISCRETE DYNAMICAL SYSTEMS

In this chapter we begin the study of a completely different type of model for processes that evolve in time, namely discrete dynamical systems, or difference equations. Unlike differential equations, these models are well suited to situations in which changes occur at specific times, rather than continuously. For example, we often measure population growth at specific times, such as the year-end of a reproductive cycle.

The study of discrete dynamical systems most often involves the process of iteration. To *iterate* means to repeat a procedure numerous times. In discrete dynamics the process that is repeated is the application of a mathematical function. We have encountered several types of iterative processes already. For example, Euler's method for solving a differential equation involves iteration.

As was the case with differential equations, we see that finding exact solutions of discrete systems is rarely possible. As usual we rely on a combination of numerical, analytic, and graphical techniques to understand these systems. Even with these sophisticated techniques, we are unable to describe even one-dimensional, nonlinear iterations completely. The reason is that many discrete nonlinear systems behave in a complex and unpredictable manner. This phenomenon, called chaos by mathematicians, is discussed at length in Sections 8.4 and 8.5.

8.1   THE DISCRETE LOGISTIC EQUATION

The population models we have studied so far all have the property that the rate of change of the population is continuous. This is a reasonable assumption for species that reproduce quickly with respect to the time scale we are considering. However, for some species, all births occur in the spring and most deaths occur during the winter; hence the assumption of continuous population change is not valid. Instead, we should think of time as discrete. If we measure the population once a summer, then we will have an accurate estimate of the population for the entire year.

We begin our study of discrete time systems by recasting several of the population models we discussed in Chapter 1 as discrete dynamical systems.

Exponential Growth Model

The **exponential growth** model is a simple but unrealistic model of population growth. We discuss it first because it illustrates the main ideas behind discrete dynamics. In this model we assume that the population of a certain species in the next generation (or other time period) is directly proportional to the population in the current generation. We let P_n denote the population of the species at the end of the nth time period. The assumption says that, for each time n, the population at the end of the $n + 1$st time step, P_{n+1}, is proportional to the population at the end of the previous time step, P_n. That is,

$$P_{n+1} = kP_n,$$

where k is the proportionality constant that determines the growth rate. This equation is an example of a **discrete dynamical system**, or a **difference equation**. Unlike the differential equation for unlimited population growth,

$$\frac{dP}{dt} = kP,$$

which tells us the rate of change of the population, this difference equation tells us directly the population P_{n+1} in the succeeding generation, provided we know P_n. Thus if we know P_0, the initial population, we can determine the population in each succeeding generation by simply computing the expression kP_n at each stage. There are no integrals to evaluate, no solution curves to approximate. All we have to do is repeatedly calculate the right-hand side of the equation, using the output of the previous calculation as the input for the next. This repetition is what we call **iteration**. Thus if we know the initial population P_0, we then compute

$$P_1 = kP_0$$
$$P_2 = kP_1 = k^2 P_0$$
$$P_3 = kP_2 = k^3 P_0$$
$$\vdots$$
$$P_n = kP_{n-1} = k^n P_0.$$

We see that the population at generation n is determined by the formula $P_n = k^n P_0$. Therefore, if $P_0 > 0$, we conclude the following: If $k > 1$, the population explodes, since

$$\lim_{n \to \infty} k^n = \infty.$$

On the other hand, if $k < 1$, the population dies out, since

$$\lim_{n \to \infty} k^n = 0.$$

Finally, if $k = 1$, the population never changes and $P_n = P_0$.

This model has the same drawbacks as the differential equation version. Although it may work well for small populations in large environments, if the population grows at all, then the model predicts unlimited growth. Thus we modify this model to account for a limited environment.

The Logistic Difference Equation

To make the exponential growth model somewhat more realistic, we add some assumptions that account for overcrowding, just as we did with the logistic differential equation in Chapter 1. The assumptions we make are:

- The population at the end of the next generation is proportional to the population at the end of the current generation when the population is very small.
- If the population is too large, then all resources will be used and the entire population will die out in the next generation and extinction will result.

This last assumption is slightly different from the one we made for the logistic differential equation. Here we assume that there is a maximum population level M that, when reached, results in extinction of the population in the next generation. Thus M is called the **annihilation parameter**. If the population ever reaches M, the species is doomed.

One model that reflects these assumptions is

$$P_{n+1} = k P_n \left(1 - \frac{P_n}{M} \right).$$

As before, P_n denotes the population at the end of generation n and P_0 is the initial population. Note that if P_n is small, the term $(1 - P_n/M)$ is approximately 1. So the difference equation becomes $P_{n+1} \approx k P_n$, which is the exponential growth model. On the other hand, if $P_n \geq M$, then $P_{n+1} \leq 0$; that is, the population is nonpositive. We interpret this to mean that the species is extinct.

Rather than deal with the large numbers that often arise in population models, we will assume that P_n represents the percentage or fraction of this maximum population alive at generation n. That is, we assume that $M = 1$ and that P_n lies between 0 and 1, with $P_n = 0$ (or P_n negative) representing extinction and $P_n = 1$ representing the maximum population level. Thus the model becomes

$$P_{n+1} = k P_n (1 - P_n),$$

which we call the **discrete logistic equation**, or **logistic difference equation**. As before, k is a parameter that depends on the specific species under investigation.

As always we must insert the caveat that this is an extremely naive model for population growth. We have neglected all kinds of other factors that affect P_n, including the effects of predators, cyclical diseases, and the variable nature of the food supply. Nevertheless this model does provide many more scenarios for changes in population than does the logistic differential equation.

Some predictions of the model

As an example of the various possibilities we encounter in the discrete logistic model, we sample the output of this model for a few k-values. Suppose we begin with a population that is exactly half the maximum population allowed, that is, $P_0 = 0.5$. Then depending on k, we find very different results when we compute successive values of P_n. Table 8.1 lists the populations (using only 4 significant digits) when $k = 0.5$, 1.5, 2, 3.2, 3.5, and 3.9. Note that these different k-values yield very different behaviors for the population. When $k = 0.5$, the population tends gradually toward extinction. When $k = 1.5$, the population seems to level out and approach an equilibrium state. When $k = 2$, the population never changes and remains fixed at 0.5. When $k = 3.2$, we see a different result: The population eventually oscillates back and forth between

Table 8.1
Successive populations of the logistic model with $P_0 = 0.5$ and $k = 0.5$, 1.5, 2, 3.2, 3.5, and 3.9

n	$k = 0.5$	$k = 1.5$	$k = 2$	$k = 3.2$	$k = 3.5$	$k = 3.9$
1	0.1250	0.3750	0.5000	0.8000	0.8750	0.9750
2	0.0546	0.3515	0.5000	0.5120	0.3828	0.0950
3	0.0258	0.3419	0.5000	0.7995	0.8269	0.3355
4	0.0125	0.3375	0.5000	0.5128	0.5008	0.8694
5	0.0062	0.3354	0.5000	0.7995	0.8749	0.4426
6	0.0030	0.3343	0.5000	0.5130	0.3828	0.9621
7	0.0015	0.3338	0.5000	0.7995	0.8269	0.1419
8	7.7×10^{-4}	0.3335	0.5000	0.5130	0.5008	0.4750
9	3.8×10^{-4}	0.3334	0.5000	0.7995	0.8749	0.9725
10	1.9×10^{-4}	0.3333	0.5000	0.5130	0.3828	0.1040
11	9.6×10^{-5}	0.3333	0.5000	0.7995	0.8269	0.3634
12	4.8×10^{-5}	0.3333	0.5000	0.5130	0.5008	0.9022
13	2.4×10^{-5}	0.3333	0.5000	0.7995	0.8749	0.3438
14	1.2×10^{-5}	0.3333	0.5000	0.5130	0.3828	0.8799
15	6.0×10^{-6}	0.3333	0.5000	0.7995	0.8269	0.4120
16	3.0×10^{-6}	0.3333	0.5000	0.5130	0.5008	0.9448
17	1.5×10^{-6}	0.3333	0.5000	0.7995	0.8749	0.2033
18	7.5×10^{-7}	0.3333	0.5000	0.5130	0.3828	0.6316
19	3.7×10^{-7}	0.3333	0.5000	0.7995	0.8269	0.9073
20	1.9×10^{-7}	0.3333	0.5000	0.5130	0.5008	0.3278

two different values. The population is high one year, approximately 0.7995, low the next, approximately 0.513, and then repeats cyclically. When $k = 3.5$, we see a similar cyclic behavior, but now the populations eventually repeat every four years instead of two. Finally, when $k = 3.9$, there is no apparent pattern to the successive populations.

Iteration

For the discrete logistic model, finding successive populations is the same as iterating a quadratic function of the form

$$L_k(x) = kx(1 - x).$$

This function depends on the parameter k and is often called the **logistic function**. To iterate this function, we begin with an initial population P_0 and then compute in succession

$$P_1 = L_k(P_0)$$
$$P_2 = L_k(P_1)$$
$$P_3 = L_k(P_2)$$
$$\vdots$$
$$P_n = L_k(P_{n-1})$$

and so forth. The list of numbers $P_0, P_1, P_2, \ldots$ that results from this iteration is called the **orbit** of P_0 under the function L_k. The initial value P_0 is sometimes called the **seed**, or initial condition, for the orbit.

In discrete dynamics the basic goal is to predict the fate of orbits for a given function. That is, the main question is: What happens to the numbers that constitute the orbit as n tends to infinity?

Sometimes predicting the fate of orbits is easy. For example, if $F(x) = x^2$, then we can easily determine what happens to all orbits. For example, the seeds $x_0 = 0$ and $x_0 = 1$ are **fixed points**, since $F(0) = 0$ and $F(1) = 1$. That is, the orbit of 0 is the constant sequence $0, 0, 0, 0, \ldots$ as is the orbit of 1: $1, 1, 1, \ldots$. The orbit of -1 under $F(x) = x^2$ is slightly different: This orbit is **eventually fixed** since $F(-1) = 1$, which is a fixed point. The orbit of -1 is $-1, 1, 1, 1, \ldots$.

For any other seed x_0, there are only two possibilities for the orbit of x_0: Either the orbit tends to the fixed point at 0, or the orbit tends to infinity. For example, if $x_0 = 1/2$, the orbit is

$$x_0 = 1/2$$
$$x_1 = 1/4$$
$$x_2 = 1/16$$
$$x_3 = 1/256$$
$$\vdots$$
$$x_n = 1/2^{2^n},$$

which tends to 0 as n tends to infinity. The fate of the orbit is the same for any other seed x_0 with $|x_0| < 1$.

If $|x_0| > 1$, successive applications of the squaring function $F(x) = x^2$ yield larger and larger results. For example, if $x_0 = 2$, we have

$$x_0 = 2$$
$$x_1 = 4$$
$$x_2 = 16$$
$$x_3 = 256$$
$$\vdots$$
$$x_n = 2^{2^n},$$

and we see that the orbit of x_0 tends to infinity.

Cycles

In a typical discrete dynamical system, there are often many different types of orbits. For example, if $G(x) = x^2 - 1$, then the orbit of 0 lies on a **cycle of period** 2, or a **periodic orbit of period** 2, since

$$x_0 = 0$$
$$x_1 = -1$$
$$x_2 = 0$$
$$x_3 = -1$$
$$\vdots$$

This orbit is the repeating sequence $0, -1, 0, -1, \ldots$.

If we choose the seed $x_0 = \sqrt{2}$, then this orbit is **eventually periodic** since we have

$$x_0 = \sqrt{2}$$
$$x_1 = 1$$
$$x_2 = 0$$
$$x_3 = -1$$
$$x_4 = 0$$
$$x_5 = -1$$
$$\vdots$$

which begins to cycle after the second iteration.

In contrast, if we choose the seed $x_0 = 0.5$, the orbit tends to the cycle of period 2 since

$$x_0 = 0.5$$
$$x_1 = (0.5)^2 - 1 = -0.75$$
$$x_2 = -0.4375$$
$$x_3 = -0.8086\ldots$$
$$\vdots$$
$$x_{20} = 0.00000\ldots$$
$$x_{21} = -1.00000\ldots$$
$$x_{22} = 0.00000\ldots$$
$$\vdots$$

It is important to realize that this orbit never actually reaches the cycle at 0 and -1. Rather, the orbit comes arbitrarily close to these two numbers, but because of round-off error, the calculator or computer eventually displays the numbers 0 and -1 in succession. Thus the orbit is not eventually periodic, it merely tends to the cycle of period 2. This behavior is technically different from that of an orbit that is eventually periodic, although in practice the fate of these orbits is essentially the same.

Discrete dynamical systems may have orbits that cycle with any period. For example, the difference equation

$$x_{n+1} = -\tfrac{3}{2}x_n^2 + \tfrac{5}{2}x_n + 1$$

admits a cycle of period 3, since for the seed $x_0 = 0$ we calculate

$$x_0 = 0$$
$$x_1 = 1$$
$$x_2 = 2$$
$$x_3 = 0.$$

Thus the orbit is the repeating sequence $0, 1, 2, 0, 1, 2, \ldots$. In general the orbit of x_0 lies on a cycle of period n if $x_n = x_0$ and n is the smallest positive integer for which this happens. A fixed point would not be regarded as a cycle of period 10, even though $x_{10} = x_0$, since 10 is not the smallest integer for which the orbit repeats. Similarly, if x_0 has period 5, we also have $x_{10} = x_0$, but we would not call this a period 10 orbit either.

Since the orbit of a fixed point is a constant sequence $x_0, x_0, x_0, \ldots$, fixed points for discrete dynamical systems can be thought of as the analogs of equilibrium points for differential equations. Both represent constant solutions in which the given system is at rest. Similarly, cycles for a discrete system are analogous to periodic solutions of differential equations, as both return to their original position after some time.

Time Series and Histograms

One convenient method to describe orbits geometrically is via a **time series**. In these diagrams we plot the iteration count on the horizontal axis and the numerical values of the orbit on the vertical axis. It often helps to draw straight lines connecting successive points in the time series. In Figure 8.1 we plot the time series corresponding to the orbit of -0.5 under $G(x) = x^2 - 1$. Note that this orbit tends to a cycle of period 2. In Figure 8.2 we plot the time series corresponding to the orbit of 0 under $H(x) = x^2 - 1.3$. Here we see that this orbit tends to a cycle of period 4. These time series are the analogs of the $x(t)$- and $y(t)$-graphs we often plot for differential equations.

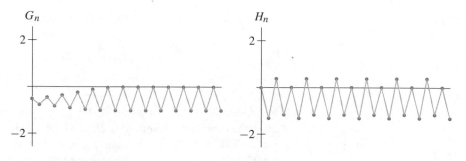

Figure 8.1
Time series for the orbit of -0.5 under $G(x) = x^2 - 1$.

Figure 8.2
Time series for the orbit of 0 under $H(x) = x^2 - 1.3$.

Figure 8.3 displays the time series for the orbit of 0.5 under iteration of the logistic function $L_{3.9}(x) = 3.9x(1 - x)$. This is neither fixed nor periodic behavior; it is difficult to see any sort of coherent pattern. We study this sort of time series in Section 8.5.

Another visual way of displaying orbits is a **histogram**. These images subdivide an interval that contains the points of an orbit into many small subintervals of equal length. Each time the orbit enters one of these subintervals, the histogram is incremented by one unit over that subinterval. Figures 8.4 and 8.5 display the histograms corresponding to the orbit of 0.123 under iteration of the logistic functions $L_{3.83}(x) = 3.83x(1 - x)$ and $L_{3.9}(x) = 3.9x(1 - x)$. In the first case the orbit quickly settles down on a period 3 cycle. In the second case we see a different view of Figure 8.3. The orbit apparently never settles down, instead visiting all of the subintervals, some more than others.

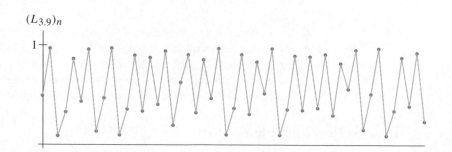

Figure 8.3
Time series for the orbit of 0.5 under $L_{3.9}(x) = 3.9x(1 - x)$.

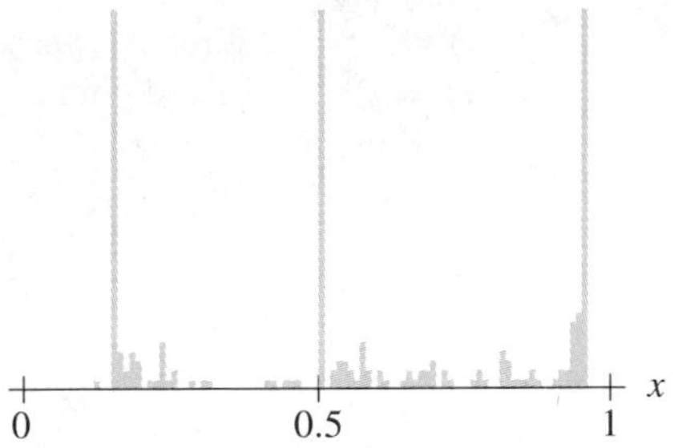

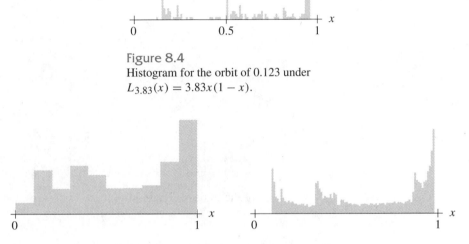

Figure 8.4
Histogram for the orbit of 0.123 under
$L_{3.83}(x) = 3.83x(1 - x)$.

Figure 8.5
Two histograms for the orbit of 0.123 under $L_{3.9}(x) = 3.9x(1 - x)$. In the first figure we subdivide the unit interval into ten subintervals of length 0.1, while in the second figure there are 100 subintervals of length 0.01.

Finding Fixed Points

As we see from the examples above, there are many different types of orbits for a discrete dynamical system. Fixed points are the simplest orbits, and these are often the most important types of orbits in such a system. As in the case of differential equations, we employ three different methods to find these special orbits: analytic, qualitative, and numerical techniques.

The analytic method involves solving equations. Given a function F, to find the fixed points for F, we need only solve the equation $F(x) = x$. For example, if $F(x) = x^2 - 2$, then we find that the fixed points are the solutions of the equation

$$x^2 - 2 = x,$$

which can be written as

$$x^2 - x - 2 = 0.$$

Factoring turns this into

$$(x + 1)(x - 2) = 0,$$

and the fixed points for F are -1 and 2.

Geometrically we can view fixed points by superimposing the graph of the diagonal line $y = x$ on the graph of F. Note that the diagonal $y = x$ meets the graph of

$F(x) = x^2 - 2$ directly over -1 and 2, as shown in Figure 8.6. As another example, $K(x) = x^3$ has 3 fixed points, which occur at the roots of $x^3 - x = 0$, or at 0, -1, and $+1$ (see Figure 8.7).

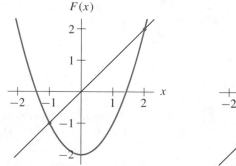

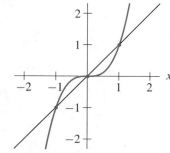

Figure 8.6
The fixed points of
$F(x) = x^2 - 2$ are $x = -1, 2$.

Figure 8.7
The fixed points of $K(x) = x^3$
are $x = 0, -1, +1$.

For the logistic equation $L_k(x) = kx(1 - x)$, we find the fixed points by solving

$$kx(1 - x) = x,$$

which yields solutions $x = 0$ and $x = (k - 1)/k$. Recall that for our population model we assume that $0 \le x \le 1$, so this second fixed point is positive only if $k > 1$. (When $k < 1$, this fixed point is negative and so is not biologically interesting.) Also, $(k - 1)/k < 1$ for all $k > 1$. The fixed point at $x = 0$ represents population extinction, and the nonzero fixed point represents a population that never changes from generation to generation. Figure 8.8 shows the graph for $k = 4$, with fixed points at 0 and 0.75.

Given a function F, finding the fixed points by solving $F(x) = x$ analytically may be difficult or impossible. If we need a more accurate value for the fixed point than we can get from graphical methods, then we can turn to numerical methods. We can use Newton's method to find values of x where $F(x) - x = 0$. For example, from Figure 8.9 we see that $C(x) = \cos x$ has a single fixed point, which we can determine numerically to be given by $0.73908\ldots$ (see the exercises).

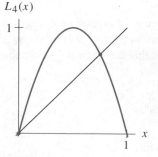

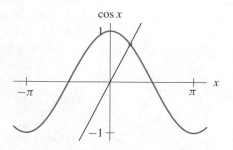

Figure 8.8
The fixed points for the logistic
model $L_4(x) = 4x(1 - x)$ are
$x = 0, 3/4$.

Figure 8.9
The fixed point for $\cos x$ occurs at
$x = 0.73908\ldots$.

Notation for Iteration

To simplify notation, we introduce the expression F^n to denote the nth iterate of F. For example, if $F(x) = x^4$, then

$$F^2(x) = F(F(x)) = F(x^4) = (x^4)^4 = x^{16},$$

and similarly $F^3(x) = x^{64}$. It is important to realize that $F^n(x)$ *does not* mean $F(x)$ raised to the nth power; rather, $F^n(x)$ means to first iterate F exactly n times, then evaluate this new function at x.

Finding Cycles

To find cycles for a discrete dynamical system, we proceed in essentially the same manner as for fixed points. For example, if $H(x) = -x^3$, the only fixed point of H is 0, since the equation for the fixed points is $-x^3 = x$ or $x(1 + x^2) = 0$. For the cycles of period 2, we must compute

$$H(H(x)) = H^2(x) = -(-x^3)^3 = x^9$$

and then solve $x^9 = x$. The solutions here are the fixed point $x = 0$ and two new solutions, $x = 1$ and $x = -1$. These latter two points form a cycle of period 2, since $H(1) = -1$ and $H(-1) = 1$.

Finding cycles is usually more difficult than finding fixed points. For example, to find cycles of period 2 for $F(x) = x^2 - 2$, we must find values of x for which $F(F(x)) = x$. We have

$$F^2(x) = (x^2 - 2)^2 - 2 = x^4 - 4x^2 + 2,$$

so that we must solve

$$x^4 - 4x^2 + 2 = x.$$

That is, we must find the roots of the fourth-degree equation

$$x^4 - 4x^2 - x + 2 = 0.$$

Luckily we know two solutions of this equation already, since we saw above that -1 and 2 are fixed points, and fixed points have orbits that repeat every two iterations as well as every iteration. So we can divide this expression by $(x + 1)(x - 2)$ to find

$$\frac{x^4 - 4x^2 - x + 2}{(x + 1)(x - 2)} = \frac{x^4 - 4x^2 - x + 2}{x^2 - x - 2} = x^2 + x - 1 = 0.$$

This quadratic equation has roots $(-1 \pm \sqrt{5})/2$. These points lie on a cycle of period 2 since we have

$$\left(\frac{-1 + \sqrt{5}}{2}\right)^2 - 2 = \frac{-1 - \sqrt{5}}{2}$$

and

$$\left(\frac{-1 - \sqrt{5}}{2}\right)^2 - 2 = \frac{-1 + \sqrt{5}}{2}.$$

In general, to find cycles of period n, we must first iterate the function F a total of n times. For example, if $L_3(x) = 3x(1 - x)$, then we compute

$$L_3^2(x) = 3[3x(1 - x)][1 - 3x(1 - x)] = -27x^4 + 54x^3 - 36x^2 + 9x$$

and

$$L_3^3(x) = L_3(L_3^2(x))$$
$$= 3\left[-27x^4 + 54x^3 - 36x^2 + 9x\right]\left[1 - (-27x^4 + 54x^3 - 36x^2 + 9x)\right],$$

which we will not bother to simplify. You should note that, when multiplied out, $L_3^3(x)$ has terms that involve x^8, so L_3^3 is an eighth-degree polynomial. Finding the solutions of $L_3^3(x) = x$ is therefore not a pleasant task.

Finding cycles geometrically

In general, finding cycles of period n involves solving the equation $F^n(x) = x$. Even for logistic functions of the form $L_k(x) = kx(1 - x)$, $L_k^n(x)$ is a polynomial of degree 2^n, and so we have little chance of finding explicit solutions. However, geometric information is relatively easy to come by. We can usually discover some information about the number of cycles of period n by sketching the graph of F^n and looking for places where this graph crosses the diagonal line $y = x$.

For example, if $H(x) = -x^3$, the graph of H shows that H has a single fixed point that occurs at 0. But $H^2(x) = x^9$ has a graph that meets the diagonal three times, at the fixed point and at a cycle of period 2, which as we saw above is given by ± 1 (see Figures 8.10 and 8.11).

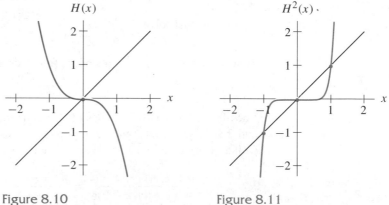

Figure 8.10
The graph of $H(x) = -x^3$.

Figure 8.11
The graph of $H^2(x) = x^9$.

In Figures 8.12–8.14 we sketch the graph of $F(x) = x^2 - 2$ as well as F^2 and F^3. Note that we see two fixed points for F and four fixed points for F^2. Two of these are the fixed points for F, and the other two lie on a cycle of period 2. For F^3 there are eight points where the graph of $y = F^3(x)$ crosses $y = x$. Two of these points are again the fixed points of F, but the other six must be periodic with period 3. The cycle of period 2 does not appear in the graph of F^3 since these orbits do not repeat after 3 iterations.

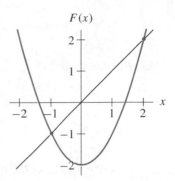

Figure 8.12
The graph of $F(x) = x^2 - 2$.

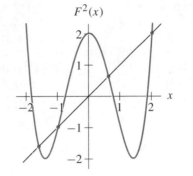

Figure 8.13
The graph of $F^2(x)$.

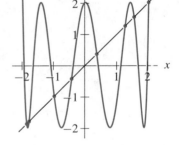

Figure 8.14
The graph of $F^3(x)$.

EXERCISES FOR SECTION 8.1

In Exercises 1–8, compute the orbit of $x_0 = 0$ for each of the given difference equations. Determine whether this orbit is fixed, cycles with some period, is eventually periodic, tends to infinity, or is none of the above. (Use a calculator as needed.)

1. $x_{n+1} = x_n^2 - 2$ **2.** $x_{n+1} = \sin(x_n)$ **3.** $x_{n+1} = e^{x_n}$

4. $x_{n+1} = x_n^2 + 1$ **5.** $x_{n+1} = -x_n^2 + x_n + 2$ **6.** $x_{n+1} = \cos(x_n)$

7. $x_{n+1} = 4x_n(1 - x_n) + 1$ **8.** $x_{n+1} = -\frac{1}{2}x_n^2 + 1$

In Exercises 9–21, find all fixed points and periodic points of period 2 for each of the given functions. If you cannot determine these values explicitly, use the graph of F or F^2 to determine how many fixed points and periodic points of period 2 F has.

9. $F(x) = -x + 2$ **10.** $F(x) = x^4$

11. $F(x) = x^2 + 1$ **12.** $F(x) = x^2 - 3$

13. $F(x) = \sin x$ **14.** $F(x) = 1/x$

15. $F(x) = -2x - x^2$ **16.** $F(x) = e^x$

17. $F(x) = -e^x$ **18.** $F(x) = x^3$

19. $F(x) = -x$ **20.** $F(x) = -2x + 1$

21. $F(x) = 2$

22. Describe the fate of the orbit of any seed under iteration of $F(x) = x^3$.

23. Describe the fate of the orbit of any seed under iteration of $F(x) = -x + 4$.

In Exercises 24–35, describe the fate of the orbit of each of the following seeds under iteration of the function

$$T(x) = \begin{cases} 2x, & \text{if } x < 1/2; \\ 2 - 2x, & \text{if } x \geq 1/2. \end{cases}$$

24. $2/3$ **25.** $1/6$ **26.** $2/5$ **27.** $2/7$

28. $3/14$ **29.** $1/8$ **30.** $1/9$ **31.** $6/11$

32. $2/9$ **33.** 0 **34.** $1/4$ **35.** $1/2$

36. Discuss the fate of the orbit of 0 under iteration of $F(x) = x^2 + c$ for each of the following c values: $c = 0.4, 0.3, 0.2, 0.1$. Use the graph and technology as necessary to explain your findings. Is there a change in the fate of the orbit of 0 as c changes? Explain in a sentence or two.

37. Consider the family of functions $F_c(x) = x^2 + c$ that depends on a parameter c. For which values of c does this function have real fixed points? How many?

38. Discuss the fate of the orbit of 0 under iteration of $F(x) = x^2 + c$ for each of the following c values: $c = -0.6, -0.7, -0.8, -0.9$. Use the graph of F and F^2 and technology as necessary to explain your findings. Is there a change in the fate of the orbit of 0 as c changes? Explain in a sentence or two.

39. Consider the family of functions $F_c(x) = x^2 + c$ that depends on a parameter c. For which values of c does this function have periodic points of period 2? How many?

8.2 FIXED POINTS AND PERIODIC POINTS

As was the case with equilibrium points for differential equations, fixed points and cycles for difference equations come in several distinct varieties, depending on what happens to nearby orbits. In this section we discuss analytic and qualitative methods for distinguishing different types of fixed points and cycles.

Graphical Iteration

As we saw in the previous section, the graph of a function and the diagonal line $y = x$ give us a geometric method for finding fixed points and cycles. In this section we see that the graph of F together with the diagonal line $y = x$ also provides a convenient way to sketch other orbits geometrically.

Given the function F and a seed x_0, we view the orbit of x_0 as follows: We begin at the point (x_0, x_0) on the diagonal and draw a vertical line from (x_0, x_0) to the graph of F; we reach the graph at (x_0, x_1), where $x_1 = F(x_0)$. Now we draw a horizontal line from this point back to the diagonal. Along this line, the y-values remain constant at $y = x_1$, so we reach the diagonal at (x_1, x_1). We interpret this point as the second point on the orbit.

Now we continue in the same fashion. First we draw a vertical line from (x_1, x_1) to the graph of F, reaching the graph at (x_1, x_2), where $x_2 = F(x_1)$. Then we draw a horizontal line back to the diagonal, reaching this line at (x_2, x_2). This is the second point on the orbit.

Continuing in this manner, the orbit of x_0 is displayed as a sequence of points along the diagonal (see Figure 8.15). The resulting picture, which sometimes resembles a staircase and sometimes a cobweb, is often called a **web diagram**. The process of drawing vertical lines to the graph and then horizontal lines to the diagonal is called **graphical analysis**, or **graphical iteration**. We sometimes add arrows to the web diagram to indicate the direction of the iteration.

Figure 8.16 depicts the web diagram for $G(x) = x^2 - 0.7$ using the seed $x_0 = 1.4$. We clearly see two fixed points for G, given by the points where the graph of G crosses the diagonal. Note that the orbit of $x_0 = 1.4$ tends away from one of these fixed points and toward the other fixed point.

Another important type of orbit for discrete systems is the cycle, or periodic orbit. Recall that for a function F, the seed x_0 lies on a cycle or periodic orbit of period n if

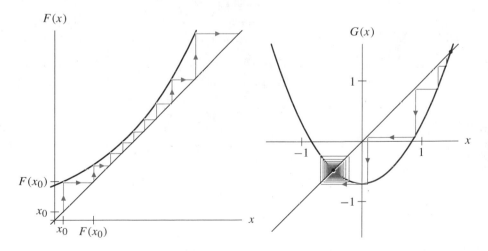

Figure 8.15
Graphical analysis showing the orbit of x_0.

Figure 8.16
Web diagram of $G(x) = x^2 - 0.7$ using the seed $x_0 = 1.4$.

$F^n(x_0) = x_0$. (Remember that F^n indicates the nth iterate of F, not the nth power.) Equivalently, x_0 lies on a cycle of period n if the orbit of x_0 is a repeating sequence of length n of the form

$$x_0, x_1, \ldots, x_{n-1}, x_0, x_1, \ldots, x_{n-1}, \ldots .$$

Graphical iteration allows us to visualize cycles without having to compute the graph of F^n. For example, 0 and -1 lie on a cycle of period 2 for $F(x) = x^2 - 1$, since $F(0) = -1$ and $F(-1) = 0$. In Figure 8.17 we see that this cycle is represented geometrically by a square in the web diagram, and we also see an orbit that tends to this cycle.

As we saw in the previous section, the logistic difference equation admits cycles of many periods. Figures 8.18 and 8.19 display some of these cycles using graphical iteration.

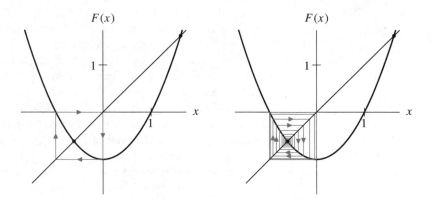

Figure 8.17
A 2-cycle for $F(x) = x^2 - 1$ and an orbit tending to the 2-cycle.

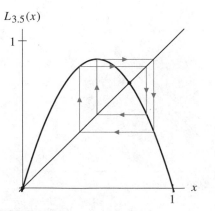

Figure 8.18
A 4-cycle for $L_{3.5}(x) = 3.5x(1 - x)$.

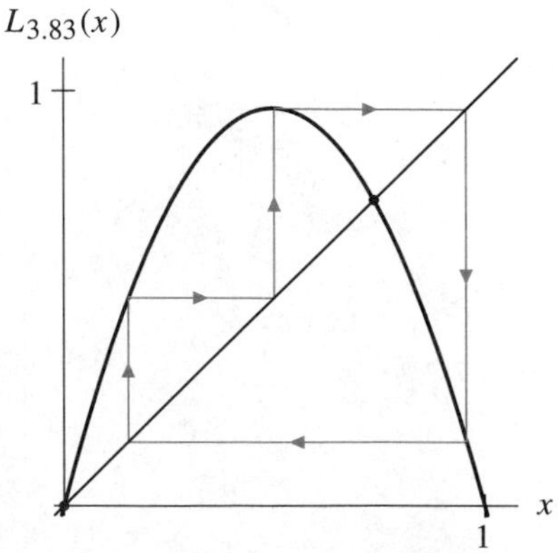

Figure 8.19
A 3-cycle for $L_{3.83}(x) = 3.83x(1 - x)$.

Attracting and Repelling Fixed Points

Figures 8.20 and 8.21 display the graphical analysis of two additional logistic functions, $L_{2.8}(x) = 2.8x(1-x)$ and $L_{3.2}(x) = 3.2x(1-x)$. In the first case observe that the displayed orbit tends to a fixed point, but when the parameter is changed to 3.2, this is no longer the case. The displayed orbit comes close to the fixed point but then tends toward a 2-cycle. This is an example of a bifurcation, a topic we discuss later. For now note that there is a dramatic difference between the behavior of orbits near the nonzero fixed point in these two cases. In the first case the orbit is attracted to the nonzero fixed point, whereas in the second the orbit seems to be repelled away. Just as with equilibrium points for a differential equation, there are different types of fixed points and cycles.

DEFINITION   A fixed point x_0 of a function F is called **attracting** if there is an interval around x_0 having the property that every seed in this interval has an orbit that remains in the interval and tends to x_0 under iteration of F. The fixed point is called **repelling** if there is an interval around x_0 having the property that every seed in this interval (except x_0) has an orbit that leaves the interval under iteration of F. A fixed point that is neither attracting nor repelling is called **neutral**.   ■

Attracting fixed points are the analogs of sinks for one-dimensional differential equations, whereas repelling fixed points correspond to sources.

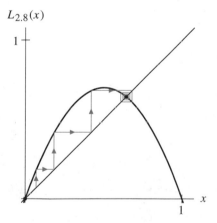

Figure 8.20
An attracting fixed point for
$L_{2.8}(x) = 2.8x(1-x)$.

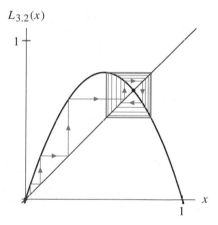

Figure 8.21
A repelling fixed point for
$L_{3.2}(x) = 3.2x(1-x)$.

Fixed points of linear functions

To see what makes certain fixed points attracting and others repelling, we first consider linear functions. For $F(x) = mx$, $x_0 = 0$ is always a fixed point. Graphical analysis shows that as long as $|m| < 1$, x_0 attracts all orbits (see Figure 8.22). Note that if $-1 < m < 0$, orbits hop back and forth about 0 as they tend to the fixed point.

When $|m| > 1$, the situation is quite different. Now all orbits tend away from 0; the origin is a repelling fixed point (see Figure 8.23). Again orbits hop from side to side as they leave the origin when $m < -1$.

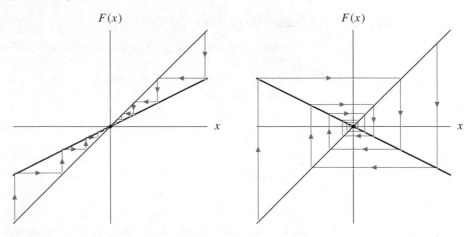

Figure 8.22
The function $F(x) = mx$ has an attracting fixed point at 0 if $|m| < 1$.

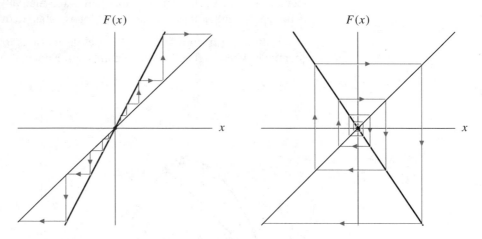

Figure 8.23
The function $F(x) = mx$ has a repelling fixed point at 0 if $|m| > 1$.

The intermediate cases $m = \pm 1$ are special. When $m = 1$, all points are fixed, whereas when $m = -1$, all nonzero points are periodic with period 2.

Classification of Fixed Points

When F is a nonlinear function, a similar story holds true, at least near fixed points. If F has a fixed point at x_0 and $|F'(x_0)| < 1$, then the graph of F is tangent to a straight line whose slope is less than 1 in absolute value. As in the case of linear functions, graphical analysis shows that nearby orbits must be attracted to x_0. Thus a fixed point x_0 for F for which $|F'(x_0)| < 1$ is attracting (see Figure 8.24). Note that this result is true only close to x_0; far away, orbits may behave very differently.

Similarly, if $|F'(x_0)| > 1$, nearby orbits are repelled away from x_0 just as in the linear case. Therefore if $|F'(x_0)| > 1$ at the fixed point, then x_0 is a repelling fixed point (see Figure 8.25).

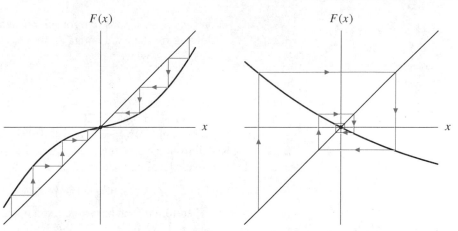

Figure 8.24
If $|F'(x_0)| < 1$, then x_0 is an attracting fixed point.

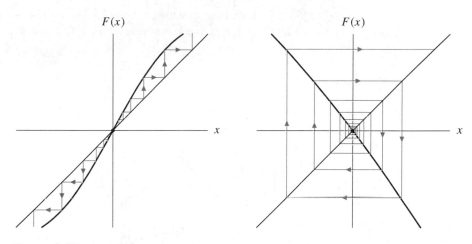

Figure 8.25
If $|F'(x_0)| > 1$, then x_0 is a repelling fixed point.

Fixed points for the logistic function

Consider the logistic function $L_{2.8}(x) = 2.8x(1 - x)$. The fixed points for $L_{2.8}$ are given by solving

$$2.8x(1 - x) = x.$$

Some algebra shows that these fixed points are $x = 0$ and $x = 1.8/2.8 \approx 0.64$. Now

$$L'_{2.8}(x) = 2.8(1 - 2x),$$

so $L'_{2.8}(0) = 2.8$ and $L'_{2.8}(0.64\ldots) = 2.8(-0.28\ldots) \approx -0.78$. Thus 0 is a repelling fixed point and $0.64\ldots$ is an attracting fixed point. We saw this qualitatively in Figure 8.20.

As a second example, suppose we consider instead the logistic function $L_{3.2}(x) = 3.2x(1 - x)$. Solving for the fixed points as above, we find fixed points at $x = 0$ and $x = 2.2/3.2 = 0.6875$. Now we have

$$L_{3.2}'(x) = 3.2(1 - 2x).$$

Thus $L_{3.2}'(0) = 3.2$ and $L_{3.2}'(0.6875) = -1.2$. In this case both fixed points are repelling. We saw this also in Figure 8.20.

These two examples are special cases of the general logistic equation $L_k(x) = kx(1 - x)$. Recall from the previous section that the fixed points of F occurred at 0 and at $(k - 1)/k$, provided $k > 1$. We compute

$$L_k'(x) = k - 2kx.$$

So $L_k'(0) = k$. Therefore 0 is an attracting fixed point when $0 \le k < 1$ and a repelling fixed point when $k > 1$. Also,

$$L_k'\left(\frac{k - 1}{k}\right) = k - 2(k - 1) = -k + 2.$$

So $(k - 1)/k$ is an attracting fixed point when $-1 < -k + 2 < 1$, that is, for $1 < k < 3$. When $k > 3$, this fixed point is repelling (see Figures 8.26 and 8.27).

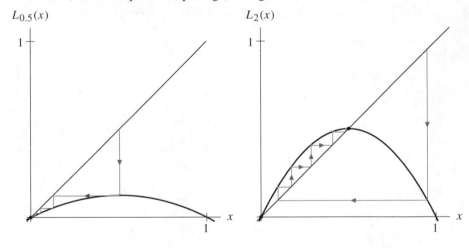

Figure 8.26
Fixed points for $L_k(x) = kx(1 - x)$ when $k = 0.5$.

Figure 8.27
Fixed points for $L_k(x) = kx(1 - x)$ when $k = 2$.

Examples of neutral fixed points

Neutral fixed points can occur only if $F'(x_0) = \pm 1$. Orbits near neutral fixed points may behave in a variety of ways. For example, consider $F(x) = -x + 4$. We have $F(2) = 2$ and $F'(2) = -1$. All other seeds have orbits that lie on cycles of period 2, since $F^2(x) = -(-x + 4) + 4 = x$.

As another example, $G(x) = x + x^2$ has a fixed point at $x = 0$. Note that $G'(0) = 1$. The graph of G shows that 0 attracts from the left but repels from the right. Thus 0 is a neutral fixed point (see Figure 8.28). If we consider instead $H(x) = x + x^3$, then again 0 is a fixed point and $H'(0) = 1$. This time, however, $x = 0$ is a repelling fixed point, as we see from graphical iteration (see Figure 8.29).

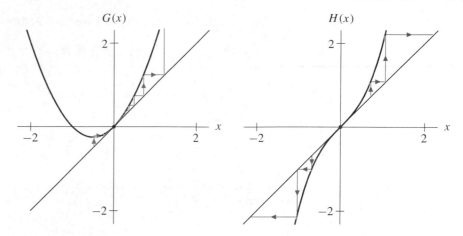

Figure 8.28

$G(x) = x + x^2$ has a neutral fixed point at $x = 0$ and $G'(0) = 1$.

Figure 8.29

$H(x) = x + x^3$ has a repelling fixed point at $x = 0$, despite the fact that $H'(0) = 1$.

Classification of Periodic Points

Periodic points can also be classified as attracting, repelling, and neutral. If x_0 lies on an n-cycle of a given function F, then the graph of F^n meets the diagonal at (x_0, x_0). That is, F^n has a fixed point at x_0. So it is natural to call the cycle attracting, repelling, or neutral depending on whether the fixed point for F^n has this property.

For example, the function $F(x) = x^2 - 1$ has a 2-cycle at 0 and -1. Since $F^2(x) = (x^2 - 1)^2 - 1 = x^4 - 2x^2$, we have

$$(F^2)'(x) = 4x^3 - 4x,$$

Therefore $(F^2)'(0) = 0$, so 0 lies on an attracting 2-cycle, as we have seen in Figure 8.17. Note also that $(F^2)'(-1) = 0$.

The fact that $(F^2)'(x_0)$ is the same at both points on the cycle in this example is no accident. The Chain Rule tells us why. Suppose x_0 and x_1 lie on a 2-cycle for F. So $F(x_0) = x_1$ and $F(x_1) = x_0$. Then we have

$$(F^2)'(x_0) = F'(F(x_0)) \cdot F'(x_0)$$
$$= F'(x_1) \cdot F'(x_0).$$

That is, the derivative of F^2 at x_0 is just the product of the derivatives along the orbit of x_0. The same is true for x_1. More generally, if $x_0 \ldots, x_{n-1}$ lies on a cycle of period n for F, then

$$(F^n)'(x_0) = F'(F^{n-1}(x_0)) \cdot F'(F^{n-2}(x_0)) \cdots F'(x_0)$$
$$= F'(x_{n-1}) \cdot F'(x_{n-2}) \cdots F'(x_0),$$

which is again the product of the derivatives along the cycle.

As a check, for $F(x) = x^2 - 1$ we have $F'(0) = 0$ and $F'(-1) = -2$. Consequently,

$$(F^2)'(0) = F'(0) \cdot F'(-1) = 0(-2) = 0$$

as before.

As a final example, consider the function

$$T(x) = \begin{cases} 2x, & \text{if } x < 1/2; \\ 2 - 2x, & \text{if } x \geq 1/2. \end{cases}$$

The seed $x_0 = 2/7$ leads to a cycle of period 3 for T with orbit $2/7, 4/7, 6/7, 2/7 \ldots$. We compute

$$T'(2/7) = 2$$
$$T'(4/7) = -2$$
$$T'(6/7) = -2.$$

Therefore we have

$$(T^3)'(2/7) = T'(2/7) \cdot T'(4/7) \cdot T'(6/7) = 8,$$

and so this cycle is repelling. Note that any cycle for this function is repelling, since the derivative will always be a product of 2's and -2's .

EXERCISES FOR SECTION 8.2

In Exercises 1–11, for each of the given functions, find all fixed points and determine whether they are attracting, repelling, or neutral.

1. $F(x) = x^2 - 2x$ **2.** $F(x) = x^5$ **3.** $F(x) = \sin x$

4. $F(x) = x^3 - x$ **5.** $F(x) = \arctan x$ **6.** $F(x) = 3x(1 - x)$

7. $F(x) = (\pi/2) \sin x$ **8.** $F(x) = x^2 - 3$ **9.** $F(x) = 1/x$

10. $F(x) = 1/x^2$ **11.** $F(x) = e^x$

In Exercises 12–17, for each of the given functions, 0 lies on a periodic orbit. First determine the period of this orbit; then decide if this cycle is attracting, repelling, or neutral.

12. $F(x) = -x^5 + 1$ **13.** $F(x) = 1 - x^2$

14. $F(x) = (\pi/2) \cos x$ **15.** $F(x) = \begin{cases} x + 1, & \text{if } x < 3.5; \\ 2x - 8, & \text{if } x \geq 3.5. \end{cases}$

16. $F(x) = 1 - x^3$ **17.** $F(x) = |x - 2| - 1$

In Exercises 18–25, each of the given functions has at least one fixed point with derivative ± 1. First find these fixed points. Then, using graphical analysis with an accurate graph or other means, determine if this fixed point is attracting, repelling, or neutral.

18. $F(x) = x - x^2$

19. $F(x) = 1/x$

20. $F(x) = \sin x$

21. $F(x) = \tan x$

22. $F(x) = -x + x^3$

23. $F(x) = -x - x^3$

24. $F(x) = e^{x-1}$ (fixed point is 1)

25. $F(x) = -e \cdot e^x$ (fixed point is -1)

26. Find all fixed points for $F_c(x) = x^2 + c$ for all values of c. Determine for which values of c each fixed point is attracting, repelling, and neutral.

27. How many fixed points does $F(x) = \tan x$ have? Are they attracting, repelling, or neutral? Why?

28. What can you say about fixed points for $F_c(x) = ce^x$ with $c > 0$? What does the graph of F_c tell you about these fixed points? Note that when $c = 1/e$, $F_c(1) = 1$.

29. Consider the function

$$
T(x) = \begin{cases} 4x, & x < 1/2; \\ 4 - 4x, & x \geq 1/2. \end{cases}
$$

Does T have any attracting cycles? Why or why not?

30. Recall from calculus that Newton's method is an iterative procedure to find the roots of a given function P. Indeed, the Newton iteration is just

$$
x_{n+1} = x_n - \frac{P(x_n)}{P'(x_n)}.
$$

(a) Show that a root x of P for which $P'(x) \neq 0$ is a fixed point for the Newton iteration.

(b) Suppose $P(x) = x^3 - x$. Determine whether the fixed points for the corresponding Newton iteration are attracting or repelling.

8.3 BIFURCATIONS

Just as with differential equations, discrete dynamical systems may undergo changes in their orbit structure as parameters vary. These changes may include the birth or death of fixed points and cycles or changes in the type of these orbits. As before, such changes are known as bifurcations.

Tangent Bifurcation

One of the simplest types of bifurcation occurs when fixed or periodic points suddenly appear or disappear. As an example, consider the family of quadratic functions $F_c(x) = x^2 + c$. These functions have fixed points when

$$
x^2 + c = x.
$$

Using the quadratic formula, we see that there are two fixed points for F_c given by

$$p_\pm = \frac{1 \pm \sqrt{1 - 4c}}{2}.$$

These fixed points are real if $1 - 4c \geq 0$, so F_c has fixed points only if $c \leq 1/4$. If $c > 1/4$, F_c has no real fixed points. If $c = 1/4$, the two fixed points are the same, since $p_+ = p_- = 1/2$. For $c < 1/4$, this fixed point splits apart into distinct fixed points p_+ and p_-. Thus a bifurcation occurs at $c = 1/4$ since a pair of fixed points is born as c decreases through $1/4$.

Qualitatively, the graph of F_c lies above the diagonal $y = x$ when $c > 1/4$, so F_c has no fixed points. When $c = 1/4$, the graph is tangent to the diagonal at one point. As we see in Figure 8.30, this is a neutral fixed point. When $c < 1/4$, the graph meets the diagonal at two distinct fixed points $p_\pm$.

To determine the character of these fixed points, we compute $F_c'(x) = 2x$, so that

$$F_c'(p_+) = 1 + \sqrt{1 - 4c}.$$

Now $\sqrt{1 - 4c} > 0$ when $c < 1/4$, so it follows that $F_c'(p_+) > 1$ for these c values. Therefore p_+ is a repelling fixed point for all $c < 1/4$. For p_-, the situation is more complicated. We have

$$F_c'(p_-) = 1 - \sqrt{1 - 4c},$$

so $F_c'(p_-) < 1$ for all values of $c < 1/4$. To guarantee that p_- is attracting, we must also have $F_c'(p_-) > -1$. This happens provided

$$-1 < 1 - \sqrt{1 - 4c} = F_c'(p_-) \, .$$
$$2 > \sqrt{1 - 4c}$$
$$-3/4 < c.$$

Therefore the fixed point p_- is attracting for $-3/4 < c < 1/4$ and repelling for $c < -3/4$.

Graphical analysis provides an alternative view of this bifurcation. Figure 8.30 shows some orbits before, at, and after the bifurcation when $c = 0.5, 0.25$, and 0.1. Of

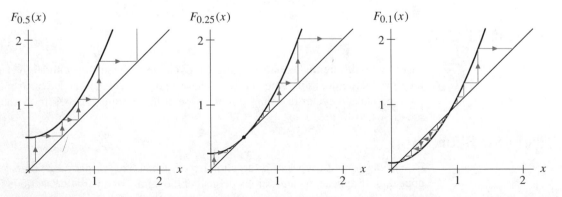

Figure 8.30
Graphical analysis of $F_c(x) = x^2 + c$ for $c = 0.5, 0.25$, and 0.1.

course, there really is no "before" and "after" the bifurcation. We could just as easily view this change as c increases, in which case this bifurcation consists of two fixed points that coalesce when $c = 1/4$ and then disappear. Note that this bifurcation is quite similar to some of the bifurcations of equilibrium points that we encountered for differential equations.

As with differential equations, a picture is a useful way to display the changes that occur in a bifurcation. Figure 8.31 shows the bifurcation diagram corresponding to the bifurcation above. In this picture the fixed points $p_\pm$ are plotted as functions of c. The parameter c is plotted on the horizontal axis, and the x-values of the fixed points are plotted on the vertical axis. As c decreases through $1/4$, we see the bifurcation: A neutral fixed point is born at the point $c = 1/4$, $x = 1/2$. This point immediately splits apart into a pair of fixed points, one attracting and one repelling. Any bifurcation of fixed points or cycles that proceeds in this manner is called a **tangent bifurcation** or **saddle node bifurcation**.

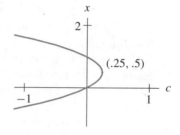

Figure 8.31
The tangent bifurcation diagram.

Pitchfork Bifurcation

The fixed point structure of a discrete dynamical system may change in other ways as well. For example, consider the family of functions $G_\alpha(x) = x^3 - \alpha x$ depending on the parameter α. The fixed points of G_α are the solutions of $x^3 - \alpha x = x$, which are $x = 0, \pm\sqrt{1 + \alpha}$. For $\alpha \leq -1$, G_α has only one fixed point, at $x = 0$, but when $\alpha > -1$, there are three fixed points; hence a bifurcation occurs at $\alpha = -1$. Note that $G'_\alpha(0) = -\alpha$, so the fixed point at 0 is attracting if $|\alpha| < 1$ and repelling if $|\alpha| > 1$. Also, $G'_\alpha(\pm\sqrt{1 + \alpha}) = 3 + 2\alpha$. Since $\alpha > -1$, it follows that both of these fixed points are repelling, and the bifurcation is as shown in Figure 8.32. Because of the shape of this diagram, this bifurcation is called a **pitchfork bifurcation**.

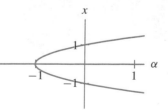

Figure 8.32
The pitchfork bifurcation diagram.

Note that two things happen at the bifurcation point. As α increases through -1, the fixed point at $x = 0$ suddenly changes from repelling to attracting. Meanwhile, a pair of new repelling fixed points is born. It is also possible for a fixed point to change from attracting to repelling and simultaneously give birth to a pair of new attracting fixed points. This type of bifurcation is also called a pitchfork bifurcation.

Period-Doubling Bifurcation

One of the most important types of bifurcations for discrete dynamical systems is the *period-doubling bifurcation*. This bifurcation produces a new cycle having twice the period of the original cycle. For example, consider again the family $F_c(x) = x^2 + c$. Recall that this function has two fixed points given by

$$p_\pm = \frac{1 \pm \sqrt{1 - 4c}}{2}.$$

The fixed point p_- is attracting if $-3/4 < c < 1/4$ and repelling if $c < -3/4$. The fact that $F_c'(p_-) = -1$ at $c = -3/4$ is our signal that a bifurcation may occur. Graphical analysis gives us a hint about what happens. In Figure 8.33, we display the web diagrams for F_c for the nearby c-values $c = -0.6$ and $c = -0.9$. Note that orbits tend to the attracting fixed point when $c = -0.6$, but they tend to an attracting cycle of period 2 when $c = -0.9$.

To find cycles of period 2 for F_c, we must solve the equation $F_c^2(x) = x$. Writing this equation out, we find

$$(x^2 + c)^2 + c = x,$$

or in simplified form,

$$x^4 + 2cx^2 - x + c^2 + c = 0.$$

This equation looks tough to solve. However, recall that any fixed point for F_c must also solve this equation, for if $F_c(x_0) = x_0$, then certainly $F_c^2(x_0) = x_0$ as well. But we know the fixed points already; they are $p_\pm$. Hence $(x - p_+)(x - p_-) = x^2 - x + c$ is a factor of $F^2(x) - x$.

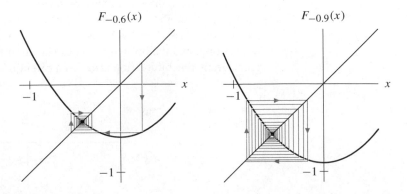

Figure 8.33
Graphical analysis of $F_c(x) = x^2 + c$ when $c = -0.6$ and $c = -0.9$.

After some long division, we find

$$\frac{x^4 + 2cx^2 - x + c^2 + c}{x^2 - x + c} = x^2 + x + c + 1.$$

Therefore to find the cycles of period 2, we need only solve the quadratic equation

$$x^2 + x + c + 1 = 0.$$

The roots of this equation are

$$q_\pm = \frac{-1 \pm \sqrt{-3 - 4c}}{2},$$

as given by the quadratic formula.

Note that $q_\pm$ are real numbers only if $c < -3/4$. Thus a cycle of period 2 appears precisely when c decreases through $c = -3/4$. We leave it to the reader to check that this cycle is attracting for c in the interval $-5/4 < c < -3/4$ and repelling for $c < -5/4$ (see the exercises). Thus we may augment the bifurcation diagram in Figure 8.31 as in Figure 8.34.

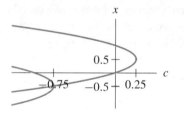

Figure 8.34
The bifurcation diagram for $F_c(x) = x^2 + c$ showing fixed points and 2-cycles.

Any bifurcation in which a given cycle changes from attracting to repelling (or vice versa) with derivative passing through -1 and that is accompanied by the birth of a new cycle having twice the original period is called a **period-doubling bifurcation**.

A qualitative view of this period-doubling bifurcation is provided by the graphs of the second iterate of F_c, namely $F_c^2(x) = x^4 + 2cx^2 + c^2 + c$. Figure 8.35 presents these graphs for $c = -0.6$, $c = -0.75$ and, $c = -0.9$. Note how the graph of F_c^2 twists through the diagonal at $c = -0.75$ to produce a pair of new fixed points for F_c^2 when $c < -0.75$.

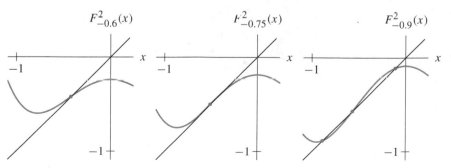

Figure 8.35
The graphs of $F_c^2(x)$ for $c = -0.6$, $c = -0.75$, and $c = -0.9$.

The Logistic Equation

The types of bifurcations discussed in this section are just the beginning of a long story of how discrete dynamical systems evolve as parameters are varied. To illustrate this in a concrete setting, we return to the logistic family of functions given by $L_k(x) = kx(1-x)$. Recall that x represents the fraction of the maximum possible population of a species, so we only consider x-values in the interval $0 \leq x \leq 1$.

We will use a combination of qualitative, analytic, and numerical methods to analyze this family of functions. However, the fate of orbits for many parameter values in this system is still an unsolved problem, so we are not be able to describe completely what happens for every parameter value and seed.

The first step in analyzing this family of functions is to find the fixed points. As we have seen, L_k has fixed points at 0 and at the point $p_k = (k-1)/k$. Note that p_k lies in the interval $0 \leq x \leq 1$ only if $k \geq 1$; otherwise, p_k is negative and so is not of interest in terms of a population model.

Differentiating L_k, we find $L'_k(x) = k(1-2x)$. Hence $L'_k(0) = k$. We conclude that 0 is an attracting fixed point for $0 \leq k < 1$ and a repelling fixed point if $k > 1$. Graphical iteration gives us much more information. We see from Figure 8.36 that, as long as $k < 1$, the graph of L_k meets the diagonal at the fixed point 0 and lies below the diagonal for all other x values. This means that all orbits tend to the fixed point. In terms of our population, we conclude that if $k < 1$, the population eventually dies out.

At the other fixed point, we compute

$$L'_k(p_k) = k\left(1 - 2\left(\frac{k-1}{k}\right)\right) = -k + 2.$$

Therefore p_k is attracting for $-1 < -k + 2 < 1$ or $1 < k < 3$. Again using graphical iteration, we see from Figure 8.36 that all seeds in the interval $0 < x < 1$ have orbits that tend to this attracting fixed point. In terms of populations, we conclude that when $1 < k < 3$, the population of our species eventually levels off at the constant value $(k-1)/k$.

If $k = 3$, we have $L'_k(p_k) = -1$, and if $k > 3$, we find $L'_k(p_k) < -1$. Thus p_k changes from an attracting to a repelling fixed point as k increases through 3, and we expect a period-doubling bifurcation at this point. In Exercise 10 we ask you to verify this fact analytically. As usual, we can use graphical iteration to see this qualitatively.

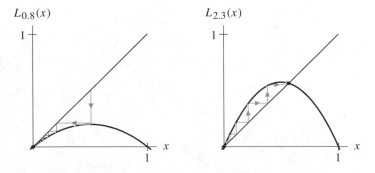

Figure 8.36

Populations die out when $k = 0.8$ but level off when $k = 2.3$.

Figure 8.37 displays orbits of L_k when $k = 2.8$ and when $k = 3.2$. Note that these orbits tend to the attracting fixed point when $k = 2.8$ but to a 2-cycle when $k = 3.2$. In terms of populations, when k is slightly larger than 3, we expect that the populations will eventually cycle in a biennial cycle: One year the population will be high; the next year it will be low.

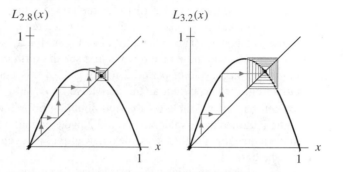

$L_{2.8}(x)$ $L_{3.2}(x)$

Figure 8.37

The period-doubling bifurcation in the logistic family when $k = 2.8$ and when $k = 3.2$.

The period-doubling route to chaos

It is impossible to use analytic methods to describe much of what occurs for the logistic family as k continues to increase, so we resort to qualitative and numerical methods. For the remainder of this section we give only a glimpse of the very complicated mathematics that arises. Figure 8.38 provides a "movie" of the graph of L_k for 6 k-values in the range $1 \le k \le 4$. As k increases, note how the slope of the tangent line to the graph of L_k at p_k decreases, eventually passing through the period-doubling point at which the slope is -1. Dynamically we see p_k change from attracting to repelling at this bifurcation point.

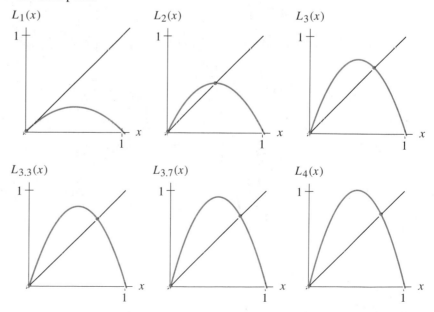

$L_1(x)$ $L_2(x)$ $L_3(x)$

$L_{3.3}(x)$ $L_{3.7}(x)$ $L_4(x)$

Figure 8.38
Graphs of $L_k(x)$ for $k = 1, 2, 3, 3.3, 3.7,$ and 4.

Figure 8.39 provides another movie, this time of the graphs of L_k^2 for 6 other k-values. The small boxes superimposed on each of these graphs are not web diagrams; rather, they are intended to draw your attention to the piece of the graph of L_k^2 inside each box. Note that these graphs look exactly like the graphs of L_k in Figure 8.38, only smaller and upside down. As we watch k change, we see these small pieces of the graph of L_k^2 behave just as the graphs of L_k do. Hence we expect L_k^2 to undergo the exact same bifurcations that L_k did in the previous pictures.

To make this precise, note that there is a fixed point for L_k^2 at the right-hand edge of the small boxes. This is actually a fixed point for L_k, not a periodic point of period 2. As k increases in these pictures, we first see the birth of a new fixed point for L_k^2 (this point lies on a cycle of period 2 for L_k). Call this point q_k. The point q_k is at first attracting, then repelling, as the derivative of L_k^2 at q_k passes through -1. Thus we expect that L_k^2 will undergo a period-doubling bifurcation as k increases. That is, for some intermediate k-value, the period 2 cycle containing q_k will cease to be attracting and become repelling. Meanwhile a new attracting cycle of twice the period, a 4-cycle, will be born.

Another way to view this would be to draw the graphs of L_k^4 over the intervals of x-values contained in each of the small boxes in Figure 8.39. If we do this, we see that the graphs of L_k^4 resemble the corresponding graphs of L_k^2 in Figure 8.39, only much smaller and turned upside down. So if we magnify these pictures and turn them around, we see that the pieces of the graphs of L_k^2 and L_k^4 look exactly like those of L_k and L_k^2 in Figures 8.38 and 8.39, respectively.

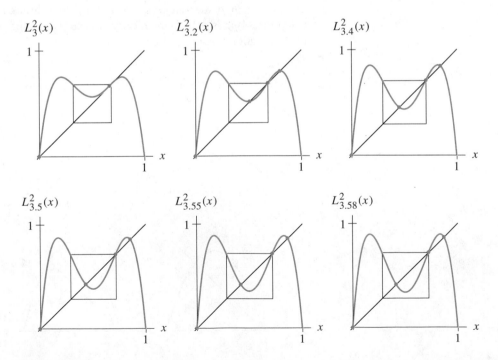

Figure 8.39
Graphs of $L_k^2(x)$ for $k = 3, 3.2, 3.4, 3.5, 3.55,$ and 3.58.

This process of iterating and zooming in to small portions of a graph is called **renormalization**. Borrowed from physics, this process allows us to recognize certain patterns that occur over and over. For example, we could now renormalize the graph of L_k^4. We would see that, over a smaller interval of k-values, the 4-cycle discussed above would itself undergo a period-doubling bifurcation. That is, the attracting 4-cycle would become repelling and a new attracting 8-cycle would be born. This scenario continues with the successive births of attracting cycles of periods 16, 32, 64, and so forth. This phenomenon has been called the "period-doubling route to chaos."

Chaos

As the parameter k continues to increase, we eventually leave the period-doubling interval and encounter a new type of behavior — chaos (see Lab 1 at the end of this chapter). For a large number of k-values in the interval $3.57 < k \leq 4$, the logistic family exhibits the type of behavior shown by the case $k = 3.9$ displayed in Figures 8.3 and 8.5. While we cannot fully explain what is happening to orbits for values in this regime, we can describe some simplified models of this chaotic behavior. This is the topic of the next section.

EXERCISES FOR SECTION 8.3

In Exercises 1–7, each function undergoes a bifurcation of fixed points at the given parameter value. In each case use analytic or qualitative methods to identify this bifurcation as a tangent, pitchfork, or period-doubling bifurcation or as none of these. Discuss the behavior of orbits near the fixed point in question at, before, and after the bifurcation.

1. $F_\alpha(x) = x + x^2 + \alpha$, $\alpha = 0$ **2.** $F_\alpha(x) = \alpha x$, $\alpha = -1$

3. $F_\alpha(x) = \alpha \sin x$, $\alpha = 1$ **4.** $F_\alpha(x) = \alpha \sin x$, $\alpha = -1$

5. $F_\alpha(x) = \alpha - x^2$, $\alpha = -1/4$ **6.** $F_\alpha(x) = \alpha \arctan x$, $\alpha = 1$

7. $F_\alpha(x) = \alpha(x + x^2)$, $\alpha = 0$

8. Consider the family of functions $F_\beta(x) = x^3 + \beta$. Find all β-values for which this family undergoes a bifurcation of fixed points. Determine the type of bifurcation that occurs at each bifurcation point, and then sketch the bifurcation diagram for this family.

9. Consider the family of functions given by

$$T_\mu(x) = \begin{cases} \mu x, & \text{if } x < 1/2; \\ \mu(1-x), & \text{if } x \geq 1/2. \end{cases}$$

Discuss in detail the bifurcation of fixed points that occurs at $\mu = 1$ for this family. Sketch the fixed-point bifurcation diagram.

10. For the logistic family $F_k(x) = kx(1 - x)$, show explicitly that there exists a cycle of period 2 for $k > 3$.

11. Consider the family of functions $F_c(x) = x^2 + c$. Sketch the graphs of F_c for $c = 0.25, 0, -0.75, -1, -1.5,$ and -2 for $-2 \le x \le 2$. Then sketch the graphs of F_c^2 for $c = -0.75, -1, -1.25, -1.3, -1.35,$ and -1.4. Is there any similarity between these two families of graphs? Discuss in a brief essay, complete with pictures.

8.4 CHAOS

In this section we begin the study of chaotic behavior. Until a few years ago most scientists and mathematicians thought that typical differential and difference equations did not exhibit the kind of behavior we now call chaos — that this kind of behavior was extremely unlikely in simple mathematical models. Now we know that this is not the case. There are many models whose behavior is quite erratic, and this behavior persists when parameters are varied. Moreover, mathematicians have begun to develop techniques to predict the onset of chaos and to determine the scope or limits of this behavior. In this chapter we use a number of qualitative and numerical techniques to explain and illustrate the concept of chaos.

An Example without Chaos

Roughly speaking, chaos means unpredictability. Thus far, most of the differential and difference equations we have encountered have been completely predictable. For example, consider the simple discrete logistic model

$$P_{n+1} = 2P_n(1 - P_n).$$

As we have seen, this system has two fixed points, at 0 and at 1/2. Graphical analysis shows that the orbit of any seed x_0 in the interval $0 < x_0 < 1$ tends to 1/2 (see Figure 8.40). The only other point, $x_0 = 1$, has an orbit that is eventually fixed. So we know the fate of all orbits.

Now in real life we rarely know the seed for our orbit with complete accuracy. For example, we most likely do not know the precise value of the population at time 0; we probably miss a few individuals in our initial count. If our model is the simple logistic model above, however, then this inaccuracy does not really matter. A small fluctuation in our initial seed in the interval $0 < x_0 < 1$ does not alter our predictions at all.

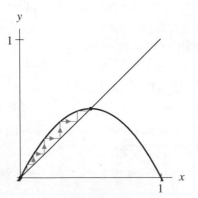

Figure 8.40
Graphical analysis of $F(x) = 2x(1 - x)$. This system has two fixed points, at 0 and at 1/2. The orbit of any seed x_0 in the interval $0 < x_0 < 1$ tends to 1/2. The only other point, $x_0 = 1$, has an orbit that is eventually fixed.

A Chaotic Logistic Function

Now let's contrast the orbits of the logistic model above with those of

$$P_{n+1} = 4P_n(1 - P_n),$$

where we have simply changed the growth parameter from 2 to 4. As always, studying this system is the same as iterating the function $L_4(x) = 4x(1 - x)$. This function has fixed points at 0 and 3/4, and eventually fixed points at 1/2 and 1. But consider the fate of virtually any other orbit. In Figure 8.41 we compute the web diagram for the first 200 points on the orbit of 0.123 for this function. It is hard to distinguish any pattern in this picture. A histogram and a time series for this orbit further confirm this erratic behavior (see Figures 8.42 and 8.43). In the histogram we see that the orbit seems to visit every small subinterval in $0 \le x \le 1$ many times.

If you repeat this experiment for a variety of seeds in the interval $0 < x < 1$, you will see essentially the same results in each case. Of course, the fixed points and eventually fixed points yield quite different diagrams. However, virtually any other seed yields figures that resemble those in Figures 8.42 and 8.43 in their complexity.

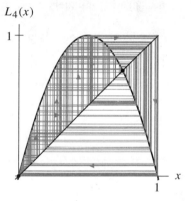

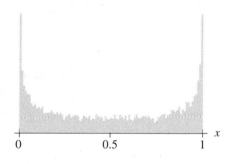

Figure 8.41
Web diagram of the orbit of 0.123
under $L_4(x) = 4x(1 - x)$.

Figure 8.42
Histogram of the orbit in Figure 8.41.

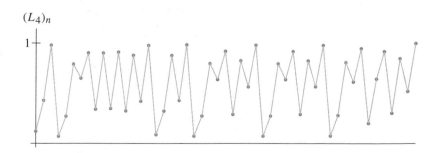

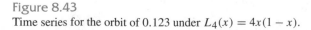

Figure 8.43
Time series for the orbit of 0.123 under $L_4(x) = 4x(1 - x)$.

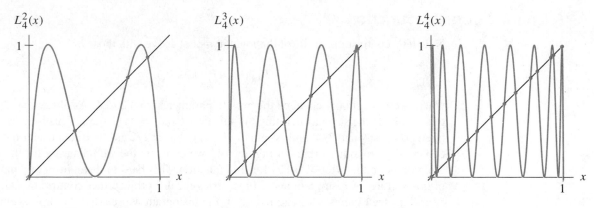

Figure 8.44
Graphs of L_4^2, L_4^3, and L_4^4.

These computer experiments suggest that most orbits of $L_4(x) = 4x(1 - x)$ behave erratically. However, there are many cycles for L_4 as well. In Figure 8.44 we plot the graphs of L_4^2, L_4^3, and L_4^4. Note that the graph of L_4^n appears to cross the diagonal $y = x$ at 2^n points, yielding infinitely many cycles of arbitrarily high period in the interval $0 \leq x \leq 1$.

Another important element in our discussion of chaotic behavior is **sensitive dependence on initial conditions**. This is illustrated in Figure 8.45, where we plot the time series for the orbits of 0.123 and 0.124. Note that these orbits initially are very close, but after very few iterations, the orbits separate and thereafter bear little resemblance to one another. Just a small change in the third decimal place of our seed has resulted in a large change in the ensuing orbit.

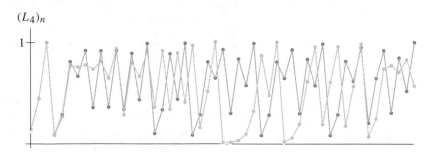

Figure 8.45
Time series for the orbit of 0.123 in dark blue; for 0.124 in light blue.

A Model Chaotic System

To characterize the chaotic behavior we observed for the logistic function $L_4(x) = 4x(1 - x)$, we turn to a discrete dynamical system that is much easier to understand. Consider the function T defined for $0 \leq x < 1$ by

$$T(x) = 10x \mod 1.$$

This notation means that to compute $T(x)$ we first multiply x by 10 and then drop the integer part, keeping only the fractional part of the result. For example

$$T\left(\tfrac{1}{3}\right) = \tfrac{10}{3} - 3 = \tfrac{1}{3},$$

$$T(0.232323\ldots) = 2.32323\ldots - 2 = 0.32323\ldots,$$

and

$$T(0.1111\ldots) = 1.111\ldots - 1 = 0.111\ldots.$$

More generally, if $0 \le x < 1$, we can write the decimal expansion of x in the form

$$x = .a_1 a_2 a_3 \ldots,$$

where the a_i are digits between 0 and 9. Then

$$10x = a_1.a_2 a_3 a_4 \ldots$$

so that

$$T(x) = 10x - a_1 = 0.a_2 a_3 a_4 \ldots;$$

that is, T simply drops or chops off the first digit in the decimal expansion of x. For this reason, T is called the **chopping function**. For example,

$$T(0.\overline{123}) = 0.\overline{231}$$

and

$$T^2(.25) = T(.5) = 0.$$

The notation $0.\overline{123}$ denotes the repeating sequence $0.123123123\ldots$.

Density of periodic points

Because of the special definition of the chopping function, periodic points of T are easy to find. Any x whose decimal expansion is repeating lies on a cycle for T. For example, $2/9$ is a fixed point, since

$$\frac{2}{9} = 0.222\ldots$$

and $T(0.222\ldots) = 0.222\ldots$. Similarly, $0.\overline{37}$ lies on a 2-cycle and $0.\overline{1234}$ lies on a 4-cycle. Clearly T has infinitely many periodic points in $[0, 1)$, and the periods of these points may be arbitrarily large. Moreover, unlike the logistic functions, we can write down all of the periodic points for T explicitly.

There is more to this story however. Given any x in the interval $[0, 1)$, we can find a periodic point of T arbitrarily close to x. This follows since, if x has the decimal expansion

$$x = 0.a_1 a_2 a_3 \ldots,$$

then the point

$$y = 0.\overline{a_1 a_2 \ldots a_n}$$

is close to x (within $1/10^n$ units) and moreover is periodic with period n.

For example, suppose we are given the number

$$x = \frac{1}{\sqrt{2}} = 0.707106781\ldots$$

and are asked to find a periodic point within 0.001 units of x. We could choose $x_1 = 0.\overline{707}$, since

$$|0.707106781\ldots - 0.\overline{707}| = 0.00060093\ldots.$$

and x_1 has period 3. But there are many other choices that work as well, including $x_2 = 0.\overline{707106}$, which is much closer to x than x_1 and has period 6. Alternatively, we could choose $0.\overline{7071999222555}$, which is also within 0.001 units of x and is periodic with period 13.

Extrapolating from this example, we see that no matter how small a subinterval we choose in $[0, 1)$, we can always find periodic points inside this subinterval. In fact we can always find infinitely many distinct periodic points inside the subinterval. A subset of $[0, 1)$, which contains points in every subinterval of $[0, 1)$, no matter how small, is said to be a **dense subset** of the interval. For example, the set of all rational numbers between 0 and 1 is a dense subset of $0 \le x < 1$. So too are the irrationals in this interval. The above argument shows that the set of periodic points for T is also a dense subset of $[0, 1)$.

Periodic orbits are by no means the only types of orbits for the chopping function; there are many eventually periodic points as well. Any point whose decimal expansion eventually begins to repeat is eventually periodic under T. For example,

$$x = 0.123\overline{4}\ldots$$

is a point whose orbit becomes fixed after 3 iterations of T. Similarly, the orbit of $0.123\overline{4567}$ is eventually periodic with period 3. Now recall that any rational number has decimal expansion that either repeats, eventually repeats, or terminates (which means it ends with a sequence of all zeroes). As a consequence, we see that any rational number in $[0, 1)$ has an orbit under T that is either periodic or eventually periodic. On the other hand, any irrational number in $[0, 1)$ has an orbit that never cycles, since the decimal expansions of irrational numbers never repeat. For example, the orbit of $1/\sqrt{2}$ is not periodic under T since $\sqrt{2}$ is an irrational number.

A dense orbit

Some of these irrational numbers have very interesting orbits. For example, consider the number x_0 whose decimal expansion begins

$$x_0 = .0123456789.$$

Suppose the next 200 terms of this expansion consist of all possible 2-blocks of digits, that is,

$$x_0 = 0.\ \underbrace{0123\ldots9}_{\text{all 1-blocks}}\ \underbrace{00\ 01\ 02\ldots10\ 11\ 12\ldots20\ 21\ldots99}_{\text{all 2-blocks}}$$

followed by all possible 3-blocks, 4-blocks, and so on, so we have

$$x_0 = 0. \quad \underbrace{0 \dots 9}_{\text{all 1-blocks}} \quad \underbrace{00 \dots 99}_{\text{all 2-blocks}} \quad \underbrace{000 \dots 999}_{\text{all 3-blocks}} \dots$$

What can we say about the orbit of x_0? Recall that each iteration of T simply drops the leading digit from this sequence. Thus given any x in $[0, 1)$, we see that there is a point on the orbit of x_0 that comes arbitrarily close to x. Indeed, we need only iterate T enough times so that the first n digits of the decimal expansion of x appear as the leading terms of this point on the orbit of x_0. Then these two points are within $1/10^n$ units of each other. For example, suppose we are given the number $1/3$ and are asked to find a point on the orbit of x_0 that lies within 0.001 of $1/3$. To do this, we simply iterate T until the 3-block that begins with 333 appears as the first entry of the sequence. Now it takes 10 iterations to remove the entries $0.012 \dots 9$, and then another 200 iterations to remove $00\ 01\ 02 \dots 98\ 99$, and then finally $3 \cdot 333$ more iterations to bring 333 to the head of the list. Thus, $T^{1209}(x_0)$ lies within the required distance of $1/3$.

Of course, there are many other points on the orbit of x_0 that lie within 0.001 units of $1/3$. All we need do is iterate T enough times until any k-block of the form $33 \dots 3$ is first in line and this point is close enough to $1/3$. As a consequence, the set of points that make up the orbit of x_0 is also a dense subset of $[0, 1)$.

Sensitive dependence on initial conditions

The final observation we wish to make about T is that this dynamical system exhibits **sensitive dependence on initial conditions**. By this we mean the following: Consider any x_0 in $[0, 1)$. Then there are points y_0 arbitrarily close to x_0 so that the orbit of y_0 is eventually "far" from that of x. For example, if we consider the fixed point $x_0 = 1/9 = 0.\overline{1}$ of T, then there are nearby points whose orbits behave very differently. For example, the seed $y_0 = 0.11 \dots 1\overline{8}$ eventually lands on the fixed point $0.\overline{8} = 8/9$. How close y_0 is to $1/9$ depends on how many initial 1's are present in the sequence. Although these two seeds are close together initially, their orbits eventually are far apart in the interval $[0, 1)$ (see Figure 8.46). As another example, we can consider a seed of the form $0.11 \dots 1\overline{35}$, which is eventually periodic with period 2, or $0.11 \dots 1\overline{23456}$, which eventually has period 5. Both of these seeds are also close to x_0, but their orbits behave quite differently. Indeed, arbitrarily close to the given seed $x_0 = 1/9$, we can find another seed whose orbit eventually has any behavior whatsoever. All we need to do is precede a given orbit by a sufficiently long string of 1's. Again, this orbit is initially close to $1/9$, but it eventually diverges from this fixed point.

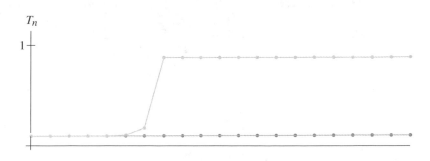

Figure 8.46
Time series for the orbit of $0.\overline{1}$ (in dark blue) and for $0.11 \dots 1\overline{8}$ (in light blue).

The general situation is similar. Given any seed x_0 whose decimal expansion is

$$x_0 = 0.a_1a_2a_3\ldots,$$

we consider the nearby seed

$$y_0 = 0.a_1a_2\ldots a_nb_1b_2b_3\ldots,$$

where $b_1b_2b_3\ldots$ are arbitrary digits. Then the seed y_0 is close to x_0 (how close depends on n, the number of decimal places to which x_0 and y_0 agree), but the fate of the two orbits can be radically different. Thus T exhibits sensitive dependence throughout the interval $[0, 1)$.

The existence of sensitive dependence in dynamical systems has profound implications for scientists and mathematicians who use difference or differential equations as mathematical models. If a given system exhibits sensitive dependence on initial conditions, then numerical predictions about the fate of orbits or solutions are to be totally distrusted. For we can never know the exact seed or initial condition for our orbit or solution because we cannot make physical measurements with infinite precision. Even if we had exact measurements, we could never carry out the necessary computations. The small numerical errors that are always introduced in such numerical procedures throw us off our original orbit and onto another whose ultimate behavior may be radically different.

Unlike T, for which we have sensitive dependence at all points in its domain, many functions exhibit sensitivity only at isolated points. Consider for example the logistic function $L_{3.2}(x) = 3.2x(1 - x)$. As shown in Figure 8.47, most orbits of this function tend to an attracting cycle of period 2. However, not all orbits share this fate. As we see in Figure 8.47, there are two repelling fixed points for $L_{3.2}$, and there are infinitely many points whose orbits eventually land on the nonzero fixed point for $L_{3.2}$. The function $L_{3.2}$ exhibits sensitive dependence at each of these seeds, since we can change the seed only slightly and find an orbit that is neither fixed nor eventually fixed, but rather tends to the attracting cycle of period 2. This behavior is much different from what we see for T; it is commonly called *unstable behavior*. Chaos occurs when we have unstable behavior everywhere, or at least on a large set.

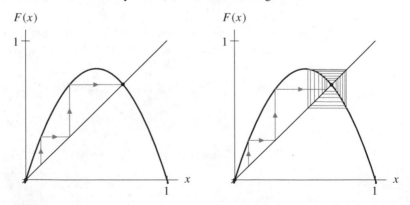

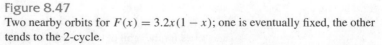

Figure 8.47
Two nearby orbits for $F(x) = 3.2x(1 - x)$; one is eventually fixed, the other tends to the 2-cycle.

Summary

We have shown that our model system T possesses an amazing amount of complexity in the interval $[0, 1)$. The periodic points of T form a dense subset of this interval, and there is also a single orbit that forms another dense subset of $[0, 1)$. Moreover, this function exhibits sensitive dependence on initial conditions, which renders precise calculations of orbits impossible. The combination of these three properties makes this system quite unpredictable, or **chaotic**.

It may seem that this type of complexity is special to our model function T because T has constant slope 10. However, over the past few years mathematicians have verified that a number of dynamical systems — both discrete systems and systems of differential equations — exhibit this type of behavior. For example, there is a large set of k-values for which the logistic function $L_k(x) = kx(1 - x)$ is chaotic. The histograms and time series we displayed earlier give qualitative evidence for this behavior.

EXERCISES FOR SECTION 8.4

1. For the logistic function $L_4(x) = 4x(1 - x)$, compute a histogram of the first 10,000 points on the orbit of $x_0 = 0.3$ by dividing the interval $[0, 1]$ into 100 subintervals, each of length 0.01, and then calculating the successive subintervals into which the orbit of x_0 falls. Then repeat this calculation using the seed $y_0 = 0.3001$. Do you see any difference? Now compute the time series for the first 100 points on the orbits of x_0 and y_0 and compare the results. Do you see any differences? Discuss your findings in a brief essay. Do not attempt this exercise without access to computing and graphing technology.

Exercises 2–11 deal with the doubling function given by

$$T(x) = \begin{cases} 2x, & \text{if } 0 \le x < 1/2; \\ 2x - 1, & \text{if } 1/2 \le x < 1. \end{cases}$$

2. Compute the orbit of the following seeds under T. Which seeds are periodic and which are eventually periodic?

(a) $x_0 = 1/5$ (b) $x_0 = 2/7$ (c) $x_0 = 3/11$ (d) $x_0 = 1/10$
(e) $x_0 = 1/6$ (f) $x_0 = 4/14$ (g) $x_0 = 4/15$

3. What can you say about the orbit of the seed $x_0 = p/2^n$ under iteration of T?

4. Sketch the graphs of T^2, T^3, and T^4. What do you expect the graph of T^n to look like? How many fixed points should T^n have?

5. Using the graphs drawn in Exercise 4, discuss the question of density of the subset of periodic points of T in the interval $0 \le x < 1$.

6. Suppose x_0 has binary representation $.a_1 a_2 a_3 \ldots$. That is,

$$x_0 = \frac{a_1}{2} + \frac{a_2}{2^2} + \frac{a_3}{2^3} + \cdots.$$

What is the binary expansion of $T(x_0)$? Of $T^n(x_0)$?

7. Using the results of Exercise 6, find all points in the interval $[0, 1)$ that are eventually fixed at 0 by T^n.

8. Using the results of Exercise 6, find all points in the interval $[0, 1)$ that are periodic with period 2, 3, or 4 under T.

9. How many points in $[0, 1)$ are fixed by T^n for each n?

10. Does T exhibit sensitive dependence on initial conditions?

11. Write down an expression for a seed x_0 in $[0, 1)$ whose orbit under T forms a dense subset of this interval.

Exercises 12–14 deal with the doubling function given by

$$T(x) = \begin{cases} 2x, & \text{if } 0 \le x < 1/2; \\ 2 - 2x, & \text{if } 1/2 \le x \le 1. \end{cases}$$

12. Sketch the graphs of T, T^2, and T^3 over the interval $[0, 1]$. How many fixed points does T^n have for each n?

13. Discuss the question of density of the set of periodic points of T.

14. Does T exhibit sensitive dependence on initial conditions?

8.5 CHAOS IN THE LORENZ SYSTEM

In this section we complete the discussion of the Lorenz system of differential equations started in Section 2.5 and continued in Section 5.5. Recall that this system is a nonlinear system of three differential equations given by

$$\frac{dx}{dt} = 10(y - x)$$
$$\frac{dy}{dt} = 28x - y - xz$$
$$\frac{dz}{dt} = -\frac{8}{3}z + xy.$$

We will first show how to reduce this problem to a two-dimensional model, and then we will further simplify the system to a one-dimensional iteration. Using techniques that we developed in the last section, we will then show how this simplified Lorenz system exhibits chaotic behavior.

The Lorenz Attractor

As we saw in Section 2.5, virtually every solution curve of the Lorenz system eventually has similar behavior: After initially meandering around three-dimensional space, the solution curve eventually winds in complicated fashion around a certain region in three-dimensional space. Our goal is to create a model of this behavior that will enable us to see more clearly how the solutions behave.

For example, Figure 8.48 shows the solution curve starting at the initial condition (0, 1, 0). After a short period of time, the solution settles down on a region where it winds around two "holes" in seemingly random manner.

The object toward which solutions of the Lorenz equation tend is called a **strange attractor**. It is an "attractor" because all solutions tend to it. It is "strange" since it is very different from the simpler attractors we have encountered thus far: equilibrium points (sinks) and certain periodic orbits.

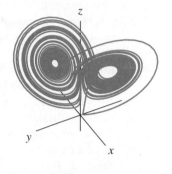

Figure 8.48
The solution of the Lorenz
equation with initial condition
(0, 1, 0).

Projections

To understand this attractor, we will try to view it from many different vantage points. The easiest way to do this is to project the solution curve onto a coordinate plane. Geometrically this means that we simply forget one of the coordinates of the solution curve. For example, to project the solution to the xy-plane, we merely plot the curve $(x(t), y(t))$ in the plane, forgetting about the third coordinate of the solution, $z(t)$. In Figure 8.49, we sketch the xy-, yz-, and xz-projections of the solution tending to the Lorenz attractor. Note that in each of these projections the solution curve apparently crosses itself. By the Uniqueness Theorem, this does not happen in the full xyz-space. When the solution crosses itself in the projection, the missing coordinate of the solution assumes different values, so the real solution curve passes "above" or "below" itself in three dimensions.

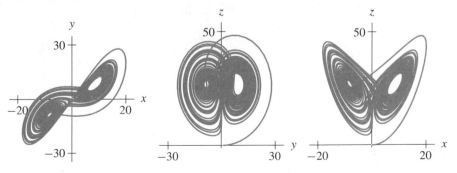

Figure 8.49
The xy-, yz-, and xz-projections of the solution in Figure 8.48.

The advantage of projections is that they can be viewed on computer screens easily, whereas three-dimensional images require some geometric intuition and perspective when viewed on a screen. The trick is to use various projections to piece together the behavior of solutions.

The Lorenz Template

The projections of the Lorenz attractor indicate that solution curves tend toward a region in the three-dimensional space that resembles the surface drawn in Figure 8.50. This image is called the **Lorenz template**, although it is sometimes called the Lorenz mask, for obvious reasons. Note that the two lobes of the template are joined along a straight line, with one lobe bending backward and the other forward of the line.

The attractor for the Lorenz system is actually a much more complicated object. It consists of infinitely many sheets packed tightly together in a complicated fashion. However, this template allows us to get a good idea of the mechanism that produces the chaotic behavior in the Lorenz system. What follows is not an exact description of the solutions of the equation; rather, it should be viewed as a simplification or model of the full system.

To understand the full phase portrait of the Lorenz system, recall from Section 5.5 that there are three equilibrium points for the Lorenz system. There is a saddle with one positive and two negative eigenvalues at the origin. There is also a pair of equilibrium points that are spiraling saddles. Each of these points have two complex eigenvalues with positive real parts and a single negative eigenvalue. Figure 8.51 indicates the positions of these equilibrium points relative to the template. Note that at each of the equilibrium points, there is a stable eigenvector pointing toward the template. This forces nearby solutions to tend toward the attractor.

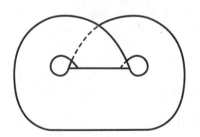

Figure 8.50
The Lorenz template.

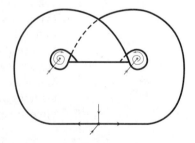

Figure 8.51
The equilibrium points for the
Lorenz system and the template.

The Poincaré return function

Now we add a typical solution curve to the template. As we have seen, solutions wind about the attractor, sometimes circling about the left lobe, at other times about the right lobe. We think of this solution as actually lying on the template, wandering back and forth between lobes (see Figure 8.52). The solution depicted in this figure is really fictitious, intended only as a representation of the behavior of a real solution of the equations.

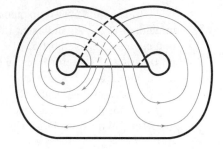

Figure 8.52
A fictitious solution lying on the Lorenz template.

Note that the typical solution on the template repeatedly crosses the line of intersection of the two lobes. We call this line L. To understand the qualitative behavior of solutions near the template, we will try to understand how this solution returns to L. Note that there is a point p on L through which the solution curve never returns to L. This is a solution that lies in both the template and the stable separatrix for the equilibrium point at $(0, 0, 0)$ (see Figure 8.53). Using the linearization near this equilibrium point, we see that solutions on either side of this solution through p evolve in different directions about the lobes (see Figure 8.54). Thus, this "disappearing" solution through p forms the boundary between those curves on the template that circle around the left lobe and those that circle around the right lobe.

Let L_ℓ denote the portion of L to the left of p (in Figure 8.53), and L_r the portion to the right of p. We wish to understand how solution curves that begin in L_ℓ or L_r return to L. To that end we will construct a function that assigns to any point in L_ℓ or L_r its next point of intersection with L. This function is called the Poincaré function, or the (first) return function on L (see Section 5.6). We denote it by ϕ. To understand how solution curves repeatedly intersect L, we need only iterate the function ϕ. It is impossible to give an analytic formula for ϕ. After all, the template itself is really fictitious. However, by observing the behavior of solutions on the xz- or yz-projections, we can give a good qualitative description of ϕ.

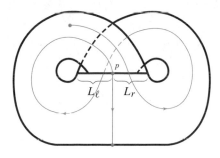

Figure 8.53
A disappearing solution on the Lorenz template.

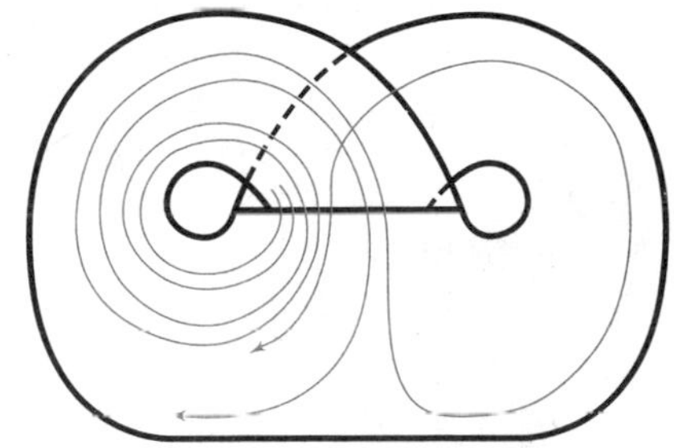

Figure 8.54
Several solutions on the Lorenz template.

In Figure 8.55 we juxtapose the yz-projection and a model for ϕ. Note that ϕ is undefined at p. As indicated in the projection, ϕ takes the interval L_ℓ to an interval that covers L_r and a portion of L_ℓ. Also, ϕ takes L_r to an interval that covers L_ℓ and a portion of L_r. Therefore we can construct an approximation of the graph of ϕ as shown in Figure 8.55.

Note that the images of both L_ℓ and L_r when they return to L are longer than L_ℓ or L_r. That is, ϕ expands these intervals. This is the reason that we have drawn the graph of ϕ with slope larger than 1.

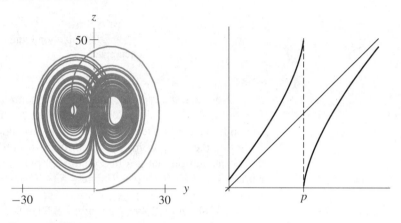

Figure 8.55
The yz-projection and a model for ϕ.

Iteration of the Poincaré Return Function

To understand the evolution of solutions as they continually reintersect L, we must iterate ϕ. As in the previous section, we see that ϕ exhibits sensitive dependence on initial conditions, since $\phi' > 1$ on L. Graphical iteration shows that there are a number of periodic points for ϕ as well (see Figure 8.56).

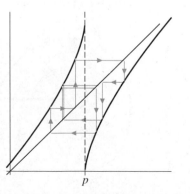

Figure 8.56
Graphical iteration of ϕ.

We need to be careful when we interpret the meaning of the observed behavior of ϕ. For example, consider the periodic orbit and the corresponding cycle for ϕ displayed in Figure 8.56. Presumably there is a real orbit for the Lorenz system that exhibits this behavior. However, such an orbit is impossible to find using numerical methods such as Euler's method, mainly because of the expansion in the direction along L. Moreover, since the Lorenz attractor is really much thicker than the template, there will be other orbits in the attractor that tend to this periodic solution. For example, note how far away solutions may eventually match up with the given periodic solution in the template. These correspond to eventually periodic points for ϕ.

Summary

In this section we have described an important technique in the study of differential equations. When we observed complicated behavior of solutions in a higher dimensional setting, we tried to describe this behavior by reducing the dimension of the system and viewing it as a discrete dynamical system. Although it is not always possible to carry out such a procedure, this type of process does work in a number of important systems of differential equations.

LAB 8.1 | The Bifurcation Diagram

When we encountered bifurcations for first-order differential equations, we saw that the bifurcation diagram (see Figures 1.82 and 1.84) provided a neat way to summarize the changes that occur in the solutions of the system. Here we introduce an analogous picture that captures the many more complicated bifurcations that occur in the logistic family $L_k(x) = kx(1 - x)$.

For a variety of k-values in the interval $1 \leq k \leq 4$, you should compute the orbit of the seed $x_0 = 0.5$. Then you will display the **asymptotic orbit** for each of these chosen k-values in the bifurcation diagram. By asymptotic orbit we mean the "tail" of the orbit. To be specific, for each chosen k-value, you should compute the first 100 points on the orbit, but you will display only the last 75 iterations. The first 25 points on the orbit should be disregarded so that you will see only the fate of the orbit.

In the bifurcation diagram you should plot the k-axis horizontally ($1 \leq k \leq 4$) and the x-axis ($0 \leq x \leq 1$) vertically. For each chosen k-value you should record all 75 points on the vertical line over the chosen k. In general there are many fewer points than 75 to record. For example, if $k = 2.8$, we show that the orbit tends quickly to an attracting fixed point that is located at approximately $x = 0.64$. So, neglecting the first 25 numbers on the orbit, you would plot the point $(2.8, 0.64)$ to indicate the presence of this fixed point.

Similarly, if $k = 3.2$, the orbit tends to a 2-cycle located at $x = 0.513$ and $x = 0.799$, so you would plot $(3.2, 0.513)$ and $(3.2, 0.799)$. Finally, if $k = 3.9$, the histogram in Figure 8.5 indicates that the orbit is distributed throughout a large subinterval of $0 \leq x \leq 1$. Over $k = 3.9$ you might sketch an interval to indicate this chaotic behavior (see Figure 8.57). The bifurcation diagram thus gives a record of the fate of the orbit of 0.5 for a collection of k-values.

In your report you should first collect and display the fate of orbits for at least 50 different k-values chosen as follows:

1. Choose 5 values in the interval $1 \leq k \leq 3$.

2. Choose 10 values in the interval $3 < k \leq 3.44$.

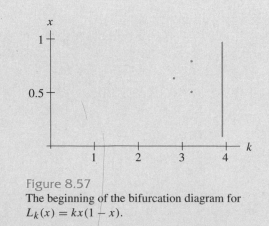

Figure 8.57
The beginning of the bifurcation diagram for
$L_k(x) = kx(1 - x)$.

3. Choose 5 values in the interval $3.44 < k \le 3.55$.

4. Choose 5 values in the interval $3.55 < k \le 3.56$.

5. Choose a number of other values just above $k = 3.56$.

6. Choose 5 values in the interval $3.57 < k \le 3.83$.

7. Choose 10 values in the interval $3.83 < k \le 3.86$.

8. Choose your remaining k-values in the regime $3.86 \le k \le 4$.

Your Report: In your report discuss the qualitative behavior of the fate of orbits of the logistic family. Can you "fill in" the diagram for other values of k with $1 \le k \le 3$? How about $3 \le k \le 3.4$? Describe a magnification of the bifurcation diagram in the interval $3.83 \le k \le 3.86$. You may need to choose additional k-values to see the structure here. Why is this interval called the "period 3 window"?

Further Projects:

1. Compute the bifurcation diagram for the quadratic family $F_c(x) = x^2 + c$. You should choose the parameter c in the interval $-2 \le c \le 0.25$ and use the seed $x_0 = 0$ in this case. The orbit should be plotted in the interval $-2 \le x \le 2$. Describe the similarity with the logistic bifurcation diagram. Can you find a period 3 window here?

2. Repeat the previous investigation using the sine family $S_\lambda(x) = \lambda \sin x$. Choose λ in the interval $1 \le \lambda \le \pi$ and use the seed $x_0 = \pi/2 \approx 1.57$. Use the interval $0 \le x \le \pi$ to plot your orbits. Now compare all three bifurcation diagrams.

LAB 8.2   Newton's Method as a Difference Equation

The **roots** of a function $f(x)$ are the values of x for which $f(x) = 0$. Given a function f, we can find approximate values for its roots using Newton's method. To use Newton's method, we first make an initial guess for the value of the root, say x_0. If this is not the root (that is, $f(x_0)$ is not 0), we can (hopefully) improve the guess by first computing the tangent line to the graph of f at x_0, then compute the point x_1, where this tangent line intersects the x-axis. The relationship between x_0 and x_1 (a good calculus review) is given by

$$x_1 = x_0 - \frac{f(x_0)}{f'(x_0)}.$$

If $f(x_1)$ is not sufficiently close to zero, then we improve the guess again by computing

$$x_2 = x_1 - \frac{f(x_1)}{f'(x_1)},$$

and so forth. What we are doing is creating the orbit of the seed x_0 under the Newton's method function

$$N(x) = x - \frac{f(x)}{f'(x)}.$$

The initial guess is the seed x_0, the first improvement is $x_1 = N(x_0)$, the second improvement is $x_2 = N(x_1)$, and so on. The hope is that the improved guesses really are improvements and that the sequence $x_0, x_1, \ldots$ approaches a root of the function f. In this lab we consider how the choice of seed affects the long-term behavior of the orbits of the Newton's method function by studying a particular example: Let

$$f(x) = x^3 - 3x^2 + 2x = x(x-1)(x-2).$$

This function has roots at $x = 0$, $x = 1$, and $x = 2$.

In your report, consider the following items:

1. For the function $f(x)$ chosen above, compute the corresponding Newton's method function $N(x) = x - f(x)/f'(x)$. Verify that this function has fixed points at the roots $x = 0$, $x = 1$ and $x = 2$. Verify that these are attracting fixed points. What does this imply for orbits of $N(x)$ for seeds x_0 near 0, 1, or 2?

2. Compute the first 20 points of the orbits under $N(x)$ for the initial seeds

$$x_0 = 0, 0.05, 0.10, 0.15, 0.20, \ldots, 0.90, 0.95, 1.$$

Sketch the segment $0 \leq x_0 \leq 1$. For each choice of x_0, if the 20th point x_{20} in its orbit under N is within 0.01 of 0, color the point red; if x_{20} is within 0.01 of 1, color it blue; if x_{20} is within 0.01 of 2, color it green; and if none of these conditions holds, color it black. Is this figure consistent with your conclusions from part 1? What implications does this picture have concerning the choice of the seed x_0 for Newton's method?

3. Repeat part 2 for the points

$$x_0 = 0.30, 0.32, 0.34, 0.36, 0.38, \ldots, 0.46, 0.48, 0.50.$$

Your report: In your report address each of the items above. Parts 2 and 3 require a computer or calculator. Do *not* hand in lists of numbers. What is important is the color-coded sketches of the x-axis and your interpretation of these sketches.

LAB 8.3   The Delayed Logistic and Iteration in Two Dimensions

In Section 8.1 the logistic equation

$$L_k(y) = ky(1-y)$$

was used to model populations that reproduce on a discrete time scale. In particular, if y_n represents the population (or population density) of some species in the nth generation, then the population in the $(n+1)$st generation is given by

$$y_{n+1} = L_k(y_n) = ky_n(1-y_n).$$

We have seen that the behavior of orbits can depend in a dramatic way on the value of the growth rate parameter k and on the seed y_0.

One hidden assumption in the discrete logistic model is that the population in the $(n + 1)$st generation depends only on the population in the nth. This may not always be the case for species that have a long maturation time or that migrate to nesting or breeding areas. For example, suppose a species returns to the same breeding area each year. The amount of food available to the nth generation in that area might depend on how much was eaten in the previous year. In this case the population of the $(n + 1)$st generation depends on the population in the nth and the $(n - 1)$st generations. This motivates the following model of population growth called the **delayed logistic** population model,

$$y_{n+1} = ky_n(1 - y_{n-1}).$$

As with the logistic model, y_n represents the population density in the nth generation. We restrict attention to the case where $0 \leq y_n \leq 1$. For the delayed logistic model, the population in the $(n + 1)$st generation is proportional to the population in the nth generation, as well as to how close the population in the $(n - 1)$st generation was to the maximum possible population density.

To study the delayed logistic model, we perform an operation very similar to the conversion of second-order differential equations into first-order systems. We introduce a new variable x_n, letting

$$x_n = y_{n-1}.$$

That is, x_n is the population density in the $(n - 1)$st generation. We then rewrite the delayed logistic equation as

$$x_{n+1} = y_n$$

$$y_{n+1} = ky_n(1 - x_n).$$

As with the conversion of second-order equations to first-order systems, the system is simpler in one way because the values x_{n+1} and y_{n+1} depend only on the values of x_n and y_n. However, it is more complicated because we have had to introduce a new variable.

We can interpret this model as an iteration by letting

$$F_k(x, y) = (y, ky(1 - x)).$$

Then

$$(x_{n+1}, y_{n+1}) = F_k(x_n, y_n);$$

that is, orbits of an initial seed (x_0, y_0) are made up of a sequence of points in the plane. Hence

$$(x_1, y_1) = F_k(x_0, y_0) = (y_0, ky_0(1 - x_0)),$$

$$(x_2, y_2) = F_k(x_1, y_1) = (y_1, ky_1(1 - x_1)),$$

and so on. In this lab we study the behavior of orbits for this two-dimensional interation. We can graph orbits in two different ways. We can form "time-series" graphs with n on the horizontal axis and x_n or y_n on the vertical axis. Alternately, we can make "phase plane" graphs, plotting the sequence of points (x_n, y_n) in the xy-plane. You should make both types of graphs for each orbit you compute.

In your report, consider the following items:

1. For several different values of k between 1.5 and 2.5, compute the first 200 iterates of the orbit with seed $(x_0, y_0) = (0.1, 0.2)$. For each of the values of k that you choose, describe the long-term behavior of the orbit. What are the biological implications of your conclusions? (For some values of k, you may need more than 200 iterates.)

2. How do your results from part 1 change if you change the initial seed? What are the biological implications of your conclusions? (Remember to keep the initial seed in the "physically meaningful" range $0 \leq x_0 \leq 1, 0 \leq y_0 \leq 1$.)

Your report: In your report describe your discoveries based on these numerical experiments. You may include a *limited* number of graphs and/or phase plane pictures of orbits to illustrate your description. You should *not* include a catalog of orbits for different initial conditions and values of k. The goal of the lab is to interpret the results of your experiments.

A FIRST-ORDER LINEAR EQUATIONS REVISITED

In this appendix we discuss an alternate approach to first-order linear differential equations. The method used here is less general than that of Section 1.8, but it involves fewer and easier calculations and allows direct access to the qualitative behavior of the solutions of certain linear equations. This method is the same as that employed in Sections 4.1–4.4 for second-order linear equations.

Linear Differential Equations

A first-order differential equation is **linear** if it can be written in the form

$$\frac{dy}{dt} + g(t)y = r(t),$$

where $g(t)$ and $r(t)$ are arbitrary functions of t. Note that we have moved all the terms involving the dependent variable y to the left-hand side of the equation. This is traditional for linear equations and helps with the nomenclature below. For more examples of linear equations, see Section 1.8.

If the right-hand side of a linear equation is zero (that is, $r(t) = 0$) then the equation is called *homogeneous* or *unforced*. Otherwise it is said to be *nonhomogeneous* or *forced*. For example

$$\frac{dy}{dt} + 2y = \sin 2t$$

is a nonhomogeneous linear equation, while

$$\frac{dy}{dt} + (\sin 2t)y = 0$$

is homogeneous.

A first-order linear differential equation has *constant coefficients* if $g(t)$ is a constant. That is, the equation has the form

$$\frac{dy}{dt} + \lambda y = r(t)$$

where λ is a constant. In this appendix we deal only with constant-coefficient equations.

Solving Constant-Coefficient Linear Differential Equations

A warm-up

Consider the constant-coefficient, homogeneous differential equation

$$\frac{dy}{dt} + 2y = 0.$$

This equation is separable and has the general solution $y(t) = ke^{-2t}$, where the constant k is determined by the initial condition. Next, consider the slightly more complicated nonhomogeneous example

$$\frac{dy}{dt} + 2y = 4.$$

Note that, if we neglect the 4 on the right-hand side, this equation has the same "homogeneous part" as the previous example. Rewriting this equation in the form

$$\frac{dy}{dt} = 2(2 - y),$$

we see that this equation is also separable and has the equilibrium solution $y(t) = 2$. If we now separate and integrate, we find

$$-\log|2 - y| = 2t + C.$$

After simplification, we have

$$y(t) = ke^{-2t} + 2.$$

The key point here is that the general solution consists of two pieces, the constant term 2, which is one particular solution of the nonhomogeneous equation as we observed above, and the exponential term, which is the general solution of the associated homogeneous equation. So the general solution of this nonhomogeneous equation is the sum of the general solution of the associated homogeneous equation and one particular solution of the nonhomogeneous equation.

Was this an accident?

No. It turns out that solutions of nonhomogeneous linear equations always behave this way. This is a very special feature of linear differential equations.

For a typical nonautonomous differential equation, there need not be any relationship between solutions with different initial conditions (except those implied by the Uniqueness Theorem). For linear differential equations, this is not the case. Our next task is to state precisely the relationship between solutions of linear equations.

LINEARITY PRINCIPLE   Suppose $y_h(t)$ is a solution of a homogeneous linear differential equation

$$\frac{dy}{dt} + g(t)y = 0,$$

then, for any constant k, the function $ky_h(t)$ is also a solution.   ∎

We can verify this statement as follows: If $y_h(t)$ ("h" for homogeneous) is a solution of the (homogeneous) differential equation

$$\frac{dy_h}{dt} + g(t)y_h(t) = 0,$$

then

$$\frac{d(ky_h)}{dt} + g(t)(ky_h(t)) = k\left(\frac{dy_h}{dt} + g(t)y_h(t)\right) = 0.$$

Hence, $ky_h(t)$ is also a solution.

The importance of the Linearity Principle is that it allows us to produce many solutions of a linear differential equation from a given solution. If, for example, $y_h(t)$ is a solution of the differential equation with $y_h(0) \neq 0$, then $ky_h(t)$ is also a solution for any k, and in fact, this is the general solution.

EXTENDED LINEARITY PRINCIPLE   Suppose $y_h(t)$ is any solution of the homogeneous linear equation differential equation

$$\frac{dy}{dt} + g(t)y = 0$$

and $y_p(t)$ ("p" for particular solution) is *any* solution of the nonhomogeneous equation

$$\frac{dy}{dt} + g(t)y = r(t).$$

Then $y_h(t) + y_p(t)$ is also a solution of the nonhomogeneous equation.   ∎

In order to check this statement, we note that we are given that

$$\frac{dy_h}{dt} + g(t)y_h(t) = 0 \quad \text{and} \quad \frac{dy_p}{dt} + g(t)y_p(t) = r(t)$$

for all t. Adding these two equations together, we find

$$\frac{d(y_h + y_p)}{dt} + g(t)(y_h(t) + y_p(t)) = \left(\frac{dy_h}{dt} + g(t)y_h(t)\right) + \left(\frac{dy_p}{dt} + g(t)y_p(t)\right)$$

$$= 0 + r(t)$$

$$= r(t)$$

for all t. Thus $y_h(t) + y_p(t)$ is a solution of the nonhomogeneous equation.

Next, suppose $ky_h(t)$ is the general solution of the homogeneous equation

$$\frac{dy}{dt} + g(t)y = 0.$$

Thus, for any number a, we can find k so that $ky_h(0) = a$. If $y_p(t)$ is a solution of the nonhomogeneous equation

$$\frac{dy}{dt} + g(t)y = r(t)$$

then, for any number y_0, we can find k such that

$$ky_h(0) = y_0 - y_p(0).$$

Hence, for any initial condition $y(0) = y_0$, we can find a solution of the nonhomogeneous equation of the form $ky_h(t) + y_p(t)$ with

$$ky_h(0) + y_p(0) = y_0.$$

In other words, $ky_h(t) + y_p(t)$ is the general solution of the nonhomogeneous equation. We often summarize this observation by saying that "the general solution of the nonhomogeneous equation is the sum of the general solution of the homogeneous equation plus one solution of the nonhomogeneous equation."

We now have a three-step algorithm for solving linear equations. We first find the general solution of the homogeneous equation. Then we find just one particular solution of the nonhomogeneous equation. Finally we add the two together. In theory, we could solve any linear differential equation using these three steps. In practice, however, this technique is usually used only for special linear equations such as those with constant coefficients. The reason is that the second step requires that we produce a particular solution of the nonhomogeneous equation. If $g(t)$ is not a constant, this step can be quite difficult. If $g(t)$ is a constant, then we can sometimes succeed using a time-honored technique in mathematics.

The Lucky Guess

One of the simplest methods to find a particular solution of a nonhomogeneous equation is simply to guess the right solution. This may seem at first like cheating, but we will see that it really is an efficient and practical method.

As a first example, consider the equation

$$\frac{dy}{dt} + 2y = e^t.$$

The homogeneous equation is $dy/dt + 2y = 0$, which is the same as

$$\frac{dy}{dt} = -2y$$

and which has general solution $y(t) = ke^{-2t}$.

To find one solution of the nonhomogeneous equation we note that the right-hand side is e^t. We need to guess a function $y_p(t)$ such that, when we insert $y_p(t)$ into the left-hand side of the equation, out comes e^t. Now we probably would not guess sines or cosines for $y_p(t)$, because the left-hand side would still involve trigonometric functions after the computation. Similarly, polynomials would not work. What we need to guess is an exponential function. So we guess

$$y_p(t) = ae^t,$$

where a is a constant that remains to be determined. (This method is called the Method of Undetermined Coefficients: We must determine a so that ae^t is actually a solution of the nonhomogeneous equation.)

Now that we have made this reasonable guess for $y_p(t)$, all that remains is to see if the guess works. So we substitute $y_p(t) = ae^t$ into the left-hand side of the differential equation, obtaining

$$\frac{dy_p}{dt} + 2y_p = ae^t + 2ae^t$$
$$= 3ae^t.$$

In order for $y_p(t)$ to be a solution, we must have $3a = 1$ or $a = 1/3$. Therefore, $y_p(t) = e^t/3$ and the general solution is

$$y(t) = ke^{-2t} + \tfrac{1}{3}e^t.$$

Another lucky guess

Suppose the right-hand side of our equation contains trigonometric functions as in

$$\frac{dy}{dt} + 2y = \cos t.$$

Then the general solution of the homogeneous equation is exactly the same as before

$$y(t) = ke^{-2t}.$$

Now, however, we must change our guess. This time we try

$$y_p(t) = a \cos t + b \sin t.$$

Note that the simpler guesses of $y_p(t) = a \cos t$ and $y_p(t) = a \sin t$ are destined to fail, since we end up with both sines and cosines when we compute the left-hand side.

To determine a and b, we substitute $y_p(t)$ into the left-hand side of the differential equation and obtain

$$\frac{dy_p}{dt} + 2y_p = \frac{d(a \cos t + b \sin t)}{dt} + 2(a \cos t + b \sin t)$$

$$= -a \sin t + b \sin t + 2a \cos t + 2b \sin t$$

$$= (-a + 2b) \sin t + (2a + b) \cos t$$

Hence, in order for $y_p(t)$ to be a solution we must choose a and b so that

$$(-a + 2b) \sin t + (2a + b) \cos t = \cos t$$

for all t. To accomplish this, we choose a and b so that $-a + 2b = 0$ and $2a + b = 1$. Solving these equations simultaneously for a and b, we obtain $a = 2/5$ and $b = 1/5$. Hence, luck is with us again and we have that

$$y_p(t) = \tfrac{2}{5} \cos t + \tfrac{1}{5} \sin t$$

is one particular solution of the nonhomogeneous equation.

Therefore, the general solution of

$$\frac{dy}{dt} + 2y = \cos t$$

is

$$y(t) = ke^{-2t} + \tfrac{2}{5} \cos t + \tfrac{1}{5} \sin t.$$

Solutions for several different initial conditions are shown in Figure 1.1.

Actually, the above guesses are not lucky. Rather, they are more like informed guesses. Given a particularly nice right-hand side of a differential equation (sines, cosines, exponentials and the like), we can usually figure out an appropriate guess. If we make an inappropriate guess (as, for example, forgetting the $b \sin t$ term above), then it will be impossible to find choices of the constants that make the guess a solution. If this happens, we simply go back and re-examine and refine the original guess.

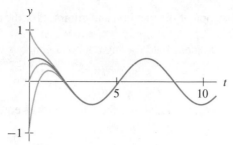

Figure 1.1

Graphs of several solutions of

$$\frac{dy}{dt} + 2y = \cos t.$$

Note that all of these graphs tend to merge relatively quickly.

Qualitative analysis

The discussion and example above give a great deal of insight into the qualitative behavior of solutions of a linear differential equation. For the equation

$$\frac{dy}{dt} + 2y = \cos t,$$

the general solution is

$$y(t) = ke^{-2t} + \tfrac{2}{5}\cos t + \tfrac{1}{5}\sin t,$$

which is the sum of the general solution of the homogeneous equation and a particular solution of the original nonhomogeneous equation. Note that the ke^{-2t} term tends quickly to zero so that, for large t, every solution is close to the particular solution

$$y_p(t) = \tfrac{2}{5}\cos t + \tfrac{1}{5}\sin t.$$

We see this clearly in Figure 1.1, where solutions with different initial conditions tend together toward the same periodic function.

Some of this behavior we could have predicted without computation. If we look at the slope field for this equation (see Figure 1.2) we see that for $y \gg 0$, the slope is negative, while for $y \ll 0$, the slope is positive. Hence, solutions that start with initial conditions far from $y = 0$ tend toward smaller $|y|$ values. The detailed behavior of solutions near zero is harder to see from the slope field. However, it is clear that solutions oscillate somehow.

Looking again at the general solution, we see that the long-term behavior of the solution is an oscillation with period 2π. This period is the same as the period of $\cos t$, the right-hand side of the differential equation. However, the amplitude and the phase

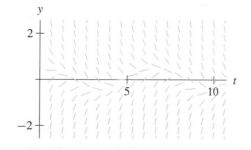

Figure 1.2

Slope field of

$$\frac{dy}{dt} = -2y + \cos t.$$

A careful examination of this field suggests that no solutions become unbounded as $t \to \infty$.

(that is, the location of the maxima and minima) for the solution is not exactly the same as $\cos t$.

These same ideas hold for any nonhomogeneous equation of the form

$$\frac{dy}{dt} + \lambda y = r(t)$$

as long as λ is positive. As above, the homogeneous equation associated with this equation is

$$\frac{dy}{dt} + \lambda y = 0,$$

which has general solution $ke^{-\lambda t}$. Because $\lambda > 0$, this solution tends to zero exponentially fast. If we let $y_p(t)$ denote one particular solution of the nonhomogeneous equation, then the general solution is

$$y(t) = ke^{-\lambda t} + y_p(t)$$

and we see that all solutions are close to $y_p(t)$ for large t. That is, the solution of the homogeneous part of the equation tends to zero, leaving only the particular solution of the nonhomogeneous part.

The discussion above relies on the fact that the solution of the homogeneous equation tends to zero (that is, $\lambda > 0$). If this is not the case, then very different behavior is possible (see the exercises).

Second Guessing

As with any guess-and-test method, we run the risk that our first guess (no matter how reasonable) might not be correct. When this happens, we simply guess again.

Consider the equation

$$\frac{dy}{dt} + 2y = 3e^{-2t}.$$

As above, because the coefficient of y is positive, solutions of the homogeneous equation tend toward $y = 0$. The slope field for the nonhomogeneous equation is given in Figure 1.3. This picture shows that all solutions seem to tend to $y = 0$.

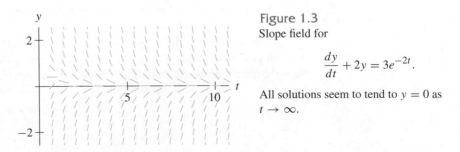

Figure 1.3
Slope field for

$$\frac{dy}{dt} + 2y = 3e^{-2t}.$$

All solutions seem to tend to $y = 0$ as $t \to \infty$.

To compute the general solution, we first note that the general solution of the homogeneous equation is $y(t) = ke^{-2t}$. To find a particular solution of the nonhomogeneous equation

$$\frac{dy}{dt} + 2y = 3e^{-2t}$$

we make the reasonable guess $y_p(t) = ae^{-2t}$, with a as the undetermined coefficient. Substituting this guess into the left-hand side of the equation yields

$$\frac{dy_p}{dt} + 2y_p = \frac{d(ae^{-2t})}{dt} + 2ae^{-2t} = -2ae^{-2t} + 2ae^{-2t} = 0.$$

This is upsetting. No matter how we pick the coefficient a, we always get zero when we substitute $y_p(t)$ into the left-hand side of the equation. There is no particular solution of the nonhomogeneous equation of the form $y_p(t) = ae^{-2t}$.

The problem is that e^{-2t} is a solution of the homogeneous equation. When we substitute it into the left-hand side, we are guaranteed to get zero. On the other hand, our guess must contain the factor e^{-2t} in order to have any hope of having the two sides agree. This leaves a wide variety of choices of possible guesses.

We need a second guess for $y_p(t)$ which contains an e^{-2t} term, is not a solution of the homogeneous equation, and is no more complicated than it needs to be. Guesses of the form $a \sin t e^{-2t}$ or ae^{bt} are clearly destined to fail. Finally, we try

$$y_p(t) = ate^{-2t},$$

where a is a coefficient to be determined; that is, we multiply our first guess by t.

Substituting this into the left-hand side of the nonhomogeneous equation

$$\frac{dy}{dt} + 2y = 3e^{-2t},$$

we obtain

$$\frac{dy_p}{dt} + 2y_p = (ae^{-2t} - 2ate^{-2t}) + 2ate^{-2t} = ae^{-2t}.$$

If we take $a = 3$, then $y_p(t) = 3te^{-2t}$ is a solution. Now we see why multiplying the first guess by t is a good idea. The derivative of $y_p(t) = ate^{-2t}$ via the Product Rule has one term that is a multiple of $y_p(t)$ and another that contains only a constant multiple of e^{-2t}.

The general solution is

$$y(t) = ke^{-2t} + 3te^{-2t},$$

and we see that all of these solutions indeed tend to zero as t increases.

Rule of thumb for second guessing

The above is a good example of what is so unsatisfying about lucky guess methods. How did we know to make the second guess equal to t times the first guess? The answer is that we have either seen a similar problem before or we can figure out at least the form of the guess by another technique. Methods for arriving at the second guess with less guesswork but more computation are given in Chapter 4 and in the exercises at the end of this appendix.

It turns out that for linear equations of any order, the examples above give a general rule of thumb for how to find particular solutions. The form of the guess depends on the right-hand side (or t term) of the equation. If the first guess doesn't yield a particular solution (which will occur if the right-hand side is a solution of the homogeneous equation), we try a second guess by multiplying the first guess by t.

EXERCISES FOR APPENDIX A

In Exercises 1–4, find the general solution of the given equation.

1. $\dfrac{dy}{dt} - y = 3e^{-2t}$

2. $\dfrac{dy}{dt} = y + \cos 2t$

3. $\dfrac{dy}{dt} - y = 3e^{-t}$

4. $\dfrac{dy}{dt} = 2y + \sin 2t$

In Exercises 5–8, find the solution of the given initial-value problem.

5. $\dfrac{dy}{dt} + 2y = e^{t/3}, \quad y(0) = 1$

6. $\dfrac{dy}{dt} + y = \cos 2t, \quad y(0) = 5$

7. $\dfrac{dy}{dt} - 2y = 3e^{-2t}, \quad y(0) = 10$

8. $\dfrac{dy}{dt} = -3y + \cos 2t, \quad y(0) = -1$

In Exercises 9–12, describe the behavior of solutions in a short paragraph. Only partial information is given about the functions in the differential equation, so only partial descriptions of some of aspects of the qualitative solutions are possible. Be sure to deal with initial conditions of different sizes and to discuss as best you can the long-term behavior of solutions.

9. $\dfrac{dy}{dt} + 2y = r(t)$, where $-1 < r(t) < 2$ for all t.

10. $\dfrac{dy}{dt} - 2y = r(t)$, where $-1 < r(t) < 2$ for all t.

11. $\dfrac{dy}{dt} + y = r(t)$, where $r(t)$ tends to 3 as $t \to \infty$.

12. $\dfrac{dy}{dt} + ay = \cos 3t + b$, where a and b are positive constants.

13. Consider the equation

$$\frac{dy}{dt} + 2y = \cos 3t.$$

To find a particular solution, it is pretty clear that our guess must contain a cosine function, but it is not quite so clear that the guess must also contain a sine function.

   (a) Guess $y_p(t) = a \cos 3t$ and substitute this guess into the equation. Is there a value of a such that $y_p(t)$ is a solution?

   (b) Write a brief paragraph explaining why the proper guess for a particular solution is $y_p(t) = a \cos 3t + b \sin 3t$.

14. Consider the equation

$$\frac{dy}{dt} + \lambda y = \cos 2t.$$

We know that, to find the general solution of this equation, we must find the general solution of the homogeneous equation $dy/dt + \lambda y = 0$ and add it to some solution of the nonhomogeneous equation. Write a brief paragraph explaining why it does not matter which solution of the original nonhomogeneous equation we use as the particular solution.

15. Consider a linear equation of the form

$$\frac{dy}{dt} + g(t)y = r_1(t) + r_2(t),$$

that is, the right-hand side is written as a sum of two functions. Suppose $y_h(t)$ is a solution of the homogeneous equation

$$\frac{dy}{dt} + g(t)y = 0,$$

suppose $y_1(t)$ is a solution of

$$\frac{dy}{dt} + g(t)y = r_1(t)$$

and suppose $y_2(t)$ is a solution of

$$\frac{dy}{dt} + g(t)y = r_2(t).$$

Show that $y_h(t) + y_1(t) + y_2(t)$ is a solution of the original equation.

16. Consider the equation

$$\frac{dy}{dt} + 2y = 3t^2 + 2t - 1.$$

In order to find the general solution, we must guess a particular solution. Since the right-hand side is a quadratic polynomial, it is reasonable to guess a quadratic for a particular solution, so let

$$y_p(t) = at^2 + bt + c.$$

Show that a, b and c can be chosen so that this is a solution.

In Exercises 17–20, find the general solution and the solution that satisfies the initial value $y(0) = 0$.

17. $\dfrac{dy}{dt} + 2y = t^2 + 2t + 1 + e^{4t}$ **18.** $\dfrac{dy}{dt} + y = t^3 + \sin 3t$

19. $\dfrac{dy}{dt} + 3y = \cos 2t + e^{3t} + e^{-4t}$ **20.** $\dfrac{dy}{dt} + y = \cos 2t + 3\sin 2t + e^{-t}$

In Exercises 21–23, we consider a very general "guess-and-test" method. From calculus, we know that many of the functions commonly encountered can be represented as power series (or Taylor series). Hence, guessing that a particular solution has the form $y_p(t) = a_0 + a_1 t + a_2 t^2 + \ldots$ is a very general guess. The disadvantage of this guess is that we must determine infinitely many coefficients $a_0, a_1, a_2, \ldots$. Also, whenever we have an infinite series, we must worry about convergence. There is a vast theory of power series techniques which can be found in older differential equations books. This theory is particularly useful in the study of partial differential equations.

21. Consider the differential equation

$$\frac{dy}{dt} + y = e^{-3t}.$$

In order to find a particular solution, we could guess $y_p(t) = ae^{-3t}$ and solve for a. Instead, suppose we replace the right-hand side of the equation with the Taylor polynomial, that is

$$\frac{dy}{dt} + y = 1 - 3t + \frac{9t^2}{2} - \frac{27t^3}{6} + \dots.$$

Now it is not so crazy to make a different guess for $y_p(t)$, namely a power series $y_p(t) = a_0 + a_1 t + a_2 t^2 + a_3 t^3 + \dots$. This guess has the advantage that we are only dealing with polynomials, but the disadvantage that we must now find infinitely many of the coefficients $a_0, a_1, \dots$. Because we do not care which particular solution we find, we may assume that $y_p(0) = 0$ ($a_0 = 0$), but this still leaves infinitely many coefficients to compute.

(a) Substitute the power series guess for $y_p(t)$ into the differential equation (assuming $a_0 = 0$) and compute a_1 and a_2.

(b) Find the solution of the differential equation with $y(0) = 0$ by methods of this appendix and verify that the first three coefficients of the Taylor polynomial of this solution are $a_0 = 0$, a_1, and a_2.

22. The method of guessing power series is clearly not the best method in the preceding problem. Sometimes, however, it is the only method available. Consider the equation

$$\frac{dy}{dt} + 2ty = e^t.$$

Note that this is not a constant-coefficient equation, so finding a particular solution is more difficult. However, we can at least get an approximation of the solution by guessing a power series.

(a) Verify that $y_p(t) = ae^t$ is not a solution for any a.

(b) Set $y_p(t) = a_0 + a_1 t + a_2 t^3 + a_3 t^3 + \dots$. Assume that $y_p(0) = 0$, so $a_0 = 0$. Substitute this guess into the equation and compute a_1 and a_2.

(c) Graph this function and compare it to the graph of the solution computed using Euler's method. Write a short paragraph describing where they are the same and where and why they are different.

23. We can use this method to motivate the "second guess" discussed in this appendix. Consider the equation

$$\frac{dy}{dt} + 2y = e^{-2t}.$$

Set $y_p(t) = a_0 + a_1 t + a_2 t^2 + a_3 t^3 + \dots$ and suppose $y_p(0) = 0$ so $a_0 = 0$.

(a) Substitute this $y_p(t)$ into the equation and solve for a_1 and a_2.

(b) Solve the equation in the usual way and verify that the first three terms of the Taylor series of the solution agree with those found in part (a).

B   COMPLEX NUMBERS AND EULER'S FORMULA

This appendix is a summary of some of the basic properties of complex numbers that we use in this text. Complex numbers and functions arise quite often when we attempt to solve differential equations, even if the equations involve only real numbers. For instance, complex numbers may appear as eigenvalues for linear systems or as roots of the characteristic equation of a second-order equation. Complex functions also appear as solutions to differential equations when we use Euler's formula.

Complex Numbers

A complex number is a number of the form $a + bi$, where a and b are real numbers and i is the "imaginary" number $\sqrt{-1}$. Equivalently, i is the number whose square is -1, that is, $i^2 = -1$. For example, $2 + 3i$, $-\pi - i$, and $\sqrt{7}\,i$ are all complex numbers. We may also regard the real number 7 as a complex number by writing it in the form $7 + 0i$.

For a complex number $z = a + bi$, the real number a is called the **real part** of z, and real number b is called the **imaginary part** of z. Note that we do not include the i in the imaginary part of a complex number. Thus, the imaginary part of a complex number is a real number.

The usual rules of arithmetic apply to complex numbers. For example, we may add two complex numbers by simply adding the corresponding real and imaginary parts. Therefore

$$(1 + 2i) + (3 + 4i) = (1 + 3) + (2 + 4)i = 4 + 6i.$$

Multiplication also obeys the usual rules of arithmetic. For example.

$$(7 + 2i) \cdot (3 + 4i) = 7 \cdot 3 + 7 \cdot 4i + 2i \cdot 3 + 2i \cdot 4i$$
$$= (7 \cdot 3 - 2 \cdot 4) + (7 \cdot 4 + 2 \cdot 3)i$$

because $i^2 = -1$, and the result is $13 + 34i$. An easier product to compute is

$$7i \cdot (1 + 2i) = -14 + 7i.$$

Division of complex numbers is a little trickier. For example, to compute the quotient

$$\frac{2 + 3i}{3 + 4i},$$

we first multiply numerator and denominator by $3 - 4i$, obtaining

$$\frac{2 + 3i}{3 + 4i} = \frac{2 + 3i}{3 + 4i} \cdot \frac{3 - 4i}{3 - 4i}.$$

The denominator becomes

$$(3 + 4i) \cdot (3 - 4i) = 3^2 + 4^2 = 25,$$

and the numerator is

$$(2 + 3i) \cdot (3 - 4i) = 18 + i.$$

Therefore

$$\frac{2+3i}{3+4i} = \frac{18+i}{25} = \frac{18}{25} + \frac{1}{25}i.$$

In general, when computing $(a + bi)/(c + di)$, we multiply numerator and denominator by $c - di$. Then the denominator becomes a real number since

$$(c + di) \cdot (c - di) = c^2 + d^2 + 0i.$$

The complex number $c - di$ is called the **complex conjugate** of $c + di$. As a second example, consider

$$\frac{1+i}{1-i} = \frac{1+i}{1-i} \cdot \frac{1+i}{1+i} = \frac{2i}{2} = i.$$

Geometry of complex numbers

Complex numbers do not lie on the real line. Instead they reside naturally in the plane. We think of the x-axis as the "real" axis in the plane and the y-axis as the "imaginary" axis. We then plot the complex number $a + bi$ at the point (a, b). For example, the imaginary number i is located at the point $(0, 1)$ in the plane, and $1 + i$ is located at $(1, 1)$.

The **magnitude** of a complex number z is the distance from z to the origin in the complex plane. We denote the magnitude of z by $|z|$, so we have

$$|z| = \sqrt{a^2 + b^2}.$$

The angle formed by the positive x-axis and a straight line from the origin to z (measured in the counterclockwise direction) is called the **polar angle** of z. For example, the number i has polar angle $\pi/2$ and magnitude 1. The complex number $-3i$ has polar angle $3\pi/2$ and magnitude 3, and $1 + i$ has polar angle $\pi/4$ and magnitude $\sqrt{2}$. Note that, to specify any complex number, all we have to do is give its real and imaginary parts. Alternatively, we can specify a complex number by giving its magnitude and polar angle.

Euler's formula

One of the most amazing formulas in all of mathematics is Euler's formula. This formula gives a surprising relationship between the exponential and the trigonometric functions. Euler's formula is

$$e^{i\theta} = \cos\theta + i \sin\theta$$

for any real number θ. We will see where this formula comes from in a moment, but before that, note that a special case of this result is

$$e^{i\pi} = \cos\pi + i \sin\pi = -1.$$

That is, when you combine three of the most "interesting" numbers in mathematics, namely e, i, and π, as above, the result is -1. Some people enjoy expressing this relationship as

$$e^{i\pi} + 1 = 0.$$

Euler's formula lets us compute the exponential of a complex number. Using the usual rules for exponentiation, we have

$$e^{a+bi} = e^a \, e^{ib} = e^a \, (\cos b + i \sin b) = e^a \cos b + i e^a \sin b.$$

For example,

$$e^{2+3i} = e^2 \cos 3 + i \, e^2 \sin 3,$$

and

$$e^{(2+3i)t} = e^{2t} \cos 3t + i \, e^{2t} \sin 3t.$$

Euler's formula is most easy to verify using the power series for the functions involved. Recall that

$$e^x = 1 + x + \frac{x^2}{2!} + \frac{x^3}{3!} + \frac{x^4}{4!} + \dots$$

$$\sin x = x - \frac{x^3}{3!} + \frac{x^5}{5!} - \frac{x^7}{7!} + \dots$$

and

$$\cos x = 1 - \frac{x^2}{2!} + \frac{x^4}{4!} - \frac{x^6}{6!} \dots.$$

If we substitute $i\theta$ for x in the power series for e^x, we obtain

$$e^{i\theta} = 1 + i\theta + \frac{i^2\theta^2}{2!} + \frac{i^3\theta^3}{3!} + \frac{i^4\theta^4}{4!} + \dots$$

$$= 1 + i\theta - \frac{\theta^2}{2!} - i\frac{\theta^3}{3!} + \frac{\theta^4}{4!} + \dots.$$

Collecting terms that comprise the real and imaginary parts, we find

$$e^{i\theta} = \left(1 - \frac{\theta^2}{2!} + \frac{\theta^4}{4!} - \frac{\theta^6}{6!} + \dots\right) + i\left(\theta - \frac{\theta^3}{3!} + \frac{\theta^5}{5!} - \frac{\theta^7}{7!} \dots\right)$$

$$= \cos\theta + i\sin\theta.$$

Geometry of complex multiplication

Using Euler's formula, we can give a geometric interpretation of multiplication of complex numbers. Suppose $z = a + bi$ has polar angle θ. Then z is located at the point in the plane whose coordinates are (a, b). Using trigonometry, we have $a = |z| \cos\theta$ and $b = |z| \sin\theta$, so we may write this complex number in **polar form** as

$$z = |z| \cos\theta + i|z| \sin\theta,$$

which with the aid of Euler's formula becomes

$$z = |z| \, e^{i\theta}.$$

Suppose we have two complex numbers z_1 and z_2 with polar angles θ_1 and θ_2, respectively. Then we may write these complex numbers in polar form as

$$z_1 = |z_1|\, e^{i\theta_1} \quad \text{and} \quad z_2 = |z_2|\, e^{i\theta_2}.$$

If we multiply z_1 and z_2, we find

$$z_1 z_2 = \left(|z_1|\, e^{i\theta_1}\right)\left(|z_2|\, e^{i\theta_2}\right)$$

$$= |z_1||z_2|\, e^{i(\theta_1+\theta_2)}.$$

Thus $z_1 z_2$ is the complex number whose magnitude is $|z_1||z_2|$ and whose polar angle is $\theta_1 + \theta_2$. Therefore the product of two complex numbers is the complex number whose magnitude is the product of the magnitudes of the factors and whose polar angle is the sum of the polar angles of the factors.

An important special case of this geometric interpretation of complex multiplication involves $w = 1/z$. First we note that $|1| = 1$ and the polar angle of 1 is 0. Then since $zw = 1$, we see that

$$|w| = \frac{1}{|z|}$$

and the polar angle of w is the negative of the polar angle of z. In other words, if

$$z = |z|\, e^{i\theta}, \quad \text{then} \quad \frac{1}{z} = \frac{1}{|z|}\, e^{-i\theta}.$$

HINTS AND ANSWERS

Hints and Answers for Section 1.1

1. (a) $P = 0$ and $P = 230$.
  (b) For $0 < P < 230$.
  (c) For $P > 230$ or the nonphysical values $P < 0$.

3. (a) $y = -3$, $y = 0$, and $y = 4$.
  (b) For $-3 < y < 0$ or $y > 4$.
  (c) For $y < -3$ or $0 < y < 4$.

5. $L = 0$.

7. (a) Beth.
  (b) Jillian.
  (c) They have the same rate.

9. (a) $\lambda = \ln(2)/5230 \approx 0.000132533$.
  (b) $\lambda = \ln(2)/8 \approx 0.0866434$.
  (c) 1/year for C-14, 1/day for I-131.
  (d) Yes.

11. Count the number of C-14 atoms that decay in a given time period. Divide this number by the time elapsed to obtain an approximation for dr/dt. Then use $dr/dt = -\lambda r$ to find r.

13. (a) $dP/dt = k(1 - P/N)P - 100$.
  (b) $dP/dt = k(1 - P/N)P - P/3$.
  (c) $dP/dt = k(1 - P/N)P - a\sqrt{P}$, where a is a positive parameter.

15. For example, one model is $dR/dt = kR(R/M - 1)$, where M is a parameter that corresponds to the threshold at which the population is too small to sustain itself over the long term. There are other reasonable models based on the stated assumptions.

17. (a) Logistic model.
  (b) Carrying capacity 64; growth-rate ≈ 0.38.
  (c) Prediction for population today is 64.

19. (a) System (i).
  (b) System (ii).
  (c) System (i).

21. (a) The species cooperate.
  (b) The species compete.

Hints and Answers for Section 1.2

1. (a) Bob and Glen.
  (b) The equilibrium solution $y = -1$.

3. $dy/dt = 3t^2 y$.

5. $y(t) = ke^{t^2/2}$, where k is any real number.

7. $y(t) = ke^{2t} - 1/2$, where k is any real number.

9. $y(t) = \ln(t + C)$, where C is any real number.

11. $y(t) = \pm\sqrt{\ln\left(k(t^2 + 1)\right)}$, where k is any positive real number. The choice of sign is determined by the initial condition.

13. $y = \left(-1 \pm \sqrt{4t + c}\right)/2$, where the sign is determined by the initial condition.

15. $y(t) = ke^t/(ke^t + 1)$, where k is any real number, and the equilibrium solution $y = 1$. Note that this is a special case of the logistic equation with growth-rate parameter 1 and carrying capacity 1.

17. $y(t) = -1 + ke^{t+t^3/3}$, where k is any real number.

19. $y^2/2 + \ln|y| = e^t + c$, where c is any real number, and the equilibrium solution $y = 0$.

21. $w = kt$ for $t > 0$ or $t < 0$, where k is any real number. (The differential equation is not defined at $t = 0$.)

23. $y^5/5 + 3y^2/2 = t^3/3 + t + c$, where c is any real number.

25. $y = 7e^{2t}/2 - 1/2$

27. $y(t) = 1/(t + 2)$

29. $y(t) = 0$

31. $y(t) = -\sqrt{4 + (2/3)\ln(t^3 + 1)}$

33. $y(t) = \tan(t^2/2 + \pi/4)$

35. **(a)** Amount of salt ≈ 0.238 lbs.
   (b) Amount of salt ≈ 1.58 lbs.
   (c) Amount of salt ≈ 2.49 lbs.
   (d) Amount of salt ≈ 2.50 lbs.
   (e) Amount of salt ≈ 2.50 lbs.

37. **(a)** The initial-value problem is
$$dT/dt = -0.2(T - 70), \quad T(0) = 170.$$
   (b) $t = (\ln 0.4)/ - 0.2 \approx 4.6$

39. $C(t) = 20/3 + ke^{-3t/100}$, where k is any real number. $C(t) = 20/3$ is an equilibrium solution.

41. **(a)** At 7%, \$278,735; at 6.85% with points, \$280,009.
   (b) Choose the 7% option.
   (c) Choose the 7% option.

Hints and Answers for Section 1.3

1.

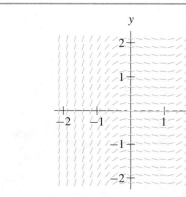

3.

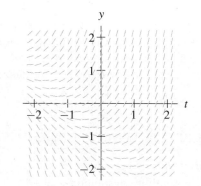

5.

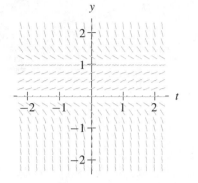

7. (a)

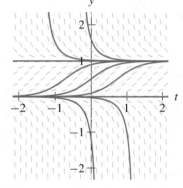

(b) The solution approaches the equilibrium value 1 from below.

9. (a)

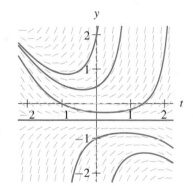

(b) The solution with $y(0) = 1/2$ tends to infinity as t increases and as t decreases.

11. (a) iv.   **(b)** v.   **(c)** viii.   **(d)** iii.

13. (a) On the line $y = 3$ in the ty-plane, the slope field is a segment with slope -1.

(b) No. Solutions with $y(0) < 3$ satisfy the inequality $y(t) < 3$ for all $t > 0$.

15.

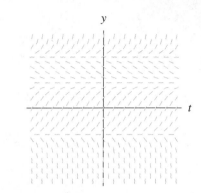

17. (a) This is enough information to give the slope field for all points (t, y) with $y > 0$.

(b)

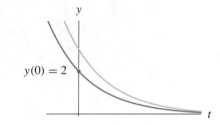

The solution with $y(0) = 2$ is a translation to the left of the given solution.

19. $v_c(t) = K + ke^{-t/RC}$

21. (a)

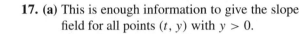

(b)

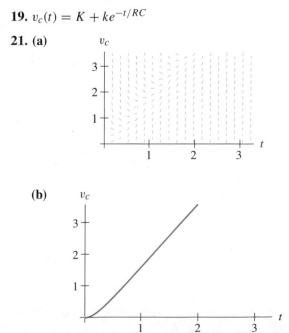

(c)

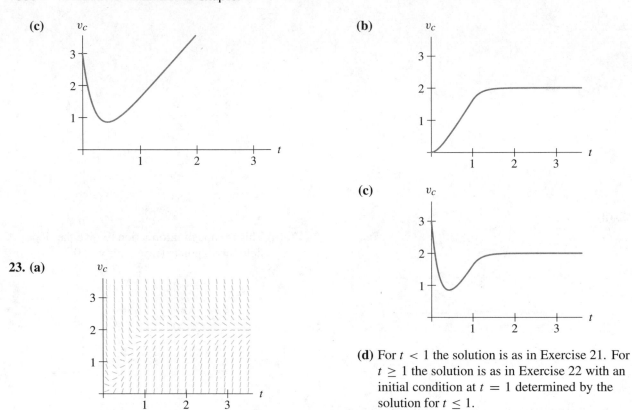

(b)

(c)

(d) For $t < 1$ the solution is as in Exercise 21. For $t \geq 1$ the solution is as in Exercise 22 with an initial condition at $t = 1$ determined by the solution for $t \leq 1$.

23. (a)

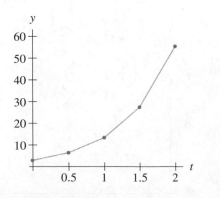

Hints and Answers for Section 1.4

1. *Hint*: When $t = 2$ the Euler approximation is $y \approx 55.5$.

3. *Hint*: When $t = 2$ the Euler approximation is $y \approx 25.49$.

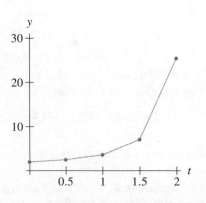

5. Selected approximate values: $w_1 = -1$, $w_2 = -1$, and $w_5 = -1$.

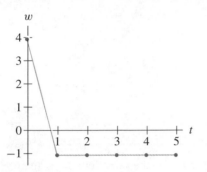

7. Selected approximate values: $y_1 = 3.36$, $y_2 = 4.27$, and $y_4 = 5.81$.

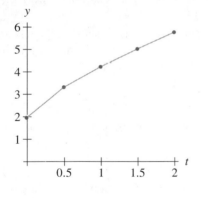

9. The answer of Exercise 8 is a translate to the right of the answer of Exercise 7. Why?

11. *Hint*: What are the constant solutions? From qualitative analysis, solutions with initial condition $w(0) = 0$ should increase until reaching the equilibrium solution at $w = 3$. However, the numerical solution indicates that the solution oscillates about $w = 3$.

13. *Hint*: What is the concavity of the solution? Are the approximations from Euler's method going to be above or below the real solution?

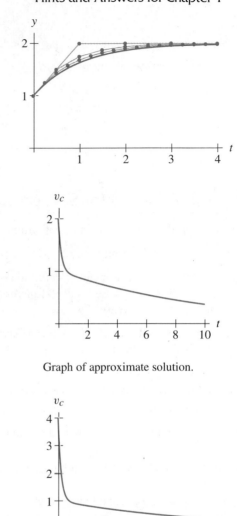

15.

Graph of approximate solution.

17.

Graph of approximate solution.

19. (a)

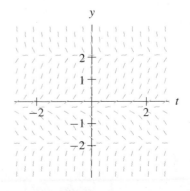

(b)

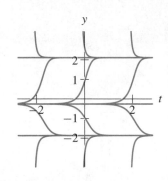

(c) The roots of the $p(y)$ are equilibria of the differential equation.

(d) $y \approx 2.115$, $y \approx -1.861$ and $y \approx -0.254$

Hints and Answers for Section 1.5

1. $y(t) < 3$ for all t in the domain of definition of $y(t)$

3. $-t^2 < y(t) < t + 2$ for all t

5. $y(t) > 3$ for all t in the domain of definition of $y(t)$, $y(t)$ tends to 3 as t decreases and $y(t)$ increases without bound as t increases

7. $0 < y(t) < 2$ for all t, and $y(t) \to 2$ as t increases and $y(t) \to 0$ as t decreases

9. (a) Substitute each solution into the differential equation and compute.

(b) Use the Existence and Uniqueness Theorem.

11. (a) *Hint*: Differentiate $y_1(t)$ at t_0 and remember that $y_1(t)$ is a solution.

(b) *Hint*: Remember that the equation is autonomous.

(c) *Hint*: Look at the slope field, but check by substituting $y_2(t)$ into both sides of the differential equation.

(d) *Hint*: Uniqueness theorem.

(e) *Hint*: Do the same four steps again.

13. (a) If $y_1(t) = 0$, then $dy_1/dt = 0 = y_1/t^2$.

(b) For any real number c, let

$$y_c(t) = \begin{cases} 0, & \text{for } t \leq 0; \\ ce^{-1/t}, & \text{for } t > 0. \end{cases}$$

The function $y_c(t)$ satisfies the equation for all $t \neq 0$. It is 0 for $t < 0$ and nonzero for $t > 0$.

(c) The differential equation, y/t^2, is not defined at $t = 0$.

15. (a) $y(t) = -1 + \sqrt{1 + \ln((1 - t/2)^2)}$

(b) Domain of solution is $t \leq 2(1 - 1/\sqrt{e})$.

(c) $y(t) \to -1$ as $t \to 2(1 - 1/\sqrt{e})$.

17. (a) $y(t) = 2 - \sqrt{t^2 + 3}$

(b) Domain of definition is all real numbers.

(c) $y(t) \to -\infty$ as $t \to \pm\infty$.

Hints and Answers for Section 1.6

1.

$y = 1$ • sink

$y = 0$ • source

3.

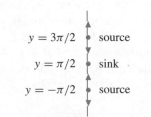

$y = 3\pi/2$ • source

$y = \pi/2$ • sink

$y = -\pi/2$ • source

... $- 3\pi/2, \pi/2, 5\pi/2, ...$ are sinks and
... $- \pi/2, 3\pi/2, 7\pi/2, ...$ are sources.

5. The sinks are $w = 0, -2\pi, -4\pi \ldots$ and $\pi, 3\pi, 5\pi$ $\ldots$. The sources are $w = 2$, $w = -\pi, -3\pi, \ldots$ and $w = 2\pi, 4\pi, 6\pi, \ldots$.

 $w = \pi$ sink

$w = 2$ source

$w = 0$ sink

7. There are no equilibrium points.

9.

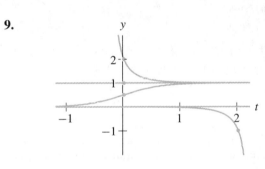

11.

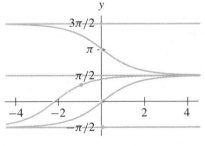

13.

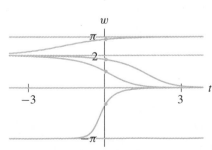

15.

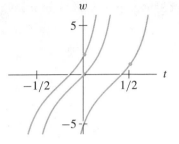

17. Since the initial condition, $y(0) = 1$, is between the roots of $y^2 - 4y + 2$, the solution $y(t)$ to the initial-value problem is always between $2 - \sqrt{2}$ and $2 + \sqrt{2}$. The limit of $y(t)$ as $t \to \infty$ is $2 - \sqrt{2}$ and the limit of $y(t)$ as $t \to -\infty$ is $2 + \sqrt{2}$.

19. The solution remains below the equilibrium point $2 - \sqrt{2}$ and is increasing for all t for which it is defined. The limit of $y(t)$ as $t \to \infty$ is $2 - \sqrt{2}$.

21. Since the initial condition, $y(3) = 1$, is between the roots of $y^2 - 4y + 2$, the solution $y(t)$ to the initial-value problem is always between $2 - \sqrt{2}$ and $2 + \sqrt{2}$. The limit of $y(t)$ as $t \to \infty$ is $2 - \sqrt{2}$ and the limit of $y(t)$ as $t \to -\infty$ is $2 + \sqrt{2}$.

23.   **25.**

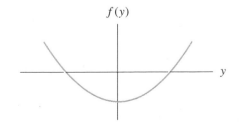

27. The following graph is one of many possible answers.

29. The following graph is one of many possible answers.

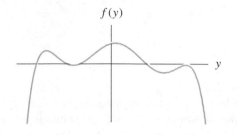

$f(y)$

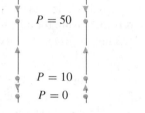

31. (a)

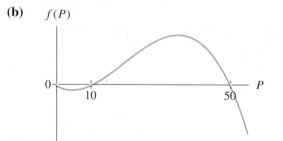

$P = 50$

$P = 10$
$P = 0$

(b) $f(P)$

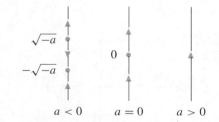

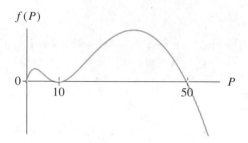

$f(P)$

(c) The functions $f(P) = P(P - 10)(50 - P)$ and $f(P) = P(P - 10)^2(50 - P)$ are two possibilities.

33. (a) *Hint*: Use the Intermediate Value Theorem.

(b) *Hint*: At a source $f(y)$ crosses the y-axis from negative to positive. Use the Intermediate Value Theorem twice.

35. (a) Source

(b) Sink

(c) Node

37. The term βx is the effect of the passengers, and $-\alpha$ is the term giving the rate of decrease of the time between trains when no passengers are present. Both α and β should be positive.

39. The only equilibrium point, $x = \alpha/\beta$, is a source. If the initial gap between trains is too large, then x will increase without bound. If it is too small, x will decrease to zero. It is very unlikely that the time between trains will remain constant for long.

Hints and Answers for Section 1.7

1.

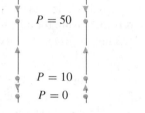

3.

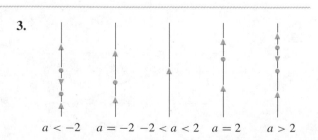

5. Only one bifurcation at $\alpha = 1$. A sink/source pair of equilibria for $\alpha < 1$ collide and form a node at $\alpha = 1$. No equilibria for $\alpha > 1$.

7. (a) $L = 16$

(b) If the population is above 40 when the licenses are issued, the population heads toward 60. If the population is below 40 when fishing begins, the fish will become extinct.

(c) As long as the fish population is well over 40 when 16 licenses are issued, unexpected perturbations in the system will not drastically effect the fish and the population will tend toward the equilibrium point near 60. (What happens if $16\frac{2}{3}$ licenses are issued?)

9. (a)

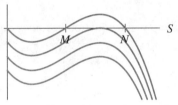

(b) Before the bifurcation, there are two positive equilibria—the smallest of which is a source while the largest is a sink. If the fox squirrel population is initially above the source, then the population will tend toward the sink. After the bifurcation, there are no positive equilibria and thus the fox squirrel population will plummet toward zero.

(c) $E = \dfrac{k}{27MN}\big((N - 2M)(M + N)(2N - M) +$

$2(M^2 - MN + N^2)^{3/2}\big)$

11. The discriminant, $D = \alpha^2 - 4\beta$. When D is positive, there are two equilibria, when D is zero, one equilibrium and when D is negative, no

equilibria. In the figure shaded region corresponds to phase lines with two equilibria. On the parabola, one. Above the parabola, none.

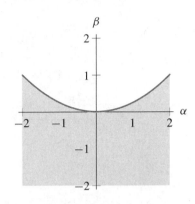

13. *Hint:* What are the signs of $f_0(0)$ and $f_0(1)$ and when (if ever) can they change as α is changed?

15. Since the sum of the indices is zero, the number of sinks and sources is the same when $\alpha = 0$. Moreover, because $f_\alpha(0) > 0$ and $f_\alpha(1) \neq 0$ for all α and f_α is continuous, there is a negative minimum value for f_α in $0 \leq y \leq 1$. Choosing a positive number M_α larger in absolute value than this minimum gives $f_\alpha + M_\alpha > 0$. Try sketching several functions which satisfy the conditions of $f_0(y)$ on the interval from 0 to 1.

17. *Hint:* There is one equilibrium for $\alpha > 0$, three equilibria for $\alpha < 0$.

19. $dy/dt = y(y + \alpha)(y^2 + \alpha)(y^4 + \alpha)$ is one of an infinite number of examples.

Hints and Answers for Section 1.8

1. $y(t) = t + c/t$

3. $y(t) = \dfrac{3e^{-t}}{2} + ce^t$

5. $y(t) = t + (1 + t^2)(\arctan(t) + c)$

7. $y(t) = t^2 - 2t + 2 + ce^{-t}$

9. $y(t) = \dfrac{t^2 + 2t + 3}{1 + t}$

11. $y(t) = t + \dfrac{2}{t}$

13. $y(t) = 2t^3 + 5t^2$

15. $y(t) = 4e^{-\cos t} \displaystyle\int e^{\cos t}\, dt$

17. $y(t) = 4e^{-1/t} \displaystyle\int e^{1/t} \cos t\, dt$

19. $a = -2$

21. (a) $dP/dt = .04P + 520$
   (b) about \$3,488.94

23. Approximately 5.08 years (61 months)

25. 4.25 parts per billion

27. (a) $dy/dt = 1/2 - y/(V_0 + t)$ with $y(0) = 0$
   (b) Note that if $V_0 = 0$, then the differential equation is undefined at $t = 0$. The amount of salt in the tank at time t is $t/4$.

Hints and Answers for Section 1.9

1. $du/dt = u^2 + u$

3. $du/dt = u/t + t(u + u^2 + \cos u)$

5. If $u = y - t$, the equation becomes $du/dt = u^2 - u - 2$.

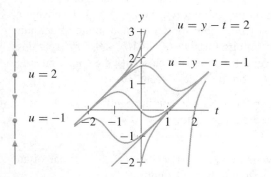

7. If $u = ty$, the equation becomes $du/dt = u \cos u$.

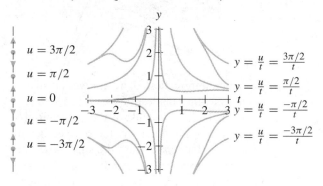

9. $y(t) = t^3(1 + t)/4 + c/t + c$

11. $dy/dt = -y^2 + 6ty + y - 9t^2 - 3t + 3$

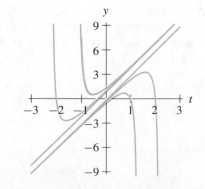

13. $dy/dt = 2y(1 - \sqrt{y})$

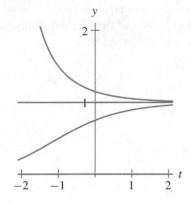

15. (a) $dS/dt = 3 - 3S/(10 + 3t)$

(b) $dC/dt = 3(1 - 2C)/(10 + 3t)$

(c) $C = 1/2$

(d) $C(5) = 0.42$

17. (a) $y = 10^{-1/3}$

(b) With $u = y - 10^{-1/3}$, the linear approximation is $du/dy = 3\sqrt[3]{10}\,u$. The origin is a source.

(c) The solution grows like $e^{3\sqrt[3]{10}\,t}$. The time necessary to double the distance from the equilibrium point is $t = (\ln 2)/(3\sqrt[3]{10})$.

19. (a) $y = -1, 3$

(b) With $u = y + 1$, the linearization is $du/dt = 4u$. $u = 0$ is a source. With $v = y - 3$, the linearization is $dv/dt = -4v$. $v = 0$ is a sink.

(c) For $y = -1$, the solution grows like e^{4t}. The time necessary to double the distance from the equilibrium point is $t = \ln 2/4$. For $y = 3$, the solution grows like e^{-4t}. The time necessary to halve the distance from the equilibrium point is $t = \ln 2/4$.

21. (a) $y - y^3/3!$

(b) $dy/dt = -3y$

(c) $y(2) = 0.04e^{-6}$

23. (a) With $u = P - 100$,
$du/dt = -0.05u(u/100 + 1) - 0.02(1 + \sin t)$.

(b) $du/dt = -0.05u - 0.02(1 + \sin t)$

(c) $u(t) = -0.4 - 2\sin t/2005 + 8\cos t/401 + ke^{-0.05t}$

(d) For large t, $ke^{-0.05t}$ becomes negligible. Also $u(t)$ fluctuates between -0.38 and -0.42. Thus P fluctuates around 99.6.

25. (a) $du/dt = f(u + y_0)$

(b) Using the Taylor expansion at $u = 0$, one obtains $f(u + y_0) = f(y_0) + f'(y_0)u + \ldots$. The linearization is $du/dt = f'(y_0)u$.

Hints and Answers for Section 2.1

1. System (i) corresponds to large predators and small prey. System (ii) corresponds to small predators and large prey.

3. *Hint*: Compute dy/dt for $y = 0$.

5. *Hint*: Compute dx/dt for $x = 0$.

7. The populations oscillate with decreasing amplitude about an equilibrium point with both populations nonzero.

9. Change only the dR/dt equation,

(i) $dR/dt = 2R - 1.2RF - \alpha$

(ii) $dR/dt = R(2 - R) - 1.2RF - \alpha$

11. In both systems, let $dF/dt = kF + 0.9RF$, where $k > 0$ is the growth-rate parameter for the predator population (ignoring any effects of overcrowding).

13. In both systems, make
$dF/dt = -F + 0.9FR + k(R - 5F)$,
where k is an immigration-rate parameter.

15. System (i) corresponds to boa constrictors and system (ii) to small cats. *Hint*: Suppose there were one predator ($y = 1$), how many units of prey are required to keep $dy/dt = 0$?

17. (a) *Hint*: Follow the behavior of a solution with initial condition near $(0, 0)$ (after application of the pesticide kills almost all of both predator and prey).

(b) After applying the pest control, you may see an explosion of the pest population due to the absence of the predator.

19. (a) $\dfrac{d^2(\sin t)}{dt^2} + \sin t = -\sin t + \sin t = 0$

(b)

(c) They are the same curve in the yv-plane.

(d) They are parameterized differently.

21. 4 lbs. per in. or 48 lbs. per ft.

23. Large

25. $da/dt = -\alpha ab$, $db/dt = -\alpha ab$, where α is the reaction-rate parameter.

27. $da/dt = k_1 - \alpha ab$, $db/dt = k_2 - \alpha ab$, where k_1 and k_2 are the rates A and B are added, respectively.

29. $da/dt = k_1 - \alpha ab + \gamma b^2/2$ and $db/dt = k_2 - \alpha ab - \gamma b^2$, where γ is the reaction-rate parameter for the reaction turning 2 B's into an A.

Hints and Answers for Section 2.2

1. (a) $\mathbf{V}(x, y) = (1, 0)$

(b) See part (c).

(c)

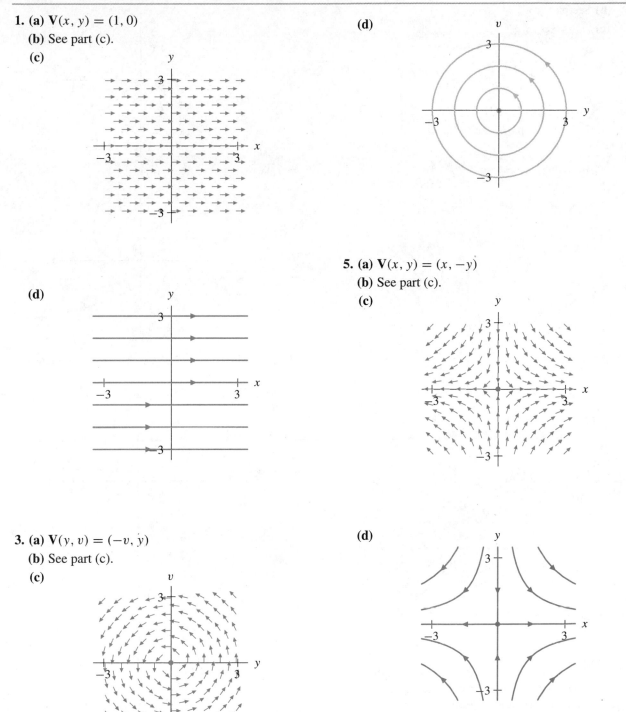

(d)

5. (a) $\mathbf{V}(x, y) = (x, -y)$

(b) See part (c).

(c)

(d)

3. (a) $\mathbf{V}(y, v) = (-v, y)$

(b) See part (c).

(c)

(d)

7. (a) $\mathbf{V}(y, v) = (v, y)$

(b) See part (c).

(c)

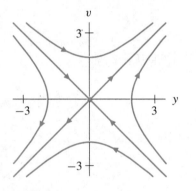

(d)

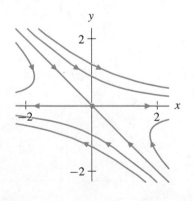

9. (a)

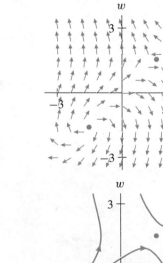

(b) The solution tends to the origin along the line $y = -x$ in the xy-phase plane. Therefore both $x(t)$ and $y(t)$ tend to zero as $t \to \infty$.

11. (a) Equilibrium is $(-1/9, 2/9)$.

(b)

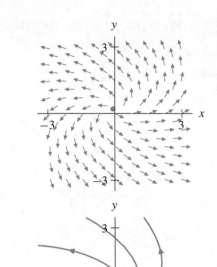

(c)

13. (a) Equilibria: $(\pi/2 + k\pi, \pi/2 + k\pi)$, where k is any integer.

(b)

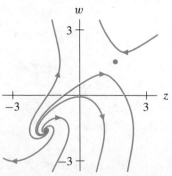

(c)

15. (a) Equilibria: $(\pi/2 + k\pi, 0)$, where k is any integer.

(b)

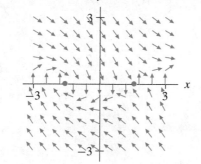

(c)

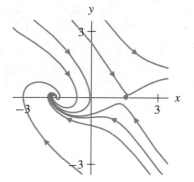

17. System ii.

19. System viii.

21.

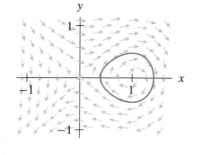

23.

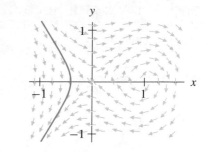

25.

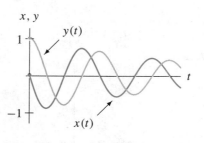

27.

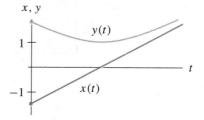

29.

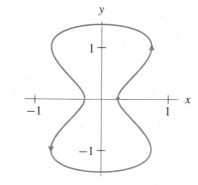

31. They will not collide.

Hints and Answers for Section 2.3

1. $(x(t), y(t))$ is a solution.

3. $(x(t), y(t))$ is not a solution.

5. $y(t) = e^{-2t}$ is not a solution.

7. $x(t) = k_2 e^{2t} - \frac{k_1}{3} e^{-t}$, $y(t) = k_1 e^{-t}$

9. (a) $x(t) = e^{2t}$, $y(t) = 0$
  (b)

  (c)

11. (a) $x(t) = \frac{1}{3} e^{2t} - \frac{1}{3} e^{-t}$, $y(t) = e^{-t}$
  (b)

(c)

13. (a) See part (c).
  (b) $y_1(t) = e^{2t}$; $y_2(t) = e^{-5t}$
  (c)

15. (a) See part (c).

(b) $y_1(t) = e^{(-2+\sqrt{3})t}$; $y_2(t) = e^{(-2-\sqrt{3})t}$

(c)

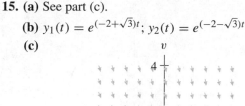

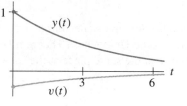

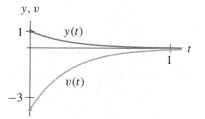

17. $m \dfrac{d^2 y}{dt^2} = -(k_1 + k_2)y + k_1 L_1 - k_2 L_2 +$

$\dfrac{k_2 - k_1}{2} - b\dfrac{dy}{dt}$, where $y = 0$ is the center,

$y < 0$ is to the left and $y > 0$ is to the right.

19. (a) $(x(t), y(t)) = (k_2 e^{-t+k_1 e^t}, -1 + k_1 e^t)$

(b) $(x, y) = (0, -1)$

(c) $(x(t), y(t)) = (\frac{1}{e} e^{-t+e^t}, -1 + e^t)$

(d)

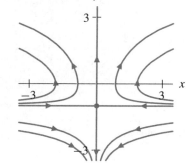

Hints and Answers for Section 2.4

1. (a) It is a solution.

(b) At $t = 10$ the Euler approximation is approximately $(-9.21, 1.41)$ and the actual solution is approximately $(-0.84, -0.54)$.

(c) At $t = 10$ the Euler approximation is approximately $(-1.41, -0.85)$ and the actual solution is approximately $(-0.84, -0.54)$.

(d) The approximation will always spiral away from the origin.

3. (a) $(x_5, y_5) \approx (0.65, -0.59)$

(b)

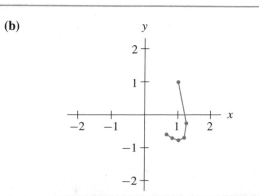

(c)

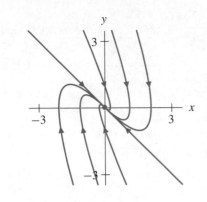

(d) It is consistent.

5. (a) $(x_5, y_5) \approx (1.94, -0.72)$

(b)

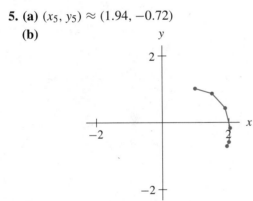

(c)

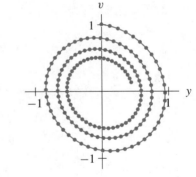

(d) It is consistent.

7. *Hint*: Substitute $\mathbf{Y}_1(t) = (e^{-t}\sin(3t), e^{-t}\cos(3t))$ into the differential equation.

9. They both give the same spiral. This does not contradict the Uniqueness Theorem because they are never in the same place at the same time.

11.

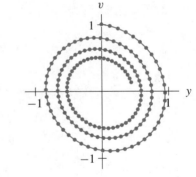

13. Assuming the vector field satisfies the hypotheses of the Uniqueness Theorem, they will not collide.

15. *Hint*: See Exercise 14

Hints and Answers for Section 2.5

1. (a) *Hint*: Substitute $(0, 0, 0)$ and $(\pm6\sqrt{2}, \pm6\sqrt{2}, 27)$ to see if all derivatives are zero.

(b) *Hint*: Set $dx/dt = dy/dt = dz/dt = 0$ and solve three simultaneous equations.

3. (a) Note that dx/dt and dy/dt vanish if $x = y = 0$.

(b) The solution is $x(t) = 0$, $y(t) = 0$, and $z(t) = e^{-8t/3}$.

(c) The solution is $x(t) = 0$, $y(t) = 0$, and $z(t) = z_0 e^{-8t/3}$.

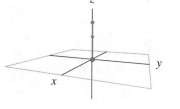

5. (a)

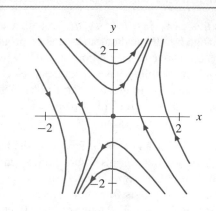

(b)

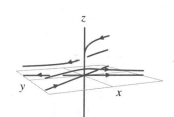

(c)

Hints and Answers for Section 3.1

1. In this case, $dx/dt = dy/dt$, so both x and y
change by identical amounts. Paul's profitability
has a positive affect on both x and y while
Bob's profitability has a negative affect. If
Paul's profits are greater than Bob's, then
the profits of both stores will increase.

3. Since $a = 1$ and $b = 0$, Paul's profits have
a positive affect on his future profits, but
Bob's profits have no affect. Since $c = 2$
and $d = 1$, profits from both stores positively
affect Bob's profits. However, the affect of
Paul's profits is twice that of Bob's profits.

5. $\mathbf{Y} = \begin{pmatrix} x \\ y \end{pmatrix}, \quad \dfrac{d\mathbf{Y}}{dt} = \begin{pmatrix} 2 & 1 \\ 1 & 1 \end{pmatrix} \mathbf{Y}$

7. $\mathbf{Y} = \begin{pmatrix} p \\ q \\ r \end{pmatrix}, \quad \dfrac{d\mathbf{Y}}{dt} = \begin{pmatrix} 3 & -2 & -7 \\ -2 & 0 & 6 \\ 0 & 7.3 & 2 \end{pmatrix} \mathbf{Y}$

9. $\dfrac{dx}{dt} = \beta y$

$\dfrac{dy}{dt} = \gamma x - y$

11. (a)

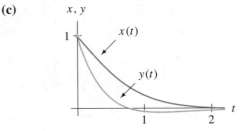

(b)

(c)

13. (a)

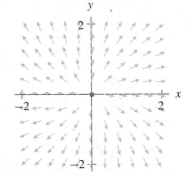

(b)

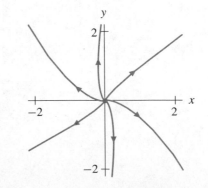

(c)

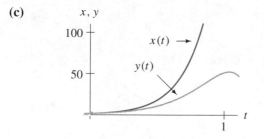

15. If $a \neq 0$, then the set of all equilibrium points (x_0, y_0) is exactly those points that lie on the line $x_0 = (-b/a)y_0$. If $a = 0$ and $b \neq 0$, then the set of all equilibrium points is the x-axis. Finally, if $a = b = 0$, then the set of all equilibrium points consists of the line defined by the equation $cx_0 + dy_0 = 0$.

17. (a) Equilibrium line $v = 0$.
   (b) Equilibrium line $v = 0$.

19. *Hint*: Let $v = dy/dt$ and $w = d^2y/dt^2$.

21. $\gamma > 0$

23. $\delta < 0$

25. (a) Substitute $\mathbf{Y}(t)$ into the differential equation and check that the left-hand side equals the right-hand side.
   (b) The solution with this initial condition is $(-2te^{2t}, (2t + 2)e^{2t})$.

27. (a) They are solutions.
   (b) They are linearly independent.

(c) $\mathbf{Y}(t) = (e^{-3t} + e^{-4t}, e^{-3t} + 2e^{-4t})$

29. (a) $\mathbf{Y}_1(t)$ is a solution, but $\mathbf{Y}_2(t)$ is not.

31. We want to show that one of the vectors is a multiple of the other.
   (a) $(x_1, y_1) = (0, 0)$ and any (x_2, y_2) are on the same line through the origin.
   (b) (x_1, y_1) is a multiple of (x_2, y_2), which implies that they are on the same line through the origin.
   (c) If $x_1 \neq 0$, we have $(x_2, y_2) = (x_2/x_1)(x_1, y_1)$. Then we can use part (b). If $x_1 = 0$ and $y_1 \neq 0$, then proceed as before. If both $x_1 = 0$ and $y_1 = 0$, then we have the situation of part (a).

33. We can give explicit solutions for (a), (c) and (d). For example, the explicit solution to part (a) is $(-2e^{-t}, 2e^{-t})$. What are the explicit solutions for parts (c) and (d)?

35. (a) $dW/dt = y_2(dx_1/dt) + x_1(dy_2/dt) - y_1(dx_2/dt) - x_2(dy_1/dt)$
   (b) Replace dx_1/dt by $ax_1 + by_1$ (and so on) in the above formula.
   (c) $W(t) = Ce^{(a+d)t}$, where C is a constant. Hence, if $C = W(0) = 0$, then $W(t) = 0$ for all t. If $W(0) \neq 0$, then $W(t) \neq 0$ for all t.
   (d) If $\mathbf{Y}_1(0)$ and $\mathbf{Y}_2(0)$ are linearly independent, then $W(0) \neq 0$ (see Exercise 32). But then $W(t) \neq 0$ for all t. Hence, $\mathbf{Y}_1(t)$ and $\mathbf{Y}_2(t)$ are linearly independent.

Hints and Answers for Section 3.2

1. (a) Eigenvalues: $\lambda_1 = -2$, $\lambda_2 = 3$
   (b) Eigenvectors: $\mathbf{V}_1 = (x_1, y_1)$ where $5x_1 = -2y_1$ and $\mathbf{V}_2 = (x_2, 0)$

(c)

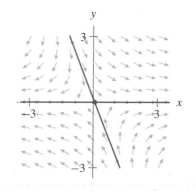

(d) $\mathbf{Y}_1(t) = e^{-2t} \begin{pmatrix} -2 \\ 5 \end{pmatrix}$, $\mathbf{Y}_2(t) = e^{3t} \begin{pmatrix} 1 \\ 0 \end{pmatrix}$

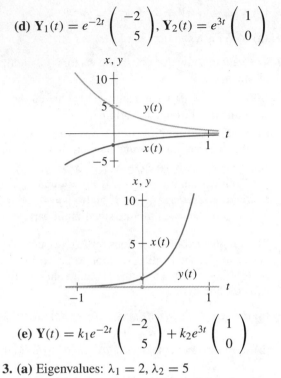

(e) $\mathbf{Y}(t) = k_1 e^{-2t} \begin{pmatrix} -2 \\ 5 \end{pmatrix} + k_2 e^{3t} \begin{pmatrix} 1 \\ 0 \end{pmatrix}$

3. (a) Eigenvalues: $\lambda_1 = 2$, $\lambda_2 = 5$

(b) Eigenvectors: $\mathbf{V}_1 = (x_1, y_1)$ where $y_1 = -x_1$
and $\mathbf{V}_2 = (x_2, y_2)$ where $x_2 = 2y_2$

(c)

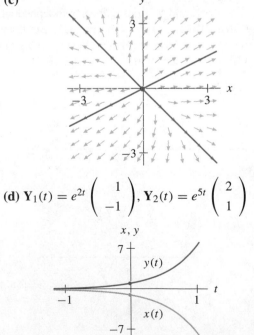

(d) $\mathbf{Y}_1(t) = e^{2t} \begin{pmatrix} 1 \\ -1 \end{pmatrix}$, $\mathbf{Y}_2(t) = e^{5t} \begin{pmatrix} 2 \\ 1 \end{pmatrix}$

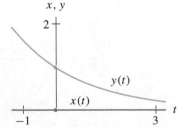

(e) $\mathbf{Y}(t) = k_1 e^{2t} \begin{pmatrix} 1 \\ -1 \end{pmatrix} + k_2 e^{5t} \begin{pmatrix} 2 \\ 1 \end{pmatrix}$.

5. (a) Eigenvalue: $\lambda_1 = -1/2$

(b) Eigenvectors: $\mathbf{V}_1 = (0, y_1)$

(c)

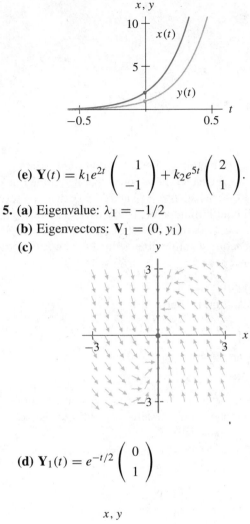

(d) $\mathbf{Y}_1(t) = e^{-t/2} \begin{pmatrix} 0 \\ 1 \end{pmatrix}$

(e) We cannot form the general solution since we do
not have two distinct eigenvalues.

7. (a) Eigenvalues: $\lambda_1 = -1$, $\lambda_2 = 4$

(b) Eigenvectors: $\mathbf{V}_1 = (x_1, y_1)$ where $y_1 = -x_1$
and $\mathbf{V}_2 = (x_2, y_2)$ where $x_2 = 4y_2$

(c)

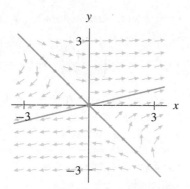

(d) $\mathbf{Y}_1(t) = e^{-t}\begin{pmatrix} 1 \\ -1 \end{pmatrix}$, $\mathbf{Y}_2(t) = e^{4t}\begin{pmatrix} 4 \\ 1 \end{pmatrix}$

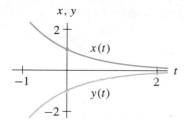

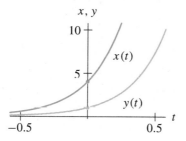

(e) $\mathbf{Y}(t) = k_1 e^{-t}\begin{pmatrix} 1 \\ -1 \end{pmatrix} + k_2 e^{4t}\begin{pmatrix} 4 \\ 1 \end{pmatrix}$

9. (a) Eigenvalues: $\lambda_1 = (3+\sqrt{5})/2$, $\lambda_2 = (3-\sqrt{5})/2$
   (b) Eigenvectors: $\mathbf{V}_1 = (x_1, y_1)$ where
   $(-1+\sqrt{5})x_1 = 2y_1$ and $\mathbf{V}_2 = (x_2, y_2)$ where
   $(1+\sqrt{5})x_2 = -2y_2$

(c)

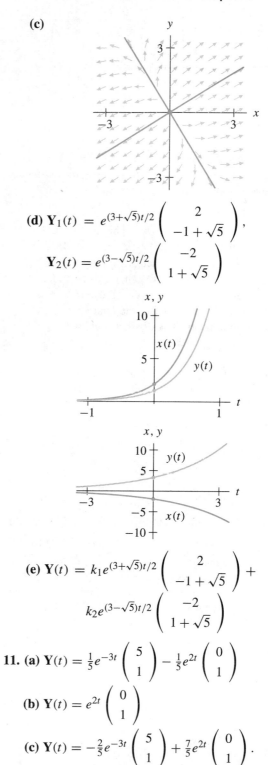

(d) $\mathbf{Y}_1(t) = e^{(3+\sqrt{5})t/2}\begin{pmatrix} 2 \\ -1+\sqrt{5} \end{pmatrix}$,

   $\mathbf{Y}_2(t) = e^{(3-\sqrt{5})t/2}\begin{pmatrix} -2 \\ 1+\sqrt{5} \end{pmatrix}$

(e) $\mathbf{Y}(t) = k_1 e^{(3+\sqrt{5})t/2}\begin{pmatrix} 2 \\ -1+\sqrt{5} \end{pmatrix} +$

   $k_2 e^{(3-\sqrt{5})t/2}\begin{pmatrix} -2 \\ 1+\sqrt{5} \end{pmatrix}$

11. (a) $\mathbf{Y}(t) = \frac{1}{5}e^{-3t}\begin{pmatrix} 5 \\ 1 \end{pmatrix} - \frac{1}{5}e^{2t}\begin{pmatrix} 0 \\ 1 \end{pmatrix}$

(b) $\mathbf{Y}(t) = e^{2t}\begin{pmatrix} 0 \\ 1 \end{pmatrix}$

(c) $\mathbf{Y}(t) = -\frac{2}{5}e^{-3t}\begin{pmatrix} 5 \\ 1 \end{pmatrix} + \frac{7}{5}e^{2t}\begin{pmatrix} 0 \\ 1 \end{pmatrix}$.

13. (a) $\mathbf{Y}(t) = \frac{2}{3}e^{-5t}\begin{pmatrix} 1 \\ -1 \end{pmatrix} + \frac{1}{3}e^{-2t}\begin{pmatrix} 1 \\ 2 \end{pmatrix}$

(b) $\mathbf{Y}(t) = e^{-5t}\begin{pmatrix} 1 \\ -1 \end{pmatrix} + e^{-2t}\begin{pmatrix} 1 \\ 2 \end{pmatrix}$

(c) $\mathbf{Y}(t) = -e^{-2t}\begin{pmatrix} 1 \\ 2 \end{pmatrix}$

15. Show that, for any vector $\mathbf{Y}_0$, $\mathbf{AY}_0 = a\mathbf{Y}_0$. Therefore, every vector is an eigenvector associated to the eigenvalue a.

17. The characteristic polynomial is $\lambda^2 - (a+d)\lambda + ad - b^2$. This quadratic has the discriminant $D = (a-d)^2 + 4b^2$. Since $D \geq 0$, the characteristic polynomial always has real roots. In other words, the matrix always has real eigenvalues. If $b \neq 0$, then $D > 0$. In this case, the characteristic polynomial always has distinct real roots.

19. (a) $\dfrac{d\mathbf{Y}}{dt} = \begin{pmatrix} 0 & 1 \\ -q & -p \end{pmatrix}\mathbf{Y}$

(b) $\lambda^2 + p\lambda + q$

(c) $(-p \pm \sqrt{p^2 - 4q})/2$

(d) The roots are distinct real numbers if $p^2 > 4q$.

(e) Note that $p^2 - 4q < p^2$.

21. (a) $dy/dt = v$ and $dv/dt = 10y - 3v$

(b) Eigenvalue $\lambda_1 = -5$ with eigenvectors $\mathbf{V}_1 = (y_1, v_1)$ where $v_1 = -5y_1$ and eigenvalue

$\lambda_2 = 2$ with eigenvectors $\mathbf{V}_2 = (y_2, v_2)$ where $v_2 = 2y_2$

(c) $\mathbf{Y}_1(t) = e^{-5t}\begin{pmatrix} 1 \\ -5 \end{pmatrix}$, $\mathbf{Y}_2(t) = e^{2t}\begin{pmatrix} 1 \\ 2 \end{pmatrix}$.

23. (a) $dy/dt = v$ and $dv/dt = -y - 4v$

(b) Eigenvalue $\lambda_1 = -2 + \sqrt{3}$ with eigenvectors $\mathbf{V}_1 = (y_1, v_1)$ where $v_1 = (-2 + \sqrt{3})y_1$ and eigenvalue $\lambda_2 = -2 - \sqrt{3}$ with eigenvectors $\mathbf{V}_2 = (y_2, v_2)$ where $v_2 = (-2 - \sqrt{3})y_2$

(c) $\mathbf{Y}_1(t) = e^{(-2+\sqrt{3})t}\begin{pmatrix} 1 \\ -2 + \sqrt{3} \end{pmatrix}$,

$\mathbf{Y}_2(t) = e^{(-2-\sqrt{3})t}\begin{pmatrix} 1 \\ -2 - \sqrt{3} \end{pmatrix}$

25. The characteristic polynomial is $\lambda^2 + \lambda + 4$, and its roots are the complex numbers $(-1 \pm \sqrt{15}\,i)/2$. Therefore there are no straight-line solutions (see the direction field).

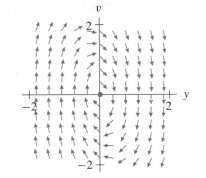

Hints and Answers for Section 3.3

1.

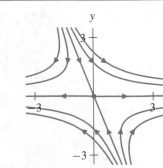

3.

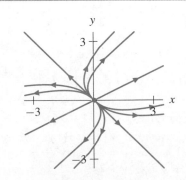

5.

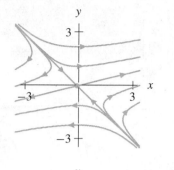

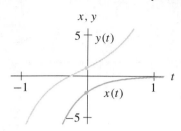

7.

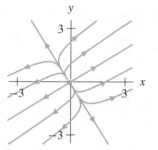

11.

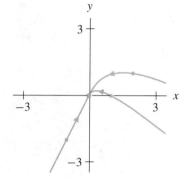

9.

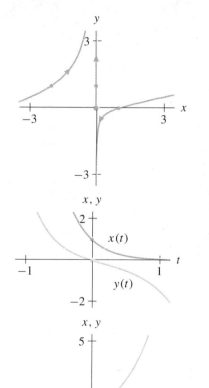

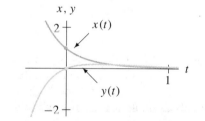

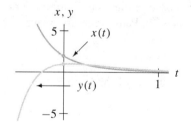

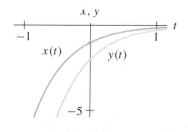

13.

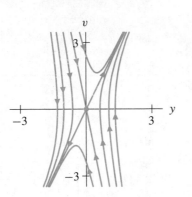

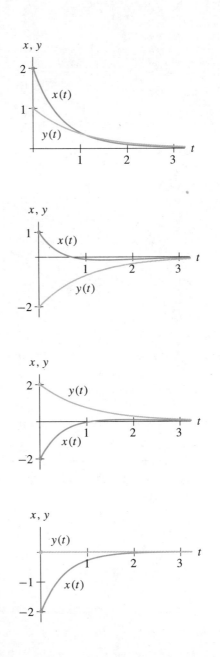

15.

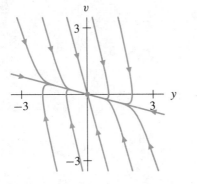

17. For any initial condition, Bob's profits y are eventually asymptotic to 0. For an initial condition on the line $y = -3x$, Paul's profits x also are eventually asymptotic to 0. For an initial condition to the left of the line $y = -3x$, Paul eventually goes broke. For an initial condition to the right of the line $y = -3x$, Paul eventually makes a fortune.

19. (a) sink

  (b) all solutions of the form $e^{-2t}(x, 0)$ and $e^{-t}(x, 2x)$

  (c)

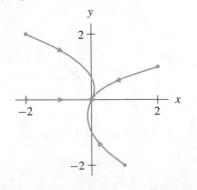

21. (a) $\dfrac{d^2 y}{dt^2} + 7\dfrac{dy}{dt} + 6y = 0$;

  As a system: $dy/dt = v,\ dv/dt = -6y - 7v$

  (b) $\lambda^2 + 7\lambda + 6 = 0$

  (c) $\lambda_1 = -6,\ \lambda_2 = -1$

  (d) The solution never crosses the rest position. It tends to the rest position like e^{-t}.

23. (a) The presence of the new species has a negative effect on the native population, but the native fish do not affect the population of the new fish.

(b) The model agrees with the premises of the system.

(c)

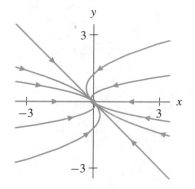

The straight-line solutions lie on the x-axis and the line $y = -x$ in the phase plane. Solutions along these lines tend toward the origin.

(d) If a small number of new fish are added to the lake, the native population drops from equilibrium. The new fish die out as the native fish return to equilibrium.

25. (a) The native fish have no affect on the population of the new fish. The introduction of new fish increases the population of the native fish.

(b) The model agrees with the system described in the problem.

(c)

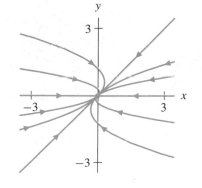

The lines containing the straight-line solutions are the x-axis and the line $y = x$. The solutions along these lines tend toward the origin. The origin is a sink.

(d) The population of the native fish initially increases and then returns toward its equilibrium value. The new fish will die out.

27. (a) Calculate the eigenvalues.

(b) Eigenvalues: $\lambda_1 = 2$, $\lambda_2 = -2$
Lines of eigenvectors: $y = 4x$, $y = 0$

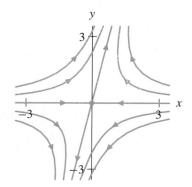

(c)

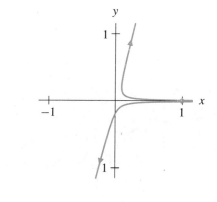

(d) $t \approx 1.96$

Hints and Answers for Section 3.4

1. $\mathbf{Y}_{\mathrm{re}}(t) = e^t \begin{pmatrix} 2\cos 3t - \sin 3t \\ \cos 3t \end{pmatrix}$

$\mathbf{Y}_{\mathrm{im}}(t) = e^t \begin{pmatrix} \cos 3t + 2\sin 3t \\ \sin 3t \end{pmatrix}$

3. (a) $\lambda_1 = 2i$, $\lambda_2 = -2i$
   (b) Center
   (c) Natural period π; Natural frequency $1/\pi$
   (d) Clockwise
   (e)

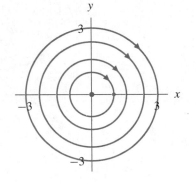

5. (a) $-1 \pm i\sqrt{2}$
   (b) Spiral sink
   (c) Natural period $= \sqrt{2}\,\pi$, natural frequency $= 1/(\sqrt{2}\,\pi)$
   (d) Clockwise

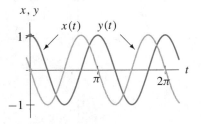

(e)

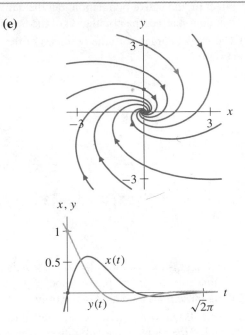

7. (a) $(3 \pm \sqrt{47}\,i)/2$
   (b) Spiral source
   (c) Natural period: $\frac{4\pi}{\sqrt{47}}$; Natural frequency: $\frac{\sqrt{47}}{4\pi}$.
   (d) Counterclockwise
   (e)

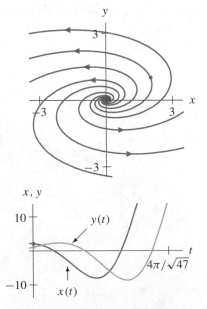

9. (a) $\mathbf{Y}(t) = k_1 \begin{pmatrix} \cos 2t \\ -\sin 2t \end{pmatrix} + k_2 \begin{pmatrix} \sin 2t \\ \cos 2t \end{pmatrix}$

(b) $\mathbf{Y}(t) = \begin{pmatrix} \cos 2t \\ -\sin 2t \end{pmatrix}$

(c) x, y

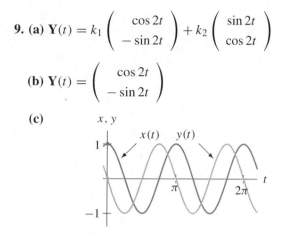

11. (a) $k_1 e^{-t} \begin{pmatrix} 2\cos\sqrt{2}\,t \\ -\sqrt{2}\sin\sqrt{2}\,t \end{pmatrix} +$

$k_2 e^{-t} \begin{pmatrix} 2\sin\sqrt{2}\,t \\ \sqrt{2}\cos\sqrt{2}\,t \end{pmatrix}$

(b) $\mathbf{Y}(t) = e^{-t} \begin{pmatrix} \sqrt{2}\sin\sqrt{2}\,t \\ \cos\sqrt{2}\,t \end{pmatrix}$

(c) x, y

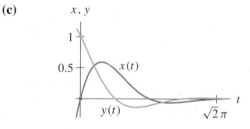

13. (a) $k_1 e^{3t/2} \begin{pmatrix} 12\cos\left(\frac{\sqrt{47}}{2}t\right) \\ \cos\left(\frac{\sqrt{47}}{2}t\right) + \sqrt{47}\sin\left(\frac{\sqrt{47}}{2}t\right) \end{pmatrix} +$

$k_2 e^{3t/2} \begin{pmatrix} 12\sin\left(\frac{\sqrt{47}}{2}t\right) \\ -\sqrt{47}\cos\left(\frac{\sqrt{47}}{2}t\right) + \sin\left(\frac{\sqrt{47}}{2}t\right) \end{pmatrix}$

(b) $e^{3t/2} \begin{pmatrix} 2\cos\left(\frac{\sqrt{47}}{2}t\right) - \frac{10}{\sqrt{47}}\sin\left(\frac{\sqrt{47}}{2}t\right) \\ \cos\left(\frac{\sqrt{47}}{2}t\right) + \frac{7}{\sqrt{47}}\sin\left(\frac{\sqrt{47}}{2}t\right) \end{pmatrix}$

(c) x, y

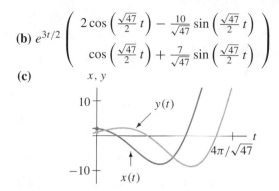

15. (a) (ii) and (v)

(b) Graph (ii) has a natural period of 1.25 and the origin is a sink. Graph (v) has a natural period of 1.25 and the origin is a source.

(c) (i) The period is not constant over time. (iii) The amplitude is not monotonically increasing nor decreasing. (iv) Oscillation stops at some t. (vi) Oscillation starts at some t. There was no prior oscillation.

17. If $\lambda_1 = \alpha + i\beta$ satisfies the equation, then $\lambda_2 = \alpha - i\beta$ does as well. A quadratic equation has at most two solutions, so they must be λ_1 and λ_2.

19. *Hint:* Suppose $\mathbf{Y}_2 = k\mathbf{Y}_1$ for some constant k. Then, $\mathbf{Y}_0 = (1 + ik)\mathbf{Y}_1$. Derive a contradiction using the fact that $\mathbf{AY}_0 = \lambda\mathbf{Y}_0$. (Note that $\mathbf{A}$ only has real entries.)

21. (a) π/β

(b) π/β

(c) $2\pi/\beta$

(d) $(\arctan(\beta/\alpha))/\beta$

23. (a) $dy/dt = v,\ dv/dt = -qy - pv$

(b) $p^2 < 4q$

(c) Spiral sink: $p^2 < 4q$ and $p > 0$; center: $p = 0$ and $q > 0$

(d) $q > 0$

25. *Hint:* Draw the phase plane and remember the Linearity Principle.

Hints and Answers for Section 3.5

1. (a) Eigenvalue: -3
 (b) Eigenvectors: $x = 0$
 (c)

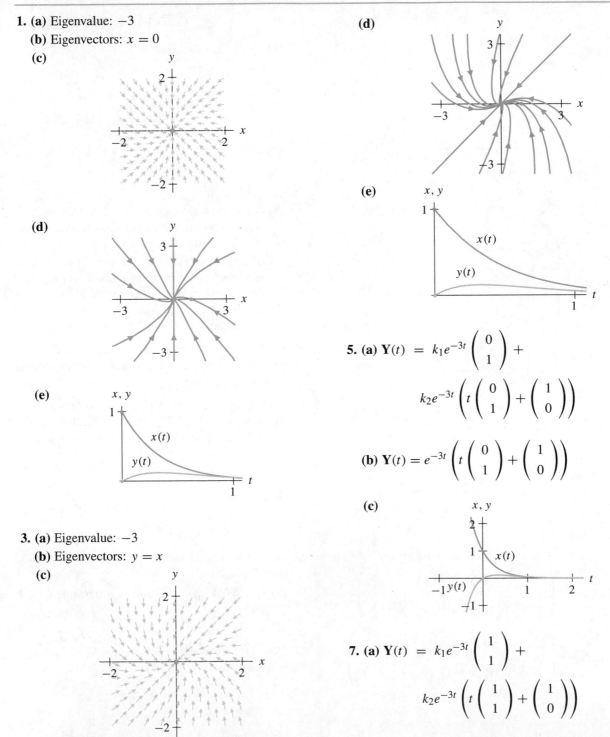

 (d)

 (e)

3. (a) Eigenvalue: -3
 (b) Eigenvectors: $y = x$
 (c)

 (d)

 (e)

5. (a) $\mathbf{Y}(t) = k_1 e^{-3t} \begin{pmatrix} 0 \\ 1 \end{pmatrix} +$

 $k_2 e^{-3t} \left(t \begin{pmatrix} 0 \\ 1 \end{pmatrix} + \begin{pmatrix} 1 \\ 0 \end{pmatrix} \right)$

 (b) $\mathbf{Y}(t) = e^{-3t} \left(t \begin{pmatrix} 0 \\ 1 \end{pmatrix} + \begin{pmatrix} 1 \\ 0 \end{pmatrix} \right)$

 (c)

7. (a) $\mathbf{Y}(t) = k_1 e^{-3t} \begin{pmatrix} 1 \\ 1 \end{pmatrix} +$

 $k_2 e^{-3t} \left(t \begin{pmatrix} 1 \\ 1 \end{pmatrix} + \begin{pmatrix} 1 \\ 0 \end{pmatrix} \right)$

(b) $\mathbf{Y}(t) = e^{-3t}\left(t\begin{pmatrix}1\\1\end{pmatrix}+\begin{pmatrix}1\\0\end{pmatrix}\right)$

(c)

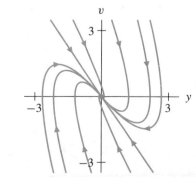

9. (a) $\alpha^2 = 4\beta$

(b) $\beta = 0$

11. (a) $p^2 - 4q > 0$

(b) $p^2 - 4q < 0$

(c) $p^2 - 4q = 0$

13. $\mathbf{A}\begin{pmatrix}1\\0\end{pmatrix} = \begin{pmatrix}a\\c\end{pmatrix} = \begin{pmatrix}\lambda\\0\end{pmatrix}$,

so we have $a = \lambda$ and $c = 0$.

Likewise, $\mathbf{A}\begin{pmatrix}0\\1\end{pmatrix} = \begin{pmatrix}b\\d\end{pmatrix} = \begin{pmatrix}0\\\lambda\end{pmatrix}$,

so we have $d = \lambda$ and $b = 0$.

15. (a) $k = 9/4$

(b) Eigenvalue: $\lambda = -3/2$
Eigenvectors: $v = -3y/2$

(c)

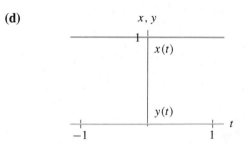

(d) $\mathbf{Y}(t) = k_1e^{-3t/2}\begin{pmatrix}2\\-3\end{pmatrix} +$

$k_2e^{-3t/2}\left(t\begin{pmatrix}2\\-3\end{pmatrix}+\begin{pmatrix}0\\2\end{pmatrix}\right)$

(e) $\mathbf{Y}(t) = e^{-3t/2}\begin{pmatrix}2\\-3\end{pmatrix} +$

$e^{-3t/2}\left(t\begin{pmatrix}3\\-9/2\end{pmatrix}+\begin{pmatrix}0\\3\end{pmatrix}\right)$

17. (a) Eigenvalues: $0, -1$.

(b) Eigenvectors: $y_1 = 0, 2y_2 = -x_2$

(c)

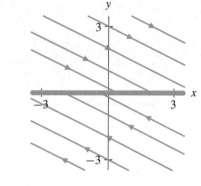

(d)

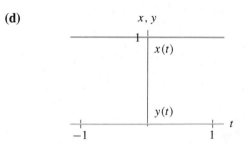

(e) $\mathbf{Y}(t) = k_1\begin{pmatrix}1\\0\end{pmatrix} + k_2e^{-t}\begin{pmatrix}2\\-1\end{pmatrix}$

(f) $\mathbf{Y}(t) = (1, 0)$

19. (a) Eigenvalues: 0, 5.

(b) Eigenvectors: $y_1 = -2x_1$, $x_2 = 2y_2$

(c)

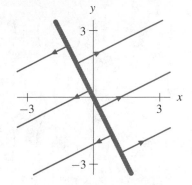

(d)

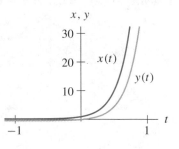

(e) $\mathbf{Y}(t) = k_1 \begin{pmatrix} 1 \\ -2 \end{pmatrix} + k_2 e^{5t} \begin{pmatrix} 2 \\ 1 \end{pmatrix}$

(f) $\mathbf{Y}(t) = \begin{pmatrix} \frac{1}{5} + \frac{4}{5}e^{5t} \\ -\frac{2}{5} + \frac{2}{5}e^{t} \end{pmatrix}$

21. (a) $\lambda = 0$ is the only eigenvalue.

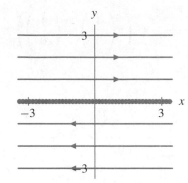

(b) $\lambda = 0$ is the only eigenvalue.

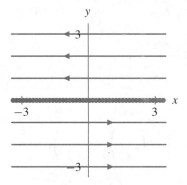

23. (a) Eigenvalues: a, d.

(b) Eigenvectors: x- and y-axes respectively if $a \neq d$

(c) *Hint*: See Exercise 14.

(d) *Hint*: Change the arrows from part (c).

Hints and Answers for Section 3.6

1. (a) $d^2y/dt^2 + 8dy/dt + 7y = 0$;
system: $dy/dt = v$, $dv/dt = -7y - 8v$

(b) $\lambda_1 = -1$ with eigenvectors $\mathbf{V}_1 = (y_1, v_1)$
where $v_1 = -y_1$; $\lambda_2 = -7$ with eigenvectors
$\mathbf{V}_2 = (y_2, v_2)$ where $v_2 = -7y_2$

(c) Overdamped

(d)

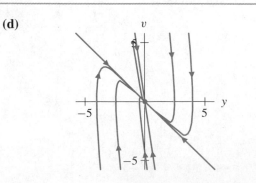

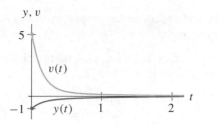

3. (a) $d^2y/dt^2 + 4dy/dt + 5y = 0$;
   system: $dy/dt = v, dv/dt = -5y - 4v$

   (b) $\lambda_1 = -2 + i$ with eigenvectors $\mathbf{V}_1 = (y_1, v_1)$
   where $v_1 = (-2 + i)y_1$; $\lambda_2 = -2 - i$
   with eigenvectors $\mathbf{V}_2 = (y_2, v_2)$ where
   $v_2 = (-2 - i)y_2$

   (c) Underdamped. Period is 2π.

   (d)

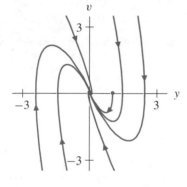

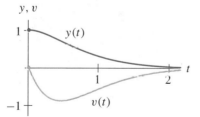

5. (a) $2d^2y/dt^2 + 3dy/dt + y = 0$;
   $dy/dt = v, dv/dt = -y/2 - 3v/2$.

   (b) $\lambda_1 = -1$ with eigenvectors $\mathbf{V}_1 = (y_1, v_1)$
   where $v_1 = -y_1$; $\lambda_2 = -1/2$ with eigenvectors
   $\mathbf{V}_2 = (y_2, v_2)$ where $v_2 = -y_2/2$

   (c) Overdamped

(d)

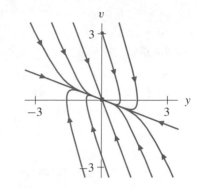

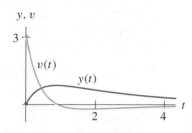

7. (a) $2d^2y/dt^2 + 3y = 0$;
   $dy/dt = v, dv/dt = -3y/2$

   (b) $\lambda = \pm i\sqrt{3/2}$ with eigenvectors $\mathbf{V} = (y, v)$ that
   satisfy $v = \pm i\sqrt{3/2}y$

   (c) Undamped with natural period $4\pi/\sqrt{6}$

   (d)

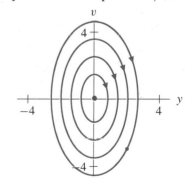

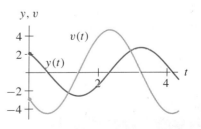

9. (a) $y(t) = k_1 e^{-7t} + k_2 e^{-t}$

(b) $y(t) = -\frac{2}{3} e^{-7t} - \frac{1}{3} e^{-t}$

(c) See the $y(t)$- and $v(t)$-graphs for Exercise 1.

11. (a) $y(t) = k_1 e^{-2t} \cos t + k_2 e^{-2t} \sin t$

(b) $y(t) = e^{-2t} \cos t + 2e^{-2t} \sin t$

(c) See the $y(t)$- and $v(t)$-graphs for Exercise 3.

13. (a) $y(t) = k_1 e^{-t} + k_2 e^{-t/2}$

(b) $y(t) = -6e^{-t} + 6e^{-t/2}$

(c) See the $y(t)$- and $v(t)$-graphs for Exercise 5.

15. (a) $y(t) = k_1 \cos\left(\sqrt{3/2}\, t\right) + k_2 \sin\left(\sqrt{3/2}\, t\right)$

(b) $y(t) = 2\cos\left(\sqrt{3/2}\, t\right) - \sqrt{6} \sin\left(\sqrt{3/2}\, t\right)$

(c) See the $y(t)$- and $v(t)$-graphs for Exercise 7.

17. *Hint*: The table will have four rows—one each for the undamped, underdamped, critically damped, and overdamped cases.

19. Use the fact that a complex number is zero if and only if both its real and imaginary part are zero.

21. The solution will approach the equilibrium no faster than $e^{-\alpha t}$, where $\alpha = b/2$ if $b^2 < 12$ and $\alpha = (b + \sqrt{b^2 - 12})/2$ if $b^2 > 12$.

23. (a) $md^2 y/dt^2 - b_{mf} dy/dt + ky = 0$

(b) $dy/dt = v$, $dv/dt = -(k/m)y + (b_{mf}/m)v$

(c) *Hint*: Modify the case-by-case analysis of this section to reflect the change in sign of the dy/dt term.

25. $m = 1/(2\pi^2)$

Hints and Answers for Section 3.7

1. *Hint*: There are three types in Sections 3.2 and 3.3, and three types in Section 3.4. These six are the most common types. The remaining types are discussed in Section 3.5.

3. (a)

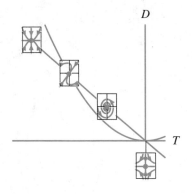

(b) Real sink: $a < -1$. Repeated negative eigenvalues: $a = -1$. Spiral sink: $-1 < a < 0$. Repeated 0 eigenvalues: $a = 0$. Saddle: $a > 0$.

(c) Bifurcations at $a = 0, -1$.

5. (a)

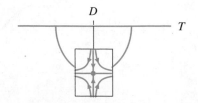

(b) One zero eigenvalue: $a = \pm 1$. Saddle: $-1 < a < 1$.

(c) Bifurcations at $a = 1, -1$.

7. (a)

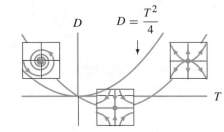

(b) Real source: $a > 1$. Zero eigenvalue: $a = 1$. Saddle: $0 < a < 1$. Repeated 0 eigenvalues: $a = 0$. Spiral sink: $a < 0$.

(c) Bifurcations at $a = 0$ and $a = 1$.

9. Eigenvalues are $a \pm |b|$. Zero eigenvalue: $a = \pm b$. Repeated eigenvalue: $b = 0$, $a \neq 0$. Repeated zero eigenvalue: $a = b = 0$. Real source: $a > |b|$. Real sink: $-a > |b|$. Saddle otherwise.

11. (a) $dy/dt = v, dv/dt = -3y - bv$

(b)

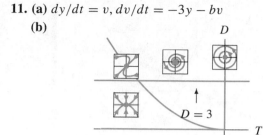

(c) Overdamped: $b > 2\sqrt{3}$. Critically damped: $b = 2\sqrt{3}$. Underdamped: $0 < b < 2\sqrt{3}$. Undamped: $b = 0$.

13. (a) $dy/dt = v, dv/dt = -\frac{2}{m}y - \frac{1}{m}v$

(b)

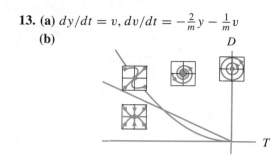

(c) Overdamped: $0 < m < 1/8$. Critically damped: $m = 1/8$. Underdamped: $m > 1/8$.

Hints and Answers for Section 3.8

1. Differentiate $\mathbf{Y}_2(t)$ and $\mathbf{Y}_3(t)$ coordinate by coordinate to see if the system of differential equations is satisfied.

3. (a) Not linearly independent
   (b) Linearly independent
   (c) Linearly independent
   (d) Linearly independent

5. (a) $\pm 1, -5$
   (b) The system decouples into a two-dimensional system in the xy-plane and a one-dimensional system in z.
   (c) There is a sink on the z-phase line and a saddle in the xy-phase plane.

(d)

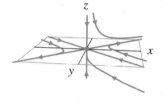

7. (a) $1, 3$
   (b) The system decouples into a two-dimensional system in the yz-plane and a one-dimensional system in x.
   (c) Source on x-axis, source in yz-plane

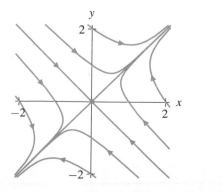

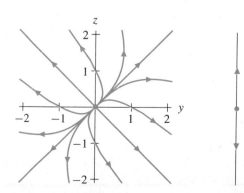

(d)

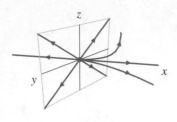

9. Using the fact that $p(a + ib) = 0$, show that the real and imaginary parts of $p(a - ib)$ are each zero.

11. (a) $-2, 1$

   (b) source on the z-axis, repeated eigenvalue sink in the xy-plane

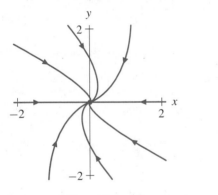

(c)

13. (a) $0, -5$

   (b) The z-axis is all equilibrium points; in the xy-plane there is a line of equilibrium points, and all solutions approach this line.

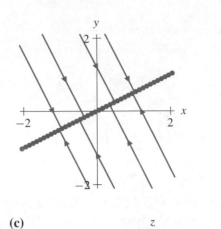

(c)

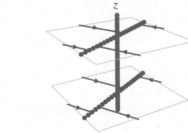

15. (a) $\lambda = 0$

   (b) the x-axis

   (c) $z(t)$ is a constant function, and the value of that constant is determined by the initial condition. For $z_0 = 0$, $y(t)$ is constant, and the solution curves lie on straight lines parallel to x-axis. For $y_0 > 0$, $x(t)$ is increasing, and for $y_0 < 0$, $x(t)$ is decreasing. For $z_0 \neq 0$, $x(t)$ is quadratic in y. Therefore, solution curves that satisfy the initial condition $z(0) = z_0$ stay on the plane $z = z_0$ and lie on a parabola.

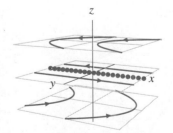

17. (a) -1

   (b) $-2 \pm i$

(c) Spiral sink

(d) *Hint*: Find the plane containing the real and imaginary parts of an eigenvector for the complex eigenvalue. This is the plane where solutions spiral. Also, on the line $x = y$ and $z = 0$, solutions tend directly toward the origin.

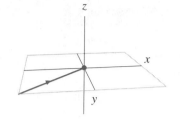

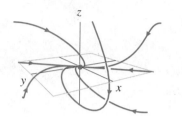

(b)

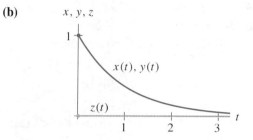

The graphs of $x(t)$ and $y(t)$ are identical, and $z(t) = 0$ for all t.

19. (a) Glen's profitability helps both Paul and Bob be profitable (and they need all the help they can get).

(b) The status of either Paul's or Bob's profits have no affect on Glen's profit.

21. (a) *Hint*: If $z = 0$, $dz/dt = 0$, so this is a two-dimensional problem.

(c) Glen continues to break even while Paul's and Bob's profits tend toward the break-even point (always staying equal).

Hints and Answers for Section 4.1

1. $y(t) = k_1 e^{-2t} + k_2 e^{-4t} + \frac{1}{3} e^{-t}$

3. $y(t) = k_1 e^{-3t} + k_2 e^{-4t} + \frac{3}{2} e^{-2t}$

5. $k_1 e^{-2t} \cos 3t + k_2 e^{-2t} \sin 3t - \frac{1}{3} e^{-2t}$

7. $y(t) = -\frac{1}{2} e^{-2t} + \frac{1}{6} e^{-4t} + \frac{1}{3} e^{-t}$

9. $y(t) = \frac{1}{3} e^{-2t} \cos 3t - \frac{1}{3} e^{-2t}$

11. (a) $y(t) = k_1 e^{-t} + k_2 e^{-3t} + \frac{4}{5} e^{-t/2}$
    (b) $y(t) = -e^{-t} + \frac{1}{5} e^{-3t} + \frac{4}{5} e^{-t/2}$
    (c) All solutions are approximately equal to
    $4e^{-t/2}/5$ for t large (so they all tend to zero).

13. (a) $y(t) = k_1 e^{-t} + k_2 e^{-3t} + \frac{1}{3} e^{-4t}$
    (b) $y(t) = \frac{1}{6} e^{-t} - \frac{1}{2} e^{-3t} + \frac{1}{3} e^{-4t}$
    (c) All solutions tend to zero, most tend to zero like e^{-t}.

15. (a) $y(t) = k_1 e^{-2t} \cos 4t + k_2 e^{-2t} \sin 4t + \frac{1}{16} e^{-2t}$
    (b) $y(t) = -\frac{1}{16} e^{-2t} \cos 4t + \frac{1}{16} e^{-2t}$
    (c) All solutions tend to zero like e^{-2t} and all but one oscillate with frequency $2/\pi$.

17. $y(t) = k_1 e^{-t} + k_2 t e^{-t} + t^2 e^{-t}/2$

19. (a) $y(t) = k_1 e^{-2t} + k_2 e^{-4t} + \frac{5}{8}$
    (b) $y(t) = -\frac{5}{4} e^{-2t} + \frac{5}{8} e^{-4t} + \frac{5}{8}$

21. (a) $y(t) = k_1 e^{-t} \cos 3t + k_2 e^{-t} \sin 3t + 1$
    (b) $y(t) = -e^{-t} \cos 3t - \frac{1}{3} e^{-t} \sin 3t + 1$

23. (a) $y(t) = k_1 \cos 3t + k_2 \sin 3t + \frac{1}{10} e^{-t}$
    (b) $y(t) = -\frac{1}{10} \cos 3t + \frac{1}{30} \sin 3t + \frac{1}{10} e^{-t}$
    (c)

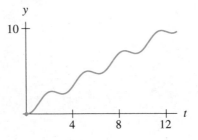

25. (a) $y(t) = k_1 \cos \sqrt{2} t + k_2 \sin \sqrt{2} t - \frac{3}{2}$
    (b) $y(t) = \frac{3}{2} \cos \sqrt{2} t - \frac{3}{2}$

(c)

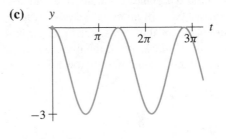

27. (a) $y(t) = k_1 \cos 3t + k_2 \sin 3t + \frac{2}{3}$
    (b) $y(t) = -\frac{2}{3} \cos 3t + \frac{2}{3}$
    (c)

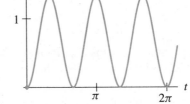

29. (a) $y(t) = k_1 \sin 2t + k_2 \cos 2t - \frac{3}{4} t^2 + \frac{1}{2} t + \frac{9}{8}$
    (b) $y(t) = -\frac{1}{4} \sin 2t + \frac{7}{8} \cos 2t - \frac{3}{4} t^2 + \frac{1}{2} t + \frac{9}{8}$

31. (a) $y(t) = k_1 \sin 2t + k_2 \cos 2t + \frac{3}{4} t + \frac{1}{2}$
    (b) $y(t) = -\frac{3}{8} \sin 2t - \frac{1}{2} \cos 2t + \frac{3}{4} t + \frac{1}{2}$
    (c) The solution tends to ∞.

33. (a) $y(t) = k_1 \sin 2t + k_2 \cos 2t - \frac{1}{80} t^2 + \frac{1}{4} t + \frac{1}{160}$
    (b) $y(t) = -\frac{1}{8} \sin 2t - \frac{1}{160} \cos 2t - \frac{1}{80} t^2 + \frac{1}{4} t + \frac{1}{160}$

(c) The solution tends to $-\infty$.

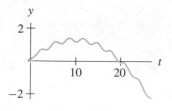

35. (a) $y(t) = k_1 e^{-3t} + k_2 e^{-2t} + \frac{1}{2}e^{-t} + \frac{2}{3}$

(b) $y(t) = \frac{11}{6}e^{-3t} - 3e^{-2t} + \frac{1}{2}e^{-t} + \frac{2}{3}$

(c) The solution tends to $2/3$.

37. (a) $y(t) = k_1 e^{-4t} + k_2 e^{-2t} - \frac{3}{16} + \frac{1}{4}t + \frac{1}{3}e^{-t}$

(b) $y(t) = \frac{5}{48}e^{-4t} - \frac{1}{4}e^{-2t} - \frac{3}{16} + \frac{1}{4}t + \frac{1}{3}e^{-t}$

(c) The solution tends to ∞.

39. (a) $y(t) = k_1 \cos 2t + k_2 \sin 2t + \frac{1}{4}t + \frac{1}{5}e^{-t}$

(b) $y(t) = -\frac{1}{5}\cos 2t - \frac{1}{40}\sin 2t + \frac{1}{4}t + \frac{1}{5}e^{-t}$

(c) The solution tends to ∞.

Hints and Answers for Section 4.2

1. $y(t) = k_1 e^{-2t} + k_2 e^{-t} + \frac{1}{10}\cos t + \frac{3}{10}\sin t$

3. $y(t) = k_1 e^{-2t} + k_2 e^{-t} - \frac{3}{10}\cos t + \frac{1}{10}\sin t$

5. $y(t) = k_1 e^{-4t} + k_2 e^{-2t} + \frac{7}{85}\cos t + \frac{6}{85}\sin t$

7. $y(t) = k_1 e^{-2t}\cos 3t + k_2 e^{-2t}\sin 3t +$

$\qquad \frac{27}{145}\cos 2t + \frac{24}{145}\sin 2t$

9. $y(t) = k_1 e^{-2t}\cos 4t + k_2 e^{-2t}\sin 4t -$

$\qquad \frac{3}{20}\sin 2t + \frac{3}{40}\cos 2t$

11. $y(t) = \frac{2}{17}e^{-4t} - \frac{1}{5}e^{-2t} + \frac{7}{85}\cos t + \frac{6}{85}\sin t$

13. $y(t) = -\frac{3}{40}e^{-2t}\cos 4t + \frac{3}{80}e^{-2t}\sin 4t -$

$\qquad \frac{3}{20}\sin 2t + \frac{3}{40}\cos 2t$

15. (a) $a = -8/145$ and $b = 9/145$

(b) $A = 1/\sqrt{145}$ and $\phi = \arctan(9/8)$

17. (ii)

19. (iv)

21. $y(t) = k_1 e^{-4t} + k_2 e^{-2t} + \frac{7}{17}\cos t + \frac{6}{17}\sin t$

23. (a) $y(t) = k_1 e^{-2t}\cos 4t + k_2 e^{-2t}\sin 4t +$

$\qquad \frac{4}{65}e^{-t}\cos t + \frac{1}{130}e^{-t}\sin t$

(b) Solutions tend to zero.

25. ϕ is the polar angle of the complex number $z = a - bi$ and $k = |z| = \sqrt{a^2 + b^2}$ (see Appendix B).

Hints and Answers for Section 4.3

1. $y(t) = k_1 \cos 3t + k_2 \sin 3t + \frac{1}{8}\cos t$

3. $y(t) = k_1 \cos 2t + k_2 \sin 2t - \frac{4}{15}\cos \frac{1}{2}t$

5. $y(t) = k_1 \cos \sqrt{3}\,t + k_2 \sin \sqrt{3}\,t - \frac{1}{6}\cos 3t$

7. $y(t) = -\frac{1}{8}\cos 3t + \frac{1}{8}\cos t$

9. $y(t) = -3\cos \sqrt{5}\,t + 3\cos 2t$

11. (a) $1/(16\pi)$

(b) $17/(16\pi)$

(c)

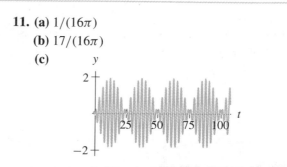

13. (a) $(\sqrt{5} - 2)/(4\pi)$

(b) $(\sqrt{5} + 2)/(4\pi)$

(c)

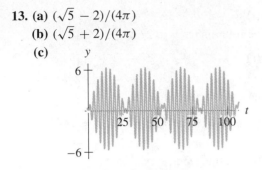

15. Equation (ii)

17. Equation (i)

19. (a) $y(t) = k_1 \cos \sqrt{15}\, t + k_2 \sin \sqrt{15}\, t -$

$\cos 4t + \frac{1}{7} \sin t$

(b) $y(t) = \cos \sqrt{15}\, t - \frac{1}{7\sqrt{15}} \sin \sqrt{15}\, t -$

$\cos 4t + \frac{1}{7} \sin t$

(c)

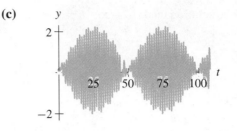

(d) *Hint*: Use the fact that, if $y_1(t)$ is a solution of $d^2y/dt^2 + 15y = \cos 4t$ and $y_2(t)$ is a solution of $d^2y/dt^2 + 15y = 2 \sin t$, then $y_1(t) + y_2(t)$ is a solution of the given differential equation.

21. The beat of "Hey, Jude" is the same (or very close to) the natural frequency of the swaying motion of the stadium.

23. *Hint*: Largest amplitude response occurs for $T = 3/2$.

Hints and Answers for Section 4.4

1. By rubbing the finger around the edge of the glass, the musician vibrates the glass with its natural frequency, and this vibration results in a sound of that frequency. By adjusting the amount of water in the glass, the frequency (and therefore the tone) can be changed. By changing the push against the glass, the amplitude of the forcing is changed. Thus, the volume of the sound changes.

3. (a) $\partial A/\partial \omega = \omega(2q - p^2 - 2\omega^2)/((q - \omega^2)^2 + p^2\omega^2)^{3/2}$

(b) *Hint*: The critical points of A as a function of ω satisfy either $\omega = 0$ or $\omega^2 = q - p^2/2$.

5. *Hint*: Use the Chain Rule applied to $y_p(t + \theta)$.

Hints and Answers for Section 4.5

1. (a) β increases.

(b) The system becomes more like a linear system.

3. (a) γ increases.

(b) *Hint*: For values of y where $c(y) \neq 0$, $|c(y)|$ is increased.

5. *Hint*: $m\, d^2y/dt^2 + \epsilon\, dy/dt + pa^2\, y = mg$

What do y, ϵ, p, a, m, and g represent?

7. $m\, d^2y/dt^2 + \epsilon\, dy/dt = mg + b(y)$ where $b(y) = -pa^2y$ if $y > 0$ and $b(y) = 0$ if $y < 0$.

What do ϵ, p, a, m, and g represent?

Hints and Answers for Section 5.1

1. Systems (i) and (iii) have the same "local picture" near $(0, 0)$.

3. (a) The linearized system at the origin is
$$dx/dt = -2x + y, \, dy/dt = -y.$$
(b) Sink.
(c)

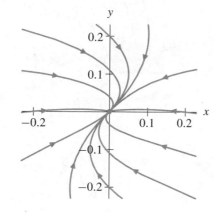

(d) The linearized system at $(2, 4)$ is
$$dx/dt = -2x + y, \, dy/dt = 4x - y. \text{ Hence,}$$
$(2, 4)$ is a saddle.

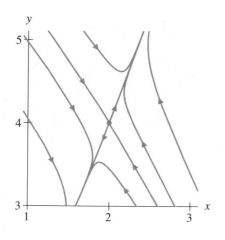

5. (a) $x(t) = x_0 e^{-t}$
(b) $y(t) = x_0^3 e^{-3t} + (y_0 - x_0^3)e^t$

(c) $x(t) = x_0 e^{-t}$ and $y(t) = x_0^3 e^{-3t} + (y_0 - x_0^3)e^t$
(d) y_0 must equal x_0^3
(e) x_0 must be 0
(f)

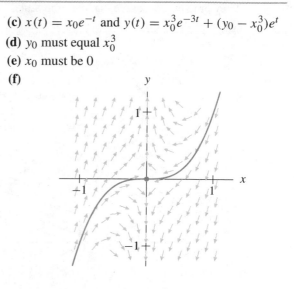

(g) For the nonlinear system, solutions tend to the origin along the curve $y = x^3$. For the linearized system, solutions tend to the origin along the x-axis.

7. (a) Equilibria: $(0, 0)$, $(0, 100)$, $(150, 0)$, $(30, 40)$. They are source, sink, sink, and saddle respectively.
(b)

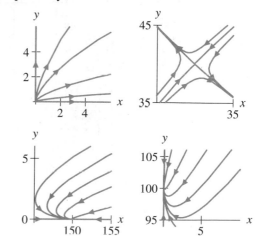

(c) See part (a).

9. (a) Equilibria: $(0, 0)$, $(0, 25)$, $(100, 0)$, $(75, 12.5)$. They are source, saddle, saddle, and sink respectively.

(b)

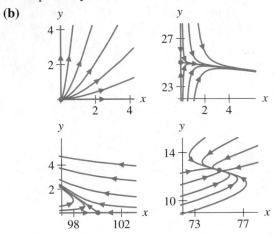

(c) See part (a).

11. (a) Equilibria: $(0, 0)$, $(0, 50)$, $(40, 0)$. They are source, sink, and saddle respectively.

(b)

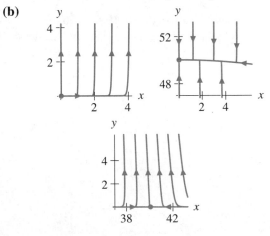

(c) See part (a).

13. (a) Equilibria: $(0, 0)$, $(0, 50)$, $(60, 0)$, $(30, 40)$, $(234/5, 88/5)$. They are source, saddle, sink, sink, saddle respectively.

(b)

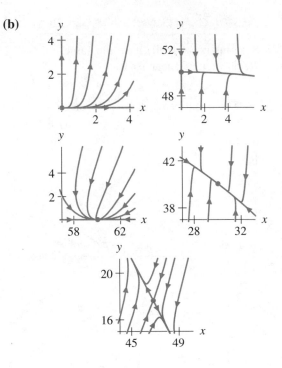

(c) See part (a).

15. (a) Equilibria: $(0, 0)$, $(1, 1)$, $(2, 0)$. The point $(0, 0)$ has a zero eigenvalue (see phase plane to determine local behavior). $(1, 1)$ is a saddle, and $(2, 0)$ is a sink.

(b)

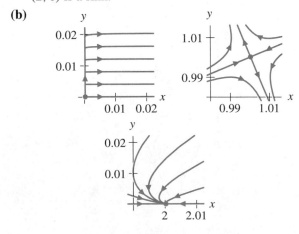

(c) See part (a).

17. (a) $dx/dt = 0$, $dy/dt = -y$

(b) Eigenvalues $0, -1$ with eigenvectors $(1, 0)$, $(0, 1)$.

(c) $dx/dt = 0, dy/dt = y$

(d) Eigenvalues 0, 1 with eigenvectors $(0, 1)$, $(1, 0)$.

(e) The equilibrium point $(0, 0)$ is a sink and the equilibrium point $(0, 1)$ looks like a saddle.

(f) The equation for dx/dt is all "higher order."

19. (a)

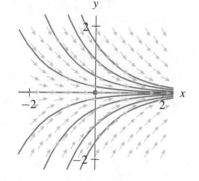

(b)

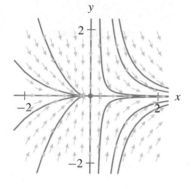

(c)

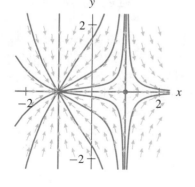

(c) If $a < 0$ all solutions decrease in the y-direction since $dy/dt < 0$. If $a > 0$ all solutions increase in the y-direction since $dy/dt > 0$. If $a = 0$ there is a curve of equilibrium points located along $y = x^2$.

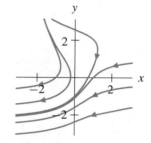

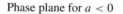
Phase plane for $a < 0$

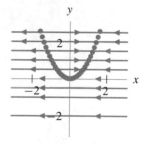

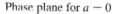
Phase plane for $a - 0$

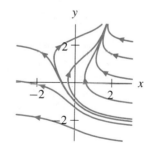

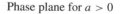
Phase plane for $a > 0$

21. (a) If $a \neq 0$, there are no equilibrium points. If $a = 0$, the curve $y = x^2$ consists entirely of equilibrium points.

(b) $a = 0$

23. (a) $(0, 0), (\pm 1/\sqrt{a}, \pm 1/\sqrt{a})$

(b) $a = 0$

(c) If $a < 0$, there are is only one equilibrium point at the origin, and this equilibrium point

is a spiral source. If $a > 0$, the system
has two additional equilibrium points, at
$(\pm 1/\sqrt{a}, \pm 1/\sqrt{a})$. These equilibrium points
come from infinity as a increases through 0.

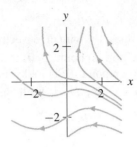

Phase plane for $a < 0$

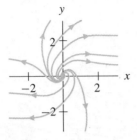

Phase plane for $a < 0$

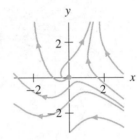

Phase plane for $a = 0$

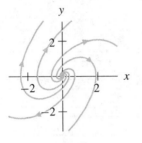

Phase plane for $a = 0$

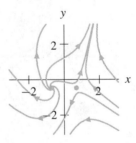

Phase plane for $a > 0$

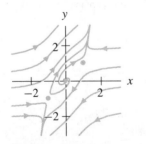

Phase plane for $a > 0$

25. (a) $(\pm\sqrt{a/2}, -a/2)$

(b) $a = 0$

(c) If $a < 0$, there are no equilibrium points. If
$a = 0$, an equilibrium point appears at the
origin. If $a > 0$, the system has two equilibrium
points, at $(\pm\sqrt{a/2}, -a/2)$.

27. (a) If X and Y are not present on the island, then
neither immigrate to the island.

(b) They are large positive.

(c) They are negative.

(d) Source or saddle.

(e) *Hint*: Note the signs of dx/dt and dy/dt for
the linearized system along the positive x- and
y-axes.

29. (a) At $(0, 0)$, $\partial f/\partial x$ is large, and $\partial g/\partial y$ is small.

(b) At $(0, 0)$, $\partial f/\partial y$ is negative with a large absolute
value and $\partial g/\partial x = 0$.

(c) Source.

(d) *Hint*: Compute the eigenvectors.

Hints and Answers for Section 5.2

1. Equilibria: $(1, 1)$ and $(-2, 4)$. All three solutions will go down and to the right without bound.

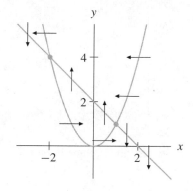

(b)

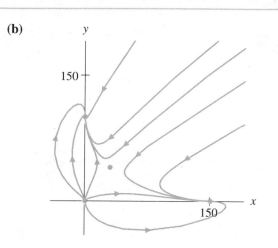

3. Equilibria: $(0, 0)$ and $(1, 1)$. Solutions with initial conditions (a) and (b) move toward the equilibrium point $(0, 0)$, while the solution with initial condition (c) travels down the line $x = 1$ toward the equilibrium point at $(1, 1)$.

(c) *Hint*: See phase plane in part (b).

7. (a)

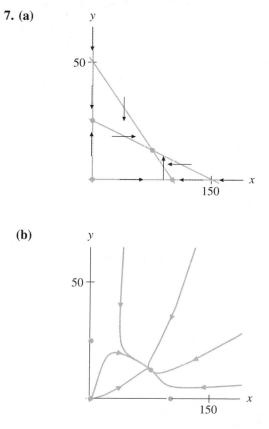

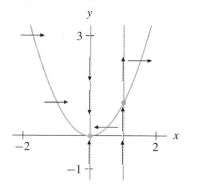

5. (a)

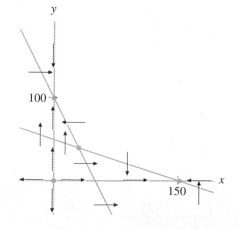

(b)

(c) *Hint*: See the phase plane in part (b).

9. (a)

(b)

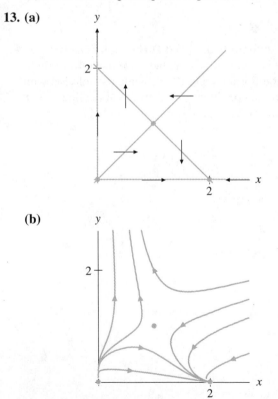

(b)

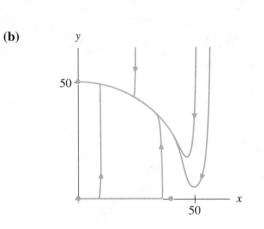

(c) *Hint*: See the phase plane in part (b).

11. (a)

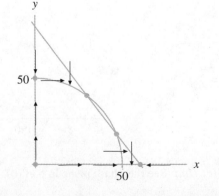

(b)

(c) *Hint*: See the phase plane in part (b).

13. (a)

(b)

(c) *Hint*: See the phase plane in part (b).

15. (a) Change the signs in front of both B and D.

(b) The nullclines are pairs of straight lines. The points $(0, 0)$, $(0, F/E)$, and $(C/A, 0)$ are equilibrium points. We need to have $AE - BD > 0$ in order for there to be an equilibrium point with both x and y positive.

17. (a)

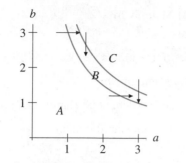

(b) In region A, the vector field points up and to the right. In region B, the vector field points down and to the right. In region C, the vector field points down and to the left.

(c) Once in region B, a solution cannot leave. As time increases, these solutions are asymptotic to the positive x-axis from above.

19. (a)

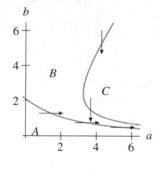

(b) In region A, the vector field points up and to the right. In region B, the vector field points down and to the right. In region C, the vector field points down and to the left.

(c) Once in region B, a solution cannot leave. As time increases, these solutions are asymptotic to the positive x-axis from above.

21.

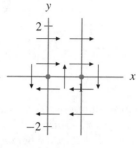

23. The linearized system at $(0, 0)$ is $dx/dt = y$ and $dy/dt = x$. The linearized system at $(1, 0)$ is $dx/dt = y$ and $dy/dt = -x$. We do not know if the equilibrium point at $(1, 0)$ is a nonlinear center. See Section 5.3.

Hints and Answers for Section 5.3

1. (a) *Hint*: compute $\partial H/\partial x$ and $\partial H/\partial y$. Then compare.

(b) $(0, 0)$ is a local minimum, $(\pm 1, 0)$ are saddles

(c) $(0, 0)$ is a center, $(\pm 1, 0)$ are saddles that are connected by separatrix orbits

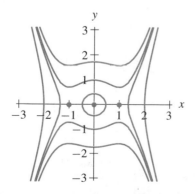

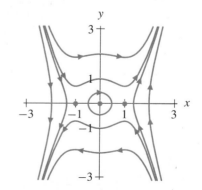

3. (a) *Hint*: compute $\partial H/\partial x$ and $\partial H/\partial y$. Then compare.

(b)

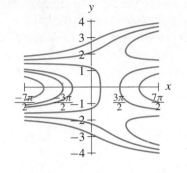

(c) The equilibrium points occur at points of the form $(\pi - 4n\pi, -\frac{\pi}{2} + 2n\pi)$ and $(\pi + 4n\pi, \frac{\pi}{2} + 2n\pi)$, where n is an integer.

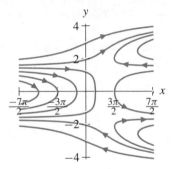

5. No, longer.

7. (a) Multiply by $\sqrt{2}$.
   (b) $\pi/\sqrt{gl}$.

9. Yes, $H(x, y) = xy - y^3$.

11. Not Hamiltonian.

13. (b) and **(c)** cannot be Hamiltonian.

15. (a) It is not Hamiltonian.
   (b)

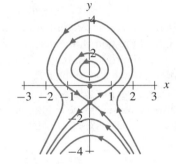

(c) Yes.
   (d) $H(x, y) = -y + 2\arctan y - x^2/2$
   (e) Changing the length of the vectors doesn't change the direction field, so the phase planes of the two systems are the same.

17. If $y > 2$, $H(x, y) = -x^2/2 + 2y + y^2/2 + 3\ln(y-2)$; if $y < 2$, $H(x, y) = -x^2/2 + 2y + y^2/2 + 3\ln(2-y)$.

19. *Hint*: For $a \neq 0$ the system has two equilibrium points, but for $a = 0$, it has only one.

Hints and Answers for Section 5.4

1. (a) Let $(x(t), y(t))$ be a solution and differentiate $L(x(t), y(t))$ with respect to t.
   (b) The level sets are circles.
   (c) All solutions tend toward the origin as t increases.

3. (a) The eigenvalues are $-0.05 \pm i\sqrt{15.99}/2$.

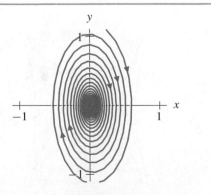

(b) Let $(x(t), y(t))$ be the solution with $x(0) = -1$, $y(0) = 1$ and compute the derivative with respect to t of $L(x(t), y(t))$ at $t = 0$.

(c) Let $(x(t), y(t))$ be any solution and compute the derivative with respect to t of $K(x(t), y(t))$.

5. It runs fast.

7. $m > b\sqrt{l}/(2\sqrt{g})$.

9. *Hint*: The amplitude of the solutions near $(0, 0)$ decreases like $e^{-tb/(2m)}$.

11. (a) $(0, 0)$ is a spiral source, $(\pi, 0)$ is a saddle.

   (b) *Hint*: The phase portrait looks like the damped pendulum except solutions spiral outward.

   (c) Oscillate with increasing amplitude.

13. (a) $dx/dt = 2x, dy/dt = -2y$.

   (b) Saddle.

   (c)

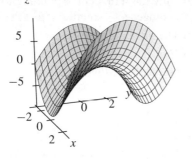

   (d) *Hint*: The axes are the lines of eigenvectors.

15. (a) The eigenvalues are 1 and -1.

   (b) The eigenvalues are -2 and -1.

   (c) *Hint*: $(0, 1)$ is an eigenvector for eigenvalue -1.

   (d) The Jacobian matrices are the same.

Hints and Answers for Section 5.5

1. (a) $\begin{pmatrix} -1 & -1 & 0 \\ 0 & 2 & 0 \\ 0 & 0 & 1 \end{pmatrix}$

   (b) Eigenvalues 1, 2, -1 and eigenvectors $(0, 0, 1)$, $(1, -3, 0)$ and $(1, 0, 0)$ respectively.

   (c) Saddle.

17. (a)
$$\frac{dx}{dt} = 2x - x^3 - 6xy^2$$
$$\frac{dy}{dt} = 2y - y^3 - 6x^2y.$$

   (b)

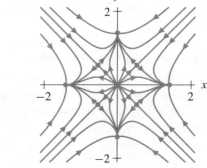

   (c) Four dead fish.

   (d)

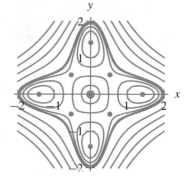

   (e) S is negative if x or y is too big.

19. (a) *Hint*: Mixed partials are equal.

   (b) $\partial f/\partial y \neq \partial g/\partial y$.

21. Let $(y(t), v(t))$ be a solution of the second system and compute $d(H(y(t), v(t))/dt$.

   (d) *Hint*: The linearized system decouples.

   (e) Solutions with $y = z = 0$ move toward the equilibrium point. All others move away.

3. (a) $\begin{pmatrix} 1 & 0 & 0 \\ 0 & 0 & 0 \\ 0 & 1 & -1 \end{pmatrix}$

(b) The eigenvalues are 1, 0 and -1 and eigenvectors are $(1, 0, 0)$, $(0, 1, 1)$ and $(0, 0, 1)$ respectively.

(c) The linearized system has a line of equilibrium points, so it does not have a name.

(d) *Hint*: The x and y, z equations decouple.

(e) Solutions with $x = 0$ move toward an equilibrium point, solutions with $x \neq 0$ move in the direction of increasing x.

5. (a) *Hint*: Set $y = 0$ and look at the system of equations, then try $x = 0$ and $z = 0$.

(b) *Hint*: What is the connection between wolves and trees if moose are extinct?

7. (a) $(0, 0, 0)$, $(1, 0, 0)$, $(0, 1, 0)$, $(0, 0, 1)$, $(1, 0, 1)$, and $((\zeta + 1)(2\zeta + 1), \zeta/(2\zeta + 1), (2\zeta + 2)/(2\zeta + 1))$.

(b) As ζ decreases, x_0 increases, y_0 decreases and z_0 increases.

(c) *Hint*: compare how ζ comes into the equation for dz/dt with the system in the section.

Hints and Answers for Section 5.6

1. Graph (ii).

3. Graph (i).

5. Graph (iii).

7. Graph (iv).

9. (a) $y(t) = -\frac{\sqrt{3}}{30} \sin \sqrt{3}\, t + \cos \sqrt{3}\, t + \sin \frac{1}{10} t$

(b) Only the terms corresponding to the homogeneous solution affect the Poincaré map.

(c) The same only larger.

(d) *Hint*: There are no saddles for the (unforced) harmonic oscillator.

Hints and Answers for Section 6.1

1. $\mathcal{L}[f] = 3/s$

3. $\mathcal{L}[h] = -10/s^3$

5. *Hint*: Use the formula $\mathcal{L}[dy/dt] = s\mathcal{L}[y] - y(0)$

7. e^{3t}

9. $(2e^{-5t/3})/3$

11. $5/3$

13. $5e^{2t} - 3e^t$

15. (a) $\mathcal{L}[dy/dt] = s\mathcal{L}[y] - y(0)$ and
   $\mathcal{L}[-y + e^{-2t}] = -\mathcal{L}[y] + 1/(s+2)$
   (b) $\mathcal{L}[y] = (2s+5)/((s+1)(s+2))$
   (c) $y(t) = 3e^{-t} - e^{-2t}$

17. (a) $\mathcal{L}[dy/dt + 7y] = s\mathcal{L}[y] - y(0) + 7\mathcal{L}[y]$
   $\mathcal{L}[1] = 1/s$
   (b) $\mathcal{L}[y] = (3s+1)/(s(s+7))$

(c) $y(t) = 20e^{-7t}/7 + 1/7$

19. (a) $\mathcal{L}[dy/dt + 9y] = s\mathcal{L}[y] - y(0) + 9\mathcal{L}[y]$
   $\mathcal{L}[2] = 2/s$
   (b) $\mathcal{L}[y] = (-2s+2)/(s(s+9))$
   (c) $y(t) = -20e^{-9t}/9 + 2/9$

21. (a) $\mathcal{L}[dy/dt + y] = s\mathcal{L}[y] - y(0) + \mathcal{L}[y]$
   $\mathcal{L}[e^{-2t}] = 1/(s+2)$
   (b) $\mathcal{L}[y] = (s+3)/((s+1)(s+2))$
   (c) $y(t) = 2e^{-t} - e^{-2t}$

23. (a) $\mathcal{L}[dy/dt] = s\mathcal{L}[y] - y(0)$ and
   $\mathcal{L}[-y + t^2] = -\mathcal{L}[y] + 2/s^3$
   (b) $\mathcal{L}[y] = (2+s^3)/(s^3(s+1))$
   (c) $y(t) = t^2 - 2t + 2 - e^{-t}$

25. $-2e^{-3t}/5 + ce^{2t}$ where $c = y(0) + 2/5$

27. *Hint*: What do you need to do with $\mathcal{L}[y^2]$?

Hints and Answers for Section 6.2

1. (a) $g_a(t) = 1 - u_a(t)$
   (b) $\mathcal{L}[g_a] = (1 - e^{-as})/s$

3. $(1 - e^{-as})/as^2$

5. $u_3(t)\left(e^{2(t-3)} - e^{(t-3)}\right)$

7. $u_1(t)\left(e^{4(t-1)} - e^{-2(t-1)/3}\right)$

9. $y(t) = -2e^{-9t} + \frac{1}{9}u_5(t) - \frac{1}{9}u_5(t)e^{-9(t-5)}$

11. $y(t) = e^{-t} + u_2(t)\left(e^{-(t-2)} - e^{-2(t-2)}\right)$

13. $y(t) = u_1(t)\left((t-2) + e^{-(t-1)}\right) + 2e^{-t}$

15. (a) $\mathcal{L}[dy/dt] = s\mathcal{L}[y] - y(0)$ and
   $\mathcal{L}[-y + u_a(t)] = -\mathcal{L}[y] + e^{-as}/s$
   (b) $\mathcal{L}[y] = y(0)/(s+1) + e^{-as}/(s(s+1))$
   (c) $y(t) = u_a(t)\left(1 - e^{-(t-a)}\right) + y(0)e^{-t}$

17. $\mathcal{L}[w] = (1 - e^{-s})/(s(1 + e^{-s}))$

19. (a) $\mathcal{L}[y] = (1 - e^{-s})/(s(s+1)(1 + e^{-s}))$
   (b) The function $w(t)$ is alternatively 1 and
   -1. While $w(t) = 1$, the solution decays
   exponentially toward $y = 1$. When $w(t)$
   changes to -1, the solution then decays toward
   $y = -1$.

Hints and Answers for Section 6.3

1. $\mathcal{L}[\sin \omega t] = \omega/(s^2 + \omega^2)$

3. $\mathcal{L}[e^{at} \sin \omega t] = \omega/(\omega^2 + (s-a)^2)$

5. *Hint*: Take the Laplace transform of both sides of

the equation and solve for $\mathcal{L}[y]$.

7. $\mathcal{L}[f] = (s^2 - \omega^2)/(\omega^2 + s^2)^2$

9. $\mathcal{L}[f] = 2/(s-a)^3$

11. $(s+1)^2 + 3^2$

13. $(s+1/2)^2 + (\sqrt{3}/2)^2$

15. $(1/3)e^{-t}\sin 3t$

17. $2e^{-t/2}\cos(\sqrt{3}t/2) + (4/\sqrt{3})e^{-t/2}\sin(\sqrt{3}t/2)$

19. Same as for real exponentials.

21. *Hint*: Recall that $e^{at}\cos\omega t$ is the real part of $e^{(a+ib)t}$.

23. $(1/3)e^{-t}\sin 3t$

25. $2e^{-t/2}\cos(\sqrt{3}t/2) + (4/\sqrt{3})e^{-t/2}\sin(\sqrt{3}t/2)$

27. (a) $\mathcal{L}[d^2y/dt^2 + 4y] = (s^2+4)\mathcal{L}[y] + 2s$
$\mathcal{L}[\cos 2t] = s/(s^2+4)$
   (b) $\mathcal{L}[y] = (-2s^3 - 7s)/(s^4 + 8s^2 + 16)$
   (c) $y(t) = -2\cos 2t + (t/4)\sin 2t$

29. (a) $\mathcal{L}[d^2y/dt^2 + 3y] = (s^2+3)\mathcal{L}[y] + 2$

$\mathcal{L}[u_4(t)\cos(5(t-4))] = se^{-4s}/(s^2+25)$
   (b) $\mathcal{L}[y] = -2/(3+s^2) + e^{-4s}s/((s^2+3)(s^2+25))$
   (c) $y(t) = -2\sin(\sqrt{3}t)/\sqrt{3} +$
$u_4(t)(\cos(\sqrt{3}(t-4)) - \cos(5(t-4)))/22$

31. (a) $\mathcal{L}[d^2y/dt^2 + 3y] = (s^2+3)\mathcal{L}[y] - 2s$
$\mathcal{L}[w] = (1 - e^{-s})/s^2$
   (b) $\mathcal{L}[y] = 2s/(s^2+3) + (1 - e^{-s})/(s^2(s^2+3))$
   (c) $y(t) = 2\cos(\sqrt{3}t) + (t/3) - \sin(\sqrt{3}t)/(3\sqrt{3}) +$
$u_1(t)((1/3) - (t/3) + \sin(\sqrt{3}(t-1))/(3\sqrt{3}))$

33. (a) $y(t) = 2\cos 3t + t(\sin 3t)/60$
   (b) *Hint*: The forcing frequency is the natural frequency and there is no damping.

35. (a) $\mathcal{L}[y] = (s+20)/(s^2 + 20s + 200) +$
$(1 - e^{-s})/((1 + e^{-s})s(s^2 + 20s + 200))$
   (b) The natural period is $\pi/5$ and the damping sends solutions to zero like e^{-5t}. The forced response has period 2 and amplitude $1/200$.

Hints and Answers for Section 6.4

1. *Hint*: Remember L'Hôpital's Rule: Differentiate numerator and denominator with respect to Δt.

3. $y(t) = e^{-t}\cos 2t + e^{-t}\sin 2t +$
$(1/2)u_3(t)e^{-(t-3)}\sin(2(t-3))$

5. $y(t) = u_1(t)e^{-(t-1)}\sin(\sqrt{2}(t-1))/\sqrt{2} -$
$(3/\sqrt{2})u_4(t)e^{-(t-4)}\sin(\sqrt{2}(t-4))$

7. (a) $\mathcal{L}[\delta_a] = e^{-as}$

$s\mathcal{L}[u_a] - u_a(0) = s(e^{-as}/s) - 0 = e^{-as}$
   (b) This relationship suggests that $du_a/dt = \delta_a$.
   (c) Approximate $u_a(t)$ with piecewise linear continuous functions and take their derivatives.

9. (a) $\mathcal{L}[y] = (1/(s^2+2))\sum_{n=1}^{\infty} e^{-ns}$
   (b) $y(t) = (1/\sqrt{2})\sum_{n=1}^{\infty} u_n(t)\sin(\sqrt{2}(t-n))$
   (c) *Hint*: It is not resonant forcing.

Hints and Answers for Section 6.5

1. $1 - e^{-t}$

3. $u_2(t)\sin(t-2)$

5. $\cos t - \cos 2t$

7. $p = 4, q = 5$

9. (a) $a(s+p)\mathcal{L}[\zeta]$

   (b) $b\mathcal{L}[\zeta]$
   (c) $(\mathcal{L}[f] + a(s+p) + b)\mathcal{L}[\zeta]$

11. (a) $\mathcal{L}[y_1] = \mathcal{L}[f_1]/(s^2 + ps + q)$
   (b) *Hint*: Calculate both and compare
   (c) *Hint*: Solve for $\mathcal{L}[y_2]$

Hints and Answers for Section 6.6

1. (a) $\mathcal{L}[y] = (2s + 2)/(s^2 + 2s + 2) +$
$4/((s^2 + 4s + 20)(s^2 + 2s + 2))$

(b) Poles are at $s = -2 \pm 4i$ and at $s = -1 \pm i$.

(c) Since the poles have negative real part, solutions decrease to zero (at a rate of e^{-t}). Since the poles are complex, the solutions oscillate (the oscillations have periods 2π and $\pi/2$).

3. (a) $\mathcal{L}[y] = (\cos(4)s + \sin(4) - se^{-4s})/((s^2 + s + 8)(s^2 + 1))$

(b) The poles are $s = \pm i$ and $s = -1/2 \pm i\sqrt{31}/2$.

(c) *Hint*: The forcing "turns off" at time $t = 4$.

5. (a) $\mathcal{L}[y] = (s + 1)/(s^2 + 16)$

(b) $s = \pm 4i$

(c) The solution is periodic and does not decay.

7. (a) $\mathcal{L}[y] = (s + 3)/(s^2 + 2s + 1)$

(b) Poles are at $s = -1$.

(c) The behavior of a second-order linear equation with a double pole should correspond to a first-order system with a double eigenvalue.

9. (a) $\mathcal{L}[y] = (s + 1)/(s^2 + 20s + 200) +$
$(1 - e^{-s})/(s(1 + e^{-s})(s^2 + 20s + 200))$

(b) The poles are $s = -10 \pm 10i$, $s = 0$ and $s = (2n + 1)\pi$ for $n = 0, \pm 1, \pm 2, \ldots$

(c) *Hint*: The natural response dies out quickly, so the solution oscillates between the forced response for forcing $+1$ and forcing -1.

Hints and Answers for Section 7.1

1. (a) $y(t) = 1/(1 + t^2)$

  (b) $y_{20} \approx 0.19334189$

  (c) $e_{20} \approx 0.0066581009$

  (d) $y_{1000} \approx 0.19987415, \ldots, y_{6000} \approx 0.19997904$

  (e) $e_{100} \approx 0.0012709461$, $e_{200} \approx 0.00063198471$,
  $e_{300} \approx 0.00042055770, \ldots,$
  $e_{6000} \approx .000020955836$

  (f) $K \approx 0.126731$

3. (a) $y(t) = e^{t^2/2}$

  (b) $y_{20} \approx 2.51066$

  (c) $e_{20} \approx 0.20762$

  (d) $y_{1000} \approx 2.71376, \ldots, y_{6000} \approx 2.71753$

  (e) $e_{100} \approx 0.0444901$, $e_{200} \approx 0.0224467, \ldots,$
  $e_{6000} \approx 0.000754848$

  (f) $K \approx 4.47023$

5. (a) $y(t) = (11/9)e^{3t} + t/3 - 2/9$

  (b) $y_{20} \approx 20.1147$

  (c) $e_{20} \approx 4.54544$

  (d) $y_{1000} \approx 24.5501, \ldots, y_{6000} \approx 24.6417$

  (e) $e_{100} \approx 1.05955$, $e_{200} \approx 0.540843, \ldots,$
  $e_{6000} \approx 0.0183987$

  (f) $K \approx 107.125$

7. (a) $M_1 = 8t^2 y^3 - 2y^2$, $M_2 = -4ty$

  (b) The estimated error is 0.00039984000 while the real error is approximately 0.00039984006.

  (c) The estimated error is 0.00079744230 while the real error is approximately 0.00079744409.

  (d) The third and fourth estimated errors are 0.0011896358 and 0.0015733200. The third and fourth real errors are 0.0011896455 and 0.0015733534.

  (e) Compare your plots with the Figures 7.5 and 7.6 in Section 7.1.

9. We cannot compare our estimate to the true error since we do not know how to calculate the true error for this differential equation. (We cannot find a closed-form solution.)

  (a) $M_1 = (y + t \sin ty) \cos ty$, $M_2 = t \cos ty$

  (b) The estimated error in the first step is 0.00134894.

  (c) The estimated error in the second step is 0.00269115.

  (d)

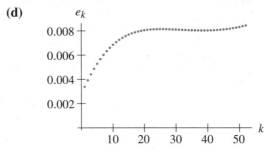

11. *Hint*: For parts (a)–(c), refer to Figures 7.3 and 7.4. For parts (d) and (e), use the inequality in part (c). For part (f), recall geometric sums from calculus. Parts (g)–(i) are just algebra. The expression $M_1(e^{M_2(t_n - t_0)} - 1)/(2M_2)$ is a constant in part (i) once the function $f(t, y)$ and the interval of time are given. In other words, $C = M_1/(2M_2)(e^{M_2(t_n - t_0)} - 1)$. The bounds that contribute to C do not depend on n.

Hints and Answers for Section 7.2

1.

Results of improved Euler's method

k	t_k	y_k
0	0.0	3.0000
1	0.5	8.2500
2	1.0	21.3750
3	1.5	54.1875
4	2.0	136.2187

3.

Results of improved Euler's method

k	t_k	y_k
0	0.0	2.000
1	0.5	2.813
2	1.0	6.618
3	1.5	129.001
4	2.0	17310268.856

5.

Results of improved Euler's method

k	t_k	y_k
0	0.0	4.00
1	1.0	1.50
2	2.0	-3.66
3	3.0	-259.96

A little qualitative analysis indicates that these results are useless (see Exercise 6).

7.

Results of improved Euler's method

k	t_k	y_k
0	0.0	2.00000
1	0.5	3.13301
2	1.0	4.01452
3	1.5	4.80396
4	2.0	5.54124

9. (a) The analytic solution is $y(t) = 1 - e^{-t}$.
   (b) $y_4 \approx 0.627471$, $e_4 \approx .00464959$
   (c) Use $n = 28$ steps. Then $e_{28} \approx 0.00008$.

11. (a) The analytic solution is $y(t) = 1/(t+2)$.
   (b) $y_4 \approx 0.252281$, $e_4 \approx 0.002281$
   (c) Use $n = 20$ steps. Then $e_{20} \approx 0.00008$.

13. (a) 53 steps
   (b) $y_{53} = 0.200095$
   (c) $e_{53} = 0.000095$

15.

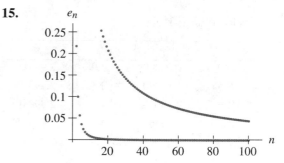

17.

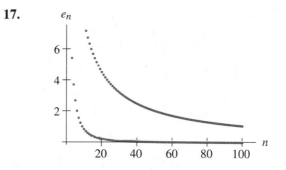

Hints and Answers for Section 7.3

1.

Results of Runge-Kutta

t_k	y_k
0.0	3.000
0.5	8.979
1.0	25.173
1.5	69.030
2.0	187.811

3.

Results of Runge-Kutta

t_k	y_k	t_k	y_k
0.0	0.00000	3.0	2.99645
0.5	1.82290	3.5	2.99882
1.0	2.70058	4.0	2.99961
1.5	2.90368	4.5	2.99987
2.0	2.96803	5.0	2.99996
2.5	2.98935		

5.

Results of Runge-Kutta

t_k	y_k
1.0	2.00000
1.5	3.10456
2.0	3.98546
2.5	4.77554
3.0	5.51352

7. (a)

Results of Runge-Kutta for the Predator-Prey System

t_k	R_k	F_k
0	1.000000	1.00000
1	1.504120	1.91806
2	0.641301	2.48192
3	0.416812	1.62154
4	0.636774	1.08434

(b)

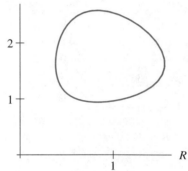

(c)

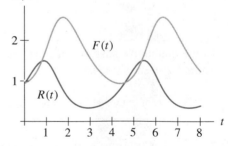

Hints and Answers for Section 7.4

1. Using single precision, we obtain $y(1) \approx 0.941274$. To make sure you are on the right track: $y_2 \approx 0.941885$, $y_4 \approx 0.924849$, $y_8 \approx 0.94101$.

3. Using single precision, we obtain $y(2) \approx 1.25938$. To make sure you are on the right track: $y_2 \approx 1.331857$, $y_4 \approx 1.256846$, $y_8 \approx 1.259110$.

Hints and Answers for Section 8.1

1. The orbit is eventually fixed.

3. The orbit tends to infinity.

5. The orbit is periodic of period 2.

7. The orbit is eventually fixed.

9. Fixed points: $x = 1$
Period 2 points: all real numbers except $x = 1$

11. There are no fixed points or period 2 points.

13. Fixed points: $x = 0$
Period 2 points: none

15. For fixed points, $x = 0, -3$. For periodic points of period 2, $x = (-1 \pm \sqrt{5})/2$.

17. There is one point which is fixed and no periodic points of period 2.

19. There is one fixed point $x = 0$ and every other point is a periodic point of period two.

21. There is one fixed point $x = 2$ and no other periodic points.

23. The orbit of any real number is a periodic orbit of period 2, except $x = 2$, which is fixed.

25. It is eventually fixed.

27. It is periodic with period 3.

29. It is eventually fixed.

31. The orbit is periodic with period 5.

33. It is a fixed point.

35. It is eventually fixed.

37. For $c = 1/4$, there is one fixed point. For $c < 1/4$, there are two fixed points. For $c > 1/4$, there are no fixed points.

39. For $c \geq -3/4$, there are no periodic points of period 2, and for $c < -3/4$, there are two periodic points of period 2.

Hints and Answers for Section 8.2

1. The fixed points $x = 0, 3$ are both repelling.

3. $x = 0$ is attracting.

5. $x = 0$ is attracting.

7. $x = 0$ is repelling, and $x = \pm\pi/2$ are attracting.

9. $x = \pm 1$ are neutral.

11. There are no fixed points.

13. The period is 2 and the orbit is attracting.

15. The period is 5 and the cycle is repelling.

17. The period is 2 and the cycle is neutral.

19. The fixed points are $x = \pm 1$ and they are neutral.

21. The fixed point is $x = 0$ and it is repelling.

23. The fixed point is $x = 0$ and it is repelling.

25. The fixed point is attracting.

27. There are infinitely many fixed points and they are all repelling.

29. All cycles are repelling. Note that $|(T^n)'(x_0)| = 4^n$.

Hints and Answers for Section 8.3

1. The bifurcation is a tangent bifurcation. For small enough $-\alpha$, the fixed point $x = \sqrt{-\alpha}$ is repelling and the other fixed point $x = -\sqrt{-\alpha}$ is attracting. For $\alpha = 0$, $x = 0$ is neutral.

3. This is a pitchfork bifurcation. For α slightly smaller than or equal to 1, the fixed point is attracting. For $\alpha > 1$, two more fixed points appear and they are attracting. The origin becomes a repelling fixed point.

5. This is a tangent bifurcation. There is one neutral fixed point for $\alpha = -1/4$, and there are two fixed points for α slightly larger than $-1/4$. One is repelling and the other is attracting.

7. The bifurcation is none of the above. If $\alpha = 0$, everything is maps to zero, which is a fixed point. If $0 < |\alpha| < 1$, then the origin is an attracting fixed point and $x = (1 - \alpha)/\alpha$ is a repelling fixed point.

9. If $\mu < 1$, the origin is the only fixed point. At $\mu = 1$ every point satisfying $x \le 1/2$ is a fixed point. If $\mu > 1$ there are two fixed points.

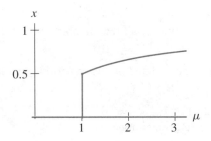

11. *Hint*: The function $F_c(x)$ is a parabola with minimum at $x = 0$ and $F_c(0) = c$. The function $F_c^2(x)$ is a quartic with local maximum at $x = 0$. Try looking at $F_c^2(x)$ upside down near $x = 0$.

Hints and Answers for Section 8.4

1.

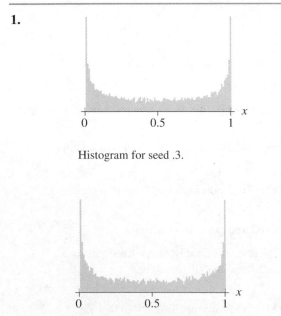

Histogram for seed .3.

Histogram for seed 3001.

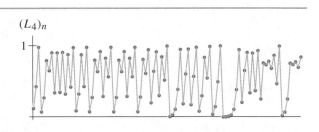

Time series for seed .3.

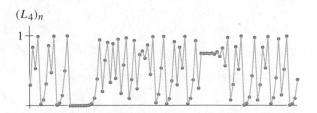

Time series for seed .3001.

3. $T^n(p/2^n) = 0$

5. Each of the intervals $[i/2^n, (i + 1)/2^n)$ for $i = 0, 1, 2, \ldots, 2^n - 2$ contains a fixed point of T^n.

7. The eventually fixed points have the form $p/2^n$ for any integer p between zero and $2^n - 1$.

9. T^n has $2^n - 1$ fixed points

11. List all possible finite strings of zeros and ones, then concatenate them to form the binary expansion of x_0.

13. Note that T^n has a fixed point in each interval of the form $[i/2^n, (i + 1)/2^n]$, so there are 2^n fixed points for T^n.

Hints and Answers for Appendix A

1. $y(t) = ke^t - e^{-2t}$

3. $y(t) = ke^t - \frac{3}{2}e^{-t}$

5. $y(t) = ke^{-2t} + \frac{3}{7}e^{t/3}$; take $k = 4/7$ for the initial-value problem.

7. $y(t) = ke^{2t} - \frac{3}{4}e^{-2t}$; take $k = 43/3$ for the initial-value problem.

9. Tends to a solution that satisfies $-1/2 \le y(t) \le 1$.

11. $y(t) \to 3$ as $t \to \infty$

13. *Hint*: The guess $y_p(t) = a\cos 3t$ leads to terms involving both $\cos 3t$ and $\sin 3t$ on the left.

15. *Hint*: Simply insert $y_h(t)$, $y_1(t)$, and $y_2(t)$ into their respective equations and add.

17. $y(t) = ke^{-2t} + \frac{1}{2}t^2 + \frac{1}{2}t + \frac{1}{4} + \frac{1}{6}e^{4t}$; take $k = -5/12$ for the initial-value problem.

19. $y(t) = ke^{-3t} + \frac{3}{13}\cos 2t + \frac{2}{13}\sin 2t + \frac{1}{6}e^{3t} - e^{-4t}$; take $k = 47/78$ for the initial-value problem.

21. (a) $a_1 = 1, a_2 = -2$
(b) $y(t) = \frac{1}{2}e^{-t} - \frac{1}{2}e^{-3t}$

23. (a) $a_1 = 1, a_2 = -2$
(b) *Hint*: The Taylor series for $y(t) = te^{-2t}$ is
$t - 2t^2 + 2t^3 + \dots$.

INDEX